Table entry for *z* is the area under the standard normal curve to the left of *z*.

Probability

| TABLE A | Standard normal probabilities (continued) |

z	.00	.01	.02	.03	.04	.05	.06	.07	.08	.09
0.0	.5000	.5040	.5080	.5120	.5160	.5199	.5239	.5279	.5319	.5359
0.1	.5398	.5438	.5478	.5517	.5557	.5596	.5636	.5675	.5714	.5753
0.2	.5793	.5832	.5871	.5910	.5948	.5987	.6026	.6064	.6103	.6141
0.3	.6179	.6217	.6255	.6293	.6331	.6368	.6406	.6443	.6480	.6517
0.4	.6554	.6591	.6628	.6664	.6700	.6736	.6772	.6808	.6844	.6879
0.5	.6915	.6950	.6985	.7019	.7054	.7088	.7123	.7157	.7190	.7224
0.6	.7257	.7291	.7324	.7357	.7389	.7422	.7454	.7486	.7517	.7549
0.7	.7580	.7611	.7642	.7673	.7704	.7734	.7764	.7794	.7823	.7852
0.8	.7881	.7910	.7939	.7967	.7995	.8023	.8051	.8078	.8106	.8133
0.9	.8159	.8186	.8212	.8238	.8264	.8289	.8315	.8340	.8365	.8389
1.0	.8413	.8438	.8461	.8485	.8508	.8531	.8554	.8577	.8599	.8621
1.1	.8643	.8665	.8686	.8708	.8729	.8749	.8770	.8790	.8810	.8830
1.2	.8849	.8869	.8888	.8907	.8925	.8944	.8962	.8980	.8997	.9015
1.3	.9032	.9049	.9066	.9082	.9099	.9115	.9131	.9147	.9162	.9177
1.4	.9192	.9207	.9222	.9236	.9251	.9265	.9279	.9292	.9306	.9319
1.5	.9332	.9345	.9357	.9370	.9382	.9394	.9406	.9418	.9429	.9441
1.6	.9452	.9463	.9474	.9484	.9495	.9505	.9515	.9525	.9535	.9545
1.7	.9554	.9564	.9573	.9582	.9591	.9599	.9608	.9616	.9625	.9633
1.8	.9641	.9649	.9656	.9664	.9671	.9678	.9686	.9693	.9699	.9706
1.9	.9713	.9719	.9726	.9732	.9738	.9744	.9750	.9756	.9761	.9767
2.0	.9772	.9778	.9783	.9788	.9793	.9798	.9803	.9808	.9812	.9817
2.1	.9821	.9826	.9830	.9834	.9838	.9842	.9846	.9850	.9854	.9857
2.2	.9861	.9864	.9868	.9871	.9875	.9878	.9881	.9884	.9887	.9890
2.3	.9893	.9896	.9898	.9901	.9904	.9906	.9909	.9911	.9913	.9916
2.4	.9918	.9920	.9922	.9925	.9927	.9929	.9931	.9932	.9934	.9936
2.5	.9938	.9940	.9941	.9943	.9945	.9946	.9948	.9949	.9951	.9952
2.6	.9953	.9955	.9956	.9957	.9959	.9960	.9961	.9962	.9963	.9964
2.7	.9965	.9966	.9967	.9968	.9969	.9970	.9971	.9972	.9973	.9974
2.8	.9974	.9975	.9976	.9977	.9977	.9978	.9979	.9979	.9980	.9981
2.9	.9981	.9982	.9982	.9983	.9984	.9984	.9985	.9985	.9986	.9986
3.0	.9987	.9987	.9987	.9988	.9988	.9989	.9989	.9989	.9990	.9990
3.1	.9990	.9991	.9991	.9991	.9992	.9992	.9992	.9992	.9993	.9993
3.2	.9993	.9993	.9994	.9994	.9994	.9994	.9994	.9995	.9995	.9995
3.3	.9995	.9995	.9995	.9996	.9996	.9996	.9996	.9996	.9996	.9997
3.4	.9997	.9997	.9997	.9997	.9997	.9997	.9997	.9997	.9997	.9998

INTRODUCTION TO THE PRACTICE OF STATISTICS

INTRODUCTION TO THE PRACTICE OF STATISTICS

FIFTH EDITION

David S. Moore
George P. McCabe
Purdue University

 W. H. Freeman and Company
New York

EXECUTIVE EDITOR: Craig Bleyer
DEVELOPMENT EDITOR: Laura Hanrahan
SENIOR MARKETING MANAGER: Krista Bettino
MEDIA EDITOR: Victoria Anderson
SUPPLEMENTS EDITOR: Amy Shaffer
PROJECT EDITOR: Mary Louise Byrd
COVER AND TEXT DESIGNER: Vicki Tomaselli
ILLUSTRATIONS: Integre Technical Publishing Co., Inc.
ILLUSTRATION COORDINATOR: Bill Page
PHOTO EDITOR: Bianca Moscatelli
PRODUCTION COORDINATOR: Paul W. Rohloff
COMPOSITION: Integre Technical Publishing Co., Inc.
MANUFACTURING: Quebecor

Library of Congress Control Number: 2004114217

ISBN: 0-7167-6400-8 (EAN: 97807167 64007)

CONTENTS

*Sections marked with an asterisk are optional.

CHAPTER | 5

Sampling Distributions   333

Introduction   334

CHAPTER | 6

Introduction to Inference   381

Introduction   382

CHAPTER | **12**

One-Way Analysis of Variance   719

CHAPTER | **13**

Two-Way Analysis of Variance   771

Additional chapters are on CD-ROM and available in a separate print supplement.

CHAPTER | **14**

Bootstrap Methods and Permutation Tests   14-1

Statistics is the science of data. *Introduction to the Practice of Statistics* (*IPS*) is an introductory text based on this principle. We present the most-used methods of basic statistics in a way that emphasizes working with data and mastering statistical reasoning. *IPS* is elementary in mathematical level but conceptually rich in statistical ideas and serious in its aim to help students think about data and use statistical methods with understanding.

Some schematic history will help place *IPS* in the universe of texts for a first course in statistics for students from a variety of disciplines. Traditional texts were almost entirely devoted to methods of inference, with quick coverage of means, medians, and histograms as a preliminary. No doubt this reflected the fact that inference is the only part of statistics that has a mathematical theory behind it. Several innovative books aimed at nontraditional audiences pioneered a broader approach that paid more attention to design of samples and experiments, the messiness of real data, and discussion of real-world statistical studies and controversies. All were written by widely known statisticians whose main business was not writing textbooks. *The Nature of Statistics* (Wallis and Roberts) has passed away, but *Statistics* (Freedman and collaborators) and *Statistics: Concepts and Controversies* (Moore) remain alive and well. None of these books tried to meet the needs of a typical first course because their audiences did not need full coverage of standard statistical methods.

IPS was the first book to successfully combine attention to broader content and reasoning with comprehensive presentation of the most-used statistical methods. It reflects the consensus among statisticians—even stronger now than when the first edition appeared—concerning the content of an introduction to our discipline. This consensus is expressed in a report from the joint curriculum committee of the American Statistical Association and the Mathematical Association of America,[1] and in discussions in leading journals.[2] *IPS* has been successful for several reasons:

1. *IPS* reflects the nature of modern statistical practice at a level suitable for beginners. Attention to data analysis and data production as well as to probability and inference is "new" only in the world of textbooks. Users of statistical methods have always paid attention to all of these. Contemporary research in statistics, driven by advances in computing, puts more emphasis on sophisticated "looking at data" and on data-analytic ways of thinking. Formal inference remains important, and receives careful treatment, but appears as part of a larger picture.

2. *IPS* has a logical overall progression, so that data analysis and data production strengthen the presentation of inference rather than stand apart from it. We stress that data analysis is an essential preliminary to inference because inference requires clean data. The most useful "goodness of fit" procedure, for example, is the normal quantile plot presented in Chapter 1 and used frequently in the inference chapters. We emphasize that when you do formal statistical inference, you are acting as if your data come from properly ran-

domized data production. We use random samples and experimental randomization to motivate the need for probability as a language for inference.

3. *IPS* presents data analysis as more than a collection of techniques for exploring data. We integrate techniques with discussion of systematic ways of thinking about data. We also work hard to make data-analytic thinking accessible to beginners by presenting a series of simple principles: always plot your data; look for overall patterns and deviations from them; when looking at the overall pattern of a distribution for one variable, consider shape, center, and spread; for relations between two variables, consider form, direction, and strength; always ask whether a relationship between variables is influenced by other variables lurking in the background. Inference is similarly treated as more than a collection of methods. We warn students about pitfalls in clear cautionary discussions—about regression and correlation (Section 2.4), experiments (page 206), sample surveys (page 222), confidence intervals (page 393), and significance tests (Section 6.3). Our goal throughout *IPS* is to present principles and techniques together in a way that is accessible to beginners and lays a foundation for those who will go on to more advanced study.

4. *IPS* integrates discussion of techniques, reasoning, and practice using real examples to drive the exposition. Students learn the technique of least-squares regression and how to interpret the regression slope. But they also learn the conceptual ties between regression and correlation, the importance of looking for influential observations (always plot your data), and to beware of averaged data and the restricted-range effect.

5. *IPS* is aware of current developments both in statistical science and in teaching statistics. For example, the first edition already favored the version of the two-sample t procedures that does not assume equal population variances and discussed the great difference in robustness between standard tests for means and for variances. In the fourth edition, we introduced the modified ("plus four") confidence intervals for proportions that are shown by both computational studies[3] and theory[4] to be superior to the standard intervals for all but very large samples. Brief optional "Beyond the Basics" sections give quick overviews of topics such as density estimation, scatterplot smoothers, nonlinear regression, and data mining. The new Chapter 14 on resampling methods offers a more extended introduction to one of the most important recent advances in statistical methodology.

The title of the book expresses our intent to introduce readers to statistics as it is used in practice. Statistics in practice is concerned with gaining understanding from data; it focuses on problem solving rather than on methods that may be useful in specific settings. A text cannot fully imitate practice, because it must teach specific methods in a logical order and must use data that are not the reader's own. Nonetheless, our interest and experience in applying statistics have influenced the nature of *IPS* in several ways.

Statistical Thinking Statistics is interesting and useful because it provides strategies and tools for using data to gain insight into real problems. As the continuing revolution in computing automates the details of doing calculations and making graphs, an emphasis on statistical concepts and on insight from data becomes both more practical for students and teachers and more important for users who must supply what is not automated. No student should complete a first statistics course, for example, without a firm grasp of the distinction between observational studies and experiments and of why randomized comparative experiments are the gold standard for evidence of causation. We have seen many statistical mistakes, but few that involved simply getting a calculation wrong. We therefore ask students to learn to explore data, always starting with plots, to think about the context of the data and the design of the study that produced the data, the possible influence of wild observations on conclusions, and the reasoning that lies behind standard methods of inference. Users of statistics who form these habits from the beginning are well prepared to learn and use more advanced methods.

Data   Data are numbers with a context, as we say in our introduction, "To Students: What Is Statistics?" A newborn who weighs 10.3 pounds is a big baby, and the birth weight could not plausibly be 10.3 ounces or 10.3 kilograms. Because context makes numbers meaningful, our examples and exercises use real data with real contexts that we briefly describe. Calculating the mean of five numbers is arithmetic, not statistics. We hope that the presence of background information, even in exercises intended for routine drill, will encourage students to always consider the meaning of their calculations as well as the calculations themselves. Note in this connection that a calculation or a graph or "reject H_0" is rarely a full answer to a statistical problem. We strongly encourage requiring students to always state a brief conclusion in the context of the problem. This helps build data sense as well as the communication skills that employers value.

Mathematics   Although statistics is a mathematical science, it is not a field of mathematics and should not be taught as if it were. A fruitful mathematical theory (based on probability, which *is* a field of mathematics) underlies some parts of basic statistics, but by no means all. The distinction between observation and experiment, for example, is a core statistical idea that is ignored by the theory.[5] Mathematically trained teachers, rightly resisting a formula-based approach, sometimes identify conceptual understanding with mathematical understanding. When teaching statistics, we must emphasize statistical ideas and recognize that mathematics is not the only vehicle for conceptual understanding. *IPS* requires only the ability to read and use equations without having each step parsed. We require no algebraic derivations, let alone calculus. Because this is a *statistics* text, it is richer in ideas and requires more thought than the low mathematical level suggests.

Calculators and Computers   Statistical calculations and graphics are in practice automated by software. We encourage instructors to use software of their choice or a graphing calculator that includes functions for both data analysis and basic inference. *IPS* includes some topics that reflect the dominance of software in practice, such as normal quantile plots and the version of the two-sample t procedures that does not require equal variances. Several times we display the output of multiple software systems for the same problem—the regression output on page 139 is an example. The point is that a student who knows the basics of regression can interpret almost any regression output. Students like this reassurance, and it helps focus their attention on understanding rather than reading output.

Nonetheless, basic statistical calculations can be done with just a "two-variable statistics" calculator with functions for correlation and the least-squares regression line as well as for the mean and standard deviation. Not all exercises are feasible for students with such a calculator, and doing graphs by hand is a bit painful, but we have taken care to make the book usable by students without access to computing.

Judgment   Statistics in practice requires judgment. It is easy to list the mathematical assumptions that justify use of a particular procedure, but not so easy to decide when the procedure can be safely used in practice. Because judgment develops through experience, an introductory course should present clear guidelines and not make unreasonable demands on the judgment of students. We have given guidelines—for example, on using the t procedures for comparing two means but avoiding the F procedures for comparing two variances—that we follow ourselves. Similarly, many exercises require students to use some judgment and (equally important) to explain their choices in words. Many students would prefer to stick to calculating, and many statistics texts allow them to. Requiring more will do them much good in the long run.

Teaching Experiences   We have successfully used *IPS* in courses taught to quite diverse student audiences. For general undergraduates from mixed disciplines, we cover Chapters 1 to 8 and one of Chapters 9, 10, and 12, omitting all optional material. For a quantitatively strong audience—sophomores planning to major in actuarial science or statistics—we move more quickly. We add Chapters 10 and 11 to the core material in Chapters 1 to 8 and include most optional content. We de-emphasize Chapter 4 (probability) because these students will take a probability course later in their program, though we make intensive use of software for simulating probabilities as well as for statistical analysis. The third group we teach contains beginning graduate students in such fields as education, family studies, and retailing. These mature but sometimes quantitatively unprepared students read the entire text (Chapters 11 and 13 lightly), again with reduced emphasis on Chapter 4 and some parts of Chapter 5. In all cases, beginning with data analysis and data production (Part I) helps students overcome their fear of statistics and builds a sound base for studying inference. We find that *IPS* can be flexibly adapted to quite varied audiences by paying attention to our clear designation of some material as optional and by varying the chapters assigned.

The Fifth Edition: What's New?

Refreshing a successful text ought to involve more than changing the edition number on the cover. In the fifth edition, regular users of *Introduction to the Practice of Statistics* (*IPS*) will find notable changes and, we believe, improvements.

NEW Topics

- **Chapter 14 on Bootstrap Methods and Permutation Tests.** We have included a new supplement chapter that examines resampling procedures, the single most important development in statistics during the past generation. Written by David Moore and a team at Insightful Corporation led by Tim Hesterberg, this chapter makes resampling methods genuinely accessible to able undergraduates who are not mathematics or statistics majors. Chapter 14 keeps *IPS* the leader among introductory texts in contact with the actual practice of statistics. This chapter can be presented at any point after Chapter 7, where students meet practical inference and the *t* procedures. Bootstrap and permutation procedures are on the way to becoming standard methods, and we hope that instructors will consider modernizing their courses by including these methods.

To easily apply these computationally intensive procedures, students require powerful but accessible software. As software packages catch up with practice, it will become possible to use most statistical software with Chapter 14. At present, we recommend S-PLUS; a version of S-PLUS is free to students. Tim Hesterberg has prepared an S-PLUS library named *IPSdata* that loads all the *IPS* data sets and the S-PLUS resampling library. For detailed instructions and links for downloading and installing this free software, please go to `www.whfreeman.com/ipsresample`. Once this is done, everything in Chapter 14 is available in menus customized for easy use. Instructors who present resampling methods will naturally consider using S-PLUS throughout the course, acquiring a personal copy or a site license to work along with the free student software.

- **Chapter 17 on Statistics for Quality** discusses process control and process capability. The chapter is notable for its emphasis on concepts and attention to problems of practice in an area often dominated by lists of recipes for control charts.

NEW Exercises and Examples

We continue our program of presenting students with new challenges by refreshing examples and exercises.

- **Half the exercises are new or revised.** Refreshing the exercises gives you and your students interesting and current problems. In addition, many of the exercises that were dropped from the third and fourth editions are available online and on the text CD as **Supplementary Exercises,** so that instructors have a large additional stock of exercises to draw from if they wish.

- **Examples.** Because the exposition is driven by real examples, new examples give us occasion to review and **refresh our presentation.** For example, Section 1.1 has been completely rewritten, in part because the familiar lead example on the passage time of light has been replaced by new data on lengths of calls to a bank's customer service center.

NEW Features

- **Statistics in Practice.** Each chapter now opens with Statistics in Practice, a brief first-person statement by a professional who uses statistics on the job. We hope these accounts by diverse (and mostly young) professionals will capture the interest of students and display the breadth of application of statistical science.

- **Caution notes.** As part of our attention to the realities of statistical practice, we have called attention to the many comments on common pitfalls that have always been a feature in *IPS*. Look for the new Caution icon in the margin.

- **Challenge exercises.** In addition to many new exercises, we have pointed out more challenging exercises with a new Challenge icon in the margin. Challenge exercises are varied: some are mathematical, some require open-ended investigation, and so on.

- **EESEE case studies.** Each chapter concludes by pointing to case studies from EESEE (the Electronic Encyclopedia of Statistical Examples and Exercises) that make use of the ideas and methods of the chapter. The explicit matching makes it easier for instructors to use EESEE to assign larger cases. EESEE is available on the text CD and online at the text Web site www.whfreeman.com/ips. William Notz, Dennis Pearl, and Elizabeth Stasny, the team at Ohio State University that produces EESEE, continue to add new case studies.

NEW applets and output screens

- **Statistical applets.** We have added to the collection of statistical applets available online at www.whfreeman.com/ips. These applets, designed for learning, are integrated throughout the text and are deliberately kept straightforward and tightly focused for easy use by students. They also work very well for classroom demonstrations. New applets energize the central limit theorem (two applets: for a continuous distribution and the approximation to binomial distributions) and significance tests (three applets: *P*-values, fixed significance levels, and power). Look for the applet icon at points in the text where applets will aid understanding and at exercises that ask students to interact with the applets.

- **Statistical output** screens are incorporated throughout the text. In addition to Minitab, SPSS, SAS, Excel, S-PLUS, and TI-83 sample screens, we have also added new CrunchIt! screens. (For more information on CrunchIt! please see the next section.)

NEW E-STAT PACK Software

E-STAT PACK software has been developed specifically to enhance *IPS*. This sophisticated and comprehensive media supplement offers four unique components to enliven your course.

StatTutor

- **StatTutor,** developed by the Brigham Young University Statistics Department, offers 84 audio-embedded statistical tutorials to help students master statistics. StatTutor lessons contain animations, applets, videos, and self-assessment quizzes and are designed to work with a variety of learning styles.

CrunchIt!

- **CrunchIt!** is a new, easy-to-use statistical software package developed by Webster West at the University of South Carolina. Like StatTutor, CrunchIt! provides more interaction and works with students to help them analyze data and derive answers completely online. A unique and superior graphing capability helps students visualize the data sets from the text.

- **iSOLVE Homework and Assessment Service** provides an automated solution for statistics homework based on problems in the text. Instructors can assign students homework online by choosing from over 500 algorithmically generated problems from the text or by creating their own questions. iSOLVE features a complete gradebook with downloadable capabilities, plus a robust Help feature and downloadable Instructor's Guide for convenient instructor support.

- **EESEE** (the Electronic Encyclopedia of Statistical Examples and Exercises), developed by a consortium of statistics teachers at Ohio State University, contains numerous case studies with real examples and real data. Each case study provides several thought-provoking questions designed to make students think carefully about statistical issues raised by the stories. EESEE includes case studies from a variety of fields, including psychology, biology, business, sociology, sports, and more, for students with different backgrounds and interests.

Acknowledgments

We are pleased that the first four editions of *Introduction to the Practice of Statistics* have helped move the teaching of introductory statistics in a direction supported by most statisticians. We are grateful to the many colleagues and students who have provided helpful comments, and we hope that they will find this new edition another step forward. In particular, we would like to thank the following colleagues who offered specific comments on the new edition:

Ilham Akhundov,
    McMaster University
Barbara Alvin,
    Eastern Washington University
Charles W. Champ,
    Georgia Southern University
Pinyuen Chen,
    Syracuse University
Smiley Cheng,
    University of Manitoba
Deann J. Christianson,
    University of the Pacific
Carolyn K. Cuff,
    Westminster College
Ginger M. Davis,
    Rice University
Alexander R. de Leon,
    University of Calgary
DeWayne R. Derryberry,
    University of Puget Sound
Brooke Fridley,
    University of
    Wisconsin—La Crosse

Brian T. Gill,
    Seattle Pacific University
Miriam Schapiro Grosof,
    Yeshiva University
Brenda Gunderson,
    University of Michigan
Christian K. Hansen,
    Eastern Washington University
Susan K. Herring,
    Sonoma State University
Chuck Hommel,
    University of Puget Sound
James R. Hughes,
    Elizabethtown College
Mark Irwin,
    Harvard University
Christopher K. Johnson,
    University of North Florida
Rudolf N. Lublinsky,
    Portland Community College
Ronald Mellado Miller,
    Brigham Young University,
    Hawaii

Kip Murray,
University of Houston
Dr. Linda M. Myers,
Harrisburg Area Community
College
Alison Paradise,
University of Puget Sound
Steve Pelikan,
University of Cincinnati
Dr. Dan Rand,
Winona State University
Larry J. Ringer,
Texas A&M University
Caren M. Rotello,
University of Massachusetts
Thomas L. Scofield,
Calvin College
Rosa Garcia Seyfried,
Harrisburg Area Community
College

Haipeng Shen,
University of North Carolina at
Chapel Hill
Angela P. Stabley,
Portland Community College
Nathan Taback,
University of Toronto
Larry Weldon,
Simon Fraser University
Nathan Wetzel,
University of
Wisconsin—Stevens Point
Vicki M. Whitledge,
University of Wisconsin—Eau
Claire
Daniel Williams,
Baruch College
Fancher E. Wolfe,
Metropolitan State University
Elaine L. Zanutto,
University of Pennsylvania

The professionals at W. H. Freeman and Company, in particular, Mary Louise Byrd, Patrick Farace, and Laura Hanrahan, have contributed greatly to the success of *IPS*. Most of all, we are grateful to the many people in varied disciplines and occupations with whom we have worked to gain understanding from data. They have provided both material for this book and the experience that enabled us to write it. What the eminent statistician John Tukey called "the real problems experience and the real data experience" has shaped our view of statistics. It has convinced us of the need for beginning instruction to focus on data and concepts, building intellectual skills that transfer to more elaborate settings and remain essential when all details are automated. We hope that users and potential users of statistical techniques will find this emphasis helpful.

A full range of supplements and media is available to help students get the most out of *IPS:*

Print Supplements

- **Study Guide** (0-7167-6358-3) prepared by Michael A. Fligner of The Ohio State University offers students explanations of crucial concepts in each section of *IPS,* plus detailed solutions to key text problems and step-through models of important statistical techniques.

- **Statistical Software Manuals** will guide students in the use of particular statistics software with *IPS.* The chapters of each manual correlate to those of *IPS* and include a set of exercises specific to each chapter's concepts. These manuals are:

  - **Excel Manual** (0-7167-6360-5) prepared by Linda Dawson of the University of Washington, Tacoma. New add-in Excel macros corresponding to the manual allow Excel users to create boxplots and normal quantile plots. These macros are available for download on the Web site www.whfreeman.com/ips.

  - **JMP Manual** (0-7167-6359-1) prepared by Thomas Devlin of Montclair State University.

  - **Minitab Manual** (0-7167-6361-3) prepared by Michael Evans of the University of Toronto.

  - **SPSS Manual** (0-7167-6363-X) prepared by Linda Sorensen of Algoma University.

  - **TI-83 Graphing Calculator Manual** (0-7167-6364-8) prepared by David K. Neal of Western Kentucky University.

  - **A free student version of S-PLUS,** with a student manual, is available on the Web. See www.whfreeman.com/ipsresample.

- **Supplement Chapters 14, 15, 16, and 17** (0-7167-6406-7) appear on the Web site and CD and are available as a printed supplement (black and white).

- **Telecourse Study Guide** (0-7167-6701-5) correlates the new edition of *IPS* to the Annenberg/CPB–produced series *Against All Odds: Inside Statistics.*

- **Activities and Projects Book** (0-7167-9809-3), prepared by Ron Millard, of Shawnee Mission South High School, and John C. Turner, of the U.S. Naval Academy, offers ideas for projects and Web-based explorations. This book asks students to write critically about statistics and is compatible with all W. H. Freeman statistics textbooks.

Media Supplements

E-STAT **PACK**

- **E-STAT PACK** (0-7167-6769-4) is the ultimate statistical package customized to the fifth edition of *IPS* with relevant examples, exercises, and case studies. The four components of E-STAT PACK are StatTutor (student tutorials), CrunchIt! (easy-to-use statistical software), iSOLVE Homework and Assessment Service (online, algorithmically generated homework problems), and EESEE (case studies). For a more detailed description of E-STAT PACK, please refer to the section "The Fifth Edition: What's New?"

- *IPS* **Web Site** `www.whfreeman.com/ips` seamlessly integrates topics from the text. On the Web site students can find:

  - **Interactive statistical applets** that allow students to manipulate data and see the corresponding results graphically. These applets automate calculations and graphics in a way that nicely complements classroom work.

  - **Data sets** in ASCII, Minitab, TI, SPSS, S-PLUS, and Excel formats.

  - **Interactive exercises and self-quizzes** to help students prepare for tests. Each quiz contains 20 multiple-choice questions that include automatic feedback for incorrect answers, reinforcement for correct answers, and text section references to aid in additional study for incorrect answers.

  - **Student version of EESEE (Electronic Encyclopedia of Statistical Examples and Exercises),** a rich repository of case studies that apply the concepts of the text to various real-world venues such as the mass media, sports, natural sciences, social sciences, and medicine. Each case study is accompanied by practice problems, and most include full data sets that are exportable to various statistical software packages. EESEE was developed by a consortium at The Ohio State University dedicated to statistical education.

  - **All tables** from the text in pdf format for quick, easy reference.

  - **Supplementary exercises for every chapter** written by David Moore and George McCabe, giving students more opportunities to make sure they understand key concepts. Solutions to odd-numbered additional exercises are also included.

  - **Activities and Projects,** the online component of a new supplement that offers ideas for Web-based explorations requiring students to write critically about statistics.

  - **Supplement Chapters 14, 15, 16, and 17** cover bootstrap methods and permutation tests, nonparametric tests, logistic regression, and statistics for quality.

- **Interactive Student CD-ROM** is included with every new copy of *IPS*. The CD contains most of the content available on the Web site, including:

  - **Interactive statistical applets.**

  - **Data sets** in ASCII, Minitab, TI, SPSS, S-PLUS, and Excel formats.

- **Interactive exercises and self-quizzes.**
- **Student version of EESEE.**
- **All tables** from the text in pdf format for quick, easy reference.
- **Supplementary exercises for every chapter.**
- **Supplement Chapters 14, 15, 16, and 17.**

- **Special Software Packages** student versions of JMP, Minitab, and SPSS are available packaged with the respective manuals. This software is not sold separately. Contact your W. H. Freeman representative or visit online at www.whfreeman.com/ips.

- **PDA Stats** (0-7167-6844-5) is a new statistical application for handheld devices designed specifically for educational use and developed by Stephen Kokoska and Shaun O'Brien of Bloomsburg University. With this easy-to-use software, students can use a Personal Digital Assistant (PDA) or Pocket PC to construct histograms and scatterplots, compute summary statistics, perform confidence interval and hypothesis test procedures, and more. PDA Stats may also be packaged with *IPS*.

FOR INSTRUCTORS

A full range of supplements and media is available to help instructors:

Media Supplements

- **E-STAT PACK** (instructor version 0-7167-6827-5) is the ultimate statistical package customized to the fifth edition of *IPS* with relevant examples, exercises, and case studies. The four components of E-STAT PACK are StatTutor (student tutorials), CrunchIt! (easy-to-use statistical software), iSOLVE Homework and Assessment Service (online, algorithmically generated homework problems), and EESEE (case studies). For a more detailed explanation of E-STAT Pack, please refer to the section "The Fifth Edition: What's New?"

- *IPS* **Instructor's Web Site,** available at www.whfreeman.com/ips after registration as an instructor, features all the student Web material plus:

  - **Instructor version of EESEE (Electronic Encyclopedia of Statistical Examples and Exercises),** with solutions to the exercises in the student version.

  - **The Instructor's Guide,** including full solutions to all exercises in pdf format.

  - **PowerPoint Slides** that can be used as is or customized. Each slide offers outlined text to match the textbook's table of contents, along with a wide range of figures and tables from *IPS*.

- **Instructor's Resource CD-ROM** (0-7167-6957-3) contains all the student CD material and the content of the Instructor's Web site and, in addition,

  - **All text figures** in an exportable presentation format, JPEG for Windows and PICT for Macintosh users.

  - **All tables** from the text in pdf format.

Assessment Tools

- **Test Bank: CD-ROM (Windows and Mac on one disk)** (0-7167-6795-3); printed (0-7167-6357-5). Prepared by Kirsten Namesnik and Brenda Gunderson of the University of Michigan, Ann Arbor, this easy-to-use CD version lets you add, edit, and resequence questions to suit your needs.

- **Online Testing,** powered by Diploma from the Brownstone Research Group, offers instructors the ability to easily create and administer secure exams over a network and over the Internet, with questions that incorporate multimedia and interactive exercises. The program lets you restrict tests to specific computers or blocks of time and includes an impressive suite of gradebook and result-analysis features.

- **Online Quizzing,** powered by QuestionMark and accessed via the *IPS* Web site, uses QuestionMark's Perception to enable instructors to easily and securely quiz students online using prewritten, multiple-choice questions for each text chapter (not from the Test Bank).

Course Management

- **Online Course Materials (WebCT, Blackboard)** can be provided as a service for adopters. We offer electronic content of *IPS*, including the complete Test Bank in either WebCT or Blackboard format.

Print

- **Instructor's Guide with Solutions** (0-7167-6356-7), prepared by Darryl Nester of Bluffton University, includes worked-out solutions to all exercises, teaching suggestions, and chapter comments.

Personal Response System (aka "Clikers")

- **Personal Response System** is the user-friendly way to make your classroom more efficient, more interactive, and more effective. This wireless, remote system (about the size of a television remote control) lets you pause to ask questions and instantly record responses, as well as take attendance, direct students through lectures, gauge your students' understanding of the material, and much more.

Introduction to the Practice of Statistics presents a wide variety of applications from diverse disciplines. The list below indicates the number of examples and exercises that relate to different fields.

Examples by Application

Biology and environmental science: 36

1.10, 1.21, 1.33, 2.5, 2.14, 3.10, 3.14, 3.27, 4.12, 4.13, 6.1, 6.27, 7.1, 7.2, 7.3, 7.10, 7.11, 10.11, 10.13, 10.14, 10.18, 10.19, 15.1, 15.2, 15.3, 15.4, 15.5, 15.6, 15.7, 15.13, 15.14, 16.5, 16.7, 16.8, 16.9, 16.10

Business and management: 46

1.1, 1.4, 1.6, 3.19, 4.9, 4.24, 4.25, 5.3, 5.4, 5.5, 5.11, 5.13, 9.9, 9.10, 9.12, 9.13, 9.14, 12.3, 12.4, 12.5, 12.6, 12.7, 12.8, 12.9, 12.10, 12.11, 12.12, 12.13, 12.14, 12.15, 12.16, 12.20, 12.22, 12.23, 12.24, 12.25, 14.1, 14.2, 14.3, 14.6, 14.9, 14.13, 17.1, 17.18, 17.19, 17.20

Demographics and population characteristics: 26

1.2, 1.5, 1.16, 1.23, 1.24, 3.1, 3.26, 4.22, 5.1, 5.8, 5.9, 5.10, 5.12, 5.17, 7.13, 7.15, 7.16, 8.10, 9.3, 9.4, 9.11, 9.15, 9.22, 9.23, 12.1, 12.2

Economics and finance: 14

1.18, 2.20, 2.25, 4.28, 6.29, 7.4, 7.5, 7.6, 7.17, 14.4, 14.5, 14.7, 14.8, 14.10

Education and child development: 50

1.7, 1.8, 1.22, 1.25, 1.26, 1.28, 1.29, 1.30, 2.2, 2.3, 2.7, 2.19, 2.22, 2.25, 2.26, 3.2, 3.3, 3.15, 4.8, 4.16, 4.19, 4.27, 4.32, 4.35, 6.2, 6.3, 6.4, 6.5, 6.6, 6.7, 6.8, 6.9, 6.10, 6.11, 6.12, 6.13, 6.15, 6.18, 6.19, 6.26, 7.14, 7.18, 11.1, 12.27, 12.28, 14.11, 14.12, 15.8, 15.9, 15.10

Fuel consumption and cars: 19

1.3, 1.11, 1.12, 1.13, 1.14, 1.15, 10.1, 10.2, 10.3, 10.4, 10.5, 10.6, 10.7, 10.8, 10.9, 10.10, 10.12, 10.21, 10.22

Manufacturing and product development: 19

1.9, 1.31, 2.4, 2.16, 3.16, 5.18, 5.20, 6.16, 6.31, 17.4, 17.5, 17.6, 17.8, 17.10, 17.11, 17.12, 17.14, 17.16, 17.17

Marketing and consumer behavior: 20

3.4, 3.8, 3.18, 3.20, 3.24, 3.25, 3.26, 5.14, 5.15, 5.16, 8.6, 9.18, 9.19, 9.20, 9.21, 11.2, 12.26, 13.1, 13.2, 17.13

Medicine and clinical studies or trials: 24

1.17, 1.32, 2.9, 2.10, 2.11, 2.12, 2.15, 2.17, 2.18, 2.28, 3.5, 3.9, 3.13, 4.15, 5.7, 6.16, 6.17, 6.30, 7.19, 7.20, 7.21, 7.22, 7.23, 9.17

Physical sciences: 6

1.19, 1.20, 2.6, 2.13, 2.27, 17.2

Psychology and human behavior: 17

2.26, 3.6, 3.7, 3.11, 3.12, 3.21, 6.24, 6.25, 7.7, 7.8, 7.12, 8.3, 8.4, 8.5, 13.10, 14.14, 17.3

Public health and nutrition: 30

2.1, 2.21, 2.24, 6.14, 6.23, 6.28, 8.1, 8.2, 8.9, 8.11, 8.12, 9.1, 9.2, 9.5, 9.6, 9.7, 9.8, 13.3, 13.4, 13.5, 13.6, 13.7, 13.8, 13.9, 13.11, 16.1, 16.2, 16.3, 16.4, 16.6

Public policy and political science: 4

3.17, 3.22, 3.23, 17.7

Sport and leisure: 18

2.8, 4.14, 4.20, 4.23, 4.26, 4.31, 4.34, 4.36, 5.2, 5.6, 5.19, 8.7, 8.8, 9.24, 9.25, 14.9, 15.11, 15.12

Technology and Internet: 3

4.33, 4.37, 4.38

Exercises by Application

Biology and environmental science: 204

1.28, 1.30, 1.58, 1.59, 1.60, 1.73, 1.91, 1.124, 1.125, 1.129, 1.138, 2.7, 2.10, 2.12, 2.13, 2.16, 2.18, 2.21, 2.22, 2.25, 2.28, 2.43, 2.44, 2.48, 2.49, 2.52, 2.53, 2.57, 2.60, 2.61, 2.64, 2.70, 2.97, 2.99, 2.103, 2.105, 2.107, 3.9, 3.30, 3.50, 4.13, 4.23, 4.28, 4.36, 4.37, 4.38, 4.39, 4.42, 4.109, 4.110, 4.111, 4.112, 4.113, 5.1, 5.35, 5.58, 5.59, 5.69, 6.5, 6.6, 6.7, 6.8, 6.12, 6.25, 6.26, 6.52, 6.53, 6.94, 6.95, 6.106, 6.107, 7.1, 7.2, 7.3, 7.4, 7.27, 7.28, 7.31, 7.32, 7.34, 7.35, 7.39, 7.44, 7.45, 7.46, 7.51, 7.68, 7.69, 7.70, 7.71, 7.75, 7.76, 7.82, 7.87, 7.88, 7.89, 7.90, 7.92, 7.99, 7.100, 7.102, 7.103, 7.104, 7.114, 7.115, 7.116, 7.126, 7.127, 7.130, 7.135, 8.8, 8.36, 8.53, 8.57, 9.24, 9.25, 9.37, 9.46, 10.4, 10.5, 10.6, 10.9, 10.10, 10.11, 10.12, 10.13, 10.15, 10.19, 10.20, 10.41, 10.44, 10.45, 11.26,

11.27, 11.28, 11.29, 11.30, 11.31, 11.32, 11.33, 11.34, 11.35, 11.36, 11.37, 11.38, 11.39, 11.40, 11.41, 11.42, 12.7, 12.9, 12.11, 12.27, 12.29, 12.31, 12.43, 12.44, 12.45, 12.60, 12.62, 13.15, 13.19, 13.20, 13.21, 13.30, 14.4, 14.5, 14.6, 14.7, 14.9, 14.33, 14.35, 14.36, 14.64, 14.69, 14.72, 14.73, 15.4, 15.5, 15.7, 15.9, 15.10, 15.12, 15.17, 15.19, 15.24, 15.25, 15.26, 15.27, 15.28, 15.29, 15.32, 15.33, 15.37, 15.38, 16.6, 16.7, 16.8, 16.28, 16.29, 16.30, 17.32, 17.34, 17.36

Business and management: 92

1.62, 1.63, 1.64, 2.11, 2.27, 2.42, 2.66, 2.86, 2.111. 3.5, 3.27, 3.53, 3.80, 4.66, 4.79, 4.80, 4.82, 4.117, 5.42, 5.57, 6.22, 6.23, 6.27, 6.81, 6.88, 6.89, 6.100, 6.109, 7.12, 7.33, 7.55, 7.56, 7.66, 7.67, 7.80, 7.81, 7.91, 7.97, 7.98, 7.117, 8.19, 8.43, 8.44, 8.56, 9.30, 9.31, 10.21, 10.22, 10.34, 14.1, 14.3, 14.15, 14.20, 14.22, 14.24, 14.25, 14.29, 14.31, 14.37, 14.38, 14.42, 14.43, 14.48, 14.49, 14.54, 14.59, 15.30, 16.2, 16.3, 16.4, 16.5, 16.13, 16.14, 16.15, 16.16, 16.23, 16.24, 17.14, 17.23, 17.43, 17.56, 17.57, 17.58, 17.59, 17.63, 17.66, 17.67, 17.68, 17.69, 17.70, 17.71, 17.72

Demographics and population characteristics: 124

1.2, 1.4, 1.5, 1.8, 1.15, 1.16, 1.31, 1.32, 1.42, 1.45, 1.47, 1.48, 1.49, 1.88, 1.89, 1.133, 1.135, 1.143, 1.146, 2.6, 2.55, 2.65, 2.68, 2.85, 3.44, 3.46, 3.47, 3.49, 3.55, 3.59, 3.62, 3.63, 3.66, 3.67, 3.68, 3.69, 3.71, 3.72, 3.73, 4.14, 4.16, 4.22, 4.24, 4.31, 4.43, 4.44, 4.45, 4.46, 4.51, 4.57, 4.61, 4.63, 4.74, 4.81, 4.86, 4.87, 4.90, 4.96, 4.97, 4.98, 4.99, 4.100, 4.101, 4.102, 4.108, 5.13, 5.14, 5.16, 5.18, 5.23, 5.51, 5.53, 5.56, 5.61, 5.62, 6.1, 6.2, 6.3, 6.4, 6.35, 6.37, 7.38, 7.59, 7.95, 7.121, 7.122, 7.123, 7.125, 7.134, 8.20, 8.37, 8.38, 8.40, 8.48, 8.49, 8.50, 8.53, 8.59, 8.60, 8.61, 8.63, 8.69, 8.71, 9.1, 9.2, 9.3, 9.4, 9.5, 9.6, 9.12, 9.13, 9.14, 9.27, 9.28, 9.35, 9.42, 9.45, 9.47, 10.38, 10.39, 10.40, 12.30, 13.29, 14.68

Economics and finance: 59

1.114, 1.144, 2.4, 2.19, 2.20, 2.24, 2.26, 2.36, 2.50, 2.51, 2.98, 2.104, 2.109, 2.110, 3.26, 3.61, 4.27, 4.33, 4.65, 4.83, 4.84, 4.85, 4.116, 5.12, 5.43, 5.48, 7.140, 7.141, 8.11, 8.12, 8.13, 8.14, 10.17, 10.27, 10.29, 10.30, 10.31, 11.16, 11.17, 11.18, 11.19, 11.20, 11.21, 11.22, 11.23, 14.10, 14.11, 14.12, 14.21, 14.34, 14.39, 14.41, 14.46, 14.52, 14.55, 14.65, 14.71, 14.74, 15.31

Education and child development: 159

1.6, 1.11, 1.14, 1.19, 1.21, 1.35, 1.37, 1.38, 1.39, 1.41, 1.52, 1.54, 1.96, 1.97, 1.98, 1.99, 1.100, 1.101, 1.102, 1.103, 1.104, 1.105, 1.106, 1.107, 1.108, 1.109, 1.113, 1.131, 1.141, 1.147, 2.2, 2.3, 2.5, 2.23, 2.39, 2.41, 2.45, 2.54, 2.58, 2.59, 2.67, 2.79, 2.84, 2.89, 2.90, 2.93, 2.112, 2.113, 3.8, 3.51, 3.64, 3.75, 3.77, 3.79, 3.89, 3.94, 4.17, 4.41, 4.59, 4.60, 4.64, 4.92, 4.93, 4.94, 4.95, 4.103, 4.118, 5.20, 5.25, 5.34, 5.36, 5.54, 5.66, 5.67, 6.9, 6.13, 6.14, 6.19, 6.30, 6.33, 6.34, 6.36, 6.50, 6.51, 6.57, 6.82, 6.83, 6.84, 6.97, 6.115, 7.8, 7.9, 7.29, 7.30, 7.41, 7.42, 7.43, 7.47, 7.48, 7.49, 7.77, 7.78, 7.79, 7.119, 7.131, 7.132, 7.133, 8.7, 8.47, 8.51, 8.66, 8.67, 8.70, 9.10, 9.15, 9.16, 9.17, 9.29, 9.38, 9.39, 10.14, 10.16, 10.18, 10.35, 10.36, 10.37, 10.43, 11.24, 12.16, 12.34, 12.35, 12.36, 12.53, 12.54, 12.58, 13.16, 13.17, 13.18, 13.39, 13.40, 13.41, 13.42, 14.2, 14.14, 14.19,

14.32, 14.45, 14.50, 14.53, 14.66, 15.1, 15.2, 15.3, 15.8, 15.16, 16.31, 16.32, 16.33, 16.34

Fuel consumption and cars: 20

1.1, 1.7, 1.9, 1.10, 2.8, 2.14, 2.32, 2.33, 2.40, 2.62, 2.73, 2.75, 4.89, 4.91, 6.16, 6.17, 6.18, 7.6, 7.7, 10.46

Manufacturing and product development: 61

1.3, 1.121, 2.46, 2.100, 2.108, 3.19, 4.69, 4.70, 4.73, 4.75, 5.29, 5.31, 5.40, 5.41, 5.44, 5.45, 5.49, 5.64, 5.65, 7.37, 7.50, 8.62, 9.34, 10.7, 10.8, 11.25, 13.25, 13.26, 17.5, 17.10, 17.11, 17.12, 17.13, 17.15, 17.16, 17.17, 17.18, 17.19, 17.24, 17.25, 17.27, 17.30, 17.31, 17.33, 17.35, 17.37, 17.38, 17.45, 17.46, 17.48, 17.49, 17.51, 17.52, 17.61, 17.64, 17.65, 17.73, 17.74, 17.75, 17.76

Marketing and consumer behavior: 85

1.12, 1.130, 3.1, 3.2, 3.11, 3.18, 3.22, 3.36, 3.39, 3.40, 3.41, 3.54, 3.56, 3.74, 3.83, 3.85, 3.86, 3.95, 4.30, 4.105, 4.107, 4.115, 4.121, 5.1, 5.2, 5.4, 5.6, 5.15, 5.22, 5.38, 5.46, 5.47, 5.50, 5.60, 6.10, 6.11, 6.24, 6.96, 6.108, 6.116, 7.26, 7.72, 7.108, 8.22, 8.24, 8.25, 8.26, 8.27, 8.28, 8.29, 8.30, 8.45, 8.46, 8.58, 8.64, 8.65, 8.68, 8.79, 9.9, 9.20, 9.21, 9.32, 9.33, 9.36, 11.7, 12.24, 12.32, 12.33, 12.55, 13.9, 13.10, 13.11, 13.12, 13.13, 13.14, 13.27, 13.28, 14.78, 14.79, 15.11, 16.10, 16.11, 16.12, 16.25, 16.26

Medicine and clinical studies or trials: 51

1.17, 1.18, 1.36, 1.50, 1.71, 1.72, 1.75, 1.77, 3.3, 3.6, 3.12, 3.13, 3.23, 3.24, 3.25, 3.31, 3.32, 3.33, 3.34, 3.65, 3.90, 3.91, 3.92, 6.34, 6.105, 7.24, 7.25, 7.36, 7.53, 7.83, 7.85, 7.105, 7.106, 7.107, 7.110, 7.111, 7.112, 7.113, 7.118, 7.128, 7.129, 8.72, 9.11, 9.43, 12.46, 12.47, 12.48, 12.49, 16.17, 16.19, 16.21

Physical sciences: 59

1.22, 1.23, 1.25, 1.27, 1.33, 1.34, 1.43, 1.46, 1.51, 1.53, 1.74, 1.76, 1.123, 1.142, 2.72, 4.72, 4.76, 5.28, 5.30, 5.32, 5.52, 6.28, 6.71, 7.40, 10.23, 10.24, 10.25, 11.43, 11.44, 11.45, 11.46, 11.47, 11.48, 11.49, 11.50, 11.51, 12.19, 12.20, 12.21, 12.22, 12.23, 12.50, 12.51, 12.52, 13.31, 13.32, 13.33, 13.34, 13.35, 13.36, 13.37, 13.38, 15.34, 15.35, 15.36

Psychology and human behavior: 46

1.29, 2.9, 2.29, 2.35, 2.37, 2.47, 2.96, 2.102, 3.4, 3.14, 3.16, 3.47, 3.93, 4.32, 5.3, 5.5, 5.7, 5.19, 5.21, 5.68, 6.55, 7.5, 7.54, 7.84, 7.86, 7.96, 7.101, 7.120, 10.28, 10.42, 11.8, 12.8, 12.10, 12.12, 12.17, 12.18, 12.61, 13.5, 13.6, 14.44, 14.77, 15.6, 15.15, 17.2, 17.8, 17.9

Public health and nutrition: 79

1.61, 1.110, 1.111, 1.112, 1.119, 2.63, 2.69, 2.81, 2.87, 2.88, 2.91, 2.92, 2.94, 2.95, 3.7, 3.10, 3.15, 3.17, 3.52, 3.70, 3.82, 4.10, 5.8, 5.17, 5.33, 5.37, 5.39, 5.63,

6.15, 6.35, 6.48, 6.49, 6.56, 6.61, 6.101, 7.10, 7.11, 7.57, 7.58, 7.109, 7.124, 8.15, 8.16, 8.17, 8.18, 10.32, 11.5, 11.6, 11.13, 11.14, 11.15, 12.15, 12.26, 12.28, 12.37, 12.38, 12.39, 12.40, 12.41, 12.42, 13.22, 13.23, 13.24, 14.56, 14.57, 14.58, 14.60, 14.61, 14.62, 14.63, 14.75, 14.76, 15.18, 15.20, 15.21, 15.22, 15.23, 16.27, 16.35

Public policy and political science: 24

3.37, 3.58, 4.52, 4.58, 4.104, 4.106, 6.90, 8.9, 8.10, 8.77, 9.40, 9.41, 14.80, 14.81, 16.18, 16.20, 16.22, 17.4, 17.26, 17.28, 17.29, 17.50, 17.60, 17.62

Sport and leisure: 53

1.20, 1.40, 1.65, 1.139, 1.140, 2.15, 2.17, 2.71, 2.74, 2.76, 2.82, 2.83, 2.101, 2.106, 3.29, 3.87, 3.88, 4.4, 4.5, 4.12, 4.19, 4.20, 4.21, 4.25, 4.26, 4.48, 4.50, 4.68, 4.71, 4.114, 4.119, 4.120, 5.11, 5.55, 5.70, 6.31, 6.54, 8.5, 8.6, 8.35, 8.39, 8.55, 8.78, 9.23, 14.43, 14.47, 14.54, 15.13, 15.14, 17.6, 17.40, 17.41, 17.42

Technology and Internet: 37

1.13, 1.132, 1.136, 4.8, 4.9, 4.18, 4.29, 4.47, 4.122, 5.4, 5.50, 5.60, 7.63, 7.64, 7.65, 8.7, 8.41, 8.42, 8.54, 8.59, 8.60, 8.61, 9.22, 9.26, 9.48, 9.49, 9.50, 9.51, 10.26, 11.16, 11.17, 11.18, 11.19, 11.20, 11.52, 12.59, 13.43

Statistics is the science of collecting, organizing, and interpreting numerical facts, which we call *data*.

We are bombarded by data in our everyday lives. The news mentions imported car sales, the latest poll of the president's popularity, and the average high temperature for today's date. Advertisements claim that data show the superiority of the advertiser's product. All sides in public debates about economics, education, and social policy argue from data. A knowledge of statistics helps separate sense from nonsense in the flood of data.

The study and collection of data are also important in the work of many professions, so that training in the science of statistics is valuable preparation for a variety of careers. The people featured at the beginning of each chapter in this book illustrate the many kinds of work that are based on statistics.

Each month, for example, government statistical offices release the latest numerical information on unemployment and inflation. Economists and financial advisors, as well as policymakers in government and business, study these data in order to make informed decisions. Doctors must understand the origin and trustworthiness of the data that appear in medical journals. Politicians rely on data from polls of public opinion. Business decisions are based on market research data that reveal consumer tastes. Engineers gather data on the quality and reliability of manufactured products. Most areas of academic study make use of numbers, and therefore also make use of the methods of statistics.

Understanding from Data

The goal of statistics is to gain understanding from data. To gain understanding, we often operate on a set of numbers—we average or graph them, for example. But we must do more, because data are not just numbers; they are numbers that have some context that helps us understand them.

You read that low birth weight is a major reason why infant mortality in the United States is higher than in most other advanced nations. The report goes on to say that 7.8% of children born in the United States have low birth weight, and that 13.4% of black infants have low birth weight.[1] To make sense of these numbers you must know what counts as low birth weight (less than 2500 grams, or 5.5 pounds) and have some feeling for the weights of babies. You probably recognize that 5.5 pounds is small, that 7.5 pounds (3400 grams) is about average, and that 10 pounds (4500 grams) is a big baby.

Another part of the context is the source of the data. How do we know that 7.8% of American babies have low birth weight or that the average weight of newborns is about 3400 grams? The data come from the National Center for Health Statistics, a government office to which the states report information from all birth certificates issued each month. These are the most complete data available about births in the United States.

When you do statistical problems—even straightforward textbook problems—don't just graph or calculate. Think about the context and state your conclusions in the specific setting of the problem. As you are learning how to do statistical calculations and graphs, remember that the goal of statistics is not calculation for its own sake but gaining understanding from numbers. The calculations and graphs can be automated by a calculator or software, but you must supply the understanding. This book presents only the most common specific procedures for statistical analysis. A thorough grasp of the principles of statistics will enable you to quickly learn more advanced methods as needed. Always keep in mind, however, that a fancy computer analysis carried out without attention to basic principles will often produce elaborate nonsense. As you read, seek to understand the principles, as well as the necessary details of methods and recipes.

The Rise of Statistics

Historically, the ideas and methods of statistics developed gradually as society grew interested in collecting and using data for a variety of applications. The earliest origins of statistics lie in the desire of rulers to count the number of inhabitants or measure the value of taxable land in their domains. As the physical sciences developed in the seventeenth and eighteenth centuries, the importance of careful measurements of weights, distances, and other physical quantities grew. Astronomers and surveyors striving for exactness had to deal with variation in their measurements. Many measurements should be better than a single measurement, even though they vary among themselves. How can we best combine many varying observations? Statistical methods that are still important were invented to analyze scientific measurements.

By the nineteenth century, the agricultural, life, and behavioral sciences also began to rely on data to answer fundamental questions. How are the heights of parents and children related? Does a new variety of wheat produce higher yields than the old, and under what conditions of rainfall and fertilizer? Can a person's mental ability and behavior be measured just as we measure height and reaction time? Effective methods for dealing with such questions developed slowly and with much debate.[2]

As methods for producing and understanding data grew in number and sophistication, the new discipline of statistics took shape in the twentieth century. Ideas and techniques that originated in the collection of government data, in the study of astronomical or biological measurements, and in the attempt to understand heredity or intelligence came together to form a unified "science of data." That science of data—statistics—is the topic of this text.

The Organization of This Book

Part I of this book, called "Looking at Data," concerns data analysis and data production. The first two chapters deal with statistical methods for organizing and describing data. These chapters progress from simpler to more complex data. Chapter 1 examines data on a single variable; Chapter 2 is devoted to relationships among two or more variables. You will learn both how to examine data produced by others and how to organize and summarize your own data. These summaries will be first graphical, then numerical, and then, when appropriate, in the form of a mathematical model that gives a compact description of the overall pattern of the data. Chapter 3 outlines arrangements (called designs) for producing data that answer specific questions. The principles presented in this chapter will help you to design proper samples and experiments and to evaluate such investigations in your field of study.

Part II, consisting of Chapters 4 to 8, introduces statistical inference—formal methods for drawing conclusions from properly produced data. Statistical inference uses the language of probability to describe how reliable its conclusions are, so some basic facts about probability are needed to understand inference. Probability is the subject of Chapters 4 and 5. Chapter 6, perhaps the most important chapter in the text, introduces the reasoning of statistical inference. Effective inference is based on good procedures for producing data (Chapter 3), careful examination of the data (Chapters 1 and 2), and an understanding of the nature of statistical inference as discussed in Chapter 6. Chapters 7 and 8 describe some of the most common specific methods of inference, for drawing conclusions about means and proportions from one and two samples.

The five, shorter chapters in Part III introduce somewhat more advanced methods of inference, dealing with relations in categorical data, regression and correlation, and analysis of variance. Supplement chapters, available bound separately and on the CD that accompanies this book, present additional statistical topics.

What Lies Ahead

Introduction to the Practice of Statistics is full of data from many different areas of life and study. Many exercises ask you to express briefly some understanding gained from the data. In practice, you would know much more about the background of the data you work with and about the questions you hope the data will answer. No textbook can be fully realistic. But it is important to form the habit of asking, "What do the data tell me?" rather than just concentrating on making graphs and doing calculations.

You should have some help in automating many of the graphs and calculations. You should certainly have a calculator with basic statistical functions. Look for keywords such as "two-variable statistics" or "regression" when you shop for a calculator. More advanced (and more expensive) calculators will do much more, including some statistical graphs. You may be asked to use software as well. There are many kinds of statistical software, from spreadsheets to large programs for advanced users of statistics. The kind of computing available to learners varies a great deal from place to place—but the big ideas of statistics don't depend on any particular level of access to computing.

Because graphing and calculating are automated in statistical practice, the most important assets you can gain from the study of statistics are an understanding of the big ideas and the beginnings of good judgment in working with data. Ideas and judgment can't (at least yet) be automated. They guide you in telling the computer what to do and in interpreting its output. This book tries to explain the most important ideas of statistics, not just teach methods. Some examples of big ideas that you will meet are "always plot your data," "randomized comparative experiments," and "statistical significance."

You learn statistics by doing statistical problems. "Practice, practice, practice." Be prepared to work problems. The basic principle of learning is persistence. Being organized and persistent is more helpful in reading this book than knowing lots of math. The main ideas of statistics, like the main ideas of any important subject, took a long time to discover and take some time to master. The gain will be worth the pain.

David S. Moore is Shanti S. Gupta Distinguished Professor of Statistics at Purdue University and was 1998 president of the American Statistical Association. He received his A.B. from Princeton and his Ph.D. from Cornell, both in mathematics. He has written many research papers in statistical theory and served on the editorial boards of several major journals. Professor Moore is an elected fellow of the American Statistical Association and of the Institute of Mathematical Statistics and an elected member of the International Statistical Institute. He has served as program director for statistics and probability at the National Science Foundation.

In recent years, Professor Moore has devoted his attention to the teaching of statistics. He was the content developer for the Annenberg/Corporation for Public Broadcasting college-level telecourse *Against All Odds: Inside Statistics* and for the series of video modules *Statistics: Decisions through Data*, intended to aid the teaching of statistics in schools. He is the author of influential articles on statistics education and of several leading texts. Professor Moore has served as president of the International Association for Statistical Education and has received the Mathematical Association of America's national award for distinguished college or university teaching of mathematics.

George P. McCabe is Associate Dean for Academic Affairs and Professor of Statistics at Purdue University. He received his B.S. in mathematics from Providence College and his Ph.D. in mathematical statistics from Columbia University. His entire professional career has been spent at Purdue, with sabbaticals at Princeton, the Commonwealth Scientific and Industrial Research Organization (CSIRO) in Melbourne, Australia, the University of Berne (Switzerland), the National Institute of Standards and Technology (NIST) in Boulder, Colorado, and the National University of Ireland in Galway. Professor McCabe is an elected fellow of the American Statistical Association and was 1998 chair of its section on Statistical Consulting. He has served on the editorial boards of several statistics journals. He has consulted with many major corporations and has testified as an expert witness on the use of statistics in several legal cases.

Professor McCabe's research interests have focused on applications of statistics. Much of his recent work has concerned nutrition, including nutrient requirements, calcium metabolism, and bone health. He is the author or coauthor of more than 140 publications in many different journals.

Looking at Data

STATISTICS IN PRACTICE

Marketing Research: A Voice for Consumers

JENNIFER KARAS *is the Marketing Research Manager at Shell Oil Products Company in Houston, Texas.*

We are all consumers. As consumers, we make decisions every day about whether or not to buy products and services and which companies to buy them from. How do consumers make these purchasing decisions? Do all consumers base their decisions on the same motivations and expectations? As a marketing researcher, I use statistics and statistical thinking to understand consumer attitudes and decisions. With this understanding, I help my company determine which types of gasoline to sell, which services to offer at our stations, and how to advertise these products and services to the public.

Marketing researchers use statistics to place consumers into only a few groups with similar purchasing motivations and behaviors. For example, there are essentially seven types of gasoline consumers, such as those motivated only by price or only by fuel quality or by speed and convenience. Segmenting consumers in this way allows companies to select the types of consumers they want to target, develop products and services that best meet their needs, and create advertising that appeals to these specific consumers.

Marketing researchers also survey the public to keep abreast of trends in behaviors and attitudes. We conduct surveys by phone, mail, and in person across the country every day. These surveys help us monitor how well we are performing, in the public's opinion, compared with our competition. I use statistics to analyze the data from these surveys to gain insight into consumers' minds. These insights help Shell decide what type of fuel we should develop next or what the gas station of the future should look like. Should we offer house calls for fill-ups? Robotic pumps at our stations? It is up to you. You are the consumer and my use of statistics gives you a voice in my company.

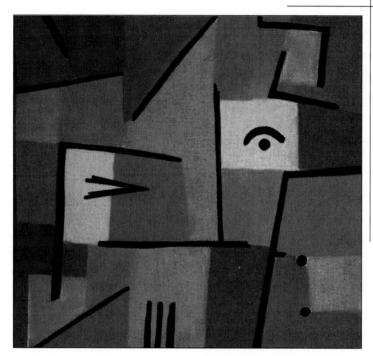

PAUL KLEE, BEWARE OF RED, 1940.
© ARS, NY (ERICH LESSING/ART RESOURCE, NY)

Looking at Data–Distributions

Introduction

Statistics is the science of learning from data. Data are numerical facts. Facts are a better basis for making decisions than just guessing, as the following example illustrates.

EXAMPLE 1.1   The size of the pension payments that a company must make to its employees after they retire has a big impact on the company's finances because funds must be put aside now to cover the future payments. If the employees retire at later ages, less money must be put aside because the fund has more years to grow before it is paid out. US Airways assumed that its pilots would retire at age 60, the age limit for pilots. That was a guess. When the airline declared bankruptcy, a look at the data showed that more than half the pilots retired earlier. The difference between the guess and the data cost the pilots and the government (which supports the pensions of bankrupt companies) a lot of money.[1]

To learn from data, we need more than just the numbers. The numbers in a medical study, for example, mean little without some knowledge of the goals of the study and of what blood pressure, heart rate, and other measurements contribute to those goals. That is, *data are numbers with a context,* and we need to understand the context if we are to make sense of the numbers. On the other hand, measurements from the study's several hundred subjects are of little value to even the most knowledgeable medical expert until the tools of statistics organize, display, and summarize them. We begin our study of statistics by mastering the art of examining data.

Variables

Any set of data contains information about some group of *individuals.* The information is organized in *variables.*

> ### INDIVIDUALS AND VARIABLES
>
> **Individuals** are the objects described by a set of data. Individuals may be people, but they may also be animals or things.
>
> A **variable** is any characteristic of an individual. A variable can take different values for different individuals.

A college's student data base, for example, includes data about every currently enrolled student. The students are the individuals described by the data set. For each individual, the data contain the values of variables such as date of birth, gender (female or male), choice of major, and grade point average. In practice, any set of data is accompanied by background information that helps us understand the data. When you plan a statistical study or explore data from someone else's work, ask yourself the following questions:

1. **Why? What purpose** do the data have? Do we hope to answer some specific questions? Do we want to draw conclusions about individuals other than those for whom we actually have data?

2. **Who?** What **individuals** do the data describe? **How many** individuals appear in the data?

3. **What?** How many **variables** do the data contain? What are the **exact definitions** of these variables? In what **unit of measurement** is each variable recorded? Weights, for example, might be recorded in pounds, in thousands of pounds, or in kilograms.

Some variables, like gender and college major, simply place individuals into categories. Others, like height and grade point average, take numerical values for which we can do arithmetic. It makes sense to give an average salary for a company's employees, but it does not make sense to give an "average" gender. We can, however, count the numbers of female and male employees and do arithmetic with these counts.

CATEGORICAL AND QUANTITATIVE VARIABLES

A **categorical variable** places an individual into one of several groups or categories.

A **quantitative variable** takes numerical values for which arithmetic operations such as adding and averaging make sense.

The **distribution** of a variable tells us what values it takes and how often it takes these values.

EXAMPLE 1.2 Here is a small part of the data set that describes the students in a large statistics class. The data come from the anonymous responses to a class questionnaire.

	A	B	C	D	E	F
1	SEX	HAND	HEIGHT	STUDY	COINS	
2	F	L	65	200	50	
3	M	L	72	30	35	
4	M	R	62	95	35	
5	F	L	64	120	0	
6	M	R	63	220	0	
7	F	R	58	60	76	
8	F	R	67	150	215	
9						

Sheet1 / Sheet2 / Sheet3

Each row records data on one individual. You will often see each row of data called **case** a **case.** Each column contains the values of one variable for all the individuals. There are 5 variables. Sex (female or male) and handedness (left-handed or right-handed) are categorical variables. The remaining three variables are quantitative: height in inches, minutes spent studying on a typical weeknight, and "How many cents in coins (not bills) are you carrying?"

Most statistical software uses this format to enter data—each row is an individual, and each column is a variable. This data set appears in a **spreadsheet** that has rows and columns ready for your use. Spreadsheets are commonly used to enter and transmit data. Most statistical software can read data from spreadsheet programs.

Knowing the context of the data—that these are student responses to a class questionnaire—helps us make sense of them. For example, one student claimed to study 30,000 minutes on a typical night. We know that this is a joke.

Measurement: know your variables

The context of data includes an understanding of the variables that are recorded. Often the variables in a statistical study are easy to understand: height in centimeters, study time in minutes, and so on. But each area of work also has its own special variables. A psychologist uses the Minnesota Multiphasic Personality Inventory (MMPI), and a physical fitness expert measures "VO2 max," the volume of oxygen consumed per minute while exercising at your maximum capacity. Both of these variables are measured with special **instruments.** VO2 max is measured by exercising while breathing into a mouthpiece connected to an apparatus that measures oxygen consumed. Scores on the MMPI are based on a long questionnaire, which is also an instrument. Part of mastering your field of work is learning what variables are important and how they are best measured. Because details of particular measurements usually require knowledge of the particular field of study, we will say little about them.

Be sure that each variable really does measure what you want it to. A poor choice of variables can lead to misleading conclusions. Often, for example, the **rate** at which something occurs is a more meaningful measure than a simple count of occurrences.

EXAMPLE 1.3   The government's Fatal Accident Reporting System says that 27,102 passenger cars were involved in fatal accidents in 2002. Only 3339 motorcycles had fatal accidents that year.[2] Does this mean that motorcycles are safer than cars? Not at all—there are many more cars than motorcycles, so we expect cars to have a higher *count* of fatal accidents.

A better measure of the dangers of driving is a *rate*, the number of fatal accidents divided by the number of vehicles on the road. In 2002, passenger cars had about 21 fatal accidents for each 100,000 vehicles registered. There were about 67 fatal accidents for each 100,000 motorcycles registered. The rate for motorcycles is more than three times the rate for cars. Motorcycles are, as we might guess, much more dangerous than cars.

1.1   Displaying Distributions with Graphs

Statistical tools and ideas help us examine data in order to describe their main features. This examination is called **exploratory data analysis.** Like an explorer crossing unknown lands, we want first to simply describe what we see. Here are two basic strategies that help us organize our exploration of a set of data:

exploratory
data analysis

- Begin by examining each variable by itself. Then move on to study the relationships among the variables.

- Begin with a graph or graphs. Then add numerical summaries of specific aspects of the data.

We will follow these principles in organizing our learning. This chapter presents methods for describing a single variable. We will study relationships among several variables in Chapter 2. Within each chapter, we will begin with graphical displays, then add numerical summaries for more complete description.

Graphs for categorical variables

The values of a categorical variable are labels for the categories, such as "female" and "male." The **distribution** of a categorical variable lists the categories and gives either the **count** or the **percent** of individuals who fall in each category. For example, how well educated are 30-something young adults? Here is the distribution of the highest level of education for people aged 25 to 34 years:[3]

Education	Count (millions)	Percent
Less than high school	4.6	11.8
High school graduate	11.6	30.6
Some college	7.4	19.5
Associate degree	3.3	8.8
Bachelor's degree	8.6	22.7
Advanced degree	2.5	6.6

Are you surprised that only 29.3% of young adults have at least a bachelor's degree?

bar graph

The graphs in Figure 1.1 display these data. The **bar graph** in Figure 1.1(a) quickly compares the sizes of the six education groups. The heights of the bars show the percents in the six categories. The **pie chart** in Figure 1.1(b) helps us see what part of the whole each group forms. For example, the "Bachelor's" slice makes up 22.7% of the pie because 22.7% of young adults have a bachelor's degree but no higher degree. We have moved that slice out to call attention to it. Because pie charts lack a scale, we have added the percents to

pie chart

(a)

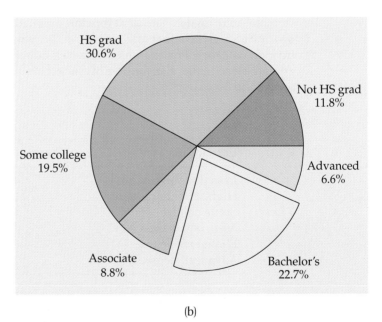

(b)

FIGURE 1.1   (a) Bar graph of the educational attainment of people aged 25 to 34 years. (b) Pie chart of the education data, with bachelor's degree holders emphasized.

the labels for the slices. *Pie charts require that you include all the categories that make up a whole. Use them only when you want to emphasize each category's relation to the whole.* Bar graphs are easier to read and are also more flexible. For example, you can use a bar graph to compare the numbers of students at your college majoring in biology, business, and political science. A pie chart cannot make this comparison because not all students fall into one of these three majors.

Bar graphs and pie charts help an audience grasp a distribution quickly. They are, however, of limited use for data analysis because it is easy to understand data on a single categorical variable, such as highest level of education, without a graph. We will move on to quantitative variables, where graphs are essential tools.

Data analysis in action: don't hang up on me

Many businesses operate call centers to serve customers who want to place an order or make an inquiry. Customers want their requests handled thoroughly. Businesses want to treat customers well, but they also want to avoid wasted time on the phone. They therefore monitor the length of calls and encourage their representatives to keep calls short. Here is an example of the difficulties this policy can cause.

EXAMPLE 1.4 We have data on the length of all 31,492 calls made to the customer service center of a small bank in a month. Table 1.1 displays the lengths of the first 80 calls. The complete data set is *eg01-004.dat,* which you can find on the text CD and Web site.[4]

Take a look at the data in Table 1.1. The numbers are meaningless without some background information. The *individuals* are calls made to the bank's call center. The *variable* recorded is the length of each call. The *units* are seconds. We see that the call lengths vary a great deal. The longest call lasted 2631 seconds, almost 44 minutes. More striking is that 8 of these 80 calls lasted less than 10 seconds. What's going on?

TABLE 1.1

Service times (seconds) for calls to a customer service center

77	289	128	59	19	148	157	203
126	118	104	141	290	48	3	2
372	140	438	56	44	274	479	211
179	1	68	386	2631	90	30	57
89	116	225	700	40	73	75	51
148	9	115	19	76	138	178	76
67	102	35	80	143	951	106	55
4	54	137	367	277	201	52	9
700	182	73	199	325	75	103	64
121	11	9	88	1148	2	465	25

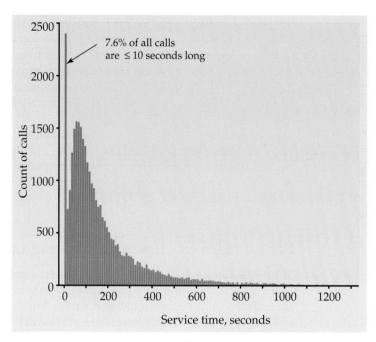

FIGURE 1.2   The distribution of call lengths for 31,492 calls to a bank's customer service center, for Example 1.4. The data show a surprising number of very short calls. These are mostly due to representatives deliberately hanging up in order to bring down their average call length.

Figure 1.2 is a histogram of the lengths of all 31,492 calls. We did not plot the few lengths greater than 1200 seconds (20 minutes). As expected, the graph shows that most calls last between about a minute and 5 minutes, with some lasting much longer when customers have complicated problems. More striking is the fact that 7.6% of all calls are no more than 10 seconds long. It turned out that the bank penalized representatives whose average call length was too long—so some representatives just hung up on customers in order to bring their average length down. Neither the customers nor the bank were happy about this. The bank changed its policy, and later data showed that calls under 10 seconds had almost disappeared.

Our examination of the call center data illustrates some important principles:

- After you understand the background of your data (individuals, variables, units of measurement), the first thing to do is almost always **plot your data.**

- When you look at a plot, look for an **overall pattern** and also for any **striking departures** from the pattern. The overall pattern in Figure 1.2 is made up of the many moderate call lengths and the long right tail of more lengthy calls. The striking departure is the surprising number of very short calls.

We now turn to the kinds of graphs that are used to describe the distribution of a quantitative variable. We will explain how to make the graphs by hand, because knowing this helps you understand what the graphs show.

However, making graphs by hand is so tedious that software is almost essential for effective data analysis unless you have just a few observations.

Stemplots

A *stemplot* (also called a stem-and-leaf plot) gives a quick picture of the shape of a distribution while including the actual numerical values in the graph. Stemplots work best for small numbers of observations that are all greater than 0.

STEMPLOT

To make a **stemplot:**

1. Separate each observation into a **stem** consisting of all but the final (rightmost) digit and a **leaf,** the final digit. Stems may have as many digits as needed, but each leaf contains only a single digit.

2. Write the stems in a vertical column with the smallest at the top, and draw a vertical line at the right of this column.

3. Write each leaf in the row to the right of its stem, in increasing order out from the stem.

EXAMPLE 1.5   The Islamic world is attracting increased attention in Europe and North America. Table 1.2 shows the percent of men and women at least 15 years old who were literate in 2002 in the major Islamic nations. We omitted countries with populations less than 3 million. Data for a few nations, such as Afghanistan and Iraq, are not available.[5]

To make a stemplot of the percents of females who are literate, use the first digits as stems and the second digits as leaves. Algeria's 60% literacy rate, for example, appears as the leaf 0 on the stem 6. Figure 1.3 shows the steps in making the plot.

TABLE 1.2

Literacy rates (percent) in Islamic nations

Country	Female percent	Male percent	Country	Female percent	Male percent
Algeria	60	78	Morocco	38	68
Bangladesh	31	50	Saudi Arabia	70	84
Egypt	46	68	Syria	63	89
Iran	71	85	Tajikistan	99	100
Jordan	86	96	Tunisia	63	83
Kazakhstan	99	100	Turkey	78	94
Lebanon	82	95	Uzbekistan	99	100
Libya	71	92	Yemen	29	70
Malaysia	85	92			

2		2	9		2	9
3		3	1 8		3	1 8
4		4	6		4	6
5		5			5	
6		6	0 3 3		6	0 3 3
7		7	1 1 0 8		7	0 1 1 8
8		8	6 2 5		8	2 5 6
9		9	9 9 9		9	9 9 9
(a)		(b)			(c)	

FIGURE 1.3 Making a stemplot of the data in Example 1.5. (a) Write the stems. (b) Go through the data and write each leaf on the proper stem. For example, the values on the 8 stem are 86, 82, and 85 in the order of the table. (c) Arrange the leaves on each stem in order out from the stem. The 8 stem now has leaves 256.

cluster   The overall pattern of the stemplot is irregular, as is often the case when there are only a few observations. There do appear to be two **clusters** of countries. The plot suggests that we might ask what explains the variation in literacy. For example, why do the three central Asian countries (Kazakhstan, Tajikistan, and Uzbekistan) have very high literacy rates?

back-to-back stemplot   When you wish to compare two related distributions, a **back-to-back stemplot** with common stems is useful. The leaves on each side are ordered out from the common stem. Here is a back-to-back stemplot comparing the distributions of female and male literacy rates in the countries of Table 1.2.

Female		Male
9	2	
8 1	3	
6	4	
	5	0
3 3 0	6	8 8
8 1 1 0	7	0 8
6 5 2	8	3 4 5 9
9 9 9	9	2 2 4 5 6
	10	0 0 0

The values on the left are the female percents, as in Figure 1.3, but ordered out from the stem, from right to left. The values on the right are the male percents. It is clear that literacy is generally higher among males than among females in these countries.

Stemplots do not work well for large data sets, where each stem must hold a large number of leaves. Fortunately, there are two modifications of the basic stemplot that are helpful when plotting the distribution of a moderate splitting stems

trimming

number of observations. You can double the number of stems in a plot by **splitting each stem** into two: one with leaves 0 to 4 and the other with leaves 5 through 9. When the observed values have many digits, it is often best to **trim** the numbers by removing the last digit or digits before making a stemplot. You must use your judgment in deciding whether to split stems and whether to trim, though statistical software will often make these choices for you. Remember that the purpose of a stemplot is to display the shape of a distribution. If a stemplot has fewer than about five stems, you should usually split the stems unless there are few observations. If there are many stems with no leaves or only one leaf, trimming will reduce the number of stems. Here is an example that makes use of both of these modifications.

EXAMPLE 1.6   Return to the 80 customer service call lengths in Table 1.1. To make a stemplot of this distribution, we first trim the call lengths to tens of seconds by dropping the last digit. For example, 56 seconds trims to 5 and 143 seconds trims to 14. (We might also round to the nearest 10 seconds, but trimming is faster than rounding if you must do it by hand.)

We can then use tens of seconds as our leaves, with the digits to the left forming stems. This gives us the single-digit leaves that a stemplot requires. For example, 56 trimmed to 5 becomes leaf 5 on the 0 stem; 143 trimmed to 14 becomes leaf 4 on the 1 stem.

Because we have 80 observations, we split the stems. Thus 56 trimmed to 5 becomes leaf 5 on the second 0 stem, along with all leaves 5 to 9. Leaves 0 to 4 go on the first 0 stem. Figure 1.4 is a stemplot of these data made by software. The software automatically did what we suggest: trimmed to tens of seconds and split stems. To save

Stem-and-Leaf Display: Call Length

```
Stem-and-leaf of C1   N  = 80
Leaf Unit = 10


   0   00000000111233444
   0   555555556667777778889
   1   00001112223344444
   1   57789
   2   0012
   2   7789
   3   2
   3   678


HI 43, 46, 47, 70, 70, 95, 114, 263
```

FIGURE 1.4   Stemplot from Minitab of the 80 call lengths in Table 1.1. The software has trimmed the data by removing the last digit. It has also split stems and listed the highest observations apart from the plot.

space, the software also listed the largest values as "HI," rather than create stems all the way up to 26. The stemplot shows the overall pattern of the distribution, with many short to moderate lengths and some very long calls.

Histograms

histogram

Stemplots display the actual values of the observations. This feature makes stemplots awkward for large data sets. Moreover, the picture presented by a stemplot divides the observations into groups (stems) determined by the number system rather than by judgment. Histograms do not have these limitations. A **histogram** breaks the range of values of a variable into classes and displays only the count or percent of the observations that fall into each class. You can choose any convenient number of classes, but you should always choose classes of equal width. Histograms are slower to construct by hand than stemplots and do not display the actual values observed. For these reasons we prefer stemplots for small data sets. The construction of a histogram is best shown by example. Any statistical software package will of course make a histogram for you.

EXAMPLE 1.7   You have probably heard that the distribution of scores on IQ tests is supposed to be roughly "bell-shaped." Let's look at some actual IQ scores. Table 1.3 displays the IQ scores of 60 fifth-grade students chosen at random from one school.[6]

1. Divide the range of the data into classes of equal width. The scores in Table 1.3 range from 81 to 145, so we choose as our classes

$$75 \le \text{IQ score} < 85$$
$$85 \le \text{IQ score} < 95$$
$$\vdots$$
$$145 \le \text{IQ score} < 155$$

Be sure to specify the classes precisely so that each individual falls into exactly one class. A student with IQ 84 would fall into the first class, but IQ 85 falls into the second.

TABLE 1.3									
IQ test scores for 60 randomly chosen fifth-grade students									
145	139	126	122	125	130	96	110	118	118
101	142	134	124	112	109	134	113	81	113
123	94	100	136	109	131	117	110	127	124
106	124	115	133	116	102	127	117	109	137
117	90	103	114	139	101	122	105	97	89
102	108	110	128	114	112	114	102	82	101

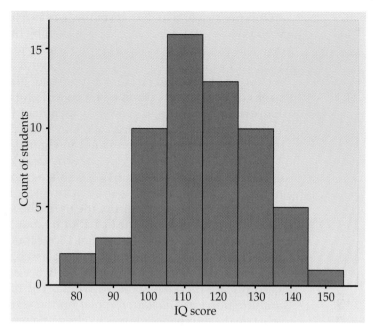

FIGURE 1.5 Histogram of the IQ scores of 60 fifth-grade students, for Example 1.7.

frequency 2. Count the number of individuals in each class. These counts are called **frequencies,** and a table of frequencies for all classes is a **frequency table.**

Class	Count	Class	Count
75 to 84	2	115 to 124	13
85 to 94	3	125 to 134	10
95 to 104	10	135 to 144	5
105 to 114	16	145 to 154	1

3. Draw the histogram. First, on the horizontal axis mark the scale for the variable whose distribution you are displaying. That's IQ score. The scale runs from 75 to 155 because that is the span of the classes we chose. The vertical axis contains the scale of counts. Each bar represents a class. The base of the bar covers the class, and the bar height is the class count. There is no horizontal space between the bars unless a class is empty, so that its bar has height zero. Figure 1.5 is our histogram. It does look roughly "bell-shaped."

Large sets of data are often reported in the form of frequency tables when it is not practical to publish the individual observations. In addition to the frequency (count) for each class, we may be interested in the fraction or percent of the observations that fall in each class. A histogram of percents looks just like a frequency histogram such as Figure 1.5. Simply relabel the vertical scale to read in percents. Use histograms of percents for comparing several distributions with different numbers of observations.

Our eyes respond to the *area* of the bars in a histogram. Because the classes are all the same width, area is determined by height and all classes are fairly represented. There is no one right choice of the classes in a histogram. Too few classes will give a "skyscraper" graph, with all values in a few classes with tall bars. Too many will produce a "pancake" graph, with most classes having one or no observations. Neither choice will give a good picture of the shape of the distribution. You must use your judgment in choosing classes to display the shape. Statistical software will choose the classes for you. The software's choice is often a good one, but you can change it if you want.

You should be aware that the appearance of a histogram can change when you change the classes. Figure 1.6 is a histogram of the customer service call lengths that are also displayed in Figure 1.2. It was produced by software with no special instructions from the user. The software's "default" histogram shows the overall shape of the distribution, but it hides the spike of very short calls by lumping all calls of less than 100 seconds into the first class. We produced Figure 1.2 by asking for smaller classes after Table 1.1 suggested that very short calls might be a problem. Software automates making graphs, but it can't replace thinking about your data. The histogram function in the *One-Variable Statistical Calculator* applet on the text CD and Web site allows you to change the number of classes by dragging with the mouse, so that it is easy to see how the choice of classes affects the histogram.

Although histograms resemble bar graphs, their details and uses are distinct. A histogram shows the distribution of counts or percents among the values of a single variable. A bar graph compares the size of different items. The horizontal axis of a bar graph need not have any measurement scale but

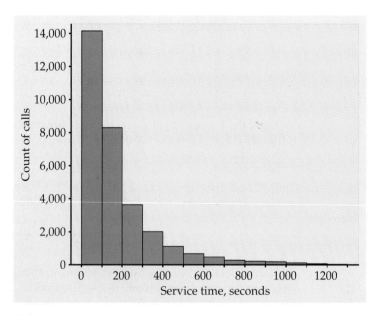

FIGURE 1.6 The "default" histogram produced by software for the call lengths in Example 1.4. This choice of classes hides the large number of very short calls that is revealed by the histogram of the same data in Figure 1.2.

simply identifies the items being compared. Draw bar graphs with blank space between the bars to separate the items being compared. Draw histograms with no space, to indicate that all values of the variable are covered. *Some spreadsheet programs, which are not primarily intended for statistics, will draw histograms as if they were bar graphs, with space between the bars. You can tell the software to eliminate the space in order to produce a proper histogram.*

Examining distributions

Making a statistical graph is not an end in itself. The purpose of the graph is to help us understand the data. After you make a graph, always ask, "What do I see?" Once you have displayed a distribution, you can see its important features as follows.

> ### EXAMINING A DISTRIBUTION
>
> In any graph of data, look for the **overall pattern** and for striking **deviations** from that pattern.
>
> You can describe the overall pattern of a distribution by its **shape, center,** and **spread.**
>
> An important kind of deviation is an **outlier,** an individual value that falls outside the overall pattern.

We will learn how to describe center and spread numerically in Section 1.2. For now, we can describe the center of a distribution by its *midpoint,* the value with roughly half the observations taking smaller values and half taking larger values. We can describe the spread of a distribution by giving the *smallest and largest values*. Stemplots and histograms display the shape of a distribution in the same way. Just imagine a stemplot turned on its side so that the larger values lie to the right. Some things to look for in describing shape are:

modes
unimodal
- Does the distribution have one or several major peaks, called **modes?** A distribution with one major peak is called **unimodal.**

symmetric
skewed
- Is it approximately symmetric or is it skewed in one direction? A distribution is **symmetric** if the values smaller and larger than its midpoint are mirror images of each other. It is **skewed to the right** if the right tail (larger values) is much longer than the left tail (smaller values).

Some variables commonly have distributions with predictable shapes. Many biological measurements on specimens from the same species and sex—lengths of bird bills, heights of young women—have symmetric distributions. Money amounts, on the other hand, usually have right-skewed distributions. There are many moderately priced houses, for example, but the few very expensive mansions give the distribution of house prices a strong right skew.

EXAMPLE 1.8   What does the histogram of IQ scores (Figure 1.5) tell us?
Shape: The distribution is *roughly symmetric* with a *single peak* in the center. We don't expect real data to be perfectly symmetric, so we are satisfied if the two sides of the histogram are roughly similar in shape and extent. **Center:** You can see from the histogram that the midpoint is not far from 110. Looking at the actual data shows that the midpoint is 114. **Spread:** The spread is from 81 to 145. There are no outliers or other strong deviations from the symmetric, unimodal pattern.

The distribution of call lengths in Figure 1.6, on the other hand, is strongly *skewed to the right*. The midpoint, the length of a typical call, is about 115 seconds, or just under 2 minutes. The spread is very large, from 1 second to 28,739 seconds.

The longest few calls are *outliers*. They stand apart from the long right tail of the distribution, though we can't see this from Figure 1.6, which omits the largest observations. The longest call lasted almost 8 hours—that may well be due to equipment failure rather than an actual customer call.

Dealing with outliers

In data sets smaller than the service call data, you can spot outliers by looking for observations that stand apart (either high or low) from the overall pattern of a histogram or stemplot. *Identifying outliers is a matter for judgment. Look for points that are clearly apart from the body of the data, not just the most extreme observations in a distribution.* You should search for an explanation for any outlier. Sometimes outliers point to errors made in recording the data. In other cases, the outlying observation may be caused by equipment failure or other unusual circumstances.

EXAMPLE 1.9   Manufacturing an electronic component requires attaching very fine wires to a semiconductor wafer. If the strength of the bond is weak, the component may fail. Here are measurements on the breaking strength (in pounds) of 23 connections:[7]

0	0	550	750	950	950	1150	1150
1150	1150	1150	1250	1250	1350	1450	1450
1450	1550	1550	1550	1850	2050	3150	

Figure 1.7 is a histogram of these data. We expect the breaking strengths of supposedly identical connections to have a roughly symmetric overall pattern, showing chance variation among the connections. Figure 1.7 does show a symmetric pattern centered at about 1250 pounds—but it also shows three *outliers* that stand apart from this pattern, two low and one high.

The engineers were able to explain all three outliers. The two low outliers had strength 0 because the bonds between the wire and the wafer were not made. The high outlier at 3150 pounds was a measurement error. Further study of the data can simply omit the three outliers. One immediate finding is that the variation in breaking strength is too large—550 pounds to 2050 pounds when we ignore the outliers. The process of bonding wire to wafer must be improved to give more consistent results.

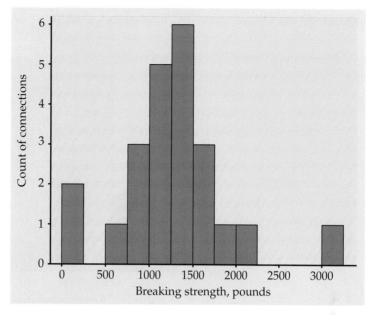

FIGURE 1.7   Histogram of a distribution with both low and high out-liers, for Example 1.9.

Time plots

Whenever data are collected over time, it is a good idea to plot the observations in time order. *Displays of the distribution of a variable that ignore time order, such as stemplots and histograms, can be misleading when there is systematic change over time.*

TIME PLOT

A **time plot** of a variable plots each observation against the time at which it was measured. Always put time on the horizontal scale of your plot and the variable you are measuring on the vertical scale. Connecting the data points by lines helps emphasize any change over time.

EXAMPLE 1.10   Table 1.4 (page 21) lists the volume of water discharged by the Mississippi River into the Gulf of Mexico for each year from 1954 to 2001.[8] The units are cubic kilometers of water—the Mississippi is a big river. Both graphs in Figure 1.8 describe these data. The histogram in Figure 1.8(a) shows the distribution of the volume discharged. The histogram is symmetric and unimodal, with center near 550 cubic kilometers. We might think that the data show just chance year-to-year fluctuation in river level about its long-term average.

   Figure 1.8(b) is a time plot of the same data. For example, the first point lies above 1954 on the "Year" scale at height 290, the volume of water discharged by the Mississippi in 1954. The time plot tells a more interesting story than the histogram. There **trend**   is a great deal of year-to-year variation, but there is also a clear increasing **trend** over

(a)

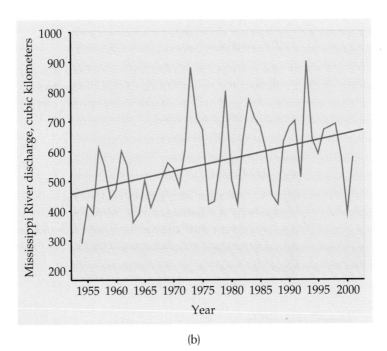

(b)

FIGURE 1.8   (a) Histogram of the volume of water discharged by the Mississippi River over the 48 years from 1954 to 2001. Data from Table 1.4. (b) Time plot of the volume of water discharged by the Mississippi River for the years 1954 to 2001. The line shows the trend toward increasing river flow, a trend that cannot be seen in the histogram in Figure 1.8(a).

TABLE 1.4

Yearly discharge of the Mississippi River, cubic kilometers of water

Year	Discharge	Year	Discharge	Year	Discharge	Year	Discharge
1954	290	1966	410	1978	560	1990	680
1955	420	1967	460	1979	800	1991	700
1956	390	1968	510	1980	500	1992	510
1957	610	1969	560	1981	420	1993	900
1958	550	1970	540	1982	640	1994	640
1959	440	1971	480	1983	770	1995	590
1960	470	1972	600	1984	710	1996	670
1961	600	1973	880	1985	680	1997	680
1962	550	1974	710	1986	600	1998	690
1963	360	1975	670	1987	450	1999	580
1964	390	1976	420	1988	420	2000	390
1965	500	1977	430	1989	630	2001	580

time. That is, there is a long-term rise in the volume of water discharged. The line on the graph is a "trend line" calculated from the data to describe this trend. The trend reflects climate change: rainfall and river flows have been increasing over most of North America.

time series

Many interesting data sets are **time series,** measurements of a variable taken at regular intervals over time. Government economic and social data are often published as time series. Some examples are the monthly unemployment rate and the quarterly gross domestic product. Weather records, the demand for electricity, and measurements on the items produced by a manufacturing process are other examples of time series. Time plots can reveal the main features of a time series.

BEYOND THE BASICS │ Decomposing time series*

When you examine a time plot, again look first for overall patterns and then for striking deviations from those patterns. Here are two important types of overall patterns to look for in a time series.

TREND AND SEASONAL VARIATION

A **trend** in a time series is a persistent, long-term rise or fall.

A pattern in a time series that repeats itself at known regular intervals of time is called **seasonal variation.**

*"Beyond the Basics" sections briefly discuss supplementary topics. Your software may make some of these topics available to you. For example, the results plotted in Figures 1.9 to 1.11 come from the Minitab statistical software.

seasonally adjusted

Because many economic time series show strong seasonal variation, government agencies often adjust for this variation before releasing economic data. The data are then said to be **seasonally adjusted.** Seasonal adjustment helps avoid misinterpretation. A rise in the unemployment rate from December to January, for example, does not mean that the economy is slipping. Unemployment almost always rises in January as temporary holiday help is laid off and outdoor employment in the north drops because of bad weather. The seasonally adjusted unemployment rate reports an increase only if unemployment rises more than normal from December to January.

EXAMPLE 1.11   Figure 1.9 is a time plot of the average retail price of regular gasoline each month for the years 1990 to 2003.[9]   The prices are *not* seasonally adjusted. You can see the upward spike in prices due to the 1990 Iraqi invasion of Kuwait, the drop in 1998 when an economic crisis in Asia reduced demand for fuel, and rapid price increases in 2000 and 2003 due to instability in the Middle East and OPEC production limits. These deviations are so large that overall patterns are hard to see.

There is nonetheless a clear *trend* of increasing price. Much of this trend just reflects inflation, the rise in the overall price level during these years. In addition, a close look at the plot shows *seasonal variation*, a regular rise and fall that recurs each year. Americans drive more in the summer vacation season, so the price of gasoline rises each spring, then drops in the fall as demand goes down.

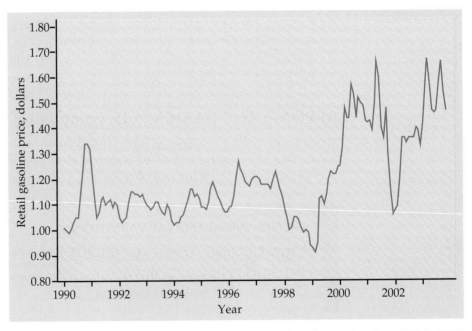

FIGURE 1.9   Time plot of the average monthly price of regular gasoline from 1990 to 2003, for Example 1.11.

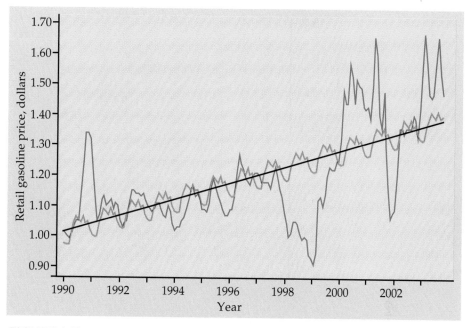

FIGURE 1.10   Time plot of gasoline prices with a trend line and seasonal variation added. These are overall patterns extracted from the data by software.

Statistical software can help us examine a time series by "decomposing" the data into systematic patterns, such as trends and seasonal variation, and the *residuals* that remain after we remove these patterns. Figure 1.10 superimposes the trend and seasonal variation on the time plot of gasoline prices. The black line shows the increasing trend. The seasonal variation appears as the blue line that regularly rises and falls each year. This is an average of the seasonal pattern for all the years in the original data, automatically extracted by software.

The trend and seasonal variation in Figure 1.10 are overall patterns in the data. Figure 1.11 is a plot of the residuals that remain when we subtract both the trend and the seasonal variation from the original data. That is, Figure 1.11 emphasizes the deviations from the pattern. In the case of gasoline prices, the deviations are large (as much as 30 cents both up and down). It is clear that we can't use trend and seasonal variation to predict gasoline prices at all accurately.

SECTION 1.1 | Summary

A data set contains information on a collection of **individuals.** Individuals may be people, animals, or things. The data for one individual make up a **case.** For each individual, the data give values for one or more **variables.** A variable describes some characteristic of an individual, such as a person's height, gender, or salary.

Some variables are **categorical** and others are **quantitative.** A categorical variable places each individual into a category, such as male or female.

FIGURE 1.11   The residuals that remain when we subtract both trend and seasonal variation from monthly gasoline prices.

A quantitative variable has numerical values that measure some characteristic of each individual, such as height in centimeters or annual salary in dollars.

Exploratory data analysis uses graphs and numerical summaries to describe the variables in a data set and the relations among them.

The **distribution** of a variable tells us what values it takes and how often it takes these values.

Bar graphs and **pie charts** display the distributions of categorical variables. These graphs use the counts or percents of the categories.

Stemplots and **histograms** display the distributions of quantitative variables. Stemplots separate each observation into a **stem** and a one-digit **leaf.** Histograms plot the **frequencies** (counts) or percents of equal-width classes of values.

When examining a distribution, look for **shape, center,** and **spread** and for clear **deviations** from the overall shape.

Some distributions have simple shapes, such as **symmetric** or **skewed.** The number of **modes** (major peaks) is another aspect of overall shape. Not all distributions have a simple overall shape, especially when there are few observations.

Outliers are observations that lie outside the overall pattern of a distribution. Always look for outliers and try to explain them.

When observations on a variable are taken over time, make a **time plot** that graphs time horizontally and the values of the variable vertically. A time plot can reveal **trends** or other changes over time.

SECTION 1.1 | Exercises

1.1   Here is a small part of a data set that describes the fuel economy (in miles per gallon) of 2004 model motor vehicles:

Make and model	Vehicle type	Transmission type	Number of cylinders	City MPG	Highway MPG
⋮					
Acura NSX	Two-seater	Automatic	6	17	24
BMW 330I	Compact	Manual	6	20	30
Cadillac Seville	Midsize	Automatic	8	18	26
Ford F150 2WD	Standard pickup truck	Automatic	6	16	19
⋮					

(a) What are the individuals in this data set?

(b) For each individual, what variables are given? Which of these variables are categorical and which are quantitative?

1.2   Here are the first lines of a professor's data set at the end of a statistics course:

```
NAME               MAJOR   POINTS   GRADE

ADVANI, SURA       COMM     397      B
BARTON, DAVID      HIST     323      C
BROWN, ANNETTE     BIOL     446      A
CHIU, SUN          PSYC     405      B
CORTEZ, MARIA      PSYC     461      A
```

What are the individuals and the variables in these data? Which variables are categorical and which are quantitative?

1.3   How can we help wood surfaces resist weathering, especially when restoring historic wooden buildings? A study of this question prepared wooden panels and then exposed them to the weather. Here are some of the variables recorded. Which of these variables are categorical, and which are quantitative?

(a) Type of wood (yellow poplar, pine, cedar)

(b) Water repellent (solvent-based, water-based)

(c) Paint thickness (millimeters)

(d) Paint color (white, gray, light blue)

(e) Weathering time (months)

1.4   You are preparing to study the television-viewing habits of college students. Describe two categorical variables and two quantitative variables that you might measure for each student. Give the units of measurement for the quantitative variables.

1.5   You want to measure the "physical fitness" of college students. Describe several variables you might use to measure fitness. What instrument or instruments does each measurement require?

1.6   Popular magazines rank colleges and universities on their "academic quality" in serving undergraduate students. Describe five variables that you would like to see measured for each college if you were choosing where to study. Give reasons for each of your choices.

1.7   Is driving becoming more dangerous? Traffic deaths declined for years, bottoming out at 39,250 killed in 1992, then began to increase again. In 2002, 42,815 people died in traffic accidents. But more vehicles drove more miles in 2002 than in 1992. In fact, the government says that motor vehicles traveled 2247 billion miles in 1992 and 2830 billion miles in 2002. Reports on transportation use deaths per 100 million miles as a measure of risk. Compare the rates for 1992 and 2002. What do you conclude?

1.8   The formal name for garbage is "municipal solid waste." Here is a breakdown of the materials that made up American municipal solid waste in 2000:[10]

Material	Weight (million tons)	Percent of total
Food scraps	25.9	11.2
Glass	12.8	5.5
Metals	18.0	7.8
Paper, paperboard	86.7	37.4
Plastics	24.7	10.7
Rubber, leather, textiles	15.8	6.8
Wood	12.7	5.5
Yard trimmings	27.7	11.9
Other	7.5	3.2
Total	231.9	100.0

roundoff error

(a) Add the weights for the nine materials given, including "Other." Each entry, including the total, is separately rounded to the nearest tenth. So the sum and the total may differ slightly because of **roundoff error.**

(b) Make a bar graph of the percents. The graph gives a clearer picture of the main contributors to garbage if you order the bars from tallest to shortest.

(c) If you use software, also make a pie chart of the percents. Comparing the two graphs, notice that it is easier to see the small differences among "Food scraps," "Plastics," and "Yard trimmings" in the bar graph.

1.9   What color is your car? Here are the most popular colors for vehicles made in North America during the 2003 model year.[11]

Color	Percent of vehicles
Silver	20.1
White	18.4
Black	11.6
Medium/dark gray	11.5
Light brown	8.8
Medium/dark blue	8.5
Medium red	6.9

What percent of vehicles had other colors? Display these data in a bar graph. Is it also correct to use a pie chart if a category for "Other" is included? If you are using software, make a pie chart if it is correct to do so.

1.10 Favorite vehicle colors differ among types of vehicle. Here are data on the most popular colors in 2003 for luxury cars and for SUVs, trucks, and vans. The entry "—" means "less than 1%." Be creative: make one bar graph that compares the two vehicle types as well as comparing colors. Arrange your graph so that it is easy to compare the two types of vehicle.

Color	Luxury car percent	SUV/truck/van percent
Black	10.9	11.6
Light brown	—	6.3
Medium/dark blue	3.8	9.3
Medium/dark gray	23.3	8.8
Medium/dark green	—	7.0
Medium red	3.9	6.2
White	30.4	22.3
Silver	18.8	17.0

1.11 The Department of Education estimates the "average unmet need" for undergraduate students—the cost of school minus estimated family contributions and financial aid. Here are the averages for full-time students at four types of institution in the 1999–2000 academic year:[12]

Public 2-year	Public 4-year	Private nonprofit 4-year	Private for-profit
$2747	$2369	$4931	$6548

Make a bar graph of these data. Write a one-sentence conclusion about the unmet needs of students. Explain clearly why it is incorrect to make a pie chart.

1.12 The JD Power Initial Quality Survey polls more than 50,000 buyers of new motor vehicles 90 days after their purchase. A two-page questionnaire asks about "things gone wrong." Here are data on problems per 100 vehicles for vehicles made by Toyota and by General Motors in recent years. Toyota has been the industry leader in quality. Make two time plots in the same graph to compare Toyota and GM. What are the most important conclusions you can draw from your graph?

Year	1998	1999	2000	2001	2002	2003	2004
GM	187	179	164	147	130	134	120
Toyota	156	134	116	115	107	115	101

1.13   Email spam is the curse of the Internet. Here is a compilation of the most common types of spam:[13]

Type of spam	Percent
Adult	14.5
Financial	16.2
Health	7.3
Leisure	7.8
Products	21.0
Scams	14.2

Make two bar graphs of these percents, one with bars ordered as in the table (alphabetical) and the other with bars in order from tallest to shortest. Comparisons are easier if you order the bars by height. A bar graph ordered from **Pareto chart**   tallest to shortest bar is sometimes called a **Pareto chart,** after the Italian economist who recommended this procedure.

1.14   Here are the percents of women among students seeking various graduate and professional degrees in the 1999–2000 academic year:[14]

Degree	Percent female
Master's in business administration	39.8
Master's in education	76.2
Other master of arts	59.6
Other master of science	53.0
Doctorate in education	70.8
Other PhD degree	54.2
Medicine (MD)	44.0
Law	50.2
Theology	20.2

(a) Explain clearly why we cannot use a pie chart to display these data.

(b) Make a bar graph of the data. (Comparisons are easier if you order the bars by height.)

1.15 The population of the United States is aging, though less rapidly than in other developed countries. Here is a stemplot of the percents of residents aged 65 and over in the 50 states, according to the 2000 census. The stems are whole percents and the leaves are tenths of a percent.

```
 5 | 7
 6 |
 7 |
 8 | 5
 9 | 6 7 9
10 | 6
11 | 0 2 2 3 3 6 7 7
12 | 0 0 1 1 1 1 3 4 4 5 7 8 9
13 | 0 0 0 1 2 2 3 3 3 4 5 5 6 8
14 | 0 3 4 5 7 9
15 | 3 6
16 |
17 | 6
```

(a) There are two outliers: Alaska has the lowest percent of older residents, and Florida has the highest. What are the percents for these two states?

(b) Ignoring Alaska and Florida, describe the shape, center, and spread of this distribution.

1.16 Make another stemplot of the percent of residents aged 65 and over in the states other than Alaska and Florida by splitting stems 8 to 15 in the plot from the previous exercise. Which plot do you prefer? Why?

1.17 People with diabetes must monitor and control their blood glucose level. The goal is to maintain "fasting plasma glucose" between about 90 and 130 milligrams per deciliter (mg/dl). Here are the fasting plasma glucose levels for 18 diabetics enrolled in a diabetes control class, five months after the end of the class:[15]

| 141 | 158 | 112 | 153 | 134 | 95 | 96 | 78 | 148 |
| 172 | 200 | 271 | 103 | 172 | 359 | 145 | 147 | 255 |

Make a stemplot of these data and describe the main features of the distribution. (You will want to trim and also split stems.) Are there outliers? How well is the group as a whole achieving the goal for controlling glucose levels?

1.18 The study described in the previous exercise also measured the fasting plasma glucose of 16 diabetics who were given individual instruction on diabetes control. Here are the data:

| 128 | 195 | 188 | 158 | 227 | 198 | 163 | 164 |
| 159 | 128 | 283 | 226 | 223 | 221 | 220 | 160 |

Make a back-to-back stemplot to compare the class and individual instruction groups. How do the distribution shapes and success in achieving the glucose control goal compare?

1.19   Figure 1.12 displays the scores of all 947 seventh-grade students in the public schools of Gary, Indiana, on the vocabulary part of the Iowa Test of Basic Skills.[16] Give a brief description of the overall pattern (shape, center, spread) of this distribution.

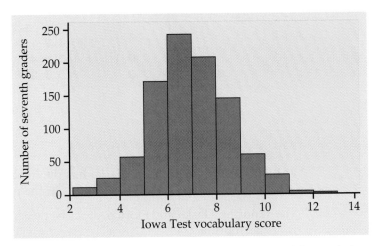

FIGURE 1.12   Histogram of the Iowa Test of Basic Skills vocabulary scores of seventh-grade students in Gary, Indiana, for Exercise 1.19.

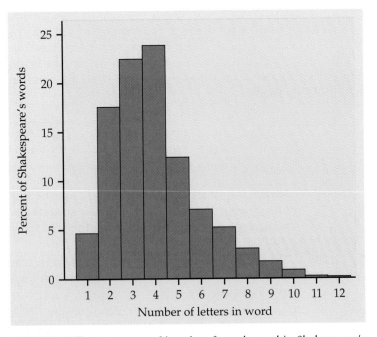

FIGURE 1.13   Histogram of lengths of words used in Shakespeare's plays, for Exercise 1.20.

1.20 Figure 1.13 is a histogram of the lengths of words used in Shakespeare's plays. Because there are so many words in the plays, we use a histogram of percents. What is the overall shape of this distribution? What does this shape say about word lengths in Shakespeare? Do you expect other authors to have word length distributions of the same general shape? Why?

1.21 Jeanna plans to attend college in her home state of Massachusetts. She looks up the tuition and fees for the 2003–2004 academic year for all 56 four-year colleges in Massachusetts (omitting art schools and other special colleges). Figure 1.14 is a histogram of the data. For state schools, Jeanna used the in-state tuition. What is the most important aspect of the overall pattern of this distribution? Why do you think this pattern appears?

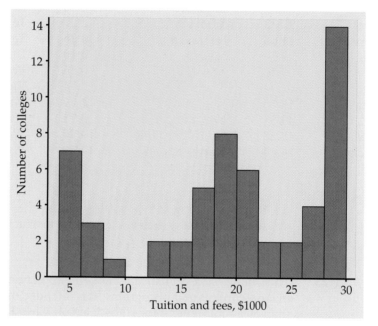

FIGURE 1.14 Histogram of the tuition and fees charged by four-year colleges in Massachusetts, for Exercise 1.21.

1.22 The states differ greatly in the kinds of severe weather that afflict them. Table 1.5 shows the average property damage caused by tornadoes per year over the period from 1950 to 1999 in each of the 50 states and Puerto Rico.[17] (To adjust for the changing buying power of the dollar over time, all damages were restated in 1999 dollars.)

(a) What are the top five states for tornado damage? The bottom five?

(b) Make a histogram of the data, by hand or using software, with classes "$0 \leq$ damage < 10," "$10 \leq$ damage < 20," and so on. Describe the shape, center, and spread of the distribution. Which states may be outliers? (To understand the outliers, note that most tornadoes in largely rural states such as Kansas cause little property damage. Damage to crops is not counted as property damage.)

TABLE 1.5

Average property damage per year due to tornadoes

State	Damage ($millions)	State	Damage ($millions)	State	Damage ($millions)
Alabama	51.88	Louisiana	27.75	Ohio	44.36
Alaska	0.00	Maine	0.53	Oklahoma	81.94
Arizona	3.47	Maryland	2.33	Oregon	5.52
Arkansas	40.96	Massachusetts	4.42	Pennsylvania	17.11
California	3.68	Michigan	29.88	Puerto Rico	0.05
Colorado	4.62	Minnesota	84.84	Rhode Island	0.09
Connecticut	2.26	Mississippi	43.62	South Carolina	17.19
Delaware	0.27	Missouri	68.93	South Dakota	10.64
Florida	37.32	Montana	2.27	Tennessee	23.47
Georgia	51.68	Nebraska	30.26	Texas	88.60
Hawaii	0.34	Nevada	0.10	Utah	3.57
Idaho	0.26	New Hampshire	0.66	Vermont	0.24
Illinois	62.94	New Jersey	2.94	Virginia	7.42
Indiana	53.13	New Mexico	1.49	Washington	2.37
Iowa	49.51	New York	15.73	West Virginia	2.14
Kansas	49.28	North Carolina	14.90	Wisconsin	31.33
Kentucky	24.84	North Dakota	14.69	Wyoming	1.78

(c) If you are using software, also display the "default" histogram that your software makes when you give it no instructions. How does this compare with your graph in (b)?

1.23 The *One-Variable Statistical Calculator* applet on the text CD and Web site will make stemplots and histograms. It is intended mainly as a learning tool rather than as a replacement for statistical software. The histogram function is particularly useful because you can change the number of classes by dragging with the mouse. The tornado damage data from Table 1.5 are available in the applet. Choose this data set and go to the "Histogram" tab.

(a) Sketch the default histogram that the applet first presents. If the default graph does not have nine classes, drag it to make a histogram with nine classes and sketch the result. This should agree with your histogram in part (b) of the previous exercise.

(b) Make a histogram with one class and also a histogram with the greatest number of classes that the applet allows. Sketch the results.

(c) Drag the graph until you find the histogram that you think best pictures the data. How many classes did you choose? Sketch your final histogram.

1.24 Burning fuels in power plants or motor vehicles emits carbon dioxide (CO_2), which contributes to global warming. Table 1.6 displays CO_2 emissions per person from countries with population at least 20 million.[18]

TABLE 1.6

Carbon dioxide emissions, metric tons per person

Country	CO$_2$	Country	CO$_2$	Country	CO$_2$
Algeria	2.3	Italy	7.3	Poland	8.0
Argentina	3.9	Iran	3.8	Romania	3.9
Australia	17.0	Iraq	3.6	Russia	10.2
Bangladesh	0.2	Japan	9.1	Saudi Arabia	11.0
Brazil	1.8	Kenya	0.3	South Africa	8.1
Canada	16.0	Korea, North	9.7	Spain	6.8
China	2.5	Korea, South	8.8	Sudan	0.2
Colombia	1.4	Malaysia	4.6	Tanzania	0.1
Congo	0.0	Mexico	3.7	Thailand	2.5
Egypt	1.7	Morocco	1.0	Turkey	2.8
Ethiopia	0.0	Myanmar	0.2	Ukraine	7.6
France	6.1	Nepal	0.1	United Kingdom	9.0
Germany	10.0	Nigeria	0.3	United States	19.9
Ghana	0.2	Pakistan	0.7	Uzbekistan	4.8
India	0.9	Peru	0.8	Venezuela	5.1
Indonesia	1.2	Philippines	0.9	Vietnam	0.5

(a) Why do you think we choose to measure emissions per person rather than total CO$_2$ emissions for each country?

(b) Display the data of Table 1.6 in a graph. Describe the shape, center, and spread of the distribution. Which countries are outliers?

1.25 Table 1.7 contains data on the mean annual temperatures (degrees Fahrenheit) for the years 1951 to 2000 at two locations in California: Pasadena and Redding.[19]   Make time plots of both time series and compare their main features. You can see why discussions of climate change often bring disagreement.

1.26 Make a histogram of the mean annual temperatures at Pasadena for the years 1951 to 2000. (Data appear in Table 1.7.) Describe the distribution of temperatures. Then explain why this histogram misses very important facts about temperatures in Pasadena.

1.27 The impression that a time plot gives depends on the scales you use on the two axes. If you stretch the vertical axis and compress the time axis, change appears to be more rapid. Compressing the vertical axis and stretching the time axis make change appear slower. Make two more time plots of the data for Pasadena in Table 1.7, one that makes mean temperature appear to increase very rapidly and one that shows only a slow increase. The moral of this exercise is: *pay close attention to the scales when you look at a time plot.*

TABLE 1.7

Mean annual temperatures (°F) in two California cities

Year	Mean temperature Pasadena	Redding	Year	Mean temperature Pasadena	Redding
1951	62.27	62.02	1976	64.23	63.51
1952	61.59	62.27	1977	64.47	63.89
1953	62.64	62.06	1978	64.21	64.05
1954	62.88	61.65	1979	63.76	60.38
1955	61.75	62.48	1980	65.02	60.04
1956	62.93	63.17	1981	65.80	61.95
1957	63.72	62.42	1982	63.50	59.14
1958	65.02	64.42	1983	64.19	60.66
1959	65.69	65.04	1984	66.06	61.72
1960	64.48	63.07	1985	64.44	60.50
1961	64.12	63.50	1986	65.31	61.76
1962	62.82	63.97	1987	64.58	62.94
1963	63.71	62.42	1988	65.22	63.70
1964	62.76	63.29	1989	64.53	61.50
1965	63.03	63.32	1990	64.96	62.22
1966	64.25	64.51	1991	65.60	62.73
1967	64.36	64.21	1992	66.07	63.59
1968	64.15	63.40	1993	65.16	61.55
1969	63.51	63.77	1994	64.63	61.63
1970	64.08	64.30	1995	65.43	62.62
1971	63.59	62.23	1996	65.76	62.93
1972	64.53	63.06	1997	66.72	62.48
1973	63.46	63.75	1998	64.12	60.23
1974	63.93	63.80	1999	64.85	61.88
1975	62.36	62.66	2000	66.25	61.58

1.28   "Recruitment," the addition of new members to a fish population, is an important measure of the health of ocean ecosystems. Here are data on the recruitment of rock sole in the Bering Sea between 1973 and 2000:[20]

Year	Recruitment (millions)	Year	Recruitment (millions)	Year	Recruitment (millions)	Year	Recruitment (millions)
1973	173	1980	1411	1987	4700	1994	505
1974	234	1981	1431	1988	1702	1995	304
1975	616	1982	1250	1989	1119	1996	425
1976	344	1983	2246	1990	2407	1997	214
1977	515	1984	1793	1991	1049	1998	385
1978	576	1985	1793	1992	505	1999	445
1979	727	1986	2809	1993	998	2000	676

(a) Make a graph to display the distribution of rock sole recruitment, then describe the pattern and any striking deviations that you see.

(b) Make a time plot of recruitment and describe its pattern. As is often the case with time series data, a time plot is needed to understand what is happening.

1.29   Asian culture does not emphasize thinness, but young Asians are often influenced by Western culture. In a study of concerns about weight among young Korean women, researchers administered the Drive for Thinness scale (a questionnaire) to 264 female college students in Seoul, South Korea. The data are in the file *ex01-029.dat* on the text CD and Web site.[21] Drive for Thinness measures excessive concern with weight and dieting and fear of weight gain. Roughly speaking, a score of 15 is typical of Western women with eating disorders but is unusually high (90th percentile) for other Western women. Graph the data and describe the shape, center, and spread of the distribution of Drive for Thinness scores for these Korean students. Are there any outliers?

1.30   Changing the choice of classes can change the appearance of a histogram. Here is an example in which a small shift in the classes, with no change in the number of classes, has an important effect on the histogram. The data are the acidity levels (measured by pH) in 105 samples of rainwater. Distilled water has pH 7.00. As the water becomes more acid, the pH goes down. The pH of rainwater is important to environmentalists because of the problem of acid rain.[22]

4.33	4.38	4.48	4.48	4.50	4.55	4.59	4.59	4.61	4.61
4.75	4.76	4.78	4.82	4.82	4.83	4.86	4.93	4.94	4.94
4.94	4.96	4.97	5.00	5.01	5.02	5.05	5.06	5.08	5.09
5.10	5.12	5.13	5.15	5.15	5.15	5.16	5.16	5.16	5.18
5.19	5.23	5.24	5.29	5.32	5.33	5.35	5.37	5.37	5.39
5.41	5.43	5.44	5.46	5.46	5.47	5.50	5.51	5.53	5.55
5.55	5.56	5.61	5.62	5.64	5.65	5.65	5.66	5.67	5.67
5.68	5.69	5.70	5.75	5.75	5.75	5.76	5.76	5.79	5.80
5.81	5.81	5.81	5.81	5.85	5.85	5.90	5.90	6.00	6.03
6.03	6.04	6.04	6.05	6.06	6.07	6.09	6.13	6.21	6.34
6.43	6.61	6.62	6.65	6.81					

(a) Make a histogram of pH with 14 classes, using class boundaries 4.2, 4.4, ..., 7.0. How many modes does your histogram show? More than one mode suggests that the data contain groups that have different distributions.

(b) Make a second histogram, also with 14 classes, using class boundaries 4.14, 4.34, ..., 6.94. The classes are those from (a) moved 0.06 to the left. How many modes does the new histogram show?

(c) Use your software's histogram function to make a histogram without specifying the number of classes or their boundaries. How does the software's default histogram compare with those in (a) and (b)?

1.31   A survey of a large college class asked the following questions:

1. Are you female or male? (In the data, male = 0, female = 1.)

2. Are you right-handed or left-handed? (In the data, right = 0, left = 1.)

3. What is your height in inches?

4. How many minutes do you study on a typical weeknight?

Figure 1.15 shows histograms of the student responses, in scrambled order and without scale markings. Which histogram goes with each variable? Explain your reasoning.

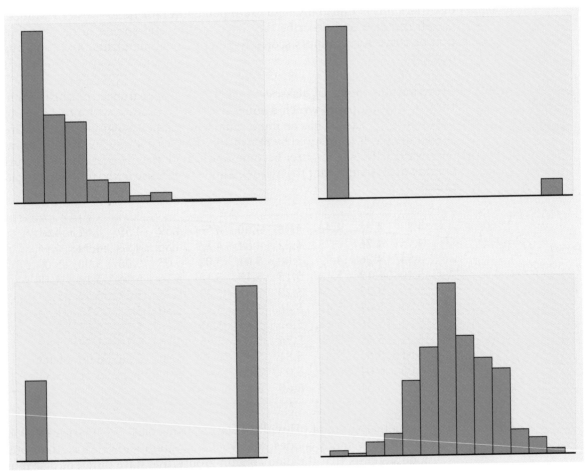

FIGURE 1.15   Match each histogram with its variable, for Exercise 1.31

1.32   Sketch a histogram for a distribution that is skewed to the left. Suppose that you and your friends emptied your pockets of coins and recorded the year marked on each coin. The distribution of dates would be skewed to the left. Explain why.

1.33   How much oil the wells in a given field will ultimately produce is key information in deciding whether to drill more wells. Here are the estimated total amounts of oil recovered from 64 wells in the Devonian Richmond Dolomite area of the Michigan basin, in thousands of barrels:[23]

21.71	53.2	46.4	42.7	50.4	97.7	103.1	51.9	43.4	69.5
156.5	34.6	37.9	12.9	2.5	31.4	79.5	26.9	18.5	14.7
32.9	196.0	24.9	118.2	82.2	35.1	47.6	54.2	63.1	69.8
57.4	65.6	56.4	49.4	44.9	34.6	92.2	37.0	58.8	21.3
36.6	64.9	14.8	17.6	29.1	61.4	38.6	32.5	12.0	28.3
204.9	44.5	10.3	37.7	33.7	81.1	12.1	20.1	30.5	7.1
10.1	18.0	3.0	2.0						

Graph the distribution and describe its main features.

1.34   In 1798 the English scientist Henry Cavendish measured the density of the earth by careful work with a torsion balance. The variable recorded was the density of the earth as a multiple of the density of water. Here are Cavendish's 29 measurements:[24]

5.50	5.61	4.88	5.07	5.26	5.55	5.36	5.29	5.58	5.65
5.57	5.53	5.62	5.29	5.44	5.34	5.79	5.10	5.27	5.39
5.42	5.47	5.63	5.34	5.46	5.30	5.75	5.68	5.85	

Present these measurements graphically by either a stemplot or a histogram and explain the reason for your choice. Then briefly discuss the main features of the distribution. In particular, what is your estimate of the density of the earth based on these measurements?

1.35   Do women study more than men? We asked the students in a large first-year college class how many minutes they studied on a typical weeknight. Here are the responses of random samples of 30 women and 30 men from the class:

Women					Men				
180	120	180	360	240	90	120	30	90	200
120	180	120	240	170	90	45	30	120	75
150	120	180	180	150	150	120	60	240	300
200	150	180	150	180	240	60	120	60	30
120	60	120	180	180	30	230	120	95	150
90	240	180	115	120	0	200	120	120	180

(a) Examine the data. Why are you not surprised that most responses are multiples of 10 minutes? We eliminated one student who claimed to study 30,000 minutes per night. Are there any other responses you consider suspicious?

(b) Make a back-to-back stemplot of these data. Report the approximate midpoints of both groups. Does it appear that women study more than men (or at least claim that they do)?

1.36   Table 1.8 gives the survival times in days of 72 guinea pigs after they were injected with tubercle bacilli in a medical experiment.[25] Make a suitable graph and describe the shape, center, and spread of the distribution of survival times. Are there any outliers?

TABLE 1.8									
Survival times (days) of guinea pigs in a medical experiment									
43	45	53	56	56	57	58	66	67	73
74	79	80	80	81	81	81	82	83	83
84	88	89	91	91	92	92	97	99	99
100	100	101	102	102	102	103	104	107	108
109	113	114	118	121	123	126	128	137	138
139	144	145	147	156	162	174	178	179	184
191	198	211	214	243	249	329	380	403	511
522	598								

1.37   Table 1.9 presents data on 78 seventh-grade students in a rural midwestern school.[26] The researcher was interested in the relationship between the students' "self-concept" and their academic performance. The data we give here include each student's grade point average (GPA), score on a standard IQ test, and gender, taken from school records. Gender is coded as F for female and M for male. The students are identified only by an observation number (OBS). The missing OBS numbers show that some students dropped out of the study. The final variable is each student's score on the Piers-Harris Children's Self-Concept Scale, a psychological test administered by the researcher.

(a) How many variables does this data set contain? Which are categorical variables and which are quantitative variables?

(b) Make a stemplot of the distribution of GPA, after rounding to the nearest tenth of a point.

(c) Describe the shape, center, and spread of the GPA distribution. Identify any suspected outliers from the overall pattern.

(d) Make a back-to-back stemplot of the rounded GPAs for female and male students. Write a brief comparison of the two distributions.

1.38   Make a graph of the distribution of IQ scores for the seventh-grade students in Table 1.9. Describe the shape, center, and spread of the distribution, as well as any outliers. IQ scores are usually said to be centered at 100. Is the midpoint for these students close to 100, clearly above, or clearly below?

1.39   Based on a suitable graph, briefly describe the distribution of self-concept scores for the students in Table 1.9. Be sure to identify any suspected outliers.

TABLE 1.9

Educational data for 78 seventh-grade students

OBS	GPA	IQ	Gender	Self-concept	OBS	GPA	IQ	Gender	Self-concept
001	7.940	111	M	67	043	10.760	123	M	64
002	8.292	107	M	43	044	9.763	124	M	58
003	4.643	100	M	52	045	9.410	126	M	70
004	7.470	107	M	66	046	9.167	116	M	72
005	8.882	114	F	58	047	9.348	127	M	70
006	7.585	115	M	51	048	8.167	119	M	47
007	7.650	111	M	71	050	3.647	97	M	52
008	2.412	97	M	51	051	3.408	86	F	46
009	6.000	100	F	49	052	3.936	102	M	66
010	8.833	112	M	51	053	7.167	110	M	67
011	7.470	104	F	35	054	7.647	120	M	63
012	5.528	89	F	54	055	0.530	103	M	53
013	7.167	104	M	54	056	6.173	115	M	67
014	7.571	102	F	64	057	7.295	93	M	61
015	4.700	91	F	56	058	7.295	72	F	54
016	8.167	114	F	69	059	8.938	111	F	60
017	7.822	114	F	55	060	7.882	103	F	60
018	7.598	103	F	65	061	8.353	123	M	63
019	4.000	106	M	40	062	5.062	79	M	30
020	6.231	105	F	66	063	8.175	119	M	54
021	7.643	113	M	55	064	8.235	110	M	66
022	1.760	109	M	20	065	7.588	110	M	44
024	6.419	108	F	56	068	7.647	107	M	49
026	9.648	113	M	68	069	5.237	74	F	44
027	10.700	130	F	69	071	7.825	105	M	67
028	10.580	128	M	70	072	7.333	112	F	64
029	9.429	128	M	80	074	9.167	105	M	73
030	8.000	118	M	53	076	7.996	110	M	59
031	9.585	113	M	65	077	8.714	107	F	37
032	9.571	120	F	67	078	7.833	103	F	63
033	8.998	132	F	62	079	4.885	77	M	36
034	8.333	111	F	39	080	7.998	98	F	64
035	8.175	124	M	71	083	3.820	90	M	42
036	8.000	127	M	59	084	5.936	96	F	28
037	9.333	128	F	60	085	9.000	112	F	60
038	9.500	136	M	64	086	9.500	112	F	70
039	9.167	106	M	71	087	6.057	114	M	51
040	10.140	118	F	72	088	6.057	93	F	21
041	9.999	119	F	54	089	6.938	106	M	56

1.40   Women were allowed to enter the Boston Marathon in 1972. Here are the times (in minutes, rounded to the nearest minute) for the winning woman from 1972 to 2003:

Year	Time	Year	Time	Year	Time	Year	Time
1972	190	1980	154	1988	145	1996	147
1973	186	1981	147	1989	144	1997	146
1974	167	1982	150	1990	145	1998	143
1975	162	1983	143	1991	144	1999	143
1976	167	1984	149	1992	144	2000	146
1977	168	1985	154	1993	145	2001	144
1978	165	1986	145	1994	142	2002	141
1979	155	1987	146	1995	145	2003	145

Make a graph that shows change over time. What overall pattern do you see? Have times stopped improving in recent years? If so, when did improvement end?

1.2   Describing Distributions with Numbers

Interested in a sporty car? Worried that it may use too much gas? The Environmental Protection Agency lists most such vehicles in its "two-seater" or "minicompact" categories. Table 1.10 gives the city and highway gas mileage for cars in these groups.[27]   (The mileages are for the basic engine and transmission combination for each car.) We want to compare two-seaters with minicompacts and city mileage with highway mileage. We can begin with graphs, but numerical summaries make the comparisons more specific.

A brief description of a distribution should include its *shape* and numbers describing its *center* and *spread*. We describe the shape of a distribution based on inspection of a histogram or a stemplot. Now we will learn specific ways to use numbers to measure the center and spread of a distribution. We can calculate these numerical measures for any quantitative variable. But to interpret measures of center and spread, and to choose among the several measures we will learn, you must think about the shape of the distribution and the meaning of the data. The numbers, like graphs, are aids to understanding, not "the answer" in themselves.

Measuring center: the mean

Numerical description of a distribution begins with a measure of its center or average. The two common measures of center are the *mean* and the *median*. The mean is the "average value" and the median is the "middle value." These are two different ideas for "center," and the two measures behave differently. We need precise recipes for the mean and the median.

TABLE 1.10

Fuel economy (miles per gallon) for model year 2004 vehicles

Two-seater cars			Minicompact cars		
Model	City	Highway	Model	City	Highway
Acura NSX	17	24	Aston Martin Vanquish	12	19
Audi TT Roadster	20	28	Audi TT Coupe	21	29
BMW Z4 Roadster	20	28	BMW 325CI	19	27
Cadillac XLR	17	25	BMW 330CI	19	28
Chevrolet Corvette	18	25	BMW M3	16	23
Dodge Viper	12	20	Jaguar XK8	18	26
Ferrari 360 Modena	11	16	Jaguar XKR	16	23
Ferrari Maranello	10	16	Lexus SC 430	18	23
Ford Thunderbird	17	23	Mini Cooper	25	32
Honda Insight	60	66	Mitsubishi Eclipse	23	31
Lamborgini Gallardo	9	15	Mitsubishi Spyder	20	29
Lamborgini Murcielago	9	13	Porsche Cabriolet	18	26
Lotus Esprit	15	22	Porsche Turbo 911	14	22
Maserati Spyder	12	17			
Mazda Miata	22	28			
Mercedes-Benz SL500	16	23			
Mercedes-Benz SL600	13	19			
Nissan 350Z	20	26			
Porsche Boxster	20	29			
Porsche Carrera 911	15	23			
Toyota MR2	26	32			

THE MEAN $\bar{x}$

To find the **mean $\bar{x}$** of a set of observations, add their values and divide by the number of observations. If the n observations are $x_1, x_2, \ldots, x_n$, their mean is

$$\bar{x} = \frac{x_1 + x_2 + \cdots + x_n}{n}$$

or, in more compact notation,

$$\bar{x} = \frac{1}{n} \sum x_i$$

The $\sum$ (capital Greek sigma) in the formula for the mean is short for "add them all up." The bar over the x indicates the mean of all the x-values. Pronounce the mean $\bar{x}$ as "x-bar." This notation is so common that writers who are discussing data use $\bar{x}$, $\bar{y}$, etc. without additional explanation. The subscripts on the observations x_i are just a way of keeping the n observations distinct. They do not necessarily indicate order or any other special facts about the data.

EXAMPLE 1.12   The mean highway mileage for the 21 two-seaters in Table 1.10 is

$$\bar{x} = \frac{x_1 + x_2 + \cdots + x_n}{n}$$

$$= \frac{24 + 28 + 28 + \cdots + 32}{21}$$

$$= \frac{518}{21} = 24.7 \text{ miles per gallon}$$

In practice, you can key the data into your calculator and hit the $\bar{x}$ key.

The data for Example 1.12 contain an outlier: the Honda Insight is a hybrid gas-electric car that doesn't belong in the same category as the 20 gasoline-powered two-seater cars. If we exclude the Insight, the mean highway mileage drops to 22.6 MPG. The single outlier adds more than 2 MPG to the mean highway mileage. This illustrates an important weakness of the mean as a measure of center: *the mean is sensitive to the influence of a few extreme observations.* These may be outliers, but a skewed distribution that has no outliers will also pull the mean toward its long tail. Because the mean cannot resist the influence of extreme observations, we say that it is not a

resistant measure **resistant measure** of center. A measure that is resistant does more than limit the influence of outliers. Its value does not respond strongly to changes in a few observations, no matter how large those changes may be. The mean fails this requirement because we can make the mean as large as we wish by making a large enough increase in just one observation.

Measuring center: the median

We used the midpoint of a distribution as an informal measure of center in the previous section. The *median* is the formal version of the midpoint, with a specific rule for calculation.

THE MEDIAN *M*

The **median *M*** is the midpoint of a distribution, the number such that half the observations are smaller and the other half are larger. To find the median of a distribution:

1. Arrange all observations in order of size, from smallest to largest.

2. If the number of observations n is odd, the median M is the center observation in the ordered list. Find the location of the median by counting $(n + 1)/2$ observations up from the bottom of the list.

3. If the number of observations n is even, the median M is the mean of the two center observations in the ordered list. The location of the median is again $(n + 1)/2$ from the bottom of the list.

Note that the formula $(n + 1)/2$ does *not* give the median, just the location of the median in the ordered list. Medians require little arithmetic, so they are easy to find by hand for small sets of data. Arranging even a moderate number of observations in order is tedious, however, so that finding the median by hand for larger sets of data is unpleasant. Even simple calculators have an $\bar{x}$ button, but you will need computer software or a graphing calculator to automate finding the median.

EXAMPLE 1.13   To find the median highway mileage for 2004 model two-seater cars, arrange the data in increasing order:

$$13\ \ 15\ \ 16\ \ 16\ \ 17\ \ 19\ \ 20\ \ 22\ \ 23\ \ 23\ \ \mathbf{23}\ \ 24\ \ 25\ \ 25\ \ 26\ \ 28\ \ 28\ \ 28\ \ 29\ \ 32\ \ 66$$

Be sure to list *all* observations, even if they repeat the same value. The median is the bold 23, the 11th observation in the ordered list. You can find the median by eye—there are 10 observations to the left and 10 to the right. Or you can use the recipe $(n + 1)/2 = 22/2 = 11$ to locate the median in the list.

What happens if we drop the Honda Insight? The remaining 20 cars have highway mileages

$$13\ \ 15\ \ 16\ \ 16\ \ 17\ \ 19\ \ 20\ \ 22\ \ 23\ \ \mathbf{23}\ \ \mathbf{23}\ \ 24\ \ 25\ \ 25\ \ 26\ \ 28\ \ 28\ \ 28\ \ 29\ \ 32$$

Because the number of observations $n = 20$ is even, there is no center observation. There is a center pair—the bold pair of 23s have 9 observations to their left and 9 to their right. The median M is the mean of the center pair, which is 23. The recipe $(n + 1)/2 = 21/2 = 10.5$ for the position of the median in the list says that the median is at location "ten and one-half," that is, halfway between the 10th and 11th observations.

You see that the median is more resistant than the mean. Removing the Honda Insight did not change the median at all. Even if we mistakenly enter the Insight's mileage as 660 rather than 66, the median remains 23. The very high value is simply one observation to the right of center. The *Mean and Median* applet on the text CD and Web site is an excellent way to compare the resistance of M and $\bar{x}$. See Exercises 1.55 to 1.57 for use of this applet.

Mean versus median

The median and mean are the most common measures of the center of a distribution. The mean and median of a symmetric distribution are close together. If the distribution is exactly symmetric, the mean and median are exactly the same. In a skewed distribution, the mean is farther out in the long tail than is the median. For example, the distribution of the sizes of the endowments of colleges and universities is strongly skewed to the right. Most institutions have modest endowments, but a few are very wealthy. The median endowment of colleges and universities in 2003 was $70 million—but the mean endowment was over $320 million. The few wealthy institutions pulled the mean up but did not affect the median. *Don't confuse the "average" value of a variable (the mean) with its "typical" value, which we might describe by the median.*

We can now give a better answer to the question of how to deal with outliers in data. First, look at the data to identify outliers and investigate their causes. You can then correct outliers if they are wrongly recorded, delete them for good reason, or otherwise give them individual attention. The three outliers in Figure 1.7 (page 19) can all be dropped from the data once we discover why they appear. If you have no clear reason to drop outliers, you may want to use resistant methods, so that outliers have little influence over your conclusions. The choice is often a matter for judgment. The government's fuel economy guide lists the Honda Insight with the other two-seaters in Table 1.10. We might choose to report median rather than mean gas mileage for all two-seaters to avoid giving too much influence to one car model. In fact, we think that the Insight doesn't belong, so we will omit it from further analysis of these data.

Measuring spread: the quartiles

A measure of center alone can be misleading. Two nations with the same median family income are very different if one has extremes of wealth and poverty and the other has little variation among families. A drug with the correct mean concentration of active ingredient is dangerous if some batches are much too high and others much too low. We are interested in the *spread* or *variability* of incomes and drug potencies as well as their centers. **The simplest useful numerical description of a distribution consists of both a measure of center and a measure of spread.**

percentile

We can describe the spread or variability of a distribution by giving several percentiles. The ***p*th percentile** of a distribution is the value such that *p* percent of the observations fall at or below it. The median is just the 50th percentile, so the use of percentiles to report spread is particularly appropriate when the median is our measure of center. The most commonly used percentiles other than the median are the *quartiles*. The first quartile is the 25th percentile, and the third quartile is the 75th percentile. (The second quartile is the median itself.) To calculate a percentile, arrange the observations in increasing order and count up the required percent from the bottom of the list. Our definition of percentiles is a bit inexact, because there is not always a value with exactly *p* percent of the data at or below it. We will be content to take the nearest observation for most percentiles, but the quartiles are important enough to require an exact rule.

THE QUARTILES Q_1 AND Q_3

To calculate the quartiles:

1. Arrange the observations in increasing order and locate the median *M* in the ordered list of observations.

2. The **first quartile Q_1** is the median of the observations whose position in the ordered list is to the left of the location of the overall median.

3. The **third quartile Q_3** is the median of the observations whose position in the ordered list is to the right of the location of the overall median.

EXAMPLE 1.14 The highway mileages of the 20 gasoline-powered two-seater cars, arranged in increasing order, are

$$13\ 15\ 16\ 16\ 17\ 19\ 20\ 22\ 23\ 23\ |\ 23\ 24\ 25\ 25\ 26\ 28\ 28\ 28\ 29\ 32$$

The median is midway between the center pair of observations. We have marked its position in the list by |. The first quartile is the median of the 10 observations to the left of the position of the median. Check that its value is $Q_1 = 18$. Similarly, the third quartile is the median of the 10 observations to the right of the |. Check that $Q_3 = 27$.

When there is an odd number of observations, the median is the unique center observation, and the rule for finding the quartiles excludes this center value. The highway mileages of the 13 minicompact cars in Table 1.10 are (in order)

$$19\ 22\ 23\ 23\ 23\ 26\ \mathbf{26}\ 27\ 28\ 29\ 29\ 31\ 32$$

The median is the bold 26. The first quartile is the median of the 6 observations falling to the left of this point in the list, $Q_1 = 23$. Similarly, $Q_3 = 29$.

We find other percentiles more informally if we are working without software. For example, we take the 90th percentile of the 13 minicompact mileages to be the 12th in the ordered list, because $0.90 \times 13 = 11.7$, which we round to 12. The 90th percentile is therefore 31 MPG.

EXAMPLE 1.15 Statistical software often provides several numerical measures in response to a single command. Figure 1.16 displays such output from the Crunch It! and Minitab software for the highway mileages of two-seater cars (without the Honda Insight).

Summary statistics

Column	n	Mean	Variance	Std. Dev.	Std. Err.	Median	Range	Min	Max	Q1	Q3
Hwy mpg	20	22.6	27.936842	5.2855315	1.1818807	23	19	13	32	18	27

(a)

(b)

FIGURE 1.16 Numerical descriptions of the highway gas mileage of two-seater cars from software, for Example 1.15. (a) Crunch It! (b) Minitab.

Both tell us that there are 20 observations and give the mean, median, quartiles, and smallest and largest data values. Both also give other measures, some of which we will meet soon. Crunch It! is basic online software that offers no choice of output. Minitab allows you to choose the descriptive measures you want from a long list.

The quartiles from Crunch It! agree with our values from Example 1.14. But Minitab's quartiles are a bit different. For example, our rule for hand calculation gives first quartile $Q_1 = 18$. Minitab's value is $Q_1 = 17.5$. *There are several rules for calculating quartiles, which often give slightly different values. The differences are always small. For describing data, just report the values that your software gives.*

The five-number summary and boxplots

In Section 1.1, we used the smallest and largest observations to indicate the spread of a distribution. These single observations tell us little about the distribution as a whole, but they give information about the tails of the distribution that is missing if we know only Q_1, M, and Q_3. To get a quick summary of both center and spread, combine all five numbers.

> ### THE FIVE-NUMBER SUMMARY
>
> The **five-number summary** of a set of observations consists of the smallest observation, the first quartile, the median, the third quartile, and the largest observation, written in order from smallest to largest. In symbols, the five-number summary is
>
> Minimum Q_1 M Q_3 Maximum

These five numbers offer a reasonably complete description of center and spread. The five-number summaries for highway gas mileages are

$$13 \ 18 \ 23 \ 27 \ 32$$

for two-seaters and

$$19 \ 23 \ 26 \ 29 \ 32$$

for minicompacts. The median describes the center of the distribution; the quartiles show the spread of the center half of the data; the minimum and maximum show the full spread of the data. The five-number summary leads to another visual representation of a distribution, the *boxplot*. Figure 1.17 shows boxplots for both city and highway gas mileages for our two groups of cars.

> ### BOXPLOT
>
> A **boxplot** is a graph of the five-number summary.
>
> - A central box spans the quartiles Q_1 and Q_3.
>
> - A line in the box marks the median M.
>
> - Lines extend from the box out to the smallest and largest observations.

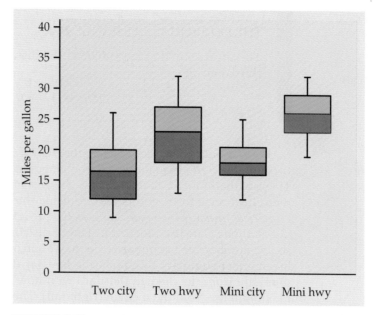

FIGURE 1.17   Boxplots of the highway and city gas mileages for cars classified as two-seaters and as minicompacts by the Environmental Protection Agency.

Because boxplots show less detail than histograms or stemplots, they are best used for side-by-side comparison of more than one distribution, as in Figure 1.17. When you look at a boxplot, first locate the median, which marks the center of the distribution. Then look at the spread. The quartiles show the spread of the middle half of the data, and the extremes (the smallest and largest observations) show the spread of the entire data set. We see at once that city mileages are lower than highway mileages. The minicompact cars have slightly higher median gas mileages than the two-seaters and their mileages are markedly less variable. In particular, the low gas mileages of the Ferraris and Lamborginis in the two-seater group pull the group minimum down.

The $1.5 \times IQR$ rule for suspected outliers

Look again at the 80 service center call lengths in Table 1.1. Figure 1.4 (page 13) is a stemplot of their distribution. You can check that the five-number summary is

$$1 \quad 54.5 \quad 103.5 \quad 200 \quad 2631$$

There is a clear outlier, a call lasting 2631 seconds, more than twice the length of any other call. How shall we describe the spread of this distribution? The smallest and largest observations are extremes that do not describe the spread of the majority of the data. The distance between the quartiles (the range of the center half of the data) is a more resistant measure of spread. This distance is called the *interquartile range*.

> ## THE INTERQUARTILE RANGE *IQR*
>
> The **interquartile range *IQR*** is the distance between the first and third quartiles,
>
> $$IQR = Q_3 - Q_1$$

For our data on service call lengths, $IQR = 200 - 54.5 = 145.5$. The quartiles and the *IQR* are not affected by changes in either tail of the distribution. They are therefore resistant, because changes in a few data points have no further effect once these points move outside the quartiles. However, *no single numerical measure of spread, such as IQR, is very useful for describing skewed distributions.* The two sides of a skewed distribution have different spreads, so one number can't summarize them. We can often detect skewness from the five-number summary by comparing how far the first quartile and the minimum are from the median (left tail) with how far the third quartile and the maximum are from the median (right tail). The interquartile range is mainly used as the basis for a rule of thumb for identifying suspected outliers.

> ## THE 1.5 × *IQR* RULE FOR OUTLIERS
>
> Call an observation a suspected outlier if it falls more than $1.5 \times IQR$ above the third quartile or below the first quartile.

EXAMPLE 1.16   For the call length data in Table 1.1,

$$1.5 \times IQR = 1.5 \times 145.5 = 218.25$$

Any values below $54.5 - 218.25 = -163.75$ or above $200 + 218.25 = 418.25$ are flagged as possible outliers. There are no low outliers, but the 8 longest calls are flagged as possible high outliers. Their lengths are

$$438 \quad 465 \quad 479 \quad 700 \quad 700 \quad 951 \quad 1148 \quad 2631$$

modified boxplot   Statistical software often uses the $1.5 \times IQR$ rule. For example, the stemplot in Figure 1.4 lists these 8 observations separately. Boxplots drawn by software are often **modified boxplots** that plot suspected outliers individually. Figure 1.18 is a modified boxplot of the call length data. The lines extend out from the central box only to the smallest and largest observations that are not flagged by the $1.5 \times IQR$ rule. The 8 largest call lengths are plotted as individual points, though 2 of them are identical and so do not appear separately.

The distribution of call lengths is very strongly skewed. We may well decide that only the longest call is truly an outlier in the sense of deviating from the overall pattern of the distribution. The other 7 calls are just part of the long right tail. The $1.5 \times IQR$ rule does not remove the need to look at the distri-

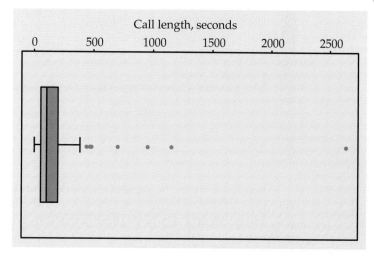

FIGURE 1.18   Modified boxplot of the call lengths in Table 1.1, for Example 1.16.

bution and use judgment. It is useful mainly to call our attention to unusual observations.

The stemplot in Figure 1.4 and the modified boxplot in Figure 1.18 tell us much more about the distribution of call lengths than the five-number summary or other numerical measures. The routine methods of statistics compute numerical measures and draw conclusions based on their values. These methods are very useful, and we will study them carefully in later chapters. But they cannot be applied blindly, by feeding data to a computer program, because *statistical measures and methods based on them are generally meaningful only for distributions of sufficiently regular shape.* This principle will become clearer as we progress, but it is good to be aware at the beginning that quickly resorting to fancy calculations is the mark of a statistical amateur. Look, think, and choose your calculations selectively.

Measuring spread: the standard deviation

The five-number summary is not the most common numerical description of a distribution. That distinction belongs to the combination of the mean to measure center and the *standard deviation* to measure spread. The standard deviation measures spread by looking at how far the observations are from their mean.

> **THE STANDARD DEVIATION *s***
>
> The **variance *s²*** of a set of observations is the average of the squares of the deviations of the observations from their mean. In symbols, the variance of n observations $x_1, x_2, \ldots, x_n$ is
>
> $$s^2 = \frac{(x_1 - \overline{x})^2 + (x_2 - \overline{x})^2 + \cdots + (x_n - \overline{x})^2}{n - 1}$$

or, in more compact notation,

$$s^2 = \frac{1}{n-1} \sum (x_i - \bar{x})^2$$

The **standard deviation** s is the square root of the variance s^2:

$$s = \sqrt{\frac{1}{n-1} \sum (x_i - \bar{x})^2}$$

The idea behind the variance and the standard deviation as measures of spread is as follows: The deviations $x_i - \bar{x}$ display the spread of the values x_i about their mean $\bar{x}$. Some of these deviations will be positive and some negative because some of the observations fall on each side of the mean. In fact, *the sum of the deviations of the observations from their mean will always be zero.* Squaring the deviations makes them all positive, so that observations far from the mean in either direction have large positive squared deviations. The variance is the average squared deviation. Therefore, s^2 and s will be large if the observations are widely spread about their mean, and small if the observations are all close to the mean.

EXAMPLE 1.17 A person's metabolic rate is the rate at which the body consumes energy. Metabolic rate is important in studies of weight gain, dieting, and exercise. Here are the metabolic rates of 7 men who took part in a study of dieting. (The units are calories per 24 hours. These are the same calories used to describe the energy content of foods.)

1792 1666 1362 1614 1460 1867 1439

Enter these data into your calculator or software and verify that

$$\bar{x} = 1600 \text{ calories} \quad s = 189.24 \text{ calories}$$

Figure 1.19 plots these data as dots on the calorie scale, with their mean marked by an asterisk (*). The arrows mark two of the deviations from the mean. If you were calculating s by hand, you would find the first deviation as

$$x_1 - \bar{x} = 1792 - 1600 = 192$$

Exercise 1.50 asks you to calculate the seven deviations, square them, and find s^2 and s directly from the deviations. Working one or two short examples by hand helps you understand how the standard deviation is obtained. In practice you will use either software or a calculator that will find s from keyed-in data. The two software outputs in

FIGURE 1.19 Metabolic rates for seven men, with the mean (*) and the deviations of two observations from the mean, for Example 1.17.

Figure 1.16 both give the variance and standard deviation for the highway mileage data.

The idea of the variance is straightforward: it is the average of the squares of the deviations of the observations from their mean. The details we have just presented, however, raise some questions.

Why do we square the deviations? Why not just average the distances of the observations from their mean? There are two reasons, neither of them obvious. First, the sum of the squared deviations of any set of observations from their mean is the smallest that the sum of squared deviations from any number can possibly be. This is not true of the unsquared distances. So squared deviations point to the mean as center in a way that distances do not. Second, the standard deviation turns out to be the natural measure of spread for a particularly important class of symmetric unimodal distributions, the *normal distributions*. We will meet the normal distributions in the next section. We commented earlier that the usefulness of many statistical procedures is tied to distributions of particular shapes. This is distinctly true of the standard deviation.

Why do we emphasize the standard deviation rather than the variance? One reason is that s, not s^2, is the natural measure of spread for normal distributions. There is also a more general reason to prefer s to s^2. Because the variance involves squaring the deviations, it does not have the same unit of measurement as the original observations. The variance of the metabolic rates, for example, is measured in squared calories. Taking the square root remedies this. The standard deviation s measures spread about the mean in the original scale.

Why do we average by dividing by n − 1 rather than n in calculating the variance? Because the sum of the deviations is always zero, the last deviation can be found once we know the other $n - 1$. So we are not averaging n unrelated numbers. Only $n - 1$ of the squared deviations can vary freely, and we average by dividing the total by $n - 1$. The number $n - 1$ is called the **degrees of freedom** of the variance or standard deviation. Many calculators offer a choice between dividing by n and dividing by $n - 1$, so be sure to use $n - 1$.

degrees of freedom

Properties of the standard deviation

Here are the basic properties of the standard deviation s as a measure of spread.

PROPERTIES OF THE STANDARD DEVIATION

- s measures spread about the mean and should be used only when the mean is chosen as the measure of center.

- $s = 0$ only when there is *no spread*. This happens only when all observations have the same value. Otherwise, $s > 0$. As the observations become more spread out about their mean, s gets larger.

- s, like the mean $\bar{x}$, is not resistant. A few outliers can make s very large.

The use of squared deviations renders s even more sensitive than x̄ to a few extreme observations. For example, dropping the Honda Insight from our list of two-seater cars reduces the mean highway mileage from 24.7 MPG to 22.6 MPG. It cuts the standard deviation more than half, from 10.8 MPG with the Insight to 5.3 MPG without it. Distributions with outliers and strongly skewed distributions have large standard deviations. The number *s* does not give much helpful information about such distributions.

Choosing measures of center and spread

How do we choose between the five-number summary and x̄ and *s* to describe the center and spread of a distribution? Because the two sides of a strongly skewed distribution have different spreads, no single number such as *s* describes the spread well. The five-number summary, with its two quartiles and two extremes, does a better job.

> ### CHOOSING A SUMMARY
>
> The five-number summary is usually better than the mean and standard deviation for describing a skewed distribution or a distribution with strong outliers. Use x̄ and *s* only for reasonably symmetric distributions that are free of outliers.

EXAMPLE 1.18   A central principle in the study of investments is that taking bigger risks is rewarded by higher returns, at least on the average over long periods of time. It is usual in finance to measure risk by the standard deviation of returns, on the grounds that investments whose returns vary a lot from year to year are less predictable and therefore more risky than those whose returns don't vary much. Compare, for example, the approximate mean and standard deviation of the annual percent returns on American common stocks and U.S. Treasury bills over the period from 1950 to 2003:

Investment	Mean return	Standard deviation
Common stocks	13.2%	17.6%
Treasury bills	5.0%	2.9%

Stocks are risky. They went up more than 13% per year on the average during this period, but they dropped almost 28% in the worst year. The large standard deviation reflects the fact that stocks have produced both large gains and large losses. When you buy a Treasury bill, on the other hand, you are lending money to the government for one year. You know that the government will pay you back with interest. That is much less risky than buying stocks, so (on the average) you get a smaller return.

```
-2 | 8 1                           0 | 9
-1 | 9 1 1 1 1 0                   1 | 0 2 5 5 6 6 6 8
-0 | 9 6 4 3                       2 | 1 5 7 7 9
 0 | 0 0 0 1 2 3 8 9 9             3 | 0 1 1 3 5 5 8 9 9
 1 | 1 3 3 4 4 6 6 6 7 8           4 | 2 4 7 7 8
 2 | 0 1 1 2 3 4 4 4 5 7 7 9 9     5 | 1 1 2 2 2 5 6 6 7 8
 3 | 0 1 1 2 3 4 6 7               6 | 2 4 5 6 9
 4 | 5                            7 | 2 7 8
 5 | 0                            8 | 0 4 8
                                  9 | 8
                                 10 | 4 5
                                 11 | 3
                                 12 |
                                 13 |
                                 14 | 7
         (a)                                (b)
```

FIGURE 1.20   Stemplots of annual returns for stocks and Treasury bills, 1950 to 2003, for Example 1.18. (a) Stock returns, in whole percents. (b) Treasury bill returns, in percents and tenths of a percent.

Are $\bar{x}$ and s good summaries for distributions of investment returns? Figure 1.20 displays stemplots of the annual returns for both investments. (Because stock returns are so much more spread out, a back-to-back stemplot does not work well. The stems in the stock stemplot are tens of percents; the stems for bills are percents. The lowest returns are −28% for stocks and 0.9% for bills.) You see that returns on Treasury bills have a right-skewed distribution. Convention in the financial world calls for $\bar{x}$ and s because some parts of investment theory use them. For describing this right-skewed distribution, however, the five-number summary would be more informative.

Remember that a graph gives the best overall picture of a distribution. Numerical measures of center and spread report specific facts about a distribution, but they do not describe its entire shape. Numerical summaries do not disclose the presence of multiple modes or gaps, for example. **Always plot your data.**

Changing the unit of measurement

The same variable can be recorded in different units of measurement. Americans commonly record distances in miles and temperatures in degrees Fahrenheit, while the rest of the world measures distances in kilometers and temperatures in degrees Celsius. Fortunately, it is easy to convert numerical descriptions of a distribution from one unit of measurement to another. This is true because a change in the measurement unit is a *linear transformation* of the measurements.

> **LINEAR TRANSFORMATIONS**
>
> A **linear transformation** changes the original variable x into the new variable x_{new} given by an equation of the form
>
> $$x_{new} = a + bx$$
>
> Adding the constant a shifts all values of x upward or downward by the same amount. In particular, such a shift changes the origin (zero point) of the variable. Multiplying by the positive constant b changes the size of the unit of measurement.

EXAMPLE 1.19   (a) If a distance x is measured in kilometers, the same distance in miles is

$$x_{new} = 0.62x$$

For example, a 10-kilometer race covers 6.2 miles. This transformation changes the units without changing the origin—a distance of 0 kilometers is the same as a distance of 0 miles.

(b) A temperature x measured in degrees Fahrenheit must be reexpressed in degrees Celsius to be easily understood by the rest of the world. The transformation is

$$x_{new} = \frac{5}{9}(x - 32) = -\frac{160}{9} + \frac{5}{9}x$$

Thus, the high of 95° F on a hot American summer day translates into 35° C. In this case

$$a = -\frac{160}{9} \quad \text{and} \quad b = \frac{5}{9}$$

This linear transformation changes both the unit size and the origin of the measurements. The origin in the Celsius scale (0° C, the temperature at which water freezes) is 32° in the Fahrenheit scale.

Linear transformations do not change the shape of a distribution. If measurements on a variable x have a right-skewed distribution, any new variable x_{new} obtained by a linear transformation $x_{new} = a + bx$ (for $b > 0$) will also have a right-skewed distribution. If the distribution of x is symmetric and unimodal, the distribution of x_{new} remains symmetric and unimodal.

Although a linear transformation preserves the basic shape of a distribution, the center and spread will change. Because linear changes of measurement scale are common, we must be aware of their effect on numerical descriptive measures of center and spread. Fortunately, the changes follow a simple pattern.

EXAMPLE 1.20 Mary and John both measure the weights of the same five newly hatched pythons. Mary measures in ounces and John uses grams. There are 28.35 grams in an ounce, so each of John's measurements is 28.35 times as large as Mary's measurement of the same python. Here are their results:

Python	1	2	3	4	5
Mary	1.13 oz	1.02 oz	1.23 oz	1.06 oz	1.16 oz
John	32 g	29 g	35 g	30 g	33 g

These two sets of numbers measure the same five weights in different units. Their means represent the same average weight in different units, so John's mean (in grams) must be 28.35 times as large as Mary's mean (in ounces). The spread of John's numbers is likewise 28.35 times the spread of Mary's, so John's standard deviation (again in grams) is 28.35 times Mary's standard deviation (in ounces). You can verify these facts with your calculator.

Here is a summary of such commonsense relations in more formal language.

EFFECT OF A LINEAR TRANSFORMATION

To see the effect of a linear transformation on measures of center and spread, apply these rules:

- Multiplying each observation by a positive number b multiplies both measures of center (mean and median) and measures of spread (interquartile range and standard deviation) by b.

- Adding the same number a (either positive or negative) to each observation adds a to measures of center and to quartiles and other percentiles but does not change measures of spread.

The measures of spread IQR and s do not change when we add the same number a to all of the observations because adding a constant changes the location of the distribution but leaves the spread unaltered. You can find the effect of a linear transformation $x_{new} = a + bx$ by combining these rules. For example, if x has mean $\bar{x}$, the transformed variable x_{new} has mean $a + b\bar{x}$.

SECTION 1.2 | Summary

A numerical summary of a distribution should report its **center** and its **spread** or **variability**.

The **mean $\bar{x}$** and the **median M** describe the center of a distribution in different ways. The mean is the arithmetic average of the observations, and the median is their midpoint.

When you use the median to describe the center of the distribution, describe its spread by giving the **quartiles**. The **first quartile Q_1** has one-fourth of the observations below it, and the **third quartile Q_3** has three-fourths of the observations below it.

The **interquartile range** is the difference between the quartiles. It is the spread of the center half of the data. The **1.5 × IQR rule** flags observations more than 1.5 × IQR beyond the quartiles as possible outliers.

The **five-number summary** consisting of the median, the quartiles, and the smallest and largest individual observations provides a quick overall description of a distribution. The median describes the center, and the quartiles and extremes show the spread.

Boxplots based on the five-number summary are useful for comparing several distributions. The box spans the quartiles and shows the spread of the central half of the distribution. The median is marked within the box. Lines extend from the box to the extremes and show the full spread of the data. In a **modified boxplot,** points identified by the 1.5 × IQR rule are plotted individually.

The **variance s^2** and especially its square root, the **standard deviation s,** are common measures of spread about the mean as center. The standard deviation s is zero when there is no spread and gets larger as the spread increases.

A **resistant measure** of any aspect of a distribution is relatively unaffected by changes in the numerical value of a small proportion of the total number of observations, no matter how large these changes are. The median and quartiles are resistant, but the mean and the standard deviation are not.

The mean and standard deviation are good descriptions for symmetric distributions without outliers. They are most useful for the normal distributions introduced in the next section. The five-number summary is a better exploratory summary for skewed distributions.

Linear transformations have the form $x_{new} = a + bx$. A linear transformation changes the origin if $a \neq 0$ and changes the size of the unit of measurement if $b > 0$. Linear transformations do not change the overall shape of a distribution. A linear transformation multiplies a measure of spread by b and changes a percentile or measure of center m into $a + bm$.

Numerical measures of particular aspects of a distribution, such as center and spread, do not report the entire shape of most distributions. In some cases, particularly distributions with multiple peaks and gaps, these measures may not be very informative.

SECTION 1.2 | Exercises

1.41   Figure 1.14 (page 31) is a histogram of the tuition and fees charged by the 56 four-year colleges in the state of Massachusetts. Here are those charges (in dollars), arranged in increasing order:

4,123	4,186	4,324	4,342	4,557	4,884	5,397	6,129
6,963	6,972	8,232	13,584	13,612	15,500	15,934	16,230
16,696	16,700	17,044	17,500	18,550	18,750	19,145	19,300
19,410	19,700	19,700	19,910	20,234	20,400	20,640	20,875
21,165	21,302	22,663	23,550	24,324	25,840	26,965	27,522
27,544	27,904	28,011	28,090	28,420	28,420	28,900	28,906
28,950	29,060	29,338	29,392	29,600	29,624	29,630	29,875

Find the five-number summary and make a boxplot. What distinctive feature of the histogram do these summaries miss? Remember that numerical summaries are not a substitute for looking at the data.

1.42   The stemplot in Exercise 1.15 (page 29) displays the distribution of the percents of residents aged 65 and over in the 50 states. Stemplots help you find the five-number summary because they arrange the observations in increasing order.

(a) Give the five-number summary of this distribution.

(b) Does the $1.5 \times IQR$ rule identify Alaska and Florida as suspected outliers? Does it also flag any other states?

1.43   Table 1.5 (page 32) shows the average property damage caused by tornadoes over a 50-year period in each of the states. The distribution is strongly skewed to the right.

(a) Give the five-number summary. Explain why you can see from these five numbers that the distribution is right-skewed.

(b) A histogram or stemplot suggests that a few states are outliers. Show that there are *no* suspected outliers according to the $1.5 \times IQR$ rule. You see once again that a rule is not a substitute for plotting your data.

(c) Find the mean property damage. Explain why the mean and median differ so greatly for this distribution.

1.44   Table 1.6 (page 33) gives carbon dioxide (CO_2) emissions per person for countries with population at least 20 million. The distribution is strongly skewed to the right. The United States and several other countries appear to be high outliers.

(a) Give the five-number summary. Explain why this summary suggests that the distribution is right-skewed.

(b) Which countries are outliers according to the $1.5 \times IQR$ rule? Make a stemplot or histogram of the data. Do you agree with the rule's suggestions about which countries are and are not outliers?

1.45   A report on the assets of American households says that the median net worth of households headed by someone aged less than 35 years is $11,600. The mean net worth of these same young households is $90,700.[28] What explains the difference between these two measures of center?

1.46   Exercise 1.33 (page 37) gives data on the total oil recovered from 64 wells. Your graph in that exercise shows that the distribution is clearly right-skewed.

(a) Find the mean and median of the amounts recovered. Explain how the relationship between the mean and the median reflects the shape of the distribution.

(b) Give the five-number summary and explain briefly how it reflects the shape of the distribution.

1.47 A small accounting firm pays each of its five clerks $25,000, two junior accountants $60,000 each, and the firm's owner $255,000. What is the mean salary paid at this firm? How many of the employees earn less than the mean? What is the median salary?

1.48 In computing the median income of any group, some federal agencies omit all members of the group who had no income. Give an example to show that the reported median income of a group can go down even though the group becomes economically better off. Is this also true of the mean income?

1.49 The firm in Exercise 1.47 gives no raises to the clerks and junior accountants, while the owner's take increases to $455,000. How does this change affect the mean? How does it affect the median?

1.50 Calculate the mean and standard deviation of the metabolic rates in Example 1.17 (page 50), showing each step in detail. First find the mean $\bar{x}$ by summing the 7 observations and dividing by 7. Then find each of the deviations $x_i - \bar{x}$ and their squares. Check that the deviations have sum 0. Calculate the variance as an average of the squared deviations (remember to divide by $n - 1$). Finally, obtain s as the square root of the variance.

1.51 A discussion of extreme weather reports that: "In most states, hurricanes occur infrequently. Yet, when a hurricane hits, the losses can be catastrophic. Average annual losses are not a meaningful measure of damage from rare but potentially catastrophic events."[29] Why is this true?

1.52 Exercise 1.35 (page 37) presented data on the nightly study time claimed by first-year college men and women. The most common methods for formal comparison of two groups use $\bar{x}$ and s to summarize the data. We wonder if this is appropriate here. Look at your back-to-back stemplot from Exercise 1.35, or make one now if you have not done so.

(a) What kinds of distributions are best summarized by $\bar{x}$ and s? It isn't easy to decide whether small data sets with irregular distributions fit the criteria. We will learn a better tool for making this decision in the next section.

(b) Each set of study times appears to contain a high outlier. Are these points flagged as suspicious by the $1.5 \times IQR$ rule? How much does removing the outlier change $\bar{x}$ and s for each group? The presence of outliers makes us reluctant to use the mean and standard deviation for these data unless we remove the outliers on the grounds that these students were exaggerating.

1.53 Many standard statistical methods that you will study in Part II of this book are intended for use with distributions that are symmetric and have no outliers. These methods start with the mean and standard deviation, $\bar{x}$ and s. Two examples of scientific data for which standard methods should work well are the pH measurements in Exercise 1.30 (page 35) and Cavendish's measurements of the density of the earth in Exercise 1.34 (page 37).

(a) Summarize each of these data sets by giving $\bar{x}$ and s.

(b) Find the median for each data set. Is the median quite close to the mean, as we expect it to be in these examples?

1.54   Many standard statistical methods that you will study in Part II of this book are intended for use with distributions that are symmetric and have no outliers. These methods start with the mean and standard deviation, $\bar{x}$ and s. For example, standard methods would typically be used for the IQ and GPA data in Table 1.9 (page 39).

(a) Find $\bar{x}$ and s for the IQ data. In large populations, IQ scores are standardized to have mean 100 and standard deviation 15. In what way does the distribution of IQ among these students differ from the overall population?

(b) Find the median IQ score. It is, as we expect, close to the mean.

(c) Find the mean and median for the GPA data. The two measures of center differ a bit. What feature of the data (see your stemplot in Exercise 1.37 or make a new stemplot) explains the difference?

1.55   The *Mean and Median* applet allows you to place observations on a line and see their mean and median visually. Place two observations on the line, by clicking below it. Why does only one arrow appear?

1.56   In the *Mean and Median* applet, place three observations on the line by clicking below it, two close together near the center of the line and one somewhat to the right of these two.

(a) Pull the single rightmost observation out to the right. (Place the cursor on the point, hold down a mouse button, and drag the point.) How does the mean behave? How does the median behave? Explain briefly why each measure acts as it does.

(b) Now drag the rightmost point to the left as far as you can. What happens to the mean? What happens to the median as you drag this point past the other two (watch carefully)?

1.57   Place five observations on the line in the *Mean and Median* applet by clicking below it.

(a) Add one additional observation *without changing the median*. Where is your new point?

(b) Use the applet to convince yourself that when you add yet another observation (there are now seven in all), the median does not change no matter where you put the seventh point. Explain why this must be true.

1.58   Different varieties of the tropical flower *Heliconia* are fertilized by different species of hummingbirds. Over time, the lengths of the flowers and the form of the hummingbirds' beaks have evolved to match each other. Here are data on the lengths in millimeters of three varieties of these flowers on the island of Dominica:[30]

H. bihai

47.12	46.75	46.81	47.12	46.67	47.43	46.44	46.64
48.07	48.34	48.15	50.26	50.12	46.34	46.94	48.36

H. caribaea red

41.90	42.01	41.93	43.09	41.47	41.69	39.78	40.57
39.63	42.18	40.66	37.87	39.16	37.40	38.20	38.07
38.10	37.97	38.79	38.23	38.87	37.78	38.01	

H. caribaea yellow

36.78	37.02	36.52	36.11	36.03	35.45	38.13	37.1
35.17	36.82	36.66	35.68	36.03	34.57	34.63	

Make boxplots to compare the three distributions. Report the five-number summaries along with your graph. What are the most important differences among the three varieties of flower?

1.59 The biologists who collected the flower length data in the previous exercise compared the three *Heliconia* varieties using statistical methods based on $\bar{x}$ and s.

(a) Find $\bar{x}$ and s for each variety.

(b) Make a stemplot of each set of flower lengths. Do the distributions appear suitable for use of $\bar{x}$ and s as summaries?

1.60 "Conservationists have despaired over destruction of tropical rainforest by logging, clearing, and burning." These words begin a report on a statistical study of the effects of logging in Borneo. Researchers compared forest plots that had never been logged (Group 1) with similar plots nearby that had been logged 1 year earlier (Group 2) and 8 years earlier (Group 3). All plots were 0.1 hectare in area. Here are the counts of trees for plots in each group:[31]

Group 1:	27	22	29	21	19	33	16	20	24	27	28	19
Group 2:	12	12	15	9	20	18	17	14	14	2	17	19
Group 3:	18	4	22	15	18	19	22	12	12			

Give a complete comparison of the three distributions, using both graphs and numerical summaries. To what extent has logging affected the count of trees? The researchers used an analysis based on $\bar{x}$ and s. Explain why this is reasonably well justified.

1.61 How does regular running affect heart rate? The RUNNERS data set, described in detail in the Data Appendix, contains heart rates for four groups of people:

Sedentary females

Sedentary males

Female runners (at least 15 miles per week)

Male runners (at least 15 miles per week)

The heart rates were measured after 6 minutes of exercise on a treadmill. There are 200 subjects in each group. Give a complete comparison of the four distributions, using both graphs and numerical summaries. How would you describe the effect of running on heart rate? Is the effect different for men and women?

The WORKERS data set, described in the Data Appendix, contains the sex, level of education, and income of 71,076 people between the ages of 25 and 64 who were employed full-time in 2001.

The boxplots in Figure 1.21 compare the distributions of income for people with five levels of education. This figure is a variation on the boxplot idea: because large data sets often contain very extreme observations, the lines extend from the central box only to the 5th and 95th percentiles. Exercises 1.62 to 1.64 concern these data.

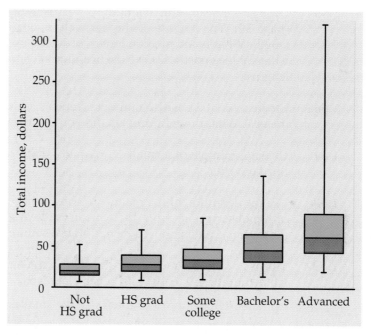

FIGURE 1.21   Boxplots comparing the distributions of income for employed people aged 25 to 64 years with five different levels of education. The lines extend from the quartiles to the 5th and 95th percentiles.

1.62   The data include 14,959 people whose highest level of education is a bachelor's degree.

(a) What is the position of the median in the ordered list of incomes (1 to 14,959)? From the boxplot, about what is the median income of people with a bachelor's degree?

(b) What is the position of the first and third quartiles in the ordered list of incomes for these people? About what are the numerical values of Q_1 and Q_3?

(c) You answered (a) and (b) from a boxplot that omits the lowest 5% and the highest 5% of incomes. Explain why leaving out these values has only a very small effect on the median and quartiles.

1.63   About what are the positions of the 5th and 95th percentiles in the ordered list of incomes of the 14,959 people with a bachelor's degree? Incomes outside this range do not appear in the boxplot. About what are the numerical values of the 5th and 95th percentiles of income? (For comparison, the largest income among all 14,959 people was $481,720. That one person made this much tells us less about the group than does the 95th percentile.)

1.64   Write a brief description of how the distribution of income changes with the highest level of education reached. Be sure to discuss center, spread, and skewness. Give some specifics read from the graph to back up your statements.

 1.65   Look at the histogram of lengths of words in Shakespeare's plays, Figure 1.13 (page 30). The heights of the bars tell us what percent of words have each length. What is the median length of words used by Shakespeare? Similarly, what are the quartiles? Give the five-number summary for Shakespeare's word lengths.

 1.66   Create a set of 5 positive numbers (repeats allowed) that have median 10 and mean 7. What thought process did you use to create your numbers?

1.67   Give an example of a small set of data for which the mean is larger than the third quartile.

 1.68   Use the definition of the mean $\bar{x}$ to show that the sum of the deviations $x_i - \bar{x}$ of the observations from their mean is always zero. This is one reason why the variance and standard deviation use squared deviations.

 1.69   This is a standard deviation contest. You must choose four numbers from the whole numbers 0 to 10, with repeats allowed.

(a) Choose four numbers that have the smallest possible standard deviation.

(b) Choose four numbers that have the largest possible standard deviation.

(c) Is more than one choice possible in either (a) or (b)? Explain.

1.70   This exercise requires a calculator with a standard deviation button or statistical software on a computer. The observations

$$10,001 \quad 10,002 \quad 10,003$$

have mean $\bar{x} = 10,002$ and standard deviation $s = 1$. Adding a 0 in the center of each number, the next set becomes

$$100,001 \quad 100,002 \quad 100,003$$

The standard deviation remains $s = 1$ as more 0s are added. Use your calculator or computer to calculate the standard deviation of these numbers, adding extra 0s until you get an incorrect answer. How soon did you go wrong? This demonstrates that calculators and computers cannot handle an arbitrary number of digits correctly.

1.71   Table 1.8 (page 38) gives the survival times of 72 guinea pigs in a medical study. Survival times—whether of cancer patients after treatment or of car batteries in everyday use—are almost always right-skewed. Make a graph to verify that this is true of these survival times. Then give a numerical summary that is appropriate for such data. Explain in simple language, to someone who knows no statistics, what your summary tells us about the guinea pigs.

1.72   A study of diet and weight gain deliberately overfed 16 volunteers for eight weeks. The mean increase in fat was $\bar{x} = 2.39$ kilograms and the standard deviation was $s = 1.14$ kilograms. What are $\bar{x}$ and s in pounds? (A kilogram is 2.2 pounds.)

1.73   Exercise 1.58 reports data on the lengths in millimeters of flowers of three varieties of *Heliconia*. In Exercise 1.59 you found the mean and standard deviation for each variety. Starting from the $\bar{x}$- and s-values in millimeters, find the means and standard deviations in inches. (A millimeter is 1/1000 of a meter. A meter is 39.37 inches.)

1.74   Henry Cavendish (see Exercise 1.34, page 37) used $\bar{x}$ to summarize his 29 measurements of the density of the earth.

(a) Find $\bar{x}$ and s for his data.

(b) Cavendish recorded the density of the earth as a multiple of the density of water. The density of water is almost exactly 1 gram per cubic centimeter, so his measurements have these units. In American units, the density of water is 62.43 pounds per cubic foot. This is the weight of a cube of water measuring 1 foot (that is, 30.48 cm) on each side. Express Cavendish's first result for the earth (5.50 g/cm³) in pounds per cubic foot. Then find $\bar{x}$ and s in pounds per cubic foot.

1.75
quintiles
deciles
Find the **quintiles** (the 20th, 40th, 60th, and 80th percentiles) of the guinea pig survival times in Table 1.8 (page 38). For quite large sets of data, the quintiles or the **deciles** (10th, 20th, 30th, etc. percentiles) give a more detailed summary than the quartiles.

1.76   Changing the unit of length from inches to centimeters multiplies each length by 2.54 because there are 2.54 centimeters in an inch. This change of units multiplies our usual measures of spread by 2.54. This is true of *IQR* and the standard deviation. What happens to the variance when we change units in this way?

1.77
trimmed mean
The **trimmed mean** is a measure of center that is more resistant than the mean but uses more of the available information than the median. To compute the 10% trimmed mean, discard the highest 10% and the lowest 10% of the observations and compute the mean of the remaining 80%. Trimming eliminates the effect of a small number of outliers. Compute the 10% trimmed mean of the guinea pig survival time data in Table 1.8 (page 38). Then compute the 20% trimmed mean. Compare the values of these measures with the median and the ordinary untrimmed mean.

1.3   Density Curves and Normal Distributions

We now have a kit of graphical and numerical tools for describing distributions. What is more, we have a clear strategy for exploring data on a single quantitative variable:

1. Always plot your data: make a graph, usually a stemplot or a histogram.

2. Look for the overall pattern and for striking deviations such as outliers.

3. Calculate an appropriate numerical summary to briefly describe center and spread.

Technology has expanded the set of graphs that we can choose for Step 1. Stemplots are intended for quick, hand-made graphs. It is possible, though painful, to make histograms by hand. Using software, clever algorithms can describe a distribution in a way that is not feasible by hand, by fitting a smooth curve to the data in addition to or instead of a histogram. The curves used are called **density curves.** Before we examine density curves in detail, here is an example of what software can do.

EXAMPLE 1.21   Figure 1.22 illustrates the use of density curves along with histograms to describe distributions.[32]   Figure 1.22(a) shows the distribution of the acidity (pH) of rainwater, from Exercise 1.30, page 35. That exercise illustrates how

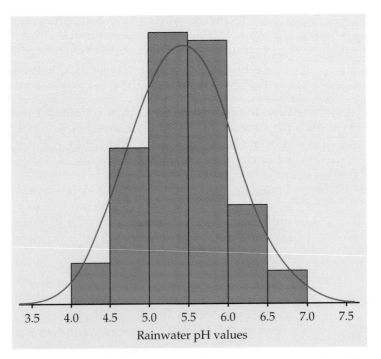

FIGURE 1.22   (a) The distribution of pH values measuring the acidity of 105 samples of rainwater, for Example 1.21. The roughly symmetric distribution is pictured by both a histogram and a density curve.

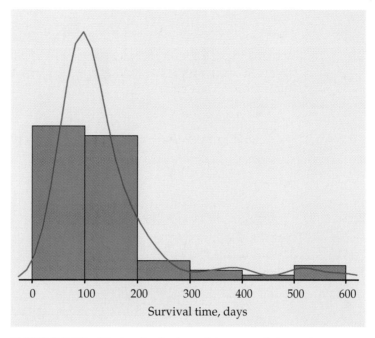

FIGURE 1.22 (b) The distribution of the survival times of 72 guinea pigs in a medical experiment, for Example 1.21. The right-skewed distribution is pictured by both a histogram and a density curve.

the choice of classes can change the shape of a histogram. The density curve and the software's default histogram agree that the distribution has a single peak and is approximately symmetric.

Figure 1.22(b) shows a strongly skewed distribution, the survival times of guinea pigs from Table 1.8, page 38. The histogram and density curve agree on the overall shape and on the "bumps" in the long right tail. The density curve shows a higher peak near the single mode of the distribution. The histogram divides the observations near the mode into two classes, thus reducing the peak.

In general, software that draws density curves describes the data in a way that is less arbitrary than choosing classes for a histogram. A smooth density curve is, however, an idealization that pictures the overall pattern of the data but ignores minor irregularities as well as any outliers. We will concentrate, not on general density curves, but on a special class, the bell-shaped normal curves.

Density curves

One way to think of a density curve is as a smooth approximation to the irregular bars of a histogram. Figure 1.23 shows a histogram of the scores of all 947 seventh-grade students in Gary, Indiana, on the vocabulary part of the Iowa Test of Basic Skills. Scores of many students on this national test have a

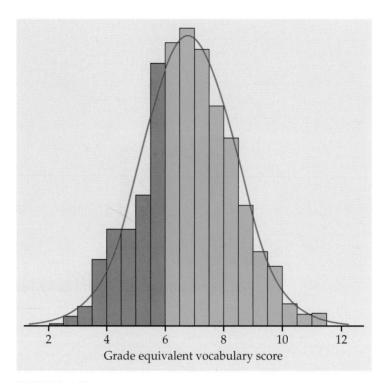

Grade equivalent vocabulary score

FIGURE 1.23 (a) The distribution of Iowa Test vocabulary scores for Gary, Indiana, seventh graders. The red bars in the histogram represent scores less than or equal to 6.0. The proportion of such scores in the data is 0.303.

very regular distribution. The histogram is symmetric, and both tails fall off quite smoothly from a single center peak. There are no large gaps or obvious outliers. The curve drawn through the tops of the histogram bars in Figure 1.23 is a good description of the overall pattern of the data.

EXAMPLE 1.22 In a histogram, the *areas* of the bars represent either counts or proportions of the observations. In Figure 1.23(a) we have colored red the bars that represent students with vocabulary scores 6.0 or lower. There are 287 such students, who make up the proportion $287/947 = 0.303$ of all Gary seventh graders. The red bars in Figure 1.23(a) make up proportion 0.303 of the total area under all the bars. If we adjust the scale so that the total area of the bars is 1, the area of the red bars will be 0.303.

In Figure 1.23(b), we have colored red the *area under the curve* to the left of 6.0. Adjust the scale so that the total area under the curve is exactly 1. Areas under the curve then represent proportions of the observations. That is, *area = proportion*. The curve is then a density curve. The red area under the density curve in Figure 1.23(b) represents the proportion of students with score 6.0 or lower. This area is 0.293, only 0.010 away from the histogram result. You can see that areas under the density curve give quite good approximations of areas given by the histogram.

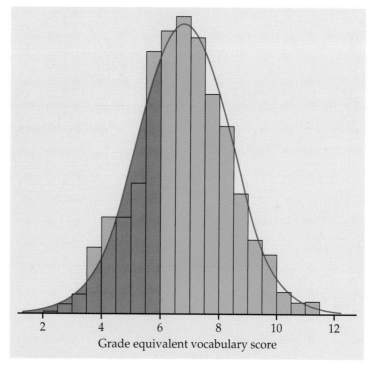

Grade equivalent vocabulary score

FIGURE 1.23 (b) The red area under the normal density curve also represents scores less than or equal to 6.0. This area is 0.293, close to the true 0.303 for the actual data.

DENSITY CURVE

A **density curve** is a curve that

- is always on or above the horizontal axis and

- has area exactly 1 underneath it.

A density curve describes the overall pattern of a distribution. The area under the curve and above any range of values is the proportion of all observations that fall in that range.

The density curve in Figure 1.23 is a *normal curve*. Density curves, like distributions, come in many shapes. Figure 1.24 shows two density curves, a symmetric normal density curve and a right-skewed curve. A density curve of an appropriate shape is often an adequate description of the overall pattern of a distribution. Outliers, which are deviations from the overall pattern, are not described by the curve.

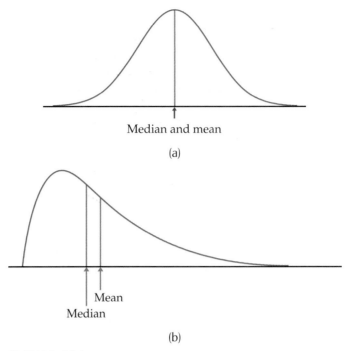

Median and mean

(a)

Mean

Median

(b)

FIGURE 1.24 (a) A symmetric density curve with its mean and median marked. (b) A right-skewed density curve with its mean and median marked.

Measuring center and spread for density curves

Our measures of center and spread apply to density curves as well as to actual sets of observations, but only some of these measures are easily seen from the curve. A **mode** of a distribution described by a density curve is a peak point of the curve, the location where the curve is highest. Because areas under a density curve represent proportions of the observations, the **median** is the point with half the total area on each side. You can roughly locate the **quartiles** by dividing the area under the curve into quarters as accurately as possible by eye. The *IQR* is then the distance between the first and third quartiles. There are mathematical ways of calculating areas under curves. These allow us to locate the median and quartiles exactly on any density curve.

What about the mean and standard deviation? The mean of a set of observations is their arithmetic average. If we think of the observations as weights strung out along a thin rod, the mean is the point at which the rod would balance. This fact is also true of density curves. The mean is the point at which the curve would balance if it were made out of solid material. Figure 1.25 illus-

FIGURE 1.25 The mean of a density curve is the point at which it would balance.

trates this interpretation of the mean. We have marked the mean and median on the density curves in Figure 1.24. A symmetric curve, such as the normal curve in Figure 1.24(a), balances at its center of symmetry. Half the area under a symmetric curve lies on either side of its center, so this is also the median. For a right-skewed curve, such as that shown in Figure 1.24(b), the small area in the long right tail tips the curve more than the same area near the center. The mean (the balance point) therefore lies to the right of the median. It is hard to locate the balance point by eye on a skewed curve. There are mathematical ways of calculating the mean for any density curve, so we are able to mark the mean as well as the median in Figure 1.24(b). The standard deviation can also be calculated mathematically, but it can't be located by eye on most density curves.

MEDIAN AND MEAN OF A DENSITY CURVE

The **median** of a density curve is the equal-areas point, the point that divides the area under the curve in half.

The **mean** of a density curve is the balance point, at which the curve would balance if made of solid material.

The median and mean are the same for a symmetric density curve. They both lie at the center of the curve. The mean of a skewed curve is pulled away from the median in the direction of the long tail.

A density curve is an idealized description of a distribution of data. For example, the symmetric density curve in Figure 1.23 is exactly symmetric, but the histogram of vocabulary scores is only approximately symmetric. We therefore need to distinguish between the mean and standard deviation of the density curve and the numbers $\bar{x}$ and s computed from the actual observations. The usual notation for the mean of an idealized distribution is μ (the Greek letter mu). We write the standard deviation of a density curve as σ (the Greek letter sigma).

mean μ
standard deviation σ

Normal distributions

One particularly important class of density curves has already appeared in Figures 1.23 and 1.24(a). These density curves are symmetric, unimodal, and bell-shaped. They are called **normal curves,** and they describe *normal distributions*. All normal distributions have the same overall shape. The exact density curve for a particular normal distribution is specified by giving its mean μ and its standard deviation σ. The mean is located at the center of the symmetric curve and is the same as the median. Changing μ without changing σ moves the normal curve along the horizontal axis without changing its spread. The standard deviation σ controls the spread of a normal curve. Figure 1.26 shows two normal curves with different values of σ. The curve with the larger standard deviation is more spread out.

normal curves

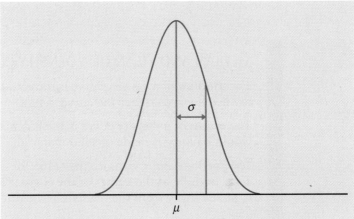

FIGURE 1.26   Two normal curves, showing the mean μ and standard deviation σ.

The standard deviation σ is the natural measure of spread for normal distributions. Not only do μ and σ completely determine the shape of a normal curve, but we can locate σ by eye on the curve. Here's how. As we move out in either direction from the center μ, the curve changes from falling ever more steeply

to falling ever less steeply

The points at which this change of curvature takes place are located at distance σ on either side of the mean μ. You can feel the change as you run your finger along a normal curve, and so find the standard deviation. Remember that μ and σ alone do not specify the shape of most distributions, and that the shape of density curves in general does not reveal σ. These are special properties of normal distributions.

There are other symmetric bell-shaped density curves that are not normal. The normal density curves are specified by a particular equation. The height of the density curve at any point x is given by

$$\frac{1}{\sigma\sqrt{2\pi}}e^{-\frac{1}{2}\left(\frac{x-\mu}{\sigma}\right)^2}$$

We will not make direct use of this fact, although it is the basis of mathematical work with normal distributions. Notice that the equation of the curve is completely determined by the mean μ and the standard deviation σ.

Why are the normal distributions important in statistics? Here are three reasons. First, normal distributions are good descriptions for some distributions of *real data*. Distributions that are often close to normal include scores on tests taken by many people (such as the Iowa Test of Figure 1.23), repeated careful measurements of the same quantity, and characteristics of biological populations (such as lengths of baby pythons and yields of corn). Second, normal distributions are good approximations to the results of many kinds of *chance outcomes*, such as tossing a coin many times. Third, and most important, we will see that many *statistical inference* procedures based on normal distributions work well for other roughly symmetric distributions. HOWEVER ... *even though many sets of data follow a normal distribution, many do not.* Most income distributions, for example, are skewed to the right and so are not normal. Nonnormal data, like nonnormal people, not only are common but are sometimes more interesting than their normal counterparts.

The 68–95–99.7 rule

Although there are many normal curves, they all have common properties. Here is one of the most important.

THE 68–95–99.7 RULE

In the normal distribution with mean μ and standard deviation σ:

- Approximately **68%** of the observations fall within σ of the mean μ.

- Approximately **95%** of the observations fall within 2σ of μ.

- Approximately **99.7%** of the observations fall within 3σ of μ.

Figure 1.27 illustrates the 68–95–99.7 rule. By remembering these three numbers, you can think about normal distributions without constantly making detailed calculations.

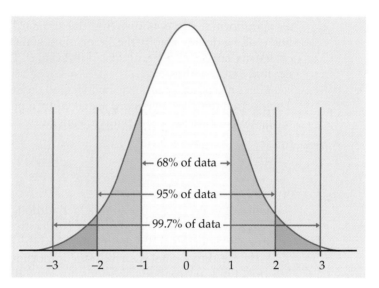

FIGURE 1.27   The 68–95–99.7 rule for normal distributions.

EXAMPLE 1.23   The distribution of heights of young women aged 18 to 24 is approximately normal with mean $\mu = 64.5$ inches and standard deviation $\sigma = 2.5$ inches. Figure 1.28 shows what the 68–95–99.7 rule says about this distribution.

Two standard deviations is 5 inches for this distribution. The 95 part of the 68–95–99.7 rule says that the middle 95% of young women are between $64.5 - 5$ and $64.5 + 5$ inches tall, that is, between 59.5 inches and 69.5 inches. This fact is exactly true for an exactly normal distribution. It is approximately true for the heights of young women because the distribution of heights is approximately normal.

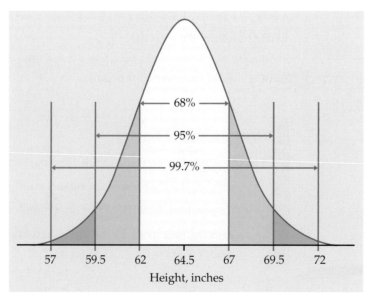

FIGURE 1.28   The 68–95–99.7 rule applied to the heights of young women, for Example 1.23.

The other 5% of young women have heights outside the range from 59.5 to 69.5 inches. Because the normal distributions are symmetric, half of these women are on the tall side. So the tallest 2.5% of young women are taller than 69.5 inches.

Because we will mention normal distributions often, a short notation is helpful. We abbreviate the normal distribution with mean μ and standard deviation σ as $N(\mu, \sigma)$. For example, the distribution of young women's heights is $N(64.5, 2.5)$.

$N(\mu, \sigma)$

Standardizing observations

As the 68–95–99.7 rule suggests, all normal distributions share many properties. In fact, all normal distributions are the same if we measure in units of size σ about the mean μ as center. Changing to these units is called *standardizing*. To standardize a value, subtract the mean of the distribution and then divide by the standard deviation.

STANDARDIZING AND z-SCORES

If x is an observation from a distribution that has mean μ and standard deviation σ, the **standardized value** of x is

$$z = \frac{x - \mu}{\sigma}$$

A standardized value is often called a **z-score**.

A z-score tells us how many standard deviations the original observation falls away from the mean, and in which direction. Observations larger than the mean are positive when standardized, and observations smaller than the mean are negative.

EXAMPLE 1.24 The heights of young women are approximately normal with $\mu = 64.5$ inches and $\sigma = 2.5$ inches. The standardized height is

$$z = \frac{\text{height} - 64.5}{2.5}$$

A woman's standardized height is the number of standard deviations by which her height differs from the mean height of all young women. A woman 68 inches tall, for example, has standardized height

$$z = \frac{68 - 64.5}{2.5} = 1.4$$

or 1.4 standard deviations above the mean. Similarly, a woman 5 feet (60 inches) tall has standardized height

$$z = \frac{60 - 64.5}{2.5} = -1.8$$

or 1.8 standard deviations less than the mean height.

We need a way to write variables, such as "height" in Example 1.24, that follow a theoretical distribution such as a normal distribution. We use capital letters near the end of the alphabet for such variables. If X is the height of a young woman, we can then shorten "the height of a young woman is less than 68 inches" to "$X < 68$." We will use lowercase x to stand for any specific value of the variable X.

We often standardize observations from symmetric distributions to express them in a common scale. We might, for example, compare the heights of two children of different ages by calculating their z-scores. The standardized heights tell us where each child stands in the distribution for his or her age group.

Standardizing is a linear transformation that transforms the data into the standard scale of z-scores. We know that a linear transformation does not change the shape of a distribution, and that the mean and standard deviation change in a simple manner. In particular, *the standardized values for any distribution always have mean 0 and standard deviation 1.*

If the variable we standardize has a normal distribution, standardizing does more than give a common scale. It makes all normal distributions into a single distribution, and this distribution is still normal. Standardizing a variable that has any normal distribution produces a new variable that has the *standard normal distribution.*

THE STANDARD NORMAL DISTRIBUTION

The **standard normal distribution** is the normal distribution $N(0, 1)$ with mean 0 and standard deviation 1.

If a variable X has any normal distribution $N(\mu, \sigma)$ with mean μ and standard deviation σ, then the standardized variable

$$Z = \frac{X - \mu}{\sigma}$$

has the standard normal distribution.

Normal distribution calculations

Areas under a normal curve represent proportions of observations from that normal distribution. There is no formula for areas under a normal curve. Calculations use either software that calculates areas or a table of areas. The table and most software calculate one kind of area: **cumulative proportions.** A cumulative proportion is the proportion of observations in a distribution that lie at or below a given value. When the distribution is given by a density curve, the cumulative proportion is the area under the curve to the left of a given value. Figure 1.29 shows the idea more clearly than words do.

The key to calculating normal proportions is to match the area you want with areas that represent cumulative proportions. Then get areas for cumulative proportions either from software or (with an extra step) from a table. The following examples show the method in pictures.

cumulative proportion *(margin note)*

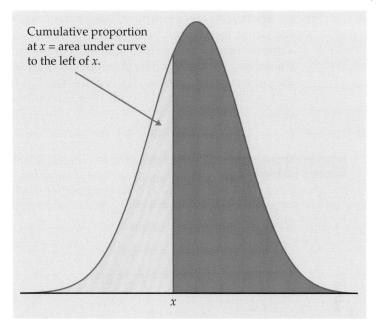

FIGURE 1.29 The *cumulative proportion* for a value x is the proportion of all observations from the distribution that are less than or equal to x. This is the area to the left of x under the normal curve.

EXAMPLE 1.25 The National Collegiate Athletic Association (NCAA) requires Division I athletes to score at least 820 on the combined mathematics and verbal parts of the SAT exam in order to compete in their first college year. (Higher scores are required for students with poor high school grades.) The scores of the 1.4 million students in the class of 2003 who took the SATs were approximately normal with mean 1026 and standard deviation 209. What proportion of all students had SAT scores of at least 820?

Here is the calculation in pictures: the proportion of scores above 820 is the area under the curve to the right of 820. That's the total area under the curve (which is always 1) minus the cumulative proportion up to 820.

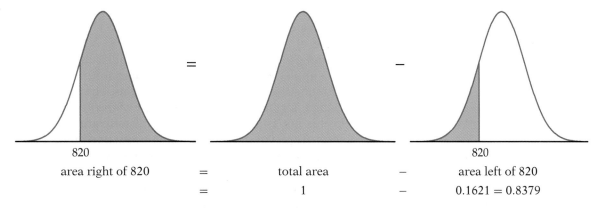

820				820
area right of 820	=	total area	−	area left of 820
	=	1	−	0.1621 = 0.8379

That is, the proportion of all SAT-takers who would be NCAA qualifiers is 0.8379, or about 84%.

There is *no* area under a smooth curve and exactly over the point 820. Consequently, the area to the right of 820 (the proportion of scores > 820) is the same as the area at or to the right of this point (the proportion of scores ≥ 820). The actual data may contain a student who scored exactly 820 on the SAT. That the proportion of scores exactly equal to 820 is 0 for a normal distribution is a consequence of the idealized smoothing of normal distributions for data.

EXAMPLE 1.26   The NCAA considers a student a "partial qualifier" eligible to practice and receive an athletic scholarship, but not to compete, if the combined SAT score is at least 720. What proportion of all students who take the SAT would be partial qualifiers? That is, what proportion have scores between 720 and 820? Here are the pictures:

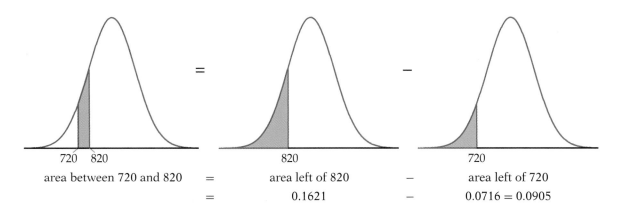

area between 720 and 820	=	area left of 820	–	area left of 720
	=	0.1621	–	0.0716 = 0.0905

About 9% of all students who take the SAT have scores between 720 and 820.

How do we find the numerical values of the areas in Examples 1.25 and 1.26? If you use software, just plug in mean 1026 and standard deviation 209. Then ask for the cumulative proportions for 820 and for 720. (Your software will probably refer to these as "cumulative probabilities." We will learn in Chapter 4 why the language of probability fits.) If you make a sketch of the area you want, you will never go wrong.

You can use the *Normal Curve* applet on the text CD and Web site to find normal proportions. The applet is more flexible than most software—it will find any normal proportion, not just cumulative proportions. The applet is an excellent way to understand normal curves. But, because of the limitations of Web browsers, the applet is not as accurate as statistical software.

If you are not using software, you can find cumulative proportions for normal curves from a table. That requires an extra step, as we now explain.

Using the standard normal table

The extra step in finding cumulative proportions from a table is that we must first standardize to express the problem in the standard scale of *z*-scores. This allows us to get by with just one table, a table of *standard normal*

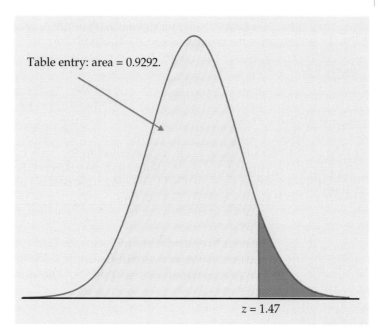

Table entry: area = 0.9292.

z = 1.47

FIGURE 1.30   The area under a standard normal curve to the left of the point $z = 1.47$ is 0.9292, for Example 1.27. Table A gives areas under the standard normal curve.

cumulative proportions. Table A in the back of the book gives cumulative proportions for the standard normal distribution. Table A also appears on the inside front cover. The pictures at the top of the table remind us that the entries are cumulative proportions, areas under the curve to the left of a value z.

EXAMPLE 1.27   What proportion of observations on a standard normal variable Z takes values less than 1.47?

Solution: To find the area to the left of 1.47, locate 1.4 in the left-hand column of Table A, then locate the remaining digit 7 as .07 in the top row. The entry opposite 1.4 and under .07 is 0.9292. This is the cumulative proportion we seek. Figure 1.30 illustrates this area.

Now that you see how Table A works, let's redo the NCAA Examples 1.25 and 1.26 using the table.

EXAMPLE 1.28   What proportion of all students who take the SAT have scores of at least 820? The picture that leads to the answer is exactly the same as in Example 1.25. The extra step is that we first standardize in order to read cumulative proportions from Table A. If X is SAT score, we want the proportion of students for which $X \geq 820$.

1. *Standardize*. Subtract the mean, then divide by the standard deviation, to transform the problem about X into a problem about a standard normal Z:

$$X \geq 820$$

$$\frac{X - 1026}{209} \geq \frac{820 - 1026}{209}$$

$$Z \geq -0.99$$

2. *Use the table*. Look at the pictures in Example 1.25. From Table A, we see that the proportion of observations less than -0.99 is 0.1611. The area to the right of -0.99 is therefore $1 - 0.1611 = 0.8389$. This is about 84%.

The area from the table in Example 1.28 (0.8389) is slightly less accurate than the area from software in Example 1.25 (0.8379) because we must round z to two places when we use Table A. The difference is rarely important in practice.

EXAMPLE 1.29   What proportion of all students who take the SAT would be partial qualifiers in the eyes of the NCAA? That is, what proportion of students have SAT scores between 720 and 820? First, sketch the areas, exactly as in Example 1.26. We again use X as shorthand for an SAT score.

1. *Standardize*.

$$720 \leq \quad X \quad < 820$$

$$\frac{720 - 1026}{209} \leq \frac{X - 1026}{209} < \frac{820 - 1026}{209}$$

$$-1.46 \leq \quad Z \quad < -0.99$$

2. *Use the table.*

$$\text{area between } -1.46 \text{ and } -0.99 = (\text{area left of } -0.99) - (\text{area left of } -1.46)$$

$$= 0.1611 - 0.0721 = 0.0890$$

As in Example 1.26, about 9% of students would be partial qualifiers.

Sometimes we encounter a value of z more extreme than those appearing in Table A. For example, the area to the left of $z = -4$ is not given directly in the table. The z-values in Table A leave only area 0.0002 in each tail unaccounted for. For practical purposes, we can act as if there is zero area outside the range of Table A.

Inverse normal calculations

Examples 1.25 to 1.29 illustrate the use of normal distributions to find the proportion of observations in a given event, such as "SAT score between 720 and 820." We may instead want to find the observed value corresponding to a given proportion.

Statistical software will do this directly. Without software, use Table A backward, finding the desired proportion in the body of the table and then reading the corresponding z from the left column and top row.

EXAMPLE 1.30   Scores on the SAT verbal test in recent years follow approximately the $N(505, 110)$ distribution. How high must a student score in order to place in the top 10% of all students taking the SAT?

Again, the key to the problem is to draw a picture. Figure 1.31 shows that we want the score x with area above it 0.10. That's the same as area below x equal to 0.90.

Statistical software has a function that will give you the x for any cumulative proportion you specify. The function often has a name such as "inverse cumulative probability." Plug in mean 505, standard deviation 110, and cumulative proportion 0.9. The software tells you that $x = 645.97$. We see that a student must score at least 646 to place in the highest 10%.

Without software, first find the standard score z with cumulative proportion 0.9, then "unstandardize" to find x. Here is the two-step process:

1. *Use the table.* Look in the body of Table A for the entry closest to 0.9. It is 0.8997. This is the entry corresponding to $z = 1.28$. So $z = 1.28$ is the standardized value with area 0.9 to its left.

2. *Unstandardize* to transform the solution from z back to the original x scale. We know that the standardized value of the unknown x is $z = 1.28$. So x itself satisfies

$$\frac{x - 505}{110} = 1.28$$

Solving this equation for x gives

$$x = 505 + (1.28)(110) = 645.8$$

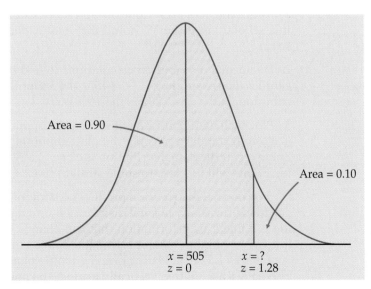

FIGURE 1.31   Locating the point on a normal curve with area 0.10 to its right, for Example 1.30. The result is $x = 646$, or $z = 1.28$ in the standard scale.

This equation should make sense: it finds the x that lies 1.28 standard deviations above the mean on this particular normal curve. That is the "unstandardized" meaning of $z = 1.28$. The general rule for unstandardizing a z-score is

$$x = \mu + z\sigma$$

Normal quantile plots

The normal distributions provide good descriptions of some distributions of real data, such as the Gary vocabulary scores. The distributions of some other common variables are usually skewed and therefore distinctly nonnormal. Examples include economic variables such as personal income and gross sales of business firms, the survival times of cancer patients after treatment, and the service lifetime of mechanical or electronic components. While experience can suggest whether or not a normal distribution is plausible in a particular case, it is risky to assume that a distribution is normal without actually inspecting the data.

A histogram or stemplot can reveal distinctly nonnormal features of a distribution, such as outliers (the breaking strengths in Figure 1.7, page 19), pronounced skewness (the survival times in Figure 1.22(b), page 65), or gaps and clusters (the Massachusetts college tuitions in Figure 1.14, page 31). If the stemplot or histogram appears roughly symmetric and unimodal, however, we need a more sensitive way to judge the adequacy of a normal model. The most useful tool for assessing normality is another graph, the **normal quantile plot.**

normal quantile plot

Here is the basic idea of a normal quantile plot. The graphs produced by software use more sophisticated versions of this idea. It is not practical to make normal quantile plots by hand.

1. Arrange the observed data values from smallest to largest. Record what percentile of the data each value occupies. For example, the smallest observation in a set of 20 is at the 5% point, the second smallest is at the 10% point, and so on.

2. Do normal distribution calculations to find the z-scores at these same percentiles. For example, $z = -1.645$ is the 5% point of the standard normal distribution, and $z = -1.282$ is the 10% point.

3. Plot each data point x against the corresponding z. If the data distribution is close to standard normal, the plotted points will lie close to the 45-degree line $x = z$. If the data distribution is close to any normal distribution, the plotted points will lie close to some straight line.

Any normal distribution produces a straight line on the plot because standardizing turns any normal distribution into a standard normal. Standardizing is a linear transformation that can change the slope and intercept of the line in our plot but cannot turn a line into a curved pattern.

Figures 1.32 to 1.34 are normal quantile plots for data we have met earlier. The data x are plotted vertically against the corresponding standard normal z-score plotted horizontally. The z-score scale extends from -3 to 3 because almost all of a standard normal curve lies between these values. These figures show how normal quantile plots behave.

USE OF NORMAL QUANTILE PLOTS

If the points on a normal quantile plot lie close to a straight line, the plot indicates that the data are normal. Systematic deviations from a straight line indicate a nonnormal distribution. Outliers appear as points that are far away from the overall pattern of the plot.

EXAMPLE 1.31 Figure 1.32 is a normal quantile plot of the breaking strengths in Example 1.9 (page 18). Lay a transparent straightedge over the center of the plot to see that most of the points lie close to a straight line. A normal distribution describes these points quite well. The only substantial deviations are short horizontal runs of points. Each run represents repeated observations having the same value—there are five measurements at 1150, for example. This phenomenon is called

granularity **granularity.** It is caused by the limited precision of the measurements and does not represent an important deviation from normality.

The high outlier at 3150 pounds lies above the line formed by the center of the data—it is farther out in the high direction than we expect normal data to be. The two low outliers at 0 lie below the line—they are suspiciously far out in the low direction. Compare Figure 1.32 with the histogram of these data in Figure 1.7 (page 19).

FIGURE 1.32 Normal quantile plot of the breaking strengths of wires bonded to a semiconductor wafer, for Example 1.31. This distribution has a normal shape except for outliers in both tails.

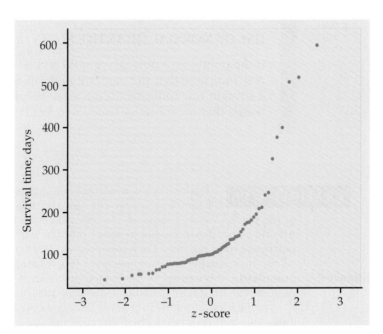

FIGURE 1.33 Normal quantile plot of the survival times of guinea pigs in a medical experiment, for Example 1.32. This distribution is skewed to the right.

EXAMPLE 1.32 Figure 1.33 is a normal quantile plot of the guinea pig survival times from Table 1.8 (page 38). Figure 1.22(b), page 65, shows that this distribution is strongly skewed to the right.

To see the right skewness in the normal quantile plot, draw a line through the left-most points, which correspond to the smaller observations. The larger observations fall systematically above this line. That is, the right-of-center observations have larger values than in a normal distribution. *In a right-skewed distribution, the largest observations fall distinctly above a line drawn through the main body of points.* Similarly, left skewness is evident when the smallest observations fall below the line. Unlike Figure 1.32, there are no individual outliers.

EXAMPLE 1.33 Figure 1.34 is a normal quantile plot of the 105 acidity (pH) measurements of rainwater from Exercise 1.30 (page 35). Histograms don't settle the question of approximate normality of these data, because their shape depends on the choice of classes. The normal quantile plot makes it clear that a normal distribution is a good description—there are only minor wiggles in a generally straight-line pattern.

As Figure 1.34 illustrates, real data almost always show some departure from the theoretical normal model. *When you examine a normal quantile plot, look for shapes that show clear departures from normality. Don't overreact to minor wiggles in the plot.* When we discuss statistical methods that are based on the normal model, we will pay attention to the sensitivity of each method

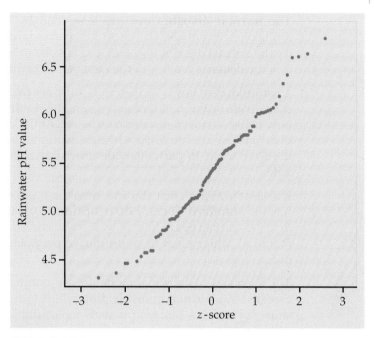

FIGURE 1.34 Normal quantile plot of the acidity (pH) values of 105 samples of rainwater, for Example 1.33. This distribution is roughly normal.

to departures from normality. Many common methods work well as long as the data are approximately normal and outliers are not present.

BEYOND THE BASICS │ Density estimation

density estimator

A density curve gives a compact summary of the overall shape of a distribution. Many distributions do not have the normal shape. There are other families of density curves that are used as mathematical models for various distribution shapes. Modern software offers a more flexible option, illustrated by the two graphs in Figure 1.22 (pages 64 and 65). A **density estimator** does not start with any specific shape, such as the normal shape. It looks at the data and draws a density curve that describes the overall shape of the data. Density estimators join stemplots and histograms as useful graphical tools for exploratory data analysis.

SECTION 1.3 │ Summary

The overall pattern of a distribution can often be described compactly by a **density curve**. A density curve has total area 1 underneath it. Areas under a density curve give proportions of observations for the distribution.

The **mean** μ (balance point), the **median** (equal-areas point), and the **quartiles** can be approximately located by eye on a density curve. The **standard deviation** σ cannot be located by eye on most density curves. The mean and median are equal for symmetric density curves, but the mean of a skewed curve is located farther toward the long tail than is the median.

The **normal distributions** are described by bell-shaped, symmetric, unimodal density curves. The mean μ and standard deviation σ completely specify the normal distribution $N(\mu, \sigma)$. The mean is the center of symmetry, and σ is the distance from μ to the change-of-curvature points on either side.

To **standardize** any observation x, subtract the mean of the distribution and then divide by the standard deviation. The resulting **z-score** $z = (x - \mu)/\sigma$ says how many standard deviations x lies from the distribution mean. All normal distributions are the same when measurements are transformed to the standardized scale. In particular, all normal distributions satisfy the **68–95–99.7 rule.**

If X has the $N(\mu, \sigma)$ distribution, then the standardized variable $Z = (X - \mu)/\sigma$ has the **standard normal distribution** $N(0, 1)$. Proportions for any normal distribution can be calculated by software or from the **standard normal table** (Table A), which gives the **cumulative proportions** of $Z < z$ for many values of z.

The adequacy of a normal model for describing a distribution of data is best assessed by a **normal quantile plot,** which is available in most statistical software packages. A pattern on such a plot that deviates substantially from a straight line indicates that the data are not normal.

SECTION 1.3 | Exercises

1.78   Sketch density curves that might describe distributions with the following shapes:

(a) Symmetric, but with two peaks (that is, two strong clusters of observations).

(b) Single peak and skewed to the left.

1.79   Remember that it is areas under a density curve, not the height of the curve, that give proportions in a distribution. To illustrate this, sketch a density curve that has its mode (peak point) at 0 on the horizontal axis but has greater area within 0.25 on either side of 1 than within 0.25 on either side of 0.

1.80

uniform distribution

If you ask a computer to generate "random numbers" between 0 and 1, you will get observations from a **uniform distribution.** Figure 1.35 graphs the density curve for a uniform distribution. Use areas under this density curve to answer the following questions.

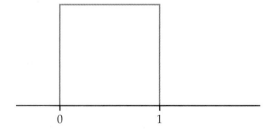

FIGURE 1.35 The density curve of a uniform distribution, for Exercise 1.80.

(a) Why is the total area under this curve equal to 1?

(b) What proportion of the observations lie above 0.75?

(c) What proportion of the observations lie between 0.25 and 0.75?

1.81   Many random number generators allow users to specify the range of the random numbers to be produced. Suppose that you specify that the outcomes are to be distributed uniformly between 0 and 2. Then the density curve of the outcomes has constant height between 0 and 2, and height 0 elsewhere.

(a) What is the height of the density curve between 0 and 2? Draw a graph of the density curve.

(b) Use your graph from (a) and the fact that areas under the curve are proportions of outcomes to find the proportion of outcomes that are less than 1.

(c) Find the proportion of outcomes that lie between 0.5 and 1.3.

1.82   What are the mean and the median of the uniform distribution in Figure 1.35? What are the quartiles?

1.83   Figure 1.36 displays three density curves, each with three points marked on it. At which of these points on each curve do the mean and the median fall?

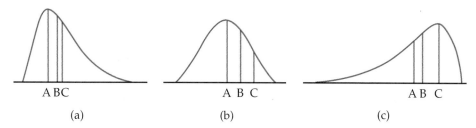

A BC   A B C   A B C

(a)   (b)   (c)

FIGURE 1.36   Three density curves, for Exercise 1.83.

1.84   The length of human pregnancies from conception to birth varies according to a distribution that is approximately normal with mean 266 days and standard deviation 16 days. Draw a density curve for this distribution on which the mean and standard deviation are correctly located.

1.85   The 68–95–99.7 rule for normal distributions is a useful approximation. You can use the *Normal Curve* applet on the text CD and Web site to see how accurate the rule is. Drag one flag across the other so that the applet shows the area under the curve between the two flags.

(a) Place the flags one standard deviation on either side of the mean. What is the area between these two values? What does the 68–95–99.7 rule say this area is?

(b) Repeat for locations two and three standard deviations on either side of the mean. Again compare the 68–95–99.7 rule with the area given by the applet.

1.86 The length of human pregnancies from conception to birth varies according to a distribution that is approximately normal with mean 266 days and standard deviation 16 days. Use the 68–95–99.7 rule to answer the following questions.

(a) Between what values do the lengths of the middle 95% of all pregnancies fall?

(b) How short are the shortest 2.5% of all pregnancies? How long do the longest 2.5% last?

1.87 Bigger animals tend to carry their young longer before birth. The length of horse pregnancies from conception to birth varies according to a roughly normal distribution with mean 336 days and standard deviation 3 days. Use the 68–95–99.7 rule to answer the following questions.

(a) Almost all (99.7%) horse pregnancies fall in what range of lengths?

(b) What percent of horse pregnancies are longer than 339 days?

1.88 One reason that normal distributions are important is that they describe how the results of an opinion poll would vary if the poll were repeated many times. About 40% of adult Americans say they are afraid to go out at night because of crime. Take many randomly chosen samples of 1050 people. The proportions of people in these samples who stay home for fear of crime will follow the normal distribution with mean 0.4 and standard deviation 0.015. Use this fact and the 68–95–99.7 rule to answer these questions.

(a) In many samples, what percent of samples give results above 0.4? Above 0.43?

(b) In a large number of samples, what range contains the central 95% of proportions of people who stay home because of crime?

1.89 The heights of women aged 20 to 29 are approximately normal with mean 64 inches and standard deviation 2.7 inches. Men the same age have mean height 69.3 inches with standard deviation 2.8 inches. What are the z-scores for a woman 6 feet tall and a man 6 feet tall? What information do the z-scores give that the actual heights do not?

 1.90 Use the *Normal Curve* applet for the standard normal distribution to say how many standard deviations above and below the mean the quartiles of any Normal distribution lie.

1.91 The normal quantile plot in Figure 1.34 (page 83) shows that the acidity (pH) measurements for rainwater samples in Exercise 1.30 are approximately normal. How well do these scores satisfy the 68–95–99.7 rule? To find out, calculate the mean $\bar{x}$ and standard deviation s of the observations. Then calculate the percent of the 105 measurements that fall between $\bar{x} - s$ and $\bar{x} + s$ and compare your result with 68%. Do the same for the intervals covering two and three standard deviations on either side of the mean. (The 68–95–99.7 rule is exact for any theoretical normal distribution. It will hold only approximately for actual data.)

1.92   Using either Table A or your calculator or software, find the proportion of observations from a standard normal distribution that satisfies each of the following statements. In each case, sketch a standard normal curve and shade the area under the curve that is the answer to the question.

(a) $Z < 1.85$ (this is a cumulative proportion)

(b) $Z > 1.85$

(c) $Z > -0.66$

(d) $-0.66 < Z < 1.85$

1.93   Using either Table A or your calculator or software, find the proportion of observations from a standard normal distribution for each of the following events. In each case, sketch a standard normal curve and shade the area representing the proportion.

(a) $Z \leq -2$ (this is a cumulative proportion)

(b) $Z \geq -2$

(c) $Z > 1.67$

(d) $-2 < Z < 1.67$

1.94   Find the value z of a standard normal variable Z that satisfies each of the following conditions. (If you use Table A, report the value of z that comes closest to satisfying the condition.) In each case, sketch a standard normal curve with your value of z marked on the axis.

(a) 20% of the observations fall below z.

(b) 30% of the observations fall above z.

1.95   The variable Z has a standard normal distribution.

(a) Find the number z that has cumulative proportion 0.8.

(b) Find the number z such that the event $Z > z$ has proportion 0.45.

1.96   The Wechsler Adult Intelligence Scale (WAIS) is the most common "IQ test." The scale of scores is set separately for each age group and is approximately normal with mean 100 and standard deviation 15. People with WAIS scores below 70 are considered mentally retarded when, for example, applying for Social Security disability benefits. What percent of adults are retarded by this criterion?

1.97   The Wechsler Adult Intelligence Scale (WAIS) is the most common "IQ test." The scale of scores is set separately for each age group and is approximately normal with mean 100 and standard deviation 15. The organization MENSA, which calls itself "the high IQ society," requires a WAIS score of 130 or higher for membership. What percent of adults would qualify for membership?

There are two major tests of readiness for college, the ACT and the SAT. ACT scores are reported on a scale from 1 to 36. The distribution of ACT scores *for more than 1 million students in a recent high school graduating class was roughly normal with mean $\mu = 20.8$ and standard deviation $\sigma = 4.8$. SAT scores are reported on a scale from 400 to 1600. The SAT scores for 1.4 million students in the same graduating class were roughly normal with mean $\mu = 1026$ and standard deviation $\sigma = 209$. Exercises 1.98 to 1.109 are based on this information.*

1.98 Tonya scores 1318 on the SAT. Jermaine scores 27 on the ACT. Assuming that both tests measure the same thing, who has the higher score?

1.99 Jacob scores 16 on the ACT. Emily scores 670 on the SAT. Assuming that both tests measure the same thing, who has the higher score?

1.100 Jose scores 1287 on the SAT. Assuming that both tests measure the same thing, what score on the ACT is equivalent to Jose's SAT score?

1.101 Maria scores 28 on the ACT. Assuming that both tests measure the same thing, what score on the SAT is equivalent to Maria's ACT score?

1.102 Reports on a student's ACT or SAT usually give the percentile as well as the actual score. The percentile is just the cumulative proportion stated as a percent: the percent of all scores that were lower than this one. Tonya scores 1318 on the SAT. What is her percentile?

1.103 Reports on a student's ACT or SAT usually give the percentile as well as the actual score. The percentile is just the cumulative proportion stated as a percent: the percent of all scores that were lower than this one. Jacob scores 16 on the ACT. What is his percentile?

1.104 It is possible to score higher than 1600 on the SAT, but scores 1600 and above are reported as 1600. What proportion of SAT scores are reported as 1600?

1.105 It is possible to score higher than 36 on the ACT, but scores 36 and above are reported as 36. What proportion of ACT scores are reported as 36?

1.106 What SAT scores make up the top 10% of all scores?

1.107 How well must Abigail do on the ACT in order to place in the top 20% of all students?

1.108 The quartiles of any distribution are the values with cumulative proportions 0.25 and 0.75. What are the quartiles of the distribution of ACT scores?

1.109 The quintiles of any distribution are the values with cumulative proportions 0.20, 0.40, 0.60, and 0.80. What are the quintiles of the distribution of SAT scores?

1.110   Too much cholesterol in the blood increases the risk of heart disease. Young women are generally less afflicted with high cholesterol than other groups. The cholesterol levels for women aged 20 to 34 follow an approximately normal distribution with mean 185 milligrams per deciliter (mg/dl) and standard deviation 39 mg/dl.[33]

(a) Cholesterol levels above 240 mg/dl demand medical attention. What percent of young women have levels above 240 mg/dl?

(b) Levels above 200 mg/dl are considered borderline high. What percent of young women have blood cholesterol between 200 and 240 mg/dl?

1.111   Middle-aged men are more susceptible to high cholesterol than the young women of Exercise 1.110. The blood cholesterol levels of men aged 55 to 64 are approximately normal with mean 222 mg/dl and standard deviation 37 mg/dl. What percent of these men have high cholesterol (levels above 240 mg/dl)? What percent have borderline high cholesterol (between 200 and 240 mg/dl)?

1.112   Osteoporosis is a condition in which the bones become brittle due to loss of minerals. To diagnose osteoporosis, an elaborate apparatus measures bone mineral density (BMD). BMD is usually reported in standardized form. The standardization is based on a population of healthy young adults. The World Health Organization (WHO) criterion for osteoporosis is a BMD 2.5 standard deviations below the mean for young adults. BMD measurements in a population of people similar in age and sex roughly follow a normal distribution.

(a) What percent of healthy young adults have osteoporosis by the WHO criterion?

(b) Women aged 70 to 79 are of course not young adults. The mean BMD in this age is about −2 on the standard scale for young adults. Suppose that the standard deviation is the same as for young adults. What percent of this older population has osteoporosis?

1.113   Changing the mean of a normal distribution by a moderate amount can greatly change the percent of observations in the tails. Suppose that a college is looking for applicants with SAT math scores 750 and above.

(a) In 2003, the scores of men on the math SAT followed a normal distribution with mean 537 and standard deviation 116. What percent of men scored 750 or better?

(b) Women's scores that year had a normal distribution with mean 503 and standard deviation 110. What percent of women scored 750 or better? You see that the percent of men above 750 is more than two and one-half times the percent of women with such high scores.

1.114   The yearly rate of return on stock indexes (which combine many individual stocks) is approximately normal. Between 1900 and 2002, U.S. common stocks had a mean yearly return of 8.3%, with a standard deviation of about

20.3%.[34] Take this normal distribution to be the distribution of yearly returns over a long period.

(a) In what range do the middle 95% of all yearly returns lie?

(b) The market is down for the year if the return is less than zero. In what percent of years is the market down?

(c) In what percent of years does the index gain 25% or more?

1.115 The length of human pregnancies from conception to birth varies according to a distribution that is approximately normal with mean 266 days and standard deviation 16 days.

(a) What percent of pregnancies last less than 240 days (that's about 8 months)?

(b) What percent of pregnancies last between 240 and 270 days (roughly between 8 months and 9 months)?

(c) How long do the longest 20% of pregnancies last?

1.116 The quartiles of any distribution are the values with cumulative proportions 0.25 and 0.75.

(a) What are the quartiles of the standard normal distribution?

(b) Using your numerical values from (a), write an equation that gives the quartiles of the $N(\mu, \sigma)$ distribution in terms of μ and σ.

(c) The length of human pregnancies from conception to birth varies according to a distribution that is approximately normal with mean 266 days and standard deviation 16 days. Apply your result from (b): what are the quartiles of the distribution of lengths of human pregnancies?

1.117 Continue your work from the previous exercise. The interquartile range IQR is the distance between the first and third quartiles of a distribution.

(a) What is the value of the IQR for the standard normal distribution?

(b) There is a constant c such that $IQR = c\sigma$ for any normal distribution $N(\mu, \sigma)$. What is the value of c?

1.118 Continue your work from the previous two exercises. The percent of the observations that are suspected outliers according to the $1.5 \times IQR$ rule is the same for any normal distribution. What is this percent?

1.119 Figure 1.37 is a normal quantile plot of the heart rates of the 200 male runners in the study described in Exercise 1.61 (page 60). The distribution is close to normal. How can you see this? Describe the nature of the small deviations from normality that are visible in the plot.

1.120 Figure 1.38 is a normal quantile plot of the emissions of carbon dioxide (CO_2) per person in 48 countries, from Table 1.6 (page 33). In what way is this distribution nonnormal? Comparing the plot with Table 1.6, which countries would you call outliers?

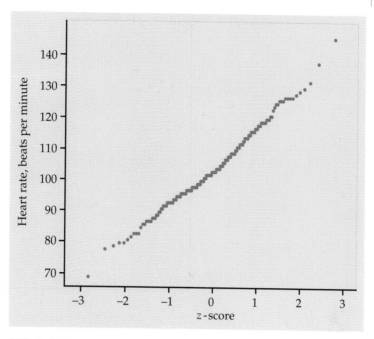

FIGURE 1.37 Normal quantile plot of the heart rates of 200 male runners, for Exercise 1.119.

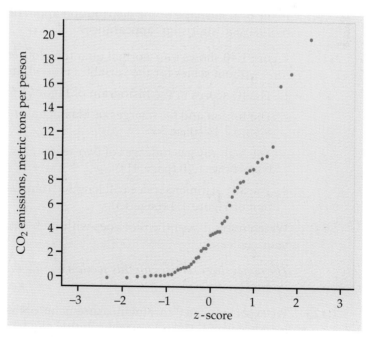

FIGURE 1.38 Normal quantile plot of CO_2 emissions in 48 countries, for Exercise 1.120.

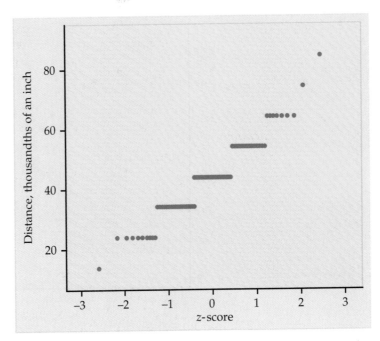

FIGURE 1.39 Normal quantile plot of distance between mounting holes, for Exercise 1.121.

1.121   The distance between two mounting holes is important to the performance of an electrical meter. The manufacturer measures this distance regularly for quality control purposes, recording the data as thousandths of an inch more than 0.600 inches. For example, 0.644 is recorded as 44. Figure 1.39 is a normal quantile plot of the distances for the last 90 electrical meters measured.[35] Is the overall shape of the distribution approximately normal? Why does the plot have a "stair-step" appearance?

1.122   Figure 1.40 shows four normal quantile plots for data that you have seen before, without scales for the variables plotted. In scrambled order, they are:

1.   The IQ scores in the histogram of Figure 1.5 (page 15).

2.   The tuition and fee charges of Massachusetts colleges in the histogram of Figure 1.14 (page 31).

3.   The highway gas mileages of two-seater cars, including the Honda Insight, from Table 1.10 (page 41).

4.   The 80 customer service call lengths from Table 1.1, displayed in the stemplot of Figure 1.4 (page 13).

Which normal quantile plot goes with each data set? Explain the reasons for your choices.

The remaining exercises for this section require the use of software that will make normal quantile plots.

1.123   We expect repeated careful measurements of the same quantity to be approximately normal. Make a normal quantile plot for Cavendish's measurements in Exercise 1.34 (page 37). Are the data approximately normal? If not, describe any clear deviations from normality.

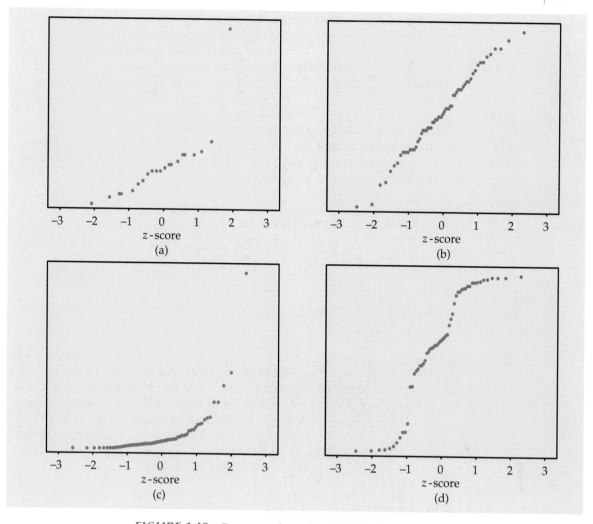

FIGURE 1.40   Four normal quantile plots, for Exercise 1.122.

1.124   The study of tropical flowers and their hummingbird pollinators (Exercise 1.58, page 59) measured lengths for three varieties of *Heliconia* flowers. We expect that such biological measurements will have roughly normal distributions.

(a) Make normal quantile plots for each of the three flower varieties. Which distribution is closest to normal?

(b) The other two distributions show the same kind of mild deviation from normality. In what way are these distributions nonnormal?

1.125   The study of the effects of logging on tree counts in the Borneo rainforest (Exercise 1.60, page 60) used statistical methods that are based on normal distributions. Make normal quantile plots for each of the three groups of forest plots. Are the three distributions roughly normal? What are the most prominent deviations from normality that you see?

1.126   Use software to generate 100 observations from the standard normal distribution. Make a histogram of these observations. How does the shape of the histogram compare with a normal density curve? Make a normal quantile plot of the data. Does the plot suggest any important deviations from normality? (Repeating this exercise several times is a good way to become familiar with how histograms and normal quantile plots look when data actually are close to normal.)

1.127 Use software to generate 100 observations from the uniform distribution described in Exercise 80 (page 84). Make a histogram of these observations. How does the histogram compare with the density curve in Figure 1.35? Make a normal quantile plot of your data. According to this plot, how does the uniform distribution deviate from normality?

CHAPTER 1 | Exercises

1.128 You want to measure the amount of "leisure time" that college students enjoy. Write a brief discussion of two issues:

(a) How will you define "leisure time"?

(b) Once you have defined leisure time, how will you measure Sally's leisure time this week?

1.129 Many plants and animals have "biological clocks" that coordinate activities with the time of day. When researchers looked at the length of the biological cycle in the plant *Arabidopsis* by measuring leaf movements, they found that the length of the cycle is not always 24 hours. Further study discovered that cycle length changes systematically with north-south location. Table 1.11 contains cycle lengths for 149 locations around the world.[36] Describe the distribution of cycle lengths with a histogram and numerical summaries. In particular, how much variation is there among locations?

TABLE 1.11

Biological clock cycle lengths for a plant species in different locations

23.89	23.72	23.74	24.35	25.05	24.56	23.69	22.33	23.79	22.12
25.39	23.08	25.64	23.98	25.84	25.46	24.37	24.13	24.40	24.74
24.44	24.82	23.56	24.96	24.21	23.85	24.57	23.44	23.64	24.23
24.01	24.58	25.57	23.73	24.11	23.21	25.08	24.03	24.62	23.51
23.21	23.41	23.69	22.97	24.65	24.65	24.29	23.89	25.08	23.89
24.95	23.09	23.21	24.66	23.88	25.33	24.38	24.68	25.34	25.22
23.45	23.39	25.43	23.16	23.95	23.25	24.72	24.89	24.88	24.71
23.58	25.98	24.28	24.25	23.16	24.19	27.22	23.77	26.21	24.33
24.34	24.89	24.32	24.14	24.00	23.48	25.81	24.99	24.18	22.73
24.18	23.95	24.48	23.89	24.24	24.96	24.58	24.29	24.31	23.64
23.87	23.68	24.87	23.00	23.48	24.26	23.34	25.11	24.69	24.97
24.64	24.49	23.61	24.07	26.60	24.91	24.76	25.09	26.56	25.13
24.81	25.63	25.63	24.69	24.41	23.79	22.88	22.00	23.33	25.12
24.00	24.31	23.03	24.51	28.55	22.96	23.61	24.72	24.04	25.18
24.30	24.22	24.39	24.73	24.68	24.14	24.57	24.42	25.62	

1.130   Product preference depends in part on the age, income, and gender of the consumer. A market researcher selects a large sample of potential car buyers. For each consumer, she records gender, age, household income, and automobile preference. Which of these variables are categorical and which are quantitative?

1.131   The 222 students enrolled in distance learning courses offered by a college ranged from 18 to 64 years of age. The mode of their ages was 19. The median age was 31.[37] Explain how this can happen.

1.132   Late in 2003, there were 77.4 million residential subscribers to Internet service in the United States. The numbers of subscribers claimed by the top 10 providers of service were as follows.[38] (There is some doubt about the accuracy of these claims.)

Service provider	Subscribers (millions)	Service provider	Subscribers (millions)
America Online	24.7	SBC	3.1
MSN	8.7	Verizon	2.1
United Online	5.2	Cox	1.8
EarthLink	5.0	Charter	1.5
Comcast	4.9	BellSouth	1.3

Display these data in a graph. How many subscribers do the many smaller providers have? Add an "Other" entry in your graph. Business people looking at this graph see an industry that offers opportunities for larger companies to take over.

1.133   The heights of people of the same sex and similar ages follow normal distributions reasonably closely. Weights, on the other hand, are not normally distributed. The weights of women aged 20 to 29 have mean 141.7 pounds and median 133.2 pounds. The first and third quartiles are 118.3 pounds and 157.3 pounds. What can you say about the shape of the weight distribution? Explain your reasoning.

1.134   What type of graph or graphs would you plan to make in a study of each of the following issues?

(a) What makes of cars do students drive? How old are their cars?

(b) How many hours per week do students study? How does the number of study hours change during a semester?

(c) Which radio stations are most popular with students?

(d) When many students measure the concentration of the same solution for a chemistry course laboratory assignment, do their measurements follow a normal distribution?

1.135   Rich and poor households differ in ways that go beyond income. Figure 1.41 displays histograms that compare the distributions of household size

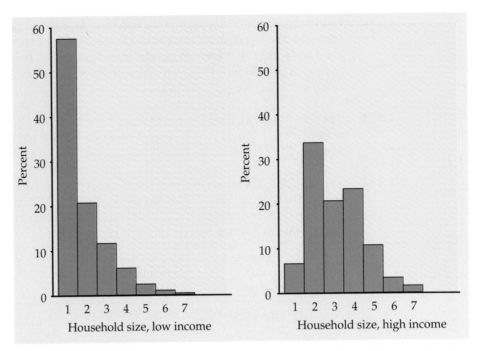

FIGURE 1.41   The distributions of household size for households with incomes less than $15,000 (*left*) and households with incomes of at least $100,000 (*right*), for Exercise 1.135.

(number of people) for low-income and high-income households in 2002.[39] Low-income households had incomes less than $15,000, and high-income households had incomes of at least $100,000.

(a) About what percent of each group of households consisted of two people?

(b) What are the important differences between these two distributions? What do you think explains these differences?

1.136   A university department installed a spam filter on its computer system. During a 21-day period, 6693 messages were tagged as spam. How much spam you get depends on what your online habits are. Here are the counts for some students and faculty in this department (with login IDs changed, of course):

ID	Count	ID	Count	ID	Count	ID	Count
AA	1818	BB	1358	CC	442	DD	416
EE	399	FF	389	GG	304	HH	251
II	251	JJ	178	KK	158	LL	103

All other department members received fewer than 100 spam messages. How many did the others receive in total? Make a graph and comment on what you learn from these data.

1.137   If two distributions have exactly the same mean and standard deviation, must their histograms have the same shape? If they have the same five-number summary, must their histograms have the same shape? Explain.

1.138 By-products from the pesticide DDT were major threats to the survival of birds of prey until use of DDT was banned at the end of 1972. Can time plots show the effect of the ban? Here are two sets of data for bald eagles nesting in the forests of northwestern Ontario.[40] The first data set gives the mean number of young per breeding area:

Year	1966	1967	1968	1969	1970	1971	1972	1973
Young	1.26	0.73	0.89	0.84	0.54	0.60	0.54	0.78

Year	1974	1975	1976	1977	1978	1979	1980	1981
Young	0.46	0.77	0.86	0.96	0.82	0.98	0.93	1.12

The second set of data are measurements of the chemical DDE (the by-product of DDT that most threatens birds of prey) from bald eagle eggs in the same area of Canada. These are in parts per million (ppm). There are often several measurements per year.

Year	1967	1967	1968	1971	1971	1972	1976	1976	1976	1976
DDE	44	95	121	125	95	87	13.3	16.4	50.4	59.8

Year	1976	1977	1977	1980	1980	1980	1981	1981	1981
DDE	56.4	0.6	23.8	16.6	14.5	24.0	7.8	48.2	53.4

Make time plots of eagle young and of mean DDE concentration in eggs. How does the effect of banning DDT appear in your plots?

1.139 Babe Ruth hit 60 home runs in 1927. Later, home-run king Mark McGwire hit 70 in 1998. A proper comparison of Ruth and McGwire should include their historical context. Here are the number of home runs by the major league leader for each year in baseball history, 1876 to 2003, in order from left to right. Make a time plot. (Be sure to add the scale of years.)

5	3	4	9	6	7	7	10	27	11	11	17	14	20	14	16	13
19	18	17	13	11	15	25	12	16	16	13	10	9	12	10	12	9
10	21	14	19	19	24	12	12	**11**	**29**	**54**	**59**	42	**41**	**46**	39	**47**
60	**54**	**46**	56	**46**	58	48	49	36	49	46	58	35	43	37	36	34
33	28	44	51	40	54	47	42	37	47	49	51	52	44	47	46	41
61	49	45	49	52	49	44	44	49	45	48	40	44	36	38	38	52
46	48	48	31	39	40	43	40	40	**49**	42	47	51	44	43	46	43
50	**52**	**58**	**70**	**65**	50	73	57	47								

(a) Describe the effect of World War II (1942 to 1945 seasons).

(b) Ruth led in the 11 years in boldface between 1918 and 1931. McGwire led in the 5 boldface years between 1987 and 1999. Briefly compare the achievements of Ruth and McGwire in the context of their times.

1.140 The single-season home run record is now held by Barry Bonds of the San Francisco Giants, who hit 73 in 2001. Here are Bonds's home run totals from

1986 (his first year) to 2003:

16 25 24 19 33 25 34 46 37 33 42 40 37 34 49 73 46 45

Make a stemplot of these data. Bonds's record year is a high outlier. How do his career mean and median number of home runs change when we drop the record 73? What general fact about the mean and median does your result illustrate?

1.141 Raw scores on behavioral tests are often transformed for easier comparison. A test of reading ability has mean 75 and standard deviation 10 when given to third graders. Sixth graders have mean score 82 and standard deviation 11 on the same test. To provide separate "norms" for each grade, we want scores in each grade to have mean 100 and standard deviation 20.

(a) What linear transformation will change third-grade scores x into new scores $x_{\text{new}} = a + bx$ that have the desired mean and standard deviation? (Use $b > 0$ to preserve the order of the scores.)

(b) Do the same for the sixth-grade scores.

(c) David is a third-grade student who scores 78 on the test. Find David's transformed score. Nancy is a sixth-grade student who scores 78. What is her transformed score? Who scores higher within his or her grade?

(d) Suppose that the distribution of scores in each grade is normal. Then both sets of transformed scores have the $N(100, 20)$ distribution. What percent of third graders have scores less than 78? What percent of sixth graders have scores less than 78?

1.142 The average damage caused by tornadoes in the states (Table 1.5, page 32) and the estimated amount of oil recovered from different oil wells (Exercise 1.33, page 37) both have right-skewed distributions. Choose one of these data sets. Make a normal quantile plot. How is the skewness of the distribution visible in the plot? Based on the plot, which observations (if any) would you call outliers?

1.143 We know that the distribution of the percents of state residents over 65 years of age has a low outlier (Alaska) and a high outlier (Florida). The stemplot in Exercise 1.15 (page 29) looks unimodal and roughly symmetric.

(a) Sketch what a normal quantile plot would look like for a distribution that is normal except for two outliers, one in each direction.

(b) If your software includes normal quantile plots, make a plot of the percent-over-65 data and discuss what you see.

1.144 Returns on common stocks are "heavy tailed." That is, they have more values far from the center in both the low and the high tails than a normal distribution would have. However, average returns for many individual stocks over longer periods of time become more normal.

(a) Sketch the appearance of a normal quantile plot for a distribution having roughly normal center and heavy tails. Explain the reasoning behind your sketch.

(b) The data file *ex01-144.dat* contains the annual returns for the years 1950 to 2003, pictured in the stemplot in Figure 1.20(a). If your software allows, make a normal quantile plot of these returns. Is the distribution clearly heavy tailed? Are there other clear deviations from normality?

1.145 Most statistical software packages have routines for generating values of variables having specified distributions. Use your statistical software to generate 25 observations from the $N(20, 5)$ distribution. Compute the mean and standard deviation $\bar{x}$ and s of the 25 values you obtain. How close are $\bar{x}$ and s to the μ and σ of the distribution from which the observations were drawn?

Repeat 19 more times the process of generating 25 observations from the $N(20, 5)$ distribution and recording $\bar{x}$ and s. Make a stemplot of the 20 values of $\bar{x}$ and another stemplot of the 20 values of s. Make normal quantile plots of both sets of data. Briefly describe each of these distributions. Are they symmetric or skewed? Are they roughly normal? Where are their centers? (The distributions of measures like $\bar{x}$ and s when repeated sets of observations are made from the same theoretical distribution will be very important in later chapters.)

1.146 Each March, the Bureau of Labor Statistics collects detailed information about more than 50,000 randomly selected households. The WORKERS data set contains data on 71,076 people from the March 2002 survey. All of these people were between 25 and 64 years of age and worked throughout the year. The Data Appendix describes this data set in detail. Describe the distribution of incomes for these people. Use graphs and numbers, and briefly state your main findings. Because this is a very large randomly selected sample, your results give a good description of incomes for all working Americans aged 25 to 64.

1.147 The CSDATA data set described in the Data Appendix contains information on 234 computer science students. We are interested in comparing the SAT mathematics scores and grade point averages of female students with those of male students. Make two sets of side-by-side boxplots to carry out these comparisons. Write a brief discussion of the male-female comparisons. Then make normal quantile plots of grade point averages and SAT math scores separately for men and women. Which students are clear outliers? Which of the four distributions are approximately normal if we ignore outliers?

EESEE CASE STUDIES

The Electronic Encyclopedia of Statistical Examples and Exercises (EESEE) is available on the text CD and Web site. These more elaborate stories, with data, provide settings for longer case studies.

1.148 Read the EESEE story "Is Old Faithful Faithful?" Write a response to Question 1 for this case study.

1.149 Read the EESEE story "Checkmating and Reading Skills." Write a report based on Questions 1 and 2 in this case study.

1.150 Read the EESEE story "Presidential Election 2000." Write a report that responds to Questions 1, 4, and 5 in this case study.

Tracking Disease in Puerto Rico

CYNTHIA PÉREZ *is Associate Professor of Biostatistics and Epidemiology at the University of Puerto Rico Graduate School of Public Health.*

Why is the burden of diabetes in Puerto Rico greater than in other states and territories of the United States? Who is at high risk for developing hepatitis C infection? Has there been a significant reduction in AIDS incidence in Puerto Rico since the beginning of the epidemic? Have the public health programs designed to reduce the spread of AIDS been effective? These questions relate to the discipline and practice of epidemiology.

After obtaining my master's degree in applied statistics, I pursued a doctoral degree in epidemiology, the basic science of public health aimed primarily at describing the patterns of disease occurrence in populations, testing hypotheses regarding disease causation, and applying these results to control health problems. For example, I recently participated in the design and conduct of a household survey to investigate the patterns of hepatitis C infection and associated factors among adults living in San Juan, Puerto Rico. Better understanding of the epidemiology of this infection will give us a broader picture of the need for hepatitis C counseling, testing, and referral and suggest new actions to fight the infection.

As an epidemiologist, I often use special regression models to compare the likelihoods that someone will develop a disease with and without being exposed to a particular risk factor. Statistical methods allow me to control for other sources of variation in the data and to judge whether the effects of the risk factor are real or due to random variation.

Complex interactions among genetics, environment, and lifestyle predispose individuals to noncommunicable and communicable health conditions. By applying rigorous epidemiological and statistical methods to investigate patterns of disease and other health conditions in populations, we can increase our understanding of their causes. Our goal is to develop effective methods for prevention and control, in order to improve the health of our communities.

WASSILY KANDINSKY, EIGHT TIMES © ARS, NY
(RÉUNION DES MUSÉES NATIONAUX/ART RESOURCE, NY)

Looking at Data—Relationships

Introduction

A medical study finds that short women are more likely to have heart attacks than women of average height, while tall women have the fewest heart attacks. An insurance group reports that heavier cars have fewer accident deaths per 100,000 vehicles registered than do lighter cars. These and many other statistical studies look at the relationship between two variables. To understand such a relationship, we must often examine other variables as well. To conclude that shorter women have higher risk from heart attacks, for example, the researchers had to eliminate the effect of other variables such as weight and exercise habits. Our topic in this chapter is relationships between variables. One of our main themes is that the relationship between two variables can be strongly influenced by other variables that are lurking in the background.

To study the relationship between two variables, we measure both variables on the same individuals. If we measure both the height and the weight of each of a large group of people, we know which height goes with each weight. These data allow us to study the connection between height and weight. A list of the heights and a separate list of the weights, two sets of single-variable data, do not show the connection between the two variables. In fact, taller people also tend to be heavier. And people who smoke more cigarettes per day tend not to live as long as those who smoke fewer. We say that pairs of variables such as height and weight or smoking and life expectancy are *associated*.

ASSOCIATION BETWEEN VARIABLES

Two variables measured on the same individuals are **associated** if some values of one variable tend to occur more often with some values of the second variable than with other values of that variable.

Statistical associations are overall tendencies, not ironclad rules. They allow individual exceptions. Although smokers on the average die earlier than nonsmokers, some people live to 90 while smoking three packs a day.

Examining relationships

When you examine the relationship between two or more variables, first ask the preliminary questions that are familiar from Chapter 1:

- What *individuals* do the data describe?

- What *variables* are present? How are they measured?

- Which variables are *quantitative* and which are *categorical?*

A medical study, for example, may record each subject's sex (female, male) and smoking status (current smoker, former smoker, never smoked) along with quantitative variables such as weight and blood pressure. We may be interested in possible associations between two quantitative variables (such as a person's weight and blood pressure), between a categorical and a quantitative variable (such as sex and blood pressure), or between two categorical variables (such as sex and smoking status).

When you examine the relationship between two variables, a new question becomes important:

- Is your purpose simply to explore the nature of the relationship, or do you hope to show that one of the variables can explain variation in the other? That is, are some of the variables *response variables* and others *explanatory variables?*

RESPONSE VARIABLE, EXPLANATORY VARIABLE

A **response variable** measures an outcome of a study. An **explanatory variable** explains or causes changes in the response variables.

It is easiest to identify explanatory and response variables when we actually set values of one variable in order to see how it affects another variable.

EXAMPLE 2.1   How does drinking beer affect the level of alcohol in our blood? The legal limit for driving in most states is 0.08%. Student volunteers at Ohio State University drank different numbers of cans of beer. Thirty minutes later, a police officer measured their blood alcohol content. Number of beers consumed is the explanatory variable, and percent of alcohol in the blood is the response variable.

When you don't set the values of either variable but just observe both variables, there may or may not be explanatory and response variables. Whether there are depends on how you plan to use the data.

EXAMPLE 2.2   A college student aid officer looks at the findings of the National Student Loan Survey. She notes data on the amount of debt of recent graduates, their current income, and how stressful they feel about college debt. She isn't interested in predictions but is simply trying to understand the situation of recent college graduates.

A sociologist looks at the same data with an eye to using amount of debt and income, along with other variables, to explain the stress caused by college debt. Now amount of debt and income are explanatory variables, and stress level is the response variable.

In many studies, the goal is to show that changes in one or more explanatory variables actually *cause* changes in a response variable. But many explanatory-response relationships do not involve direct causation. The SAT scores of high school students help predict the students' future college grades, but high SAT scores certainly don't cause high college grades.

Some of the statistical techniques in this chapter require us to distinguish explanatory from response variables; others make no use of this distinction.

independent variable   You will often see explanatory variables called **independent variables** and

dependent variable   response variables called **dependent variables.** The idea behind this language is that response variables depend on explanatory variables. Because the words "independent" and "dependent" have other meanings in statistics that are unrelated to the explanatory-response distinction, we prefer to avoid those words.

Most statistical studies examine data on more than one variable. Fortunately, statistical analysis of several-variable data builds on the tools used for examining individual variables. The principles that guide our work also remain the same:

- Start with a graphical display of the data.

- Look for overall patterns and deviations from those patterns.

- Based on what you see, use numerical summaries to describe specific aspects of the data.

2.1   Scatterplots

The most useful graph for displaying the relationship between two quantitative variables is a *scatterplot.*

SCATTERPLOT

A **scatterplot** shows the relationship between two quantitative variables measured on the same individuals. The values of one variable appear on the horizontal axis, and the values of the other variable appear on the vertical axis. Each individual in the data appears as the point in the plot fixed by the values of both variables for that individual.

Always plot the explanatory variable, if there is one, on the horizontal axis (the x axis) of a scatterplot. As a reminder, we usually call the explanatory variable x and the response variable y. If there is no explanatory-response distinction, either variable can go on the horizontal axis.

EXAMPLE 2.3   More than a million high school seniors take the SAT college entrance examination each year. We sometimes see the states "rated" by the average SAT scores of their seniors. For example, Illinois students average 1179 on the SAT, which looks better than the 1038 average of Massachusetts students. Rating states by SAT scores makes little sense, because average SAT score is largely explained by what percent of a state's students take the SAT. The scatterplot in Figure 2.1 allows us to see how the mean SAT score in each state is related to the percent of that state's high school seniors who take the SAT.[1]

Each point on the plot represents a single individual—that is, a single state. Because we think that the percent taking the exam influences mean score, percent taking is the explanatory variable and we plot it horizontally. For example, 20% of West

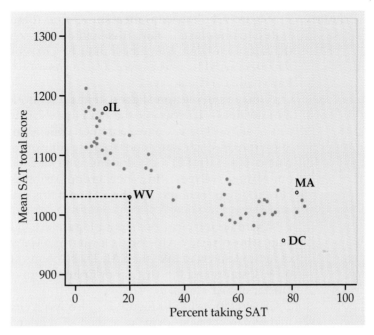

FIGURE 2.1   State mean SAT scores plotted against the percent of high school seniors in each state who take the SAT exams, for Example 2.3. The point for West Virginia (20% take the SAT, mean score 1032) is high-lighted.

Virginia high school seniors take the SAT, and their mean score is 1032. West Virginia appears as the point (20, 1032) in the scatterplot, above 20 on the x axis and to the right of 1032 on the y axis.

We see at once that state average SAT score is closely related to the percent of students who take the SAT. Illinois has a high mean score, but only 11% of Illinois seniors take the SAT. In Massachusetts, on the other hand, 82% of seniors take the exam.

Interpreting scatterplots

To look more closely at a scatterplot such as Figure 2.1, apply the strategies of exploratory analysis learned in Chapter 1.

EXAMINING A SCATTERPLOT

In any graph of data, look for the **overall pattern** and for striking **deviations** from that pattern.

You can describe the overall pattern of a scatterplot by the **form, direction,** and **strength** of the relationship.

An important kind of deviation is an **outlier,** an individual value that falls outside the overall pattern of the relationship.

clusters

Figure 2.1 shows an interesting *form:* there are two distinct **clusters** of states. In one cluster, more than half of high school seniors take the SAT, and the mean scores are low. Fewer than 40% of seniors in states in the other cluster take the SAT—fewer than 20% in most of these states—and these states have higher mean scores.

Clusters in a graph suggest that the data describe several distinct kinds of individuals. The two clusters in Figure 2.1 do in fact describe two distinct sets of states. There are two common college entrance examinations, the SAT and the ACT. Each state tends to prefer one or the other. In ACT states (the left cluster in Figure 2.1) most students who take the SAT are applying to selective out-of-state colleges. This select group performs well. In SAT states (the right cluster), many seniors take the SAT, and this broader group has a lower mean score.

There are no clear *outliers* in Figure 2.1, but each cluster does include a state whose mean SAT score is lower than we would expect from the percent of its students who take the SAT. These points are West Virginia in the cluster of ACT states and the District of Columbia (a city rather than a state) in the cluster of SAT states.

The relationship in Figure 2.1 also has a clear *direction:* states in which a higher percent of students take the SAT tend to have lower mean scores. This is true both between the clusters and within each cluster. This is a *negative association* between the two variables.

POSITIVE ASSOCIATION, NEGATIVE ASSOCIATION

Two variables are **positively associated** when above-average values of one tend to accompany above-average values of the other and below-average values also tend to occur together.

Two variables are **negatively associated** when above-average values of one accompany below-average values of the other, and vice versa.

---

linear relationship

When a scatterplot shows distinct clusters, it is often useful to describe the overall pattern separately within each cluster. The *form* of the relationship in the ACT states is roughly **linear.** That is, the points roughly follow a straight line. The *strength* of a relationship in a scatterplot is determined by how closely the points follow a clear form. The linear pattern among the ACT states is moderately strong because the points show only modest scatter about the straight-line pattern. In summary, the ACT states in Figure 2.1 show a moderately strong negative linear relationship. The cluster of SAT states shows a much weaker relationship between percent taking the SAT and mean SAT score.

Adding categorical variables to scatterplots

The Census Bureau groups the states into four broad regions, named Midwest, Northeast, South, and West. We might ask about regional patterns in SAT exam scores. Figure 2.2 repeats part of Figure 2.1, with an important difference. We have plotted only the Northeast and Midwest groups of states,

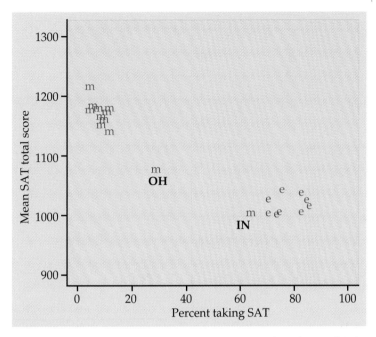

FIGURE 2.2   State mean SAT scores and percent taking the SAT for the northeastern states (plot symbol "e") and the midwestern states (plot symbol "m").

using the plot symbol "e" for the northeastern states and the symbol "m" for the midwestern states.

The regional comparison is striking. The 9 northeastern states are all SAT states—in fact, at least 70% of high school graduates in each of these states take the SAT. The 12 midwestern states are mostly ACT states. In 10 of these states, the percent taking the SAT is between 4% and 11%. One midwestern state is clearly an outlier within the region. Indiana is an SAT state (63% take the SAT) that falls close to the northeastern cluster. Ohio, where 28% take the SAT, also lies outside the midwestern cluster.

In dividing the states into regions, we introduced a third variable into the scatterplot. "Region" is a categorical variable that has four values, although we plotted data from only two of the four regions. The two regions are displayed by the two different plotting symbols.[2]

> ## CATEGORICAL VARIABLES IN SCATTERPLOTS
>
> To add a categorical variable to a scatterplot, use a different plot color or symbol for each category.

More examples of scatterplots

Experience in examining scatterplots is the foundation for more detailed study of relationships among quantitative variables. Here is an example with a pattern different from that in Figure 2.1.

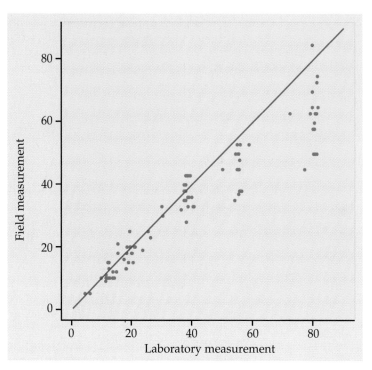

FIGURE 2.3 Depths of small defects in pipe for the Trans–Alaska Oil Pipeline, measured in the field and in the laboratory, for Example 2.4. If the two measurements were the same, the points would lie on the $y = x$ line that is drawn on the graph.

EXAMPLE 2.4 The Trans-Alaska Oil Pipeline is a tube formed from 1/2-inch thick steel that carries oil across 800 miles of sensitive arctic and subarctic terrain. The pipe and the welds that join pipe segments were carefully examined before installation. How accurate are field measurements of the depth of small defects? Figure 2.3 compares the results of measurements on 100 defects made in the field with measurements of the same defects made in the laboratory.[3] We plot the laboratory results on the x axis because they are a standard against which we compare the field results.

What is the overall pattern of this scatterplot? There is a positive linear association between the two variables. This is what we expect from two measurements of the same quantity. If field and laboratory measurements agree, the points will all fall on the $y = x$ line drawn on the plot, except for small random variations in the measurements. In fact, we see that the points for larger defects fall systematically below this line. That is, the field measurements are too small compared with the laboratory results as a standard. This is an important finding that can be used to adjust future field measurements.

Another part of the overall pattern is that the strength of the linear relationship decreases as the size of the defects increases. Field data show more variation (vertical spread in the plot) for large defect sizes than for small sizes. An increase in the spread in a response variable as the size of the response increases is a common pattern. It implies that predictions of the response based on the overall pattern will be less accurate for large responses.

Did you notice a fine point of graphing technique? Because both x and y measure the same thing, the graph is square and the same scales appear on both axes.

Some scatterplots appear quite different from the cloud of points in Figure 2.1 and the linear pattern in Figure 2.3. This is true, for example, in experiments in which measurements of a response variable are taken at a few selected levels of the explanatory variable. The following example illustrates the use of scatterplots in this setting.

EXAMPLE 2.5 Here is one way in which nature regulates the size of animal populations: high population density attracts predators, who remove a higher proportion of the population than when the density of the prey is low. One study looked at kelp perch and their common predator, the kelp bass. The researcher set up four large circular pens on sandy ocean bottom in southern California. He chose young perch at random from a large group and placed 10, 20, 40, and 60 perch in the four pens. Then he dropped the nets protecting the pens, allowing bass to swarm in, and counted the perch left after 2 hours. Here are data on the proportions of perch eaten in four repetitions of this setup:[4]

Perch	Proportion killed			
10	0.0	0.1	0.3	0.3
20	0.2	0.3	0.3	0.6
40	0.075	0.3	0.6	0.725
60	0.517	0.55	0.7	0.817

The scatterplot in Figure 2.4 displays the results of this experiment. Because number of perch in a pen is the explanatory variable, we plot it horizontally as the

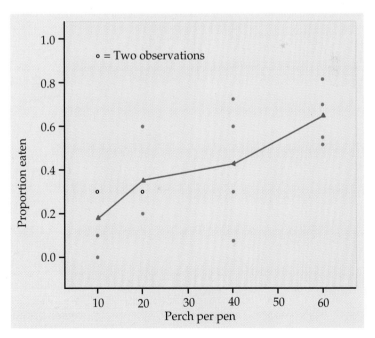

FIGURE 2.4 Data from an experiment in ecology: proportion of perch eaten by bass plotted against the number of perch present, for Example 2.5. The lines connect the mean responses (triangles) for each group.

x variable. The proportion of perch eaten by bass is the response variable *y*. Notice that there are two identical responses in the 10-perch group and also in the 20-perch group. These pairs of observations occupy the same points on the plot, so we use a different symbol for points that represent two observations. *Most software does not alert you to repeated values in your data when making scatterplots.* This can affect the impression the plot creates, especially when there are just a few points.

The vertical spread of points above each pen size shows the variation in proportions of perch eaten by bass. To see the overall pattern behind this variation, plot the mean response for each pen size. In Figure 2.4, these means are marked by small triangles and joined by line segments. There is a clear positive association between number of prey present and proportion eaten by predators. Moreover, the relationship is not far from linear.

BEYOND THE BASICS | Scatterplot smoothers

A scatterplot provides a complete picture of the relationship between two quantitative variables. A complete picture is often too detailed for easy interpretation, so we try to describe the plot in terms of an overall pattern and deviations from that pattern. Though we can often do this by eye, more systematic methods of extracting the overall pattern are helpful. This is called **smoothing** a scatterplot. Example 2.5 suggests how to proceed when we are plotting a response variable *y* against an explanatory variable *x*. We smoothed Figure 2.4 by averaging the *y*-values separately for each *x*-value. Though not all scatterplots have many *y*-values at the same value of *x*, as did Figure 2.4,

smoothing

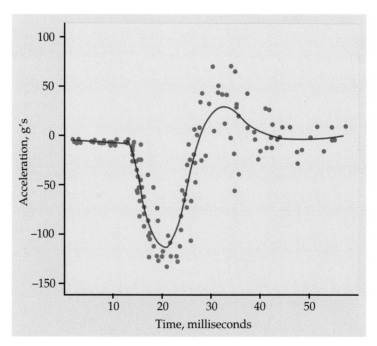

FIGURE 2.5 Smoothing a scatterplot, for Example 2.5. Time plot of the acceleration of the head of a crash test dummy as a motorcycle hits a wall, with the overall pattern calculated by a scatterplot smoother.

modern software provides scatterplot smoothers that form an overall pattern by looking at the *y*-values for points in the neighborhood of each *x*-value. Smoothers use *resistant* calculations, so they are not affected by outliers in the plot.

EXAMPLE 2.6   Crash a motorcycle into a wall. The rider, fortunately, is a dummy with an instrument to measure acceleration (change of velocity) mounted in its head. Figure 2.5 plots the acceleration of the dummy's head against time in milliseconds.[5] Acceleration is measured in g's, or multiples of the acceleration due to gravity at the earth's surface. The motorcycle approaches the wall at a constant speed (acceleration near 0). As it hits, the dummy's head snaps forward and decelerates violently (negative acceleration reaching more than 100 g's), then snaps back again (up to 75 g's) and wobbles a bit before coming to rest.

The scatterplot has a clear overall pattern, but it does not obey a simple form such as linear. Moreover, the strength of the pattern varies, from quite strong at the left of the plot to weaker (much more scatter) at the right. A scatterplot smoother deals with this complexity quite effectively and draws a line on the plot to represent the overall pattern.

Categorical explanatory variables

Scatterplots display the association between two quantitative variables. To display a relationship between a categorical explanatory variable and a quantitative response variable, make a side-by-side comparison of the distributions of the response for each category. We have already met some tools for such comparisons:

- A back-to-back stemplot compares two distributions. See the comparison of literacy rates (the quantitative response) for females and males (two categories) on page 12.

- Side-by-side boxplots compare any number of distributions. See the comparison of gas mileage (the quantitative response) for minicompact and two-seater cars on the highway and in the city (four categories) in Figure 1.17, page 47.

You can also use a type of scatterplot to display the association between a categorical explanatory variable and a quantitative response. Suppose, for example, that the prey-predator study of Example 2.5 had compared four species of prey rather than four densities of prey. The plot in Figure 2.4 remains helpful if we mark the prey species as A, B, C, and D at equal intervals on the horizontal axis in place of the count of perch per pen. A graph of the mean or median responses at the four locations still shows the overall nature of the relationship.

Many categorical variables, like prey species or type of car, have no natural order from smallest to largest. In such situations we cannot speak of a positive or negative association with the response variable. If the mean responses in our plot increase as we go from left to right, we could make them decrease by writing the categories in the opposite order. The plot simply presents a side-by-side comparison of several distributions. The categorical variable labels the distributions. Some categorical variables do have a least-to-most order, however. We can then speak of the direction of the association between the categorical explanatory variable and the quantitative response. Look again at the boxplots of income by level of education in Figure 1.21, page 61. Although the Census Bureau records education in categories, such as "did not graduate from high school," the categories have an order from less education to more education. The boxes in Figure 1.21 are arranged in order of increasing education. They show a positive association between education and income: people with more education tend to have higher incomes.

SECTION 2.1 | Summary

To study relationships between variables, we must measure the variables on the same group of individuals.

If we think that a variable x may explain or even cause changes in another variable y, we call x an **explanatory variable** and y a **response variable.**

A **scatterplot** displays the relationship between two quantitative variables. Mark values of one variable on the horizontal axis (x axis) and values of the other variable on the vertical axis (y axis). Plot each individual's data as a point on the graph.

Always plot the explanatory variable, if there is one, on the x axis of a scatterplot. Plot the response variable on the y axis.

Plot points with different colors or symbols to see the effect of a categorical variable in a scatterplot.

In examining a scatterplot, look for an overall pattern showing the **form, direction,** and **strength** of the relationship, and then for **outliers** or other deviations from this pattern.

Form: Linear relationships, where the points show a straight-line pattern, are an important form of relationship between two variables. Curved relationships and **clusters** are other forms to watch for.

Direction: If the relationship has a clear direction, we speak of either **positive association** (high values of the two variables tend to occur together) or **negative association** (high values of one variable tend to occur with low values of the other variable).

Strength: The **strength** of a relationship is determined by how close the points in the scatterplot lie to a simple form such as a line.

To display the relationship between a categorical explanatory variable and a quantitative response variable, make a graph that compares the distributions of the response for each category of the explanatory variable.

SECTION 2.1 | Exercises

2.1   In each of the following situations, is it more reasonable to simply explore the relationship between the two variables or to view one of the variables as an explanatory variable and the other as a response variable? In the latter case, which is the explanatory variable and which is the response variable?

(a) The amount of time spent studying for a statistics exam and the grade on the exam.

(b) The weight in kilograms and height in centimeters of a person.

(c) Inches of rain in the growing season and the yield of corn in bushels per acre.

(d) A student's scores on the SAT math exam and the SAT verbal exam.

(e) A family's income and the years of education their eldest child completes.

2.2   How well does the income of a college student's parents predict how much the student will borrow to pay for college? We have data on parents' income and college debt for a sample of 1200 recent college graduates. What are the explanatory and response variables? Are these variables categorical or quantitative? Do you expect a positive or negative association between these variables? Why?

2.3   A study of reading ability in schoolchildren chose 60 fifth-grade children at random from a school. The researchers had the children's scores on an IQ test and on a test of reading ability.[6] Figure 2.6 plots reading test score (response) against IQ score (explanatory).

(a) Explain why we should expect a positive association between IQ and reading score for children in the same grade. Does the scatterplot show a positive association?

(b) A group of four points appear to be outliers. In what way do these children's IQ and reading scores deviate from the overall pattern?

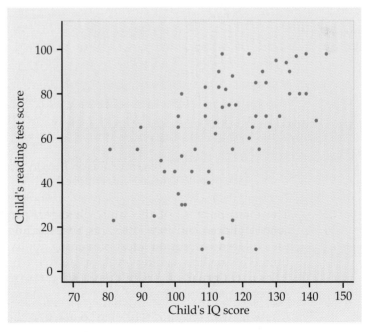

FIGURE 2.6   IQ and reading test scores for 60 fifth-grade children, for Exercise 2.3.

(c) Ignoring the outliers, is the association between IQ and reading scores roughly linear? Is it very strong? Explain your answers.

2.4 What is the relationship between returns from buying Treasury bills and returns from buying common stocks? The stemplots in Figure 1.20 (page 53) show the two individual distributions of percent returns. To see the relationship, we need a scatterplot. Figure 2.7 plots the annual returns on stocks for the years 1950 to 2003 against the returns on Treasury bills for the same years.

(a) The best year for stocks during this period was 1954. The worst year was 1974. About what were the returns on stocks in those two years?

(b) Treasury bills are a measure of the general level of interest rates. The years around 1980 saw very high interest rates. Treasury bill returns peaked in 1981. About what was the percent return that year?

(c) Some people say that high Treasury bill returns tend to go with low returns on stocks. Does such a pattern appear clearly in Figure 2.7? Does the plot have any clear pattern?

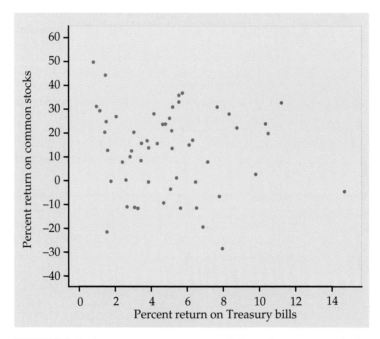

FIGURE 2.7   Percent returns on Treasury bills and common stocks for the years 1950 to 2003, for Exercise 2.4.

2.5 The main purpose of the study cited in Exercise 2.3 was to ask whether schoolchildren can estimate their own reading ability. The researchers had the children's scores on a test of reading ability. They asked each child to estimate his or her reading level, on a scale from 1 (low) to 5 (high). Figure 2.8 is a scatterplot of the children's estimates (response) against their reading scores (explanatory).

(a) What explains the "stair-step" pattern in the plot?

(b) Is there an overall positive association between reading score and self-estimate?

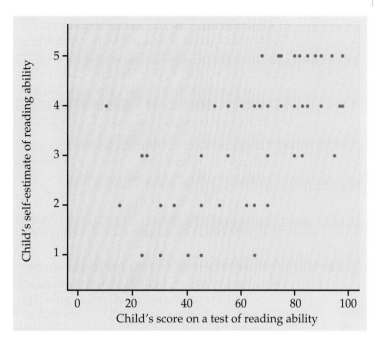

FIGURE 2.8 Reading test scores for 60 fifth-grade children and the children's estimates of their own reading levels, for Exercise 2.5.

(c) There is one clear outlier. What is this child's self-estimated reading level? Does this appear to over- or underestimate the level as measured by the test?

2.6 Table 1.2 (page 11) shows the percent of men and women at least 15 years old who were literate in 2002 in the major Islamic nations for which data were available. Make a scatterplot of these data, taking male literacy as the explanatory variable. Describe the direction, form, and strength of the relationship. Are there any identical observations that plot as the same point? Are there any clear outliers?

2.7 Often the percent of an animal species in the wild that survive to breed again is lower following a successful breeding season. This is part of nature's self-regulation, tending to keep population size stable. A study of merlins (small falcons) in northern Sweden observed the number of breeding pairs in an isolated area and the percent of males (banded for identification) who returned the next breeding season. Here are data for nine years:[7]

Pairs	28	29	29	29	30	32	33	38	38
Percent	82	83	70	61	69	58	43	50	47

(a) Why is the response variable the *percent* of males that return rather than the *number* of males that return?

(b) Make a scatterplot. To emphasize the pattern, also plot the mean response for years with 29 and 38 breeding pairs and draw lines connecting the mean responses for the six values of the explanatory variable.

(c) Describe the pattern. Do the data support the theory that a smaller percent of birds survive following a successful breeding season?

2.8   Table 1.10 (page 41) gives the city and highway gas mileages for minicompact and two-seater cars. We expect a positive association between the city and highway mileages of a group of vehicles. We have already seen that the Honda Insight is a different type of car, so omit it as you work with these data.

(a) Make a scatterplot that shows the relationship between city and highway mileage, using city mileage as the explanatory variable. Use different plotting symbols for the two types of cars.

(b) Interpret the plot. Is there a positive association? Is the form of the plot roughly linear? Is the form of the relationship similar for the two types of car? What is the most important difference between the two types?

2.9   We often describe our emotional reaction to social rejection as "pain." A clever study asked whether social rejection causes activity in areas of the brain that are known to be activated by physical pain. If it does, we really do experience social and physical pain in similar ways. Subjects were first included and then deliberately excluded from a social activity while increases in blood flow in their brains were measured. After each activity, the subjects filled out questionnaires that assessed how excluded they felt.

Here are data for 13 subjects.[8]   The explanatory variable is "social distress" measured by each subject's questionnaire score after exclusion relative to the score after inclusion. (So values greater than 1 show the degree of distress caused by exclusion.) The response variable is activity in the anterior cingulate cortex, a region of the brain that is activated by physical pain.

Subject	Social distress	Brain activity	Subject	Social distress	Brain activity
1	1.26	−0.055	8	2.18	0.025
2	1.85	−0.040	9	2.58	0.027
3	1.10	−0.026	10	2.75	0.033
4	2.50	−0.017	11	2.75	0.064
5	2.17	−0.017	12	3.33	0.077
6	2.67	0.017	13	3.65	0.124
7	2.01	0.021			

Plot brain activity against social distress. Describe the direction, form, and strength of the relationship, as well as any outliers. Do the data suggest that brain activity in the "pain" region really is directly related to the distress from social exclusion?

2.10   Many plants and animals have "biological clocks" that coordinate activities with the time of day. When researchers looked at the length of the biological cycles in the plant *Arabidopsis* by measuring leaf movements, they found that the length of the cycle is not always 24 hours. The researchers suspected that the plants adapt their clocks to their north-south position. Plants don't know geography, but they do respond to light, so the researchers looked at the relationship between the plants' cycle lengths and the length of the day on June 21 at their locations. The data file *ex02-010.dat* has data on cycle length and day length, both in hours, for 146 plants.[9] Plot cycle length as the response variable against day length as the explanatory variable. Does there appear to be a positive association? Is it a strong association? Explain your answers.

2.11   Management theory says that the value of a business should depend on its operating income, the income produced by the business after taxes. (Operating income excludes income from sales of assets and investments, which don't reflect the actual business.) Total revenue, which ignores costs, should be less important. Table 2.1 shows the values, operating incomes, and revenues of an unusual group of businesses: the teams in the National Basketball Association (NBA).[10] Professional sports teams are generally privately owned, often

TABLE 2.1

NBA teams as businesses

Team	Value ($millions)	Revenue ($millions)	Income ($millions)
Los Angeles Lakers	447	149	22.8
New York Knicks	401	160	13.5
Chicago Bulls	356	119	49.0
Dallas Mavericks	338	117	−17.7
Philadelphia 76ers	328	109	2.0
Boston Celtics	290	97	25.6
Detroit Pistons	284	102	23.5
San Antonio Spurs	283	105	18.5
Phoenix Suns	282	109	21.5
Indiana Pacers	280	94	10.1
Houston Rockets	278	82	15.2
Sacramento Kings	275	102	−16.8
Washington Wizards	274	98	28.5
Portland Trail Blazers	272	97	−85.1
Cleveland Cavaliers	258	72	3.8
Toronto Raptors	249	96	10.6
New Jersey Nets	244	94	−1.6
Utah Jazz	239	85	13.8
Miami Heat	236	91	7.9
Minnesota Timberwolves	230	85	6.9
Memphis Grizzlies	227	63	−19.7
Denver Nuggets	218	75	7.9
New Orleans Hornets	216	80	21.9
Los Angeles Clippers	208	72	15.9
Atlanta Hawks	202	78	−8.4
Orlando Magic	199	80	13.1
Seattle Supersonics	196	70	2.4
Golden State Warriors	188	70	7.8
Milwaukee Bucks	174	70	−15.1

by very wealthy individuals who may treat their team as a source of prestige rather than as a business.

(a) Plot team value against revenue. There are several outliers. Which teams are these, and in what way are they outliers? Is there a positive association between value and revenue? Is the pattern roughly linear?

(b) Now plot value against operating income. Are the same teams outliers? Does revenue or operating income do a better job of predicting the value of an NBA team?

2.12 Metatarsus adductus (call it MA) is a turning in of the front part of the foot that is common in adolescents and usually corrects itself. Hallux abducto valgus (call it HAV) is a deformation of the big toe that is not common in youth and often requires surgery. Perhaps the severity of MA can help predict the severity of HAV. Table 2.2 gives data on 38 consecutive patients who came to a medical center for HAV surgery.[11] Using X-rays, doctors measured the angle of deformity for both MA and HAV. They speculated that there is a positive association—more serious MA is associated with more serious HAV.

(a) Make a scatterplot of the data in Table 2.2. (Which is the explanatory variable?)

(b) Describe the form, direction, and strength of the relationship between MA angle and HAV angle. Are there any clear outliers in your graph?

(c) Do you think the data confirm the doctors' speculation?

TABLE 2.2

Two measurements of foot deformities

HAV angle	MA angle	HAV angle	MA angle	HAV angle	MA angle
28	18	21	15	16	10
32	16	17	16	30	12
25	22	16	10	30	10
34	17	21	7	20	10
38	33	23	11	50	12
26	10	14	15	25	25
25	18	32	12	26	30
18	13	25	16	28	22
30	19	21	16	31	24
26	10	22	18	38	20
28	17	20	10	32	37
13	14	18	15	21	23
20	20	26	16		

2.13 Metabolic rate, the rate at which the body consumes energy, is important in studies of weight gain, dieting, and exercise. The table below gives data on the lean body mass and resting metabolic rate for 12 women and 7 men who are subjects in a study of dieting. Lean body mass, given in kilograms, is a person's weight leaving out all fat. Metabolic rate is measured in calories burned per 24 hours, the same calories used to describe the energy content of foods. The researchers believe that lean body mass is an important influence on metabolic rate.

Subject	Sex	Mass	Rate	Subject	Sex	Mass	Rate
1	M	62.0	1792	11	F	40.3	1189
2	M	62.9	1666	12	F	33.1	913
3	F	36.1	995	13	M	51.9	1460
4	F	54.6	1425	14	F	42.4	1124
5	F	48.5	1396	15	F	34.5	1052
6	F	42.0	1418	16	F	51.1	1347
7	M	47.4	1362	17	F	41.2	1204
8	F	50.6	1502	18	M	51.9	1867
9	F	42.0	1256	19	M	46.9	1439
10	M	48.7	1614				

(a) Make a scatterplot of the data, using different symbols or colors for men and women.

(b) Is the association between these variables positive or negative? What is the form of the relationship? How strong is the relationship? Does the pattern of the relationship differ for women and men? How do the male subjects as a group differ from the female subjects as a group?

2.14 How does the fuel consumption of a car change as its speed increases? Here are data for a British Ford Escort. Speed is measured in kilometers per hour, and fuel consumption is measured in liters of gasoline used per 100 kilometers traveled.[12]

Speed (km/h)	Fuel used (liters/100 km)	Speed (km/h)	Fuel used (liter/100 km)
10	21.00	90	7.57
20	13.00	100	8.27
30	10.00	110	9.03
40	8.00	120	9.87
50	7.00	130	10.79
60	5.90	140	11.77
70	6.30	150	12.83
80	6.95		

(a) Make a scatterplot. (Which variable should go on the x axis?)

(b) Describe the form of the relationship. In what way is it not linear? Explain why the form of the relationship makes sense.

(c) It does not make sense to describe the variables as either positively associated or negatively associated. Why not?

(d) Is the relationship reasonably strong or quite weak? Explain your answer.

TABLE 2.3

World record times for the 10,000-meter run

Men				Women	
Record year	Time (seconds)	Record year	Time (seconds)	Record year	Time (seconds)
1912	1880.8	1962	1698.2	1967	2286.4
1921	1840.2	1963	1695.6	1970	2130.5
1924	1835.4	1965	1659.3	1975	2100.4
1924	1823.2	1972	1658.4	1975	2041.4
1924	1806.2	1973	1650.8	1977	1995.1
1937	1805.6	1977	1650.5	1979	1972.5
1938	1802.0	1978	1642.4	1981	1950.8
1939	1792.6	1984	1633.8	1981	1937.2
1944	1775.4	1989	1628.2	1982	1895.3
1949	1768.2	1993	1627.9	1983	1895.0
1949	1767.2	1993	1618.4	1983	1887.6
1949	1761.2	1994	1612.2	1984	1873.8
1950	1742.6	1995	1603.5	1985	1859.4
1953	1741.6	1996	1598.1	1986	1813.7
1954	1734.2	1997	1591.3	1993	1771.8
1956	1722.8	1997	1587.8		
1956	1710.4	1998	1582.7		
1960	1698.8	2004	1580.3		

2.15 Table 2.3 shows the progress of world record times (in seconds) for the 10,000-meter run up to mid-2004.[13] Concentrate first on the women's world record. Make a scatterplot with year as the explanatory variable. Describe the pattern of improvement over time that your plot displays.

2.16 How fast do icicles grow? Japanese researchers measured the growth of icicles in a cold chamber under various conditions of temperature, wind, and water flow.[14] Table 2.4 contains data produced under two sets of conditions. In both cases, there was no wind and the temperature was set at −11° C. Water flowed over the icicle at a higher rate (29.6 milligrams per second) in run 8905 and at a slower rate (11.9 mg/s) in run 8903.

(a) Make a scatterplot of the length of the icicle in centimeters versus time in minutes, using separate symbols for the two runs.

(b) Write a careful explanation of what your plots show about the growth of icicles.

2.17 Table 2.3 shows the progress of world record times (in seconds) for the 10,000-meter run for both men and women.

TABLE 2.4

Growth of icicles over time

	Run 8903				Run 8905		
Time (min)	Length (cm)	Time (min)	Length (cm)	Time (min)	Length (cm)	Time (min)	Length (cm)
10	0.6	130	18.1	10	0.3	130	10.4
20	1.8	140	19.9	20	0.6	140	11.0
30	2.9	150	21.0	30	1.0	150	11.9
40	4.0	160	23.4	40	1.3	160	12.7
50	5.0	170	24.7	50	3.2	170	13.9
60	6.1	180	27.8	60	4.0	180	14.6
70	7.9			70	5.3	190	15.8
80	10.1			80	6.0	200	16.2
90	10.9			90	6.9	210	17.9
100	12.7			100	7.8	220	18.8
110	14.4			110	8.3	230	19.9
120	16.6			120	9.6	240	21.1

(a) Make a scatterplot of world record time against year, using separate symbols for men and women. Describe the pattern for each sex. Then compare the progress of men and women.

(b) Women began running this long distance later than men, so we might expect their improvement to be more rapid. Moreover, it is often said that men have little advantage over women in distance running as opposed to sprints, where muscular strength plays a greater role. Do the data appear to support these claims?

2.18   To demonstrate the effect of nematodes (microscopic worms) on plant growth, a botanist introduces different numbers of nematodes into 16 planting pots. He then transplants a tomato seedling into each pot. Here are data on the increase in height of the seedlings (in centimeters) 14 days after planting:[15]

Nematodes	Seedling growth			
0	10.8	9.1	13.5	9.2
1,000	11.1	11.1	8.2	11.3
5,000	5.4	4.6	7.4	5.0
10,000	5.8	5.3	3.2	7.5

(a) Make a scatterplot of the response variable (growth) against the explanatory variable (nematode count). Then compute the mean growth for each group of seedlings, plot the means against the nematode counts, and connect these four points with line segments.

(b) Briefly describe the conclusions about the effects of nematodes on plant growth that these data suggest.

2.19 Fidelity Investments, like other large mutual-fund companies, offers many "sector funds" that concentrate their investments in narrow segments of the stock market. These funds often rise or fall by much more than the market as a whole. We can group them by broader market sector to compare returns. Here are percent total returns for 23 Fidelity "Select Portfolios" funds for the year 2003, a year in which stocks rose sharply:[16]

Market sector	Fund returns (percent)						
Consumer	23.9	14.1	41.8	43.9	31.1		
Financial services	32.3	36.5	30.6	36.9	27.5		
Technology	26.1	62.7	68.1	71.9	57.0	35.0	59.4
Natural resources	22.9	7.6	32.1	28.7	29.5	19.1	

(a) Make a plot of total return against market sector (space the four market sectors equally on the horizontal axis). Compute the mean return for each sector, add the means to your plot, and connect the means with line segments.

(b) Based on the data, which of these market sectors were the best places to invest in 2003? Hindsight is wonderful.

(c) Does it make sense to speak of a positive or negative association between market sector and total return?

2.20 The data for 2003 in the previous exercise make sector funds look attractive. Stocks rose sharply in 2003, after falling sharply in 2002 (and also in 2001 and 2000). Let's look at the percent returns for both 2003 and 2002 for these same 23 funds. Here they are:

2002 return	2003 return	2002 return	2003 return	2002 return	2003 return
−17.1	23.9	−0.7	36.9	−37.8	59.4
−6.7	14.1	−5.6	27.5	−11.5	22.9
−21.1	41.8	−26.9	26.1	−0.7	36.9
−12.8	43.9	−42.0	62.7	64.3	32.1
−18.9	31.1	−47.8	68.1	−9.6	28.7
−7.7	32.3	−50.5	71.9	−11.7	29.5
−17.2	36.5	−49.5	57.0	−2.3	19.1
−11.4	30.6	−23.4	35.0		

Do a careful graphical analysis of these data: side-by-side comparison of the distributions of returns in 2002 and 2003 and also a description of the relationship between the returns of the same funds in these two years. What are your most important findings? (The outlier is Fidelity Gold Fund.)

2.2 Correlation

A scatterplot displays the form, direction, and strength of the relationship between two quantitative variables. Linear (straight-line) relations are particularly important because a straight line is a simple pattern that is quite common. We say a linear relationship is strong if the points lie close to a straight line, and weak if they are widely scattered about a line. Our eyes are not good judges of how strong a relationship is. The two scatterplots in Figure 2.9 depict exactly the same data, but the lower plot is drawn smaller in a large field. The lower plot seems to show a stronger relationship. Our eyes can be fooled by changing the plotting scales or the amount of white space around the cloud of points in a scatterplot.[17] We need to follow our strategy for data analysis by using a numerical measure to supplement the graph. *Correlation* is the measure we use.

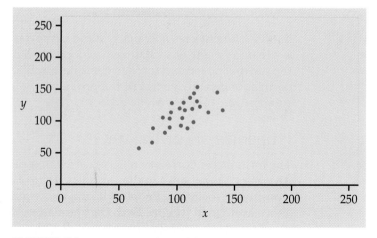

FIGURE 2.9 Two scatterplots of the same data. The linear pattern in the lower plot appears stronger because of the surrounding white space.

The correlation r

We have data on variables x and y for n individuals. Think, for example, of measuring height and weight for n people. Then x_1 and y_1 are your height and your weight, x_2 and y_2 are my height and my weight, and so on. For the ith individual, height x_i goes with weight y_i. Here is the definition of correlation.

> ### CORRELATION
>
> The **correlation** measures the direction and strength of the linear relationship between two quantitative variables. Correlation is usually written as r.
>
> Suppose that we have data on variables x and y for n individuals. The means and standard deviations of the two variables are $\bar{x}$ and s_x for the x-values, and $\bar{y}$ and s_y for the y-values. The correlation r between x and y is
>
> $$ r = \frac{1}{n-1} \sum \left(\frac{x_i - \bar{x}}{s_x} \right) \left(\frac{y_i - \bar{y}}{s_y} \right) $$

As always, the summation sign $\sum$ means "add these terms for all the individuals." The formula for the correlation r is a bit complex. It helps us see what correlation is but is not convenient for actually calculating r. In practice you should use software or a calculator that finds r from keyed-in values of two variables x and y. Exercise 2.21 asks you to calculate a correlation step-by-step from the definition to solidify its meaning.

The formula for r begins by standardizing the observations. Suppose, for example, that x is height in centimeters and y is weight in kilograms and that we have height and weight measurements for n people. Then $\bar{x}$ and s_x are the mean and standard deviation of the n heights, both in centimeters. The value

$$ \frac{x_i - \bar{x}}{s_x} $$

is the standardized height of the ith person, familiar from Chapter 1. The standardized height says how many standard deviations above or below the mean a person's height lies. Standardized values have no units—in this example, they are no longer measured in centimeters. Standardize the weights also. The correlation r is an average of the products of the standardized height and the standardized weight for the n people.

Properties of correlation

The formula for correlation helps us see that r is positive when there is a positive association between the variables. Height and weight, for example, have a positive association. People who are above average in height tend to also be above average in weight. Both the standardized height and the standardized weight for such a person are positive. People who are below average in height tend also to have below-average weight. Then both standardized height and standardized weight are negative. In both cases, the products in the formula for r are mostly positive and so r is positive. In the same way, we can see that r is negative when the association between x and y is negative. More detailed study of the formula gives more detailed properties of r. Here is what you need to know in order to interpret correlation:

- Correlation makes no use of the distinction between explanatory and response variables. It makes no difference which variable you call x and which you call y in calculating the correlation.

- *Correlation requires that both variables be quantitative, so that it makes sense to do the arithmetic indicated by the formula for r.* We cannot calculate a correlation between the incomes of a group of people and what city they live in, because city is a categorical variable.

- Because r uses the standardized values of the observations, r does not change when we change the units of measurement of x, y, or both. Measuring height in inches rather than centimeters and weight in pounds rather than kilograms does not change the correlation between height and weight. The correlation r itself has no unit of measurement; it is just a number.

- Positive r indicates positive association between the variables, and negative r indicates negative association.

- The correlation r is always a number between -1 and 1. Values of r near 0 indicate a very weak linear relationship. The strength of the relationship increases as r moves away from 0 toward either -1 or 1. Values of r close to -1 or 1 indicate that the points lie close to a straight line. The extreme values $r = -1$ and $r = 1$ occur only when the points in a scatterplot lie exactly along a straight line.

- Correlation measures the strength of only the linear relationship between two variables. *Correlation does not describe curved relationships between variables, no matter how strong they are.*

- *Like the mean and standard deviation, the correlation is not resistant: r is strongly affected by a few outlying observations.* Use r with caution when outliers appear in the scatterplot.

The scatterplots in Figure 2.10 illustrate how values of r closer to 1 or -1 correspond to stronger linear relationships. To make the essential meaning of r clear, the standard deviations of both variables in these plots are equal and the horizontal and vertical scales are the same. In general, it is not so easy to guess the value of r from the appearance of a scatterplot. Remember that changing the plotting scales in a scatterplot may mislead our eyes, but it does not change the standardized values of the variables and therefore cannot change the correlation. To explore how extreme observations can influence r, use the *Correlation and Regression* applet available on the text CD and Web site.

EXAMPLE 2.7 The real data we have examined also illustrate the behavior of correlation. Figure 2.1 (page 105), despite the clusters, shows a quite strong negative linear association between the percent of a state's high school seniors who take the SAT exam and their mean SAT score. The correlation is $r = -0.877$.

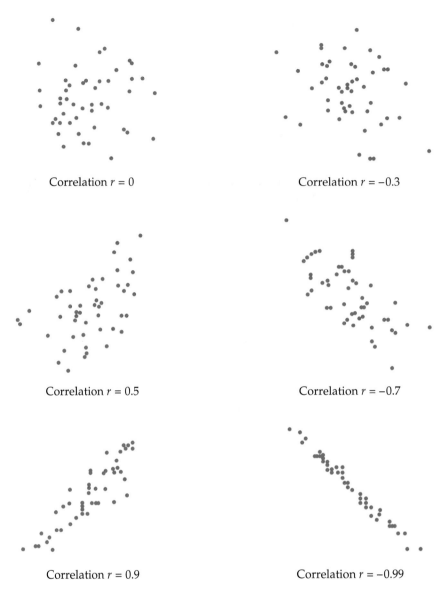

FIGURE 2.10   How the correlation r measures the direction and strength of linear association.

Figure 2.3 (page 108) shows a strong positive linear association between the two measurements of defect depth. The correlation is $r = 0.944$. That the pattern doesn't follow the $y = x$ line drawn on the graph doesn't matter—correlation measures closeness to whatever line describes the data, not to a specific line that we specify in advance.

Figure 2.7 (page 114) shows a very weak relationship between returns on Treasury bills and on common stocks. We expect a small negative r, and calculation gives $r = -0.113$.

The correlation between time and acceleration for the motorcycle crash data graphed in Figure 2.5 (page 110) is $r = 0.296$. Because the relationship is not at all linear, r provides no useful information. *Always plot your data before calculating common statistical measures such as correlation.*

Finally, remember that **correlation is not a complete description of two-variable data,** even when the relationship between the variables is linear. You should give the means and standard deviations of both x and y along with the correlation. (Because the formula for correlation uses the means and standard deviations, these measures are the proper choices to accompany a correlation.) Conclusions based on correlations alone may require rethinking in the light of a more complete description of the data.

EXAMPLE 2.8   Until a scandal at the 2002 Olympics brought change, figure skating was scored by judges on a scale from 0.0 to 6.0. The scores were often controversial. We have the scores awarded by two judges, Pierre and Elena, to many skaters. How well do they agree? We calculate that the correlation between their scores is $r = 0.9$. But the mean of Pierre's scores is 0.8 point lower than Elena's mean.

These facts do not contradict each other. They are simply different kinds of information. The mean scores show that Pierre awards lower scores than Elena. But because Pierre gives *every* skater a score about 0.8 point lower than Elena, the correlation remains high. Adding the same number to all values of either x or y does not change the correlation. If both judges score the same skaters, the competition is scored consistently because Pierre and Elena agree on which performances are better than others. The high r shows their agreement. But if Pierre scores some skaters and Elena others, we must add 0.8 points to Pierre's scores to arrive at a fair comparison.

SECTION 2.2 | Summary

The **correlation r** measures the direction and strength of the linear (straight line) association between two quantitative variables x and y. Although you can calculate a correlation for any scatterplot, r measures only linear relationships.

Correlation indicates the direction of a linear relationship by its sign: $r > 0$ for a positive association and $r < 0$ for a negative association.

Correlation always satisfies $-1 \leq r \leq 1$ and indicates the strength of a relationship by how close it is to -1 or 1. Perfect correlation, $r = \pm 1$, occurs only when the points lie exactly on a straight line.

Correlation ignores the distinction between explanatory and response variables. The value of r is not affected by changes in the unit of measurement of either variable. Correlation is not resistant, so outliers can greatly change the value of r.

SECTION 2.2 | Exercises

2.21   Coffee is a leading export from several developing countries. When coffee prices are high, farmers often clear forest to plant more coffee trees. Here are data on prices paid to coffee growers in Indonesia and the rate of deforestation in a national park that lies in a coffee-producing region for five years:[18]

Price (cents per pound)	Deforestation (percent)
29	0.49
40	1.59
54	1.69
55	1.82
72	3.10

(a) Make a scatterplot. Which is the explanatory variable? What kind of pattern does your plot show?

(b) Find the correlation r step-by-step. That is, find the mean and standard deviation of the two variables. Then find the five standardized values for each variable and use the formula for r. Explain how your value for r matches your graph in (a).

(c) Now enter these data into your calculator or software and use the correlation function to find r. Check that you get the same result as in (b).

2.22 Figure 2.4 (page 109) displays the positive association between number of prey (perch) present in an area and the proportion eaten by predators (bass).

(a) Do you think the correlation between these variables is closest to $r = 0.1$, $r = 0.6$, or $r = 0.9$? Explain the reason for your guess.

(b) Calculate the correlation. Was your guess correct?

2.23 Figure 2.6 (page 113) displays the positive association between the IQ scores of fifth-grade students and their reading scores. Do you think the correlation between these variables is closest to $r = 0.1$, $r = 0.6$, or $r = 0.9$? Explain the reason for your guess.

2.24 Mutual-fund reports often give correlations to describe how the prices of different investments are related. You look at the correlations between three Fidelity funds and the Standard & Poor's 500-Stock Index, which describes stocks of large U.S. companies. The three funds are Dividend Growth (stocks of large U.S. companies), Small Cap Stock (stocks of small U.S. companies), and Emerging Markets (stocks in developing countries). For 2003, the three correlations are $r = 0.35$, $r = 0.81$, and $r = 0.98$.[19]

(a) Which correlation goes with each fund? Explain your answer.

(b) The correlations of the three funds with the index are all positive. Does this tell you that stocks went up in 2003? Explain your answer.

2.25 Coffee is currently priced in dollars. If it were priced in euros, and the dollar prices in Exercise 2.21 were translated into the equivalent prices in euros, would the correlation between coffee price and percent deforestation change? Explain your answer.

2.26 Exercise 2.20 (page 122) gives data on the returns from 23 Fidelity "sector funds" in 2002 (a down year for stocks) and 2003 (an up year).

(a) Make a scatterplot if you did not do so in Exercise 2.20. Fidelity Gold Fund, the only fund with a positive return in both years, is an extreme outlier.

(b) To demonstrate that correlation is not resistant, find r for all 23 funds and then find r for the 22 funds other than Gold. Explain from Gold's position in your plot why omitting this point makes r more negative.

2.27 Table 2.1 (page 117) gives the values of the 29 teams in the National Basketball Association, along with their total revenues and operating incomes. You made scatterplots of value against both explanatory variables in Exercise 2.11.

(a) Find the correlations of team value with revenue and with operating income. Do you think that the two values of r provide a good first comparison of what the plots show about predicting value?

(b) Portland is an outlier in the plot of value against income. How does r change when you remove Portland? Explain from the position of this point why the change has the direction it does.

2.28 Your scatterplots for Exercises 2.10 (page 116) and 2.16 (Table 2.4, page 121) illustrate a quite weak linear association and a very strong linear association. Find the correlations that go with these plots. It isn't surprising that a laboratory experiment on physical behavior (the icicles) gives a much stronger correlation than field data on living things (the biological clock). How strong a correlation must be to interest scientists depends on the field of study.

2.29 A student wonders if tall women tend to date taller men than do short women. She measures herself, her dormitory roommate, and the women in the adjoining rooms; then she measures the next man each woman dates. Here are the data (heights in inches):

Women (x)	66	64	66	65	70	65
Men (y)	72	68	70	68	71	65

(a) Make a scatterplot of these data. Based on the scatterplot, do you expect the correlation to be positive or negative? Near ± 1 or not?

(b) Find the correlation r between the heights of the men and women.

(c) How would r change if all the men were 6 inches shorter than the heights given in the table? Does the correlation tell us whether women tend to date men taller than themselves?

(d) If heights were measured in centimeters rather than inches, how would the correlation change? (There are 2.54 centimeters in an inch.)

(e) If every woman dated a man exactly 3 inches taller than herself, what would be the correlation between male and female heights?

2.30   Make a scatterplot of the following data:

x	1	2	3	4	10	10
y	1	3	3	5	1	11

Use your calculator to show that the correlation is about 0.5. What feature of the data is responsible for reducing the correlation to this value despite a strong straight-line association between x and y in most of the observations?

2.31   You are going to use the *Correlation and Regression* applet to make different scatterplots with 10 points that have correlation close to 0.7. *Many patterns can have the same correlation. Always plot your data before you trust a correlation.*

(a) Stop after adding the first two points. What is the value of the correlation? Why does it have this value no matter where the two points are located?

(b) Make a lower-left to upper-right pattern of 10 points with correlation about $r = 0.7$. (You can drag points up or down to adjust r after you have 10 points.) Make a rough sketch of your scatterplot.

(c) Make another scatterplot with 9 points in a vertical stack at the left of the plot. Add one point far to the right and move it until the correlation is close to 0.7. Make a rough sketch of your scatterplot.

(d) Make yet another scatterplot with 10 points in a curved pattern that starts at the lower left, rises to the right, then falls again at the far right. Adjust the points up or down until you have a quite smooth curve with correlation close to 0.7. Make a rough sketch of this scatterplot also.

2.32   Exercise 2.14 (page 119) gives data on gas mileage against speed for a small car. Make a scatterplot if you have not already done so, then find the correlation r. Explain why r is close to zero despite a strong relationship between speed and gas used.

2.33   Table 1.10 (page 41) gives the city and highway gas mileages for 21 two-seater cars, including the Honda Insight gas-electric hybrid car.

(a) Make a scatterplot of highway mileage y against city mileage x for all 21 cars. There is a strong positive linear association. The Insight lies far from the other points. Does the Insight extend the linear pattern of the other cars, or is it far from the line they form?

(b) Find the correlation between city and highway mileages both without and with the Insight. Based on your answer to (a), explain why r changes in this direction when you add the Insight.

2.34   Go to the *Correlation and Regression* applet. Click on the scatterplot to create a group of 10 points in the lower-left corner of the scatterplot with a strong straight-line pattern (correlation about 0.9).

(a) Add one point at the upper right that is in line with the first 10. How does the correlation change?

(b) Drag this last point down until it is opposite the group of 10 points. How small can you make the correlation? Can you make the correlation negative? *A single outlier can greatly strengthen or weaken a correlation. Always plot your data to check for outlying points.*

2.35 Suppose that women always married men 2 years older than themselves. Draw a scatterplot of the ages of 5 married couples, with the wife's age as the explanatory variable. What is the correlation r for your data? Why?

2.36 Investment reports often include correlations. Following a table of correlations among mutual funds, a report adds, "Two funds can have perfect correlation, yet different levels of risk. For example, Fund A and Fund B may be perfectly correlated, yet Fund A moves 20% whenever Fund B moves 10%." Write a brief explanation, for someone who knows no statistics, of how this can happen. Include a sketch to illustrate your explanation.

2.37 A college newspaper interviews a psychologist about student ratings of the teaching of faculty members. The psychologist says, "The evidence indicates that the correlation between the research productivity and teaching rating of faculty members is close to zero." The paper reports this as "Professor McDaniel said that good researchers tend to be poor teachers, and vice versa." Explain why the paper's report is wrong. Write a statement in plain language (don't use the word "correlation") to explain the psychologist's meaning.

2.38 Each of the following statements contains a blunder. Explain in each case what is wrong.

(a) "There is a high correlation between the gender of American workers and their income."

(b) "We found a high correlation ($r = 1.09$) between students' ratings of faculty teaching and ratings made by other faculty members."

(c) "The correlation between planting rate and yield of corn was found to be $r = 0.23$ bushel."

2.39 Table 1.9 (page 39) reports data on 78 seventh-grade students. We expect a positive association between IQ and GPA. Moreover, some people think that self-concept is related to school performance. Examine in detail the relationships between GPA and the two explanatory variables IQ and self-concept. Are the relationships roughly linear? How strong are they? Are there unusual points? What is the effect of removing these points?

2.40 Consider again the correlation r between the speed of a car and its gas consumption, from the data in Exercise 2.14 (page 119).

(a) Transform the data so that speed is measured in miles per hour and fuel consumption in gallons per mile. (There are 1.609 kilometers in a mile and 3.785 liters in a gallon.) Make a scatterplot and find the correlation for both the original and the transformed data. How did the change of units affect your results?

(b) Now express fuel consumption in miles per gallon. (So each value is $1/x$ if x is gallons per mile.) Again make a scatterplot and find the correlation. How did this change of units affect your results?

(*Lesson:* The effects of a linear transformation of the form $x_{\text{new}} = a + bx$ are simple. The effects of a nonlinear transformation are more complex.)

2.3   Least-Squares Regression

Correlation measures the direction and strength of the linear (straight-line) relationship between two quantitative variables. If a scatterplot shows a linear relationship, we would like to summarize this overall pattern by drawing a line on the scatterplot. A *regression line* summarizes the relationship between two variables, but only in a specific setting: when one of the variables helps explain or predict the other. That is, regression describes a relationship between an explanatory variable and a response variable.

REGRESSION LINE

A **regression line** is a straight line that describes how a response variable y changes as an explanatory variable x changes. We often use a regression line to **predict** the value of y for a given value of x. Regression, unlike correlation, requires that we have an explanatory variable and a response variable.

EXAMPLE 2.9   Does fidgeting keep you slim? Some people don't gain weight even when they overeat. Perhaps fidgeting and other "nonexercise activity" (NEA) explains why—the body might spontaneously increase nonexercise activity when fed more. Researchers deliberately overfed 16 healthy young adults for 8 weeks. They measured fat gain (in kilograms) and, as an explanatory variable, increase in energy use (in calories) from activity other than deliberate exercise—fidgeting, daily living, and the like. Here are the data:[20]

NEA increase (cal)	−94	−57	−29	135	143	151	245	355
Fat gain (kg)	4.2	3.0	3.7	2.7	3.2	3.6	2.4	1.3

NEA increase (cal)	392	473	486	535	571	580	620	690
Fat gain (kg)	3.8	1.7	1.6	2.2	1.0	0.4	2.3	1.1

Figure 2.11 is a scatterplot of these data. The plot shows a moderately strong negative linear association with no outliers. The correlation is $r = -0.7786$. People with larger increases in nonexercise activity do indeed gain less fat. A line drawn through the points will describe the overall pattern well.

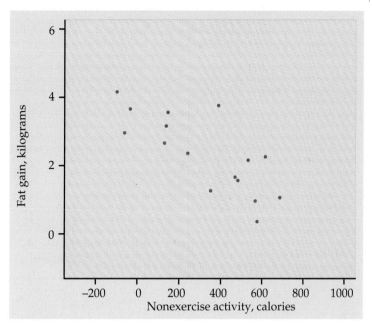

FIGURE 2.11   Weight gain after 8 weeks of overeating, plotted against increase in nonexercise activity over the same period, for Example 2.9.

Fitting a line to data

fitting a line   When a scatterplot displays a linear pattern, we can describe the overall pattern by drawing a straight line through the points. Of course, no straight line passes exactly through all of the points. **Fitting a line** to data means drawing a line that comes as close as possible to the points. The equation of a line fitted to the data gives a compact description of the dependence of the response variable y on the explanatory variable x.

> **STRAIGHT LINES**
>
> Suppose that y is a response variable (plotted on the vertical axis) and x is an explanatory variable (plotted on the horizontal axis). A straight line relating y to x has an equation of the form
>
> $$y = a + bx$$
>
> In this equation, b is the **slope,** the amount by which y changes when x increases by one unit. The number a is the **intercept,** the value of y when $x = 0$.

Any straight line describing the nonexercise activity data has the form

$$\text{fat gain} = a + (b \times \text{NEA increase})$$

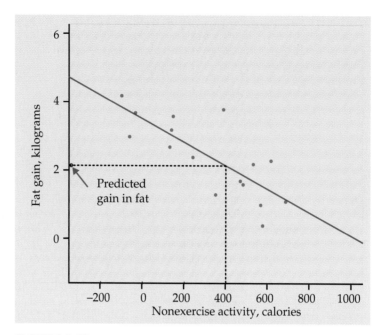

FIGURE 2.12 A regression line fitted to the nonexercise activity data and used to predict fat gain for an NEA increase of 400 calories.

In Figure 2.12 we have drawn the regression line with the equation

$$\text{fat gain} = 3.505 - (0.00344 \times \text{NEA increase})$$

The figure shows that this line fits the data well. The slope $b = -0.00344$ tells us that fat gained goes down by 0.00344 kilogram for each added calorie of NEA. The slope b of a line $y = a + bx$ is the *rate of change* in the response y as the explanatory variable x changes. The slope of a regression line is an important numerical description of the relationship between the two variables. The intercept, $a = 3.505$ kilograms, is the estimated fat gain if NEA does not change when a person overeats.

Prediction

prediction · We can use a regression line to **predict** the response y for a specific value of the explanatory variable x.

EXAMPLE 2.10 · Based on the linear pattern, we want to predict the fat gain for an individual whose NEA increases by 400 calories when she overeats. To use the fitted line to predict fat gain, go "up and over" on the graph in Figure 2.12. From 400 calories on the x axis, go up to the fitted line and over to the y axis. The graph shows that the predicted gain in fat is a bit more than 2 kilograms.

If we have the equation of the line, it is faster and more accurate to substitute $x = 400$ in the equation. The predicted fat gain is

$$\text{fat gain} = 3.505 - (0.00344 \times 400) = 2.13 \text{ kilograms}$$

The accuracy of predictions from a regression line depends on how much scatter about the line the data show. In Figure 2.12, fat gains for similar increases in NEA show a spread of 1 or 2 kilograms. The regression line summarizes the pattern but gives only roughly accurate predictions.

EXAMPLE 2.11   Can we predict the fat gain for someone whose nonexercise activity increases by 1500 calories when she overeats? We can certainly substitute 1500 calories into the equation of the line. The prediction is

$$\text{fat gain} = 3.505 - (0.00344 \times 1500) = -1.66 \text{ kilograms}$$

That is, we predict that this individual loses fat when she overeats. This prediction is not trustworthy. Look again at Figure 2.12. An NEA increase of 1500 calories is far outside the range of our data. We can't say whether increases this large ever occur, or whether the relationship remains linear at such extreme values. Predicting fat gain when NEA increases by 1500 calories *extrapolates* the relationship beyond what the data show.

EXTRAPOLATION

Extrapolation is the use of a regression line for prediction far outside the range of values of the explanatory variable x used to obtain the line. Such predictions are often not accurate.

Least-squares regression

Different people might draw different lines by eye on a scatterplot. This is especially true when the points are widely scattered. We need a way to draw a regression line that doesn't depend on our guess as to where the line should go. No line will pass exactly through all the points, but we want one that is as close as possible. We will use the line to predict y from x, so we want a line that is as close as possible to the points in the *vertical* direction. That's because the prediction errors we make are errors in y, which is the vertical direction in the scatterplot.

The line in Figure 2.12 predicts 2.13 kilograms of fat gain for an increase in nonexercise activity of 400 calories. If the actual fat gain turns out to be 2.3 kilograms, the error is

$$\text{error} = \text{observed gain} - \text{predicted gain}$$

$$= 2.3 - 2.13 = 0.17 \text{ kilograms}$$

Errors are positive if the observed response lies above the line, and negative if the response lies below the line. We want a regression line that makes these prediction errors as small as possible. Figure 2.13 illustrates the idea. For clarity, the plot shows only three of the points from Figure 2.12, along with the line, on an expanded scale. The line passes below two of the points and above one of them. The vertical distances of the data points from the line appear as vertical line segments. A "good" regression line makes these distances as small

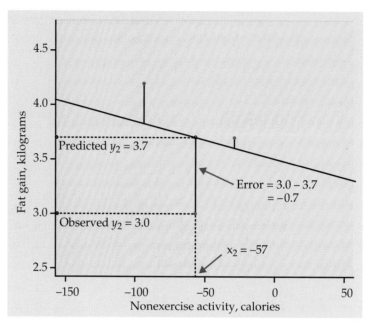

FIGURE 2.13 The least-squares idea: make the errors in predicting y as small as possible by minimizing the sum of their squares.

as possible. There are many ways to make "as small as possible" precise. The most common is the *least-squares* idea. The line in Figures 2.12 and 2.13 is in fact the least-squares regression line.

LEAST-SQUARES REGRESSION LINE

The **least-squares regression line of y on x** is the line that makes the sum of the squares of the vertical distances of the data points from the line as small as possible.

Here is the least-squares idea expressed as a mathematical problem. We represent n observations on two variables x and y as

$$(x_1, y_1), \ (x_2, y_2), \ \ldots, \ (x_n, y_n)$$

If we draw a line $y = a + bx$ through the scatterplot of these observations, the line predicts the value of y corresponding to x_i as $\hat{y}_i = a + bx_i$. We write $\hat{y}$ (read "y-hat") in the equation of a regression line to emphasize that the line gives a *predicted* response $\hat{y}$ for any x. The predicted response will usually not be exactly the same as the actually *observed* response y. The method of least squares chooses the line that makes the sum of the squares of these errors as small as possible. To find this line, we must find the values of the intercept a and the slope b that minimize

$$\sum (\text{error})^2 = \sum (y_i - a - bx_i)^2$$

for the given observations x_i and y_i. For the NEA data, for example, we must find the a and b that minimize

$$(-94 - a - 4.2b)^2 + (-57 - a - 3.0b)^2 + \cdots + (690 - a - 1.1b)^2$$

These values are the intercept and slope of the least-squares line.

You will use software or a calculator with a regression function to find the equation of the least-squares regression line from data on x and y. We will therefore give the equation of the least-squares line in a form that helps our understanding but is not efficient for calculation.

EQUATION OF THE LEAST-SQUARES REGRESSION LINE

We have data on an explanatory variable x and a response variable y for n individuals. The means and standard deviations of the sample data are $\bar{x}$ and s_x for x and $\bar{y}$ and s_y for y, and the correlation between x and y is r. The equation of the least-squares regression line of y on x is

$$\hat{y} = a + bx$$

with **slope**

$$b = r\frac{s_y}{s_x}$$

and **intercept**

$$a = \bar{y} - b\bar{x}$$

EXAMPLE 2.12   Verify from the data in Example 2.9 that the mean and standard deviation of the 16 increases in NEA are

$$\bar{x} = 324.8 \text{ calories} \quad \text{and} \quad s_x = 257.66 \text{ calories}$$

The mean and standard deviation of the 16 fat gains are

$$\bar{y} = 2.388 \text{ kg} \quad \text{and} \quad s_y = 1.1389 \text{ kg}$$

The correlation between height and age is $r = -0.7786$. The least-squares regression line of height y on age x therefore has slope

$$b = r\frac{s_y}{s_x} = -0.7786\frac{1.1389}{257.66}$$

$$= -0.00344 \text{ kg per calorie}$$

and intercept

$$a = \bar{y} - b\bar{x} = 2.388 + (0.00344)(324.8)$$

$$= 3.505 \text{ kg}$$

The equation of the least-squares line is

$$\hat{y} = 3.505 - 0.00344x$$

When doing calculations like this by hand, you may need to carry extra decimal places in the preliminary calculations to get accurate values of the slope and intercept. Using software or a calculator with a regression function eliminates this worry.

Interpreting the regression line

The slope $b = -0.00344$ kilograms per calorie in Example 2.12 is the rate of change in fat gain as NEA increases. The units "kilograms of fat gained per calorie of NEA" come from the units of y (kilograms) and x (calories). Although the correlation does not change when we change the units of measurement, the equation of the least-squares line does change. The slope in grams per calorie is would be 1000 times as large as the slope in kilograms per calorie, because there are 1000 grams in a kilogram. The small value of the slope, $b = -0.00344$, does not mean that the effect of increased NEA on fat gain is small—it just reflects the choice of kilograms as the unit for fat gain. *The slope and intercept of the least-squares line depend on the units of measurement—you can't conclude anything from their size.*

The expression $b = rs_y/s_x$ for the slope says that along the regression line, **a change of one standard deviation in x corresponds to a change of r standard deviations in y.** When the variables are perfectly correlated ($r = 1$ or $r = -1$), the change in the predicted response $\hat{y}$ is the same (in standard deviation units) as the change in x. Otherwise, because $-1 \le r \le 1$, the change in $\hat{y}$ is less than the change in x. As the correlation grows less strong, the prediction $\hat{y}$ moves less in response to changes in x.

The least-squares regression line always passes through the point $(\bar{x}, \bar{y})$ on the graph of y against x. Check that when you substitute $\bar{x} = 324.8$ into the equation of the regression line in Example 2.12, the result is $\hat{y} = 2.388$, equal to the mean of y. So the least-squares regression line of y on x is the line with slope rs_y/s_x that passes through the point $(\bar{x}, \bar{y})$. We can describe regression entirely in terms of the basic descriptive measures $\bar{x}$, s_x, $\bar{y}$, s_y, and r. If both x and y are standardized variables, so that their means are 0 and their standard deviations are 1, then the regression line has slope r and passes through the origin.

Figure 2.14 displays the basic regression output for the nonexercise activity data from two statistical software packages. Other software produces very similar output. You can find the slope and intercept of the least-squares line, calculated to more decimal places than we need, in both outputs. The software also provides information that we do not yet need, including some that we trimmed from Figure 2.14. Part of the art of using software is to ignore the extra information that is almost always present. Look for the results that you need. Once you understand a statistical method, you can read output from almost any software.

Correlation and regression

Least-squares regression looks at the distances of the data points from the line only in the y direction. So the two variables x and y play different roles in regression.

CrunchIt!

Minitab

FIGURE 2.14 Regression results for the nonexercise activity data from two statistical software packages. Other software produces similar output.

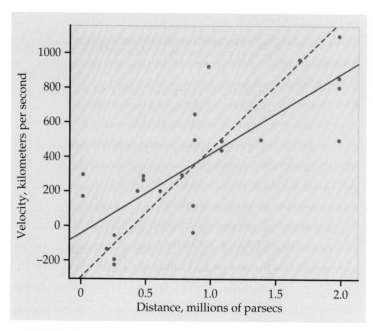

FIGURE 2.15   Hubble's data on the velocity and distance of 24 galaxies, for Example 2.13. The lines are least-squares regression lines of velocity on distance (solid) and of distance on velocity (dashed).

EXAMPLE 2.13   Figure 2.15 is a scatterplot of data that played a central role in the discovery that the universe is expanding. They are the distances from Earth of 24 spiral galaxies and the speed at which these galaxies are moving away from us, reported by the astronomer Edwin Hubble in 1929.[21] There is a positive linear relationship, $r = 0.7842$. More distant galaxies are moving away more rapidly. Astronomers believe that there is in fact a perfect linear relationship, and that the scatter is caused by imperfect measurements.

The two lines on the plot are the two least-squares regression lines. The regression line of velocity on distance is solid. The regression line of distance on velocity is dashed. *Although there is only one correlation between velocity and distance, regression of velocity on distance and regression of distance on velocity give different lines. In doing regression, you must choose which variable is explanatory.*

Even though the correlation r ignores the distinction between explanatory and response variables, there is a close connection between correlation and regression. We saw that the slope of the least-squares line involves r. Another connection between correlation and regression is even more important. In fact, the numerical value of r as a measure of the strength of a linear relationship is best interpreted by thinking about regression. Here is the fact we need.

The correlation between NEA increase and fat gain for the 16 subjects in Example 2.9 is $r = -0.7786$. Because $r^2 = 0.606$, the straight-line relationship between NEA and fat gain explains about 61% of the vertical scatter in fat gains in Figure 2.12. When you report a regression, give r^2 as a measure of how successfully the regression explains the response. Both software outputs in Figure 2.14 include r^2, either in decimal form or as a percent. When you see a correlation, square it to get a better feel for the strength of the association. Perfect correlation ($r = -1$ or $r = 1$) means the points lie exactly on a line. Then $r^2 = 1$ and all of the variation in one variable is accounted for by the linear relationship with the other variable. If $r = -0.7$ or $r = 0.7$, $r^2 = 0.49$ and about half the variation is accounted for by the linear relationship. In the r^2 scale, correlation ± 0.7 is about halfway between 0 and ± 1.

The use of r^2 to describe the success of regression in explaining the response y is very common. It rests on the fact that there are two sources of variation in the responses y in a regression setting. Figure 2.16 gives a rough visual picture of the two sources. One reason the fat gains of the subjects vary

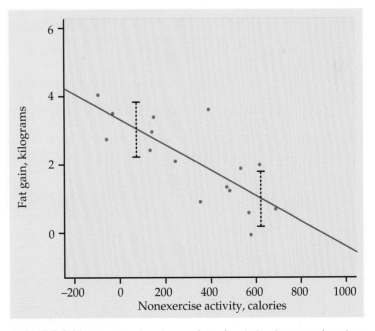

FIGURE 2.16   Explained and unexplained variation in regression. As x increases, it pulls y with it along the line. That is the variation explained by the regression. The scatter of the data points above and below the line, suggested by the dashed segments, is not explained by the regression.

is that there is a relationship between fat gain y and increase in NEA x. As x increases from −94 calories to 690 calories among the 16 subjects, it pulls fat gain y with it along the regression line in the figure. The linear relationship explains this part of the variation in fat gains.

The fat gains do not lie exactly on the line, however, but are scattered above and below it. This is the second source of variation in y, and the regression line tells us nothing about how large it is. The vertical dashed lines in Figure 2.16 show a rough average for the spread in y when we fix a value of x. We use r^2 to measure variation along the line as a fraction of the total variation in the fat gains. In Figure 2.16, about 61% of the variation in fat gains among the 16 subjects is due to the straight-line tie between y and x. The remaining 39% is vertical scatter in the observed responses remaining after the line has fixed the predicted responses.

*Understanding r^2

Here is a more specific interpretation of r^2. The fat gains y in Figure 2.16 range from 0.4 kilograms to 4.2 kilograms. The variance of these responses, a measure of how variable they are, is

$$\text{variance of observed values } y = 1.297$$

Much of this variability is due to the fact that as x increases from −94 calories to 690 calories it pulls height y along with it. If the only variability in the observed responses were due to the straight-line dependence of fat gain on NEA, the observed gains would lie exactly on the regression line. That is, they would be the same as the predicted gains $\hat{y}$. We can compute the predicted gains by substituting the NEA values for each subject into the equation of the least-squares line. Their variance describes the variability in the predicted responses. The result is

$$\text{variance of predicted values } \hat{y} = 0.786$$

This is what the variance would be if the responses fell exactly on the line, that is, if the linear relationship explained 100% of the observed variation in y. Because the responses don't fall exactly on the line, the variance of the predicted values is smaller than the variance of the observed values. Here is the fact we need:

$$r^2 = \frac{\text{variance of predicted values } \hat{y}}{\text{variance of observed values } y}$$
$$= \frac{0.786}{1.297} = 0.606$$

This fact is always true. The squared correlation gives the variance the responses would have if there were no scatter about the least-squares line as a fraction of the variance of the actual responses. This is the exact meaning of "fraction of variation explained" as an interpretation of r^2.

These connections with correlation are special properties of least-squares regression. They are not true for other methods of fitting a line to data. One

*This explanation is optional reading.

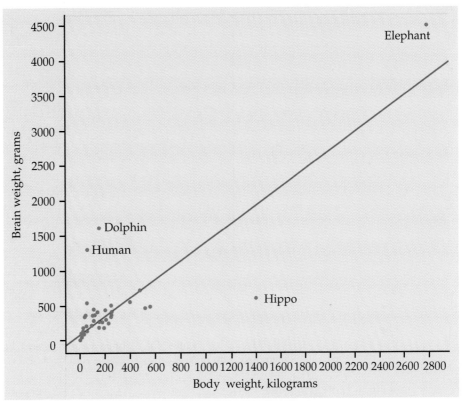

FIGURE 2.17 Scatterplot of brain weight against body weight for 96 species of mammals.

reason that least squares is the most common method for fitting a regression line to data is that it has many convenient special properties.

BEYOND THE BASICS | Transforming relationships

How is the weight of an animal's brain related to the weight of its body? Figure 2.17 is a scatterplot of brain weight against body weight for 96 species of mammals.[22] The line is the least-squares regression line for predicting brain weight from body weight. The outliers are interesting. We might say that dolphins and humans are smart, hippos are dumb, and African elephants are just big. That's because dolphins and humans have larger brains than their body weights suggest, hippos have smaller brains, and the elephant is much heavier than any other mammal in both body and brain.

EXAMPLE 2.14   The plot in Figure 2.17 is not very satisfactory. Most mammals are so small relative to elephants and hippos that their points overlap to form a blob in the lower-left corner of the plot. The correlation between brain weight and body weight is $r = 0.86$, but this is misleading. If we remove the elephant, the correlation for the other 95 species is $r = 0.50$. Figure 2.18 is a scatterplot of the data with the four outliers removed to allow a closer look at the other 92 observations. We can now see that the relationship is not linear. It bends to the right as body weight increases.

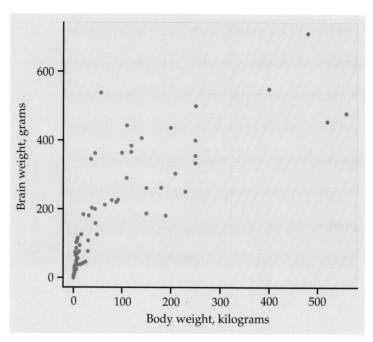

FIGURE 2.18   Scatterplot of brain weight against body weight for mammals, with outliers removed.

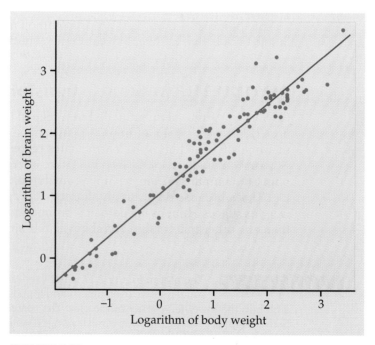

FIGURE 2.19   Scatterplot of the logarithm of brain weight against the logarithm of body weight for 96 species of mammals.

Biologists know that data on sizes often behave better if we take *logarithms* before doing more analysis. Figure 2.19 plots the logarithm of brain weight against the logarithm of body weight for all 96 species. The effect is almost magical. There are no longer any extreme outliers or very influential observations. The pattern is very linear, with correlation $r = 0.96$. The vertical spread about the least-squares line is similar everywhere, so that predictions of brain weight from body weight will be about equally precise for any body weight (in the log scale).

transforming

Example 2.14 shows that **transforming** data by applying a function such as the logarithm can greatly simplify statistical analysis. Transforming data is common in statistical practice. There are systematic principles that describe how transformations behave and guide the search for transformations that will, for example, make a distribution more normal or a curved relationship more linear. You can read about these principles in the extra material entitled *Transforming Relationships* available on the text CD and Web site.

SECTION 2.3 | Summary

A **regression line** is straight line that describes how a response variable y changes as an explanatory variable x changes.

The most common method of fitting a line to a scatterplot is least squares. The **least-squares regression line** is the straight line $\hat{y} = a + bx$ that minimizes the sum of the squares of the vertical distances of the observed y-values from the line.

You can use a regression line to **predict** the value of y for any value of x by substituting this x into the equation of the line. **Extrapolation** beyond the range of x-values spanned by the data is risky.

The **slope** b of a regression line $\hat{y} = a + bx$ is the rate at which the predicted response $\hat{y}$ changes along the line as the explanatory variable x changes. Specifically, b is the change in $\hat{y}$ when x increases by 1. The numerical value of the slope depends on the units used to measure x and y.

The **intercept** a of a regression line $\hat{y} = a + bx$ is the predicted response $\hat{y}$ when the explanatory variable $x = 0$. This prediction is of no statistical use unless x can actually take values near 0.

The least-squares regression line of y on x is the line with slope $b = rs_y/s_x$ and intercept $a = \bar{y} - b\bar{x}$. This line always passes through the point $(\bar{x}, \bar{y})$.

Correlation and regression are closely connected. The correlation r is the slope of the least-squares regression line when we measure both x and y in standardized units. The square of the correlation r^2 is the fraction of the variance of one variable that is explained by least-squares regression on the other variable.

SECTION 2.3 | Exercises

2.41   Figure 2.6 (page 113) plots the reading test scores of 60 fifth-grade students against their IQ scores. There are four outliers, children whose reading scores are very low for their IQs. If we eliminate these four children, there is a positive linear association. The least-squares regression line for predicting reading score from IQ, omitting the outliers, is

$$\text{reading score} = -33.4 + (0.882 \times \text{IQ})$$

(a) Draw a graph of this line for IQs between 90 and 140. (*Hint:* It is enough to find the predicted reading score for IQ 90 and IQ 140.)

(b) An eccentric professor believes that IQ 100 should produce reading score 50, and that reading score should increase by 1 point for every additional point of IQ. What is the equation of the professor's line?

2.42 Table 2.1 (page 117) gives the values of the 29 teams in the National Basketball Association, along with their operating incomes and revenues. Plots and correlations show that revenue predicts team value much better than does operating income. The least-squares regression line for predicting value from revenue is

$$\text{value} = 21.4 + (2.59 \times \text{revenue})$$

(a) What is the slope of this line? Say in simple language what the slope says about the relationship of value to revenue.

(b) The Los Angeles Lakers are the NBA's most valuable team, valued at $447 million, with $149 million in revenue. Use the line to predict the value of the Lakers from their revenue. What is the error in this prediction?

(c) The correlation between revenue and team value is $r = 0.9265$. What does the correlation say about the success of the regression line in predicting the values of the 29 teams?

2.43 Figure 1.8(b) (page 20) is a time plot of the volume of water discharged by the Mississippi River for the years 1954 to 2001. Water volume is recorded in cubic kilometers. The trend line on the plot is the least-squares regression line. The equation of this line is

$$\text{water discharged} = -7792 + (4.2255 \times \text{year})$$

(a) How much (on the average) does the volume of water increase with each passing year?

(b) What does the equation say about the volume of water flowing out of the Mississippi in the year 1780? Why is this extrapolation clearly nonsense?

(c) What is the predicted volume discharged in 1990 (round to the nearest cubic kilometer)? What is the prediction error for 1990?

(d) Can you see evidence of the great floods of 1973 and 1993, even on the plot of annual water discharged? Explain.

2.44 Example 2.5 (page 109) gives data from an experiment in ecology. Figure 2.4 is the scatterplot of proportion of perch eaten by bass against the number of perch in a pen before the bass were let in. There is a roughly linear pattern. The least-squares line for predicting proportion eaten from initial count of perch is

$$\text{proportion eaten} = 0.120 + (0.0086 \times \text{count})$$

(a) When 10 more perch are added to a pen, what happens to the proportion that are eaten (according to the line)? Explain your answer.

(b) If there are no perch in a pen, what proportion does the line predict will be eaten? Explain why this prediction is nonsense. What is wrong with using the regression line to predict *y* when *x* = 0? *You see that the intercept, though it is needed to draw the line, may have no statistical interpretation if x = 0 is outside the range of the data.*

2.45   Every few years, the National Assessment of Educational Progress asks a national sample of 17-year-olds to perform the same math tasks. The goal is to get an honest picture of progress in math. Here are the last few national mean scores, on a scale of 0 to 500:[23]

Year	1973	1978	1982	1986	1990	1992	1994	1996	1999
Score	304	300	298	302	305	307	306	307	308

(a) Make a time plot of the mean scores, by hand. This is just a scatterplot of score against year. There is a slow linear increasing trend.

(b) Find the regression line of mean score on time step-by-step. First calculate the mean and standard deviation of each variable and their correlation (use a calculator with these functions). Then find the equation of the least-squares line from these. Draw the line on your scatterplot. What percent of the year-to-year variation in scores is explained by the linear trend?

(c) Now use software or the regression function on your calculator to verify your regression line.

2.46   Figure 2.3 (page 108) plots field measurements on the depth of 100 small defects in the Trans-Alaska Oil Pipeline against laboratory measurements of the same defects. Drawing the *y* = *x* line on the graph shows that field measurements tend to be too low for larger defect depths. The data appear in the file *ex02-046.dat*.

(a) Find the equation of the least-squares regression line for predicting field measurement from laboratory measurement. Make a scatterplot with this line drawn on it. How does the least-squares line differ from the *y* = *x* line?

(b) What is the slope of the *y* = *x* line? What is the slope of the regression line? Say in simple language what these slopes mean.

2.47   Exercise 2.9 (page 116) gives data from a study that shows that social exclusion causes "real pain." That is, activity in the area of the brain that responds to physical pain goes up as distress from social exclusion goes up. Your scatterplot in Exercise 2.9 shows a moderately strong linear relationship.

(a) What is the equation of the least-squares regression line for predicting brain activity from social distress score? Make a scatterplot with this line drawn on it.

(b) On your plot, show the "up and over" lines that predict brain activity for social distress score 2.0. Use the equation of the regression line to get the predicted brain activity level. Verify that it agrees with your plot.

(c) What percent of the variation in brain activity among these subjects is explained by the straight-line relationship with social distress score?

2.48 Your scatterplot in Exercise 2.12 (page 118) suggests that the severity of the mild foot deformity called MA can help predict the severity of the more serious deformity called HAV. Table 2.2 (page 118) gives data for 38 young patients.

(a) Find the equation of the least-squares regression line for predicting HAV angle from MA angle. Add this line to the scatterplot you made in Exercise 2.12.

(b) A new patient has MA angle 25 degrees. What do you predict this patient's HAV angle to be?

(c) Does knowing MA angle allow doctors to predict HAV angle accurately? Explain your answer from the scatterplot, then calculate a numerical measure to support your finding.

2.49 Table 2.4 (page 121) gives data on the growth of icicles at two rates of water flow. You examined these data in Exercise 2.16. Use least-squares regression to estimate the rate (centimeters per minute) at which icicles grow at these two flow rates. How does flow rate affect growth?

2.50 Exercise 2.20 (page 122) gives the returns of 23 Fidelity "sector funds" for the years 2002 and 2003. These mutual funds invest in narrow segments of the stock market. They often rise faster than the overall market in up years such as 2003 and fall faster than the market in down years such as 2002. A scatterplot shows that Fidelity Gold Fund—the only fund that went up in 2002—is an outlier. In Exercise 2.26, you showed that this outlier has a strong effect on the correlation. The least-squares line, like the correlation, is not resistant.

(a) Find the equations of two least-squares lines for predicting 2003 return from 2002 return, one for all 23 funds and one omitting Fidelity Gold Fund. Make a scatterplot with both lines drawn on it. The two lines are very different.

(b) Starting with the least-squares idea, explain why adding Fidelity Gold Fund to the other 22 funds moves the line in the direction that your graph shows.

2.51 The scatterplot in Figure 2.7 (page 114) suggests that returns on common stocks may be somewhat lower in years with high interest rates. Here is part of the output from software for the regression of stock returns on the Treasury bill returns for the same years:

```
Stock = 16.639318 - 0.67974913 Tbill
Sample size: 54
R (correlation coefficient) = -0.113
R-sq = 0.01275773
Estimate of error standard deviation: 17.680649
```

If you knew the return on Treasury bills for next year, do you think you could predict the return on stocks quite accurately? Use both the scatterplot in Figure 2.7 and a number from the regression output to justify your answer.

2.52   Find the mean and standard deviation of the times and icicle lengths for the data on run 8903 in Table 2.4 (page 121). Find the correlation between the two variables. Use these five numbers to find the equation of the regression line for predicting length from time. Verify that your result agrees with that in Exercise 2.49. Use the same five numbers to find the equation of the regression line for predicting the time an icicle has been growing from its length. What units does the slope of each of these lines have?

2.53   Compute the mean and the standard deviation of the metabolic rates and lean body masses in Exercise 2.13 (page 118) and the correlation between these two variables. Use these values to find the slope of the regression line of metabolic rate on lean body mass. Also find the slope of the regression line of lean body mass on metabolic rate. What are the units for each of the two slopes?

2.54   Table 1.9 (page 39) reports data on 78 seventh-grade students. We want to know how well each of IQ score and self-concept score predicts GPA using least-squares regression. We also want to know which of these explanatory variables predicts GPA better. Give numerical measures that answer these questions, and explain your answers.

2.55   The mean height of American women in their early twenties is about 64.5 inches and the standard deviation is about 2.5 inches. The mean height of men the same age is about 68.5 inches, with standard deviation about 2.7 inches. If the correlation between the heights of husbands and wives is about $r = 0.5$, what is the equation of the regression line of the husband's height on the wife's height in young couples? Draw a graph of this regression line. Predict the height of the husband of a woman who is 67 inches tall.

 2.56   Use the equation for the least-squares regression line to show that this line always passes through the point $(\bar{x}, \bar{y})$.

2.57   The data for run 8903 in Table 2.4 (page 121) describe how the length y in centimeters of an icicle increases over time x. Time is measured in minutes.

(a) What are the numerical values and units of measurement for each of $\bar{x}$, s_x, $\bar{y}$, s_y, and the correlation r between x and y?

(b) There are 2.54 centimeters in an inch. If we measure length y in inches rather than in centimeters, what are the new values of $\bar{y}$, s_y, and the correlation r?

(c) If we measure length y in inches rather than in centimeters, what is the new value of the slope b of the least-squares line for predicting length from time?

2.58   In Professor Friedman's economics course the correlation between the students' total scores before the final examination and their final examination scores is $r = 0.6$. The pre-exam totals for all students in the course have mean 280 and standard deviation 30. The final exam scores have mean 75 and standard deviation 8. Professor Friedman has lost Julie's final exam but knows that her total before the exam was 300. He decides to predict her final exam score from her pre-exam total.

(a) What is the slope of the least-squares regression line of final exam scores on pre-exam total scores in this course? What is the intercept?

(b) Use the regression line to predict Julie's final exam score.

(c) Julie doesn't think this method accurately predicts how well she did on the final exam. Calculate r^2 and use the value you get to argue that her actual score could have been much higher or much lower than the predicted value.

2.59 A study of class attendance and grades among first-year students at a state university showed that in general students who attended a higher percent of their classes earned higher grades. Class attendance explained 16% of the variation in grade index among the students. What is the numerical value of the correlation between percent of classes attended and grade index?

2.60 Fenthion is a pesticide used to control the olive fruit fly. There are government limits on the amount of pesticide residue that can be present in olive products. Because the pesticide decays over time, producers of olive oil might simply store the oil until the fenthion has decayed. The simple exponential decay model says that the concentration C of pesticide remaining after time t is

$$C = C_0 e^{-kt}$$

where C_0 is the initial concentration and k is a constant that determines the rate of decay. This model is a straight line if we take the logarithm of the concentration:

$$\log C = \log C_0 - kt$$

(The logarithm here is the natural logarithm, not the common logarithm with base 10.) Here are data on the concentration (milligrams of fenthion per kilogram of oil) in specimens of Greek olive oil:[24]

Days stored	Concentration				
28	0.99	0.99	0.96	0.95	0.93
84	0.96	0.94	0.91	0.91	0.90
183	0.89	0.87	0.86	0.85	0.85
273	0.87	0.86	0.84	0.83	0.83
365	0.83	0.82	0.80	0.80	0.79

(a) Plot the natural logarithm of concentration against days stored. Notice that there are several pairs of identical data points. Does the pattern suggest that the model of simple exponential decay describes the data reasonably well, at least over this interval of time? Explain your answer.

(b) Regress the logarithm of concentration on time. Use your result to estimate the value of the constant k.

2.61 Unfortunately, the main product of the decay of the pesticide fenthion is fenthion sulfoxide, which is also toxic. Here are data on the total concentration of fenthion and fenthion sulfoxide in the same specimens of olive oil described in the previous exercise:

Days stored	Concentration				
28	1.03	1.03	1.01	0.99	0.99
84	1.05	1.04	1.00	0.99	0.99
183	1.03	1.02	1.01	0.98	0.98
273	1.07	1.06	1.03	1.03	1.02
365	1.06	1.02	1.01	1.01	0.99

(a) Plot concentration against days stored. Your software may fill the available space in the plot, which in this case hides the pattern. Try a plot with vertical scale from 0.8 to 1.2. Be sure your plot takes note of the pairs of identical data points.

(b) What is the slope of the least-squares line for predicting concentration of fenthion and fenthion sulfoxide from days stored? Explain why this value agrees with the graph.

(c) What do the data say about the idea of reducing fenthion in olive oil by storing the oil before selling it?

2.4   Cautions about Correlation and Regression

Correlation and regression are among the most common statistical tools. They are used in more elaborate form to study relationships among many variables, a situation in which we cannot see the essentials by studying a single scatterplot. We need a firm grasp of the use and limitations of these tools, both now and as a foundation for more advanced statistics.

Residuals

A regression line describes the overall pattern of a linear relationship between an explanatory variable and a response variable. Deviations from the overall pattern are also important. In the regression setting, we see deviations by looking at the scatter of the data points about the regression line. The vertical distances from the points to the least-squares regression line are as small as possible in the sense that they have the smallest possible sum of squares. Because they represent "left-over" variation in the response after fitting the regression line, these distances are called *residuals*.

> **RESIDUALS**
>
> A **residual** is the difference between an observed value of the response variable and the value predicted by the regression line. That is,
>
> $$\text{residual} = \text{observed } y - \text{predicted } y$$
> $$= y - \hat{y}$$

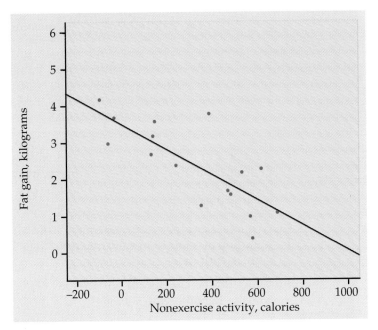

FIGURE 2.20   (a) Scatterplot of fat gain versus increase in nonexercise activity, with the least-squares line, for Example 2.15.

EXAMPLE 2.15   Example 2.9 (page 132) describes measurements on 16 young people who volunteered to overeat for 8 weeks. Those whose nonexercise activity (NEA) spontaneously rose substantially gained less fat than others. Figure 2.20(a) is a scatterplot of these data. The pattern is linear. The least-squares line is

$$\text{fat gain} = 3.505 - (0.00344 \times \text{NEA increase})$$

One subject's NEA rose by 135 calories. That subject gained 2.7 kilograms of fat. The predicted gain for 135 calories is

$$\hat{y} = 3.505 - (0.00344 \times 135) = 3.04 \text{ kg}$$

The residual for this subject is therefore

$$\text{residual} = \text{observed } y - \text{predicted } y$$
$$= y - \hat{y}$$
$$= 2.7 - 3.04 = -0.34 \text{ kg}$$

The 16 data points used in fitting the least-squares line produce 16 residuals. Rounded to two decimal places, they are

0.37	−0.70	0.10	−0.34	0.19	0.61	−0.26	−0.98
1.64	−0.18	−0.23	0.54	−0.54	−1.11	0.93	−0.03

Most regression software will calculate and store these residuals for you.

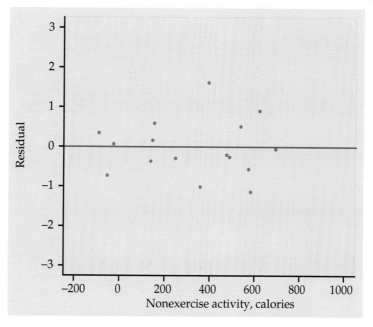

FIGURE 2.20   (b) Residual plot for the regression displayed in Figure 2.20(a). The line at $y = 0$ marks the mean of the residuals.

Because the residuals show how far the data fall from our regression line, examining the residuals helps assess how well the line describes the data. Although residuals can be calculated from any model fitted to the data, the residuals from the least-squares line have a special property: **the mean of the least-squares residuals is always zero.** The residuals in Example 2.15 sum to 0.01 because we rounded to two decimal places.

You can see the residuals in the scatterplot of Figure 2.20(a) by looking at the vertical deviations of the points from the line. The *residual plot* in Figure 2.20(b) makes it easier to study the residuals by plotting them against the explanatory variable, increase in NEA.

RESIDUAL PLOTS

A **residual plot** is a scatterplot of the regression residuals against the explanatory variable. Residual plots help us assess the fit of a regression line.

Because the mean of the residuals is always zero, the horizontal line at zero in Figure 2.20(b) helps orient us. This line (residual = 0) corresponds to the fitted line in Figure 2.20(a). The residual plot magnifies the deviations from the line to make patterns easier to see. If the regression line catches the overall pattern of the data, there should be *no pattern* in the residuals. That is, the residual plot should show an unstructured horizontal band centered at zero. The residuals in Figure 2.20(b) do have this irregular scatter.

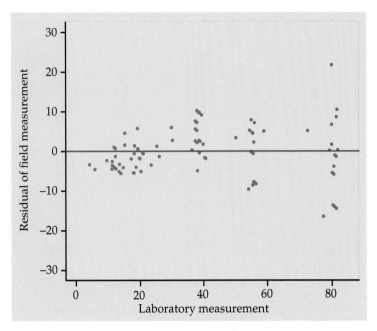

FIGURE 2.21   Residual plot for the regression of field measurements of Alaska pipeline defects on laboratory measurements of the same defects, for Example 2.16.

You can see the same thing in the scatterplot of Figure 2.20(a) and the residual plot of Figure 2.20(b). It's just a bit easier in the residual plot. Deviations from an irregular horizontal pattern point out ways in which the regression line fails to catch the overall pattern. For example, if the overall pattern in the scatterplot is curved rather than straight, the residuals will magnify the curved pattern, moving up and down rather than straight across. Exercise 2.62 is an example of this. Here is a different kind of example.

EXAMPLE 2.16   Figure 2.3 (page 108) plots field measurements on the depth of 100 small defects in the Trans-Alaska Oil Pipeline against laboratory measurements of the same defects. The $y = x$ line on the graph shows that field measurements tend to be too low for larger defect depths. The least-squares regression line for predicting field result from lab result, unlike the $y = x$ line, goes through the center of the points. Figure 2.21 is the residual plot for these data.

Although the horizontal line at zero does go through the middle of the points, the residuals are more spread out both above and below the line as we move to the right. The field measurements are more variable as the true defect depth measured in the lab increases. There is indeed a straight-line pattern, but the regression line doesn't catch the important fact that the variability of field measurements increases with defect depth. The scatterplot makes this clear, and the residual plot magnifies the picture.

Outliers and influential observations

When you look at scatterplots and residual plots, look for striking individual points as well as for an overall pattern. Here is an example of data that contain some unusual cases.

TABLE 2.5

Two measures of glucose level in diabetics

Subject	HbA (%)	FPG (mg/ml)	Subject	HbA (%)	FPG (mg/ml)	Subject	HbA (%)	FPG (mg/ml)
1	6.1	141	7	7.5	96	13	10.6	103
2	6.3	158	8	7.7	78	14	10.7	172
3	6.4	112	9	7.9	148	15	10.7	359
4	6.8	153	10	8.7	172	16	11.2	145
5	7.0	134	11	9.4	200	17	13.7	147
6	7.1	95	12	10.4	271	18	19.3	255

EXAMPLE 2.17   People with diabetes must manage their blood sugar levels carefully. They measure their fasting plasma glucose (FPG) several times a day with a glucose meter. Another measurement, made at regular medical checkups, is called HbA. This is roughly the percent of red blood cells that have a glucose molecule attached. It measures average exposure to glucose over a period of several months. Table 2.5 gives data on both HbA and FPG for 18 diabetics five months after they had completed a diabetes education class.[25]

Because both FPG and HbA measure blood glucose, we expect a positive association. The scatterplot in Figure 2.22 shows a surprisingly weak relationship, with correlation $r = 0.4819$. The line on the plot is the least-squares regression line for predicting

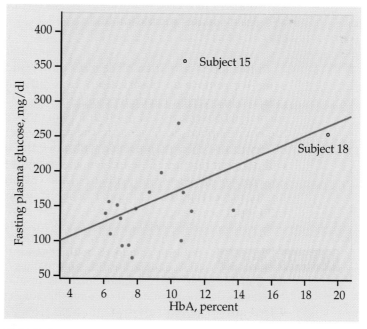

FIGURE 2.22   Scatterplot of fasting plasma glucose against HbA (which measures long-term blood glucose), with the least-squares line, for Example 2.17.

FPG from HbA. Its equation is

$$\hat{y} = 66.4 + 10.41x$$

It appears that one-time measurements of FPG can vary quite a bit among people with similar long-term levels, as measured by HbA.

Two unusual cases are marked in Figure 2.22. Subjects 15 and 18 are unusual in different ways. Subject 15 has dangerously high FPG and lies far from the regression line in the y direction. Subject 18 is close to the line but far out in the x direction. The residual plot in Figure 2.23 confirms that Subject 15 has a large residual and that Subject 18 does not.

Points that are outliers in the x direction, like Subject 18, can have a strong influence on the position of the regression line. Least-squares lines make the sum of squares of the vertical distances to the points as small as possible. A point that is extreme in the x direction with no other points near it pulls the line toward itself.

OUTLIERS AND INFLUENTIAL OBSERVATIONS IN REGRESSION

An **outlier** is an observation that lies outside the overall pattern of the other observations. Points that are outliers in the y direction of a scatterplot have large regression residuals, but other outliers need not have large residuals.

An observation is **influential** for a statistical calculation if removing it would markedly change the result of the calculation. Points that are outliers in the x direction of a scatterplot are often influential for the least-squares regression line.

Influence is a matter of degree—how much does a calculation change when we remove an observation? It is difficult to assess influence on a regression line without actually doing the regression both with and without the suspicious observation. A point that is an outlier in x is often influential. But if the point happens to lie close to the regression line calculated from the other observations, then its presence will move the line only a little and the point will not be influential. The influence of a point that is an outlier in y depends on whether there are many other points with similar values of x that hold the line in place. Figures 2.22 and 2.23 identify two unusual observations. How influential are they?

EXAMPLE 2.18   Subjects 15 and 18 both influence the correlation between FPG and HbA, in opposite directions. Subject 15 weakens the linear pattern; if we drop this point, the correlation increases from $r = 0.4819$ to $r = 0.5684$. Subject 18 extends the linear pattern; if we omit this subject, the correlation drops from $r = 0.4819$ to $r = 0.3837$.

To assess influence on the least-squares line, we recalculate the line leaving out a suspicious point. Figure 2.24 shows three least-squares lines. The solid line is the regression line of FPG on HbA based on all 18 subjects. This is the same line that appears in Figure 2.22. The dotted line is calculated from all subjects except Subject 18. You

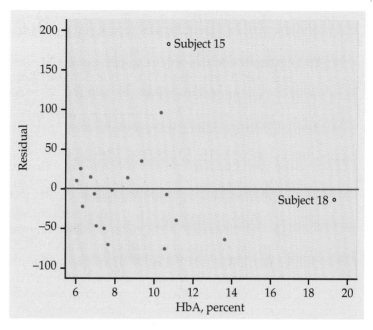

FIGURE 2.23 Residual plot for the regression of FPG on HbA. Subject 15 is an outlier in y. Subject 18 is an outlier in x that may be influential but does not have a large residual.

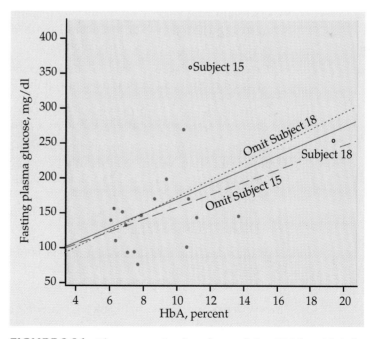

FIGURE 2.24 Three regression lines for predicting FPG from HbA, for Example 2.18. The solid line uses all 18 subjects. The dotted line leaves out Subject 18. The dashed line leaves out Subject 15. "Leaving one out" calculations are the surest way to assess influence.

see that point 18 does pull the line down toward itself. But the influence of Subject 18 is not very large—the dotted and solid lines are close together for HbA values between 6 and 14, the range of all except Subject 18.

The dashed line similarly omits Subject 15, the outlier in y. Comparing the solid and dashed lines, we see that Subject 15 pulls the regression line up. The influence is again not large, but it exceeds the influence of Subject 18.

The best way to see how points that are extreme in x can influence the regression line is to use the *Correlation and Regression* applet on the text CD and Web site. As Exercise 2.78 demonstrates, moving one point can pull the line to almost any position on the graph.

We did not need the distinction between outliers and influential observations in Chapter 1. A single large salary that pulls up the mean salary $\bar{x}$ for a group of workers is an outlier because it lies far above the other salaries. It is also influential because the mean changes when it is removed. In the regression setting, however, not all outliers are influential. Because influential observations draw the regression line toward themselves, we may not be able to spot them by looking for large residuals.

Beware the lurking variable

Correlation and regression are powerful tools for measuring the association between two variables and for expressing the dependence of one variable on the other. These tools must be used with an awareness of their limitations. We have seen that:

- Correlation measures *only linear association*, and fitting a straight line makes sense only when the overall pattern of the relationship is linear. Always plot your data before calculating.

- *Extrapolation* (using a fitted model far outside the range of the data that we used to fit it) often produces unreliable predictions.

- Correlation and least-squares regression are *not resistant*. Always plot your data and look for potentially influential points.

Another caution is even more important: the relationship between two variables can often be understood only by taking other variables into account. *Lurking variables* can make a correlation or regression misleading.

> ### LURKING VARIABLE
>
> A **lurking variable** is a variable that is not among the explanatory or response variables in a study and yet may influence the interpretation of relationships among those variables.

EXAMPLE 2.19 Is high school math the key to success in college? A College Board study of 15,941 high school graduates found a strong correlation between how much math minority students took in high school and their later success in college. News articles quoted the head of the College Board as saying that "math is the gatekeeper for success in college."[26] Maybe so, but we should also think about lurking variables. Minority students from middle-class homes with educated parents no doubt take more high school math courses. They also are more likely to have a stable family, parents who emphasize education and can pay for college, and so on. These students would succeed in college even if they took fewer math courses. The family background of the students is a lurking variable that probably explains much of the relationship between math courses and college success.

EXAMPLE 2.20   Figure 2.25 displays a strong positive linear association. The correlation between these variables is $r = 0.9749$. Because $r^2 = 0.9504$, regression of y on x will explain 95% of the variation in the values of y.

The explanatory variable in Figure 2.25 is the dollar value of goods imported into the United States in the years between 1990 and 2001. The response variable is private spending on health in the same years. There is no economic relationship between these variables. The strong association is due entirely to the fact that both imports and health spending grew rapidly in these years. The common year for each point is a lurking variable. Any two variables that both increase over time will show a strong association. This does not mean that one variable explains or influences the other. In this example, the scatterplot and correlation are correct as exercises in following recipes, but they shed no light on any real situation.

Correlations such as that in Example 2.20 are sometimes called "nonsense correlations." The correlation is real. What is nonsense is the suggestion that the variables are directly related so that changing one of the variables *causes*

FIGURE 2.25   The relationship between private spending on health and the value of goods imported in the same year, for Example 2.20.

changes in the other. The question of causation is important enough to merit separate treatment in Section 2.5. For now, just remember that an association between two variables x and y can reflect many types of relationship among x, y, and one or more lurking variables.

ASSOCIATION DOES NOT IMPLY CAUSATION

An association between an explanatory variable x and a response variable y, even if it is very strong, is not by itself good evidence that changes in x actually cause changes in y.

Lurking variables sometimes create a correlation between x and y, as in Examples 2.19 and 2.20. They can also hide a true relationship between x and y, as the following example illustrates.

EXAMPLE 2.21   A study of housing conditions and health in the city of Hull, England, measured a large number of variables for each of the wards into which the city is divided. Two of the variables were an index x of overcrowding and an index y of the lack of indoor toilets. Because x and y are both measures of inadequate housing, we expect a high correlation. Yet the correlation was only $r = 0.08$. How can this be? Investigation disclosed that some poor wards were dominated by public housing. These wards had high values of x but low values of y because public housing always includes indoor toilets. Other poor wards lacked public housing, and in these wards high values of x were accompanied by high values of y. Because the relationship between x and y differed in the two types of wards, analyzing all wards together obscured the nature of the relationship.[27]

Figure 2.26 shows in simplified form how groups formed by a categorical lurking variable, as in the housing example, can make the correlation r

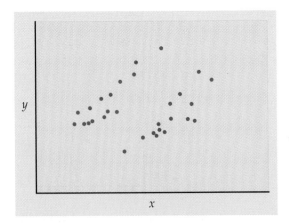

FIGURE 2.26   This scatterplot has a low r even though there is a strong correlation within each of the two clusters.

misleading. The groups appear as clusters of points in the scatterplot. There is a strong relationship between x and y within each of the clusters. In fact, $r = 0.85$ and $r = 0.91$ in the two clusters. However, because similar values of x correspond to quite different values of y in the two clusters, x alone is of little value for predicting y. The correlation for all points displayed is therefore low: $r = 0.14$. This example is another reminder to plot the data rather than simply calculate numerical measures such as the correlation.

Beware correlations based on averaged data

Regression or correlation studies sometimes work with averages or other measures that combine information from many individuals. For example, if we plot the average height of young children against their age in months, we will see a very strong positive association with correlation near 1. But individual children of the same age vary a great deal in height. A plot of height against age for individual children will show much more scatter and lower correlation than the plot of average height against age.

A correlation based on averages over many individuals is usually higher than the correlation between the same variables based on data for individuals. This fact reminds us again of the importance of noting exactly what variables a statistical study involves.

The restricted-range problem

A regression line is often used to predict the response y to a given value x of the explanatory variable. Successful prediction does not require a cause-and-effect relationship. If both x and y respond to the same underlying unmeasured variables, x may help us predict y even though x has no direct influence on y. For example, the scores of SAT exams taken in high school help predict college grades. There is no cause-and-effect tie between SAT scores and college grades. Rather, both reflect a student's ability and knowledge.

How well do SAT scores, perhaps with the help of high school grades, predict college GPA? We can use the correlation r and its square to get a rough answer. There is, however, a subtle difficulty.

EXAMPLE 2.22   Combining several studies for students graduating from college since 1980, the College Board reports these correlations between explanatory variables and the overall GPA of college students:

SAT math and verbal	High school grades	SAT plus grades
$r = 0.36$	$r = 0.42$	$r = 0.52$

Because $0.52^2 = 0.27$, we see that SAT scores plus students' high school records explain about 27% of the variation in GPA among college students.

The subtle problem? Colleges differ greatly in the range of students they attract. Almost all students at Princeton have high SAT scores and did well in high school. At

Generic State College, most students are in the middle range of SAT scores and high school performance. Both sets of students receive the full spread of grades. We suspect that if Princeton admitted weaker students they would get lower grades, and that the typical Princeton student would get very high grades at Generic State. This is the *restricted-range problem:* the data do not contain information on the full range of both explanatory and response variables. *When data suffer from restricted range, r and r^2 are lower than they would be if the full range could be observed.*

Thus $r = 0.52$ understates the actual ability of SAT scores and high school grades to predict college GPA. One investigator found 21 colleges that enrolled the full range of high school graduates. Sure enough, at these colleges, $r = 0.65$.[28]

Did you notice that the correlations in Example 2.22 involve more than one explanatory variable? It is common to use several explanatory variables together to predict a response. This is called *multiple regression*. Each r in the example is a *multiple correlation coefficient*, whose square is the proportion of variation in the response explained by the multiple regression. Chapter 11 introduces multiple regression.

BEYOND THE BASICS | Data mining

Chapters 1 and 2 of this book are devoted to the important aspect of statistics called *exploratory data analysis* (EDA). We use graphs and numerical summaries to examine data, searching for patterns and paying attention to striking deviations from the patterns we find. In discussing regression, we advanced to using the pattern we find (in this case, a linear pattern) for prediction.

Suppose now that we have a truly enormous data base, such as all purchases recorded by the cash register scanners of a national retail chain during the past week. Surely this trove of data contains patterns that might guide business decisions. If we could see clearly the types of activewear preferred in large California cities and compare the preferences of small Midwest cities—right now, not at the end of the season—we might improve profits in both parts of the country by matching stock with demand. This sounds much like EDA, and indeed it is. There is, however, a saying in computer science that a big enough difference of scale amounts to a difference of kind. Exploring really large data bases in the hope of finding useful patterns is called **data mining.** Here are some distinctive features of data mining:

data mining

- When you have 100 gigabytes of data, even straightforward calculations and graphics become impossibly time-consuming. So efficient algorithms are very important.

- The structure of the data base and the process of storing the data (the fashionable term is *data warehousing*), perhaps by unifying data scattered across many departments of a large corporation, require careful consideration.

- Data mining requires automated tools that work based on only vague queries by the user. The process is too complex to do step-by-step as we have done in EDA.

All of these features point to the need for sophisticated computer science as a basis for data mining. Indeed, data mining is often thought of as a part of computer science. Yet many statistical ideas and tools—mostly tools for dealing with multidimensional data, not the sort of thing that appears in a first statistics course—are very helpful. Like many modern developments, data mining crosses the boundaries of traditional fields of study.

Do remember that the perils we associate with blind use of correlation and regression are yet more perilous in data mining, where the fog of an immense data base prevents clear vision. Extrapolation, ignoring lurking variables, and confusing association with causation are traps for the unwary data miner.

SECTION 2.4 | Summary

You can examine the fit of a regression line by plotting the **residuals,** which are the differences between the observed and predicted values of y. Be on the lookout for points with unusually large residuals and also for nonlinear patterns and uneven variation about the line.

Also look for **influential observations,** individual points that substantially change the regression line. Influential observations are often outliers in the x direction, but they need not have large residuals.

Correlation and regression must be **interpreted with caution.** Plot the data to be sure that the relationship is roughly linear and to detect outliers and influential observations.

Lurking variables may explain the relationship between the explanatory and response variables. Correlation and regression can be misleading if you ignore important lurking variables.

We cannot conclude that there is a cause-and-effect relationship between two variables just because they are strongly associated. **High correlation does not imply causation.**

A correlation based on averages is usually higher than if we used data for individuals. A correlation based on data with a **restricted range** is often lower than would be the case if we could observe the full range of the variables.

SECTION 2.4 | Exercises

2.62 Exercise 2.14 (page 119) gives data on the fuel consumption y of a car at various speeds x. The relationship is strongly curved: fuel used decreases with increasing speed at low speeds, then increases again as higher speeds are reached. The equation of the least-squares regression line for these data is

$$\hat{y} = 11.058 - 0.01466x$$

The residuals, in the same order as the observations, are

10.09	2.24	−0.62	−2.47	−3.33	−4.28	−3.73	−2.94
−2.17	−1.32	−0.42	0.57	1.64	2.76	3.97	

(a) Make a scatterplot of the observations and draw the regression line on your plot. The line is a poor description of the curved relationship.

(b) Check that the residuals have sum zero (up to roundoff error).

(c) Make a plot of the residuals against the values of x. Draw a horizontal line at height zero on your plot. The residuals show the same pattern about this line as the data points show about the regression line in the scatterplot in (a).

2.63 A study of nutrition in developing countries collected data from the Egyptian village of Nahya. Here are the mean weights (in kilograms) for 170 infants in Nahya who were weighed each month during their first year of life:[29]

Age (months)	1	2	3	4	5	6	7	8	9	10	11	12
Weight (kg)	4.3	5.1	5.7	6.3	6.8	7.1	7.2	7.2	7.2	7.2	7.5	7.8

(a) Plot weight against time.

(b) A hasty user of statistics enters the data into software and computes the least-squares line without plotting the data. The result is

```
The regression equation is
weight = 4.88 + 0.267 age
```

Plot this line on your graph. Is it an acceptable summary of the overall pattern of growth? Remember that you can calculate the least-squares line for *any* set of two-variable data. It's up to you to decide if it makes sense to fit a line.

(c) Fortunately, the software also prints out the residuals from the least-squares line. In order of age along the rows, they are

−0.85	−0.31	0.02	0.35	0.58	0.62
0.45	0.18	−0.08	−0.35	−0.32	−0.28

Verify that the residuals have sum zero (except for roundoff error). Plot the residuals against age and add a horizontal line at zero. Describe carefully the pattern that you see.

2.64 Exercise 2.60 (page 150) gives data on the concentration of the pesticide fenthion in Greek olive oil that has been stored for various lengths of time. The exponential decay model used to describe how concentration decreases over time proposes a curved relationship between storage time and concentration. Do the residuals from fitting a regression line show a curved pattern? The least-squares line for predicting concentration is

$$\hat{y} = 0.965 - 0.00045x$$

(a) The first batch of olive oil was stored for 28 days and had fenthion concentration 0.99 mg/kg. What is the predicted concentration for this batch? What is the residual?

(b) The residuals, arranged as in the data table in Exercise 2.60, are:

Days stored	Residual				
28	0.0378	0.0378	0.0078	−0.0022	−0.0222
84	0.0329	0.0129	−0.0171	−0.0171	−0.0271
183	0.0072	−0.0128	−0.0228	−0.0328	−0.0328
273	0.0275	0.0175	−0.0025	−0.0125	−0.0125
365	0.0286	0.0186	−0.0014	−0.0014	−0.0114

Check that your residual from (a) agrees (up to roundoff error) with the value 0.0378 given here. Verify that the residuals sum to zero (again up to roundoff error).

(c) Make a residual plot. Is a curved pattern visible? Is the curve very strong? (Software often makes the pattern hard to see because it fills the entire plot area. Try a plot with vertical scale from −0.1 to 0.1.)

2.65 Your plot for Exercise 2.63 shows that increase of the mean weight of children in Nahya is very linear during the first 5 months of life. The correlation of age and weight is $r = 0.9964$ for the first 5 months. Weight in these data is the mean for 170 children. Explain why the correlation between age and weight for the 170 individual children would surely be much smaller.

2.66 Your company gives a test of cognitive ability to job applicants before deciding whom to hire. Your boss has asked you to use company records to see if this test really helps predict the performance ratings of employees. Explain carefully to your boss why the restricted-range problem may make it difficult to see a strong relationship between test scores and performance ratings.

 2.67 The effect of a lurking variable can be surprising when individuals are divided into groups. In recent years, the mean SAT score of all high school seniors has increased. But the mean SAT score has decreased for students at each level of high school grades (A, B, C, and so on). Explain how grade inflation in high school (the lurking variable) can account for this pattern. *A relationship that holds for each group within a population need not hold for the population as a whole. In fact, the relationship can even change direction.*

 2.68 Here is another example of the group effect cautioned about in the previous exercise. Explain how as a nation's population grows older mean income can go down for workers in each age group, yet still go up for all workers.

2.69 Careful statistical studies often include examination of potential lurking variables. This was true of the study of the effect of nonexercise activity (NEA) on fat gain (Example 2.9, page 132), our lead example in Section 2.3. Overeating may lead our bodies to spontaneously increase NEA (fidgeting and the like). Our bodies might also spontaneously increase their basal metabolic rate (BMR), which measures energy use while resting. If both energy uses increased, regressing fat gain on NEA alone would be misleading. Here are data on BMR and fat gain for the same 16 subjects whose NEA we examined earlier:

BMR increase (cal)	117	352	244	−42	−3	134	136	−32
Fat gain (kg)	4.2	3.0	3.7	2.7	3.2	3.6	2.4	1.3

BMR increase (cal)	−99	9	−15	−70	165	172	100	35
Fat gain (kg)	3.8	1.7	1.6	2.2	1.0	0.4	2.3	1.1

The correlation between NEA and fat gain is $r = -0.7786$. The slope of the regression line for predicting fat gain from NEA is $b = -0.00344$ kilograms per calorie. What are the correlation and slope for BMR and fat gain? Explain why these values show that BMR has much less effect on fat gain than does NEA.

2.70   Gas chromatography is a technique used to detect very small amounts of a substance, for example, a contaminant in drinking water. Laboratories use regression to calibrate such techniques. The data below show the results of five measurements for each of four amounts of the substance being investigated.[30] The explanatory variable x is the amount of substance in the specimen, measured in nanograms (ng), units of 10^{-9} gram. The response variable y is the reading from the gas chromatograph.

Amount (ng)	Response				
0.25	6.55	7.98	6.54	6.37	7.96
1.00	29.7	30.0	30.1	29.5	29.1
5.00	211	204	212	213	205
20.00	929	905	922	928	919

(a) Make a scatterplot of these data. The relationship appears to be approximately linear, but the wide variation in the response values makes it hard to see detail in this graph.

(b) Compute the least-squares regression line of y on x, and plot this line on your graph.

(c) Now compute the residuals and make a plot of the residuals against x. It is much easier to see deviations from linearity in the residual plot. Describe carefully the pattern displayed by the residuals.

2.71   Here are the golf scores of 11 members of a women's golf team in two rounds of tournament play:

Player	1	2	3	4	5	6	7	8	9	10	11
Round 1	89	90	87	95	86	81	105	83	88	91	79
Round 2	94	85	89	89	81	76	89	87	91	88	80

(a) Plot the data with the Round 1 scores on the x axis and the Round 2 scores on the y axis. There is a generally linear pattern except for one potentially influential observation. Circle this observation on your graph.

(b) Here are the equations of two least-squares lines. One of the equations is calculated from all 11 data points and the other omits the influential observation.

$$\hat{y} = 20.49 + 0.754x$$
$$\hat{y} = 50.01 + 0.410x$$

Draw both lines on your scatterplot. Which line omits the influential observation? How do you know this?

2.72 Drilling down beneath a lake in Alaska yields chemical evidence of past changes in climate. Biological silicon, left by the skeletons of single-celled creatures called diatoms, measures the abundance of life in the lake. A rather complex variable based on the ratio of certain isotopes relative to ocean water gives an indirect measure of moisture, mostly from snow. As we drill down, we look farther into the past. Here are data from 2300 to 12,000 years ago:[31]

Isotope (%)	Silicon (mg/g)	Isotope (%)	Silicon (mg/g)	Isotope (%)	Silicon (mg/g)
−19.90	97	−20.71	154	−21.63	224
−19.84	106	−20.80	265	−21.63	237
−19.46	118	−20.86	267	−21.19	188
−20.20	141	−21.28	296	−19.37	337

(a) Make a scatterplot of silicon (response) against isotope (explanatory). Ignoring the outlier, describe the direction, form, and strength of the relationship. The researchers say that this and relationships among other variables they measured are evidence for cyclic changes in climate that are linked to changes in the sun's activity.

(b) The researchers single out one point: "The open circle in the plot is an outlier that was excluded in the correlation analysis." Circle this outlier on your graph. What is the correlation with and without this point? The point strongly influences the correlation.

(c) Is the outlier also strongly influential for the regression line? Calculate and draw on your graph two regression lines, and discuss what you see.

2.73 Table 1.10 (page 41) gives the city and highway gas mileages for 21 two-seater cars, including the Honda Insight gas-electric hybrid car. In Exercise 2.33 you investigated the influence of the Insight on the correlation between city and highway mileage.

(a) Make a scatterplot of highway mileage (response) against city mileage (explanatory) for all 21 cars.

(b) Use software or a graphing calculator to find the regression line for predicting highway mileage from city mileage and also the 21 residuals for this regression. Make a residual plot with a horizontal line at zero. (The "stacks" in the plot are due to the fact that mileage is measured only to the nearest mile per gallon.)

(c) Which car has the largest positive residual? The largest negative residual?

(d) The Honda Insight, an extreme outlier, does not have the largest residual in either direction. Why is this not surprising?

2.74   Runners are concerned about their form when racing. One measure of form is the stride rate, the number of steps taken per second. As running speed increases, the stride rate should also increase. In a study of 21 of the best American female runners, researchers measured the stride rate for different speeds. The following table gives the speeds (in feet per second) and the mean stride rates for these runners:[32]

Speed	15.86	16.88	17.50	18.62	19.97	21.06	22.11
Stride rate	3.05	3.12	3.17	3.25	3.36	3.46	3.55

(a) Plot the data with speed on the x axis and stride rate on the y axis. Does a straight line adequately describe these data?

(b) Find the equation of the regression line of stride rate on speed. Draw this line on your plot.

(c) For each of the speeds given, obtain the predicted value of the stride rate and the residual. Verify that the residuals add to zero.

(d) Plot the residuals against speed. Describe the pattern. What does the plot indicate about the adequacy of the linear fit? Are there any potentially influential observations?

2.75   Continue your work in Exercise 2.73. Find the regression line for predicting highway mileage from city mileage for the 20 two-seater cars other than the Honda Insight. Draw both regression lines on your scatterplot. Is the Insight very influential for the least-squares line? (Look at the position of the lines for city mileages between 10 and 30 MPG, values that cover most cars.) What explains your result?

2.76   Exercise 2.74 gives data on the mean stride rate of a group of 21 elite female runners at various running speeds. Find the correlation between speed and stride rate. Would you expect this correlation to increase or decrease if we had data on the individual stride rates of all 21 runners at each speed? Why?

2.77   It isn't easy to guess the position of the least-squares line by eye. Use the *Correlation and Regression* applet to compare a line you draw with the least-squares line. Click on the scatterplot to create a group of 15 to 20 points from lower left to upper right with a clear positive straight-line pattern (correlation around 0.7). Click the "Draw line" button and use the mouse to draw a line through the middle of the cloud of points from lower left to upper right. Note the "thermometer" that appears above the plot. The red portion is the sum of the squared vertical distances from the points in the plot to the least-squares line. The green portion is the "extra" sum of squares for your line—it shows by how much your line misses the smallest possible sum of squares.

(a) You drew a line by eye through the middle of the pattern. Yet the right-hand part of the bar is probably almost entirely green. What does that tell you?

(b) Now click the "Show least-squares line" box. Is the slope of the least-squares line smaller (the new line is less steep) or larger (line is steeper) than that of your line? If you repeat this exercise several times, you will consistently get the same result. *The least-squares line minimizes the vertical distances of the points from the line. It is not the line through the "middle" of the cloud of points.* This is one reason why it is hard to draw a good regression line by eye.

2.78   Go to the *Correlation and Regression* applet. Click on the scatterplot to create a group of 10 points in the lower-left corner of the scatterplot with a strong straight-line pattern (correlation about 0.9). In Exercise 2.34 you started here to see that correlation r is not resistant. Now click the "Show least-squares line" box to display the regression line.

(a) Add one point at the upper right that is far from the other 10 points but exactly on the regression line. Why does this outlier have no effect on the line even though it changes the correlation?

(b) Now drag this last point down until it is opposite the group of 10 points. You see that one end of the least-squares line chases this single point, while the other end remains near the middle of the original group of 10. What makes the last point so influential?

2.79   There is a strong positive correlation between years of education and income for economists employed by business firms. (In particular, economists with doctorates earn more than economists with only a bachelor's degree.) There is also a strong positive correlation between years of education and income for economists employed by colleges and universities. But when all economists are considered, there is a *negative* correlation between education and income. The explanation for this is that business pays high salaries and employs mostly economists with bachelor's degrees, while colleges pay lower salaries and employ mostly economists with doctorates. Sketch a scatterplot with two groups of cases (business and academic) that illustrates how a strong positive correlation within each group and a negative overall correlation can occur together. (*Hint:* Begin by studying Figure 2.26.)

2.80   Table 2.6 presents four sets of data prepared by the statistician Frank Anscombe to illustrate the dangers of calculating without first plotting the data.[33]

(a) Without making scatterplots, find the correlation and the least-squares regression line for all four data sets. What do you notice? Use the regression line to predict y for $x = 10$.

TABLE 2.6										
Four data sets for exploring correlation and regression										
Data Set A										
x	10	8	13	9	11	14	6	4	12	7	5
y	8.04	6.95	7.58	8.81	8.33	9.96	7.24	4.26	10.84	4.82	5.68
Data Set B										
x	10	8	13	9	11	14	6	4	12	7	5
y	9.14	8.14	8.74	8.77	9.26	8.10	6.13	3.10	9.13	7.26	4.74
Data Set C										
x	10	8	13	9	11	14	6	4	12	7	5
y	7.46	6.77	12.74	7.11	7.81	8.84	6.08	5.39	8.15	6.42	5.73
Data Set D										
x	8	8	8	8	8	8	8	8	8	8	19
y	6.58	5.76	7.71	8.84	8.47	7.04	5.25	5.56	7.91	6.89	12.50

(b) Make a scatterplot for each of the data sets and add the regression line to each plot.

(c) In which of the four cases would you be willing to use the regression line to describe the dependence of y on x? Explain your answer in each case.

2.81 If you need medical care, should you go to a hospital that handles many cases like yours? Figure 2.27 presents some data for heart attacks. The figure plots mortality rate (the proportion of patients who died) against the number of heart attack patients treated for a large number of hospitals in a recent year. The line on the plot is the least-squares regression line for predicting mortality from number of patients.

(a) Do the plot and regression generally support the thesis that mortality is lower at hospitals that treat more heart attacks? Is the relationship very strong?

(b) In what way is the pattern of the plot nonlinear? Does the nonlinearity strengthen or weaken the conclusion that heart attack patients should avoid hospitals that treat few heart attacks? Why?

2.82 A multimedia statistics learning system includes a test of skill in using the computer's mouse. The software displays a circle at a random location on the computer screen. The subject tries to click in the circle with the mouse as quickly as possible. A new circle appears as soon as the subject clicks the old one. Table 2.7 gives data for one subject's trials, 20 with each hand. Distance is the distance from the cursor location to the center of the new circle, in units whose actual size depends on the size of the screen. Time is the time required to click in the new circle, in milliseconds.[34]

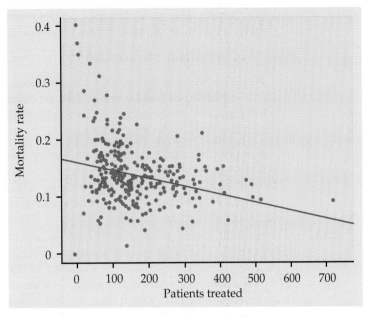

FIGURE 2.27 Mortality of heart attack patients and number of heart attack cases treated for a large group of hospitals, for Exercise 2.81.

TABLE 2.7					
Reaction times in a computer game					
Time	Distance	Hand	Time	Distance	Hand
115	190.70	right	240	190.70	left
96	138.52	right	190	138.52	left
110	165.08	right	170	165.08	left
100	126.19	right	125	126.19	left
111	163.19	right	315	163.19	left
101	305.66	right	240	305.66	left
111	176.15	right	141	176.15	left
106	162.78	right	210	162.78	left
96	147.87	right	200	147.87	left
96	271.46	right	401	271.46	left
95	40.25	right	320	40.25	left
96	24.76	right	113	24.76	left
96	104.80	right	176	104.80	left
106	136.80	right	211	136.80	left
100	308.60	right	238	308.60	left
113	279.80	right	316	279.80	left
123	125.51	right	176	125.51	left
111	329.80	right	173	329.80	left
95	51.66	right	210	51.66	left
108	201.95	right	170	201.95	left

(a) We suspect that time depends on distance. Make a scatterplot of time against distance, using separate symbols for each hand.

(b) Describe the pattern. How can you tell that the subject is right-handed?

(c) Find the regression line of time on distance separately for each hand. Draw these lines on your plot. Which regression does a better job of predicting time from distance? Give numerical measures that describe the success of the two regressions.

(d) It is possible that the subject got better in later trials due to learning. It is also possible that he got worse due to fatigue. Plot the residuals from each regression against the time order of the trials (down the columns in Table 2.7). Is either of these systematic effects of time visible in the data?

2.83 Are baseball players paid according to their performance? To study this question, a statistician analyzed the salaries of over 260 Major League hitters along with such explanatory variables as career batting average, career home runs per time at bat, and years in the Major Leagues.[35] This is a *multiple regression* with several explanatory variables. More detail on multiple regression appears in Chapter 11, but the fit of the model is assessed just as we have done in this chapter, by calculating and plotting the residuals:

$$\text{residual} = \text{observed } y - \text{predicted } y$$

(a) Figure 2.28(a) is a plot of the residuals against the predicted salary. When points are too close together to plot separately, the plot uses letters of the

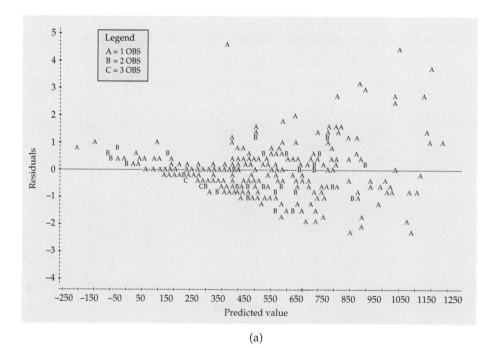

(a)

FIGURE 2.28 (a) Plot of the residuals for predicting baseball players' salaries against the predicted salaries.

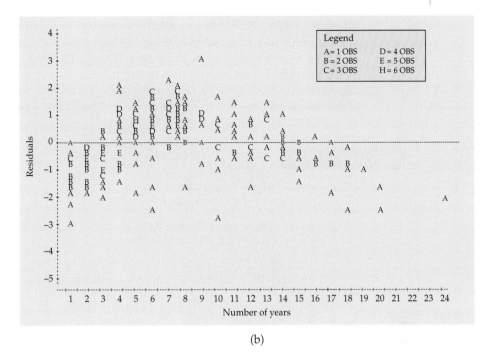

(b)

FIGURE 2.28 (b) Plot of the residuals for predicting the logarithms of salaries against years in the Major Leagues.

alphabet to show how many points there are at each position. Describe the pattern that appears on this residual plot. Will the regression model predict high or low salaries more precisely?

(b) After studying the residuals in more detail, the statistician decided to predict the logarithm of salary rather than the salary itself. One reason was that while salaries are not normally distributed (the distribution is skewed to the right), their logarithms are nearly normal. When the response variable is the logarithm of salary, a plot of the residuals against the predicted value is satisfactory. Figure 2.28(b) is a plot of the residuals against the number of years the player has been in the Major Leagues. Describe the pattern that you see. Will the model overestimate or underestimate the salaries of players who are new to the Majors? Of players who have been in the Major Leagues about 8 years? Of players with more than 15 years in the Majors?

2.5 The Question of Causation

In many studies of the relationship between two variables, the goal is to establish that changes in the explanatory variable *cause* changes in the response variable. Even when a strong association is present, the conclusion that this association is due to a causal link between the variables is often elusive. What ties between two variables (and others lurking in the background) can explain an observed association? What constitutes good evidence for causation? We begin our consideration of these questions with a set of examples. In each case, there is a clear association between an explanatory variable x and a response variable y. Moreover, the association is positive whenever the direction makes sense.

EXAMPLE 2.23   Here are some examples of observed association between x and y:

1. x = mother's body mass index
   y = daughter's body mass index

2. x = amount of the artificial sweetener saccharin in a rat's diet
   y = count of tumors in the rat's bladder

3. x = a student's SAT score as a high school senior
   y = the student's first-year college grade point average

4. x = monthly flow of money into stock mutual funds
   y = monthly rate of return for the stock market

5. x = whether a person regularly attends religious services
   y = how long the person lives

6. x = the number of years of education a worker has
   y = the worker's income

Explaining association: causation

Figure 2.29 shows in outline form how a variety of underlying links between variables can explain association. The dashed double-arrow line represents an observed association between the variables x and y. Some associations are explained by a direct cause-and-effect link between these variables. The first diagram in Figure 2.29 shows "x causes y" by a solid arrow running from x to y.

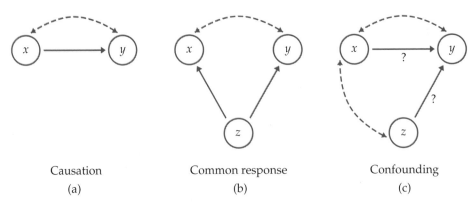

Causation	Common response	Confounding
(a)	(b)	(c)

FIGURE 2.29   Some possible explanations for an observed association. The dashed double-arrow lines show an association. The solid arrows show a cause-and-effect link. The variable x is explanatory, y is a response variable, and z is a lurking variable.

EXAMPLE 2.24   Items 1 and 2 in Example 2.23 are examples of direct causation. Thinking about these examples, however, shows that "causation" is not a simple idea.

1. A study of Mexican American girls aged 9 to 12 years recorded body mass index (BMI), a measure of weight relative to height, for both the girls and their mothers. People with high BMI are overweight or obese. The study also measured hours of television, minutes of physical activity, and intake of several kinds of food. The strongest correlation ($r = 0.506$) was between the BMI of daughters and the BMI of their mothers.[36]

Body type is in part determined by heredity. Daughters inherit half their genes from their mothers. There is therefore a direct causal link between the BMI of mothers and daughters. Yet the mothers' BMIs explain only 25.6% (that's r^2 again) of the variation among the daughters' BMIs. Other factors, such as diet and exercise, also influence BMI. *Even when direct causation is present, it is rarely a complete explanation of an association between two variables.*

2. The best evidence for causation comes from experiments that actually change x while holding all other factors fixed. If y changes, we have good reason to think that x caused the change in y. Experiments show that large amounts of saccharin—about 5% of the entire diet—cause bladder tumors in rats. Should we avoid saccharin as a replacement for sugar in food? Rats are not people. Although we can't experiment with people, studies of people who consume different amounts of saccharin fail to find an association between saccharin and bladder tumors.[37] *Even well-established causal relations may not generalize to other settings.*

Explaining association: common response

"Beware the lurking variable" is good advice when thinking about an association between two variables. The second diagram in Figure 2.29 illustrates **common response.** The observed association between the variables x and y is explained by a lurking variable z. Both x and y change in response to changes in z. This common response creates an association even though there may be no direct causal link between x and y.

common response

EXAMPLE 2.25   The third and fourth items in Example 2.23 illustrate how common response can create an association.

3. Students who are smart and who have learned a lot tend to have both high SAT scores and high college grades. The positive correlation is explained by this common response to students' ability and knowledge.

4. There is a strong positive correlation between how much money individuals add to mutual funds each month and how well the stock market does the same month. Is the new money driving the market up? The correlation may be explained in part by common response to underlying investor sentiment: when optimism reigns, individuals send money to funds and large institutions also invest more. The institutions would drive up prices even if individuals did nothing. In addition, what causation there is may operate in the other direction: when the market is doing well, individuals rush to add money to their mutual funds.[38]

Explaining association: confounding

We noted in Example 2.24 that inheritance no doubt explains part of the association between the body mass indexes (BMIs) of daughters and their mothers. Can we use r or r^2 to say how much inheritance contributes to the daughters' BMIs? No. It may well be that mothers who are overweight also set an example of little exercise, poor eating habits, and lots of television. Their daughters pick up these habits to some extent, so the influence of heredity is mixed up with influences from the girls' environment. We call this mixing of influences *confounding*.

> ## CONFOUNDING
>
> Two variables are **confounded** when their effects on a response variable cannot be distinguished from each other. The confounded variables may be either explanatory variables or lurking variables.

When many variables interact with each other, confounding of several variables often prevents us from drawing conclusions about causation. The third diagram in Figure 2.29 illustrates confounding. Both the explanatory variable x and the lurking variable z may influence the response variable y. Because x is confounded with z, we cannot distinguish the influence of x from the influence of z. We cannot say how strong the direct effect of x on y is. In fact, it can be hard to say if x influences y at all.

EXAMPLE 2.26   The last two associations in Example 2.23 (Items 5 and 6) are explained in part by confounding.

5. Many studies have found that people who are active in their religion live longer than nonreligious people. But people who attend church or mosque or synagogue also take better care of themselves than nonattenders. They are less likely to smoke, more likely to exercise, and less likely to be overweight. The effects of these good habits are confounded with the direct effects of attending religious services.

6. It is likely that more education is a cause of higher income—many highly paid professions require advanced education. However, confounding is also present. People who have high ability and come from prosperous homes are more likely to get many years of education than people who are less able or poorer. Of course, people who start out able and rich are more likely to have high earnings even without much education. We can't say how much of the higher income of well-educated people is actually caused by their education.

Many observed associations are at least partly explained by lurking variables. Both common response and confounding involve the influence of a lurking variable (or variables) z on the response variable y. The distinction between these two types of relationship is less important than the common element, the influence of lurking variables. The most important lesson of these examples is one we have already emphasized: **even a very strong association between two variables is not by itself good evidence that there is a cause-and-effect link between the variables.**

Establishing causation

How can a direct causal link between x and y be established? The best method—indeed, the only fully compelling method—of establishing causation is to conduct a carefully designed experiment in which the effects of possible lurking variables are controlled. Chapter 3 explains how to design convincing experiments.

Many of the sharpest disputes in which statistics plays a role involve questions of causation that cannot be settled by experiment. Does gun control reduce violent crime? Does living near power lines cause cancer? Has "outsourcing" work to overseas locations reduced overall employment in the United States? All of these questions have become public issues. All concern associations among variables. And all have this in common: they try to pinpoint cause and effect in a setting involving complex relations among many interacting variables. Common response and confounding, along with the number of potential lurking variables, make observed associations misleading. Experiments are not possible for ethical or practical reasons. We can't assign some people to live near power lines or compare the same nation with and without strong gun controls.

EXAMPLE 2.27 Electric currents generate magnetic fields. So living with electricity exposes people to magnetic fields. Living near power lines increases exposure to these fields. Really strong fields can disturb living cells in laboratory studies. Some people claim that the weaker fields we experience if we live near power lines cause leukemia in children.

It isn't ethical to do experiments that expose children to magnetic fields. It's hard to compare cancer rates among children who happen to live in more and less exposed locations because leukemia is rare and locations vary in many ways other than magnetic fields. We must rely on studies that compare children who have leukemia with children who don't.

A careful study of the effect of magnetic fields on children took five years and cost $5 million. The researchers compared 638 children who had leukemia and 620 who did not. They went into the homes and actually measured the magnetic fields in the children's bedrooms, in other rooms, and at the front door. They recorded facts about nearby power lines for the family home and also for the mother's residence when she was pregnant. Result: no evidence of more than a chance connection between magnetic fields and childhood leukemia.[39]

"No evidence" that magnetic fields are connected with childhood leukemia doesn't prove that there is no risk. It says only that a careful study could not find any risk that stands out from the play of chance that distributes leukemia cases across the landscape. Critics continue to argue that the study failed to measure some lurking variables, or that the children studied don't fairly represent all children. Nonetheless, a carefully designed study comparing children with and without leukemia is a great advance over haphazard and sometimes emotional counting of cancer cases.

EXAMPLE 2.28 Despite the difficulties, it is sometimes possible to build a strong case for causation in the absence of experiments. The evidence that smoking causes lung cancer is about as strong as nonexperimental evidence can be.

Doctors had long observed that most lung cancer patients were smokers. Comparison of smokers and similar nonsmokers showed a very strong association between smoking and death from lung cancer. Could the association be due to common response? Might there be, for example, a genetic factor that predisposes people both to nicotine addiction and to lung cancer? Smoking and lung cancer would then be positively associated even if smoking had no direct effect on the lungs. Or perhaps confounding is to blame. It might be that smokers live unhealthy lives in other ways (diet, alcohol, lack of exercise) and that some other habit confounded with smoking is a cause of lung cancer. How were these objections overcome?

Let's answer this question in general terms: What are the criteria for establishing causation when we cannot do an experiment?

- *The association is strong.* The association between smoking and lung cancer is very strong.

- *The association is consistent.* Many studies of different kinds of people in many countries link smoking to lung cancer. That reduces the chance that a lurking variable specific to one group or one study explains the association.

- *Higher doses are associated with stronger responses.* People who smoke more cigarettes per day or who smoke over a longer period get lung cancer more often. People who stop smoking reduce their risk.

- *The alleged cause precedes the effect in time.* Lung cancer develops after years of smoking. The number of men dying of lung cancer rose as smoking became more common, with a lag of about 30 years. Lung cancer kills more men than any other form of cancer. Lung cancer was rare among women until women began to smoke. Lung cancer in women rose along with smoking, again with a lag of about 30 years, and has now passed breast cancer as the leading cause of cancer death among women.

- *The alleged cause is plausible.* Experiments with animals show that tars from cigarette smoke do cause cancer.

Medical authorities do not hesitate to say that smoking causes lung cancer. The U.S. Surgeon General states that cigarette smoking is "the largest avoidable cause of death and disability in the United States."[40] The evidence for causation is overwhelming—but it is not as strong as the evidence provided by well-designed experiments.

SECTION 2.5 | Summary

Some observed associations between two variables are due to a **cause-and-effect** relationship between these variables, but others are explained by **lurking variables.**

The effect of lurking variables can operate through **common response** if changes in both the explanatory and response variables are caused by changes in lurking variables. **Confounding** of two variables (either explanatory or lurking variables) means that we cannot distinguish their effects on the response variable.

That an association is due to causation is best established by an **experiment** that changes the explanatory variable while controlling other influences on the response.

In the absence of experimental evidence, be cautious in accepting claims of causation. Good evidence of causation requires a strong association that appears consistently in many studies, a clear explanation for the alleged causal link, and careful examination of possible lurking variables.

SECTION 2.5 | Exercises

2.84 Many colleges offer online versions of some courses that are also taught in the classroom. It often happens that the students who enroll in the online version do better than the classroom students on the course exams. This does not show that online instruction is more effective than classroom teaching, because the kind of people who sign up for online courses are often quite different from the classroom students. Suggest some student characteristics that you think could be confounded with online versus classroom. Use a diagram like Figure 2.29(c) to illustrate your ideas.

2.85 Data show that men who are married, and also divorced or widowed men, earn quite a bit more than men who have never been married. This does not mean that a man can raise his income by getting married. Suggest several lurking variables that you think are confounded with marital status and that help explain the association between marital status and income. Use a diagram like Figure 2.29(c) to illustrate your ideas.

2.86 "Based on an examination of twenty-two companies that announced large layoffs during 1994, Downs found a strong (.31) correlation between the size of the layoffs and the compensation of the CEOs."[41] This correlation is probably explained by common response to a lurking variable, the size of the company as measured by its number of employees. Explain how common response could create the observed correlation. Use a diagram like Figure 2.29(b) to illustrate your explanation.

2.87 A college fitness center offers an exercise program for staff members who choose to participate. The program assesses each participant's fitness, using a treadmill test, and also administers a personality questionnaire. There is a moderately strong positive correlation between fitness score and score for self-confidence. Is this good evidence that improving fitness increases self-confidence? Explain why or why not.

2.88 An article entitled "The Health and Wealth of Nations" says: "The positive correlation between health and income per capita is one of the best-known relations in international development. This correlation is commonly thought to reflect a causal link running from income to health. ... Recently, however, another intriguing possibility has emerged: that the health-income correlation is partly explained by a causal link running the other way—from health to income."[42]

Explain how higher income in a nation can cause better health. Then explain how better health can cause higher national income. There is no simple way to determine the direction of the link.

2.89 The Kalamazoo (Michigan) Symphony once advertised a "Mozart for Minors" program with this statement: "Question: Which students scored 51 points higher in verbal skills and 39 points higher in math? Answer: Students who had experience in music."[43] In fact, good academic performance and early exposure to classical music are in part common responses to lurking variables. What background information about students could explain the association? Use a diagram like Figure 2.29(b) to show the situation.

2.90 A study finds that high school students who take the SAT, enroll in an SAT coaching course, and then take the SAT a second time raise their SAT mathematics scores from a mean of 521 to a mean of 561.[44] What factors other than "taking the course causes higher scores" might explain this improvement?

2.91 A study showed that women who work in the production of computer chips have abnormally high numbers of miscarriages. The union claimed that exposure to chemicals used in production caused the miscarriages. Another possible explanation is that these workers spend most of their work time standing up. Illustrate these relationships in a diagram like those in Figure 2.29.

2.92 A study shows that there is a positive correlation between the size of a hospital (measured by its number of beds x) and the median number of days y that patients remain in the hospital. Does this mean that you can shorten a hospital stay by choosing a small hospital? Use a diagram like those in Figure 2.29 to explain the association.

2.93 Children who watch many hours of television get lower grades in school on the average than those who watch less TV. Explain clearly why this fact does not show that watching TV *causes* poor grades. In particular, suggest some other variables that may be confounded with heavy TV viewing and may contribute to poor grades.

2.94 People who use artificial sweeteners in place of sugar tend to be heavier than people who use sugar. Does this mean that artificial sweeteners cause weight gain? Give a more plausible explanation for this association.

2.95 A sign in a fitness center says, "Mortality is halved for men over 65 who walk at least 2 miles a day."

(a) Mortality is eventually 100% for everyone. What do you think "mortality is halved" means?

(b) Assuming that the claim is true, explain why this fact does not show that exercise *causes* lower mortality.

2.96 People who do well tend to feel good about themselves. Perhaps helping people feel good about themselves will help them do better in their jobs and in life. Raising self-esteem became for a time a goal in many schools and companies. Can you think of explanations for the association between high self-esteem and good performance other than "self-esteem causes better work"?

CHAPTER 2 | Exercises

2.97 Endangered animal species often live in isolated patches of habitat. If the population in a patch varies a lot (due to weather, for example), the species is more likely to disappear from that patch in a bad year. Here is a general question: Is there less variation in population size when a patch of habitat has more diverse vegetation? If so, maintaining habitat diversity can help protect endangered species.

A researcher measured the variation over time in the population of a cricket species in 45 habitat patches. He also measured the diversity of each patch.[45] He reported his results by giving the least-squares equation

$$\text{population variation} = 84.4 - 0.13 \times \text{diversity}$$

along with the fact that $r^2 = 0.34$. Do these results support the idea that more diversity goes with less variation in population size? Is the relationship very strong or only moderately strong?

2.98 In the long run, the price of a company's stock ought to parallel changes in the company's earnings. Table 2.8 gives data on the annual growth rates in earnings and in stock prices (both in percent) for major industry groups as set by Standard & Poor's.[46]

(a) Make a graph showing how earnings growth explains growth in stock price. Does it appear to be true that (on the average in the long run) stock price growth parallels earnings growth?

(b) What percent of the variation in stock price growth among industry groups can be explained by the linear relationship with earnings growth?

TABLE 2.8

Percent growth in stock price and earnings for industry groups

Industry	Earnings growth (%)	Price growth (%)	Industry	Earnings growth (%)	Price growth (%)
Auto	3.3	2.9	Oil: international	7.7	7.7
Banks	8.6	6.5	Oil equipment/		
Chemicals	6.6	3.1	services	10.1	10.8
Computers	10.2	5.3	Railroad	6.6	6.6
Drugs	11.3	10.0	Retail: department		
Electrical			stores	10.1	9.5
equipment	8.5	8.2	Retail: food	6.9	6.9
Food	7.6	6.5	Soft drinks	12.7	12.0
Household			Steel	−1.0	−1.6
products	9.7	10.1	Tobacco	12.3	11.7
Machinery	5.1	4.7	Utilities: electric	2.8	1.4
Oil: domestic	7.4	7.3	Utilities: gas	5.2	6.2

(c) If stock prices exactly followed earnings, the slope of the least-squares line for predicting price growth from earnings growth would be 1. Explain why. What is the slope of the least-squares line for these data?

(d) What is the correlation between earnings growth and price growth? If we had data on all of the individual companies in these 20 industries, would the correlation be higher or lower? Why?

2.99 The usual way to study the brain's response to sounds is to have subjects listen to "pure tones." The response to recognizable sounds may differ. To compare responses, researchers anesthetized macaque monkeys. They fed pure tones and also monkey calls directly to their brains by inserting electrodes. Response to the stimulus was measured by the firing rate (electrical spikes per second) of neurons in various areas of the brain. Table 2.9 contains the responses for 37 neurons.[47]

(a) One notable finding is that responses to monkey calls are generally stronger than responses to pure tones. Give a numerical measure that supports this finding.

(b) Make a scatterplot of monkey call response against pure tone response (explanatory variable). Find the least-squares line and add it to your plot. Mark on your plot the point with the largest residual (positive or negative) and also a point that is an outlier in the x direction.

(c) How influential are each of these points for the correlation r?

(d) How influential are each of these points for the regression line?

TABLE 2.9

Neuron response to tones and monkey calls

Tone	Call	Tone	Call	Tone	Call	Tone	Call
474	500	145	42	71	134	35	103
256	138	141	241	68	65	31	70
241	485	129	194	59	182	28	192
226	338	113	123	59	97	26	203
185	194	112	182	57	318	26	135
174	159	102	141	56	201	21	129
176	341	100	118	47	279	20	193
168	85	74	62	46	62	20	54
161	303	72	112	41	84	19	66
150	208						

2.100 How strong is a building material such as plywood? To be specific, support a 24-inch by 2-inch strip of plywood at both ends and apply force in the middle until the strip breaks. The modulus of rupture (MOR) is the force needed to break the strip. We would like to be able to predict MOR without actually breaking the wood. The modulus of elasticity (MOE) is found by bending the wood without breaking it. Both MOE and MOR are measured in pounds per square inch. Here are data for 32 specimens of the same type of plywood:[48]

MOE	MOR	MOE	MOR	MOE	MOR	MOE	MOR
2,005,400	11,591	1,774,850	10,541	2,181,910	12,702	1,747,010	11,794
1,166,360	8,542	1,457,020	10,314	1,559,700	11,209	1,791,150	11,413
1,842,180	12,750	1,959,590	11,983	2,372,660	12,799	2,535,170	13,920
2,088,370	14,512	1,720,930	10,232	1,580,930	12,062	1,355,720	9,286
1,615,070	9,244	1,355,960	8,395	1,879,900	11,357	1,646,010	8,814
1,938,440	11,904	1,411,210	10,654	1,594,750	8,889	1,472,310	6,326
2,047,700	11,208	1,842,630	10,223	1,558,770	11,565	1,488,440	9,214
2,037,520	12,004	1,984,690	13,499	2,212,310	15,317	2,349,090	13,645

Can we use MOE to predict MOR accurately? Use the data to write a discussion of this question.

2.101 Some statistical methods require that the residuals from a regression line have a normal distribution. The residuals in the nonexercise activity example, which we found in Example 2.15 (page 152), behave well in other ways. Is their distribution close to normal? Make a normal quantile plot to find out.

2.102 Asian culture does not emphasize thinness, but young Asians are often influenced by Western culture. In a study of concerns about weight among young Korean women, researchers administered the Drive for Thinness scale (a questionnaire) to 264 female college students in Seoul, South Korea. This scale measures excessive concern with weight and dieting and fear of weight gain. In Exercise 1.29, you examined the distribution of Drive for Thinness scores among these college women. The study looked at several explanatory variables. One was "Body Dissatisfaction," also measured by a questionnaire. The data are in the file *ex02-102.dat* on the text CD and Web site.

(a) Make a scatterplot of Drive for Thinness (response) against Body Dissatisfaction. The appearance of the plot is a result of the fact that both variables take only whole-number values. Such variables are common in the social and behavioral sciences.

(b) Add the least-squares line to your plot. The line shows a linear relationship. How strong is this relationship? Body Dissatisfaction was more strongly correlated with Drive for Thinness than any of the other explanatory variables examined. Rather weak relationships are common in social and behavioral sciences, because individuals vary a great deal. Using several explanatory variables together improves prediction of the response. This is *multiple regression,* discussed in Chapter 11.

2.103 To study the energy savings due to adding solar heating panels to a house, researchers measured the natural gas consumption of the house for more than a year, then installed solar panels and observed the natural gas consumption for almost two years. The explanatory variable *x* is degree-days per day during the several weeks covered by each observation, and the response variable *y* is gas

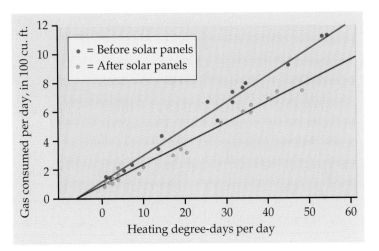

FIGURE 2.30 The regression of residential natural gas consumption on heating degree-days before and after installation of solar heating panels, for Exercise 2.103.

consumption (in hundreds of cubic feet) per day during the same period. Figure 2.30 plots y against x, with different colors for observations taken before and after the installation of the solar panels.[49] The least-squares regression lines were computed separately for the before and after data and are drawn on the plot. The regression lines are

$$\text{Before: } \hat{y} = 1.089 + 0.189x$$

$$\text{After: } \hat{y} = 0.853 + 0.157x$$

(a) Does the scatterplot suggest that a straight line is an appropriate description of the relationship between degree-days and natural gas consumption? Do any individual observations appear to have large residuals or to be highly influential?

(b) About how much additional natural gas was consumed per day for each additional degree-day before the panels were added? After the panels were added?

(c) The daily average temperature during January in this location is about 30°, which corresponds to 35 degree-days per day. Use the regression lines to predict daily gas usage for a day with 35 degree-days before and after installation of the panels.

(d) The Energy Information Agency says that natural gas cost consumers about $1.20 per 100 cubic feet in the fall of 2003. At this rate, how much money do the solar panels save in the 31 days of January?

2.104 Investment theory uses the standard deviation of returns to describe the volatility (risk) of an investment. To describe how the risk of a specific security is related to that of the market as a whole, we use least-squares regression. Figure 2.31 plots the monthly percent total return y on Philip Morris common

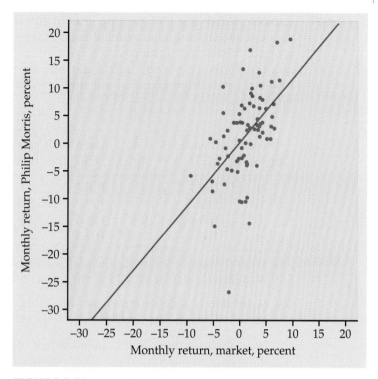

FIGURE 2.31   Monthly returns on an individual stock plotted against the returns for the stock market as a whole, with the least-squares line, for Exercise 2.104.

stock against the monthly return x on the Standard & Poor's 500-Stock Index, which represents the market, for the period between August 1990 and June 1997. The one clear outlier turns out not to be very influential. Here are the basic descriptive measures:

$$\bar{x} = 1.304 \qquad s_x = 3.392 \qquad r = 0.5251$$
$$\bar{y} = 1.878 \qquad s_y = 7.554$$

(a) Find the equation of the least-squares line from this information. What percent of the volatility in Philip Morris stock is explained by the linear relationship with the market as a whole?

(b) Explain carefully what the slope of the line tells us about how Philip Morris stock responds to changes in the market. This slope is called "beta" in investment theory.

(c) Returns on most individual stocks have a positive correlation with returns on the entire market. That is, when the market goes up, an individual stock tends to also go up. Explain why an investor should prefer stocks with beta > 1 when the market is rising and stocks with beta < 1 when the market is falling.

2.105 Lamb's-quarter is a common weed that interferes with the growth of corn. An agriculture researcher planted corn at the same rate in 16 small plots of ground, then weeded the plots by hand to allow a fixed number of lamb's-quarter plants to grow in each meter of corn row. No other weeds were allowed to grow. Here are the yields of corn (bushels per acre) in each of the plots:[50]

Weeds per meter	Corn yield	Weeds per meter	Corn yield	Weeds per meter	Corn yield	Weeds per meter	Corn yield
0	166.7	1	166.2	3	158.6	9	162.8
0	172.2	1	157.3	3	176.4	9	142.4
0	165.0	1	166.7	3	153.1	9	162.8
0	176.9	1	161.1	3	156.0	9	162.4

(a) What are the explanatory and response variables in this experiment?

(b) Make side-by-side stemplots of the yields, after rounding to the nearest bushel. What do you conclude about the effect of this weed on corn yield?

(c) Make a scatterplot of corn yield against weeds per meter. Find the least-squares regression line and add it to your plot. The advantage of regression over the side-by-side comparison in (b) is that we can use the fitted model to draw conclusions for counts of weeds other than the ones the researcher actually used. What does the slope of the fitted line tell us about the effect of lamb's-quarter on corn yield?

(d) Predict the yield for corn grown under these conditions with 6 lamb's-quarter plants per meter of row.

2.106 The table below gives data on the relationship between running speed (feet per second) and stride rate (steps taken per second) for elite female runners:[51]

Speed	15.86	16.88	17.50	18.62	19.97	21.06	22.11
Stride rate	3.05	3.12	3.17	3.25	3.36	3.46	3.55

Here are the corresponding data from the same source for male runners:

Speed	15.86	16.88	17.50	18.62	19.97	21.06	22.11
Stride rate	2.92	2.98	3.03	3.11	3.22	3.31	3.41

(a) Plot the data for both groups on one graph using different symbols to distinguish between the points for females and those for males.

(b) Suppose now that the data came to you without identification as to gender. Compute the least-squares line from all of the data and plot it on your graph.

(c) Compute the residuals from this line for each observation. Make a plot of the residuals against speed. How does the fact that the data come from two distinct groups show up in the residual plot?

2.107 Expose marine bacteria to X-rays for time periods from 1 to 15 minutes. Here are the number of surviving bacteria (in hundreds) on a culture plate after each exposure time:[52]

Time	Count	Time	Count	Time	Count
1	355	6	106	11	36
2	211	7	104	12	32
3	197	8	60	13	21
4	166	9	56	14	19
5	142	10	38	15	15

Theory suggests that the relationship between time and the *logarithm* of the count of surviving bacteria is linear.

(a) Find the regression line of count on time and make plots of the data and the residuals. In what way is this relationship nonlinear?

(b) Repeat your work for the regression of the logarithm of count on time. Make a numerical comparison of the linearity of the two relationships.

2.108 Wood scientists are interested in replacing solid-wood building material by less expensive products made from wood flakes. They collected the following data to examine the relationship between the length (in inches) and the strength (in pounds per square inch) of beams made from wood flakes:[53]

Length	5	6	7	8	9	10	11	12	13	14
Strength	446	371	334	296	249	254	244	246	239	234

(a) Make a scatterplot that shows how the length of a beam affects its strength.

(b) Describe the overall pattern of the plot. Are there any outliers?

(c) Fit a least-squares line to the entire set of data. Graph the line on your scatterplot. Does a straight line adequately describe these data?

(d) The scatterplot suggests that the relation between length and strength can be described by *two* straight lines, one for lengths less than 9 inches and another for lengths 9 inches or greater. Fit least-squares lines to these two subsets of the data, and draw the lines on your plot. Do they describe the data adequately? What question would you now ask the wood experts?

2.109 One reason to invest abroad is that markets in different countries don't move in step. When American stocks go down, foreign stocks may go up. So an investor who holds both bears less risk. That's the theory. Now we read: "The correlation between changes in American and European share prices has risen from 0.4 in the mid-1990s to 0.8 in 2000."[54] Explain to an investor who knows no statistics why this fact reduces the protection provided by buying European stocks.

2.110 The same article that claims that the correlation between changes in stock prices in Europe and the United States was 0.8 in 2000 goes on to say: "Crudely, that means that movements on Wall Street can explain 80% of price movements in Europe." Is this true? What is the correct percent explained if $r = 0.8$?

2.111 Someone says, "There is a strong positive correlation between the number of firefighters at a fire and the amount of damage the fire does. So sending lots of firefighters just causes more damage." Explain why this reasoning is wrong.

2.112 We expect that students who do well on the midterm exam in a course will usually also do well on the final exam. Gary Smith of Pomona College looked at the exam scores of all 346 students who took his statistics class over a 10-year period.[55] The least-squares line for predicting final exam score from midterm exam score was $\hat{y} = 46.6 + 0.41x$.

Octavio scores 10 points above the class mean on the midterm. How many points above the class mean do you predict that he will score on the final? (*Hint:* What is the predicted final score for the class mean midterm score $\bar{x}$?) This is an example of **regression to the mean**, the phenomenon that gave "regression" its name: students who do well on the midterm will on the average do less well on the final, but still above the class mean.

regression to the mean

2.113 Can we predict college grade point average from SAT scores and high school grades? The CSDATA data set described in the Data Appendix contains information on this issue for a large group of computer science students. We will look only at SAT mathematics scores as a predictor of later college GPA, using the variables SATM and GPA from CSDATA. Make a scatterplot, obtain r and r^2, and draw on your plot the least-squares regression line for predicting GPA from SATM. Then write a brief discussion of the ability of SATM alone to predict GPA. (In Chapter 11 we will see how combining several explanatory variables improves our ability to predict.)

EESEE CASE STUDIES

The Electronic Encyclopedia of Statistical Examples and Exercises (EESEE) is available on the text CD and Web site. These more elaborate stories, with data, provide settings for longer case studies.

2.114 Read the EESEE story "Is Old Faithful Faithful?" Write a response to Questions 2 and 3 for this case study.

2.115 Read the EESEE story " Will Women Outrun Men?" Write a report based on Questions 1, 2, and 4 for this case study.

2.116 Read the EESEE story "Surviving the *Titanic*." Write a report based on Questions 1, 2, and 3 in this case study.

2.117 Read the EESEE story "The Summer Rally—The Bear Facts." Write a report that responds to all of the questions posed in this case study.

Sampling the Nation

TOMMY WRIGHT *is Chief of the Statistical Research Division at the U.S. Bureau of the Census.*

The Bureau of the Census is best known for the every-ten-years census that attempts to gather information on every household in America. But much of our work involves sampling. In the words of the sociologist Frederick Stephan, "All scientific observation, whether statistical or not, is based on sampling. . . . The common practice of taking a small part or portion for tasting or testing to determine the characteristics of the whole precedes recorded history."

Throughout my career, I have focused on statistical sampling, which uses random selection of a small part of some large group to draw conclusions about the whole. At the Census Bureau, we select samples from all businesses and from all households on a monthly, quarterly, and annual basis. Many of the facts reported about the people and economy of the United States—the proportion of people unemployed, the number of new housing starts, the number of people in poverty—come from Census Bureau samples.

As a mathematical statistician, I think about both practical and theoretical problems. My colleagues and I must deal with practical issues of data collection such as how to ask a question, how to get responses, what to do when some units don't respond, how to match records in one data base with those in another, and how to protect and keep secret the responses from individual units.

These practical issues motivate theoretical problems that require mathematical skills. One of my main interests is the important and controversial issue of using samples to measure any undercounting or overcounting in the every-ten-years census. I also enjoy, almost as recreation, developing simple (but efficient) sampling methods and finding simple ways to explain well-known statistical results. For example, choosing a random sample of size 1000 usually requires generating 1000 random numbers—I found a simple algorithm that does the job for a sample of any size with just one randomization. The combination of mathematical challenges and practical importance makes doing statistics at the Census Bureau a fascinating career.

WASSILY KANDINSKY, ON THE POINTS © ARS, NY
(GIRAUDON/ART RESOURCE, NY)

Producing Data

Introduction

Numerical data are raw material for the growth of knowledge. Statistics provides tools that help data produce knowledge rather than confusion. These tools include methods for producing data as well as for interpreting already available data.

Chapters 1 and 2 explored the art of *data analysis*. They showed how to use graphs and numbers to uncover the nature of a set of data. We can distinguish two different attitudes toward data. We may ask, "What do we see in the data?" Or we may ask, "What answer do the data give to this specific question?" This distinction becomes important as we turn to producing data, because trustworthy answers to specific questions often require data produced for exactly that purpose.

exploratory data analysis

"What do we see?" is the attitude of **exploratory data analysis.** Sometimes we meet data that seem interesting and that deserve careful exploration. We do not know what the data may show, and we may have no specific questions in mind before we begin our examination. We have explored the relationship between SAT scores in the states and the percent of students who take the SAT, looked for explanations of obesity in girls, and examined the relationship between education and income. Exploratory analysis relies heavily on plotting the data. It seeks patterns that suggest interesting conclusions or questions for further study. However, *exploratory analysis alone can rarely provide convincing evidence for its conclusions, because striking patterns in data can arise from many sources.*

We may also use data to provide clear answers to specific questions such as "What car colors are most popular among young adults?" or "Does social rejection trigger the same reactions in the brain as physical pain?" or "Do recent college graduates with higher college debt tend to also have higher incomes?" Data that answer such questions are the product of intelligent effort, just like hybrid tomatoes and DVD players. This chapter is devoted to developing the skills needed to produce trustworthy data and to judge the quality of data produced by others. The techniques for producing data discussed in this chapter involve no formulas or graphs, but they are among the most important ideas in statistics.

statistical inference

Statistical techniques for producing data open the door to a further advance in data analysis, **formal statistical inference,** which answers specific questions with a known degree of confidence. Statistical inference uses the descriptive tools we have already met, in combination with new kinds of reasoning. It is primarily numerical rather than graphical. The later chapters of this book are devoted to inference. We will see that careful design of data production is the most important prerequisite for trustworthy inference.

3.1   First Steps

The first questions you must answer when you plan to produce data are familiar: "What individuals shall I study?" "What variables shall I measure?" Answering these questions gives a clear statement of your interests. We will take for granted in this chapter that all the variables we work with are clearly defined and satisfactorily measured—recall from page 6 that understanding your variables can require serious effort.

Statistics begins to play an essential role when the focus turns from measurements on individuals (whether ball bearings, rats, or people) to arrangements for collecting data from many individuals. These arrangements or patterns for producing data are called **designs.** Some questions that a design must address are: How many individuals shall we collect data from? How shall we select the individuals to be studied? If, as in many experiments, several groups of individuals are to receive different treatments, how shall we form the groups? If there is no planned design for producing data, we may be misled by haphazard or incomplete data or by confounding.

designs

Where to find data: the library and the Internet

You want data on a question of interest to you. Perhaps you want to know what causes of death are most common among young people. Or whether the mathematics performance of American schoolchildren is getting better.

It is tempting to simply draw conclusions from our own experience, making no use of more broadly representative data. We think (without really thinking) that the students at our college are typical. We hear a lot about AIDS, so we assume that it must be a leading cause of death among young people. Or we recall an unusual incident that sticks in our memory exactly because it is unusual. An airplane crash that kills several hundred people has more impact than data on all flights showing that flying is much safer than driving. We are relying on *anecdotes* rather than data.

ANECDOTAL EVIDENCE

Anecdotal evidence is based on haphazardly selected individual cases, which often come to our attention because they are striking in some way. These cases need not be representative of any larger group of cases.

A better tactic is to head for the library or the Internet. There you will find abundant data, not gathered specifically to answer your questions but available for your use. The most recent data can be found online, but locating them can be challenging. Government statistical offices are the primary source for demographic, economic, and social data. Many nations have a single national statistical office, such as Statistics Canada (www.statcan.ca) or Mexico's INEGI (www.inegi.gob.mx). The United States does not have a national statistical office. More than 70 different federal agencies collect data. Fortunately, you can reach most of them through the government's handy Fed-Stats site (www.fedstats.gov).

EXAMPLE 3.1   If you visit the National Center for Health Statistics, www.cdc.gov/nchs, you will learn that accidents are much the most common cause of death among people aged 20 to 24, accounting for over 40% of all deaths. Homicide is next, followed by suicide. AIDS ranks seventh, behind heart disease and cancer, at 1% of all deaths. The data also show that it is dangerous to be a young man: the overall death rate for men aged 20 to 24 is three times that for women, and the death rate from homicide is more than five times higher among men.

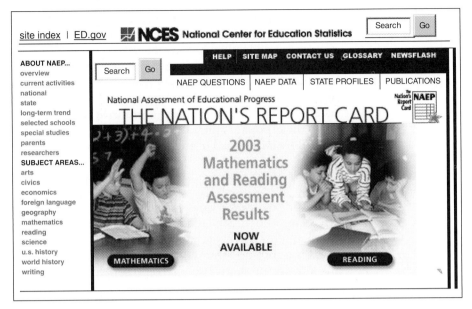

FIGURE 3.1   The Web sites of government statistical offices are prime sources of data. Here is the home page of the National Assessment of Educational Progress.

If you select the National Center for Education Statistics, `nces.ed.gov`, you will find in the latest National Assessment of Educational Progress (Figure 3.1) full detail about the math skills of schoolchildren. Mathematics scores have slowly but steadily increased since 1990. All racial/ethnic groups, both men and women, and students in most states are getting better in math.

The library and the Internet are sources of *available data*.

> ### AVAILABLE DATA
>
> **Available data** are data that were produced in the past for some other purpose but that may help answer a present question.

Available data are the only data used in most student reports. Because producing new data is expensive, we all use available data whenever possible. However, the clearest answers to present questions often require data produced to answer those specific questions. Statistical designs for producing data rely on either *sampling* or *experiments*.

Sampling

How have the attitudes of Americans, on issues ranging from abortion to work, changed over time? The most important source of such data is the University of Chicago's General Social Survey (GSS). The GSS interviews about

sample
population

3000 adult residents of the United States every second year. That is, the GSS selects a **sample** of adults to represent the larger **population** of all English-speaking adults living in the United States. The idea of *sampling* is to study a part in order to gain information about the whole. Data are often produced by sampling a population of people or things. Opinion polls, for example, report the views of the entire country based on interviews with a sample of about 1000 people. Government reports on employment and unemployment are produced from a monthly sample of about 60,000 households. The quality of manufactured items is monitored by inspecting small samples each hour or each shift.

census

In all of these examples, the expense of examining every item in the population makes sampling a practical necessity. Timeliness is another reason for preferring a sample to a **census,** which is an attempt to contact every individual in the entire population. We want information on current unemployment and public opinion next week, not next year. Moreover, a carefully conducted sample is often more accurate than a census. Accountants, for example, sample a firm's inventory to verify the accuracy of the records. Attempting to count every last item in the warehouse would be not only expensive but inaccurate. Bored people do not count carefully.

If conclusions based on a sample are to be valid for the entire population, a sound design for selecting the sample is required. Sampling designs are the topic of Section 3.3.

Experiments

A sample survey collects information about a population by selecting and measuring a sample from the population. The goal is a picture of the population, disturbed as little as possible by the act of gathering information. Sample surveys are one kind of *observational study*.

> **OBSERVATION VERSUS EXPERIMENT**
>
> An **observational study** observes individuals and measures variables of interest but does not attempt to influence the responses.
>
> An **experiment** deliberately imposes some treatment on individuals in order to observe their responses.

An observational study, even one based on a statistical sample, is a poor way to gauge the effect of an intervention. To see the response to a change, we must actually impose the change. When our goal is to understand cause and effect, experiments are the only source of fully convincing data.

EXAMPLE 3.2 A study of child care enrolled 1364 infants in 1991 and plans to follow them through their sixth year in school. In 2003, the researchers published an article finding that "the more time children spent in child care from birth to age four-and-a-half, the more adults tended to rate them, both at age four-and-a-half and at kindergarten, as less likely to get along with others, as more assertive, as disobedient, and as aggressive."[1]

This is an observational study. Parents made all child care decisions and the study did not attempt to influence them. "The study authors noted that their study was not designed to prove a cause and effect relationship. That is, the study cannot prove whether spending more time in child care causes children to have more problem behaviors." Perhaps employed parents who use child care are under stress and the children react to their parents' stress. Perhaps single parents are more likely to use child care. Perhaps parents are more likely to place in child care children who already have behavior problems. We can imagine an experiment that would remove these difficulties. From a large group of young children, choose some to be placed in child care and others to remain at home. This is an experiment because the treatment (child care or not) is imposed on the children. Of course, this particular experiment is neither practical nor ethical.

In Example 3.2, the effect of child care on behavior is *confounded* with (mixed up with) the characteristics of families who use child care. Observational studies of the effect of one variable on another often fail because the explanatory variable is confounded with lurking variables. Because experiments allow us to pin down the effects of specific variables of interest to us, they are the preferred method for gaining knowledge in science, medicine, and industry.

Experiments, like samples, require careful design. Because of the unique importance of experiments as a method for producing data, we begin the discussion of statistical designs for data collection in Section 3.2 with the principles underlying the design of experiments.

SECTION 3.1 | Summary

Data analysis is sometimes **exploratory** in nature. Exploratory analysis asks what the data tell us about the variables and their relations to each other. The conclusions of an exploratory analysis may not generalize beyond the specific data studied.

Statistical inference produces answers to specific questions, along with a statement of how confident we can be that the answer is correct. The conclusions of statistical inference are usually intended to apply beyond the individuals actually studied. Successful statistical inference usually requires **production of data** intended to answer the specific questions posed.

Anecdotal evidence based on a few individual cases is rarely trustworthy. **Available data** collected for other purposes, such as the data produced by government statistical offices, are often helpful.

We can produce data intended to answer specific questions by sampling or experimentation. **Sampling** selects a part of a population of interest to represent the whole. **Experiments** are distinguished from **observational studies** such as sample surveys by the active imposition of some treatment on the subjects of the experiment.

SECTION 3.1 | Exercises

3.1   Jamie is a hard-core computer programmer. He and all his friends prefer Jolt Cola (caffeine equivalent to two cups of coffee) to either Coke or Pepsi (caffeine equivalent to less than one cup of coffee). Explain why Jamie's experience is not good evidence that most young people prefer Jolt to Coke or Pepsi.[2]

3.2   When the discussion turns to the pros and cons of wearing automobile seat belts, Herman always brings up the case of a friend who survived an accident because he was not wearing a seat belt. The friend was thrown out of the car and landed on a grassy bank, suffering only minor injuries, while the car burst into flames and was destroyed. Explain briefly why this anecdote does not provide good evidence that it is safer not to wear seat belts.

3.3   One study of cell phones and the risk of brain cancer looked at a group of 469 people who have brain cancer. The investigators matched each cancer patient with a person of the same sex, age, and race who did not have brain cancer, then asked about use of cell phones.[3]   Result: "Our data suggest that use of handheld cellular telephones is not associated with risk of brain cancer." Is this an observational study or an experiment? Why? What are the explanatory and response variables?

3.4   A typical hour of prime-time television shows three to five violent acts. Linking family interviews and police records shows a clear association between time spent watching TV as a child and later aggressive behavior.[4]

(a) Explain why this is an observational study rather than an experiment. What are the explanatory and response variables?

(b) Suggest several lurking variables describing a child's home life that may be confounded with how much TV he or she watches. Explain why confounding makes it difficult to conclude that more TV *causes* more aggressive behavior.

3.5   An educational software company wants to compare the effectiveness of its computer animation for teaching cell biology with that of a textbook presentation. The company tests the biological knowledge of each of a group of first-year college students, then divides them into two groups. One group uses the animation, and the other studies the text. The company retests all the students and compares the increase in understanding of cell biology in the two groups. Is this an experiment? Why or why not? What are the explanatory and response variables?

3.6   Several large observational studies suggested that women who take hormones such as estrogen after menopause have lower risk of a heart attack than women who do not take hormones. Hormone replacement became popular. But in 2002, several careful experiments showed that hormone replacement does not reduce heart attacks. The National Institutes of Health, after reviewing the evidence, concluded that the observational studies were wrong. Taking hormones after menopause quickly fell out of favor.

(a) Explain the difference between an observational study and an experiment to compare women who do and don't take hormones after menopause.

(b) Suggest some characteristics of women who choose to take hormones that might affect the rate of heart attacks. In an observational study, these characteristics are confounded with taking hormones.

3.7   Moderate use of alcohol is associated with better health. Some studies suggest that drinking wine rather than beer or spirits confers added health benefits.

(a) Explain the difference between an observational study and an experiment to compare people who drink wine with people who drink beer.

(b) Suggest some characteristics of wine drinkers that might benefit their health. In an observational study, these characteristics are confounded with drinking wine.

3.8   Try to find information on this question: what percent of college undergraduates work part-time or full-time while they are taking classes? Start with the Web site of the National Center for Education Statistics nces.ed.gov. Keep a log of your search for this information.

3.2   Design of Experiments

A study is an experiment when we actually do something to people, animals, or objects in order to observe the response. Here is the basic vocabulary of experiments.

> ## EXPERIMENTAL UNITS, SUBJECTS, TREATMENT
>
> The individuals on which the experiment is done are the **experimental units.** When the units are human beings, they are called **subjects.** A specific experimental condition applied to the units is called a **treatment.**

Because the purpose of an experiment is to reveal the response of one variable to changes in other variables, the distinction between explanatory and response variables is important. The explanatory variables in an experiment are often called **factors.** Many experiments study the joint effects of several factors. In such an experiment, each treatment is formed by combining a specific value (often called a **level**) of each of the factors.

factors

level of a factor

EXAMPLE 3.3   Do smaller classes in elementary school really benefit students in areas such as scores on standard tests, staying in school, and going on to college? We might do an observational study that compares students who happened to be in smaller and larger classes in their early school years. Small classes are expensive, so they are more common in schools that serve richer communities. Students in small classes tend to also have other advantages: their schools have more resources, their parents are better educated, and so on. Confounding makes it impossible to isolate the effects of small classes.

The Tennessee STAR program was an experiment on the effects of class size. It has been called "one of the most important educational investigations ever carried out."

The *subjects* were 6385 students who were beginning kindergarten. Each student was assigned to one of three *treatments:* regular class (22 to 25 students) with one teacher, regular class with a teacher and a full-time teacher's aide, and small class (13 to 17 students). These treatments are levels of a single *factor,* the type of class. The students stayed in the same type of class for four years, then all returned to regular classes. In later years, students from the small classes had higher scores on standard tests, were less likely to fail a grade, had better high school grades, and so on. The benefits of small classes were greatest for minority students.[5]

Example 3.3 illustrates the big advantage of experiments over observational studies. **In principle, experiments can give good evidence for causation.** In an experiment, we study the specific factors we are interested in, while controlling the effects of lurking variables. All the students in the Tennessee STAR program followed the usual curriculum at their schools. Because students were assigned to different class types within their schools, school resources and family backgrounds were not confounded with class type. The only systematic difference was the type of class. When students from the small classes did better than those in the other two types, we can be confident that class size made the difference.

EXAMPLE 3.4   What are the effects of repeated exposure to an advertising message? The answer may depend both on the length of the ad and on how often it is repeated. An experiment investigated this question using undergraduate students as *subjects.* All subjects viewed a 40-minute television program that included ads for a digital camera. Some subjects saw a 30-second commercial; others, a 90-second version. The same commercial was shown either 1, 3, or 5 times during the program.

This experiment has two *factors:* length of the commercial, with 2 levels, and repetitions, with 3 levels. The 6 combinations of one level of each factor form 6 *treatments.* Figure 3.2 shows the layout of the treatments. After viewing, all of the subjects answered questions about their recall of the ad, their attitude toward the camera, and their intention to purchase it. These are the *response variables.*[6]

FIGURE 3.2   The treatments in the study of advertising, for Example 3.4. Combining the levels of the two factors forms six treatments.

Example 3.4 shows how experiments allow us to study the combined effects of several factors. The interaction of several factors can produce effects that could not be predicted from looking at the effects of each factor alone. Perhaps longer commercials increase interest in a product, and more commercials also increase interest, but if we both make a commercial longer and show it more often, viewers get annoyed and their interest in the product drops. The two-factor experiment in Example 3.4 will help us find out.

Comparative experiments

Laboratory experiments in science and engineering often have a simple design with only a single treatment, which is applied to all of the experimental units. The design of such an experiment can be outlined as

Treatment $\longrightarrow$ Observe response

For example, we may subject a beam to a load (treatment) and measure its deflection (observation). We rely on the controlled environment of the laboratory to protect us from lurking variables. When experiments are conducted in the field or with living subjects, such simple designs often yield invalid data. That is, we cannot tell whether the response was due to the treatment or to lurking variables. A medical example will show what can go wrong.

EXAMPLE 3.5   "Gastric freezing" is a clever treatment for ulcers in the upper intestine. The patient swallows a deflated balloon with tubes attached, then a refrigerated liquid is pumped through the balloon for an hour. The idea is that cooling the stomach will reduce its production of acid and so relieve ulcers. An experiment reported in the *Journal of the American Medical Association* showed that gastric freezing did reduce acid production and relieve ulcer pain. The treatment was safe and easy and was widely used for several years. The design of the experiment was

Gastric freezing $\longrightarrow$ Observe pain relief

placebo effect   The gastric freezing experiment was poorly designed. The patients' response may have been due to the **placebo effect.** A placebo is a dummy treatment. Many patients respond favorably to any treatment, even a placebo. This may be due to trust in the doctor and expectations of a cure or simply to the fact that medical conditions often improve without treatment. The response to a dummy treatment is the placebo effect.

A later experiment divided ulcer patients into two groups. One group was treated by gastric freezing as before. The other group received a placebo treatment in which the liquid in the balloon was at body temperature rather than freezing. The results: 34% of the 82 patients in the treatment group improved, but so did 38% of the 78 patients in the placebo group. This and other properly designed experiments showed that gastric freezing was no better than a placebo, and its use was abandoned.[7]

The first gastric freezing experiment gave misleading results because the effects of the explanatory variable were confounded with the placebo effect. We can defeat confounding by *comparing* two groups of patients, as in the second gastric freezing experiment. The placebo effect and other lurking variables now operate on both groups. The only difference between the groups is the actual effect of gastric freezing. The group of patients who received a sham

control group   treatment is called a **control group,** because it enables us to control the effects of outside variables on the outcome. Control is the first basic principle of statistical design of experiments. Comparison of several treatments in the same environment is the simplest form of control.

Uncontrolled experiments in medicine and the behavioral sciences can be dominated by such influences as the details of the experimental arrangement, the selection of subjects, and the placebo effect. The result is often *bias.*

> ### BIAS
>
> The design of a study is **biased** if it systematically favors certain outcomes.

An uncontrolled study of a new medical therapy, for example, is biased in favor of finding the treatment effective because of the placebo effect. It should not surprise you to learn that uncontrolled studies in medicine give new therapies a much higher success rate than proper comparative experiments. Well-designed experiments usually compare several treatments.

Randomization

experimental design   The **design of an experiment** first describes the response variable or variables, the factors (explanatory variables), and the layout of the treatments, with comparison as the leading principle. Figure 3.2 illustrates this aspect of the design of a study of response to advertising. The second aspect of design is the rule used to assign the experimental units to the treatments. Comparison of the effects of several treatments is valid only when all treatments are applied to similar groups of experimental units. If one corn variety is planted on more fertile ground, or if one cancer drug is given to more seriously ill patients, comparisons among treatments are meaningless. Systematic differences among the groups of experimental units in a comparative experiment cause bias. How can we assign experimental units to treatments in a way that is fair to all of the treatments?

Experimenters often attempt to match groups by elaborate balancing acts. Medical researchers, for example, try to match the patients in a "new drug" experimental group and a "standard drug" control group by age, sex, physical condition, smoker or not, and so on. Matching is helpful but not adequate—there are too many lurking variables that might affect the outcome. The experimenter is unable to measure some of these variables and will not think of others until after the experiment. Some important variables, such as how advanced a cancer patient's disease is, are so subjective that an experimenter might bias the study by, for example, assigning more advanced cancer cases to a promising new treatment in the unconscious hope that it will help them.

The statistician's remedy is to rely on chance to make an assignment that does not depend on any characteristic of the experimental units and that does not rely on the judgment of the experimenter in any way. The use of chance can be combined with matching, but the simplest design creates groups by chance alone. Here is an example.

FIGURE 3.3 Outline of a randomized comparative experiment, for Example 3.6.

EXAMPLE 3.6   Does talking on a hands-free cell phone distract drivers? Undergraduate students "drove" in a high-fidelity driving simulator equipped with a hands-free cell phone. The car ahead brakes: how quickly does the subject respond? Twenty students (the control group) simply drove. Another 20 (the experimental group) talked on the cell phone while driving.

This experiment has a single factor (cell phone use) with two levels. The researchers must divide the 40 student subjects into two groups of 20. To do this in a completely unbiased fashion, put the names of the 40 students in a hat, mix them up, and draw 20. These students form the experimental group and the remaining 20 make up the control group. Figure 3.3 outlines the design of this experiment.[8]

---

randomization   The use of chance to divide experimental units into groups is called **randomization.** The design in Figure 3.3 combines comparison and randomization to arrive at the simplest randomized comparative design. This "flowchart" outline presents all the essentials: randomization, the sizes of the groups and which treatment they receive, and the response variable. There are, as we will see later, statistical reasons for generally using treatment groups about equal in size.

Randomized comparative experiments

The logic behind the randomized comparative design in Figure 3.3 is as follows:

- Randomization produces two groups of subjects that we expect to be similar in all respects before the treatments are applied.

- Comparative design helps ensure that influences other than the cell phone operate equally on both groups.

- Therefore, differences in average brake reaction time must be due either to talking on the cell phone or to the play of chance in the random assignment of subjects to the two groups.

That "either-or" deserves more comment. We cannot say that *any* difference in the average reaction times of the experimental and control groups is caused by talking on the cell phone. There would be some difference even if both groups were treated the same, because the natural variability among people means that some react faster than others. Chance assigns the faster-reacting students to one group or the other, so that there is a chance difference between the groups. We would not trust an experiment with just one subject in each group, for example. The results would depend too much on which group got lucky and received the subject with quicker reactions. If we assign many students to each group, however, the effects of chance will average out. There will be little difference in the average reaction times in the two groups unless talking on the cell phone causes a difference. "Use enough subjects to reduce chance variation" is the third big idea of statistical design of experiments.

PRINCIPLES OF EXPERIMENTAL DESIGN

The basic principles of statistical design of experiments are

1. Control the effects of lurking variables on the response, most simply by comparing two or more treatments.

2. Randomize—use impersonal chance to assign experimental units to treatments.

3. Repeat each treatment on many units to reduce chance variation in the results.

We hope to see a difference in the responses so large that it is unlikely to happen just because of chance variation. We can use the laws of probability, which give a mathematical description of chance behavior, to learn if the treatment effects are larger than we would expect to see if only chance were operating. If they are, we call them *statistically significant.*

STATISTICAL SIGNIFICANCE

An observed effect so large that it would rarely occur by chance is called **statistically significant.**

You will often see the phrase "statistically significant" in reports of investigations in many fields of study. It tells you that the investigators found good evidence for the effect they were seeking. The cell phone study, for example, reported statistically significant evidence that talking on a cell phone increases the mean reaction time of drivers when the car in front of them brakes.

How to randomize

The idea of randomization is to assign subjects to treatments by drawing names from a hat. In practice, experimenters use software to carry out randomization. Most statistical software will choose 20 out of a list of 40 at random, for example. The list might contain the names of 40 human subjects. The 20 chosen form one group, and the 20 that remain form the second group. The *Simple Random Sample* applet on the text CD and Web site makes it particularly easy to choose treatment groups at random.

You can randomize without software by using a *table of random digits.* Thinking about random digits helps you to understand randomization even if you will use software in practice. Table B at the back of the book and on the back endpaper is a table of random digits.

> **RANDOM DIGITS**
>
> A **table of random digits** is a list of the digits 0, 1, 2, 3, 4, 5, 6, 7, 8, 9 that has the following properties:
>
> **1.** The digit in any position in the list has the same chance of being any one of 0, 1, 2, 3, 4, 5, 6, 7, 8, 9.
>
> **2.** The digits in different positions are independent in the sense that the value of one has no influence on the value of any other.

You can think of Table B as the result of asking an assistant (or a computer) to mix the digits 0 to 9 in a hat, draw one, then replace the digit drawn, mix again, draw a second digit, and so on. The assistant's mixing and drawing saves us the work of mixing and drawing when we need to randomize. Table B begins with the digits 19223950340575628713. To make the table easier to read, the digits appear in groups of five and in numbered rows. The groups and rows have no meaning—the table is just a long list of digits having the properties 1 and 2 described above.

Our goal is to use random digits for experimental randomization. We need the following facts about random digits, which are consequences of the basic properties 1 and 2:

- Any *pair* of random digits has the same chance of being any of the 100 possible pairs: 00, 01, 02, . . . , 98, 99.

- Any *triple* of random digits has the same chance of being any of the 1000 possible triples: 000, 001, 002, . . . , 998, 999.

- . . . and so on for groups of four or more random digits.

EXAMPLE 3.7   In the cell phone experiment of Example 3.6, we must divide 40 students at random into two groups of 20 students each.

Step 1: Label. Give each student a numerical label, using as few digits as possible. Two digits are needed to label 40 students, so we use labels

$$01, 02, 03, \ldots, 39, 40$$

It is also correct to use labels 00 to 39 or some other choice of 40 two-digit labels.

Step 2: Table. Start anywhere in Table B and read two-digit groups. Suppose we begin at line 130, which is

$$69051 \quad 64817 \quad 87174 \quad 09517 \quad 84534 \quad 06489 \quad 87201 \quad 97245$$

The first 10 two-digit groups in this line are

$$69 \quad 05 \quad 16 \quad 48 \quad 17 \quad 87 \quad 17 \quad 40 \quad 95 \quad 17$$

Each of these two-digit groups is a label. The labels 00 and 41 to 99 are not used in this example, so we ignore them. The first 20 labels between 01 and 40 that we encounter in the table choose students for the experimental group. Of the first 10 labels in line 130, we ignore four because they are too high (over 40). The others are 05, 16, 17, 17, 40, and 17. The students labeled 05, 16, 17, and 40 go into the experimental group. Ignore the second and third 17s because that student is already in the group. Run your finger across line 130 (and continue to the following lines) until you have chosen 20 students. They are the students labeled

05, 16, 17, 40, 20, 19, 32, 04, 25, 29, 37, 39, 31, 18, 07, 13, 33, 02, 36, 23

You should check at least the first few of these. These students form the experimental group. The remaining 20 are the control group.

As Example 3.7 illustrates, randomization requires two steps: assign labels to the experimental units and then use Table B to select labels at random. Be sure that all labels are the same length so that all have the same chance to be chosen. Use the shortest possible labels—one digit for 9 or fewer individuals, two digits for 10 to 100 individuals, and so on. Don't try to scramble the labels as you assign them. Table B will do the required randomizing, so assign labels in any convenient manner, such as in alphabetical order for human subjects. You can read digits from Table B in any order—along a row, down a column, and so on—because the table has no order. As an easy standard practice, we recommend reading along rows.

completely randomized design   When all experimental units are allocated at random among all treatments, as in Example 3.7, the experimental design is **completely randomized.** Completely randomized designs can compare any number of treatments. The treatments can be formed by levels of a single factor or by more than one factor.

EXAMPLE 3.8   Figure 3.2 (page 199) displays six treatments formed by the two factors in an experiment on response to a TV commercial. Suppose that we have 150 students who are willing to serve as subjects. We must assign 25 students at random to each group. Figure 3.4 outlines the completely randomized design.

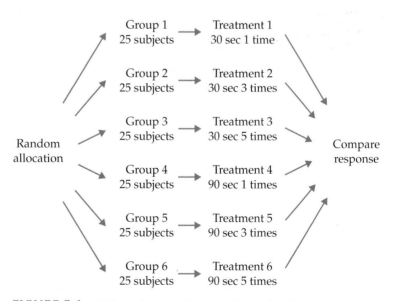

FIGURE 3.4   Outline of a completely randomized design comparing six treatments, for Example 3.8.

To carry out the random assignment, label the 150 students 001 to 150. (Three digits are needed to label 150 subjects.) Enter Table B and read three-digit groups until you have selected 25 students to receive Treatment 1 (a 30-second ad shown once). If you start at line 140, the first few labels for Treatment 1 subjects are 129, 048, and 003.

Continue in Table B to select 25 more students to receive Treatment 2 (a 30-second ad shown 3 times). Then select another 25 for Treatment 3 and so on until you have assigned 125 of the 150 students to Treatments 1 through 5. The 25 students who remain get Treatment 6. The randomization is straightforward, but very tedious to do by hand. We recommend the *Simple Random Sample* applet. Exercise 3.21 shows how to use the applet to do the randomization for this example.

Cautions about experimentation

The logic of a randomized comparative experiment depends on our ability to treat all the experimental units identically in every way except for the actual treatments being compared. Good experiments therefore require careful attention to details. For example, the subjects in the second gastric freezing experiment (Example 3.5) all got the same medical attention during the study. **double-blind** Moreover, the study was **double-blind**—neither the subjects themselves nor the medical personnel who worked with them knew which treatment any subject had received. The double-blind method avoids unconscious bias by, for example, a doctor who doesn't think that "just a placebo" can benefit a patient.

Many—perhaps most—experiments have some weaknesses in detail. The environment of an experiment can influence the outcomes in unexpected ways. Although experiments are the gold standard for evidence of cause and effect, really convincing evidence usually requires that a number of studies in different places with different details produce similar results. Here are some brief examples of what can go wrong.

EXAMPLE 3.9 A study of the effects of marijuana recruited young men who used marijuana. Some were randomly assigned to smoke marijuana cigarettes, while others were given placebo cigarettes. This failed: the control group recognized that their cigarettes were phony and complained loudly. It may be quite common for blindness to fail because the subjects can tell which treatment they are receiving.[9]

EXAMPLE 3.10 To study genetic influence on behavior, experimenters "knock out" a gene in one group of mice and compare their behavior with that of a control group of normal mice. The results of these experiments often don't agree as well as hoped, so investigators did exactly the same experiment with the same genetic strain of mice in Oregon, Alberta (Canada), and New York. Many results were very different. It appears that small differences in the lab environments have big effects on the behavior of the mice. Remember this the next time you read that our genes control our behavior.[10]

lack of realism The most serious potential weakness of experiments is **lack of realism.** The subjects or treatments or setting of an experiment may not realistically duplicate the conditions we really want to study. Here is an example.

EXAMPLE 3.11   How do layoffs at a workplace affect the workers who remain on the job? Psychologists asked student subjects to proofread text for extra course credit, then "let go" some of the workers (who were actually accomplices of the experimenters). Some subjects were told that those let go had performed poorly (Treatment 1). Others were told that not all could be kept and that it was just luck that they were kept and others let go (Treatment 2). We can't be sure that the reactions of the students are the same as those of workers who survive a layoff in which other workers lose their jobs. Many behavioral science experiments use student subjects in a campus setting. Do the conclusions apply to the real world?

Lack of realism can limit our ability to apply the conclusions of an experiment to the settings of greatest interest. Most experimenters want to generalize their conclusions to some setting wider than that of the actual experiment. *Statistical analysis of an experiment cannot tell us how far the results will generalize to other settings.* Nonetheless, the randomized comparative experiment, because of its ability to give convincing evidence for causation, is one of the most important ideas in statistics.

Matched pairs designs

Completely randomized designs are the simplest statistical designs for experiments. They illustrate clearly the principles of control, randomization, and repetition. However, completely randomized designs are often inferior to more elaborate statistical designs. In particular, matching the subjects in various ways can produce more precise results than simple randomization.

matched pairs design   The simplest use of matching is a **matched pairs design,** which compares just two treatments. The subjects are matched in pairs. For example, an experiment to compare two advertisements for the same product might use pairs of subjects with the same age, sex, and income. The idea is that matched subjects are more similar than unmatched subjects, so that comparing responses within a number of pairs is more efficient than comparing the responses of groups of randomly assigned subjects. Randomization remains important: which one of a matched pair sees the first ad is decided at random. One common variation of the matched pairs design imposes both treatments on the same subjects, so that each subject serves as his or her own control. Here is an example.

EXAMPLE 3.12   Example 3.6 describes an experiment on the effects of talking on a cell phone while driving. The experiment compared two treatments, driving in a simulator and driving in the simulator while talking on a hands-free cell phone. The response variable is the time the driver takes to apply the brake when the car in front brakes suddenly. In Example 3.6, 40 student subjects were assigned at random, 20 students to each treatment. This is a completely randomized design, outlined in Figure 3.3. Subjects differ in driving skill and reaction times. The completely randomized design relies on chance to create two similar groups of subjects.

In fact, the experimenters used a matched pairs design in which all subjects drove both with and without using the cell phone. They compared each individual's reaction times with and without the phone. If all subjects drove first with the phone and then without it, the effect of talking on the cell phone would be confounded with the fact that this is the first run in the simulator. The proper procedure requires that all subjects first be trained in using the simulator, that the *order* in which a subject drives with and without the phone be random, and that the two drives be on separate days to reduce carryover effects.

The completely randomized design uses chance to decide which 20 subjects will drive with the cell phone. The other 20 drive without it. The matched pairs design uses chance to decide which 20 subjects will drive first with and then without the cell phone. The other 20 drive first without and then with the phone.

Block designs

The matched pairs design of Example 3.12 uses the principles of comparison of treatments, randomization, and repetition on several experimental units. However, the randomization is not complete (all subjects randomly assigned to treatment groups) but restricted to assigning the order of the treatments for each subject. *Block designs* extend the use of "similar subjects" from pairs to larger groups.

BLOCK DESIGN

A **block** is a group of experimental units or subjects that are known before the experiment to be similar in some way that is expected to affect the response to the treatments. In a **block design,** the random assignment of units to treatments is carried out separately within each block.

Block designs can have blocks of any size. A block design combines the idea of creating equivalent treatment groups by matching with the principle of forming treatment groups at random. Blocks are another form of *control*. They control the effects of some outside variables by bringing those variables into the experiment to form the blocks. Here are some typical examples of block designs.

EXAMPLE 3.13   The progress of a type of cancer differs in women and men. A clinical experiment to compare three therapies for this cancer therefore treats sex as a blocking variable. Two separate randomizations are done, one assigning the female subjects to the treatments and the other assigning the male subjects. Figure 3.5 outlines the design of this experiment. Note that there is no randomization involved in making up the blocks. They are groups of subjects who differ in some way (sex in this case) that is apparent before the experiment begins.

EXAMPLE 3.14   The soil type and fertility of farmland differ by location. Because of this, a test of the effect of tillage type (two types) and pesticide application (three application schedules) on soybean yields uses small fields as blocks. Each block is divided into six plots, and the six treatments are randomly assigned to plots separately within each block.

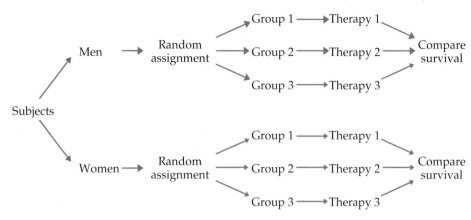

FIGURE 3.5   Outline of a block design, for Example 3.13. The blocks consist of male and female subjects. The treatments are three therapies for cancer.

EXAMPLE 3.15   The Tennessee STAR class size experiment (Example 3.3) used a block design. It was important to compare different class types in the same school because the children in a school come from the same neighborhood, follow the same curriculum, and have the same school environment outside class. In all, 79 schools across Tennessee participated in the program. That is, there were 79 blocks. New kindergarten students were randomly placed in the three types of class separately within each school.

Blocks allow us to draw separate conclusions about each block, for example, about men and women in the cancer study in Example 3.13. Blocking also allows more precise overall conclusions because the systematic differences between men and women can be removed when we study the overall effects of the three therapies. The idea of blocking is an important additional principle of statistical design of experiments. A wise experimenter will form blocks based on the most important unavoidable sources of variability among the experimental units. Randomization will then average out the effects of the remaining variation and allow an unbiased comparison of the treatments.

SECTION 3.2 | Summary

In an experiment, one or more **treatments** are imposed on the **experimental units** or **subjects.** Each treatment is a combination of **levels** of the explanatory variables, which we call **factors.**

The **design** of an experiment refers to the choice of treatments and the manner in which the experimental units or subjects are assigned to the treatments.

The basic principles of statistical design of experiments are **control, randomization,** and **repetition.**

The simplest form of control is **comparison.** Experiments should compare two or more treatments in order to prevent **confounding** the effect of a treatment with other influences, such as lurking variables.

Randomization uses chance to assign subjects to the treatments. Randomization creates treatment groups that are similar (except for chance variation) before the treatments are applied. Randomization and comparison together prevent **bias,** or systematic favoritism, in experiments.

You can carry out randomization by giving numerical labels to the experimental units and using a **table of random digits** to choose treatment groups.

Repetition of the treatments on many units reduces the role of chance variation and makes the experiment more sensitive to differences among the treatments.

Good experiments require attention to detail as well as good statistical design. Many behavioral and medical experiments are **double-blind. Lack of realism** in an experiment can prevent us from generalizing its results.

In addition to comparison, a second form of control is to restrict randomization by forming **blocks** of experimental units that are similar in some way that is important to the response. Randomization is then carried out separately within each block.

Matched pairs are a common form of blocking for comparing just two treatments. In some matched pairs designs, each subject receives both treatments in a random order. In others, the subjects are matched in pairs as closely as possible, and one subject in each pair receives each treatment.

SECTION 3.2 | Exercises

For each of the experimental situations described in Exercises 3.9 to 3.11, identify the experimental units or subjects, the factors, the treatments, and the response variables.

3.9   Ability to grow in shade may help pines found in the dry forests of Arizona resist drought. How well do these pines grow in shade? Investigators planted pine seedlings in a greenhouse in either full light or light reduced to 5% of normal by shade cloth. At the end of the study, they dried the young trees and weighed them.

3.10   Most American adolescents don't eat well and don't exercise enough. Can middle schools increase physical activity among their students? Can they persuade students to eat better? Investigators designed a "physical activity intervention" to increase activity in physical education classes and during leisure periods throughout the school day. They also designed a "nutrition intervention" that improved school lunches and offered ideas for healthy home-packed lunches. Each participating school was randomly assigned to one of the interventions, both interventions, or no intervention. The investigators observed physical activity and lunchtime consumption of fat.

3.11   How can we reduce the rate of refusals in telephone surveys? Most people who answer at all listen to the interviewer's introductory remarks and then decide whether to continue. One study made telephone calls to randomly selected households to ask opinions about the next election. In some calls, the interviewer gave her name, in others she identified the university she was representing, and in still others she identified both herself and the university. For each type of call, the interviewer either did or did not offer to send a copy of the final survey results to the person interviewed. Do these differences in the introduction affect whether the interview is completed?

3.12 The Bayer Aspirin Web site claims that "Nearly five decades of research now link aspirin to the prevention of stroke and heart attacks." The most important evidence for this claim comes from the Physicians' Health Study, a large medical experiment involving 22,000 male physicians. One group of about 11,000 physicians took an aspirin every second day, while the rest took a placebo. After several years the study found that subjects in the aspirin group had significantly fewer heart attacks than subjects in the placebo group.

(a) Identify the experimental subjects, the factor and its levels, and the response variable in the Physicians' Health Study.

(b) Use a diagram to outline a completely randomized design for the Physicians' Health Study.

(c) What does it mean to say that the aspirin group had "significantly fewer heart attacks"?

3.13 Doctors identify "chronic tension-type headaches" as headaches that occur almost daily for at least six months. Can antidepressant medications or stress management training reduce the number and severity of these headaches? Are both together more effective than either alone? Investigators compared four treatments: antidepressant alone, placebo alone, antidepressant plus stress management, and placebo plus stress management. Outline the design of the experiment. The headache sufferers named below have agreed to participate in the study. Use software or Table B at line 130 to randomly assign the subjects to the treatments.

Acosta	Duncan	Han	Liang	Padilla	Valasco
Asihiro	Durr	Howard	Maldonado	Plochman	Vaughn
Bennett	Edwards	Hruska	Marsden	Rosen	Wei
Bikalis	Farouk	Imrani	Montoya	Solomon	Wilder
Chen	Fleming	James	O'Brian	Trujillo	Willis
Clemente	George	Kaplan	Ogle	Tullock	Zhang

3.14 How does smoking marijuana affect willingness to work? Canadian researchers persuaded young adult men who used marijuana to live for 98 days in a "planned environment." The men earned money by weaving belts. They used their earnings to pay for meals and other consumption and could keep any money left over. One group smoked two potent marijuana cigarettes every evening. The other group smoked two weak marijuana cigarettes. All subjects could buy more cigarettes but were given strong or weak cigarettes, depending on their group. Did the weak and strong groups differ in work output and earnings?[11]

(a) Outline the design of this experiment.

(b) The following table lists the names of the 20 subjects. Use software or Table B at line 131 to carry out the randomization your design requires.

Abbott	Decker	Gutierrez	Lucero	Rosen
Afifi	Engel	Hwang	McNeill	Thompson
Brown	Fluharty	Iselin	Morse	Travers
Chen	Gerson	Kaplan	Quinones	Ullmann

3.15   Eye cataracts are responsible for over 40% of blindness around the world. Can drinking tea regularly slow the growth of cataracts? We can't experiment on people, so we use rats as subjects. Researchers injected 18 young rats with a substance that causes cataracts. One group of the rats also received black tea extract; a second group received green tea extract; and a third got a placebo, a substance with no effect on the body. The response variable was the growth of cataracts over the next six weeks. Yes, both tea extracts did slow cataract growth.[12]

(a) Outline the design of this experiment.

(b) Use software or Table B, starting at line 142, to assign rats to treatments.

3.16   Workers who survive a layoff of other employees at their location may suffer from "survivor guilt." A study of survivor guilt and its effects used as subjects 120 students who were offered an opportunity to earn extra course credit by doing proofreading. Each subject worked in the same cubicle as another student, who was an accomplice of the experimenters. At a break midway through the work, one of three things happened:

Treatment 1: The accomplice was told to leave; it was explained that this was because she performed poorly.

Treatment 2: It was explained that unforeseen circumstances meant there was only enough work for one person. By "chance," the accomplice was chosen to be laid off.

Treatment 3: Both students continued to work after the break.

The subjects' work performance after the break was compared with performance before the break.[13]

(a) Outline the design of this completely randomized experiment.

(b) If you are using software, choose the subjects for Treatment 1. If not, use Table B at line 123 to choose the first four subjects for Treatment 1.

3.17   Twenty-four public middle schools agree to participate in the experiment described in Exercise 3.10. Use a diagram to outline a completely randomized design for this experiment. Then do the randomization required to assign schools to treatments. If you use Table B, start at line 105.

3.18   Stores advertise price reductions to attract customers. What type of price cut is most attractive? Market researchers prepared ads for athletic shoes announcing different levels of discounts (20%, 40%, 60%, or 80%). The student subjects who read the ads were also given "inside information" about the fraction of shoes on sale (25%, 50%, 75%, or 100%). Each subject then rated the attractiveness of the sale on a scale of 1 to 7.[14]

(a) There are two factors. Make a sketch like Figure 3.2 (page 199) that displays the treatments formed by all combinations of levels of the factors.

(b) Outline a completely randomized design using 80 student subjects. Use software or Table B at line 133 to choose the subjects for the first treatment.

3.19 A maker of fabric for clothing is setting up a new line to "finish" the raw fabric. The line will use either metal rollers or natural bristle rollers to raise the surface of the fabric; a dyeing cycle time of either 30 minutes or 40 minutes; and a temperature of either 150° or 175° Celsius. An experiment will compare all combinations of these choices. Three specimens of fabric will be subjected to each treatment and scored for quality.

(a) What are the factors and the treatments? How many individuals (fabric specimens) does the experiment require?

(b) Outline a completely randomized design for this experiment. (You need not actually do the randomization.)

3.20 You can use the *Simple Random Sample* applet to choose a treatment group at random once you have labeled the subjects. Example 3.7 (page 204) uses Table B to choose 20 students from a group of 40 for the treatment group in a study of the effect of cell phones on driving. Use the applet to choose the 20 students for the experimental group. Which students did you choose? The remaining 20 students make up the control group.

3.21 The *Simple Random Sample* applet allows you to randomly assign experimental units to more than two groups without difficulty. Example 3.8 (page 205) describes a randomized comparative experiment in which 150 students are randomly assigned to six groups of 25.

(a) Use the applet to randomly choose 25 out of 150 students to form the first group. Which students are in this group?

(b) The "Population hopper" now contains the 125 students that were not chosen, in scrambled order. Click "Sample" again to choose 25 of these remaining students to make up the second group. Which students were chosen?

(c) Click "Sample" three more times to choose the third, fourth, and fifth groups. Don't take the time to write down these groups. Check that there are only 25 students remaining in the "Population hopper." These subjects get Treatment 6. Which students are they?

3.22 Experiments with more than one factor allow insight into interactions between the factors. A study of the attractiveness of advertised price discounts had two factors: percent of all goods on sale (25%, 50%, 75%, or 100%) and whether the discount was stated precisely as 60% off or as a range, 50% to 70% off. Subjects rated the attractiveness of the sale on a scale of 1 to 7. Figure 3.6 shows the mean ratings for the eight treatments formed from the two factors.[15] Based on these results, write a careful description of how percent on sale and precise discount versus range of discounts influence the attractiveness of a sale.

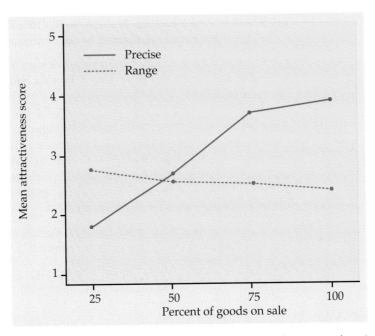

FIGURE 3.6 Mean responses to eight treatments in an experiment with two factors, showing interaction between the factors, for Exercise 3.22.

3.23 "Bee pollen is effective for combating fatigue, depression, cancer, and colon disorders." So says a Web site that offers the pollen for sale. We wonder if bee pollen really does prevent colon disorders. Here are two ways to study this question. Explain why the second design will produce more trustworthy data.

1. Find 200 women who take bee pollen regularly. Match each with a woman of the same age, race, and occupation who does not take bee pollen. Follow both groups for 5 years.

2. Find 400 women who do not have colon disorders. Assign 200 to take bee pollen capsules and the other 200 to take placebo capsules that are identical in appearance. Follow both groups for 5 years.

3.24 Health care providers are giving more attention to relieving the pain of cancer patients. An article in the journal *Cancer* surveyed a number of studies and concluded that controlled-release morphine tablets, which release the painkiller gradually over time, are more effective than giving standard morphine when the patient needs it.[16] The "methods" section of the article begins: "Only those published studies that were controlled (i.e., randomized, double blind, and comparative), repeated-dose studies with CR morphine tablets in cancer pain patients were considered for this review." Explain the terms in parentheses to someone who knows nothing about medical trials.

3.25 Does the herb Saint-John's-wort relieve major depression? Here are some excerpts from the report of a study of this issue.[17] The study concluded that the herb is no more effective than a placebo.

(a) "Design: Randomized, double-blind, placebo-controlled clinical trial . . . ." Explain the meaning of each of the terms in this description.

(b) "Participants . . . were randomly assigned to receive either Saint-John's-wort extract ($n = 98$) or placebo ($n = 102$). . . . The primary outcome measure was the rate of change in the Hamilton Rating Scale for Depression over the treatment period." Based on this information, use a diagram to outline the design of this clinical trial.

3.26   Puzzling but true: stocks tend to go down on Mondays. There is no convincing explanation for this fact. A recent study looked at this "Monday effect" in more detail, using data on the daily returns of stocks on several U.S. exchanges over a 30-year period. Here are some of the findings:

> *To summarize, our results indicate that the well-known Monday effect is caused largely by the Mondays of the last two weeks of the month. The mean Monday return of the first three weeks of the month is, in general, not significantly different from zero and is generally significantly higher than the mean Monday return of the last two weeks. Our finding seems to make it more difficult to explain the Monday effect.*[18]

A friend thinks that "significantly" in this article has its plain English meaning, roughly "I think this is important." Explain in simple language what "significantly higher" and "not significantly different from zero" actually tell us here.

3.27   Does adding the five-digit postal ZIP code to an address really speed up delivery of letters? Does adding the four more digits that make up "ZIP + 4" speed delivery yet more? What about mailing a letter on Monday, Thursday, or Saturday? Describe the design of an experiment on the speed of first-class mail delivery. For simplicity, suppose that all letters go from you to a friend, so that the sending and receiving locations are fixed.

3.28   The *Simple Random Sample* applet can demonstrate how randomization works to create similar groups for comparative experiments. Suppose that (unknown to the experimenters) the 20 even-numbered students among the 40 subjects for the cell phone study in Example 3.7 (page 204) have fast reactions, and that the odd-numbered students have slow reactions. We would like the experimental and control groups to contain similar numbers of the fast reactors. Use the applet to choose 10 samples of size 20 from the 40 students. (Be sure to click "Reset" after each sample.) Record the counts of even-numbered students in each of your 10 samples. You see that there is considerable chance variation but no systematic bias in favor of one or the other group in assigning the fast-reacting students. Larger samples from larger populations will on the average do a better job of making the two groups equivalent.

3.29   We often see players on the sidelines of a football game inhaling oxygen. Their coaches think this will speed their recovery. We might measure recovery from intense exercise as follows: Have a football player run 100 yards three times in quick succession. Then allow three minutes to rest before running 100 yards again. Time the final run. Because players vary greatly in speed, you plan a matched pairs experiment using 20 football players as subjects. Describe the design of such an experiment to investigate the effect of inhaling oxygen during the rest period. Why should each player's two trials be on different days? Use Table B at line 170 to decide which players will get oxygen on their first trial.

3.30 The concentration of carbon dioxide (CO_2) in the atmosphere is increasing rapidly due to our use of fossil fuels. Because plants use CO_2 to fuel photosynthesis, more CO_2 may cause trees and other plants to grow faster. An elaborate apparatus allows researchers to pipe extra CO_2 to a 30-meter circle of forest. We want to compare the growth in base area of trees in treated and untreated areas to see if extra CO_2 does in fact increase growth. We can afford to treat three circular areas.[19]

(a) Describe the design of a completely randomized experiment using 6 well-separated 30-meter circular areas in a pine forest. Sketch the forest area with the 6 circles and carry out the randomization your design calls for.

(b) Regions within the forest may differ in soil fertility. Describe a matched pairs design using three pairs of circles that will reduce the extra variation due to different fertility. Sketch the forest area with the new arrangement of circles and carry out the randomization your design calls for.

3.31 Calcium is important to the development of young girls. To study how the bodies of young girls process calcium, investigators used the setting of a summer camp. Calcium was given in Hawaiian Punch at either a high or a low level. The camp diet was otherwise the same for all girls. Suppose that there are 60 campers.

(a) Outline a completely randomized design for this experiment.

(b) Describe a matched pairs design in which each girl receives both levels of calcium (with a "washout period" between). What is the advantage of the matched pairs design over the completely randomized design?

(c) The same randomization can be used in different ways for both designs. Label the subjects 01 to 60. You must choose 30 of the 60. Use Table B at line 160 to choose just the first 5 of the 30. How are the 30 subjects chosen treated in the completely randomized design? How are they treated in the matched pairs design?

3.32 Twenty overweight females have agreed to participate in a study of the effectiveness of four reducing regimens, A, B, C, and D. The researcher first calculates how overweight each subject is by comparing the subject's actual weight with her "ideal" weight. Below are the subjects and their excess weights in pounds.

Birnbaum	35	Hernandez	25	Moses	25	Smith	29
Brown	34	Jackson	33	Nevesky	39	Stall	33
Brunk	30	Kendall	28	Obrach	30	Tran	35
Dixon	34	Loren	32	Rodriguez	30	Wilansky	42
Festinger	24	Mann	28	Santiago	27	Williams	22

The response variable is the weight lost after eight weeks of treatment. Because the initial amount overweight will influence the response variable, a block design is appropriate.

(a) Arrange the subjects in order of increasing excess weight. Form five blocks by grouping the four least overweight, then the next four, and so on.

(b) Use Table B to do the required random assignment of subjects to the four reducing regimens separately within each block. Be sure to explain exactly how you used the table.

3.33 Fractures of the spine are common and serious among women with advanced osteoporosis (low mineral density in the bones). Can taking strontium renelate help? A large medical trial assigned 1649 women to take either strontium renelate or a placebo each day. All of the subjects had osteoporosis and had had at least one fracture. All were taking calcium supplements and receiving standard medical care. The response variables were measurements of bone density and counts of new fractures over three years. The subjects were treated at 10 medical centers in 10 different countries.[20] Outline an appropriate design for this experiment. Explain why this is the proper design.

3.34 An ultramarathon, as you might guess, is a footrace longer than the 26.2 miles of a marathon. Runners commonly develop respiratory infections after an ultramarathon. Will taking 600 milligrams of vitamin C daily reduce these infections? Researchers randomly assigned ultramarathon runners to receive either vitamin C or a placebo. Separately, they also randomly assigned these treatments to a group of nonrunners the same age as the runners. All subjects were watched for 14 days after the big race to see if infections developed.[21]

(a) What is the name for this experimental design?

(b) Use a diagram to outline the design.

(c) The report of the study said:

> *Sixty-eight percent of the runners in the placebo group reported the development of symptoms of upper respiratory tract infection after the race; this was significantly more than that reported by the vitamin C–supplemented group (33%).*

Explain to someone who knows no statistics why "significantly more" means there is good reason to think that vitamin C works.

3.35 Table B is a table of random digits. Which of the following statements are true of a table of random digits, and which are false? Explain your answers.

(a) There are exactly four 0s in each row of 40 digits.

(b) Each pair of digits has chance 1/100 of being 00.

(c) The digits 0000 can never appear as a group, because this pattern is not random.

3.3   Sampling Design

A political scientist wants to know what percent of college-age adults consider themselves conservatives. An automaker hires a market research firm to learn what percent of adults aged 18 to 35 recall seeing television advertisements for a new sport utility vehicle. Government economists inquire about average household income. In all these cases, we want to gather information about a large group of individuals. We will not, as in an experiment, impose a treatment in order to observe the response. Time, cost, and inconvenience forbid contacting every individual. In such cases, we gather information about only part of the group—a *sample*—in order to draw conclusions about the whole. **sample survey** **Sample surveys** are an important kind of observational study.

> ## POPULATION AND SAMPLE
>
> The entire group of individuals that we want information about is called the **population.**
>
> A **sample** is a part of the population that we actually examine in order to gather information.

Notice that "population" is defined in terms of our desire for knowledge. If we wish to draw conclusions about all U.S. college students, that group is our population even if only local students are available for questioning. The sample is the part from which we draw conclusions about the whole. The **de-** **sample design** **sign** of a sample survey refers to the method used to choose the sample from the population. Poor sample designs can produce misleading conclusions, as the following examples illustrate.

EXAMPLE 3.16   A mill produces large coils of thin steel for use in manufacturing home appliances. The quality engineer wants to submit a sample of 5-centimeter squares to detailed laboratory examination. She asks a technician to cut a sample of 10 such squares. Wanting to provide "good" pieces of steel, the technician carefully avoids the visible defects in the coil material when cutting the sample. The laboratory results are wonderful but the customers complain about the material they are receiving.

EXAMPLE 3.17   The American Family Association (AFA) is a conservative group that claims to stand for "traditional family values." It regularly posts online poll questions on its Web site—just click on a response to take part. Because the respondents are people who visit this site, the poll results always support AFA's positions. Well, almost always. In 2004, AFA's online poll asked about the heated issue of allowing same-sex marriage. Soon, email lists and social-network sites favored mostly by young liberals pointed to the AFA poll. Almost 850,000 people responded, and 60% of them favored legalization of same-sex marriage. AFA claimed that homosexual rights groups had skewed its poll.

Online polls are now everywhere—some sites will even provide help in conducting your own online poll. As the AFA poll illustrates, you can't trust the results. People who take the trouble to respond to an open invitation are not representative of the entire adult population. That's true of regular visitors to AFA's site, of the activists who made a special effort to vote in the marriage poll, and of the people who bother to respond to write-in, call-in, or online polls in general.

In both Examples 3.16 and 3.17, the sample was selected in a manner that guaranteed that it would not be representative of the entire population. These sampling schemes display *bias*, or systematic error, in favoring some parts of the population over others. Online polls use *voluntary response samples*, a particularly common form of biased sample.

> ## VOLUNTARY RESPONSE SAMPLE
>
> A **voluntary response sample** consists of people who choose themselves by responding to a general appeal. Voluntary response samples are biased because people with strong opinions, especially negative opinions, are most likely to respond.

The remedy for bias in choosing a sample is to allow impersonal chance to do the choosing, so that there is neither favoritism by the sampler (as in Example 3.16) nor voluntary response (as in Example 3.17). Random selection of a sample eliminates bias by giving all individuals an equal chance to be chosen, just as randomization eliminates bias in assigning experimental subjects.

Simple random samples

The simplest sampling design amounts to placing names in a hat (the population) and drawing out a handful (the sample). This is *simple random sampling*.

> ## SIMPLE RANDOM SAMPLE
>
> A **simple random sample (SRS)** of size n consists of n individuals from the population chosen in such a way that every set of n individuals has an equal chance to be the sample actually selected.

Each treatment group in a completely randomized experimental design is an SRS drawn from the available experimental units. We select an SRS by labeling all the individuals in the population and using software or a table of random digits to select a sample of the desired size, just as in experimental randomization. Notice that an SRS not only gives each individual an equal chance to be chosen (thus avoiding bias in the choice) but gives every possible sample an equal chance to be chosen. There are other random sampling designs that give each individual, but not each sample, an equal chance. One such design, systematic random sampling, is described in Exercise 3.47.

EXAMPLE 3.18   A campus newspaper plans a major article on spring break destinations. The authors intend to call a few randomly chosen resorts at each destination to ask about their attitudes toward groups of students as guests. Here are the resorts listed in one city. The first step is to label the members of this population as shown.

01	Aloha Kai	08	Captiva	15	Palm Tree	22	Sea Shell
02	Anchor Down	09	Casa del Mar	16	Radisson	23	Silver Beach
03	Banana Bay	10	Coconuts	17	Ramada	24	Sunset Beach
04	Banyan Tree	11	Diplomat	18	Sandpiper	25	Tradewinds
05	Beach Castle	12	Holiday Inn	19	Sea Castle	26	Tropical Breeze
06	Best Western	13	Lime Tree	20	Sea Club	27	Tropical Shores
07	Cabana	14	Outrigger	21	Sea Grape	28	Veranda

Now enter Table B, and read two-digit groups until you have chosen three resorts. If you enter at line 185, the authors will call Banana Bay (03), Palm Tree (15), and Cabana (07). Most statistical software will select an SRS for you, eliminating the need for Table B. The *Simple Random Sample* applet on the text CD and Web site is a convenient way to automate this task.

Stratified samples

The general framework for designs that use chance to choose a sample is a *probability sample*.

> ### PROBABILITY SAMPLE
>
> A **probability sample** is a sample chosen by chance. We must know what samples are possible and what chance, or probability, each possible sample has.

Some probability sampling designs (such as an SRS) give each member of the population an *equal* chance to be selected. This may not be true in more elaborate sampling designs. In every case, however, the use of chance to select the sample is the essential principle of statistical sampling.

Designs for sampling from large populations spread out over a wide area are usually more complex than an SRS. For example, it is common to sample important groups within the population separately, then combine these samples. This is the idea of a *stratified sample*.

> ### STRATIFIED RANDOM SAMPLE
>
> To select a **stratified random sample,** first divide the population into groups of similar individuals, called **strata.** Then choose a separate SRS in each stratum and combine these SRSs to form the full sample.

Choose the strata based on facts known before the sample is taken. For example, a population of election districts might be divided into urban, suburban, and rural strata. A stratified design can produce more exact information than an SRS of the same size by taking advantage of the fact that individuals in the same stratum are similar to one another. Think of the extreme case in which all individuals in each stratum are identical: just one individual from each stratum is then enough to completely describe the population. Strata for sampling are similar to blocks in experiments. We have two names because the idea of grouping similar units before randomizing arose separately in sampling and in experiments.

EXAMPLE 3.19 A dentist is suspected of defrauding insurance companies by describing some dental procedures incorrectly on claim forms and overcharging for them. An investigation begins by examining a sample of his bills for the past three years. Because there are five suspicious procedures, the investigators take a stratified sample. That is, they randomly select bills for each of the five procedures separately.

Multistage samples

Another common means of restricting random selection is to choose the sample in stages. This is common practice for national samples of households or people. For example, data on employment and unemployment are gathered by the government's Current Population Survey, which conducts interviews in about 60,000 households each month. The cost of sending interviewers to the widely scattered households in an SRS would be too high. Moreover, the government wants data broken down by states and large cities. The Current Population Survey therefore uses a **multistage sampling design.** The final *multistage sample* sample consists of clusters of nearby households that an interviewer can easily visit. Most opinion polls and other national samples are also multistage, though interviewing in most national samples today is done by telephone rather than in person, eliminating the economic need for clustering. The Current Population Survey sampling design is roughly as follows:[22]

Stage 1. Divide the United States into 2007 geographical areas called Primary Sampling Units, or PSUs. PSUs do not cross state lines. Select a sample of 754 PSUs. This sample includes the 428 PSUs with the largest population and a stratified sample of 326 of the others.

Stage 2. Divide each PSU selected into smaller areas called "blocks." Stratify the blocks using ethnic and other information and take a stratified sample of the blocks in each PSU.

Stage 3. Sort the housing units in each block into clusters of four nearby units. Interview the households in a probability sample of these clusters.

Analysis of data from sampling designs more complex than an SRS takes us beyond basic statistics. But the SRS is the building block of more elaborate designs, and analysis of other designs differs more in complexity of detail than in fundamental concepts.

Cautions about sample surveys

Random selection eliminates bias in the choice of a sample from a list of the population. Sample surveys of large human populations, however, require much more than a good sampling design.[23] To begin, we need an accurate and complete list of the population. Because such a list is rarely available, most samples suffer from some degree of *undercoverage*. A sample survey of households, for example, will miss not only homeless people but prison inmates and students in dormitories. An opinion poll conducted by telephone will miss the 6% of American households without residential phones. The results of national sample surveys therefore have some bias if the people not covered—who most often are poor people—differ from the rest of the population.

A more serious source of bias in most sample surveys is *nonresponse*, which occurs when a selected individual cannot be contacted or refuses to cooperate. Nonresponse to sample surveys often reaches 50% or more, even with careful planning and several callbacks. Because nonresponse is higher in urban areas, most sample surveys substitute other people in the same area to avoid favoring rural areas in the final sample. If the people contacted differ from those who are rarely at home or who refuse to answer questions, some bias remains.

UNDERCOVERAGE AND NONRESPONSE

Undercoverage occurs when some groups in the population are left out of the process of choosing the sample.

Nonresponse occurs when an individual chosen for the sample can't be contacted or does not cooperate.

EXAMPLE 3.20   How bad is nonresponse? The Current Population Survey (CPS) has the lowest nonresponse rate of any poll we know: only about 4% of the households in the CPS sample refuse to take part and another 3% or 4% can't be contacted. People are more likely to respond to a government survey such as the CPS, and the CPS contacts its sample in person before doing later interviews by phone.

The General Social Survey (Figure 3.7) is the nation's most important social science research survey. The GSS also contacts its sample in person, and it is run by a university. Despite these advantages, its most recent survey had a 30% rate of nonresponse.

What about polls done by the media and by market research and opinion polling firms? We don't know their rates of nonresponse, because they won't say. That itself is a bad sign. The Pew Research Center imitated a careful telephone survey and published the results: out of 2879 households called, 1658 were never at home, refused, or would not finish the interview. That's a nonresponse rate of 58%.[24]

Most sample surveys, and almost all opinion polls, are now carried out by telephone. This and other details of the interview method can affect the results.

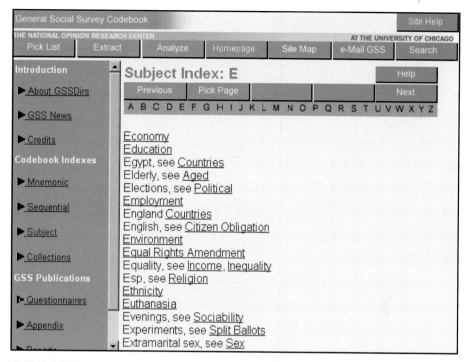

FIGURE 3.7 Part of the subject index for the General Social Survey (GSS). The GSS has assessed attitudes on a wide variety of topics since 1972. Its continuity over time makes the GSS a valuable source for studies of changing attitudes.

EXAMPLE 3.21   A Pew Research Center poll has asked about belief in God for many years. In response to the statement "I never doubt the existence of God," subjects are asked to choose from the responses

completely agree   mostly agree   mostly disagree   completely disagree

In 1990, subjects were interviewed in person and were handed a card with the four responses on it. In 1991, the poll switched to telephone interviews. In 1990, 60% said "completely agree," in line with earlier years. In 1991, 71% completely agreed. The increase is probably explained by the effect of hearing "completely agree" read first by the interviewer.[25]

response bias   The behavior of the respondent or of the interviewer can cause **response bias** in sample results. Respondents may lie, especially if asked about illegal or unpopular behavior. The race or sex of the interviewer can influence responses to questions about race relations or attitudes toward feminism. Answers to questions that ask respondents to recall past events are often inaccurate because of faulty memory. For example, many people "telescope" events in the past, bringing them forward in memory to more recent time periods. "Have you visited a dentist in the last 6 months?" will often elicit a "Yes" from someone who last visited a dentist 8 months ago.[26]

EXAMPLE 3.22   "One of the most frequently observed survey measurement errors is the overreporting of voting behavior."[27] People know they should vote, so those who didn't vote tend to save face by saying that they did. Here are the data from a typical sample of 663 people after an election:

		What they said	
		I voted	I didn't
What they did	Voted	358	13
	Didn't vote	120	172

You can see that 478 people (72%) said that they voted, but only 371 people (56%) actually did vote.

---

wording of questions

The **wording of questions** is the most important influence on the answers given to a sample survey. Confusing or leading questions can introduce strong bias, and even minor changes in wording can change a survey's outcome. Here are some examples.

EXAMPLE 3.23   In response to the question "Are you heterosexual, homosexual, or bisexual?" in a social science research survey, one woman answered, "It's just me and my husband, so bisexual." The issue is serious, even if the example seems silly: reporting about sexual behavior is difficult because people understand and misunderstand sexual terms in many ways.

How do Americans feel about government help for the poor? Only 13% think we are spending too much on "assistance to the poor," but 44% think we are spending too much on "welfare." How do the Scots feel about the movement to become independent from England? Well, 51% would vote for "independence for Scotland," but only 34% support "an independent Scotland separate from the United Kingdom." It seems that "assistance to the poor" and "independence" are nice, hopeful words. "Welfare" and "separate" are negative words.[28]

The statistical design of sample surveys is a science, but this science is only part of the art of sampling. Because of nonresponse, response bias, and the difficulty of posing clear and neutral questions, you should hesitate to fully trust reports about complicated issues based on surveys of large human populations. *Insist on knowing the exact questions asked, the rate of nonresponse, and the date and method of the survey before you trust a poll result.*

SECTION 3.3 | Summary

A sample survey selects a **sample** from the **population** of all individuals about which we desire information. We base conclusions about the population on data about the sample.

The **design** of a sample refers to the method used to select the sample from the population. **Probability sampling designs** use impersonal chance to select a sample.

The basic probability sample is a **simple random sample (SRS).** An SRS gives every possible sample of a given size the same chance to be chosen.

Choose an SRS by labeling the members of the population and using a **table of random digits** to select the sample. Software can automate this process.

To choose a **stratified random sample,** divide the population into **strata,** groups of individuals that are similar in some way that is important to the response. Then choose a separate SRS from each stratum and combine them to form the full sample.

Multistage samples select successively smaller groups within the population in stages, resulting in a sample consisting of clusters of individuals. Each stage may employ an SRS, a stratified sample, or another type of sample.

Failure to use probability sampling often results in **bias,** or systematic errors in the way the sample represents the population. **Voluntary response** samples, in which the respondents choose themselves, are particularly prone to large bias.

In human populations, even probability samples can suffer from bias due to **undercoverage** or **nonresponse,** from **response bias** due to the behavior of the interviewer or the respondent, or from misleading results due to **poorly worded questions.**

SECTION 3.3 | Exercises

3.36   A committee on community relations in a college town plans to survey local businesses about the importance of students as customers. From telephone book listings, the committee chooses 150 businesses at random. Of these, 73 return the questionnaire mailed by the committee. What is the population for this sample survey? What is the sample? What is the rate (percent) of nonresponse?

3.37   A Gallup poll asked, "Do you think the U.S. should take the leading role in world affairs, take a major role but not the leading role, take a minor role, or take no role at all in world affairs?" Gallup's report said, " Results are based on telephone interviews with 1,002 national adults, aged 18 and older, conducted Feb. 9–12, 2004."[29]

(a)   What is the population for this sample survey? What was the sample size?

(b)   Gallup notes that the order of the four possible responses was rotated when the question was read over the phone. Why was this done?

3.38   For each of the following sampling situations, identify the population as exactly as possible. That is, say what kind of individuals the population consists of and say exactly which individuals fall in the population. If the information given is not complete, complete the description of the population in a reasonable way.

(a)   An opinion poll contacts 1161 adults and asks them, "Which political party do you think has better ideas for leading the country in the twenty-first century?"

(b)   A furniture maker buys hardwood in large lots. The supplier is supposed to dry the wood before shipping—wood that is not dry won't hold its size and shape. The furniture maker chooses 5 pieces of wood from each lot and tests their moisture content. If any piece exceeds 12% moisture content, the entire lot is sent back.

(c) The American Community Survey (ACS) will replace the census "long form" starting with the 2010 census. The main part of the ACS contacts 250,000 addresses by mail each month, with follow-up by phone and in person if there is no response. Each household answers questions about their housing, economic, and social status.

3.39 An opinion poll calls 1800 randomly chosen residential telephone numbers, then asks to speak with an adult member of the household. The interviewer asks, "How many movies have you watched in a movie theater in the past 12 months?"

(a) What population do you think the poll has in mind?

(b) In all, 1131 people respond. What is the rate (percent) of nonresponse?

(c) What source of response error is likely for the question asked?

3.40 You are planning a report on apartment living in a college town. You decide to select 5 apartment complexes at random for in-depth interviews with residents. Select a simple random sample of 5 of the following apartment complexes. If you use Table B, start at line 117.

Ashley Oaks	Country View	Mayfair Village
Bay Pointe	Country Villa	Nobb Hill
Beau Jardin	Crestview	Pemberly Courts
Bluffs	Del-Lynn	Peppermill
Brandon Place	Fairington	Pheasant Run
Briarwood	Fairway Knolls	Richfield
Brownstone	Fowler	Sagamore Ridge
Burberry	Franklin Park	Salem Courthouse
Cambridge	Georgetown	Village Manor
Chauncey Village	Greenacres	Waterford Court
Country Squire	Lahr House	Williamsburg

3.41 You want to ask a sample of college students the question "How much do you trust information about health that you find on the Internet—a great deal, somewhat, not much, or not at all?" You try out this and other questions on a pilot group of 10 students chosen from your class. The class members are

Anderson	Eckstein	Johnson	Puri
Arroyo	Fernandez	Kim	Richards
Batista	Fullmer	Molina	Rodriguez
Bell	Gandhi	Morgan	Samuels
Burke	Garcia	Nguyen	Shen
Calloway	Glauser	Palmiero	Velasco
Delluci	Helling	Percival	Washburn
Drasin	Husain	Prince	Zhao

Choose an SRS. If you use Table B, start at line 139.

3.42 After you have labeled the individuals in a population, the *Simple Random Sample* applet automates the task of choosing an SRS. Use the applet to choose the sample in Exercise 3.40.

3.43 There are approximately 371 active telephone area codes covering Canada, the United States, and some Caribbean areas. (More are created regularly.) You want to choose an SRS of 25 of these area codes for a study of available telephone numbers. Label the codes 001 to 371 and use the *Simple Random Sample* applet to choose your sample. (If you use Table B, start at line 149 and choose only the first 5 codes in the sample.)

3.44 The Census Bureau divides the entire country into "census tracts" that contain about 4000 people. Each tract is in turn divided into small "blocks," which in urban areas are bounded by local streets. An SRS of blocks from a census tract is often the next-to-last stage in a multistage sample. Figure 3.8 shows part of census tract 8051.12, in Cook County, Illinois, west of Chicago. The 44 blocks in this tract are divided into three "block groups." Group 1 contains 6 blocks numbered 1000 to 1005; Group 2 (outlined in Figure 3.8) contains 12 blocks numbered 2000 to 2011; Group 3 contains 26 blocks numbered 3000 to 3025. Use Table B, beginning at line 125, to choose an SRS of 5 of the 44 blocks in this census tract. Explain carefully how you labeled the blocks.

FIGURE 3.8 Census blocks in Cook County, Illinois, for Exercises 3.44 and 3.46. The outlined area is a block group.

3.45   In using Table B repeatedly to choose samples or do randomization for experiments, you should not always begin at the same place, such as line 101. Why not?

3.46   Exercise 3.44 asks you to choose an SRS of blocks from the census tract pictured in Figure 3.8. You might instead choose a stratified sample of one block from the 6 blocks in Group 1, two from the 12 blocks in Group 2, and three from the 26 blocks in Group 3. Choose such a sample, explaining carefully how you labeled blocks and used Table B.

3.47

systematic random sample

Systematic random samples are often used to choose a sample of apartments in a large building or dwelling units in a block at the last stage of a multistage sample. An example will illustrate the idea of a systematic sample. Suppose that we must choose 4 addresses out of 100. Because 100/4 = 25, we can think of the list as four lists of 25 addresses. Choose 1 of the first 25 at random, using Table B. The sample contains this address and the addresses 25, 50, and 75 places down the list from it. If 13 is chosen, for example, then the systematic random sample consists of the addresses numbered 13, 38, 63, and 88.

(a) A study of dating among college students wanted a sample of 200 of the 9000 single male students on campus. The sample consisted of every 45th name from a list of the 9000 students. Explain why the survey chooses every 45th name.

(b) Use Table B at line 145 to choose the starting point for this systematic sample.

3.48   The previous exercise introduces systematic random samples. Explain carefully why a systematic random sample *does* give every individual the same chance to be chosen but is *not* a simple random sample.

3.49   A poll of opinion in California uses random digit dialing to choose telephone numbers at random. Numbers are selected separately within each California area code. The size of the sample in each area code is proportional to the population living there.

(a) What is the name for this kind of sampling design?

(b) California area codes, in rough order from north to south, are

530	707	916	209	415	925	510	650	408	831	805	559	760
661	818	213	626	323	562	709	310	949	909	858	619	

Another California survey does not call numbers in all area codes but starts with an SRS of 10 area codes. Choose such an SRS. If you use Table B, start at line 111.

3.50   Stratified samples are widely used to study large areas of forest. Based on satellite images, a forest area in the Amazon basin is divided into 14 types. Foresters studied the four most commercially valuable types: alluvial climax forests of quality levels 1, 2, and 3, and mature secondary forest. They divided the area of each type into large parcels, chose parcels of each type at random, and counted tree species in a 20- by 25-meter rectangle randomly placed within each parcel selected. Here is some detail:

Forest type	Total parcels	Sample size
Climax 1	36	4
Climax 2	72	7
Climax 3	31	3
Secondary	42	4

Choose the stratified sample of 18 parcels. Be sure to explain how you assigned labels to parcels. If you use Table B, start at line 162.

3.51 A club has 30 student members and 10 faculty members. The students are

Abel	Fisher	Huber	Moran	Reinmann
Carson	Golomb	Jimenez	Moskowitz	Santos
Chen	Griswold	Jones	Neyman	Shaw
David	Hein	Kiefer	O'Brien	Thompson
Deming	Hernandez	Klotz	Pearl	Utts
Elashoff	Holland	Liu	Potter	Vlasic

and the faculty members are

Andrews	Fernandez	Kim	Moore	Rabinowitz
Besicovitch	Gupta	Lightman	Phillips	Yang

The club can send 4 students and 2 faculty members to a convention and decides to choose those who will go by random selection. Use Table B to choose a stratified random sample of 4 students and 2 faculty members.

3.52 At a party there are 30 students over age 21 and 20 students under age 21. You choose at random 3 of those over 21 and separately choose at random 2 of those under 21 to interview about attitudes toward alcohol. You have given every student at the party the same chance to be interviewed: what is that chance? Why is your sample not an SRS?

3.53 Accountants use stratified samples during audits to verify a company's records of such things as accounts receivable. The stratification is based on the dollar amount of the item and often includes 100% sampling of the largest items. One company reports 5000 accounts receivable. Of these, 100 are in amounts over $50,000; 500 are in amounts between $1000 and $50,000; and the remaining 4400 are in amounts under $1000. Using these groups as strata, you decide to verify all of the largest accounts and to sample 5% of the midsize accounts and 1% of the small accounts. How would you label the two strata from which you will sample? Use Table B, starting at line 115, to select the first 5 accounts from each of these strata.

3.54   A common form of nonresponse in telephone surveys is "ring-no-answer." That is, a call is made to an active number but no one answers. The Italian National Statistical Institute looked at nonresponse to a government survey of households in Italy during the periods January 1 to Easter and July 1 to August 31. All calls were made between 7 and 10 p.m., but 21.4% gave "ring-no-answer" in one period versus 41.5% "ring-no-answer" in the other period.[30] Which period do you think had the higher rate of no answers? Why? Explain why a high rate of nonresponse makes sample results less reliable.

3.55   The list of individuals from which a sample is actually selected is called the **sampling frame.** Ideally, the frame should list every individual in the population, but in practice this is often difficult. A frame that leaves out part of the population is a common source of undercoverage.

sampling frame

(a) Suppose that a sample of households in a community is selected at random from the telephone directory. What households are omitted from this frame? What types of people do you think are likely to live in these households? These people will probably be underrepresented in the sample.

(b) It is usual in telephone surveys to use random digit dialing equipment that selects the last four digits of a telephone number at random after being given the area code and the exchange (the first three digits). Which of the households that you mentioned in your answer to (a) will be included in the sampling frame by random digit dialing?

3.56   The Excite Poll can be found online at `poll.excite.com`. The question appears on the screen, and you simply click buttons to vote "Yes," "No," "Not sure," or "Don't care." On February 17, 2004, the question was "Do you think that beer advertisements are targeted towards minors?" In all, 4316 (29%) said "Yes," another 8986 (62%) said "No," and the remaining 1182 were not sure or didn't care.

(a) What is the sample size for this poll?

(b) That's a much larger sample than standard sample surveys. In spite of this, we can't trust the result to give good information about any clearly defined population. Why?

3.57   Comment on each of the following as a potential sample survey question. Is the question clear? Is it slanted toward a desired response?

(a) "Some cell phone users have developed brain cancer. Should all cell phones come with a warning label explaining the danger of using cell phones?"

(b) "Do you agree that a national system of health insurance should be favored because it would provide health insurance for everyone and would reduce administrative costs?"

(c) "In view of escalating environmental degradation and incipient resource depletion, would you favor economic incentives for recycling of resource-intensive consumer goods?"

3.58   In 2000, when the federal budget showed a large surplus, the Pew Research Center asked two questions of random samples of adults. Both questions stated that Social Security would be "fixed." Here are the uses suggested for the remaining surplus:

> *Should the money be used for a tax cut, or should it be used to fund new government programs?*

> *Should the money be used for a tax cut, or should it be spent on programs for education, the environment, health care, crime-fighting and military defense?*

One of these questions drew 60% favoring a tax cut; the other, only 22%. Which wording pulls respondents toward a tax cut? Why?

3.59   A teacher asks her class, "How many children are there in your family, including yourself?" The mean response is about 3 children. According to the 2000 census, families that have children average 1.86 children. Why is a sample like this biased toward higher outcomes?

3.60   Write your own examples of bad sample survey questions.

(a) Write a biased question designed to get one answer rather than another.

(b) Write a question that is confusing, so that it is hard to answer.

3.61   Spain's Centro de Investigaciones Sociológicos carried out a sample survey on the economic attitudes of Spaniards.[31]   Of the 2496 adults interviewed, 72% agreed that "Employees with higher performance must get higher pay." On the other hand, 71% agreed that "Everything a society produces should be distributed among its members as equally as possible and there should be no major differences." Use these conflicting results as an example in a short explanation of why opinion polls often fail to reveal public attitudes clearly.

3.4   Toward Statistical Inference

A market research firm interviews a random sample of 2500 adults. Result: 66% find shopping for clothes frustrating and time-consuming. That's the truth about the 2500 people in the sample. What is the truth about the almost 220 million American adults who make up the population? Because the sample was chosen at random, it's reasonable to think that these 2500 people represent the entire population fairly well. So the market researchers turn the *fact* that 66% of the *sample* find shopping frustrating into an *estimate* that about 66% of *all adults* feel this way. That's a basic move in statistics: use a fact about a sample to estimate the truth about the whole population. We call

statistical inference   this **statistical inference** because we infer conclusions about the wider population from data on selected individuals. To think about inference, we must keep straight whether a number describes a sample or a population. Here is the vocabulary we use.

> ## PARAMETERS AND STATISTICS
>
> A **parameter** is a number that describes the **population.** A parameter is a fixed number, but in practice we do not know its value.
>
> A **statistic** is a number that describes a **sample.** The value of a statistic is known when we have taken a sample, but it can change from sample to sample. We often use a statistic to estimate an unknown parameter.

EXAMPLE 3.24   Are attitudes toward shopping changing? Sample surveys show that fewer people enjoy shopping than in the past. A survey by the market research firm Yankelovich Clancy Shulman asked a nationwide random sample of 2500 adults if they agreed or disagreed that "I like buying new clothes, but shopping is often frustrating and time-consuming." Of the respondents, 1650, or 66%, said they agreed.[32] The proportion of the sample who agree is

$$\hat{p} = \frac{1650}{2500} = 0.66 = 66\%$$

The number $\hat{p} = 0.66$ is a *statistic.* The corresponding *parameter* is the proportion (call it p) of all adult U.S. residents who would have said "Agree" if asked the same question. We don't know the value of the parameter p, so we use the statistic $\hat{p}$ to estimate it.

Sampling variability

If Yankelovich took a second random sample of 2500 adults, the new sample would have different people in it. It is almost certain that there would not be exactly 1650 positive responses. That is, the value of the statistic $\hat{p}$ will *sampling variability* vary from sample to sample. This basic fact is called **sampling variability:** the value of a statistic varies in repeated random sampling. Could it happen that one random sample finds that 66% of adults find clothes shopping frustrating and a second random sample finds that only 42% feel this way? Random samples eliminate *bias* from the act of choosing a sample, but they can still be wrong because of the *variability* that results when we choose at random. If the variation when we take repeat samples from the same population is too great, we can't trust the results of any one sample.

We are saved by the second great advantage of random samples. The first advantage is that choosing at random eliminates favoritism. That is, random sampling attacks bias. The second advantage is that if we take lots of random samples of the same size from the same population, the variation from sample to sample will follow a predictable pattern. **All of statistical inference is based on one idea: to see how trustworthy a procedure is, ask what would happen if we repeated it many times.**

To understand why sampling variability is not fatal, we ask, "What would happen if we took many samples?" Here's how to answer that question:

- Take a large number of samples from the same population.

- Calculate the sample proportion $\hat{p}$ for each sample.

- Make a histogram of the values of $\hat{p}$.

- Examine the distribution displayed in the histogram for shape, center, and spread, as well as outliers or other deviations.

In practice it is too expensive to take many samples from a large population such as all adult U.S. residents. But we can imitate many samples by using random digits. Using random digits from a table or computer software to imitate chance behavior is called **simulation.**

simulation

EXAMPLE 3.25   We will simulate drawing simple random samples (SRSs) of size 100 from the population of all adult U.S. residents. Suppose that in fact 60% of the population find clothes shopping time-consuming and frustrating. Then the true value of the parameter we want to estimate is $p = 0.6$. (Of course, we would not sample in practice if we already knew that $p = 0.6$. We are sampling here to understand how sampling behaves.)

We can imitate the population by a table of random digits, with each entry standing for a person. Six of the ten digits (say 0 to 5) stand for people who find shopping frustrating. The remaining four digits, 6 to 9, stand for those who do not. Because all digits in a random number table are equally likely, this assignment produces a population proportion of frustrated shoppers equal to $p = 0.6$. We then imitate an SRS of 100 people from the population by taking 100 consecutive digits from Table B. The statistic $\hat{p}$ is the proportion of 0s to 5s in the sample.

Here are the first 100 entries in Table B with digits 0 to 5 highlighted:

19223	95034	05756	28713	96409	12531	42544	82853
73676	47150	99400	01927	27754	42648	82425	36290
45467	71709	77558	00095				

There are 64 digits between 0 and 5, so $\hat{p} = 64/100 = 0.64$. A second SRS based on the second 100 entries in Table B gives a different result, $\hat{p} = 0.55$. The two sample results are different, and neither is equal to the true population value $p = 0.6$. That's sampling variability.

Sampling distributions

Simulation is a powerful tool for studying chance. Now that we see how simulation works, it is faster to abandon Table B and to use a computer programmed to produce random digits.

EXAMPLE 3.26   Figure 3.9 illustrates the process of choosing many samples and finding the sample proportion $\hat{p}$ for each one. Follow the flow of the figure from the population at the left, to choosing an SRS and finding the $\hat{p}$ for this sample, to collecting together the $\hat{p}$'s from many samples. The histogram at the right of the figure shows the distribution of the values of $\hat{p}$ from 1000 separate SRSs of size 100 drawn from a population with $p = 0.6$.

Of course, Yankelovich interviewed 2500 people, not just 100. Figure 3.10 is parallel to Figure 3.9. It shows the process of choosing 1000 SRSs, each of size 2500, from

FIGURE 3.9   The results of many SRSs have a regular pattern. Here, we draw 1000 SRSs of size 100 from the same population. The population proportion is $p = 0.60$. The histogram shows the distribution of the 1000 sample proportions $\hat{p}$.

FIGURE 3.10   The distribution of sample proportions $\hat{p}$ for 1000 SRSs of size 2500 drawn from the same population as in Figure 3.9. The two histograms have the same scale. The statistic from the larger sample is less variable.

a population in which the true proportion is $p = 0.6$. The 1000 values of $\hat{p}$ from these samples form the histogram at the right of the figure. Figures 3.9 and 3.10 are drawn on the same scale. Comparing them shows what happens when we increase the size of our samples from 100 to 2500. These histograms display the *sampling distribution* of the statistic $\hat{p}$ for two sample sizes.

SAMPLING DISTRIBUTION

The **sampling distribution** of a statistic is the distribution of values taken by the statistic in all possible samples of the same size from the same population.

Strictly speaking, the sampling distribution is the ideal pattern that would emerge if we looked at all possible samples of size 100 from our population. A distribution obtained from a fixed number of trials, like the 1000 trials in Figure 3.9, is only an approximation to the sampling distribution. We will see that probability theory, the mathematics of chance behavior, can sometimes describe sampling distributions exactly. The interpretation of a sampling distribution is the same, however, whether we obtain it by simulation or by the mathematics of probability.

We can use the tools of data analysis to describe any distribution. Let's apply those tools to Figures 3.9 and 3.10.

- **Shape:** The histograms look normal. Figure 3.11 is a normal quantile plot of the values of $\hat{p}$ for our samples of size 100. It confirms that the distribution in Figure 3.9 is close to normal. The 1000 values for samples of size 2500 in Figure 3.10 are even closer to normal. The normal curves drawn through the histograms describe the overall shape quite well.

- **Center:** In both cases, the values of the sample proportion $\hat{p}$ vary from sample to sample, but the values are centered at 0.6. Recall that $p = 0.6$ is the true population parameter. Some samples have a $\hat{p}$ less than 0.6 and some greater, but there is no tendency to be always low or always high. That is, $\hat{p}$ has no **bias** as an estimator of p. This is true for both large and small samples. (Want the details? The mean of the 1000 values of $\hat{p}$ is 0.598

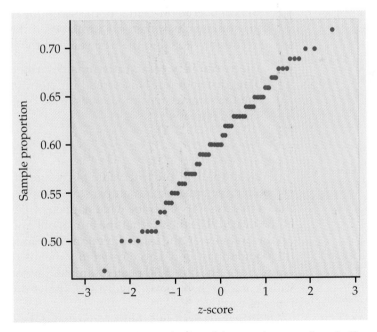

FIGURE 3.11   Normal quantile plot of the sample proportions in Figure 3.9. The distribution is close to normal except for some granularity due to the fact that sample proportions from a sample of size 100 can take only values that are multiples of 0.01. Because a plot of 1000 points is hard to read, this plot presents only every 10th value.

for samples of size 100 and 0.6002 for samples of size 2500. The median value of $\hat{p}$ is exactly 0.6 for samples of both sizes.)

• **Spread:** The values of $\hat{p}$ from samples of size 2500 are much less spread out than the values from samples of size 100. In fact, the standard deviations are 0.051 for Figure 3.9 and 0.0097, or about 0.01, for Figure 3.10.

Although these results describe just two sets of simulations, they reflect facts that are true whenever we use random sampling.

Bias and variability

Our simulations show that a sample of size 2500 will almost always give an estimate $\hat{p}$ that is close to the truth about the population. Figure 3.10 illustrates this fact for just one value of the population proportion, but it is true for any population. Samples of size 100, on the other hand, might give an estimate of 50% or 70% when the truth is 60%.

Thinking about Figures 3.9 and 3.10 helps us restate the idea of bias when we use a statistic like $\hat{p}$ to estimate a parameter like p. It also reminds us that variability matters as much as bias.

BIAS AND VARIABILITY

Bias concerns the center of the sampling distribution. A statistic used to estimate a parameter is **unbiased** if the mean of its sampling distribution is equal to the true value of the parameter being estimated.

The **variability of a statistic** is described by the spread of its sampling distribution. This spread is determined by the sampling design and the sample size n. Statistics from larger probability samples have smaller spreads.

We can think of the true value of the population parameter as the bull's-eye on a target, and of the sample statistic as an arrow fired at the bull's-eye. Bias and variability describe what happens when an archer fires many arrows at the target. *Bias* means that the aim is off, and the arrows land consistently off the bull's-eye in the same direction. The sample values do not center about the population value. Large *variability* means that repeated shots are widely scattered on the target. Repeated samples do not give similar results but differ widely among themselves. Figure 3.12 shows this target illustration of the two types of error.

Notice that small variability (repeated shots are close together) can accompany large bias (the arrows are consistently away from the bull's-eye in one direction). And small bias (the arrows center on the bull's-eye) can accompany large variability (repeated shots are widely scattered). A good sampling scheme, like a good archer, must have both small bias and small variability. Here's how we do this.

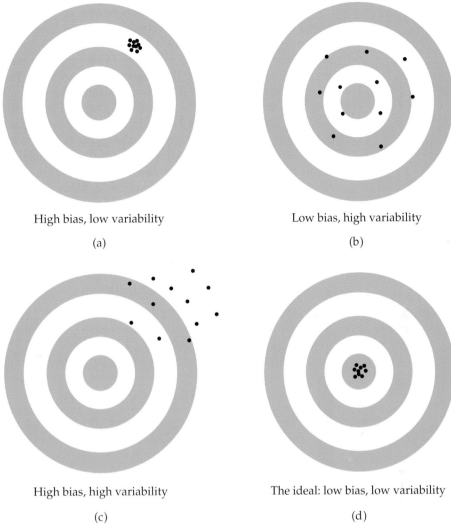

High bias, low variability

(a)

Low bias, high variability

(b)

High bias, high variability

(c)

The ideal: low bias, low variability

(d)

FIGURE 3.12   Bias and variability in shooting arrows at a target. Bias means the archer systematically misses in the same direction. Variability means that the arrows are scattered.

MANAGING BIAS AND VARIABILITY

To reduce bias, use random sampling. When we start with a list of the entire population, simple random sampling produces unbiased estimates—the values of a statistic computed from an SRS neither consistently overestimate nor consistently underestimate the value of the population parameter.

To reduce the variability of a statistic from an SRS, use a larger sample. You can make the variability as small as you want by taking a large enough sample.

In practice, Yankelovich takes only one sample. We don't know how close to the truth an estimate from this one sample is because we don't know what the truth about the population is. But *large random samples almost always give an estimate that is close to the truth.* Looking at the pattern of many samples shows that we can trust the result of one sample. The Current Population Survey's sample of 60,000 households estimates the national unemployment rate very accurately. Of course, only probability samples carry this guarantee. The American Family Association's voluntary response sample (Example 3.17, page 218) is worthless even though 850,000 people responded. Using a probability sampling design and taking care to deal with practical difficulties reduce bias in a sample. The size of the sample then determines how close to the population truth the sample result is likely to fall. Results from a sample survey usually come with a **margin of error** that sets bounds on the size of the likely error. The margin of error directly reflects the variability of the sample statistic, so it is smaller for larger samples. We will describe the details in later chapters.

margin of error

Sampling from large populations

Yankelovich's sample of 2500 adults is only about 1 out of every 90,000 adults in the United States. Does it matter whether we sample 1-in-100 individuals in the population or 1-in-90,000?

> ### POPULATION SIZE DOESN'T MATTER
>
> The variability of a statistic from a random sample does not depend on the size of the population, as long as the population is at least 100 times larger than the sample.

Why does the size of the population have little influence on the behavior of statistics from random samples? To see why this is plausible, imagine sampling harvested corn by thrusting a scoop into a lot of corn kernels. The scoop doesn't know whether it is surrounded by a bag of corn or by an entire truckload. As long as the corn is well mixed (so that the scoop selects a random sample), the variability of the result depends only on the size of the scoop.

The fact that the variability of sample results is controlled by the size of the sample has important consequences for sampling design. An SRS of size 2500 from the 220 million adult residents of the United States gives results as precise as an SRS of size 2500 from the 665,000 adult inhabitants of San Francisco. This is good news for designers of national samples but bad news for those who want accurate information about the citizens of San Francisco. If both use an SRS, both must use the same size sample to obtain equally trustworthy results.

Why randomize?

Why randomize? The act of randomizing guarantees that the results of analyzing our data are subject to the laws of probability. The behavior of statistics is described by a sampling distribution. The form of the distribution is known, and in many cases is approximately normal. Often the center of the distribution lies at the true parameter value, so that the notion that randomization eliminates bias is made more precise. The spread of the distribution describes the variability of the statistic and can be made as small as we wish by choosing a large enough sample. In a randomized experiment, we can reduce variability by choosing larger groups of subjects for each treatment.

These facts are at the heart of formal statistical inference. Later chapters will have much to say in more technical language about sampling distributions and the way statistical conclusions are based on them. What any user of statistics must understand is that all the technical talk has its basis in a simple question: *What would happen if the sample or the experiment were repeated many times?* The reasoning applies not only to an SRS but also to the complex sampling designs actually used by opinion polls and other national sample surveys. The same conclusions hold as well for randomized experimental designs. The details vary with the design but the basic facts are true whenever randomization is used to produce data.

Remember that proper statistical design is not the only aspect of a good sample or experiment. *The sampling distribution shows only how a statistic varies due to the operation of chance in randomization. It reveals nothing about possible bias due to undercoverage or nonresponse in a sample, or to lack of realism in an experiment.* The actual error in estimating a parameter by a statistic can be much larger than the sampling distribution suggests. What is worse, there is no way to say how large the added error is. The real world is less orderly than statistics textbooks imply.

BEYOND THE BASICS | Capture-recapture sampling

Sockeye salmon return to reproduce in the river where they were hatched four years earlier. How many salmon survived natural perils and heavy fishing to make it back this year? How many mountain sheep are there in Colorado? Are migratory songbird populations in North America decreasing or holding their own? These questions concern the size of animal populations. Biologists address them with a special kind of repeated sampling, called *capture-recapture sampling.*

EXAMPLE 3.27   You are interested in the number of least flycatchers migrating along a major route in the north-central United States. You set up "mist nets" that capture the birds but do not harm them. The birds caught in the net are fitted with a small aluminum leg band and released. Last year you banded and released 200 least flycatchers. This year you repeat the process. Your net catches 120 least flycatchers, 12 of which have tags from last year's catch.

The proportion of your second sample that have bands should estimate the proportion in the entire population that are banded. So if N is the unknown number of least flycatchers, we should have approximately

proportion banded in sample = proportion banded in population

$$\frac{12}{120} = \frac{200}{N}$$

Solve for N to estimate that the total number of flycatchers migrating while your net was up this year is approximately

$$N = 200 \times \frac{120}{12} = 2000$$

The capture-recapture idea extends the use of a sample proportion to estimate a population proportion. The idea works well if both samples are SRSs from the population and the population remains unchanged between samples. In practice, complications arise because, for example, some of the birds tagged last year died before this year's migration. Variations on capture-recapture samples are widely used in wildlife studies and are now finding other applications. One way to estimate the census undercount in a district is to consider the census as "capturing and marking" the households that respond. Census workers then visit the district, take an SRS of households, and see how many of those counted by the census show up in the sample. Capture-recapture estimates the total count of households in the district. As with estimating wildlife populations, there are many practical pitfalls. Our final word is as before: the real world is less orderly than statistics textbooks imply.

SECTION 3.4 | Summary

A number that describes a population is a **parameter.** A number that can be computed from the data is a **statistic.** The purpose of sampling or experimentation is usually **inference:** use sample statistics to make statements about unknown population parameters.

A statistic from a probability sample or randomized experiment has a **sampling distribution** that describes how the statistic varies in repeated data production. The sampling distribution answers the question "What would happen if we repeated the sample or experiment many times?" Formal statistical inference is based on the sampling distributions of statistics.

A statistic as an estimator of a parameter may suffer from **bias** or from high **variability.** Bias means that the center of the sampling distribution is not equal to the true value of the parameter. The variability of the statistic is described by the spread of its sampling distribution. Variability is usually reported by giving a **margin of error** for conclusions based on sample results.

Properly chosen statistics from randomized data production designs have no bias resulting from the way the sample is selected or the way the experimental units are assigned to treatments. We can reduce the variability of the statistic by increasing the size of the sample or the size of the experimental groups.

SECTION 3.4 | Exercises

State whether each boldface number in Exercises 3.62 to 3.65 is a parameter or a statistic.

3.62   An opinion poll uses random digit dialing equipment to dial 2000 randomly chosen residential telephone numbers. Of these, **621** are unlisted numbers. This isn't surprising, because **35%** of all residential numbers are unlisted.

3.63   A study of voting chose 663 registered voters at random shortly after an election. Of these, **72%** said they had voted in the election. Election records show that only **56%** of registered voters voted in the election.

3.64   The Tennessee STAR experiment randomly assigned children to regular or small classes during their first four years of school. When these children reached high school, **40.2%** of blacks from small classes took the ACT or SAT college entrance exams. Only **31.7%** of blacks from regular classes took one of these exams.

3.65   How does caffeine affect our bodies? In a matched pairs experiment, subjects pushed a button as quickly as they could after taking a caffeine pill and again after taking a placebo pill. The mean pushes per minute were $\bar{x} = 283$ for the placebo and $\bar{x} = 311$ for caffeine.

3.66   Figure 3.13 shows histograms of four sampling distributions of statistics intended to estimate the same parameter. Label each distribution relative to the others as high or low bias and as high or low variability.

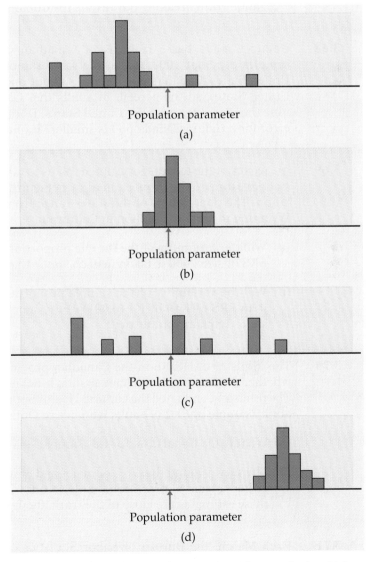

FIGURE 3.13   Which of these sampling distributions displays high or low bias and high or low variability? (For Exercise 3.66.)

3.67 A New York Times News Service article on a poll concerned with the opinions of Hispanics includes this paragraph:[33]

> *The poll was conducted by telephone from July 13 to 27, with 3,092 adults nationwide, 1,074 of whom described themselves as Hispanic. It has a margin of sampling error of plus or minus three percentage points for the entire poll and plus or minus four percentage points for Hispanics. Sample sizes for most Hispanic nationalities, like Cubans or Dominicans, were too small to break out the results separately.*

(a) Why is the "margin of sampling error" larger for Hispanics than for all 3092 respondents?

(b) Why would a very small sample size prevent a responsible news organization from breaking out results for Cubans?

3.68 Gallup Canada bases its polls of Canadian public opinion on telephone samples of about 1000 adults, the same sample size as Gallup uses in the United States. Canada's population is about one-ninth as large as that of the United States, so the percent of adults that Gallup interviews in Canada is nine times as large as in the United States. Does this mean that the margin of error for a Gallup Canada poll is smaller? Explain your answer.

3.69 An agency of the federal government plans to take an SRS of residents in each state to estimate the proportion of owners of real estate in each state's population. The populations of the states range from less than 500,000 people in Wyoming to about 35 million in California.

(a) Will the variability of the sample proportion vary from state to state if an SRS of size 2000 is taken in each state? Explain your answer.

(b) Will the variability of the sample proportion change from state to state if an SRS of 1/10 of 1% (0.001) of the state's population is taken in each state? Explain your answer.

3.70 The Ministry of Health in the Canadian province of Ontario wants to know whether the national health care system is achieving its goals in the province. The ministry conducted the Ontario Health Survey, which interviewed a probability sample of 61,239 adults who live in Ontario.[34]

(a) What is the population for this sample survey? What is the sample?

(b) The survey found that 76% of males and 86% of females in the sample had visited a general practitioner at least once in the past year. Do you think these estimates are close to the truth about the entire population? Why?

3.71 Each March, the Bureau of Labor Statistics carries out an Annual Demographic Supplement to its monthly Current Population Survey. The WORKERS data set contains the incomes of all 71,076 people in the March 2002

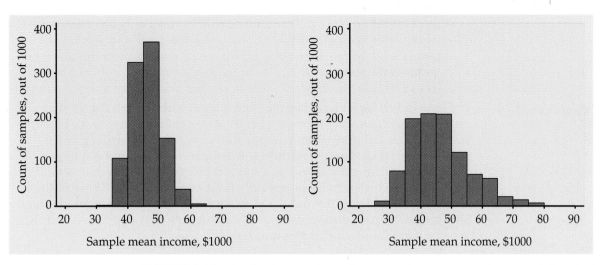

FIGURE 3.14   The distribution of sample means $\bar{x}$ for 1000 SRSs of size 100 (*left*) and of size 25 (*right*) from the same population, for Exercise 3.71.

sample who were between 25 and 64 years of age and had been employed during all of 2001. We will treat these 71,076 people as a population.

(a) If you have suitable software, choose and save an SRS of size 25 from the 71,076 incomes in the WORKERS file. What is the mean income $\bar{x}$ for your sample?

(b) We chose 1000 SRSs of size 25 and another 1000 SRSs of size 100. We calculated the mean income $\bar{x}$ for each of these samples. Figure 3.14 shows histograms of the two sets of 1000 sample means, using the same classes and drawn to the same scale for easy comparison. These represent the two sampling distributions. About what is the range (from smallest to largest) of the sample means for samples of size 25? For samples of size 100? What general fact about sampling distributions does this comparison illustrate?

(c) Based on the sample means for samples of size 100, about what is the value of the mean income for this entire population?

The remaining exercises demonstrate the idea of a sampling distribution. Sampling distributions are the basis for statistical inference. We strongly recommend doing some of these exercises.

3.72   Coin tossing can illustrate the idea of a sampling distribution. The population is all outcomes (heads or tails) we would get if we tossed a coin forever. The parameter p is the proportion of heads in this population. We suspect that p is close to 0.5. That is, we think the coin will show about one-half heads in the long run. The sample is the outcomes of 20 tosses, and the statistic $\hat{p}$ is the proportion of heads in these 20 tosses (count of heads divided by 20).

(a) Toss a coin 20 times and record the value of $\hat{p}$.

(b) Repeat this sampling process 10 times. Make a stemplot of the 10 values of $\hat{p}$. You are constructing the sampling distribution of $\hat{p}$. Is the center of this distribution close to 0.5? (Ten repetitions give only a crude approximation to the sampling distribution. If possible, pool your work with that of other students to obtain several hundred repetitions and make a histogram of the values of $\hat{p}$.)

3.73 The *Probability* applet simulates tossing a coin, with the advantage that you can choose the true long-term proportion, or probability, of a head. Example 3.25 discusses sampling from a population in which proportion $p = 0.6$ (the parameter) find shopping frustrating. Tossing a coin with probability $p = 0.6$ of a head simulates this situation: each head is a person who finds shopping frustrating, and each tail is a person who does not. Set the "Probability of heads" in the applet to 0.6 and the number of tosses to 25. This simulates an SRS of size 25 from this population. By alternating between "Toss" and "Reset" you can take many samples quickly.

(a) Take 50 samples, recording the number of heads in each sample. Make a histogram of the 50 sample proportions (count of heads divided by 25). You are constructing the sampling distribution of this statistic.

(b) Another population contains only 20% who approve of legal gambling. Take 50 samples of size 25 from this population, record the number in each sample who approve, and make a histogram of the 50 sample proportions. How do the centers of your two histograms reflect the differing truths about the two populations?

3.74 Statistical software can speed simulations. We are interested in the sampling distribution of the proportion $\hat{p}$ of people who find shopping frustrating in an SRS from a population in which proportion p find shopping frustrating. Here, p is a parameter and $\hat{p}$ is a statistic used to estimate p. We will see in Chapter 5 that "binomial" is the key word to look for in the software menus. For example, in CrunchIt! go to "Simulate data" in the "Data" menu, and choose "Binomial."

(a) Set $n = 50$ and $p = 0.6$ and generate 100 binomial observations. These are the counts for 100 SRSs of size 50 when 60% of the population finds shopping frustrating. Save these counts and divide them by 50 to get values of $\hat{p}$ from 100 SRSs. Make a stemplot of the 100 values of $\hat{p}$.

(b) Repeat this process with $p = 0.3$, representing a population in which only 30% of people find shopping frustrating. Compare your two stemplots. How does changing the parameter p affect the center and spread of the sampling distribution?

(c) Now generate 100 binomial observations with $n = 200$ and $p = 0.6$. This simulates 100 SRSs, each of size 200. Obtain the 100 sample proportions $\hat{p}$ and make a stemplot. Compare this with your stemplot from (a). How does changing the sample size n affect the center and spread of the sampling distribution?

3.75 We can construct a sampling distribution by hand in the case of a very small sample from a very small population. The population contains 10 students. Here are their scores on an exam:

Student	0	1	2	3	4	5	6	7	8	9
Score	82	62	80	58	72	73	65	66	74	62

The parameter of interest is the mean score, which is 69.4. The sample is an SRS of $n = 4$ students drawn from this population. The students are labeled 0 to 9 so that a single random digit from Table B chooses one student for the sample.

(a) Use Table B to draw an SRS of size 4 from this population. Write the four scores in your sample and calculate the mean $\bar{x}$ of the sample scores. This statistic is an estimate of the population parameter.

(b) Repeat this process 10 times. Make a histogram of the 10 values of $\bar{x}$. You are constructing the sampling distribution of $\bar{x}$. Is the center of your histogram close to 69.4? (Ten repetitions give only a crude approximation to the sampling distribution. If possible, pool your work with that of other students—using different parts of Table B—to obtain several hundred repetitions and make a histogram of the values of $\bar{x}$. This histogram is a better approximation to the sampling distribution.)

3.76 The *Simple Random Sample* applet can illustrate the idea of a sampling distribution. Form a population labeled 1 to 100. We will choose an SRS of 10 of these numbers. That is, in this exercise, the numbers themselves are the population, not just labels for 100 individuals. The mean of the whole numbers 1 to 100 is 50.5. This is the parameter, the mean of the population.

(a) Use the applet to choose an SRS of size 10. Which 10 numbers were chosen? What is their mean? This is a statistic, the sample mean $\bar{x}$.

(b) Although the population and its mean 50.5 remain fixed, the sample mean changes as we take more samples. Take another SRS of size 10. (Use the "Reset" button to return to the original population before taking the second sample.) What are the 10 numbers in your sample? What is their mean? This is another value of $\bar{x}$.

(c) Take 8 more SRSs from this same population and record their means. You now have 10 values of the sample mean $\bar{x}$ from 10 SRSs of the same size from the same population. Make a histogram of the 10 values and mark the population mean 50.5 on the horizontal axis. Are your 10 sample values roughly centered at the population value? (If you kept going forever, your $\bar{x}$-values would form the sampling distribution of the sample mean; the population mean would indeed be the center of this distribution.)

3.77 The CSDATA data file contains the college grade point averages (GPAs) of all 224 students in a university entering class who planned to major in computer science. This is our population. Statistical software can take repeated samples to illustrate sampling variability.

(a) Using software, describe this population with a histogram and with numerical summaries. In particular, what is the mean GPA in the population? This is a parameter.

(b) Choose an SRS of 20 members from this population. Make a histogram of the GPAs in the sample and find their mean. The sample mean is a statistic. Briefly compare the distributions of GPA in the sample and in the population.

(c) Repeat the process of choosing an SRS of size 20 four more times (five in all). Record the five histograms of your sample GPAs. Does it seem reasonable to you from this small trial that an SRS will usually produce a sample that is generally representative of the population?

3.78 Continue the previous exercise, using software to illustrate the idea of a sampling distribution.

(a) Choose 20 more SRSs of size 20 in addition to the 5 you have already chosen. Don't make histograms of these latest samples—just record the mean GPA for each sample. Make a histogram of the 25 sample means. This histogram is a rough approximation to the sampling distribution of the mean.

(b) One sign of bias would be that the distribution of the sample means was systematically on one side of the true population mean. Mark the population mean GPA on your histogram of the 25 sample means. Is there a clear bias?

(c) Find the mean and standard deviation of your 25 sample means. We expect that the mean will be close to the true mean of the population. Is it? We also expect that the standard deviation of the sampling distribution will be smaller than the standard deviation of the population. Is it?

CHAPTER 3 | Exercises

3.79 The WORKERS data set contains information about 14,959 people aged 25 to 64 whose highest level of education is a bachelor's degree.

(a) In order to select an SRS of these people, how would you assign labels?

(b) Use Table B at line 185 to choose the first three members of the SRS.

3.80 Will cash bonuses speed the return to work of unemployed people? The Illinois Department of Employment Security designed an experiment to find out. The subjects were 10,065 people aged 20 to 54 who were filing claims for unemployment insurance. Some were offered $500 if they found a job within 11 weeks and held it for at least 4 months. Others could tell potential employers that the state would pay the employer $500 for hiring them. A control group got neither kind of bonus.[35]

(a) Suggest a few response variables of interest to the state and outline the design of the experiment.

(b) How will you label the subjects for random assignment? Use Table B at line 167 to choose the first 3 subjects for the first treatment.

3.81   What is the name for each of these study designs?

(a) A study to compare two methods of preserving wood started with boards of southern white pine. Each board was ripped from end to end to form two edge-matched specimens. One was assigned to Method A, the other to Method B.

(b) A survey on youth and smoking contacted by telephone 300 smokers and 300 nonsmokers, all 14 to 22 years of age.

(c) Does air pollution induce DNA mutations in mice? Starting with 40 male and 40 female mice, 20 of each sex were housed in a polluted industrial area downwind from a steel mill. The other 20 of each sex were housed at an unpolluted rural location 30 kilometers away.

3.82   A large observational study used records from Canada's national health care system to compare the effectiveness of two ways to treat prostate disease. The two treatments are traditional surgery and a new method that does not require surgery. The records described many patients whose doctors had chosen one or the other method. The study found that patients treated by the new method were significantly more likely to die within 8 years.[36]

(a) Further study of the data showed that this conclusion was wrong. The extra deaths among patients who received the new treatment could be explained by lurking variables. What lurking variables might be confounded with a doctor's choice of surgical or nonsurgical treatment?

(b) You have 300 prostate patients who are willing to serve as subjects in an experiment to compare the two methods. Use a diagram to outline the design of a randomized comparative experiment.

3.83   A researcher studying the effect of price promotions on consumers' expectations makes up two different histories of the store price of a hypothetical brand of laundry detergent for the past year. Students in a marketing course view one or the other price history on a computer. Some students see a steady price, while others see regular promotions that temporarily cut the price. Then the students are asked what price they would expect to pay for the detergent. Is this study an experiment? Why? What are the explanatory and response variables?

3.84   What is the best way to answer each of the questions below: an experiment, a sample survey, or an observational study that is not a sample survey? Explain your choices.

(a) Are people generally satisfied with how things are going in the country right now?

(b) Do college students learn basic accounting better in a classroom or using an online course?

(c) How long do your teachers wait on the average after they ask their class a question?

3.85   Give an example of a question about college students, their behavior, or their opinions that would best be answered by

(a) a sample survey.

(b) an observational study that is not a sample survey.

(c) an experiment.

3.86   Do consumers prefer the taste of a cheeseburger from McDonald's or from Wendy's in a blind test in which neither burger is identified? Describe briefly the design of a matched pairs experiment to investigate this question. How will you use randomization?

3.87   How does the time it takes a bicycle rider to travel 100 meters depend on which gear is used and how steep the course is? It may be, for example, that higher gears are faster on the level but lower gears are faster on steep inclines. Discuss the design of a two-factor experiment to investigate this issue, using one bicycle with three gears and one rider. How will you use randomization?

3.88   The previous two exercises illustrate the use of statistically designed experiments to answer questions that arise in everyday life. Select a question of interest to you that an experiment might answer and carefully discuss the design of an appropriate experiment.

3.89   You want to investigate the attitudes of students at your school about the faculty's commitment to teaching. The student government will pay the costs of contacting about 500 students.

(a) Specify the exact population for your study; for example, will you include part-time students?

(b) Describe your sample design. Will you use a stratified sample?

(c) Briefly discuss the practical difficulties that you anticipate; for example, how will you contact the students in your sample?

3.90   A drug manufacturer is studying how a new drug behaves in patients. Investigators compare two doses: 5 milligrams (mg) and 10 mg. The drug can be administered by injection, by a skin patch, or by intravenous drip. Concentration in the blood after 30 minutes (the response variable) may depend both on the dose and on the method of administration.

(a) Make a sketch that describes the treatments formed by combining dosage and method. Then use a diagram to outline a completely randomized design for this two-factor experiment.

(b) "How many subjects?" is a tough issue. We will explain the basic ideas in Chapter 6. What can you say now about the advantage of using larger groups of subjects?

3.91   Few people want to eat discolored french fries. Potatoes are kept refrigerated before being cut for french fries to prevent spoiling and preserve flavor. But immediate processing of cold potatoes causes discoloring due to complex chemical reactions. The potatoes must therefore be brought to room temperature before processing. Design an experiment in which tasters will rate the color and flavor of french fries prepared from several groups of potatoes. The potatoes will be fresh picked or stored for a month at room temperature or stored for a month refrigerated. They will then be sliced and cooked either immediately or after an hour at room temperature.

(a) What are the factors and their levels, the treatments, and the response variables?

(b) Describe and outline the design of this experiment.

(c) It is efficient to have each taster rate fries from all treatments. How will you use randomization in presenting fries to the tasters?

3.92   The drug that is the subject of the experiment in Exercise 3.90 may behave differently in men and women. How would you modify your experimental design to take this into account?

3.93   As the millennium approached, *Time* magazine launched a Person of the Century Poll on the Internet. Web users could choose the person "who most influenced the course of history" during the 20th century. The top choice was Elvis Presley, with 625,045 votes. Further down the list, Linus Torvalds (originator of the Linux operating system for PCs) beat out Nelson Mandela and Princess Diana. *Time* ignored an online campaign to get out the vote for Jesus, leaving him off the list. Use this example in a brief discussion of the weaknesses of voluntary response samples.

3.94   The requirement that human subjects give their informed consent to participate in an experiment can greatly reduce the number of available subjects. For example, a study of new teaching methods asks the consent of parents for their children to be taught by either a new method or the standard method. Many parents do not return the forms, so their children must continue to follow the standard curriculum. Why is it not correct to consider these children as part of the control group along with children who are randomly assigned to the standard method?

3.95   Sample survey questions are usually read from a computer screen. In a Computer Aided Personal Interview (CAPI), the interviewer reads the questions and enters the responses. In a Computer Aided Self Interview (CASI), the interviewer stands aside and the respondent reads the questions and enters responses. One method almost always shows a higher percent of subjects admitting use of illegal drugs. Which method? Explain why.

EESEE CASE STUDIES

The Electronic Encyclopedia of Statistical Examples and Exercises (EESEE) is available on the text CD and Web site. These more elaborate stories, with data, provide settings for longer case studies.

3.96   Read the EESEE story "Checkmating and Reading Skills." Respond to Question 2 for this story.

3.97   Read the EESEE story "Anecdotes of Bias" and answer all of the questions posed about these incidents.

3.98   Read the EESEE story "Is Caffeine Dependence Real?" Write a report based on Questions 1, 2, 3, and 4.

3.99   Read the EESEE story "Does Use of an Incentive Increase Response Rates for Mailed Surveys?" Answer all of the questions for this case study.

Probability and Inference

Research in Probability Theory

JAIME SAN MARTÍN *is Professor of Mathematical Engineering, Universidad de Chile, Santiago, Chile.*

I was born in Santiago, Chile, and studied at Universidad de Chile, my country's most important university. My subject was mathematical engineering, which is mathematics with an orientation toward the mathematical aspects of engineering problems. After earning a doctorate in the United States, I returned to an academic career at Universidad de Chile.

My main work, in addition to teaching, is research in probability theory. That is, I work to discover new mathematical facts about systems governed by the laws of chance. One of my interests is Brownian motion. In physics, Brownian motion is the rapid jittery movement of small particles, such as pollen grains suspended in water—that's the setting in which Robert Brown first observed it in 1828. Einstein showed that the movement results from constant collisions with randomly moving water molecules. In probability theory, Brownian motion loses its physical meaning and becomes a very important class of mathematical models for random behavior. These models can describe phenomena far from physics, such as the movement of prices on a stock exchange. The random paths traced out by Brownian motion have unusual and subtle properties. I use advanced mathematics to learn more about these properties.

The mathematics of probability is a powerful tool for describing real-world phenomena. I have complemented my theoretical research with applied work in the area of forestry. In 1986, the pine shoot moth entered Chile. This moth attacks our pine plantations, one of the most important sectors of our economy. Researchers at Universidad de Chile work with one of Chile's largest forestry companies to model different aspects of this plague. At first, we were interested in modeling the way the moth infestation evolves in time and space. Now we are concentrating on giving a mathematical description of how the deformations caused by the moth in the trunks of pine trees evolve, in order to decide whether a tree will recover from the attack or should be harvested immediately. The mathematics of probability theory is the basis for our models, but we also use statistics to design experiments and to fit our models to data.

Probability: The Study of Randomness

Introduction

The reasoning of statistical inference rests on asking, "How often would this method give a correct answer if I used it very many times?" When we produce data by random sampling or randomized comparative experiments, the laws of probability answer the question "What would happen if we did this many times?"

4.1   Randomness

Toss a coin, or choose an SRS. The result can't be predicted in advance, because the result will vary when you toss the coin or choose the sample repeatedly. But there is nonetheless a regular pattern in the results, a pattern that emerges clearly only after many repetitions. This remarkable fact is the basis for the idea of probability.

EXAMPLE 4.1   When you toss a coin, there are only two possible outcomes, heads or tails. Figure 4.1 shows the results of tossing a coin 5000 times twice. For each number of tosses from 1 to 5000, we have plotted the proportion of those tosses that gave a head. Trial A (solid line) begins tail, head, tail, tail. You can see that the proportion of heads for Trial A starts at 0 on the first toss, rises to 0.5 when the second toss gives a head, then falls to 0.33 and 0.25 as we get two more tails. Trial B, on the other hand, starts with five straight heads, so the proportion of heads is 1 until the sixth toss.

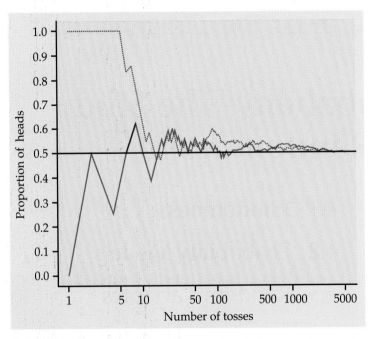

FIGURE 4.1   The proportion of tosses of a coin that give a head varies as we make more tosses. Eventually, however, the proportion approaches 0.5, the probability of a head. This figure shows the results of two trials of 5000 tosses each.

The proportion of tosses that produce heads is quite variable at first. Trial A starts low and Trial B starts high. As we make more and more tosses, however, the proportions of heads for both trials get close to 0.5 and stay there. If we made yet a third trial at tossing the coin a great many times, the proportion of heads would again settle down to 0.5 in the long run. We say that 0.5 is the *probability* of a head. The probability 0.5 appears as a horizontal line on the graph.

The *Probability* applet on the text CD and Web site animates Figure 4.1. It allows you to choose the probability of a head and simulate any number of tosses of a coin with that probability. Try it. You will see that the proportion of heads gradually settles down close to the probability. Equally important, you will also see that the proportion in a small or moderate number of tosses can be far from the probability. *Probability describes only what happens in the long run. Most people expect chance outcomes to show more short-term regularity than is actually true.*

The language of probability

"Random" in statistics is not a synonym for "haphazard" but a description of a kind of order that emerges in the long run. We often encounter the unpredictable side of randomness in our everyday experience, but we rarely see enough repetitions of the same random phenomenon to observe the long-term regularity that probability describes. You can see that regularity emerging in Figure 4.1. In the very long run, the proportion of tosses that give a head is 0.5. This is the intuitive idea of probability. Probability 0.5 means "occurs half the time in a very large number of trials."

RANDOMNESS AND PROBABILITY

We call a phenomenon **random** if individual outcomes are uncertain but there is nonetheless a regular distribution of outcomes in a large number of repetitions.

The **probability** of any outcome of a random phenomenon is the proportion of times the outcome would occur in a very long series of repetitions.

We might suspect that a coin has probability 0.5 of coming up heads just because the coin has two sides. As Exercises 4.1 and 4.2 illustrate, such suspicions are not always correct. The idea of probability is empirical. That is, it is based on observation rather than theorizing. Probability describes what happens in very many trials, and we must actually observe many trials to pin down a probability. In the case of tossing a coin, some diligent people have in fact made thousands of tosses.

EXAMPLE 4.2   The French naturalist Count Buffon (1707–1788) tossed a coin 4040 times. Result: 2048 heads, or proportion 2048/4040 = 0.5069 for heads.
Around 1900, the English statistician Karl Pearson heroically tossed a coin 24,000 times. Result: 12,012 heads, a proportion of 0.5005.

While imprisoned by the Germans during World War II, the South African statistician John Kerrich tossed a coin 10,000 times. Result: 5067 heads, proportion of heads 0.5067.

Thinking about randomness

That some things are random is an observed fact about the world. The outcome of a coin toss, the time between emissions of particles by a radioactive source, and the sexes of the next litter of lab rats are all random. So is the outcome of a random sample or a randomized experiment. Probability theory is the branch of mathematics that describes random behavior. Of course, we can never observe a probability exactly. We could always continue tossing the coin, for example. Mathematical probability is an idealization based on imagining what would happen in an indefinitely long series of trials.

The best way to understand randomness is to observe random behavior—not only the long-run regularity but the unpredictable results of short runs. You can do this with physical devices, as in Exercises 4.1 to 4.3, but software simulations (imitations) of random behavior allow faster exploration. Exercises 4.5 to 4.9 suggest some simulations of random behavior, mostly using the applets on the text CD and Web site. As you explore randomness, remember:

independence   • You must have a long series of **independent** trials. That is, the outcome of one trial must not influence the outcome of any other. Imagine a crooked gambling house where the operator of a roulette wheel can stop it where she chooses—she can prevent the proportion of "red" from settling down to a fixed number. These trials are not independent.

• The idea of probability is empirical. Simulations start with given probabilities and imitate random behavior, but we can estimate a real-world probability only by actually observing many trials.

• Nonetheless, simulations are very useful because we need long runs of trials. In situations such as coin tossing, the proportion of an outcome often requires several hundred trials to settle down to the probability of that outcome. The kinds of physical random devices suggested in the exercises are too slow for this. Short runs give only rough estimates of a probability.

The uses of probability

Probability theory originated in the study of games of chance. Tossing dice, dealing shuffled cards, and spinning a roulette wheel are examples of deliberate randomization. In that respect, they are similar to random sampling. Although games of chance are ancient, they were not studied by mathematicians until the sixteenth and seventeenth centuries. It is only a mild simplification to say that probability as a branch of mathematics arose when seventeenth-century French gamblers asked the mathematicians Blaise Pascal and Pierre de Fermat for help. Gambling is still with us, in casinos and state lotteries. We will make use of games of chance as simple examples that illustrate the principles of probability.

Careful measurements in astronomy and surveying led to further advances in probability in the eighteenth and nineteenth centuries because the results of repeated measurements are random and can be described by distributions much like those arising from random sampling. Similar distributions appear in data on human life span (mortality tables) and in data on lengths or weights in a population of skulls, leaves, or cockroaches.[1] Now, we employ the mathematics of probability to describe the flow of traffic through a highway system, the Internet, or a computer processor; the genetic makeup of individuals or populations; the energy states of subatomic particles; the spread of epidemics or rumors; and the rate of return on risky investments. Although we are interested in probability because of its usefulness in statistics, the mathematics of chance is important in many fields of study.

SECTION 4.1 | Summary

A **random phenomenon** has outcomes that we cannot predict but that nonetheless have a regular distribution in very many repetitions.

The **probability** of an event is the proportion of times the event occurs in many repeated trials of a random phenomenon.

SECTION 4.1 | Exercises

4.1   Hold a penny upright on its edge under your forefinger on a hard surface, then snap it with your other forefinger so that it spins for some time before falling. Based on 50 spins, what is the probability of heads?

4.2   You may feel that it is obvious that the probability of a head in tossing a coin is about 1/2 because the coin has two faces. Such opinions are not always correct. The previous exercise asked you to spin a penny rather than toss it—that changes the probability of a head. Now try another variation. Stand a nickel on edge on a hard, flat surface. Pound the surface with your hand so that the nickel falls over. What is the probability that it falls with heads upward? Make at least 50 trials to estimate the probability of a head.

4.3   It is reasonable to think that packages of M&M's Milk Chocolate Candies are filled at the factory with candies chosen at random from the very large number produced. So a package of M&M's contains a number of repetitions of a random phenomenon: choosing a candy at random and noting its color. What is the probability that an M&M's Milk Chocolate Candy is blue? To find out, buy one or more packs. How many candies did you examine? How many were blue? What is your estimate of the probability that a randomly chosen candy is blue?

4.4   Many Internet sites give the probabilities of being dealt various five-card poker hands. For example, the probability of being dealt two pairs is approximately 1/21. Explain is simple language what "probability 1/21" means. Also explain why it does *not* mean that in 21 deals you will get exactly one two-pair hand.

4.5   The basketball player Shaquille O'Neal makes about half of his free throws over an entire season. Use the *Probability* applet to simulate 100 free throws shot independently by a player who has probability 0.5 of making each shot.

(a) What percent of the 100 shots did he hit?

(b) Examine the sequence of hits and misses. How long was the longest run of shots made? Of shots missed? (Sequences of random outcomes often show runs longer than our intuition thinks likely.)

4.6   The idea of probability is that the *proportion* of heads in many tosses of a balanced coin eventually gets close to 0.5. But does the actual *count* of heads get close to one-half the number of tosses? Let's find out. Set the "Probability of heads" in the *Probability* applet to 0.5 and the number of tosses to 40. You can extend the number of tosses by clicking "Toss" again to get 40 more. Don't click "Reset" during this exercise.

(a) After 40 tosses, what is the proportion of heads? What is the count of heads? What is the difference between the count of heads and 20 (one-half the number of tosses)?

(b) Keep going to 120 tosses. Again record the proportion and count of heads and the difference between the count and 60 (half the number of tosses).

(c) Keep going. Stop at 240 tosses and again at 480 tosses to record the same facts. Although it may take a long time, the laws of probability say that the proportion of heads will always get close to 0.5 and also that the difference between the count of heads and half the number of tosses will always grow without limit.

4.7   Here is a question that a French gambler asked the mathematicians Fermat and Pascal at the very beginning of probability theory: what is the probability of getting at least one six in rolling four dice? The *Law of Large Numbers* applet allows you to roll several dice and watch the outcomes. (Ignore the title of the applet for now.) Because simulation—just like real random phenomena—often takes very many trials to estimate a probability accurately, let's simplify the question: is this probability clearly greater than 0.5, clearly less than 0.5, or quite close to 0.5? Use the applet to roll four dice until you can confidently answer this question. You will have to set "Rolls" to 1 so that you have time to look at the four up-faces. Keep clicking "Roll dice" to roll again and again. How many times did you roll four dice? What percent of your rolls produced at least one six?

4.8   About 30% of adult Internet users are between 18 and 29 years of age. Suppose the probability that a randomly chosen Internet user is in this age group is exactly 0.3. Use the *Probability* applet to make a study of short-term variability and long-term regularity as follows.

(a) Set the probability of heads to 0.3. Each head stands for an Internet user who is between 18 and 29 and each tail is a user who is not. Set the number of tosses to 20. Click "Toss" to make 20 tosses. What was the proportion of heads? Do this 25 times, keep a record of the 25 proportions of heads, and make a stemplot of these numbers. *Lesson:* In the short run (20 repetitions) proportions are quite variable and are often not close to the probability.

(b) With the probability of heads still set to 0.3, make 200 tosses. (Set tosses to 40 and click "Toss" five times without a reset.) What was the proportion of heads? Do this 25 times and make a stemplot of the 25 proportions of heads. *Lesson:* More repetitions make proportions less variable and generally closer to the probability. Of course, 200 repetitions is not "the long run," but you see the idea.

4.9 If you use statistical software, you can continue the exploration begun in Exercise 4.8. Software allows you to simulate many independent "Yes/No" trials more quickly if all you want to save is the count of "Yes" outcomes. The keyword "Binomial" simulates n independent Yes/No trials, each with probability p of a "Yes," and records just the count of "Yes" outcomes.

(a) Simulate 100 draws of 20 Internet users from the population in Exercise 4.8. (That is, ask the software to generate 100 binomial observations, each with $n = 20$ trials and probability $p = 0.3$ of a "Yes.") Record the count in the 18 to 29 age group on each draw. Convert the counts into percents of the 20 Internet users in each trial who are 18 to 29. Make a histogram of these 100 percents. Describe the shape, center, and spread of this distribution.

(b) Now simulate drawing 320 Internet users. (That is, set $n = 320$ and $p = 0.3$.) Do this 100 times and record the percent in the 18 to 29 age group for each of the 100 draws. Make a histogram of the percents and describe the shape, center, and spread of the distribution.

(c) In what ways are the distributions in parts (a) and (b) alike? In what ways do they differ? (Because regularity emerges in the long run, we expect the results of drawing 320 subjects to be less variable than the results of drawing 20 subjects.)

4.2 Probability Models

The idea of probability as a proportion of outcomes in very many repeated trials guides our intuition but is hard to express in mathematical form. A description of a random phenomenon in the language of mathematics is called a **probability model.** To see how to proceed, think first about a very simple random phenomenon, tossing a coin once. When we toss a coin, we cannot know the outcome in advance. What do we know? We are willing to say that the outcome will be either heads or tails. Because the coin appears to be balanced, we believe that each of these outcomes has probability 1/2. This description of coin tossing has two parts:

probability model

- A list of possible outcomes

- A probability for each outcome

This two-part description is the starting point for a probability model. We will begin by describing the outcomes of a random phenomenon and then learn how to assign probabilities to the outcomes.

Sample spaces

A probability model first tells us what outcomes are possible.

> **SAMPLE SPACE**
>
> The **sample space** *S* of a random phenomenon is the set of all possible outcomes.

The name "sample space" is natural in random sampling, where each possible outcome is a sample and the sample space contains all possible samples. To specify *S*, we must state what constitutes an individual outcome and then state which outcomes can occur. We often have some freedom in defining the sample space, so the choice of *S* is a matter of convenience as well as correctness. The idea of a sample space, and the freedom we may have in specifying it, are best illustrated by examples.

EXAMPLE 4.3　Toss a coin. There are only two possible outcomes, and the sample space is

$$S = \{\text{heads, tails}\}$$

or, more briefly, $S = \{\text{H, T}\}$.

EXAMPLE 4.4　Let your pencil point fall blindly into Table B of random digits and record the value of the digit it lands on. The possible outcomes are

$$S = \{0, 1, 2, 3, 4, 5, 6, 7, 8, 9\}$$

EXAMPLE 4.5　Toss a coin four times and record the results. That's a bit vague. To be exact, record the results of each of the four tosses in order. A typical outcome is then HTTH. Counting shows that there are 16 possible outcomes. The sample space *S* is the set of all 16 strings of four H's and T's.

Suppose that our only interest is the number of heads in four tosses. Now we can be exact in a simpler fashion. The random phenomenon is to toss a coin four times and count the number of heads. The sample space contains only five outcomes:

$$S = \{0, 1, 2, 3, 4\}$$

This example illustrates the importance of carefully specifying what constitutes an individual outcome.

Although these examples seem remote from the practice of statistics, the connection is surprisingly close. Suppose that in conducting an opinion poll you select four people at random from a large population and ask each if he or she favors reducing federal spending on low-interest student loans. The answers are "Yes" or "No." The possible outcomes—the sample space—are exactly as in Example 4.5 if we replace heads by "Yes" and tails by "No." Similarly, the possible outcomes of an SRS of 1500 people are the same in principle as the possible outcomes of tossing a coin 1500 times. One of the great advantages of mathematics is that the essential features of quite different phenomena can be described by the same probability model.

EXAMPLE 4.6 Most statistical software has a function that will generate a random number between 0 and 1. The sample space is

$$S = \{\text{all numbers between 0 and 1}\}$$

This S is a mathematical idealization. Any specific random number generator produces numbers with some limited number of decimal places so that, strictly speaking, not all numbers between 0 and 1 are possible outcomes. For example, Minitab generates random numbers like 0.736891, with six decimal places. The entire interval from 0 to 1 is easier to think about. It also has the advantage of being a suitable sample space for different software systems that produce random numbers with different numbers of digits.

Intuitive probability

A sample space S lists the possible outcomes of a random phenomenon. To complete a mathematical description of the random phenomenon, we must also give the probabilities with which these outcomes occur.

The true long-term proportion of any outcome—say, "exactly 2 heads in four tosses of a coin"—can be found only empirically, and then only approximately. How then can we describe probability mathematically? Rather than immediately attempting to give "correct" probabilities, let's confront the easier task of laying down rules that any assignment of probabilities must satisfy. We need to assign probabilities not only to single outcomes but also to sets of outcomes.

> **EVENT**
>
> An **event** is an outcome or a set of outcomes of a random phenomenon. That is, an event is a subset of the sample space.

EXAMPLE 4.7 Take the sample space S for four tosses of a coin to be the 16 possible outcomes in the form HTHH. Then "exactly 2 heads" is an event. Call this event A. The event A expressed as a set of outcomes is

$$A = \{\text{HHTT, HTHT, HTTH, THHT, THTH, TTHH}\}$$

In a probability model, events have probabilities. What properties must any assignment of probabilities to events have? Here are some basic facts about any probability model. These facts follow from the idea of probability as "the long-run proportion of repetitions on which an event occurs."

1. **Any probability is a number between 0 and 1.** Any proportion is a number between 0 and 1, so any probability is also a number between 0 and 1. An event with probability 0 never occurs, and an event with probability 1 occurs on every trial. An event with probability 0.5 occurs in half the trials in the long run.

2. **All possible outcomes together must have probability 1.** Because some outcome must occur on every trial, the sum of the probabilities for all possible outcomes must be exactly 1.

3. **If two events have no outcomes in common, the probability that one or the other occurs is the sum of their individual probabilities.** If one event occurs in 40% of all trials, a different event occurs in 25% of all trials, and the two can never occur together, then one or the other occurs on 65% of all trials because 40% + 25% = 65%.

4. **The probability that an event does not occur is 1 minus the probability that the event does occur.** If an event occurs in (say) 70% of all trials, it fails to occur in the other 30%. The probability that an event occurs and the probability that it does not occur always add to 100%, or 1.

Probability rules

Formal probability uses mathematical notation to state Facts 1 to 4 more concisely. We use capital letters near the beginning of the alphabet to denote events. If A is any event, we write its probability as $P(A)$. Here are our probability facts in formal language. As you apply these rules, remember that they are just another form of intuitively true facts about long-run proportions.

PROBABILITY RULES

Rule 1. The probability $P(A)$ of any event A satisfies $0 \le P(A) \le 1$.

Rule 2. If S is the sample space in a probability model, then $P(S) = 1$.

Rule 3. Two events A and B are **disjoint** if they have no outcomes in common and so can never occur together. If A and B are disjoint,

$$P(A \text{ or } B) = P(A) + P(B)$$

This is the **addition rule for disjoint events.**

Rule 4. The **complement** of any event A is the event that A does not occur, written as A^c. The **complement rule** states that

$$P(A^c) = 1 - P(A)$$

You may find it helpful to draw a picture to remind yourself of the meaning of complements and disjoint events. A picture like Figure 4.2 that shows the sample space S as a rectangular area and events as areas within S is called

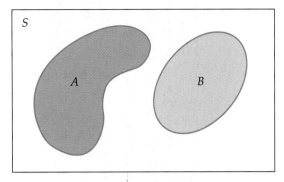

FIGURE 4.2   Venn diagram showing disjoint events *A* and *B*. Disjoint events have no common outcomes.

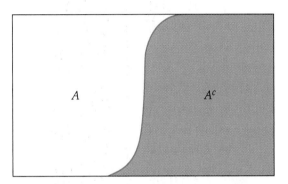

FIGURE 4.3   Venn diagram showing the complement A^c of an event *A*. The complement consists of all outcomes that are not in *A*.

Venn diagram   a **Venn diagram.** The events *A* and *B* in Figure 4.2 are disjoint because they do not overlap. As Figure 4.3 shows, the complement A^c contains exactly the outcomes that are not in *A*.

EXAMPLE 4.8   Distance learning courses are rapidly gaining popularity among college students. Choose at random an undergraduate taking a distance learning course for credit, and record the student's age. "At random" means that we give every such student the same chance to be the one we choose. This is an SRS of size 1. The probability of any age group is just the proportion of all distance learners in that age group—if we drew many students, this is the proportion we would get. Here is the probability model:[2]

Age group	18 to 23 years	24 to 29 years	30 to 39 years	40 years or over
Probability	0.57	0.17	0.14	0.12

Each probability is between 0 and 1. The probabilities add to 1 because these outcomes together make up the sample space *S*.

The probability that the student we draw is not in the traditional undergraduate age range of 18 and 23 years is, by the complement rule,

$$P(\text{not 18 to 23 years}) = 1 - P(\text{18 to 23 years})$$
$$= 1 - 0.57 = 0.43$$

That is, if 57% of distance learners are between 18 and 23 years old, then the remaining 43% are not in this age group.

The events "30 to 39 years" and "40 years or over" are disjoint because no student can be in both age groups. So the addition rule says that

$$P(\text{30 years or over}) = P(\text{30 to 39 years}) + P(\text{40 years or over})$$
$$= 0.14 + 0.12 = 0.26$$

That is, 26% of undergraduates in distance learning courses are at least 30 years old.

Assigning probabilities: finite number of outcomes

The individual outcomes of a random phenomenon are always disjoint. So the addition rule provides a way to assign probabilities to events with more than one outcome: start with probabilities for individual outcomes and add to get probabilities for events. This idea works well when there are only a finite (fixed and limited) number of outcomes.

> **PROBABILITIES IN A FINITE SAMPLE SPACE**
>
> Assign a probability to each individual outcome. These probabilities must be numbers between 0 and 1 and must have sum 1.
>
> The probability of any event is the sum of the probabilities of the outcomes making up the event.

EXAMPLE 4.9   Faked numbers in tax returns, payment records, invoices, expense account claims, and many other settings often display patterns that aren't present in legitimate records. Some patterns, like too many round numbers, are obvious and easily avoided by a clever crook. Others are more subtle. It is a striking fact that the first digits of numbers in legitimate records often follow a distribution known as **Benford's law.** Here it is (note that a first digit can't be 0):[3]

Benford's law

First digit	1	2	3	4	5	6	7	8	9
Probability	0.301	0.176	0.125	0.097	0.079	0.067	0.058	0.051	0.046

Benford's law usually applies to the first digits of the sizes of similar quantities, such as invoices, expense account claims, and county populations. Investigators can detect fraud by comparing the first digits in records such as invoices paid by a business with these probabilities.

Consider the events

$$A = \{\text{first digit is 1}\}$$

$$B = \{\text{first digit is 6 or greater}\}$$

From the table of probabilities,

$$P(A) = P(1) = 0.301$$

$$P(B) = P(6) + P(7) + P(8) + P(9)$$

$$= 0.067 + 0.058 + 0.051 + 0.046 = 0.222$$

Note that $P(B)$ is not the same as the probability that a first digit is strictly greater than 6. The probability $P(6)$ that a first digit is 6 is included in "6 or greater" but not in "greater than 6."

The probability that a first digit is anything other than 1 is, by the complement rule,

$$P(A^c) = 1 - P(A)$$

$$= 1 - 0.301 = 0.699$$

The events A and B are disjoint, so the probability that a first digit is either 1 or 6 or greater is, by the addition rule,

$$P(A \text{ or } B) = P(A) + P(B)$$

$$= 0.301 + 0.222 = 0.523$$

Be careful to apply the addition rule only to disjoint events. Check that the probability of the event C that a first digit is odd is

$$P(C) = P(1) + P(3) + P(5) + P(7) + P(9) = 0.609$$

The probability

$$P(B \text{ or } C) = P(1) + P(3) + P(5) + P(6) + P(7) + P(8) + P(9) = 0.727$$

is *not* the sum of $P(B)$ and $P(C)$, because events B and C are not disjoint. Outcomes 7 and 9 are common to both events.

Assigning probabilities: equally likely outcomes

Assigning correct probabilities to individual outcomes often requires long observation of the random phenomenon. In some circumstances, however, we are willing to assume that individual outcomes are equally likely because of some balance in the phenomenon. Ordinary coins have a physical balance that should make heads and tails equally likely, for example, and the table of random digits comes from a deliberate randomization.

EXAMPLE 4.10   You might think that first digits are distributed "at random" among the digits 1 to 9. The 9 possible outcomes would then be equally likely. The sample space for a single digit is

$$S = \{1, 2, 3, 4, 5, 6, 7, 8, 9\}$$

Because the total probability must be 1, the probability of each of the 9 outcomes must be 1/9. That is, the assignment of probabilities to outcomes is

First digit	1	2	3	4	5	6	7	8	9
Probability	1/9	1/9	1/9	1/9	1/9	1/9	1/9	1/9	1/9

The probability of the event B that a randomly chosen first digit is 6 or greater is

$$P(B) = P(6) + P(7) + P(8) + P(9)$$
$$= \frac{1}{9} + \frac{1}{9} + \frac{1}{9} + \frac{1}{9} = \frac{4}{9} = 0.444$$

Compare this with the Benford's law probability in Example 4.9. A crook who fakes data by using "random" digits will end up with too many first digits 6 or greater and too few 1s and 2s.

In Example 4.10 all outcomes have the same probability. Because there are 9 equally likely outcomes, each must have probability 1/9. Because exactly 4 of the 9 equally likely outcomes are 6 or greater, the probability of this event is 4/9. In the special situation where all outcomes are equally likely, we have a simple rule for assigning probabilities to events.

EQUALLY LIKELY OUTCOMES

If a random phenomenon has k possible outcomes, all equally likely, then each individual outcome has probability $1/k$. The probability of any event A is

$$P(A) = \frac{\text{count of outcomes in } A}{\text{count of outcomes in } S}$$
$$= \frac{\text{count of outcomes in } A}{k}$$

Most random phenomena do not have equally likely outcomes, so the general rule for finite sample spaces is more important than the special rule for equally likely outcomes.

Independence and the multiplication rule

Rule 3, the addition rule for disjoint events, describes the probability that *one or the other* of two events A and B will occur in the special situation when A and B cannot occur together because they are disjoint. Our final rule describes the probability that *both* events A and B occur, again only in a special situation. More general rules appear in Section 4.5, but in our study of statistics we will need only the rules that apply to special situations.

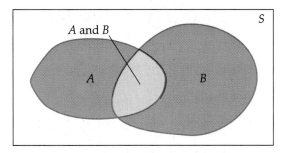

FIGURE 4.4 Venn diagram showing the event {A and B}. This event consists of outcomes common to A and B.

Suppose that you toss a balanced coin twice. You are counting heads, so two events of interest are

$$A = \{\text{first toss is a head}\}$$

$$B = \{\text{second toss is a head}\}$$

The events A and B are not disjoint. They occur together whenever both tosses give heads. We want to compute the probability of the event {A and B} that *both* tosses are heads. The Venn diagram in Figure 4.4 illustrates the event {A and B} as the overlapping area that is common to both A and B.

The coin tossing of Buffon, Pearson, and Kerrich described at the beginning of this chapter makes us willing to assign probability 1/2 to a head when we toss a coin. So

$$P(A) = 0.5$$

$$P(B) = 0.5$$

What is $P(A \text{ and } B)$? Our common sense says that it is 1/4. The first coin will give a head half the time and then the second will give a head on half of those trials, so both coins will give heads on $1/2 \times 1/2 = 1/4$ of all trials in the long run. This reasoning assumes that the second coin still has probability 1/2 of a head after the first has given a head. This is true—we can verify it by tossing two coins many times and observing the proportion of heads on the second toss after the first toss has produced a head. We say that the events "head on the first toss" and "head on the second toss" are *independent*. Here is our final probability rule.

THE MULTIPLICATION RULE FOR INDEPENDENT EVENTS

Rule 5. Two events A and B are **independent** if knowing that one occurs does not change the probability that the other occurs. If A and B are independent,

$$P(A \text{ and } B) = P(A)P(B)$$

This is the **multiplication rule for independent events.**

Our definition of independence is rather informal. We will make this informal idea precise in Section 4.5. In practice, though, we rarely need a precise definition of independence, because independence is usually *assumed* as part of a probability model when we want to describe random phenomena that seem to be physically unrelated to each other.

EXAMPLE 4.11   Because a coin has no memory and most coin-tossers cannot influence the fall of the coin, it is safe to assume that successive coin tosses are independent. For a balanced coin this means that after we see the outcome of the first toss, we still assign probability 1/2 to heads on the second toss.

On the other hand, the colors of successive cards dealt from the same deck are not independent. A standard 52-card deck contains 26 red and 26 black cards. For the first card dealt from a shuffled deck, the probability of a red card is 26/52 = 0.50 because the 52 possible cards are equally likely. Once we see that the first card is red, we know that there are only 25 reds among the remaining 51 cards. The probability that the second card is red is therefore only 25/51 = 0.49. Knowing the outcome of the first deal changes the probabilities for the second.

If a doctor measures your blood pressure twice, it is reasonable to assume that the two results are independent because the first result does not influence the instrument that makes the second reading. But if you take an IQ test or other mental test twice in succession, the two test scores are not independent. The learning that occurs on the first attempt influences your second attempt.

When independence is part of a probability model, the multiplication rule applies. Here is an example.

EXAMPLE 4.12   Gregor Mendel used garden peas in some of the experiments that revealed that inheritance operates randomly. The seed color of Mendel's peas can be either green or yellow. Two parent plants are "crossed" (one pollinates the other) to produce seeds. Each parent plant carries two genes for seed color, and each of these genes has probability 1/2 of being passed to a seed. The two genes that the seed receives, one from each parent, determine its color. The parents contribute their genes independently of each other.

Suppose that both parents carry the G and the Y genes. The seed will be green if both parents contribute a G gene; otherwise it will be yellow. If M is the event that the male contributes a G gene and F is the event that the female contributes a G gene, then the probability of a green seed is

$$P(M \text{ and } F) = P(M)P(F)$$

$$= (0.5)(0.5) = 0.25$$

In the long run, 1/4 of all seeds produced by crossing these plants will be green.

The multiplication rule applies only to independent events; you cannot use it if events are not independent. Here is a distressing example of misuse of the multiplication rule.

EXAMPLE 4.13   Sudden infant death syndrome (SIDS) causes babies to die suddenly (often in their cribs) with no explanation. Deaths from SIDS have been greatly reduced by placing babies on their backs, but as yet no cause is known.

When more than one SIDS death occurs in a family, the parents are sometimes accused. One "expert witness" popular with prosecutors in England told juries that there is only a 1 in 73 million chance that two children in the same family could have died naturally. Here's his calculation: the rate of SIDS in a nonsmoking middle-class family is 1 in 8500. So the probability of two deaths is

$$\frac{1}{8500} \times \frac{1}{8500} = \frac{1}{72,250,000}$$

Several women were convicted of murder on this basis, without any direct evidence that they harmed their children.

As the Royal Statistical Society said, this reasoning is nonsense. It assumes that SIDS deaths in the same family are independent events. The cause of SIDS is unknown: "There may well be unknown genetic or environmental factors that predispose families to SIDS, so that a second case within the family becomes much more likely."[4] The British government decided to review the cases of 258 parents convicted of murdering their babies.

The multiplication rule $P(A \text{ and } B) = P(A)P(B)$ holds if A and B are independent but not otherwise. The addition rule $P(A \text{ or } B) = P(A) + P(B)$ holds if A and B are disjoint but not otherwise. Resist the temptation to use these simple formulas when the circumstances that justify them are not present. *You must also be certain not to confuse disjointness and independence. Disjoint events cannot be independent.* If A and B are disjoint, then the fact that A occurs tells us that B cannot occur—look again at Figure 4.2. Unlike disjointness or complements, independence cannot be pictured by a Venn diagram, because it involves the probabilities of the events rather than just the outcomes that make up the events.

Applying the probability rules

If two events A and B are independent, then their complements A^c and B^c are also independent and A^c is independent of B. Suppose, for example, that 75% of all registered voters in a suburban district are Republicans. If an opinion poll interviews two voters chosen independently, the probability that the first is a Republican and the second is not a Republican is $(0.75)(0.25) = 0.1875$. The multiplication rule also extends to collections of more than two events, provided that all are independent. Independence of events A, B, and C means that no information about any one or any two can change the probability of the remaining events. The formal definition is a bit messy. Fortunately, independence is usually assumed in setting up a probability model. We can then use the multiplication rule freely, as in this example.

EXAMPLE 4.14   The "six degrees of separation" idea says that any two people on earth are linked by a chain of friends of friends of friends no more than 6 steps long. Does this work online? That is, if we give you the name and location of someone you don't know, can you get a message to that person by sending email to an acquaintance who you think is closer to your target and asking him or her to pass on the message?

As you might guess, success depends on how willing the people in the chain are to take the next step. Here's a model based on a large study: after the first step (you send the message to someone you know), the probability that the message is passed on is 0.37 and steps are independent.[5] Take A_i to be the event that the ith person after you in the chain passes on the message.

The probability that the next two links pass on the message is

$$P(A_1 \text{ and } A_2) = P(A_1)P(A_2)$$

$$= 0.37 \times 0.37 = 0.1369$$

If there really are six degrees of separation between you and a specific person you don't know, you need at the very least five people after you to pass on the message. The probability that five people will all pass the message is

$$P(A_1 \text{ and } A_2 \text{ and } A_3 \text{ and } A_4 \text{ and } A_5) = P(A_1)P(A_2)P(A_3)P(A_4)(P(A_5)$$

$$= 0.37 \times 0.37 \times 0.37 \times 0.37 \times 0.37$$

$$= 0.37^5 = 0.0069$$

Even if five people in a row keep your message moving, they may not succeed in reaching your target. In the study, only 384 of 24,163 chains reached the target. It's very unlikely that you can pass a message by way of friends of friends of friends successfully.

By combining the rules we have learned, we can compute probabilities for rather complex events. Here is an example.

EXAMPLE 4.15   Many people who come to clinics to be tested for HIV, the virus that causes AIDS, don't come back to learn the test results. Clinics now use "rapid HIV tests" that give a result in a few minutes. Applied to people who have no HIV antibodies, one rapid test has probability about 0.004 of producing a false positive (that is, of falsely indicating that antibodies are present).[6] If a clinic tests 200 people who are free of HIV antibodies, what is the probability that at least one false positive will occur?

It is reasonable to assume as part of the probability model that the test results for different individuals are independent. The probability that the test is positive for a single person is 0.004, so the probability of a negative result is $1 - 0.004 = 0.996$ by the complement rule. The probability of at least one false positive among the 200 people tested is therefore

$$P(\text{at least one positive}) = 1 - P(\text{no positives})$$

$$= 1 - P(200 \text{ negatives})$$

$$= 1 - 0.996^{200}$$

$$= 1 - 0.4486 = 0.5514$$

The probability is greater than 1/2 that at least one of the 200 people will test positive for HIV, even though no one has the virus.

SECTION 4.2 | Summary

A **probability model** for a random phenomenon consists of a sample space S and an assignment of probabilities P.

(b) Describe the event $X \leq 2$ in words. What is its probability? What is the probability that $X < 2$?

4.42 The length of human pregnancies from conception to birth varies according to a distribution that is approximately normal with mean 266 days and standard deviation 16 days. Call the length of a randomly chosen pregnancy Y. What is $P(Y > 300)$?

4.43 How do rented housing units differ from units occupied by their owners? Here are the distributions of the number of rooms for owner-occupied units and renter-occupied units in San Jose, California:[16]

Rooms	1	2	3	4	5	6	7	8	9	10
Owned	0.003	0.002	0.023	0.104	0.210	0.224	0.197	0.149	0.053	0.035
Rented	0.008	0.027	0.287	0.363	0.164	0.093	0.039	0.013	0.003	0.003

Make probability histograms of these two distributions, using the same scales. What are the most important differences between the distributions for owner-occupied and rented housing units?

4.44 In government data, a household consists of all occupants of a dwelling unit, while a family consists of two or more persons who live together and are related by blood or marriage. So all families form households, but some households are not families. Here are the distributions of household size and of family size in the United States:

Number of persons	1	2	3	4	5	6	7
Household probability	0.27	0.33	0.16	0.14	0.06	0.03	0.01
Family probability	0	0.44	0.22	0.20	0.09	0.03	0.02

Make probability histograms for these two discrete distributions, using the same scales. What are the most important differences between the sizes of households and families?

4.45 Let the random variable X be the number of rooms in a randomly chosen owner-occupied housing unit in San Jose, California. Exercise 4.43 gives the distribution of X.

(a) Express "the unit has 5 or more rooms" in terms of X. What is the probability of this event?

(b) Express the event $\{X > 5\}$ in words. What is its probability?

(c) What important fact about discrete random variables does comparing your answers to (a) and (b) illustrate?

4.46 Choose an American household at random and let the random variable Y be the number of persons living in the household. Exercise 4.44 gives the distribution of Y.

The **sample space S** is the set of all possible outcomes of the random phenomenon. Sets of outcomes are called **events.** *P* assigns a number *P(A)* to an event *A* as its probability.

The **complement A^c** of an event *A* consists of exactly the outcomes that are not in *A*. Events *A* and *B* are **disjoint** if they have no outcomes in common. Events *A* and *B* are **independent** if knowing that one event occurs does not change the probability we would assign to the other event.

Any assignment of probability must obey the rules that state the basic properties of probability:

Rule 1. $0 \le P(A) \le 1$ for any event *A*.

Rule 2. $P(S) = 1$.

Rule 3. **Addition rule:** If events *A* and *B* are **disjoint,** then $P(A \text{ or } B) = P(A) + P(B)$.

Rule 4. **Complement rule:** For any event *A*, $P(A^c) = 1 - P(A)$.

Rule 5. **Multiplication rule:** If events *A* and *B* are **independent,** then $P(A \text{ and } B) = P(A)P(B)$.

SECTION 4.2 | Exercises

4.10 A randomly chosen subject arrives for a study of exercise and fitness. Describe a sample space for each of the following. (In some cases, you may have some freedom in your choice of *S*.)

(a) The subject is either female or male.

(b) After 10 minutes on an exercise bicycle, you ask the subject to rate his or her effort on the Rate of Perceived Exertion (RPE) scale. RPE ranges in whole-number steps from 6 (no exertion at all) to 20 (maximal exertion).

(c) You measure VO2, the maximum volume of oxygen consumed per minute during exercise. VO2 is generally between 2.5 liters per minute and 6 liters per minute.

(d) You measure the maximum heart rate (beats per minute).

4.11 Choose a student at random from a large statistics class. Give a reasonable sample space *S* for answers to each of these questions. (In some cases you may have some freedom in specifying *S*.)

(a) Are you right-handed or left-handed?

(b) What is your height in centimeters? (One inch is 2.54 centimeters.)

(c) How much money in coins (not bills) are you carrying?

(d) How many minutes did you study last night?

4.12 Role-playing games like Dungeons & Dragons use many different types of dice. Suppose that a four-sided die has faces marked 1, 2, 3, 4. The intelligence of a character is determined by rolling this die twice and adding 1 to the sum of the spots.

(a) What is the sample space for rolling the die twice (spots on first and second rolls)?

(b) What is the sample space for the character's intelligence?

4.13   All human blood can be "ABO-typed" as one of O, A, B, or AB, but the distribution of the types varies a bit among groups of people. Here is the distribution of blood types for a randomly chosen person in the United States:

Blood type	O	A	B	AB
U.S. probability	0.45	0.40	0.11	?

(a) What is the probability of type AB blood in the United States?

(b) Maria has type B blood. She can safely receive blood transfusions from people with blood types O and B. What is the probability that a randomly chosen American can donate blood to Maria?

4.14   The distribution of blood types in China differs from the U.S. distribution given in the previous exercise:

Blood type	O	A	B	AB
China probability	0.35	0.27	0.26	0.12

Choose an American and a Chinese at random, independently of each other. What is the probability that both have type O blood? What is the probability that both have the same blood type?

4.15   In each of the following situations, state whether or not the given assignment of probabilities to individual outcomes is legitimate, that is, satisfies the rules of probability. If not, give specific reasons for your answer.

(a) Roll a die and record the count of spots on the up-face: $P(1) = 0$, $P(2) = 1/6$, $P(3) = 1/3$, $P(4) = 1/3$, $P(5) = 1/6$, $P(6) = 0$.

(b) Choose a college student at random and record sex and enrollment status: P(female full-time) = 0.56, P(female part-time) = 0.24, P(male full-time) = 0.44, P(male part-time) = 0.17.

(c) Deal a card from a shuffled deck: P(clubs) = 12/52, P(diamonds) = 12/52, P(hearts) = 12/52, P(spades) = 16/52.

4.16   Canada has two official languages, English and French. Choose a Canadian at random and ask, "What is your mother tongue?" Here is the distribution of responses, combining many separate languages from the broad Asian/Pacific region:[7]

Language	English	French	Asian/Pacific	Other
Probability	0.59	0.23	0.07	?

(a) What probability should replace "?" in the distribution?

(b) What is the probability that a Canadian's mother tongue is not English?

4.17 Choose a young adult (age 25 to 34 years) at random. The probability is 0.12 that the person chosen did not complete high school, 0.31 that the person has a high school diploma but no further education, and 0.29 that the person has at least a bachelor's degree.

(a) What must be the probability that a randomly chosen young adult has some education beyond high school but does not have a bachelor's degree?

(b) What is the probability that a randomly chosen young adult has at least a high school education?

4.18 A majority of email messages are now "spam." Choose a spam email message at random. Here is the distribution of topics:[8]

Topic	Adult	Financial	Health	Leisure	Products	Scams
Probability	0.145	0.162	0.073	0.078	0.210	0.142

(a) What is the probability that a spam email does not concern one of these topics?

(b) Corinne is particularly annoyed by spam offering "adult" content (that is, pornography) and scams. What is the probability that a randomly chosen spam email falls into one or the other of these categories?

4.19 Dugout Lou thinks that the probabilities for the American League baseball champion are as follows. The Yankees have probability 0.6 of winning. The Red Sox and Angels have equal probabilities of winning. The Athletics and White Sox also have equal probabilities, but their probabilities are one-third that of the Red Sox and Angels. No other team has a chance. What is Lou's assignment of probabilities to teams?

4.20 Role-playing games like Dungeons & Dragons use many different types of dice. Suppose that a four-sided die has faces marked 1, 2, 3, 4. The intelligence of a character is determined by rolling this die twice and adding 1 to the sum of the spots. The faces are equally likely and the two rolls are independent. In Exercise 4.12 you gave the sample space S for the character's intelligence. Now give the assignment of probabilities to the outcomes in S.

4.21   There are many ways to produce crooked dice. To *load* a die so that 6 comes up too often and 1 (which is opposite 6) comes up too seldom, add a bit of lead to the filling of the spot on the 1 face. Because the spot is solid plastic, this works even with transparent dice. If a die is loaded so that 6 comes up with probability 0.2 and the probabilities of the 2, 3, 4, and 5 faces are not affected, what is the assignment of probabilities to the six faces?

4.22   The 2000 census allowed each person to choose from a long list of races. That is, in the eyes of the Census Bureau, you belong to whatever race you say you belong to. "Hispanic/Latino" is a separate category; Hispanics may be of any race. If we choose a resident of the United States at random, the 2000 census gives these probabilities:

	Hispanic	Not Hispanic
Asian	0.000	0.036
Black	0.003	0.121
White	0.060	0.691
Other	0.062	0.027

Let A be the event that a randomly chosen American is Hispanic, and let B be the event that the person chosen is white.

(a) Verify that the table gives a legitimate assignment of probabilities.

(b) What is $P(A)$?

(c) Describe B^c in words and find $P(B^c)$ by the complement rule.

(d) Express "the person chosen is a non-Hispanic white" in terms of events A and B. What is the probability of this event?

4.23   Human blood is typed as O, A, B, or AB and also as Rh-positive or Rh-negative. ABO type and Rh-factor type are independent because they are governed by different genes. In the American population, 84% of people are Rh-positive. Use the information about ABO type in Exercise 4.13 to give the probability distribution of blood type (ABO and Rh) for a randomly chosen person.

4.24   Exercise 4.22 assigns probabilities for the ethnic background of a randomly chosen resident of the United States. Let A be the event that the person chosen is Hispanic, and let B be the event that he or she is white. Are events A and B independent? How do you know?

4.25   A roulette wheel has 38 slots, numbered 0, 00, and 1 to 36. The slots 0 and 00 are colored green, 18 of the others are red, and 18 are black. The dealer spins the wheel and at the same time rolls a small ball along the wheel in the opposite direction. The wheel is carefully balanced so that the ball is equally likely to land in any slot when the wheel slows. Gamblers can bet on various combinations of numbers and colors.

(a) What is the probability that the ball will land in any one slot?

(b) If you bet on "red," you win if the ball lands in a red slot. What is the probability of winning?

(c) The slot numbers are laid out on a board on which gamblers place their bets. One column of numbers on the board contains all multiples of 3, that is, 3, 6, 9, . . . , 36. You place a "column bet" that wins if any of these numbers comes up. What is your probability of winning?

4.26   A state lottery's "Pick 3" game asks players to choose a three-digit number, 000 to 999. The state chooses the winning three-digit number at random, so that each number has probability 1/1000. You win if the winning number contains the digits in your number, in any order.

(a) Your number is 123. What is your probability of winning?

(b) Your number is 112. What is your probability of winning?

4.27   The personal identification numbers (PINs) for automatic teller machines usually consist of four digits. You notice that most of your PINs have at least one 0, and you wonder if the issuers use lots of 0s to make the numbers easy to remember. Suppose that PINs are assigned at random, so that all four-digit numbers are equally likely.

(a) How many possible PINs are there?

(b) What is the probability that a PIN assigned at random has at least one 0?

4.28   People with type O-negative blood are universal donors. That is, any patient can receive a transfusion of O-negative blood. Only 7% of the American population have O-negative blood. If 10 people appear at random to give blood, what is the probability that at least 1 of them is a universal donor?

4.29   Internet sites often vanish or move, so that references to them can't be followed. In fact, 13% of Internet sites referenced in major scientific journals are lost within two years after publication.[9] If a paper contains seven Internet references, what is the probability that all seven are still good two years later? What specific assumptions did you make in order to calculate this probability?

4.30   Most sample surveys use random digit dialing equipment to call residential telephone numbers at random. The telephone polling firm Zogby International reports that the probability that a call reaches a live person is 0.2.[10] Calls are independent.

(a) A polling firm places 5 calls. What is the probability that none of them reaches a person?

(b) When calls are made to New York City, the probability of reaching a person is only 0.08. What is the probability that none of 5 calls made to New York City reaches a person?

4.31   Government data show that 6% of the American population are at least 75 years of age and that about 51% are women. Explain why it is wrong to conclude that because $(0.06)(0.51) = 0.0306$ about 3% of the population are women aged 75 or over.

4.32   Here's more evidence that our intuition about chance behavior is not very accurate. A six-sided die has four green and two red faces, all equally probable. Psychologists asked students to say which of these color sequences is most likely to come up at the beginning of a long set of rolls of this die:

RGRRR

RGRRRG

GRRRRR

More than 60% chose the second sequence.[11] What is the correct probability of each sequence?

4.33   The "random walk" theory of securities prices holds that price movements in disjoint time periods are independent of each other. Suppose that we record only whether the price is up or down each year, and that the probability that our portfolio rises in price in any one year is 0.65. (This probability is approximately correct for a portfolio containing equal dollar amounts of all common stocks listed on the New York Stock Exchange.)

(a) What is the probability that our portfolio goes up for 3 consecutive years?

(b) If you know that the portfolio has risen in price 2 years in a row, what probability do you assign to the event that it will go down next year?

(c) What is the probability that the portfolio's value moves in the same direction in both of the next 2 years?

4.34   Show that any assignment of probabilities to events that obeys Rules 2 and 3 on page 262 automatically obeys the complement rule (Rule 4). This implies that a mathematical treatment of probability can start from just Rules 1, 2, and 3. These rules are sometimes called *axioms* of probability.

4.35   Show that if events A and B obey the multiplication rule, $P(A \text{ and } B) = P(A)P(B)$, then A and the complement B^c of B also obey the multiplication rule, $P(A \text{ and } B^c) = P(A)P(B^c)$. That is, if events A and B are independent, then A and B^c are also independent. (*Hint:* Start by drawing a Venn diagram and noticing that the events "A and B" and "A and B^c" are disjoint.)

Mendelian inheritance. Some traits of plants and animals depend on inheritance of a single gene. This is called Mendelian inheritance, after Gregor Mendel (1822–1884). Exercises 4.36 to 4.39 are based on the following information about Mendelian inheritance of blood type.

Each of us has an ABO blood type, which describes whether two characteristics called A and B are present. Every human being has two blood type alleles (gene forms), one inherited from our mother and one from our father. Each of these alleles can be A, B, or O. Which two we inherit determines our blood type.

Here is a table that shows what our blood type is for each combination of two alleles:

Alleles inherited	Blood type
A and A	A
A and B	AB
A and O	A
B and B	B
B and O	B
O and O	O

We inherit each of a parent's two alleles with probability 0.5. We inherit independently from our mother and father.

4.36 Hannah and Jacob both have alleles A and B.

(a) What blood types can their children have?

(b) What is the probability that their next child has each of these blood types?

4.37 Nancy and David both have alleles B and O.

(a) What blood types can their children have?

(b) What is the probability that their next child has each of these blood types?

4.38 Jennifer has alleles A and O. José has alleles A and B. They have two children. What is the probability that both children have blood type A? What is the probability that both children have the same blood type?

4.39 Jasmine has alleles A and O. Joshua has alleles B and O.

(a) What is the probability that a child of these parents has blood type O?

(b) If Jasmine and Joshua have three children, what is the probability that all three have blood type O? What is the probability that the first child has blood type O and the next two do not?

4.3 Random Variables

Sample spaces need not consist of numbers. When we toss a coin four times, we can record the outcome as a string of heads and tails, such as HTTH. In statistics, however, we are most often interested in numerical outcomes such as the count of heads in the four tosses. It is convenient to use a shorthand notation: Let X be the number of heads. If our outcome is HTTH, then $X = 2$. If the next outcome is TTTH, the value of X changes to $X = 1$. The possible values of X are 0, 1, 2, 3, and 4. Tossing a coin four times will give X one of these possible values. Tossing four more times will give X another and probably different value. We call X a *random variable* because its values vary when the coin tossing is repeated.

> ### RANDOM VARIABLE
>
> A **random variable** is a variable whose value is a numerical outcome of a random phenomenon.

We usually denote random variables by capital letters near the end of the alphabet, such as X or Y. Of course, the random variables of greatest interest to us are outcomes such as the mean $\bar{x}$ of a random sample, for which we will keep the familiar notation.[12] As we progress from general rules of probability toward statistical inference, we will concentrate on random variables. When a random variable X describes a random phenomenon, the sample space S just lists the possible values of the random variable. We usually do not mention S separately. There remains the second part of any probability model, the assignment of probabilities to events. There are two main ways of assigning probabilities to the values of a random variable. The two types of probability models that result will dominate our application of probability to statistical inference.

Discrete random variables

We have learned several rules of probability but only one method of assigning probabilities: state the probabilities of the individual outcomes and assign probabilities to events by summing over the outcomes. The outcome probabilities must be between 0 and 1 and have sum 1. When the outcomes are numerical, they are values of a random variable. We will now attach a name to random variables having probability assigned in this way.[13]

> ### DISCRETE RANDOM VARIABLE
>
> A **discrete random variable** X has a finite number of possible values. The **probability distribution** of X lists the values and their probabilities:
>
> | Value of X | x_1 | x_2 | x_3 | $\cdots$ | x_k |
> |---|---|---|---|---|---|
> | Probability | p_1 | p_2 | p_3 | $\cdots$ | p_k |
>
> The probabilities p_i must satisfy two requirements:
>
> **1.** Every probability p_i is a number between 0 and 1.
>
> **2.** $p_1 + p_2 + \cdots + p_k = 1$.
>
> Find the probability of any event by adding the probabilities p_i of the particular values x_i that make up the event.

EXAMPLE 4.16   North Carolina State University posts the grade distributions for its courses online.[14] Students in Statistics 101 in the fall 2003 semester received 21% A's, 43% B's, 30% C's, 5% D's, and 1% F's. Choose a Statistics 101 student at random. To "choose at random" means to give every student the same chance to be chosen. The student's grade on a four-point scale (with A = 4) is a random variable X.

The value of X changes when we repeatedly choose students at random, but it is always one of 0, 1, 2, 3, or 4. Here is the distribution of X:

Value of X	0	1	2	3	4
Probability	0.01	0.05	0.30	0.43	0.21

The probability that the student got a B or better is the sum of the probabilities of an A and a B. In the language of random variables,

$$P(X \geq 3) = P(X = 3) + P(X = 4)$$
$$= 0.43 + 0.21 = 0.64$$

We can use histograms to show probability distributions as well as distributions of data. Figure 4.5 displays **probability histograms** that compare the probability model for random digits (Example 4.10) with the model given by

probability histogram

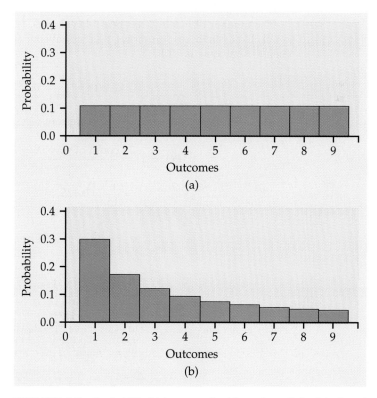

FIGURE 4.5   Probability histograms for (a) random digits 1 to 9 and (b) Benford's law. The height of each bar shows the probability assigned to a single outcome.

Benford's law (Example 4.9). The height of each bar shows the probability of the outcome at its base. Because the heights are probabilities, they add to 1. As usual, all the bars in a histogram have the same width. So the areas also display the assignment of probability to outcomes. Think of these histograms as idealized pictures of the results of very many trials. The histograms make it easy to quickly compare the two distributions.

EXAMPLE 4.17 What is the probability distribution of the discrete random variable X that counts the number of heads in four tosses of a coin? We can derive this distribution if we make two reasonable assumptions:

- The coin is balanced, so each toss is equally likely to give H or T.

- The coin has no memory, so tosses are independent.

The outcome of four tosses is a sequence of heads and tails such as HTTH. There are 16 possible outcomes in all. Figure 4.6 lists these outcomes along with the value of X for each outcome. The multiplication rule for independent events tells us that, for example,

$$P(\text{HTTH}) = \frac{1}{2} \times \frac{1}{2} \times \frac{1}{2} \times \frac{1}{2} = \frac{1}{16}$$

Each of the 16 possible outcomes similarly has probability 1/16. That is, these outcomes are equally likely.

The number of heads X has possible values 0, 1, 2, 3, and 4. These values are *not* equally likely. As Figure 4.6 shows, there is only one way that $X = 0$ can occur: namely, when the outcome is TTTT. So

$$P(X = 0) = \frac{1}{16} = 0.0625$$

The event $\{X = 2\}$ can occur in six different ways, so that

$$P(X = 2) = \frac{\text{count of ways } X = 2 \text{ can occur}}{16}$$

$$= \frac{6}{16} = 0.375$$

TTTT	HTTT THTT TTHT TTTH	HTTH HTHT THTH HHTT THHT TTHH	HHHT HHTH HTHH THHH	HHHH
$X = 0$	$X = 1$	$X = 2$	$X = 3$	$X = 4$

FIGURE 4.6 Possible outcomes in four tosses of a coin, for Example 4.17. The outcomes are arranged by the values of the random variable X, the number of heads.

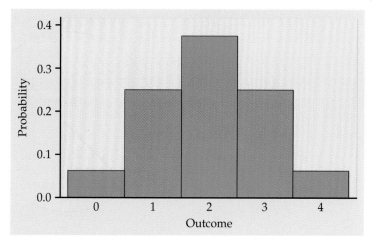

FIGURE 4.7   Probability histogram for the number of heads in four tosses of a coin, for Example 4.17.

We can find the probability of each value of X from Figure 4.6 in the same way. Here is the result:

Value of X	0	1	2	3	4
Probability	0.0625	0.25	0.375	0.25	0.0625

Figure 4.7 is a probability histogram for this distribution. The probability distribution is exactly symmetric. The probabilities (box heights) are idealizations of the proportions after many tosses of four coins. The distribution of proportions would be nearly symmetric but is unlikely to be exactly symmetric.

Any event involving the number of heads observed can be expressed in terms of X, and its probability can be found from the distribution of X. For example, the probability of tossing at least two heads is

$$P(X \geq 2) = 0.375 + 0.25 + 0.0625 = 0.6875$$

The probability of at least one head is most simply found by use of the complement rule:

$$P(X \geq 1) = 1 - P(X = 0)$$
$$= 1 - 0.0625 = 0.9375$$

Recall that tossing a coin n times is similar to choosing an SRS of size n from a large population and asking a yes-or-no question. We will extend the results of Example 4.17 when we return to sampling distributions in the next chapter.

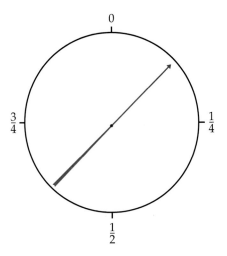

FIGURE 4.8 A spinner that generates a random number between 0 and 1.

Continuous random variables

When we use the table of random digits to select a digit between 0 and 9, the result is a discrete random variable. The probability model assigns probability 1/10 to each of the 10 possible outcomes. Suppose that we want to choose a number at random between 0 and 1, allowing *any* number between 0 and 1 as the outcome. Software random number generators will do this. You can visualize such a random number by thinking of a spinner (Figure 4.8) that turns freely on its axis and slowly comes to a stop. The pointer can come to rest anywhere on a circle that is marked from 0 to 1. The sample space is now an entire interval of numbers:

$$S = \{\text{all numbers } x \text{ such that } 0 \leq x \leq 1\}$$

How can we assign probabilities to events such as $\{0.3 \leq x \leq 0.7\}$? As in the case of selecting a random digit, we would like all possible outcomes to be equally likely. But we cannot assign probabilities to each individual value of x and then sum, because there are infinitely many possible values. Instead we use a new way of assigning probabilities directly to events—as *areas under a density curve*. Any density curve has area exactly 1 underneath it, corresponding to total probability 1.

EXAMPLE 4.18

uniform distribution

The random number generator will spread its output uniformly across the entire interval from 0 to 1 as we allow it to generate a long sequence of numbers. The results of many trials are represented by the density curve of a **uniform distribution.** This density curve appears in blue in Figure 4.9. It has height 1 over the interval from 0 to 1, and height 0 everywhere else. The area under the density curve is 1, the area of a square with base 1 and height 1. The probability of any event is the area under the density curve and above the event in question.

As Figure 4.9(a) illustrates, the probability that the random number generator produces a number X between 0.3 and 0.7 is

$$P(0.3 \leq X \leq 0.7) = 0.4$$

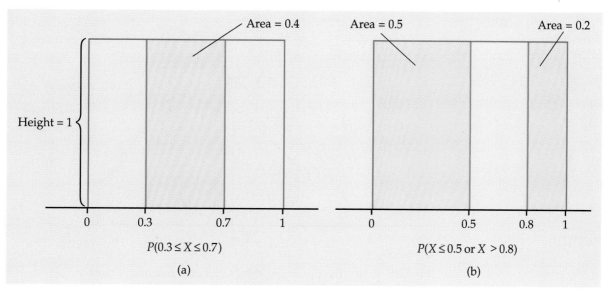

FIGURE 4.9 Assigning probabilities for generating a random number between 0 and 1, for Example 4.18. The probability of any interval of numbers is the area above the interval and under the density curve.

because the area under the density curve and above the interval from 0.3 to 0.7 is 0.4. The height of the density curve is 1, and the area of a rectangle is the product of height and length, so the probability of any interval of outcomes is just the length of the interval.

Similarly,

$$P(X \leq 0.5) = 0.5$$

$$P(X > 0.8) = 0.2$$

$$9(X \leq 0.5 \text{ or } X > 0.8) = 0.7$$

Notice that the last event consists of two nonoverlapping intervals, so the total area above the event is found by adding two areas, as illustrated by Figure 4.9(b). This assignment of probabilities obeys all of our rules for probability.

Probability as area under a density curve is a second important way of assigning probabilities to events. Figure 4.10 illustrates this idea in general form. We call X in Example 4.18 a *continuous random variable* because its values are not isolated numbers but an entire interval of numbers.

CONTINUOUS RANDOM VARIABLE

A **continuous random variable** X takes all values in an interval of numbers. The **probability distribution** of X is described by a density curve. The probability of any event is the area under the density curve and above the values of X that make up the event.

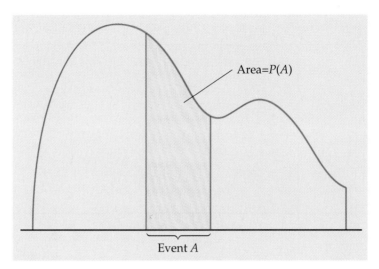

Area=$P(A)$

Event A

FIGURE 4.10 The probability distribution of a continuous random variable assigns probabilities as areas under a density curve. The total area under any density curve is 1.

The probability model for a continuous random variable assigns probabilities to intervals of outcomes rather than to individual outcomes. In fact, **all continuous probability distributions assign probability 0 to every individual outcome.** Only intervals of values have positive probability. To see that this is true, consider a specific outcome such as $P(X = 0.8)$ in Example 4.18. The probability of any interval is the same as its length. The point 0.8 has no length, so its probability is 0.

Although this fact may seem odd, it makes intuitive, as well as mathematical, sense. The random number generator produces a number between 0.79 and 0.81 with probability 0.02. An outcome between 0.799 and 0.801 has probability 0.002. A result between 0.799999 and 0.800001 has probability 0.000002. You see that as we home in on 0.8 the probability gets closer to 0. To be consistent, the probability of outcome *exactly* equal to 0.8 must be 0. Because there is no probability exactly at $X = 0.8$, the two events $\{X > 0.8\}$ and $\{X \geq 0.8\}$ have the same probability. We can ignore the distinction between > and ≥ when finding probabilities for continuous (but not discrete) random variables.

Normal distributions as probability distributions

The density curves that are most familiar to us are the normal curves. Because any density curve describes an assignment of probabilities, *normal distributions are probability distributions*. Recall that $N(\mu, \sigma)$ is our shorthand for the normal distribution having mean μ and standard deviation σ. In the language of random variables, if X has the $N(\mu, \sigma)$ distribution, then the standardized variable

$$Z = \frac{X - \mu}{\sigma}$$

is a standard normal random variable having the distribution $N(0, 1)$.

EXAMPLE 4.19   Students are reluctant to report cheating by other students. A sample survey puts this question to an SRS of 400 undergraduates: "You witness two students cheating on a quiz. Do you go to the professor?" Suppose that if we could ask all undergraduates, 12% would answer "Yes."[15]

The proportion $p = 0.12$ is a *parameter* that describes the population of all undergraduates. The proportion $\hat{p}$ of the sample who answer "Yes" is a *statistic* used to estimate p. The statistic $\hat{p}$ is a random variable, because repeating the SRS would give a different sample of 400 undergraduates and a different value of $\hat{p}$. We will see in the next chapter that $\hat{p}$ has approximately the $N(0.12, 0.016)$ distribution. The mean 0.12 of this distribution is the same as the population parameter because $\hat{p}$ is an unbiased estimate of p. The standard deviation is controlled mainly by the size of the sample.

What is the probability that the survey result differs from the truth about the population by more than 2 percentage points? Because $p = 0.12$, the survey misses by more than two percentage points if $\hat{p} < 0.10$ or $\hat{p} > 0.14$. Figure 4.11 shows this probability as an area under a normal density curve. You can find it by software or by standardizing and using Table A. Let's start with the complement rule,

$$P(\hat{p} < 0.10 \text{ or } \hat{p} > 0.14) = 1 - P(0.10 \leq \hat{p} \leq 0.14)$$

From Table A,

$$P(0.10 \leq \hat{p} \leq 0.14) = P\left(\frac{0.10 - 0.12}{0.016} \leq \frac{\hat{p} - 0.12}{0.016} \leq \frac{0.14 - 0.12}{0.016}\right)$$

$$= P(-1.25 \leq Z \leq 1.25)$$

$$= 0.8944 - 0.1056 = 0.7888$$

The probability we seek is therefore $1 - 0.7888 = 0.2112$. About 21% of sample results will be off by more than 2 percentage points. The arrangement of this calculation is familiar from our earlier work with normal distributions. Only the language of probability is new.

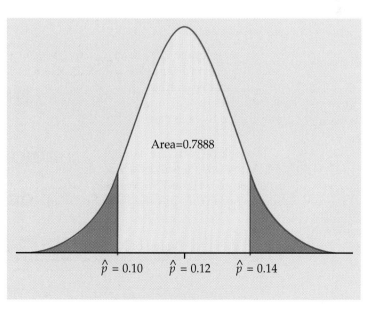

FIGURE 4.11   Probability in Example 4.19 as area under a normal density curve.

We began this chapter with a general discussion of the idea of probability and the properties of probability models. Two very useful specific types of probability models are distributions of discrete and continuous random variables. In our study of statistics we will employ only these two types of probability models.

SECTION 4.3 | Summary

A **random variable** is a variable taking numerical values determined by the outcome of a random phenomenon. The **probability distribution** of a random variable X tells us what the possible values of X are and how probabilities are assigned to those values.

A random variable X and its distribution can be **discrete** or **continuous.**

A **discrete random variable** has finitely many possible values. The probability distribution assigns each of these values a probability between 0 and 1 such that the sum of all the probabilities is exactly 1. The probability of any event is the sum of the probabilities of all the values that make up the event.

A **continuous random variable** takes all values in some interval of numbers. A **density curve** describes the probability distribution of a continuous random variable. The probability of any event is the area under the curve above the values that make up the event.

Normal distributions are one type of continuous probability distribution.

You can picture a probability distribution by drawing a **probability histogram** in the discrete case or by graphing the density curve in the continuous case.

SECTION 4.3 | Exercises

4.40 Exercise 4.10 (page 271) asks you to give sample spaces for several kinds of measurement on a randomly chosen person. Some of these measurements are random variables, although we did not use the term. Is the random variable in each of parts (b), (c), and (d) of Exercise 4.10 discrete or continuous? Explain your choice in each case.

4.41 Spell-checking software catches "nonword errors" that result in a string of letters that is not a word, as when "the" is typed as "teh." When undergraduates are asked to write a 250-word essay (without spell-checking), the number X of nonword errors has the following distribution:

Value of X	0	1	2	3	4
Probability	0.1	0.2	0.3	0.3	0.1

(a) Write the event "at least one nonword error" in terms of X. What is the probability of this event?

(a) Express "more than one person lives in this household" in terms of Y. What is the probability of this event?

(b) What is $P(2 < Y \leq 4)$?

(c) What is $P(Y \neq 2)$?

4.47   Example 4.14 (page 269) gives a model for an attempt to send an email message to someone you don't know by sending email to an acquaintance who you think is closer to your target and asking him or her to pass on the message. At each step after you send the initial email, the probability is 0.37 that the recipient will pass on your message. Steps are independent. Let T be the total length of the chain. That is, if your friend does not relay your message, $T = 2$ because the chain stops at the 2nd person. If your friend passes it on but the next person does not, $T = 3$, and so on.

(a) What is $P(T = 2)$? What is $P(T = 3)$?

(b) Express in words the meaning of $P(T \leq 4)$. What is the value of this probability?

4.48   Some games of chance rely on tossing two dice. Each die has six faces, marked with 1, 2, …, 6 spots called pips. The dice used in casinos are carefully balanced so that each face is equally likely to come up. When two dice are tossed, each of the 36 possible pairs of faces is equally likely to come up. The outcome of interest to a gambler is the sum of the pips on the two up-faces. Call this random variable X.

(a) Write down all 36 possible pairs of faces.

(b) If all pairs have the same probability, what must be the probability of each pair?

(c) Write the value of X next to each pair of faces and use this information with the result of (b) to give the probability distribution of X. Draw a probability histogram to display the distribution.

(d) One bet available in craps wins if a 7 or an 11 comes up on the next roll of two dice. What is the probability of rolling a 7 or an 11 on the next roll?

(e) Several bets in craps lose if a 7 is rolled. If any outcome other than 7 occurs, these bets either win or continue to the next roll. What is the probability that anything other than a 7 is rolled?

4.49   Nonstandard dice can produce interesting distributions of outcomes. You have two balanced, six-sided dice. One is a standard die, with faces having 1, 2, 3, 4, 5, and 6 spots. The other die has three faces with 0 spots and three faces with 6 spots. Find the probability distribution for the total number of spots Y on the up-faces when you roll these two dice.

4.50   Role-playing games like Dungeons & Dragons use many different types of dice, usually having either 4, 6, 8, 10, 12, or 20 sides. Roll a balanced 8-sided die and a balanced 6-sided die and add the spots on the up-faces. Call the sum X. What is the probability distribution of the random variable X?

4.51   The Census Bureau reports that 27% of California residents are foreign-born. Suppose that you choose three Californians at random, so that each has probability 0.27 of being foreign-born and the three are independent of each other. Let the random variable W be the number of foreign-born people you chose.

(a) What are the possible values of W?

(b) Look at your three people in order. There are eight possible arrangements of foreign (F) and domestic (D) birth. For example, FFD means the first two are foreign-born and the third is not. What is the probability of each arrangement? (Use the multiplication rule.)

(c) What is the value of W for each arrangement in (b)? What is the probability of each possible value of W? (This is the distribution of a Yes/No response for an SRS of size 3. In principle the same idea works for an SRS of any size.)

4.52   Weary of the low turnout in student elections, a college administration decides to choose an SRS of three students to form an advisory board that represents student opinion. Suppose that 40% of all students oppose the use of student fees to fund student interest groups, and that the opinions of the three students on the board are independent. Then the probability is 0.4 that each opposes the funding of interest groups.

(a) Call the three students A, B, and C. What is the probability that A and B support funding and C opposes it?

(b) List all possible combinations of opinions that can be held by students A, B, and C. (*Hint:* There are eight possibilities.) Then give the probability of each of these outcomes. Note that they are not equally likely.

(c) Let the random variable X be the number of student representatives who oppose the funding of interest groups. Give the probability distribution of X.

(d) Express the event "a majority of the advisory board opposes funding" in terms of X and find its probability.

4.53   Let X be a random number between 0 and 1 produced by the idealized uniform random number generator described in Example 4.18 and Figure 4.9. Find the following probabilities:

(a) $P(X < 0.5)$

(b) $P(X \leq 0.5)$

(c) What important fact about continuous random variables does comparing your answers to (a) and (b) illustrate?

4.54   Let the random variable X be a random number with the uniform density curve in Figure 4.9. Find the following probabilities:

(a) $P(X \geq 0.27)$

(b) $P(X = 0.27)$

(c) $P(0.27 < X < 1.27)$

(d) $P(0.1 \le X \le 0.2$ or $0.8 \le X \le 0.9)$

(e) The probability that X is not in the interval 0.3 to 0.8.

4.55   Many random number generators allow users to specify the range of the random numbers to be produced. Suppose that you specify that the range is to be all numbers between 0 and 2. Call the random number generated Y. Then the density curve of the random variable Y has constant height between 0 and 2, and height 0 elsewhere.

(a) What is the height of the density curve between 0 and 2? Draw a graph of the density curve.

(b) Use your graph from (a) and the fact that probability is area under the curve to find $P(Y \le 1)$.

(c) Find $P(0.5 < Y < 1.3)$.

(d) Find $P(Y \ge 0.8)$.

4.56   Generate *two* random numbers between 0 and 1 and take Y to be their sum. Then Y is a continuous random variable that can take any value between 0 and 2. The density curve of Y is the triangle shown in Figure 4.12.

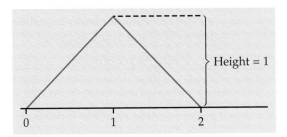

FIGURE 4.12   The density curve for the sum Y of two random numbers, for Exercise 4.56.

(a) Verify by geometry that the area under this curve is 1.

(b) What is the probability that Y is less than 1? (Sketch the density curve, shade the area that represents the probability, then find that area. Do this for (c) also.)

(c) What is the probability that Y is less than 0.5?

4.57   How many close friends do you have? Suppose that the number of close friends adults claim to have varies from person to person with mean $\mu = 9$ and standard deviation $\sigma = 2.5$. An opinion poll asks this question of an SRS of 1100 adults. We will see in the next chapter that in this situation the sample mean response $\bar{x}$ has approximately the normal distribution with mean 9 and standard deviation 0.075. What is $P(8 \le \bar{x} \le 10)$, the probability that the statistic $\bar{x}$ estimates the parameter μ to within ± 1?

4.58   A sample survey contacted an SRS of 663 registered voters in Oregon shortly after an election and asked respondents whether they had voted. Voter records show that 56% of registered voters had actually voted. We will see in the next chapter that in this situation the proportion $\hat{p}$ of the sample who voted has approximately the normal distribution with mean $\mu = 0.56$ and standard deviation $\sigma = 0.019$.

(a) If the respondents answer truthfully, what is $P(0.52 \leq \hat{p} \leq 0.60)$? This is the probability that the statistic $\hat{p}$ estimates the parameter 0.56 within plus or minus 0.04.

(b) In fact, 72% of the respondents said they had voted ($\hat{p} = 0.72$). If respondents answer truthfully, what is $P(\hat{p} \geq 0.72)$? This probability is so small that it is good evidence that some people who did not vote claimed that they did vote.

4.4   Means and Variances of Random Variables

The probability histograms and density curves that picture the probability distributions of random variables resemble our earlier pictures of distributions of data. In describing data, we moved from graphs to numerical measures such as means and standard deviations. Now we will make the same move to expand our descriptions of the distributions of random variables. We can speak of the mean winnings in a game of chance or the standard deviation of the randomly varying number of calls a travel agency receives in an hour. In this section we will learn more about how to compute these descriptive measures and about the laws they obey.

The mean of a random variable

The mean $\bar{x}$ of a set of observations is their ordinary average. The mean of a random variable X is also an average of the possible values of X, but with an essential change to take into account the fact that not all outcomes need be equally likely. An example will show what we must do.

EXAMPLE 4.20   Most states and Canadian provinces have government-sponsored lotteries. Here is a simple lottery wager, from the Tri-State Pick 3 game that New Hampshire shares with Maine and Vermont. You choose a three-digit number, 000 to 999. The state chooses a three-digit winning number at random and pays you $500 if your number is chosen. Because there are 1000 three-digit numbers, you have probability 1/1000 of winning. Taking X to be the amount your ticket pays you, the probability distribution of X is

Payoff X	$0	$500
Probability	0.999	0.001

What is your average payoff from many tickets? The ordinary average of the two possible outcomes $0 and $500 is $250, but that makes no sense as the average because $500 is much less likely than $0. In the long run you receive $500 once in every 1000 tickets and $0 on the remaining 999 of 1000 tickets. The long-run average payoff is

$$\$500\frac{1}{1000} + \$0\frac{999}{1000} = \$0.50$$

or fifty cents. That number is the mean of the random variable X. (Tickets cost $1, so in the long run the state keeps half the money you wager.)

If you play Tri-State Pick 3 several times, we would as usual call the mean of the actual amounts you win $\bar{x}$. The mean in Example 4.20 is a different quantity—it is the long-run average winnings you expect if you play a very large number of times. Just as probabilities are an idealized description of long-run proportions, the mean of a probability distribution describes the long-run average outcome. We can't call this mean $\bar{x}$, so we need a different symbol. The common symbol for the **mean of a probability distribution** is μ, the Greek letter mu. We used μ in Chapter 1 for the mean of a normal distribution, so this is not a new notation. We will often be interested in several random variables, each having a different probability distribution with a different mean. To remind ourselves that we are talking about the mean of X, we often write μ_X rather than simply μ. In Example 4.20, $\mu_X = \$0.50$. Notice that, as often happens, the mean is not a possible value of X. You will often find the mean of a random variable X called the **expected value** of X. This term can be misleading, for we don't necessarily expect one observation on X to be close to its expected value.

The mean of any discrete random variable is found just as in Example 4.20. It is an average of the possible outcomes, but a weighted average in which each outcome is weighted by its probability. Because the probabilities add to 1, we have total weight 1 to distribute among the outcomes. An outcome that occurs half the time has probability one-half and gets one-half the weight in calculating the mean. Here is the general definition.

mean μ

expected value

MEAN OF A DISCRETE RANDOM VARIABLE

Suppose that X is a discrete random variable whose distribution is

Value of X	x_1	x_2	x_3	$\cdots$	x_k
Probability	p_1	p_2	p_3	$\cdots$	p_k

To find the **mean** of X, multiply each possible value by its probability, then add all the products:

$$\mu_X = x_1p_1 + x_2p_2 + \cdots + x_kp_k$$
$$= \sum x_ip_i$$

EXAMPLE 4.21 If first digits in a set of data appear "at random," the nine possible digits 1 to 9 all have the same probability. The probability distribution of the first digit X is then

First digit X	1	2	3	4	5	6	7	8	9
Probability	1/9	1/9	1/9	1/9	1/9	1/9	1/9	1/9	1/9

The mean of this distribution is

$$\mu_X = 1 \times \frac{1}{9} + 2 \times \frac{1}{9} + 3 \times \frac{1}{9} + 4 \times \frac{1}{9} + 5 \times \frac{1}{9} + 6 \times \frac{1}{9} + 7 \times \frac{1}{9} + 8 \times \frac{1}{9} + 9 \times \frac{1}{9}$$

$$= 45 \times \frac{1}{9} = 5$$

If, on the other hand, the data obey Benford's law, the distribution of the first digit V is

First digit V	1	2	3	4	5	6	7	8	9
Probability	0.301	0.176	0.125	0.097	0.079	0.067	0.058	0.051	0.046

The mean of V is

$$\mu_V = (1)(0.301) + (2)(0.176) + (3)(0.125) + (4)(0.097) + (5)(0.079) + (6)(0.067) +$$
$$(7)(0.058) + (8)(0.051) + (9)(0.046)$$
$$= 3.441$$

The means reflect the greater probability of smaller first digits under Benford's law.

Figure 4.13 locates the means of X and V on the two probability histograms. Because the discrete uniform distribution of Figure 4.13(a) is symmetric, the mean lies at the center of symmetry. We can't locate the mean of the right-skewed distribution of Figure 4.13(b) by eye—calculation is needed.

What about continuous random variables? The probability distribution of a continuous random variable X is described by a density curve. Chapter 1 (page 68) showed how to find the mean of the distribution: it is the point at which the area under the density curve would balance if it were made out of solid material. The mean lies at the center of symmetric density curves such as the normal curves. Exact calculation of the mean of a distribution with a skewed density curve requires advanced mathematics.[17] The idea that the mean is the balance point of the distribution applies to discrete random variables as well, but in the discrete case we have a formula that gives us this point.

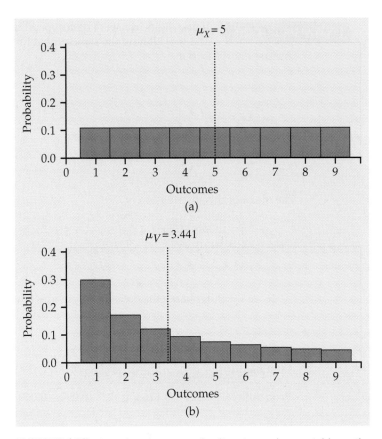

FIGURE 4.13   Locating the mean of a discrete random variable on the probability histogram for (a) digits between 1 and 9 chosen at random; (b) digits between 1 and 9 chosen from records that obey Benford's law.

Statistical estimation and the law of large numbers

We would like to estimate the mean height μ of the population of all American women between the ages of 18 and 24 years. This μ is the mean μ_X of the random variable X obtained by choosing a young woman at random and measuring her height. To estimate μ, we choose an SRS of young women and use the sample mean $\bar{x}$ to estimate the unknown population mean μ. In the language of Section 3.4 (page 232), μ is a *parameter* and $\bar{x}$ is a *statistic*. Statistics obtained from probability samples are random variables, because their values would vary in repeated sampling. The sampling distributions of statistics are just the probability distributions of these random variables.

It seems reasonable to use $\bar{x}$ to estimate μ. An SRS should fairly represent the population, so the mean $\bar{x}$ of the sample should be somewhere near the mean μ of the population. Of course, we don't expect $\bar{x}$ to be exactly equal to μ, and we realize that if we choose another SRS, the luck of the draw will probably produce a different $\bar{x}$.

If $\bar{x}$ is rarely exactly right and varies from sample to sample, why is it nonetheless a reasonable estimate of the population mean μ? We gave one answer in Section 3.4: $\bar{x}$ is unbiased and we can control its variability by choosing the sample size. Here is another answer: if we keep on adding observations to our random sample, the statistic $\bar{x}$ is *guaranteed* to get as close as we wish to the parameter μ and then stay that close. We have the comfort of knowing that if we can afford to keep on measuring more women, eventually we will estimate the mean height of all young women very accurately. This remarkable fact is called the *law of large numbers*. It is remarkable because it holds for *any* population, not just for some special class such as normal distributions.

LAW OF LARGE NUMBERS

Draw independent observations at random from any population with finite mean μ. Decide how accurately you would like to estimate μ. As the number of observations drawn increases, the mean $\bar{x}$ of the observed values eventually approaches the mean μ of the population as closely as you specified and then stays that close.

The behavior of $\bar{x}$ is similar to the idea of probability. In the long run, the *proportion* of outcomes taking any value gets close to the *probability* of that value, and the *average outcome* gets close to the distribution *mean*. Figure 4.1 (page 254) shows how proportions approach probability in one example. Here is an example of how sample means approach the distribution mean.

EXAMPLE 4.22   The distribution of the heights of all young women is close to the normal distribution with mean 64.5 inches and standard deviation 2.5 inches. Suppose that $\mu = 64.5$ were exactly true. Figure 4.14 shows the behavior of the mean height $\bar{x}$ of n women chosen at random from a population whose heights

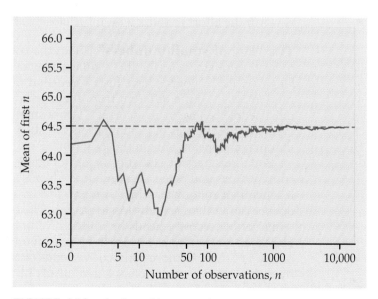

FIGURE 4.14   The law of large numbers in action. As we take more observations, the sample mean $\bar{x}$ always approaches the mean μ of the population.

follow the $N(64.5, 2.5)$ distribution. The graph plots the values of $\bar{x}$ as we add women to our sample. The first woman drawn had height 64.21 inches, so the line starts there. The second had height 64.35 inches, so for $n = 2$ the mean is

$$\bar{x} = \frac{64.21 + 64.35}{2} = 64.28$$

This is the second point on the line in the graph.

At first, the graph shows that the mean of the sample changes as we take more observations. Eventually, however, the mean of the observations gets close to the population mean $\mu = 64.5$ and settles down at that value. The law of large numbers says that this *always* happens.

The *Law of Large Numbers* applet on the text CD and Web site animates a graph like Figure 4.14. Use it to better understand the law of large numbers.

The mean μ of a random variable is the average value of the variable in two senses. By its definition, μ is the average of the possible values, weighted by their probability of occurring. The law of large numbers says that μ is also the long-run average of many independent observations on the variable. The law of large numbers can be proved mathematically starting from the basic laws of probability.

Thinking about the law of large numbers

The law of large numbers says broadly that the average results of many independent observations are stable and predictable. The gamblers in a casino may win or lose, but the casino will win in the long run because the law of large numbers says what the average outcome of many thousands of bets will be. An insurance company deciding how much to charge for life insurance and a fast-food restaurant deciding how many beef patties to prepare also rely on the fact that averaging over many individuals produces a stable result. It is worth the effort to think a bit more closely about so important a fact.

The "law of small numbers"   Both the rules of probability and the law of large numbers describe the regular behavior of chance phenomena *in the long run*. Psychologists have discovered that our intuitive understanding of randomness is quite different from the true laws of chance.[18] For example, most people believe in an incorrect "law of small numbers." That is, we expect even short sequences of random events to show the kind of average behavior that in fact appears only in the long run.

Some teachers of statistics begin a course by asking students to toss a coin 50 times and bring the sequence of heads and tails to the next class. The teacher then announces which students just wrote down a random-looking sequence rather than actually tossing a coin. The faked tosses don't have enough "runs" of consecutive heads or consecutive tails. Runs of the same outcome don't look random to us but are in fact common. For example, the probability of a run of three or more consecutive heads or tails in just 10 tosses is greater than 0.8.[19] The runs of consecutive heads or consecutive tails that appear in real coin tossing (and that are predicted by the mathematics of probability) seem surprising to us. Because we don't expect to see long runs, we may conclude that the coin tosses are not independent or that some influence is disturbing the random behavior of the coin.

EXAMPLE 4.23   Belief in the law of small numbers influences behavior. If a basketball player makes several consecutive shots, both the fans and her teammates believe that she has a "hot hand" and is more likely to make the next shot. This is doubtful. Careful study suggests that runs of baskets made or missed are no more frequent in basketball than would be expected if each shot were independent of the player's previous shots. Baskets made or missed are just like heads and tails in tossing a coin. (Of course, some players make 30% of their shots in the long run and others make 50%, so a coin-toss model for basketball must allow coins with different probabilities of a head.) Our perception of hot or cold streaks simply shows that we don't perceive random behavior very well.[20]

Our intuition doesn't do a good job of distinguishing random behavior from systematic influences. This is also true when we look at data. We need statistical inference to supplement exploratory analysis of data because probability calculations can help verify that what we see in the data is more than a random pattern.

How large is a large number?   The law of large numbers says that the actual mean outcome of many trials gets close to the distribution mean μ as more trials are made. It doesn't say how many trials are needed to guarantee a mean outcome close to μ. That depends on the *variability* of the random outcomes. The more variable the outcomes, the more trials are needed to ensure that the mean outcome $\bar{x}$ is close to the distribution mean μ. Casinos understand this: the outcomes of games of chance are variable enough to hold the interest of gamblers. Only the casino plays often enough to rely on the law of large numbers. Gamblers get entertainment; the casino has a business.

BEYOND THE BASICS | More laws of large numbers

The law of large numbers is one of the central facts about probability. It helps us understand the mean μ of a random variable. It explains why gambling casinos and insurance companies make money. It assures us that statistical estimation will be accurate if we can afford enough observations. The basic law of large numbers applies to independent observations that all have the same distribution. Mathematicians have extended the law to many more general settings. Here are two of these.

Is there a winning system for gambling?   Serious gamblers often follow a system of betting in which the amount bet on each play depends on the outcome of previous plays. You might, for example, double your bet on each spin of the roulette wheel until you win—or, of course, until your fortune is exhausted. Such a system tries to take advantage of the fact that you have a memory even though the roulette wheel does not. Can you beat the odds with a system based on the outcomes of past plays? No. Mathematicians have established a stronger version of the law of large numbers that says that if you do not have an infinite fortune to gamble with, your long-run average winnings μ remain the same as long as successive trials of the game (such as spins of the roulette wheel) are independent.

What if observations are not independent?   You are in charge of a process that manufactures video screens for computer monitors. Your equipment measures the tension on the metal mesh that lies behind each screen and is critical to its image quality. You want to estimate the mean tension μ for the process by the average $\bar{x}$ of the measurements. Alas, the tension measurements are not independent. If the tension on one screen is a bit too high, the tension on the next is more likely to also be high. Many real-world processes are like this—the process stays stable in the long run, but observations made close together are likely to be both above or both below the long-run mean. Again the mathematicians come to the rescue: as long as the dependence dies out fast enough as we take measurements farther and farther apart in time, the law of large numbers still holds.

Rules for means

You are studying flaws in the painted finish of refrigerators made by your firm. Dimples and paint sags are two kinds of surface flaw. Not all refrigerators have the same number of dimples: many have none, some have one, some two, and so on. You ask for the average number of imperfections on a refrigerator. The inspectors report finding an average of 0.7 dimples and 1.4 sags per refrigerator. How many total imperfections of both kinds (on the average) are there on a refrigerator? That's easy: if the average number of dimples is 0.7 and the average number of sags is 1.4, then counting both gives an average of $0.7 + 1.4 = 2.1$ flaws.

In more formal language, the number of dimples on a refrigerator is a random variable X that varies as we inspect one refrigerator after another. We know only that the mean number of dimples is $\mu_X = 0.7$. The number of paint sags is a second random variable Y having mean $\mu_Y = 1.4$. (As usual, the subscripts keep straight which variable we are talking about.) The total number of both dimples and sags is another random variable, the sum $X + Y$. Its mean μ_{X+Y} is the average number of dimples and sags together. It is just the sum of the individual means μ_X and μ_Y. That's an important rule for how means of random variables behave.

Here's another rule. The crickets living in a field have mean length 1.2 inches. What is the mean in centimeters? There are 2.54 centimeters in an inch, so the length of a cricket in centimeters is 2.54 times its length in inches. If we multiply every observation by 2.54, we also multiply their average by 2.54. The mean in centimeters must be 2.54×1.2, or about 3.05 centimeters. More formally, the length in inches of a cricket chosen at random from the field is a random variable X with mean μ_X. The length in centimeters is $2.54X$, and this new random variable has mean $2.54\mu_X$.

The point of these examples is that means behave like averages. Here are the rules we need.

RULES FOR MEANS

Rule 1. If X is a random variable and a and b are fixed numbers, then

$$\mu_{a+bX} = a + b\mu_X$$

Rule 2. If X and Y are random variables, then

$$\mu_{X+Y} = \mu_X + \mu_Y$$

EXAMPLE 4.24   Linda is a sales associate at a large auto dealership. At her commission rate of 25% of gross profit on each vehicle she sells, Linda expects to earn $350 for each car sold and $400 for each truck or SUV sold. Linda motivates herself by using probability estimates of her sales. For a sunny Saturday in April, she estimates her car sales as follows:

Cars sold	0	1	2	3
Probability	0.3	0.4	0.2	0.1

Linda's estimate of her truck or SUV sales is

Vehicles sold	0	1	2
Probability	0.4	0.5	0.1

Take X to be the number of cars Linda sells and Y the number of trucks or SUVs. The means of these random variables are

$$\mu_X = (0)(0.3) + (1)(0.4) + (2)(0.2) + (3)(0.1)$$
$$= 1.1 \text{ cars}$$
$$\mu_Y = (0)(0.4) + (1)(0.5) + (2)(0.1)$$
$$= 0.7 \text{ trucks or SUVs}$$

Linda's earnings, at $350 per car and $400 per truck or SUV, are

$$Z = 350X + 400Y$$

Combining Rules 1 and 2, her mean earnings are

$$\mu_Z = 350\mu_X + 400\mu_Y$$
$$= (350)(1.1) + (400)(0.7) = \$665$$

This is Linda's best estimate of her earnings for the day. It's a bit unusual for individuals to use probability estimates, but they are a common tool for business planners.

---

personal probability   The probabilities in Example 4.24 are **personal probabilities** that describe Linda's informed opinion about her sales in the coming weekend. Although personal probabilities need not be based on observing many repetitions of a random phenomenon, they must obey the rules of probability if they are to make sense. Personal probability extends the usefulness of probability models to one-time events, but remember that they are subject to the follies of human opinion. Overoptimism is common: 40% of college students think that they will eventually reach the top 1% in income.

The variance of a random variable

The mean is a measure of the center of a distribution. A basic numerical description requires in addition a measure of the spread or variability of the distribution. The variance and the standard deviation are the measures of spread that accompany the choice of the mean to measure center. Just as for the mean, we need a distinct symbol to distinguish the variance of a random variable from the variance s^2 of a data set. We write the variance of a random variable X as σ_X^2. Once again the subscript reminds us which variable we have in mind. The definition of the variance σ_X^2 of a random variable is similar to the definition of the sample variance s^2 given in Chapter 1. That is, the variance is an average value of the squared deviation $(X - \mu_X)^2$ of the variable X from its mean μ_X. As for the mean, the average we use is a weighted average in which each outcome is weighted by its probability in order to take account of outcomes that are not equally likely. Calculating this weighted average is straightforward for discrete random variables but requires advanced mathematics in the continuous case. Here is the definition.

> **VARIANCE OF A DISCRETE RANDOM VARIABLE**
>
> Suppose that X is a discrete random variable whose distribution is
>
> | Value of X | x_1 | x_2 | x_3 | $\cdots$ | x_k |
> |---|---|---|---|---|---|
> | Probability | p_1 | p_2 | p_3 | $\cdots$ | p_k |
>
> and that μ_X is the mean of X. The **variance** of X is
>
> $$\sigma_X^2 = (x_1 - \mu_X)^2 p_1 + (x_2 - \mu_X)^2 p_2 + \cdots + (x_k - \mu_X)^2 p_k$$
> $$= \sum (x_i - \mu_X)^2 p_i$$
>
> The **standard deviation** σ_X of X is the square root of the variance.

EXAMPLE 4.25 In Example 4.24 we saw that the number X of cars that Linda hopes to sell has distribution

Cars sold	0	1	2	3
Probability	0.3	0.4	0.2	0.1

We can find the mean and variance of X by arranging the calculation in the form of a table. Both μ_X and σ_X^2 are sums of columns in this table.

x_i	p_i	$x_i p_i$	$(x_i - \mu_X)^2 p_i$
0	0.3	0.0	$(0 - 1.1)^2(0.3) = 0.363$
1	0.4	0.4	$(1 - 1.1)^2(0.4) = 0.004$
2	0.2	0.4	$(2 - 1.1)^2(0.2) = 0.162$
3	0.1	0.3	$(3 - 1.1)^2(0.1) = 0.361$

$$\mu_X = 1.1 \qquad\qquad \sigma_X^2 = 0.890$$

We see that $\sigma_X^2 = 0.89$. The standard deviation of X is $\sigma_X = \sqrt{0.89} = 0.943$. The standard deviation is a measure of the variability of the number of cars Linda sells. As in the case of distributions for data, the standard deviation of a probability distribution is easiest to understand for normal distributions.

Rules for variances

What are the facts for variances that parallel Rules 1 and 2 for means? *The mean of a sum of random variables is always the sum of their means, but this addition rule is true for variances only in special situations.* To understand why, take X to be the percent of a family's after-tax income that is spent and Y the percent that is saved. When X increases, Y decreases by the same amount. Though X and Y may vary widely from year to year, their sum $X + Y$ is always 100% and does not vary at all. It is the association between the variables X and Y that prevents their variances from adding. If random variables are independent, this kind of association between their values is ruled out and their variances do add. Two random variables X and Y are **independent** if knowing that any event involving X alone did or did not occur tells us nothing about the occurrence of any event involving Y alone. Probability models often assume independence when the random variables describe outcomes that appear unrelated to each other. You should ask in each instance whether the assumption of independence seems reasonable.

independence

When random variables are not independent, the variance of their sum depends on the **correlation** between them as well as on their individual variances. In Chapter 2, we met the correlation r between two observed variables measured on the same individuals. We defined (page 124) the correlation r as an average of the products of the standardized x and y observations. The correlation between two random variables is defined in the same way, once again using a weighted average with probabilities as weights. We won't give the details—it is enough to know that the correlation between two random variables has the same basic properties as the correlation r calculated from data. We use ρ, the Greek letter rho, for the correlation between two random variables. The correlation ρ is a number between -1 and 1 that measures the direction and strength of the linear relationship between two variables. **The correlation between two independent random variables is zero.**

correlation

Returning to family finances, if X is the percent of a family's after-tax income that is spent and Y the percent that is saved, then $Y = 100 - X$. This is a perfect linear relationship with a negative slope, so the correlation between X and Y is $\rho = -1$. With the correlation at hand, we can state the rules for manipulating variances.

> ## RULES FOR VARIANCES
>
> **Rule 1.** If X is a random variable and a and b are fixed numbers, then
>
> $$\sigma^2_{a+bX} = b^2\sigma^2_X$$
>
> **Rule 2.** If X and Y are independent random variables, then
>
> $$\sigma^2_{X+Y} = \sigma^2_X + \sigma^2_Y$$
>
> $$\sigma^2_{X-Y} = \sigma^2_X + \sigma^2_Y$$
>
> This is the **addition rule for variances of independent random variables.**
>
> **Rule 3.** If X and Y have correlation ρ, then
>
> $$\sigma^2_{X+Y} = \sigma^2_X + \sigma^2_Y + 2\rho\sigma_X\sigma_Y$$
>
> $$\sigma^2_{X-Y} = \sigma^2_X + \sigma^2_Y - 2\rho\sigma_X\sigma_Y$$
>
> This is the **general addition rule for variances of random variables**.

Because a variance is the average of squared deviations from the mean, multiplying X by a constant b multiplies σ^2_X by the square of the constant. Adding a constant a to a random variable changes its mean but does not change its variability. The variance of $X + a$ is therefore the same as the variance of X. Because the square of -1 is 1, the addition rule says that the variance of a difference of independent random variables is the *sum* of the variances. For independent random variables, the difference $X - Y$ is more variable than either X or Y alone because variations in both X and Y contribute to variation in their difference.

As with data, we prefer the standard deviation to the variance as a measure of the variability of a random variable. *Rule 2 for variances implies that standard deviations of independent random variables do not add. To combine standard deviations, use the rules for variances.* For example, the standard deviations of $2X$ and $-2X$ are both equal to $2\sigma_X$ because this is the square root of the variance $4\sigma^2_X$.

EXAMPLE 4.26   The payoff X of a \$1 ticket in the Tri-State Pick 3 game is \$500 with probability 1/1000 and 0 the rest of the time. Here is the combined calculation of mean and variance:

x_i	p_i	x_ip_i	$(x_i - \mu_X)^2 p_i$	
0	0.999	0	$(0 - 0.5)^2(0.999) =$	0.24975
500	0.001	0.5	$(500 - 0.5)^2(0.001) =$	249.50025
		$\mu_X = 0.5$	$\sigma^2_X = 249.75$	

The mean payoff is 50 cents. The standard deviation is $\sigma_X = \sqrt{249.75} = \$15.80$. It is usual for games of chance to have large standard deviations because large variability makes gambling exciting.

If you buy a Pick 3 ticket, your winnings are $W = X - 1$ because the dollar you paid for the ticket must be subtracted from the payoff. By the rules for means, the mean amount you win is

$$\mu_W = \mu_X - 1 = -\$0.50$$

That is, you lose an average of 50 cents on a ticket. The rules for variances remind us that the variance and standard deviation of the winnings $W = X - 1$ are the same as those of X. Subtracting a fixed number changes the mean but not the variance.

Suppose now that you buy a \$1 ticket on each of two different days. The payoffs X and Y on the two tickets are independent because separate drawings are held each day. Your total payoff $X + Y$ has mean

$$\mu_{X+Y} = \mu_X + \mu_Y = \$0.50 + \$0.50 = \$1.00$$

Because X and Y are independent, the variance of $X + Y$ is

$$\sigma^2_{X+Y} = \sigma^2_X + \sigma^2_Y = 249.75 + 249.75 = 499.5$$

The standard deviation of the total payoff is

$$\sigma_{X+Y} = \sqrt{499.5} = \$22.35$$

This is not the same as the sum of the individual standard deviations, which is $\$15.80 + \$15.80 = \$31.60$. Variances of independent random variables add; standard deviations do not.

EXAMPLE 4.27   Scores on the mathematics part of the SAT college entrance exam in a recent year had mean 519 and standard deviation 115. Scores on the verbal part of the SAT had mean 507 and standard deviation 111. What are the mean and standard deviation of total SAT score?[21]

Think of choosing one student's scores at random. Expressed in the language of random variables,

$$\text{SAT math score } X \qquad \mu_X = 519 \qquad \sigma_X = 115$$
$$\text{SAT verbal score } Y \qquad \mu_Y = 507 \qquad \sigma_Y = 111$$

The total score is $X + Y$. The mean is easy:

$$\mu_{X+Y} = \mu_X + \mu_Y = 519 + 507 = 1026$$

The variance and standard deviation of the total *cannot be computed* from the information given. SAT verbal and math scores are not independent, because students who score high on one exam tend to score high on the other also. Therefore, Rule 2 does not apply and we need to know ρ, the correlation between X and Y, to apply Rule 3.

The correlation between SAT math and verbal scores was $\rho = 0.71$. By Rule 3,

$$\sigma^2_{X+Y} = \sigma^2_X + \sigma^2_Y + 2\rho\sigma_X\sigma_Y$$
$$= (115)^2 + (111)^2 + (2)(0.71)(115)(111)$$
$$= 43,672$$

The variance of the sum $X + Y$ is greater than the sum of the variances $\sigma^2_X + \sigma^2_Y$ because of the positive correlation between SAT math scores and SAT verbal scores.

That is, X and Y tend to move up together and down together, which increases the variability of their sum. Find the standard deviation from the variance,

$$\sigma_{X+Y} = \sqrt{43{,}672} = 209$$

Total SAT scores had mean 1026 and standard deviation 209.

EXAMPLE 4.28   Zadie has invested 20% of her funds in Treasury bills and 80% in an "index fund" that represents all U.S. common stocks. The rate of return of an investment over a time period is the percent change in the price during the time period, plus any income received. If X is the annual return on T-bills and Y the annual return on stocks, the portfolio rate of return is

$$R = 0.2X + 0.8Y$$

The returns X and Y are random variables because they vary from year to year. Based on annual returns between 1950 and 2003, we have

X = annual return on T-bills	$\mu_X = 5.0\%$	$\sigma_X = 2.9\%$
Y = annual return on stocks	$\mu_Y = 13.2\%$	$\sigma_Y = 17.6\%$
Correlation between X and Y	$\rho = -0.11$	

Stocks had higher returns than T-bills on the average, but the standard deviations show that returns on stocks varied much more from year to year. That is, the risk of investing in stocks is greater than the risk for T-bills because their returns are less predictable. For the return R on Zadie's portfolio of 20% T-bills and 80% stocks,

$$R = 0.2X + 0.8Y$$

$$\mu_R = 0.2\mu_X + 0.8\mu_Y$$

$$= (0.2 \times 5.0) + (0.8 \times 13.2) = 11.56\%$$

To find the variance of the portfolio return, combine Rules 1 and 3. Use the fact that, for example, the variance of $0.2X$ is $(0.2)^2$ times the variance of X. Also use the fact that changing scales does not change the correlation, so that the correlation between $0.2X$ and $0.8Y$ is the same as the correlation between X and Y.

$$\sigma_R^2 = \sigma_{0.2X}^2 + \sigma_{0.8Y}^2 + 2\rho\sigma_{0.2X}\sigma_{0.8Y}$$

$$= (0.2)^2\sigma_X^2 + (0.8)^2\sigma_Y^2 + 2\rho(0.2 \times \sigma_X)(0.8 \times \sigma_Y)$$

$$= (0.2)^2(2.9)^2 + (0.8)^2(17.6)^2 + (2)(-0.11)(0.2 \times 2.9)(0.8 \times 17.6)$$

$$= 196.786$$

$$\sigma_R = \sqrt{196.786} = 14.03\%$$

The portfolio has a smaller mean return than an all-stock portfolio, but it is also less risky. That's why Zadie put some funds into Treasury bills.

SECTION 4.4 | Summary

The probability distribution of a random variable X, like a distribution of data, has a **mean μ_X** and a **standard deviation σ_X.**

The **law of large numbers** says that the average of the values of X observed in many trials must approach μ.

The **mean** μ is the balance point of the probability histogram or density curve. If X is discrete with possible values x_i having probabilities p_i, the mean is the average of the values of X, each weighted by its probability:

$$\mu_X = x_1 p_1 + x_2 p_2 + \cdots + x_k p_k$$

The **variance** σ_X^2 is the average squared deviation of the values of the variable from their mean. For a discrete random variable,

$$\sigma_X^2 = (x_1 - \mu)^2 p_1 + (x_2 - \mu)^2 p_2 + \cdots + (x_k - \mu)^2 p_k$$

The **standard deviation** σ_X is the square root of the variance. The standard deviation measures the variability of the distribution about the mean. It is easiest to interpret for normal distributions.

The mean and variance of a continuous random variable can be computed from the density curve, but to do so requires more advanced mathematics.

The means and variances of random variables obey the following rules. If a and b are fixed numbers, then

$$\mu_{a+bX} = a + b\mu_X$$
$$\sigma_{a+bX}^2 = b^2 \sigma_X^2$$

If X and Y are any two random variables having correlation ρ, then

$$\mu_{X+Y} = \mu_X + \mu_Y$$
$$\sigma_{X+Y}^2 = \sigma_X^2 + \sigma_Y^2 + 2\rho\sigma_X\sigma_Y$$
$$\sigma_{X-Y}^2 = \sigma_X^2 + \sigma_Y^2 - 2\rho\sigma_X\sigma_Y$$

If X and Y are **independent,** then $\rho = 0$. In this case,

$$\sigma_{X+Y}^2 = \sigma_X^2 + \sigma_Y^2$$
$$\sigma_{X-Y}^2 = \sigma_X^2 + \sigma_Y^2$$

SECTION 4.4 | Exercises

4.59 Example 4.16 gives the distribution of grades (A = 4, B = 3, and so on) in Statistics 101 at North Carolina State University as

Value of X	0	1	2	3	4
Probability	0.01	0.05	0.30	0.43	0.21

Find the average (that is, the mean) grade in this course.

4.60 Typographical and spelling errors can be either "nonword errors" or "word errors." A nonword error is not a real word, as when "the" is typed as "teh." A word error is a real word, but not the right word, as when "lose" is typed as "loose." When undergraduates are asked to write a 250-word essay (without spell-checking), the number of nonword errors has the following distribution:

Errors	0	1	2	3	4
Probability	0.1	0.2	0.3	0.3	0.1

The number of word errors has this distribution:

Errors	0	1	2	3
Probability	0.4	0.3	0.2	0.1

What are the mean numbers of nonword errors and word errors in an essay?

4.61 How do rented housing units differ from units occupied by their owners? Exercise 4.43 (page 287) gives the distributions of the number of rooms for owner-occupied units and renter-occupied units in San Jose, California. Find the mean number of rooms for both types of housing unit. How do the means reflect the differences between the distributions that you found in Exercise 4.43?

4.62 Figure 4.12 (page 290) displays the density curve of the sum $Y = X_1 + X_2$ of two independent random numbers, each uniformly distributed between 0 and 1.

(a) The mean of a continuous random variable is the balance point of its density curve. Use this fact to find the mean of Y from Figure 4.12.

(b) Use the same fact to find the means of X_1 and X_2. (They have the density curve pictured in Figure 4.9 on page 283.) Verify that the mean of Y is the sum of the mean of X_1 and the mean of X_2.

4.63 Which of the two distributions of room counts appears more spread out in the probability histograms you made in Exercise 4.43 (page 287)? Why? Find the standard deviation for both distributions. The standard deviation provides a numerical measure of spread.

4.64 Find the mean and standard deviation of the total number of errors (nonword errors plus word errors) in an essay if the error counts have the distributions given in Exercise 4.60 and

(a) The counts of nonword and word errors are independent.

(b) Students who make many nonword errors also tend to make many word errors, so that the correlation between the two error counts is 0.5.

4.65 You buy a hot stock for $1000. The stock either gains 30% or loses 25% each day, each with probability 0.5. Its returns on consecutive days are independent of each other. You plan to sell the stock after two days.

(a) What are the possible values of the stock after two days, and what is the probability for each value? What is the probability that the stock is worth more after two days than the $1000 you paid for it?

(b) What is the mean value of the stock after two days? You see that these two criteria give different answers to the question, "Should I invest?"

4.66 To use Benford's law (Example 4.9, page 264) to detect faked invoice amounts, compare the distribution of first digits among the invoices with the Benford's law distribution. The process is more effective when you have very many invoices from the vendor in question, rather than just a few. Explain why this is true.

4.67 The rules for means and variances allow you to find the mean and variance of a sum of random variables without first finding the distribution of the sum, which is usually much harder to do.

(a) A single toss of a balanced coin has either 0 or 1 head, each with probability 1/2. What are the mean and standard deviation of the number of heads?

(b) Toss a coin four times. Use the rules for means and variances to find the mean and standard deviation of the total number of heads.

(c) Example 4.17 (page 280) finds the distribution of the number of heads in four tosses. Find the mean and standard deviation from this distribution. Your results in (b) and (c) should agree.

4.68 In Exercise 4.50 (page 288) you found the distribution of the intelligence of a character in a role-playing game, assigned by adding the spots on a roll of an 8-sided and a 6-sided die. The outcomes of the two dice are independent. Use this fact to answer these questions: What is the average (mean) intelligence for such characters? How spread out are their intelligences, as measured by the standard deviation of the distribution? (Note that you don't need the distribution from Exercise 4.40 to answer these questions.)

4.69 A mechanical assembly (Figure 4.15) consists of a rod with a bearing on each end. The three parts are manufactured independently, and all vary a bit from part to part. The length of the rod has mean 10 centimeters (cm) and standard deviation 0.005 millimeters (mm). The length of a bearing has mean 2 cm and standard deviation 0.001 mm. What are the mean and standard deviation of the total length of the assembly?

Bearing   Rod   Bearing

FIGURE 4.15 Sketch of a mechanical assembly, for Exercise 4.69.

4.70 Continue your work in the previous exercise. Dimensions of mechanical parts are often roughly normal. According to the 68–95–99.7 rule, 95% of rods have lengths within $\pm d_1$ of 10 cm and 95% of bearings have lengths within $\pm d_2$ of 2 cm.

(a) What are the values of d_1 and d_2? These are often called the "natural tolerances" of the parts.

(b) Statistical theory says that any sum of independent normal random variables has a normal distribution. So the total length of the assembly is roughly normal. What is the natural tolerance for the total length? It is *not* $d_1 + 2d_2$, because standard deviations don't add.

4.71 In which of the following games of chance would you be willing to assume independence of X and Y in making a probability model? Explain your answer in each case.

(a) In blackjack, you are dealt two cards and examine the total points X on the cards (face cards count 10 points). You can choose to be dealt another card and compete based on the total points Y on all three cards.

(b) In craps, the betting is based on successive rolls of two dice. X is the sum of the faces on the first roll, and Y the sum of the faces on the next roll.

4.72 For each of the following situations, would you expect the random variables X and Y to be independent? Explain your answers.

(a) X is the rainfall (in inches) on November 6 of this year, and Y is the rainfall at the same location on November 6 of next year.

(b) X is the amount of rainfall today, and Y is the rainfall at the same location tomorrow.

(c) X is today's rainfall at the Orlando, Florida, airport, and Y is today's rainfall at Disney World just outside Orlando.

4.73 A time and motion study measures the time required for an assembly-line worker to perform a repetitive task. The data show that the time required to bring a part from a bin to its position on an automobile chassis varies from car to car with mean 11 seconds and standard deviation 2 seconds. The time required to attach the part to the chassis varies with mean 20 seconds and standard deviation 8 seconds.

(a) What is the mean time required for the entire operation of positioning and attaching the part?

(b) Industry quality programs strive to reduce variation in processes. A training program reduces the standard deviation for attaching the part from 8 seconds to 4 seconds. Will this change your result in (a) if the mean times don't change? Why is reducing the variation nonetheless worthwhile to the automaker?

(c) The study finds that the times required for the two steps are independent. A part that takes a long time to position, for example, does not take more or less time to attach than other parts. How would your answer in (a) change if the two variables were dependent with correlation 0.3?

4.74   A report of the National Center for Health Statistics says that the heights of 20-year-old men have mean 176.8 centimeters (cm) and standard deviation 7.2 cm. There are 2.54 centimeters in an inch. What are the mean and standard deviation in inches?

4.75   Find the standard deviation of the time required for the two-step assembly operation studied in Exercise 4.73, assuming that the study shows the two times to be independent. Redo the calculation assuming that the two times are dependent, with correlation 0.3. Explain in nontechnical language why positive correlation increases the variability of the total time.

4.76   You have two instruments with which to measure the height of a tower. If the true height is 100 meters, measurements with the first instrument vary with mean 100 meters and standard deviation 1.2 meters. Measurements with the second instrument vary with mean 100 meters and standard deviation 0.85 meter. You make one measurement with each instrument. Your results are X_1 for the first and X_2 for the second, and are independent.

(a) To combine the two measurements, you might average them, $Y = (X_1 + X_2)/2$. What are the mean and standard deviation of Y?

(b) It makes sense to give more weight to the less variable measurement because it is more likely to be close to the truth. Statistical theory says that to make the standard deviation as small as possible you should weight the two measurements inversely proportional to their variances. The variance of X_2 is very close to half the variance of X_1, so X_2 should get twice the weight of X_1. That is, use

$$W = \frac{1}{3}X_1 + \frac{2}{3}X_2$$

What are the mean and standard deviation of W?

4.77   We know that variances add if the random variables involved are uncorrelated ($\rho = 0$), but not otherwise. The opposite extreme is perfect positive correlation ($\rho = 1$). Show by using the general addition rule for variances that in this case the standard deviations add. That is, $\sigma_{X+Y} = \sigma_X + \sigma_Y$ if $\rho_{XY} = 1$.

4.78   Here is a simple way to create a random variable X that has mean μ and standard deviation σ: X takes only the two values $\mu - \sigma$ and $\mu + \sigma$, each with probability 0.5. Use the definition of the mean and variance for discrete random variables to show that X does have mean μ and standard deviation σ.

Insurance. *The business of selling insurance is based on probability and the law of large numbers. Consumers buy insurance because we all face risks that are unlikely but carry high cost. Think of a fire destroying your home. So we form a group to share the risk: we all pay a small amount, and the insurance policy pays a large amount to those few of us whose homes burn down. The insurance company sells many policies, so it can rely on the law of large numbers. Exercises 4.79 to 4.82 explore aspects of insurance.*

4.79   An insurance company looks at the records for millions of homeowners and sees that the mean loss from fire in a year is $\mu = \$250$ per person. (Most of us have no loss, but a few lose their homes. The $250 is the average loss.) The company plans to sell fire insurance for $250 plus enough to cover its costs and profit. Explain clearly why it would be stupid to sell only 12 policies. Then explain why selling thousands of such policies is a safe business.

4.80 In fact, the insurance company sees that in the entire population of home-owners, the mean loss from fire is $\mu = \$250$ and the standard deviation of the loss is $\sigma = \$300$. What are the mean and standard deviation of the total loss for 12 policies? (Losses on separate policies are independent.) What are the mean and standard deviation of the average loss for 12 policies?

4.81 According to the current Commissioners' Standard Ordinary mortality table, adopted by state insurance regulators in December 2002, a 25-year-old man has these probabilities of dying during the next five years:[22]

Age at death	25	26	27	28	29
Probability	0.00039	0.00044	0.00051	0.00057	0.00060

(a) What is the probability that the man does not die in the next five years?

(b) An online insurance site offers a term insurance policy that will pay $100,000 if a 25-year-old man dies within the next 5 years. The cost is $175 per year. So the insurance company will take in $875 from this policy if the man does not die within five years. If he does die, the company must pay $100,000. Its loss depends on how many premiums were paid, as follows:

Age at death	25	26	27	28	29
Loss	$99,825	$99,650	$99,475	$99,300	$99,125

What is the insurance company's mean cash intake from such polices?

4.82 It would be quite risky for you to insure the life of a 25-year-old friend under the terms of Exercise 4.81. There is a high probability that your friend would live and you would gain $875 in premiums. But if he were to die, you would lose almost $100,000. Explain carefully why selling insurance is not risky for an insurance company that insures many thousands of 25-year-old men.

Portfolio analysis. Here are the means, standard deviations, and correlations for the annual returns from three Fidelity mutual funds for the 10 years ending in February 2004.[23] Because there are three random variables, there are three correlations. We use subscripts to show which pair of random variables a correlation refers to.

W = annual return on 500 Index Fund $\mu_W = 11.12\%$ $\sigma_W = 17.46\%$

X = annual return on Investment Grade Bond Fund $\mu_X = 6.46\%$ $\sigma_X = 4.18\%$

Y = annual return on Diversified International Fund $\mu_Y = 11.10\%$ $\sigma_Y = 15.62\%$

Correlations

$$\rho_{WX} = -0.22 \qquad \rho_{WY} = 0.56 \qquad \rho_{XY} = -0.12$$

Exercises 4.83 to 4.85 make use of these historical data.

4.83   Many advisors recommend using roughly 20% foreign stocks to diversify port-folios of U.S. stocks. You see that the 500 Index (U.S. stocks) and Diversified International (foreign stocks) Funds had almost the same mean returns. A portfolio of 80% 500 Index and 20% Diversified International will deliver this mean return with less risk. Verify this by finding the mean and standard deviation of returns on this portfolio. (Example 4.28, on page 304, shows how to find the mean and standard deviation for the portfolio.)

4.84   Diversification works better when the investments in a portfolio have small correlations. To demonstrate this, suppose that returns on 500 Index Fund and Diversified International Fund had the means and standard deviations we have given but were uncorrelated ($\rho_{WY} = 0$). Show that the standard deviation of a portfolio that combines 80% 500 Index with 20% Diversified International is then smaller than your result from the previous exercise. What happens to the mean return if the correlation is 0?

4.85   Portfolios often contain more than two investments. The rules for means and variances continue to apply, though the arithmetic gets messier. A portfolio containing proportions a of 500 Index Fund, b of Investment Grade Bond Fund, and c of Diversified International Fund has return $R = aW + bX + cY$. Because a, b, and c are the proportions invested in the three funds, $a + b + c = 1$. The mean and variance of the portfolio return R are

$$\mu_R = a\mu_W + b\mu_X + c\mu_Y$$
$$\sigma_R^2 = a^2\sigma_W^2 + b^2\sigma_X^2 + c^2\sigma_Y^2 + 2ab\rho_{WX}\sigma_W\sigma_X + 2ac\rho_{WY}\sigma_W\sigma_Y + 2bc\rho_{XY}\sigma_X\sigma_Y$$

A basic well-diversified portfolio has 60% in 500 Index, 20% in Investment Grade Bond, and 20% in Diversified International. What are the (historical) mean and standard deviation of the annual returns for this portfolio? What does an investor gain by choosing this diversified portfolio over 100% U.S. stocks? What does the investor lose (at least in this time period)?

4.5   General Probability Rules*

Our study of probability has concentrated on random variables and their dis-tributions. Now we return to the laws that govern any assignment of proba-bilities. The purpose of learning more laws of probability is to be able to give probability models for more complex random phenomena. We have already met and used five rules.

---

*This section extends the rules of probability discussed in Section 4.2. This material is not needed for understanding the statistical methods in later chapters. It can therefore be omitted if desired.

RULES OF PROBABILITY

Rule 1. $0 \le P(A) \le 1$ for any event A

Rule 2. $P(S) = 1$

Rule 3. Addition rule: If A and B are **disjoint** events, then

$$P(A \text{ or } B) = P(A) + P(B)$$

Rule 4. Complement rule: For any event A,

$$P(A^c) = 1 - P(A)$$

Rule 5. Multiplication rule: If A and B are **independent** events, then

$$P(A \text{ and } B) = P(A)P(B)$$

General addition rules

Probability has the property that if A and B are disjoint events, then $P(A \text{ or } B) = P(A) + P(B)$. What if there are more than two events, or if the events are not disjoint? These circumstances are covered by more general addition rules for probability.

UNION

The **union** of any collection of events is the event that at least one of the collection occurs.

For two events A and B, the union is the event $\{A \text{ or } B\}$ that A or B or both occur. From the addition rule for two disjoint events we can obtain rules for more general unions. Suppose first that we have several events—say A, B, and C—that are disjoint in pairs. That is, no two can occur together. The Venn diagram in Figure 4.16 illustrates three disjoint events. The addition rule for two disjoint events extends to the following law:

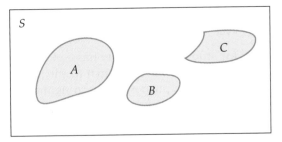

FIGURE 4.16   The addition rule for disjoint events: $P(A \text{ or } B \text{ or } C) = P(A) + P(B) + P(C)$ when events A, B, and C are disjoint.

ADDITION RULE FOR DISJOINT EVENTS

If events A, B, and C are disjoint in the sense that no two have any outcomes in common, then

$$P(\text{one or more of } A, B, C) = P(A) + P(B) + P(C)$$

This rule extends to any number of disjoint events.

EXAMPLE 4.29 Generate a random number X between 0 and 1. What is the probability that the first digit after the decimal point will be odd? The random number X is a continuous random variable whose density curve has constant height 1 between 0 and 1 and is 0 elsewhere. The event that the first digit of X is odd is the union of five disjoint events. These events are

$$0.10 \leq X < 0.20$$
$$0.30 \leq X < 0.40$$
$$0.50 \leq X < 0.60$$
$$0.70 \leq X < 0.80$$
$$0.90 \leq X < 1.00$$

Figure 4.17 illustrates the probabilities of these events as areas under the density curve. Each area is 0.1. The union of the five therefore has probability equal to the sum, or

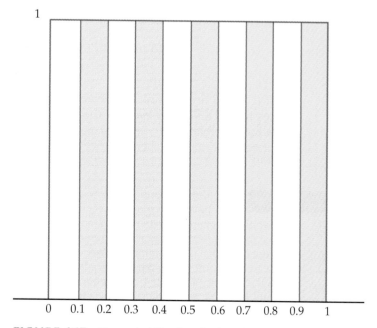

FIGURE 4.17 The probability that the first digit of a random number is odd is the sum of the probabilities of the 5 disjoint events shown. See Example 4.29.

0.5. As we should expect, a random number is equally likely to begin with an odd or an even digit.

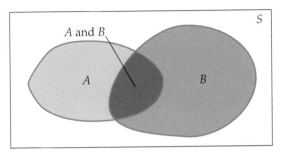

FIGURE 4.18 The union of two events that are not disjoint. The general addition rule says that $P(A \text{ or } B) = P(A) + P(B) - P(A \text{ and } B)$.

If events A and B are not disjoint, they can occur together. The probability of their union is then *less* than the sum of their probabilities. As Figure 4.18 suggests, the outcomes common to both are counted twice when we add probabilities, so we must subtract this probability once. Here is the addition rule for the union of any two events, disjoint or not.

> **GENERAL ADDITION RULE FOR UNIONS OF TWO EVENTS**
>
> For any two events A and B,
>
> $$P(A \text{ or } B) = P(A) + P(B) - P(A \text{ and } B)$$

If A and B are disjoint, the event $\{A \text{ and } B\}$ that both occur has no outcomes in it. This *empty event* is the complement of the sample space S and must have probability 0. So the general addition rule includes Rule 3, the addition rule for disjoint events.

EXAMPLE 4.30   Deborah and Matthew are anxiously awaiting word on whether they have been made partners of their law firm. Deborah guesses that her probability of making partner is 0.7 and that Matthew's is 0.5. (These are personal probabilities reflecting Deborah's assessment of chance.) This assignment of probabilities does not give us enough information to compute the probability that at least one of the two is promoted. In particular, adding the individual probabilities of promotion gives the impossible result 1.2. If Deborah also guesses that the probability that *both* she and Matthew are made partners is 0.3, then by the addition rule for unions

$$P(\text{at least one is promoted}) = 0.7 + 0.5 - 0.3 = 0.9$$

The probability that *neither* is promoted is then 0.1 by the complement rule.

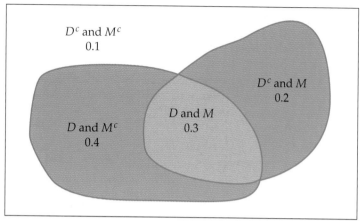

D = Deborah is made partner
M = Matthew is made partner

FIGURE 4.19   Venn diagram and probabilities for Example 4.30.

Venn diagrams are a great help in finding probabilities for unions because you can just think of adding and subtracting areas. Figure 4.19 shows some events and their probabilities for Example 4.30. What is the probability that Deborah is promoted and Matthew is not? The Venn diagram shows that this is the probability that Deborah is promoted minus the probability that both are promoted, $0.7 - 0.3 = 0.4$. Similarly, the probability that Matthew is promoted and Deborah is not is $0.5 - 0.3 = 0.2$. The four probabilities that appear in the figure add to 1 because they refer to four disjoint events whose union is the entire sample space.

Conditional probability

The probability we assign to an event can change if we know that some other event has occurred. This idea is the key to many applications of probability.

EXAMPLE 4.31   Slim is a professional poker player. He stares at the dealer, who prepares to deal. What is the probability that the card dealt to Slim is an ace? There are 52 cards in the deck. Because the deck was carefully shuffled, the next card dealt is equally likely to be any of the cards that Slim has not seen. Four of the 52 cards are aces. So

$$P(\text{ace}) = \frac{4}{52} = \frac{1}{13}$$

This calculation assumes that Slim knows nothing about any cards already dealt. Suppose now that he is looking at 4 cards already in his hand, and that one of them is an ace. He knows nothing about the other 48 cards except that exactly 3 aces are among them. Slim's probability of being dealt an ace *given what he knows* is now

$$P(\text{ace} \mid 1 \text{ ace in 4 visible cards}) = \frac{3}{48} = \frac{1}{16}$$

Knowing that there is 1 ace among the 4 cards Slim can see changes the probability that the next card dealt is an ace.

conditional
probability
The new notation $P(A \mid B)$ is a **conditional probability.** That is, it gives the probability of one event (the next card dealt is an ace) under the condition that we know another event (exactly 1 of the 4 visible cards is an ace). You can read the bar | as "given the information that."

In Example 4.31 we could find probabilities because we are willing to use an equally likely probability model for a shuffled deck of cards. Here is an example based on data.

EXAMPLE 4.32 Students at the University of New Harmony received 10,000 course grades last semester. Table 4.1 breaks down these grades by which school of the university taught the course. The schools are Liberal Arts, Engineering and Physical Sciences, and Health and Human Services. (Table 4.1 is based closely on grade distributions at an actual university, simplified a bit for clarity.[24])

It is common knowledge that college grades are lower in engineering and the physical sciences (EPS) than in liberal arts and social sciences. Consider the two events

$$A = \text{the grade comes from an EPS course}$$

$$B = \text{the grade is below a B}$$

There are 10,000 grades, of which 3656 are below B. Choosing at random gives each grade an equal chance, so the probability of choosing a grade below a B is

$$P(B) = \frac{3,656}{10,000} = 0.3656$$

To find the *conditional* probability that a grade is below a B, *given the information* that it comes from the EPS school, look only at the "Engineering and Physical Sciences" row. The EPS grades are all in this row, so the information given says that only this row is relevant. The conditional probability is

$$P(B \mid A) = \frac{800}{1600} = 0.5$$

The conditional probability that a grade is below a B, given that we know it comes from an EPS course, is much higher than the probability for a randomly chosen grade.

TABLE 4.1

Grades awarded at a university, by school

	Grade level			
	A	B	Below B	Total
Liberal Arts	2,142	1,890	2,268	6,300
Engineering and Physical Sciences	368	432	800	1,600
Health and Human Services	882	630	588	2,100
Total	3,392	2,952	3,656	10,000

CAUTION

It is easy to confuse probabilities and conditional probabilities involving the same events. For example, Table 4.1 says that

$$P(A) = \frac{1600}{10,000} = 0.16$$

$$P(A \text{ and } B) = \frac{800}{10,000} = 0.08$$

$$P(B \mid A) = \frac{800}{1600} = 0.5$$

Be sure you understand how we found these three results. There is a relationship among these three probabilities. The probability that a grade is both from EPS *and* below a B is the product of the probabilities that it is from EPS and that it is below a B *given* that it is from EPS. That is,

$$P(A \text{ and } B) = P(A) \times P(B \mid A)$$

$$= \frac{1600}{10,000} \times \frac{800}{1600}$$

$$= \frac{800}{10,000} = 0.08 \quad \text{(as before)}$$

Try to think your way through this in words: First, the grade is from EPS; then, given that it is from EPS, it is below B. We have just discovered the fundamental multiplication rule of probability.

MULTIPLICATION RULE

The probability that both of two events A and B happen together can be found by

$$P(A \text{ and } B) = P(A)P(B \mid A)$$

Here $P(B \mid A)$ is the conditional probability that B occurs given the information that A occurs.

EXAMPLE 4.33 The multiplication rule is just common sense made formal. For example, 29% of Internet users download music files, and 67% of downloaders say they don't care if the music is copyrighted.[25] So the percent of Internet users who download music (event A) *and* don't care about copyright (event B) is 67% of the 29% who download, or

$$(0.67)(0.29) = 0.1943 = 19.43\%$$

The multiplication rule expresses this as

$$P(A \text{ and } B) = P(A) \times P(B \mid A)$$

$$= (0.29)(0.67) = 0.1943$$

EXAMPLE 4.34 Slim is still at the poker table. At the moment, he wants very much to draw two diamonds in a row. As he sits at the table looking at his hand and at the upturned cards on the table, Slim sees 11 cards. Of these, 4 are diamonds. The full deck contains 13 diamonds among its 52 cards, so 9 of the 41 unseen cards are diamonds. To find Slim's probability of drawing two diamonds, first calculate

$$P(\text{first card diamond}) = \frac{9}{41}$$

$$P(\text{second card diamond} \mid \text{first card diamond}) = \frac{8}{40}$$

Slim finds both probabilities by counting cards. The probability that the first card drawn is a diamond is 9/41 because 9 of the 41 unseen cards are diamonds. If the first card is a diamond, that leaves 8 diamonds among the 40 remaining cards. So the *conditional* probability of another diamond is 8/40. The multiplication rule now says that

$$P(\text{both cards diamonds}) = \frac{9}{41} \times \frac{8}{40} = 0.044$$

Slim will need luck to draw his diamonds.

If $P(A)$ and $P(A \text{ and } B)$ are given, we can rearrange the multiplication rule to produce a *definition* of the conditional probability $P(B \mid A)$ in terms of unconditional probabilities.

DEFINITION OF CONDITIONAL PROBABILITY

When $P(A) > 0$, the **conditional probability** of B given A is

$$P(B \mid A) = \frac{P(A \text{ and } B)}{P(A)}$$

Be sure to keep in mind the distinct roles in $P(B \mid A)$ of the event B whose probability we are computing and the event A that represents the information we are given. The conditional probability $P(B \mid A)$ makes no sense if the event A can never occur, so we require that $P(A) > 0$ whenever we talk about $P(B \mid A)$.

EXAMPLE 4.35 What is the conditional probability that a grade at the University of New Harmony is an A, given that it comes from a liberal arts course? We see from Table 4.1 that

$$P(\text{liberal arts course}) = \frac{6300}{10{,}000} = 0.63$$

$$P(\text{A grade } and \text{ liberal arts course}) = \frac{2142}{10{,}000} = 0.2142$$

The definition of conditional probability therefore says that

$$P(\text{A grade} \mid \text{liberal arts course}) = \frac{P(\text{A grade } and \text{ liberal arts course})}{P(\text{liberal arts course})}$$

$$= \frac{0.2142}{0.63} = 0.34$$

Check that this agrees with the result obtained from the "Liberal Arts" row of Table 4.1:

$$P(\text{A grade} \mid \text{liberal arts course}) = \frac{2142}{6300} = 0.34$$

General multiplication rules

The definition of conditional probability reminds us that in principle all probabilities, including conditional probabilities, can be found from the assignment of probabilities to events that describes a random phenomenon. More often, however, conditional probabilities are part of the information given to us in a probability model, and the multiplication rule is used to compute $P(A \text{ and } B)$. This rule extends to more than two events.

The union of a collection of events is the event that *any* of them occur. Here is the corresponding term for the event that *all* of them occur.

INTERSECTION

The **intersection** of any collection of events is the event that *all* of the events occur.

To extend the multiplication rule to the probability that all of several events occur, the key is to condition each event on the occurrence of *all* of the preceding events. For example, the intersection of three events A, B, and C has probability

$$P(A \text{ and } B \text{ and } C) = P(A)P(B \mid A)P(C \mid A \text{ and } B)$$

EXAMPLE 4.36   Only 5% of male high school basketball, baseball, and football players go on to play at the college level. Of these, only 1.7% enter major league professional sports. About 40% of the athletes who compete in college and then reach the pros have a career of more than 3 years.[26] Define these events:

$$A = \{\text{competes in college}\}$$

$$B = \{\text{competes professionally}\}$$

$$C = \{\text{pro career longer than 3 years}\}$$

What is the probability that a high school athlete competes in college and then goes on to have a pro career of more than 3 years? We know that

$$P(A) = 0.05$$

$$P(B \mid A) = 0.017$$

$$P(C \mid A \text{ and } B) = 0.4$$

The probability we want is therefore

$$P(A \text{ and } B \text{ and } C) = P(A)P(B \mid A)P(C \mid A \text{ and } B)$$

$$= 0.05 \times 0.017 \times 0.4 = 0.00034$$

Only about 3 of every 10,000 high school athletes can expect to compete in college and have a professional career of more than 3 years. High school students would be wise to concentrate on studies rather than on unrealistic hopes of fortune from pro sports.

Tree diagrams

Probability problems often require us to combine several of the basic rules into a more elaborate calculation. Here is an example that illustrates how to solve problems that have several stages.

EXAMPLE 4.37   Online chat rooms are dominated by the young. Teens are the biggest users. If we look only at adult Internet users (age 18 and over), 47% of the 18 to 29 age group chat, as do 21% of those aged 30 to 49 and just 7% of those 50 and over. To learn what percent of all Internet users participate in chat, we also need the age breakdown of users. Here it is: 29% of adult Internet users are age 18 to 29 (event A_1), another 47% are 30 to 49 years old (event A_2), and the remaining 24% are age 50 and over (event A_3).[27]

tree diagram   What is the probability that a randomly chosen user of the Internet participates in chat rooms (event C)? To find out, use the **tree diagram** in Figure 4.20 to organize your thinking. Each segment in the tree is one stage of the problem. Each complete branch shows a path through the two stages. The probability written on each segment is the conditional probability of an Internet user following that segment given that he or she has reached the node from which it branches.

Starting at the left, an Internet user falls into one of the three age groups. The probabilities of these groups

$$P(A_1) = 0.29 \qquad P(A_2) = 0.47 \qquad P(A_3) = 0.24$$

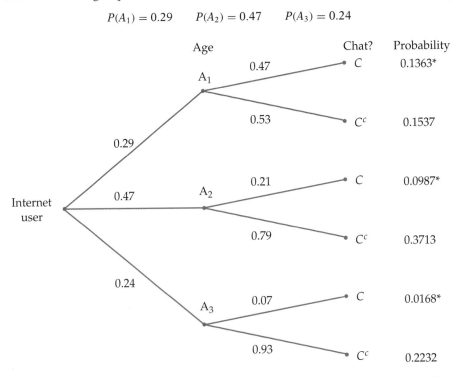

FIGURE 4.20   Tree diagram for Example 4.37. The probability $P(C)$ is the sum of the probabilities of the three branches marked with asterisks (*).

mark the leftmost branches in the tree. Conditional on being 18 to 29 years old, the probability of participating in chat is $P(C \mid A_1) = 0.47$. So the conditional probability of *not* participating is

$$P(C^c \mid A_1) = 1 - 0.47 = 0.53$$

These conditional probabilities mark the paths branching out from the A_1 node in Figure 4.20. The other two age group nodes similarly lead to two branches marked with the conditional probabilities of chatting or not. The probabilities on the branches from any node add to 1 because they cover all possibilities given that this node was reached.

There are three disjoint paths to C, one for each age group. By the addition rule, $P(C)$ is the sum of their probabilities. The probability of reaching C through the 18 to 29 age group is

$$P(C \text{ and } A_1) = P(A_1)P(C \mid A_1)$$

$$= 0.29 \times 0.47 = 0.1363$$

Follow the paths to C through the other two age groups. The probabilities of these paths are

$$P(C \text{ and } A_2) = P(A_2)P(C \mid A_2) = (0.47)(0.21) = 0.0987$$

$$P(C \text{ and } A_3) = P(A_3)P(C \mid A_3) = (0.24)(0.07) = 0.0168$$

The final result is

$$P(C) = 0.1363 + 0.0987 + 0.0168 = 0.2518$$

About 25% of all adult Internet users take part in chat rooms.

It takes longer to explain a tree diagram than it does to use it. Once you have understood a problem well enough to draw the tree, the rest is easy. Tree diagrams combine the addition and multiplication rules. The multiplication rule says that the probability of reaching the end of any complete branch is the product of the probabilities written on its segments. The probability of any outcome, such as the event C that an adult Internet user takes part in chat rooms, is then found by adding the probabilities of all branches that are part of that event.

Bayes's rule

There is another kind of probability question that we might ask in the context of thinking about online chat. What percent of adult chat room participants are age 18 to 29?

EXAMPLE 4.38   In the notation of Example 4.37 this is the conditional probability $P(A_1 \mid C)$. Start from the definition of conditional probability and then apply the results of Example 4.37:

$$P(A_1 \mid C) = \frac{P(A_1 \text{ and } C)}{P(C)}$$

$$= \frac{0.1363}{0.2518} = 0.5413$$

Over half of adult chat room participants are between 18 and 29 years old. Compare this conditional probability with the original information (unconditional) that 29% of adult Internet users are between 18 and 29 years old. Knowing that a person chats increases the probability that he or she is young.

We know the probabilities $P(A_1)$, $P(A_2)$, and $P(A_3)$ that give the age distribution of adult Internet users. We also know the conditional probabilities $P(C \mid A_1)$, $P(C \mid A_2)$, and $P(C \mid A_3)$ that a person from each age group chats. Example 4.37 shows how to use this information to calculate $P(C)$. The method can be summarized in a single expression that adds the probabilities of the three paths to C in the tree diagram:

$$P(C) = P(A_1)P(C \mid A_1) + P(A_2)P(C \mid A_2) + P(A_3)P(C \mid A_3)$$

In Example 4.38 we calculated the "reverse" conditional probability $P(A_1 \mid C)$. The denominator 0.2518 in that example came from the expression just above. Put in this general notation, we have another probability law.

> ## BAYES'S RULE
>
> Suppose that $A_1, A_2, \ldots, A_k$ are disjoint events whose probabilities are not 0 and add to exactly 1. That is, any outcome is in exactly one of these events. Then if C is any other event whose probability is not 0 or 1,
>
> $$P(A_i \mid C) = \frac{P(C \mid A_i)P(A_i)}{P(C \mid A_1)P(A_1) + P(C \mid A_2)P(A_2) + \cdots + P(A_k)P(C \mid A_k)}$$

The numerator in Bayes's rule is always one of the terms in the sum that makes up the denominator. The rule is named after Thomas Bayes, who wrestled with arguing from outcomes like C back to antecedents like the A_i in a book published in 1763. It is far better to think your way through problems like Examples 4.37 and 4.38 rather than memorize these formal expressions.

Independence again

The conditional probability $P(B \mid A)$ is generally not equal to the unconditional probability $P(B)$. That is because the occurrence of event A generally gives us some additional information about whether or not event B occurs. If knowing that A occurs gives no additional information about B, then A and B are independent events. The formal definition of independence is expressed in terms of conditional probability.

> ## INDEPENDENT EVENTS
>
> Two events A and B that both have positive probability are **independent** if
>
> $$P(B \mid A) = P(B)$$

This definition makes precise the informal description of independence given in Section 4.2. We now see that the multiplication rule for independent events, $P(A \text{ and } B) = P(A)P(B)$, is a special case of the general multiplication rule, $P(A \text{ and } B) = P(A)P(B \mid A)$, just as the addition rule for disjoint events is a special case of the general addition rule.

SECTION 4.5 | Summary

The **complement** A^c of an event A contains all outcomes that are not in A. The **union** {A or B} of events A and B contains all outcomes in A, in B, or in both A and B. The **intersection** {A and B} contains all outcomes that are in both A and B, but not outcomes in A alone or B alone.

The **conditional probability** $P(B \mid A)$ of an event B given an event A is defined by

$$P(B \mid A) = \frac{P(A \text{ and } B)}{P(A)}$$

when $P(A) > 0$. In practice, conditional probabilities are most often found from directly available information.

The essential general rules of elementary probability are

Legitimate values: $0 \leq P(A) \leq 1$ for any event A

Total probability 1: $P(S) = 1$

Complement rule: $P(A^c) = 1 - P(A)$

Addition rule: $P(A \text{ or } B) = P(A) + P(B) - P(A \text{ and } B)$

Multiplication rule: $P(A \text{ and } B) = P(A)P(B \mid A)$

If A and B are **disjoint**, then $P(A \text{ and } B) = 0$. The general addition rule for unions then becomes the special addition rule, $P(A \text{ or } B) = P(A) + P(B)$.

A and B are **independent** when $P(B \mid A) = P(B)$. The multiplication rule for intersections then becomes $P(A \text{ and } B) = P(A)P(B)$.

In problems with several stages, draw a **tree diagram** to organize use of the multiplication and addition rules.

SECTION 4.5 | Exercises

4.86 Call a household prosperous if its income exceeds $100,000. Call the household educated if the householder completed college. Select an American household at random, and let A be the event that the selected household is prosperous and B the event that it is educated. According to the Current Population Survey, $P(A) = 0.138$, $P(B) = 0.261$, and the probability that a household is both prosperous and educated is $P(A \text{ and } B) = 0.082$. What is the probability $P(A \text{ or } B)$ that the household selected is either prosperous or educated?

4.87 In the setting of the previous exercise, what is the conditional probability that a household is prosperous, given that it is educated? Explain why your result shows that events A and B are not independent.

4.88   Draw a Venn diagram that shows the relation between the events A and B in Exercise 4.86. Indicate each of the following events on your diagram and use the information in Exercise 4.86 to calculate the probability of each event. Finally, describe in words what each event is.

(a) $\{A \text{ and } B\}$

(b) $\{A \text{ and } B^c\}$

(c) $\{A^c \text{ and } B\}$

(d) $\{A^c \text{ and } B^c\}$

4.89   Motor vehicles sold to individuals are classified as either cars or light trucks (including SUVs) and as either domestic or imported. In early 2004, 69% of vehicles sold were light trucks, 78% were domestic, and 55% were domestic light trucks. Let A be the event that a vehicle is a light truck and B the event that it is imported. Write each of the following events in set notation and give its probability.

(a) The vehicle is a car.

(b) The vehicle is an imported car.

4.90   In 2000, the Internal Revenue Service received 129,075,000 individual tax returns. Of these, 10,855,000 reported an adjusted gross income of at least $100,000, and 240,000 reported at least $1 million.[28]   If you know that a randomly chosen return shows an income of $100,000 or more, what is the conditional probability that the income is at least $1 million?

4.91   Using the information in Exercise 4.89, answer these questions.

(a) Given that a vehicle is imported, what is the conditional probability that it is a car?

(b) Are the events "vehicle is a car" and "vehicle is imported" independent? Justify your answer.

4.92   Julie is graduating from college. She has studied biology, chemistry, and computing and hopes to work as a forensic scientist applying her science background to crime investigation. Late one night she thinks about some jobs she has applied for. Let A, B, and C be the events that Julie is offered a job by

$A =$ the Connecticut Office of the Chief Medical Examiner

$B =$ the New Jersey Division of Criminal Justice

$C =$ the federal Disaster Mortuary Operations Response Team

Julie writes down her personal probabilities for being offered these jobs:

$P(A) = 0.6$ $P(B) = 0.4$ $P(C) = 0.2$

$P(A \text{ and } B) = 0.1$ $P(A \text{ and } C) = 0.05$ $P(B \text{ and } C) = 0.05$

$P(A \text{ and } B \text{ and } C) = 0$

Make a Venn diagram of the events A, B, and C. As in Figure 4.19 (page 315), mark the probabilities of every intersection involving these events and their complements. Use this diagram for Exercises 4.93 to 4.95.

4.93   What is the probability that Julie is offered at least one of the three jobs?

4.94   What is the probability that Julie is offered both the Connecticut and New Jersey jobs, but not the federal job?

4.95   If Julie is offered the federal job, what is the conditional probability that she is also offered the New Jersey job? If Julie is offered the New Jersey job, what is the conditional probability that she is also offered the federal job?

4.96   Here are the counts (in thousands) of earned degrees in the United States in the 2005–2006 academic year, classified by level and by the sex of the degree recipient:[29]

	Bachelor's	Master's	Professional	Doctorate	Total
Female	784	276	39	20	1119
Male	559	197	44	25	825
Total	1343	473	83	45	1944

(a) If you choose a degree recipient at random, what is the probability that the person you choose is a woman?

(b) What is the conditional probability that you choose a woman, given that the person chosen received a professional degree?

(c) Are the events "choose a woman" and "choose a professional degree recipient" independent? How do you know?

4.97   The previous exercise gives the counts (in thousands) of earned degrees in the United States in the 2005–2006 academic year. Use these data to answer the following questions.

(a) What is the probability that a randomly chosen degree recipient is a man?

(b) What is the conditional probability that the person chosen received a bachelor's degree, given that he is a man?

(c) Use the multiplication rule to find the probability of choosing a male bachelor's degree recipient. Check your result by finding this probability directly from the table of counts.

Working. *In the language of government statistics, you are "in the labor force" if you are available for work and either working or actively seeking work. The unemployment rate is the proportion of the labor force (not of the entire population) who are unemployed. Here are data from the Current Population Survey for the civilian population aged 25 years and over at the end of 2003. The table entries are counts in thousands of people.[30] Exercises 4.98 to 4.101 concern these data.*

Highest education	Total population	In labor force	Employed
Did not finish high school	28,021	12,623	11,552
High school but no college	59,844	38,210	36,249
Some college, but no bachelor's degree	46,777	33,928	32,429
College graduate	51,568	40,414	39,250

4.98   Find the unemployment rate for people with each level of education. How does the unemployment rate change with education? Explain carefully why your results show that level of education and being employed are not independent.

4.99   (a) What is the probability that a randomly chosen person 25 years of age or older is in the labor force?

(b) If you know that the person chosen is a college graduate, what is the conditional probability that he or she is in the labor force?

(c) Are the events "in the labor force" and "college graduate" independent? How do you know?

4.100   You know that a person is employed. What is the conditional probability that he or she is a college graduate? You know that a second person is a college graduate. What is the conditional probability that he or she is employed?

4.101   Beware the lurking variable. The low labor force participation rate of people who did not finish high school is explained by the confounding of education level with a variable that lurks behind the "aged 25 years and over" restriction for these data. Explain this confounding.

4.102   The probability that a randomly chosen student at the University of New Harmony is a woman is 0.6. The probability that the student is studying education is 0.15. The conditional probability that the student is a woman, given that the student is studying education, is 0.8. What is the conditional probability that the student is studying education, given that she is a woman?

4.103   As explained in Exercise 4.60 (page 305), spelling errors in a text can be either nonword errors or word errors. Nonword errors make up 25% of all errors. A human proofreader will catch 90% of nonword errors and 70% of word errors. What percent of all errors will the proofreader catch? (Draw a tree diagram to organize the information given.)

4.104   The voters in a large city are 40% white, 40% black, and 20% Hispanic. (Hispanics may be of any race in official statistics, but in this case we are speaking of political blocks.) A black mayoral candidate anticipates attracting 30% of the white vote, 90% of the black vote, and 50% of the Hispanic vote. Draw a tree diagram with probabilities for the race (white, black, or Hispanic) and vote (for or against the candidate) of a randomly chosen voter. What percent of the overall vote does the candidate expect to get?

4.105 At a self-service gas station, 40% of the customers pump regular gas, 35% pump midgrade, and 25% pump premium gas. Of those who pump regular, 30% pay at least $20. Of those who pump midgrade, 50% pay at least $20. And of those who pump premium, 60% pay at least $20. What is the probability that the next customer pays at least $20?

4.106 In the election described in Exercise 4.104, what percent of the candidate's votes come from black voters? (Write this as a conditional probability.)

4.107 In the setting of Exercise 4.105, what percent of customers who pay at least $20 pump premium? (Write this as a conditional probability.)

4.108 Of the 16,071 degrees in mathematics given by U.S. colleges and universities in the most recent year, 73% were bachelor's degrees, 21% were master's degrees, and the rest were doctorates. Moreover, women earned 48% of the bachelor's degrees, 42% of the master's degrees, and 29% of the doctorates.[31] You choose a mathematics degree at random and find that it was awarded to a woman. What is the probability that it is a bachelor's degree?

Genetic counseling. Conditional probabilities and Bayes's rule are a basis for counseling people who may have genetic defects that can be passed to their children. Exercises 4.109 to 4.113 concern genetic counseling settings.

4.109 **Albinism.** People with albinism have little pigment in their skin, hair, and eyes. The gene that governs albinism has two forms (called alleles), which we denote by *a* and *A*. Each person has a pair of these genes, one inherited from each parent. A child inherits one of each parent's two alleles, independently with probability 0.5. Albinism is a recessive trait, so a person is albino only if the inherited pair is *aa*.

(a) Beth's parents are not albino but she has an albino brother. This implies that both of Beth's parents have type *Aa*. Why?

(b) Which of the types *aa*, *Aa*, *AA* could a child of Beth's parents have? What is the probability of each type?

(c) Beth is not albino. What are the conditional probabilities for Beth's possible genetic types, given this fact? (Use the definition of conditional probability.)

4.110 **Albinism, continued.** Beth knows the probabilities for her genetic types from part (c) of the previous exercise. She marries Bob, who is albino. Bob's genetic type must be *aa*.

(a) What is the conditional probability that a child of Beth and Bob is non-albino if Beth has type *Aa*? What is the conditional probability of a non-albino child if Beth has type *AA*?

(b) Beth and Bob's first child is non-albino. What is the conditional probability that Beth is a carrier, type *Aa*?

4.111   **Cystic fibrosis.** Cystic fibrosis is a lung disorder that often results in death. It is inherited but can be inherited only if both parents are carriers of an abnormal gene. In 1989, the CF gene that is abnormal in carriers of cystic fibrosis was identified. The probability that a randomly chosen person of European ancestry carries an abnormal CF gene is 1/25. (The probability is less in other ethnic groups.) The CF20m test detects most but not all harmful mutations of the CF gene. The test is positive for 90% of people who are carriers. It is (ignoring human error) never positive for people who are not carriers. Jason tests positive. What is the probability that he is a carrier?

4.112   **Cystic fibrosis, continued.** Jason knows that he is a carrier of cystic fibrosis. His wife, Julianne, has a brother with cystic fibrosis, which means the probability is 2/3 that she is a carrier. If Julianne is a carrier, each child she has with Jason has probability 1/4 of having cystic fibrosis. If she is not a carrier, her children cannot have the disease. Jason and Julianne have one child, who does not have cystic fibrosis. This information reduces the probability that Julianne is a carrier. Use Bayes's rule to find the conditional probability that Julianne is a carrier, given that she and Jason have one child who does not have cystic fibrosis.

4.113   **Muscular dystrophy.** Muscular dystrophy is an incurable muscle-wasting disease. The most common and serious type, called DMD, is caused by a sex-linked recessive mutation. Specifically: women can be carriers but do not get the disease; a son of a carrier has probability 0.5 of having DMD; a daughter has probability 0.5 of being a carrier. As many as 1/3 of DMD cases, however, are due to spontaneous mutations in sons of mothers who are not carriers. Toni has one son, who has DMD.

In the absence of other information, the probability is 1/3 that the son is the victim of a spontaneous mutation and 2/3 that Toni is a carrier. There is a screening test called the CK test that is positive with probability 0.7 if a woman is a carrier and with probability 0.1 if she is not. Toni's CK test is positive. What is the probability that she is a carrier?

CHAPTER 4 | Exercises

4.114   An ancient Korean drinking game involves a 14-sided die. The players roll the die in turn and must submit to whatever humiliation is written on the up-face: something like "Keep still when tickled on face." Six of the 14 faces are squares. Let's call them A, B, C, D, E, and F for short. The other eight faces are triangles, which we will call 1, 2, 3, 4, 5, 6, 7, and 8. Each of the squares is equally likely. Each of the triangles is also equally likely, but the triangle probability differs from the square probability. The probability of getting a square is 0.72. Give the probability model for the 14 possible outcomes.

4.115   Two wine tasters rate each wine they taste on a scale of 1 to 5. From data on their ratings of a large number of wines, we obtain the following probabilities for both tasters' ratings of a randomly chosen wine:

Taster 1	Taster 2				
	1	2	3	4	5
1	0.03	0.02	0.01	0.00	0.00
2	0.02	0.08	0.05	0.02	0.01
3	0.01	0.05	0.25	0.05	0.01
4	0.00	0.02	0.05	0.20	0.02
5	0.00	0.01	0.01	0.02	0.06

(a) Why is this a legitimate assignment of probabilities to outcomes?

(b) What is the probability that the tasters agree when rating a wine?

(c) What is the probability that Taster 1 rates a wine higher than 3? What is the probability that Taster 2 rates a wine higher than 3?

4.116 Rotter Partners is planning a major investment. The amount of profit X is uncertain but a probabilistic estimate gives the following distribution (in millions of dollars):

Profit	1	1.5	2	4	10
Probability	0.1	0.2	0.4	0.2	0.1

(a) Find the mean profit μ_X and the standard deviation σ_X of the profit.

(b) Rotter Partners owes its source of capital a fee of $200,000 plus 10% of the profits X. So the firm actually retains

$$Y = 0.9X - 0.2$$

from the investment. Find the mean and standard deviation of Y.

4.117 A grocery store gives its customers cards that may win them a prize when matched with other cards. The back of the card announces the following probabilities of winning various amounts if a customer visits the store 10 times:

Amount	$1000	$200	$50	$10
Probability	1/10,000	1/1000	1/100	1/20

(a) What is the probability of winning nothing?

(b) What is the mean amount won?

(c) What is the standard deviation of the amount won?

4.118   The College Board finds that the distribution of students' SAT scores depends on the level of education their parents have. Children of parents who did not finish high school have SAT math scores X with mean 442 and standard deviation 108. Scores Y of children of parents with graduate degrees have mean 566 and standard deviation 109. Perhaps we should standardize to a common scale for equity. Find positive numbers a, b, c, and d such that $a + bX$ and $c + dY$ both have mean 500 and standard deviation 100.

4.119   Joe buys a ticket in the Tri-State Pick 3 lottery every day, always betting on 856. He will win something if the winning number contains 8, 5, and 6 in any order. Each day, Joe has probability 0.006 of winning, and he wins (or not) independently of other days because a new drawing is held each day. What is the probability that Joe's first winning ticket comes on the 20th day?

4.120   Slot machines are now video games, with winning determined by electronic random number generators. In the old days, slot machines were like this: you pull the lever to spin three wheels; each wheel has 20 symbols, all equally likely to show when the wheel stops spinning; the three wheels are independent of each other. Suppose that the middle wheel has 9 bells among its 20 symbols, and the left and right wheels have 1 bell each.

(a)   You win the jackpot if all three wheels show bells. What is the probability of winning the jackpot?

(b)   What is the probability that the wheels stop with exactly 2 bells showing?

The following exercises require familiarity with the material presented in the optional Section 4.5.

4.121   In the setting of Exercise 4.115, Taster 1's rating for a wine is 3. What is the conditional probability that Taster 2's rating is higher than 3?

4.122   Students have different patterns of Internet use than other adults. Among adult Internet users, 4.1% are full-time students and another 2.9% are part-time students. Students are much more likely to access the Internet from someplace other than work or home: 58% of full-time students do so, as do 30% of part-time students, but only 21% of other users.[32] What percent of all adult users reach the Internet from someplace other than home or work?

4.123   Independence of events is not always obvious. Toss two balanced coins independently. The four possible combinations of heads and tails in order each have probability 0.25. The events

$$A = \text{head on the first toss}$$

$$B = \text{both tosses have the same outcome}$$

may seem intuitively related. Show that $P(B \mid A) = P(B)$ so that A and B are in fact independent.

4.124   Choose a point at random in the square with sides $0 \le x \le 1$ and $0 \le y \le 1$. This means that the probability that the point falls in any region within the square is the area of that region. Let X be the x coordinate and Y the y coordinate of the point chosen. Find the conditional probability $P(Y < 1/2 \mid Y > X)$. (*Hint:* Sketch the square and the events $Y < 1/2$ and $Y > X$.)

CHALLENGE

randomized response

4.125 It is difficult to conduct sample surveys on sensitive issues because many people will not answer questions if the answers might embarrass them. **Randomized response** is an effective way to guarantee anonymity while collecting information on topics such as student cheating or sexual behavior. Here is the idea. To ask a sample of students whether they have plagiarized a term paper while in college, have each student toss a coin in private. If the coin lands heads *and* they have not plagiarized, they are to answer "No." Otherwise they are to give "Yes" as their answer. Only the student knows whether the answer reflects the truth or just the coin toss, but the researchers can use a proper random sample with follow-up for nonresponse and other good sampling practices.

Suppose that in fact the probability is 0.3 that a randomly chosen student has plagiarized a paper. Draw a tree diagram in which the first stage is tossing the coin and the second is the truth about plagiarism. The outcome at the end of each branch is the answer given to the randomized-response question. What is the probability of a "No" answer in the randomized-response poll? If the probability of plagiarism were 0.2, what would be the probability of a "No" response on the poll? Now suppose that you get 39% "No" answers in a randomized-response poll of a large sample of students at your college. What do you estimate to be the percent of the population who have plagiarized a paper?

EESEE CASE STUDIES

The Electronic Encyclopedia of Statistical Examples and Exercises (EESEE) is available on the text CD and Web site. These more elaborate stories, with data, provide settings for longer case studies.

4.126 Read the EESEE story "Anecdotes of Probability" and answer all of the questions.

4.127 Read the EESEE story "On-Time Flights" and answer all of the questions asked.

4.128 Read the EESEE story "Baby Hearing Screening." Respond to all the questions posed in this case study. (This story makes use of the optional material in Section 4.5.)

STATISTICS IN PRACTICE

Statistical Software for Solving Business Problems

JING SHYR *is Chief Statistician and Vice President of Research and Development at SPSS, Inc.*

SPSS is one of the world's leading providers of statistical software. As chief statistician, my job blends statistics and computing. I lead a group of statisticians and software engineers who design tools and applications for customers who apply statistics in varied settings. We place special emphasis on helping our customers solve business problems.

Many business users of SPSS software are concerned with predicting the right product or the right offer for each of their customers: What kinds of people are the most profitable customers? At what kinds of people should we aim a new advertising campaign? How can we retain more of our current customers? Other business problems have an engineering flavor: Which parts in a complex product fail most often in use? Is it more economical to repair or replace a failed part? These business problems translate into statistical problems concerning what data to collect and what statistical models to apply. Good software can help users obtain business insight in a timely fashion.

One of the biggest challenges today in software development is handling model building for very large data sets. Think, for example, of the enormous volume of data produced by the checkout scanners of a national retail chain. We look at several approaches to this problem, such as developing efficient algorithms for large-scale computation, dividing the data into subsets for analysis, and integrating analysis algorithms into the data base software. As a statistician, I work with software engineers to develop a robust system to handle large quantities of data more efficiently.

Since joining SPSS 17 years ago, I've found myself more fulfilled by my work as my experience has grown. As I've studied statistics, I've always had an interest in how statistics can benefit our society. Seeing software that I have helped develop solve the business problems of our customers brings the feeling of satisfaction that comes with knowing that a job is well done.

Sampling Distributions

Introduction

Statistical inference draws conclusions about a population or process on the basis of data. The data are summarized by *statistics* such as means, proportions, and the slopes of least-squares regression lines. When the data are produced by random sampling or randomized experimentation, a statistic is a random variable that obeys the laws of probability theory. The link between probability and data is formed by the *sampling distributions* of statistics. A sampling distribution shows how a statistic would vary in repeated data production. That is, a sampling distribution is a probability distribution that answers the question "What would happen if we did this many times?" In Section 3.4 we looked at several sampling distributions empirically by simulating 1000 random samples. Considered as a probability distribution, a sampling distribution is an idealized mathematical description of the results of an indefinitely large number of samples rather than the results of a particular 1000 samples.

THE DISTRIBUTION OF A STATISTIC

A statistic from a random sample or randomized experiment is a random variable. The probability distribution of the statistic is its **sampling distribution.**

Probability distributions also play a second role in statistical inference. Any quantity that can be measured for each member of a population is described by the distribution of its values for all members of the population. This is the context in which we first met distributions, as models for the overall pattern of data. Imagine choosing one individual at random from the population. The results of repeated choices have a probability distribution that is the same as the distribution for the population.

EXAMPLE 5.1
The distribution of heights of women between the ages of 18 and 24 is approximately normal with mean 64.5 inches and standard deviation 2.5 inches. Select a woman at random and measure her height. The result is a random variable X. We don't know the height of a randomly chosen woman, but we do know that in repeated sampling X will have the same $N(64.5, 2.5)$ distribution that describes the pattern of heights in the entire population. We call $N(64.5, 2.5)$ the *population distribution*.

POPULATION DISTRIBUTION

The **population distribution** of a variable is the distribution of its values for all members of the population. The population distribution is also the probability distribution of the variable when we choose one individual at random from the population.

The population of all women between the ages of 18 and 24 actually exists, so that we can in principle draw an SRS from it. Sometimes the population that we want to draw conclusions about does not actually exist in the physical world. For example, we might observe an industrial process producing rods with lengths that vary according to a normal distribution. The population contains all rods that would be produced if the process continued forever in its present state. That population does not actually exist, so we cannot draw an SRS from it. Yet we are often justified in treating the rods as if they were independent observations randomly chosen from this hypothetical population. Once again we describe the population (or the process, if you prefer) by the normal distribution that governs individual observations.

To progress from discussing probability as a topic in itself to probability as a foundation for inference, we must study the sampling distributions of some common statistics. We will meet other sampling distributions later in our study of inference. In each case, the nature of the sampling distribution depends on both the nature of the population distribution and the way we collect the data from the population.

5.1   Sampling Distributions for Counts and Proportions

A sample survey asks 2500 adults whether they agree that "I like buying new clothes, but shopping is often frustrating and time-consuming." The number who say "Agree" is a random variable X. A state tax auditor selects a sample of 150 sales records from a company to examine whether or not the sales are subject to sales tax. A number X are wrongly classified. In both of these examples, the random variable X is a **count** of the occurrences of some outcome in a fixed number of observations. There are 2500 observations in the first example and 150 observations in the second. If the number of observations is n, then the **sample proportion** is $\hat{p} = X/n$. For example, if 1650 of the 2500 shoppers say "Agree," the sample proportion is

count

sample proportion

$$\hat{p} = \frac{1650}{2500} = 0.66$$

Sample counts and sample proportions are common statistics. This section describes their sampling distributions.

The binomial distributions for sample counts

The distribution of a count X depends on how the data are produced. Here is a simple but common situation.

THE BINOMIAL SETTING

1. There are a fixed number n of observations.

2. The n observations are all independent.

3. Each observation falls into one of just two categories, which for convenience we call "success" and "failure."

4. The probability of a success, call it p, is the same for each observation.

Think of tossing a coin n times as an example of the binomial setting. Each toss gives either heads or tails. The outcomes of successive tosses are independent. If we call heads a success, then p is the probability of a head and remains the same as long as we toss the same coin. The number of heads we count is a random variable X. The distribution of X, and more generally of the count of successes in any binomial setting, is completely determined by the number of observations n and the success probability p.

BINOMIAL DISTRIBUTIONS

The distribution of the count X of successes in the binomial setting is called the **binomial distribution** with parameters n and p. The parameter n is the number of observations, and p is the probability of a success on any one observation. The possible values of X are the whole numbers from 0 to n. As an abbreviation, we say that X is $B(n, p)$.

The binomial distributions are an important class of discrete probability distributions. Later in this section we will learn how to assign probabilities to outcomes and how to find the mean and standard deviation of binomial distributions. *The most important skill for using binomial distributions is the ability to recognize situations to which they do and don't apply.*

EXAMPLE 5.2   **(a)** Toss a balanced coin 10 times and count the number X of heads. There are $n = 10$ observations. Successive tosses are independent. If the coin is balanced, the probability of a head is $p = 0.5$ on each toss. The number of heads we observe has the binomial distribution $B(10, 0.5)$.

(b) Deal 10 cards from a shuffled deck and count the number X of red cards. There are 10 observations, and each gives either a red or a black card. But the observations are *not* independent. If the first card is black, the second is more likely to be red because there are now more red cards than black cards remaining in the deck. The count X does *not* have a binomial distribution.

(c) Genetics says that children receive genes from their parents independently. Each child of a particular pair of parents has probability 0.25 of having type O blood. If these parents have 5 children, the number who have type O blood is the count X of successes in 5 independent trials with probability 0.25 of a success on each trial. So X has the $B(5, 0.25)$ distribution.

(d) Engineers define reliability as the probability that an item will perform its function under specific conditions for a specific period of time. Replacement heart valves made of animal tissue, for example, have probability 0.77 of performing well for 15 years.[1] The probability of failure is therefore 0.23. It is reasonable to assume that valves in different patients fail (or not) independently of each other. The number of patients in a group of 500 who will need another valve replacement within 15 years has the $B(500, 0.23)$ distribution.

Binomial distributions in statistical sampling

The binomial distributions are important in statistics when we wish to make inferences about the proportion p of "successes" in a population. Here is a typical example.

EXAMPLE 5.3   The financial records of businesses may be audited by state tax authorities to test compliance with tax laws. It is too time-consuming to examine, for example, all sales and purchases made by a company during the period covered by the audit. The auditor will examine samples, sometimes chosen by judgment but now often chosen at random. Random sampling is fast and easy when records are kept in electronic form. It is usual to stratify the population of records by dollar value and to give more attention to high-value items.

An audit examines an SRS of 150 sales records out of 10,000 available. One issue is whether each sale was correctly classified as subject to state sales tax or not. Suppose that in fact 800 of the 10,000 sales are incorrectly classified. Is the count X of misclassified records in the sample a binomial random variable?

Choosing an SRS from a population is not quite a binomial setting. Removing one record in Example 5.3 changes the proportion of bad records in the remaining population, so the state of the second record chosen is not independent of the first. Because the population is large, however, removing a few items has a very small effect on the composition of the remaining population. Successive inspection results are very nearly independent. The population proportion of misclassified records is

$$p = \frac{800}{10,000} = 0.08$$

If the first record chosen is bad, the proportion of bad records remaining is $799/9999 = 0.079908$. If the first record is good, the proportion of bad records left is $800/9999 = 0.080008$. These proportions are so close to 0.08 that for practical purposes we can act as if removing one record has no effect on the proportion of misclassified records remaining. We act as if the count X of misclassified sales records in the audit sample has the binomial distribution $B(150, 0.08)$.

> ## SAMPLING DISTRIBUTION OF A COUNT
>
> A population contains proportion p of successes. If the population is much larger than the sample, the count X of successes in an SRS of size n has approximately the binomial distribution $B(n, p)$.
>
> The accuracy of this approximation improves as the size of the population increases relative to the size of the sample. As a rule of thumb, we will use the binomial sampling distribution for counts when the population is at least 20 times as large as the sample.

Finding binomial probabilities: software and tables

We will later give a formula for the probability that a binomial random variable takes any of its values. In practice, you will rarely have to use this formula for calculations. Some calculators and most statistical software packages calculate binomial probabilities.

FIGURE 5.1   Binomial probabilities: output from the Minitab statistical software, for Example 5.4.

EXAMPLE 5.4   In the audit setting of Example 5.3, what is the probability that the audit finds exactly 10 misclassified sales records? What is the probability that the audit finds no more than 10 misclassified records? Figure 5.1 shows the output from one statistical software system. You see that if the count X has the $B(150, 0.08)$ distribution,

$$P(X = 10) = 0.106959$$

$$P(X \leq 10) = 0.338427$$

It was easy to request these calculations in the software's menus. The output supplies more decimal places than we need and uses labels that may not be helpful (for example, "Probability Density Function" when the distribution is discrete, not continuous). But, as usual with software, we can ignore distractions and find the results we need.

If you do not have suitable computing facilities, you can still shorten the work of calculating binomial probabilities for some values of n and p by looking up probabilities in Table C in the back of this book. The entries in the table are the probabilities $P(X = k)$ of individual outcomes for a binomial random variable X.

EXAMPLE 5.5   Suppose that the audit in Example 5.3 chose just 15 sales records. What is the probability that no more than 1 of the 15 is misclassified? The count X of misclassified records in the sample has approximately the $B(15, 0.08)$ dis-

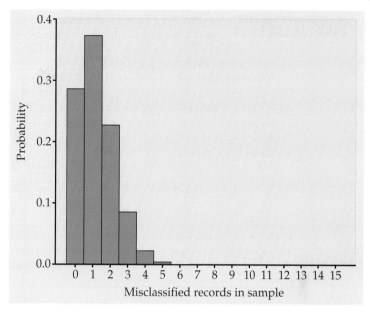

FIGURE 5.2   Probability histogram for the binomial distribution with $n = 15$ and $p = 0.08$, for Example 5.5.

tribution. Figure 5.2 is a probability histogram for this distribution. The distribution is strongly skewed. Although X can take any whole-number value from 0 to 15, the probabilities of values larger than 5 are so small that they do not appear in the histogram.

We want to calculate

$$P(X \leq 1) = P(X = 0) + P(X = 1)$$

when X is $B(15, 0.08)$. To use Table C for this calculation, look opposite $n = 15$ and under $p = 0.08$. This part of the table appears at the left. The entry opposite each k is $P(X = k)$. Blank entries are 0 to four decimal places, so we have omitted most of them here. You see that

$$P(X \leq 1) = P(X = 0) + P(X = 1)$$
$$= 0.2863 + 0.3734 = 0.6597$$

About two-thirds of all samples will contain no more than 1 bad record. In fact, almost 29% of the samples will contain no bad records. A sample of size 15 cannot be trusted to provide adequate evidence of the presence of misclassified sales records. Most state audit manuals require larger samples, such as the $n = 150$ used in Example 5.4.

		p
n	k	.08
15	0	.2863
	1	.3734
	2	.2273
	3	.0857
	4	.0223
	5	.0043
	6	.0006
	7	.0001
	8	
	9	

The values of p that appear in Table C are all 0.5 or smaller. When the probability of a success is greater than 0.5, restate the problem in terms of the number of failures. The probability of a failure is less than 0.5 when the probability of a success exceeds 0.5. When using the table, always stop to ask whether you must count successes or failures.

EXAMPLE 5.6   Corinne is a basketball player who makes 75% of her free throws over the course of a season. In a key game, Corinne shoots 12 free throws and misses 5 of them. The fans think that she failed because she was nervous. Is it unusual for Corinne to perform this poorly?

To answer this question, assume that free throws are independent with probability 0.75 of a success on each shot. (Studies of long sequences of free throws have found no evidence that they are dependent, so this is a reasonable assumption.) Because the probability of making a free throw is greater than 0.5, we count misses in order to use Table C. The probability of a miss is $1 - 0.75$, or 0.25. The number X of misses in 12 attempts has the $B(12, 0.25)$ distribution.

We want the probability of missing 5 or more. This is

$$P(X \geq 5) = P(X = 5) + P(X = 6) + \cdots + P(X = 12)$$

$$= 0.1032 + 0.0401 + \cdots + 0.0000 = 0.1576$$

Corinne will miss 5 or more out of 12 free throws about 16% of the time, or roughly one of every six games. While below her average level, this performance is well within the range of the usual chance variation in her shooting.

Binomial mean and standard deviation

If a count X is $B(n, p)$, what are the mean μ_X and the standard deviation σ_X? We can guess the mean. If Corinne makes 75% of her free throws, the mean number made in 12 tries should be 75% of 12, or 9. That's μ_X when X is $B(12, 0.75)$. Intuition suggests more generally that the mean of the $B(n, p)$ distribution should be np. Can we show that this is correct and also obtain a short formula for the standard deviation? Because binomial distributions are discrete probability distributions, we could find the mean and variance by using the definitions in Section 4.4. For example, we could obtain the mean by multiplying each outcome by its probability and adding over all outcomes. Here is an easier way.

A binomial random variable X is the count of successes in n independent observations that each have the same probability p of success. Let the random variable S_i indicate whether the ith observation is a success or failure by taking the values $S_i = 1$ if a success occurs and $S_i = 0$ if the outcome is a failure. The S_i are independent because the observations are, and each S_i has the same simple distribution:

Outcome	1	0
Probability	p	$1 - p$

From the definition of the mean of a discrete random variable (page 292), we know that the mean of each S_i is

$$\mu_S = (1)(p) + (0)(1 - p) = p$$

Similarly, the definition of the variance (page 300) shows that $\sigma_S^2 = p(1 - p)$. Because each S_i is 1 for a hit and 0 for a miss, the total number of hits X is just the sum of the S_i's:

$$X = S_1 + S_2 + \cdots + S_n$$

Apply the addition rules for means and variances to this sum. The mean of X is the sum of the means of the S_i's:

$$\mu_X = \mu_{S_1} + \mu_{S_2} + \cdots + \mu_{S_n}$$
$$= n\mu_S = np$$

Similarly, the variance is n times the variance of a single S, so that $\sigma_X^2 = np(1-p)$. The standard deviation σ_X is the square root of the variance. Here is the result.

BINOMIAL MEAN AND STANDARD DEVIATION

If a count X has the binomial distribution $B(n,p)$, then

$$\mu_X = np$$
$$\sigma_X = \sqrt{np(1-p)}$$

EXAMPLE 5.7   The Helsinki Heart Study asked whether the anticholesterol drug gemfibrozil reduces heart attacks. In planning such an experiment, the researchers must be confident that the sample sizes are large enough to enable them to observe enough heart attacks. The Helsinki study planned to give gemfibrozil to about 2000 men aged 40 to 55 and a placebo to another 2000. The probability of a heart attack during the five-year period of the study for men this age is about 0.04. What are the mean and standard deviation of the number of heart attacks that will be observed in one group if the treatment does not change this probability?

There are 2000 independent observations, each having probability $p = 0.04$ of a heart attack. The count X of heart attacks is $B(2000, 0.04)$, so that

$$\mu_X = np = (2000)(0.04) = 80$$
$$\sigma_X = \sqrt{np(1-p)} = \sqrt{(2000)(0.04)(0.96)} = 8.76$$

The expected number of heart attacks is large enough to permit conclusions about the effectiveness of the drug. In fact, there were 84 heart attacks among the 2035 men actually assigned to the placebo, quite close to the mean. The gemfibrozil group of 2046 men suffered only 56 heart attacks. This is evidence that the drug does reduce the chance of a heart attack.

Sample proportions

What proportion of a company's sales records have an incorrect sales tax classification? What percent of adults favor stronger laws restricting firearms? In statistical sampling we often want to estimate the **proportion** p of "successes" in a population. Our estimator is the sample proportion of successes:

$$\hat{p} = \frac{\text{count of successes in sample}}{\text{size of sample}}$$
$$= \frac{X}{n}$$

Be sure to distinguish between the proportion $\hat{p}$ and the count X. The count takes whole-number values between 0 and n, but a proportion is always a number between 0 and 1. In the binomial setting, the count X has a binomial distribution. The proportion $\hat{p}$ does *not* have a binomial distribution. We can, however, do probability calculations about $\hat{p}$ by restating them in terms of the count X and using binomial methods. As an example, we return to the sample survey that we studied by simulation in Example 3.26 (page 233).

EXAMPLE 5.8   A sample survey asks a nationwide random sample of 2500 adults if they agree or disagree that "I like buying new clothes, but shopping is often frustrating and time-consuming." Suppose that 60% of all adults would agree if asked this question. What is the probability that the sample proportion who agree is at least 58%?

The count X who agree has the binomial distribution $B(2500, 0.6)$. The sample proportion $\hat{p} = X/2500$ does *not* have a binomial distribution, because it is not a count. We therefore translate any question about the sample proportion $\hat{p}$ into a question about the count X. Because 58% of 2500 is 1450,

$$P(\hat{p} \geq 0.58) = P(X \geq 1450)$$

$$= P(X = 1450) + P(X = 1451) + \cdots + P(X = 2500)$$

This is a rather elaborate calculation. We must add more than 1000 binomial probabilities. Software tells us that $P(\hat{p} \geq 0.58) = 0.9802$. Because some software packages cannot handle an n as large as 2500, we need another way to do this calculation.

As a first step, find the mean and standard deviation of a sample proportion. We know the mean and standard deviation of a sample count, so apply the rules from Section 4.4 for the mean and variance of a constant times a random variable. Here is the result.

> ## MEAN AND STANDARD DEVIATION OF A SAMPLE PROPORTION
>
> Let $\hat{p}$ be the sample proportion of successes in an SRS of size n drawn from a large population having population proportion p of successes. The mean and standard deviation of $\hat{p}$ are
>
> $$\mu_{\hat{p}} = p$$
>
> $$\sigma_{\hat{p}} = \sqrt{\frac{p(1-p)}{n}}$$

The formula for $\sigma_{\hat{p}}$ is exactly correct in the binomial setting. It is approximately correct for an SRS from a large population. We will use it when the population is at least 20 times as large as the sample.

EXAMPLE 5.9   The mean and standard deviation of the proportion of the survey respondents in Example 5.8 who find shopping frustrating are

$$\mu_{\hat{p}} = p = 0.6$$

$$\sigma_{\hat{p}} = \sqrt{\frac{p(1-p)}{n}} = \sqrt{\frac{(0.6)(0.4)}{2500}} = 0.0098$$

The fact that the mean of $\hat{p}$ is p states in statistical language that the sample proportion $\hat{p}$ in an SRS is an *unbiased estimator* of the population proportion p. When a sample is drawn from a new population having a different value of the population proportion p, the sampling distribution of the unbiased estimator $\hat{p}$ changes so that its mean moves to the new value of p. We observed this fact empirically in Section 3.4 and have now verified it from the laws of probability.

The variability of $\hat{p}$ about its mean, as described by the variance or standard deviation, decreases as the sample size increases. So a sample proportion from a large sample will usually lie quite close to the population proportion p. We observed this in the simulation experiment on page 233 in Section 3.4. Now we have discovered exactly how the variability decreases: the standard deviation is $\sqrt{p(1-p)/n}$. *The $\sqrt{n}$ in the denominator means that the sample size must be multiplied by 4 if we wish to divide the standard deviation in half.*

Normal approximation for counts and proportions

We discovered empirically in Section 3.4 that the sampling distribution of a sample proportion $\hat{p}$ is close to normal. Now we know that the distribution of $\hat{p}$ is that of a binomial count divided by the sample size n. This seems at first to be a contradiction. To clear up the matter, look at Figure 5.3. This is a probability histogram of the exact distribution of the proportion of frustrated shoppers $\hat{p}$, based on the binomial distribution $B(2500, 0.6)$. There are hundreds of narrow bars, one for each of the 2501 possible values of $\hat{p}$. Most have probabilities too small to show in a graph. *The probability histogram looks very normal!* In fact, both the count X and the sample proportion $\hat{p}$ are approximately normal in large samples.

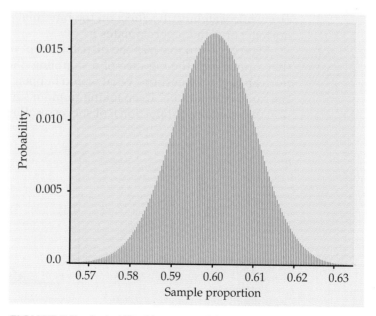

FIGURE 5.3   Probability histogram of the sample proportion $\hat{p}$ based on a binomial count with $n = 2500$ and $p = 0.6$. The distribution is very close to normal.

NORMAL APPROXIMATION FOR COUNTS AND PROPORTIONS

Draw an **SRS** of size n from a large population having population proportion p of successes. Let X be the count of successes in the sample and $\hat{p} = X/n$ the sample proportion of successes. When n is large, the sampling distributions of these statistics are approximately normal:

$$X \text{ is approximately } N\left(np, \sqrt{np(1-p)}\right)$$

$$\hat{p} \text{ is approximately } N\left(p, \sqrt{\frac{p(1-p)}{n}}\right)$$

As a rule of thumb, we will use this approximation for values of n and p that satisfy $np \geq 10$ and $n(1-p) \geq 10$.

These normal approximations are easy to remember because they say that $\hat{p}$ and X are normal with their usual means and standard deviations. Whether or not you use the normal approximations should depend on how accurate your calculations need to be. For most statistical purposes great accuracy is not required. Our "rule of thumb" for use of the normal approximations reflects this judgment.

The accuracy of the normal approximations improves as the sample size n increases. They are most accurate for any fixed n when p is close to $1/2$, and least accurate when p is near 0 or 1. You can compare binomial distributions with their normal approximations by using the *Normal Approximation to Binomial* applet on the text CD and Web site. The applet allows you to change n or p while watching the effect on the binomial probability histogram and the normal curve that approximates it.

Figure 5.4 summarizes the distribution of a sample proportion in a form that emphasizes the big idea of a sampling distribution. Sampling distributions answer the question "What would happen if we took many samples from the same population?" Keep taking random samples of size n from a population that contains proportion p of successes. Find the sample proportion $\hat{p}$ for

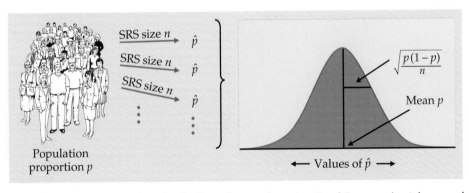

FIGURE 5.4 The sampling distribution of a sample proportion $\hat{p}$ is approximately normal with mean p and standard deviation $\sqrt{p(1-p)/n}$.

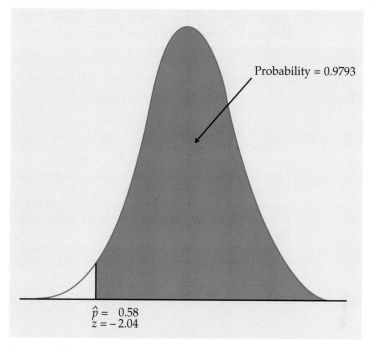

FIGURE 5.5   The normal probability calculation for Example 5.10.

each sample. Collect all the $\hat{p}$'s and display their distribution. That's the sampling distribution of $\hat{p}$.

EXAMPLE 5.10   Let's compare the normal approximation for the calculation of Example 5.8 with the exact calculation from software. We want to calculate $P(\hat{p} \geq 0.58)$ when the sample size is $n = 2500$ and the population proportion is $p = 0.6$. Example 5.9 shows that

$$\mu_{\hat{p}} = p = 0.6$$

$$\sigma_{\hat{p}} = \sqrt{\frac{p(1-p)}{n}} = 0.0098$$

Act as if $\hat{p}$ were normal with mean 0.6 and standard deviation 0.0098. The approximate probability, as illustrated in Figure 5.5, is

$$P(\hat{p} \geq 0.58) = P\left(\frac{\hat{p} - 0.6}{0.0098} \geq \frac{0.58 - 0.6}{0.0098}\right)$$

$$\doteq P(Z \geq -2.04) = 0.9793$$

That is, about 98% of all samples have a sample proportion that is at least 0.58. Because the sample was large, this normal approximation is quite accurate. It misses the software value 0.9802 by only 0.0009.

EXAMPLE 5.11   The audit described in Example 5.3 examined an SRS of 150 sales records for compliance with sales tax laws. In fact, 8% of all the company's sales records have an incorrect sales tax classification. The count X of bad records in the sample has approximately the $B(150, 0.08)$ distribution. The distribution of the count of bad records in a sample of 15 is distinctly nonnormal, as Figure 5.2 showed. When the sample size is increased to 150, however, the shape of the binomial distribution becomes roughly normal.

According to the normal approximation to the binomial distributions, the count X is approximately normal with mean and standard deviation

$$\mu_X = np = (150)(0.08) = 12$$

$$\sigma_X = \sqrt{np(1-p)} = \sqrt{(150)(0.08)(0.92)} = 3.3226$$

Figure 5.6 displays the probability histogram of the binomial distribution with the density curve of the approximating normal distribution superimposed. Both distributions have the same mean and standard deviation, and both the area under the histogram and the area under the curve are 1. The normal curve fits the histogram reasonably well. Look closely: the histogram is slightly skewed to the right, a property that the symmetric normal curve can't match. The normal approximation is less accurate for $n = 150$ than for $n = 2500$ in the previous example.

The normal approximation for the probability of no more than 10 misclassified records is the area to the left of $X = 10$ under the normal curve. Using Table A,

$$P(X \leq 10) = P\left(\frac{X-12}{3.3226} \leq \frac{10-12}{3.3226}\right)$$

$$\doteq P(Z \leq -0.60) = 0.2743$$

Software tells us that the actual binomial probability that no more than 10 of the records in the sample are misclassified is $P(X \leq 10) = 0.3384$. The normal approximation is only roughly accurate. Because $np = 12$, this combination of n and p is close to the border of the values for which we are willing to use the approximation.

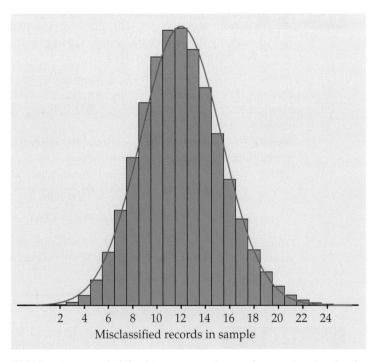

FIGURE 5.6   Probability histogram and normal approximation for the binomial distribution with $n = 150$ and $p = 0.08$, for Example 5.11.

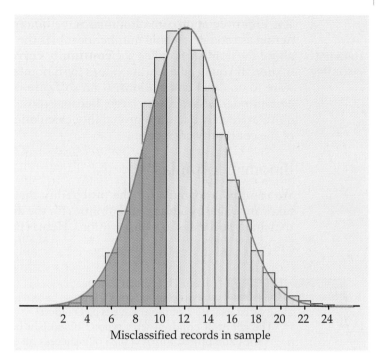

Misclassified records in sample

FIGURE 5.7   Area under the normal approximation curve for the probability in Example 5.11.

The continuity correction*

Figure 5.7 illustrates an idea that greatly improves the accuracy of the normal approximation to binomial probabilities. The binomial probability $P(X \leq 10)$ is the area of the histogram bars for values 0 to 10. The bar for $X = 10$ actually extends from 9.5 to 10.5. Because the discrete binomial distribution puts probability only on whole numbers, the probabilities $P(X \leq 10)$ and $P(X \leq 10.5)$ are the same. The normal distribution spreads probability continuously, so these two normal probabilities are different. The normal approximation is more accurate if we consider $X = 10$ to extend from 9.5 to 10.5, matching the bar in the probability histogram.

The event $\{X \leq 10\}$ includes the outcome $X = 10$. Figure 5.7 shades the area under the normal curve that matches all the histogram bars for outcomes 0 to 10, bounded on the right not by 10, but by 10.5. So $P(X \leq 10)$ is calculated as $P(X \leq 10.5)$. On the other hand, $P(X < 10)$ excludes the outcome $X = 10$, so we exclude the entire interval from 9.5 to 10.5 and calculate $P(X \leq 9.5)$ from the normal table. Here is the result of the normal calculation in Example 5.11 improved in this way:

$$P(X \leq 10) = P(X \leq 10.5)$$

$$= P\left(\frac{X - 12}{3.3226} \leq \frac{10.5 - 12}{3.3226}\right)$$

$$\doteq P(Z \leq -0.45) = 0.3264$$

*This material can be omitted if desired.

The improved approximation misses the binomial probability by only 0.012. Acting as though a whole number occupies the interval from 0.5 below to 0.5 above the number is called the **continuity correction** to the normal approximation. If you need accurate values for binomial probabilities, try to use software to do exact calculations. If no software is available, use the continuity correction unless n is very large. Because most statistical purposes do not require extremely accurate probability calculations, we do not emphasize use of the continuity correction.

continuity correction

Binomial formulas*

We can find a formula for the probability that a binomial random variable takes any value by adding probabilities for the different ways of getting exactly that many successes in n observations. Here is the example we will use to show the idea.

EXAMPLE 5.12 Each child born to a particular set of parents has probability 0.25 of having blood type O. If these parents have 5 children, what is the probability that exactly 2 of them have type O blood?

The count of children with type O blood is a binomial random variable X with $n = 5$ tries and probability $p = 0.25$ of a success on each try. We want $P(X = 2)$.

Because the method doesn't depend on the specific example, we will use "S" for success and "F" for failure. In Example 5.12, "S" would stand for type O blood. Do the work in two steps.

Step 1: Find the probability that a specific 2 of the 5 tries give successes, say the first and the third. This is the outcome SFSFF. The multiplication rule for independent events tells us that

$$P(\text{SFSFF}) = P(S)P(F)P(S)P(F)P(F)$$

$$= (0.25)(0.75)(0.25)(0.75)(0.75)$$

$$= (0.25)^2(0.75)^3$$

Step 2: Observe that the probability of *any one* arrangement of 2 S's and 3 F's has this same probability. That's true because we multiply together 0.25 twice and 0.75 three times whenever we have 2 S's and 3 F's. The probability that $X = 2$ is the probability of getting 2 S's and 3 F's in any arrangement whatsoever. Here are all the possible arrangements:

SSFFF   SFSFF   SFFSF   SFFFS   FSSFF
FSFSF   FSFFS   FFSSF   FFSFS   FFFSS

There are 10 of them, all with the same probability. The overall probability of 2 successes is therefore

$$P(X = 2) = 10(0.25)^2(0.75)^3 = 0.2637$$

The pattern of this calculation works for any binomial probability. To use it, we need to be able to count the number of arrangements of k successes in

---

*The formula for binomial probabilities is useful in many settings, but we will not need it in our study of statistical inference. This section can therefore be omitted if desired.

n observations without actually listing them. We use the following fact to do the counting.

> ## BINOMIAL COEFFICIENT
>
> The number of ways of arranging k successes among n observations is given by the **binomial coefficient**
>
> $$\binom{n}{k} = \frac{n!}{k!\,(n-k)!}$$
>
> for $k = 0, 1, 2, \ldots, n$.

factorial   The formula for binomial coefficients uses the **factorial** notation. The factorial $n!$ for any positive whole number n is

$$n! = n \times (n-1) \times (n-2) \times \cdots \times 3 \times 2 \times 1$$

Also, $0! = 1$. Notice that the larger of the two factorials in the denominator of a binomial coefficient will cancel much of the $n!$ in the numerator. For example, the binomial coefficient we need for Example 5.12 is

$$\binom{5}{2} = \frac{5!}{2!\,3!}$$

$$= \frac{(5)(4)(3)(2)(1)}{(2)(1) \times (3)(2)(1)}$$

$$= \frac{(5)(4)}{(2)(1)} = \frac{20}{2} = 10$$

This agrees with our earlier enumeration.

The notation $\binom{n}{k}$ is not related to the fraction $\frac{n}{k}$. A helpful way to remember its meaning is to read it as "binomial coefficient n choose k." Binomial coefficients have many uses in mathematics, but we are interested in them only as an aid to finding binomial probabilities. The binomial coefficient $\binom{n}{k}$ counts the number of ways in which k successes can be distributed among n observations. The binomial probability $P(X = k)$ is this count multiplied by the probability of any specific arrangement of the k successes. Here is the formula we seek.

> ## BINOMIAL PROBABILITY
>
> If X has the binomial distribution $B(n, p)$ with n observations and probability p of success on each observation, the possible values of X are $0, 1, 2, \ldots, n$. If k is any one of these values, the **binomial probability** is
>
> $$P(X = k) = \binom{n}{k} p^k (1-p)^{n-k}$$

Here is an example of the use of the binomial probability formula.

EXAMPLE 5.13   The number X of misclassified sales records in the auditor's sample in Example 5.5 has the $B(15, 0.08)$ distribution. The probability of finding no more than 1 misclassified record is

$$P(X \leq 1) = P(X = 0) + P(X = 1)$$

$$= \binom{15}{0}(0.08)^0(0.92)^{15} + \binom{15}{1}(0.08)^1(0.92)^{14}$$

$$= \frac{15!}{0!\,15!}(1)(0.2863) + \frac{15!}{1!\,14!}(0.08)(0.3112)$$

$$= (1)(1)(0.2863) + (15)(0.08)(0.3112)$$

$$= 0.2863 + 0.3734 = 0.6597$$

The calculation used the facts that $0! = 1$ and that $a^0 = 1$ for any number $a \neq 0$. The result agrees with that obtained from Table C in Example 5.5.

SECTION 5.1 | Summary

A **count** X of successes has the **binomial distribution** $B(n, p)$ in the **binomial setting:** there are n trials, all independent, each resulting in a success or a failure, and each having the same probability p of a success.

Binomial probabilities are most easily found by software. There is an exact formula that is practical for calculations when n is small. Table C contains binomial probabilities for some values of n and p. For large n, you can use the normal approximation.

The binomial distribution $B(n, p)$ is a good approximation to the **sampling distribution of the count of successes** in an SRS of size n from a large population containing proportion p of successes. We will use this approximation when the population is at least 20 times larger than the sample.

The mean and standard deviation of a **binomial count** X and a **sample proportion** of successes $\hat{p} = X/n$ are

$$\mu_X = np \qquad\qquad \mu_{\hat{p}} = p$$

$$\sigma_X = \sqrt{np(1-p)} \qquad \sigma_{\hat{p}} = \sqrt{\frac{p(1-p)}{n}}$$

The sample proportion $\hat{p}$ is therefore an unbiased estimator of the population proportion p.

The **normal approximation** to the binomial distribution says that if X is a count having the $B(n, p)$ distribution, then when n is large,

$$X \text{ is approximately } N\left(np, \sqrt{np(1-p)}\right)$$

$$\hat{p} \text{ is approximately } N\left(p, \sqrt{\frac{p(1-p)}{n}}\right)$$

We will use these approximations when $np \geq 10$ and $n(1 - p) \geq 10$. The **continuity correction** improves the accuracy of the normal approximations.

The exact **binomial probability formula** is

$$P(X = k) = \binom{n}{k} p^k (1 - p)^{n-k}$$

where the possible values of X are $k = 0, 1, \ldots, n$. The binomial probability formula uses the **binomial coefficient**

$$\binom{n}{k} = \frac{n!}{k! \, (n - k)!}$$

Here the **factorial** $n!$ is

$$n! = n \times (n - 1) \times (n - 2) \times \cdots \times 3 \times 2 \times 1$$

for positive whole numbers n and $0! = 1$. The binomial coefficient counts the number of ways of distributing k successes among n trials.

SECTION 5.1 | Exercises

Most binomial probability calculations required in these exercises can be done by using Table C or the normal approximation. Your instructor may request that you use the binomial probability formula or software. In exercises requiring the normal approximation, you should use the continuity correction if you studied that topic.

5.1   In each situation below, is it reasonable to use a binomial distribution for the random variable X? Give reasons for your answer in each case. If a binomial distribution applies, give the values of n and p.

(a) Most calls made at random by sample surveys don't succeed in talking with a live person. Of calls to New York City, only 1/12 succeed. A survey calls 500 randomly selected numbers in New York City. X is the number that reach a live person.

(b) At peak periods, 25% of attempted logins to an Internet service provider fail. Login attempts are independent and each has the same probability of failing. Darci logs in repeatedly until she succeeds. X is the number of the login attempt that finally succeeds.

(c) On a bright October day, Canada geese arrive to foul the pond at an apartment complex at the average rate of 12 geese per hour; X is the number of geese that arrive in the next three hours.

5.2   In each situation below, is it reasonable to use a binomial distribution for the random variable X? Give reasons for your answer in each case.

(a) An auto manufacturer chooses one car from each hour's production for a detailed quality inspection. One variable recorded is the count X of finish defects (dimples, ripples, etc.) in the car's paint.

(b) The pool of potential jurors for a murder case contains 100 persons chosen at random from the adult residents of a large city. Each person in the pool is asked whether he or she opposes the death penalty; X is the number who say "Yes."

(c) Joe buys a ticket in his state's "Pick 3" lottery game every week; X is the number of times in a year that he wins a prize.

5.3   Typographic errors in a text are either nonword errors (as when "the" is typed as "teh") or word errors that result in a real but incorrect word. Spell-checking software will catch nonword errors but not word errors. Human proofreaders catch 70% of word errors. You ask a fellow student to proofread an essay in which you have deliberately made 20 word errors.

(a) If the student matches the usual 70% rate, what is the distribution of the number of errors caught? What is the distribution of the number of errors missed?

(b) Missing 9 or more out of 20 errors seems a poor performance. What is the probability that a proofreader who catches 70% of word errors misses 9 or more out of 20?

5.4   What kinds of Web sites do males aged 18 to 34 visit most often? Pornographic sites take first place, but about 50% of male Internet users in this age group visit an auction site such as eBay at least once a month.[2] Interview a random sample of 12 male Internet users aged 18 to 34.

(a) What is the distribution of the number who have visited an online auction site in the past month?

(b) What is the probability that at least 8 of the 12 have visited an auction site in the past month?

5.5   Return to the proofreading setting of Exercise 5.3.

(a) What is the mean number of errors caught? What is the mean number of errors missed? You see that these two means must add to 20, the total number of errors.

(b) What is the standard deviation σ of the number of errors caught?

(c) Suppose that a proofreader catches 90% of word errors, so that $p = 0.9$. What is σ in this case? What is σ if $p = 0.99$? What happens to the standard deviation of a binomial distribution as the probability of a success gets close to 1?

5.6   Suppose that 50% of male Internet users aged 18 to 34 have visited an auction site at least once in the past month.

(a) If you interview 12 at random, what is the mean of the count X who have visited an auction site? What is the mean of the proportion $\hat{p}$ in your sample who have visited an auction site?

(b) Repeat the calculations in (a) for samples of size 120 and 1200. What happens to the mean count of successes as the sample size increases? What happens to the mean proportion of successes?

5.7   In the proofreading setting of Exercise 5.3, what is the smallest number of misses m with $P(X \geq m)$ no larger than 0.05? You might consider m or more misses as evidence that a proofreader actually catches fewer than 70% of word errors.

5.8   Some of the methods in this section are approximations rather than exact probability results. We have given rules of thumb for safe use of these approximations.

(a) You are interested in attitudes toward drinking among the 75 members of a fraternity. You choose 25 members at random to interview. One question is "Have you had five or more drinks at one time during the last week?" Suppose that in fact 20% of the 75 members would say "Yes." Explain why you *cannot* safely use the $B(25, 0.2)$ distribution for the count X in your sample who say "Yes."

(b) The National AIDS Behavioral Surveys found that 0.2% (that's 0.002 as a decimal fraction) of adult heterosexuals had both received a blood transfusion and had a sexual partner from a group at high risk of AIDS. Suppose that this national proportion holds for your region. Explain why you *cannot* safely use the normal approximation for the sample proportion who fall in this group when you interview an SRS of 500 adults.

5.9   Each entry in a table of random digits like Table B has probability 0.1 of being a 0, and digits are independent of each other.

(a) What is the probability that a group of five digits from the table will contain at least one 0?

(b) What is the mean number of 0s in lines 40 digits long?

5.10   The *Probability* applet on the text CD and Web site simulates tosses of a coin. You can choose the number of tosses n and the probability p of a head. You can therefore use the applet to simulate binomial random variables.

   The count of misclassified sales records in Example 5.5 (page 338) has the binomial distribution with $n = 15$ and $p = 0.08$. Set these values for the number of tosses and probability of heads in the applet. Table C shows that the probability of getting a sample with exactly 1 misclassified record is 0.3734. This is the long-run proportion of samples with exactly 1 bad record. Click "Toss" and "Reset" repeatedly to simulate 20 samples. Record the number of bad records (the count of heads) in each of the 20 samples. What proportion of the 20 samples had exactly 1 bad record? Remember that probability tells us only what happens in the long run

5.11   In 1998, Mark McGwire of the St. Louis Cardinals hit 70 home runs, a new Major League record. Was this feat as surprising as most of us thought? In the three seasons before 1998, McGwire hit a home run in 11.6% of his times at bat. He went to bat 509 times in 1998. If he continues his past performance, McGwire's home run count in 509 times at bat has approximately the binomial distribution with $n = 509$ and $p = 0.116$.

(a) What is the mean number of home runs McGwire will hit in 509 times at bat?

(b) What is the probability that he hits 70 or more home runs?

(c) In 2001, Barry Bonds of the San Francisco Giants hit 73 home runs, breaking McGwire's record. This was surprising. In the three previous seasons, Bonds hit a home run in 8.65% of his times at bat. He batted 476 times in 2001. Considering his home run count as a binomial random variable with $n = 476$ and $p = 0.0865$, what is the probability of 73 or more home runs?

5.12 A believer in the "random walk" theory of the behavior of stock prices thinks that an index of stock prices has probability 0.65 of increasing in any year. Moreover, the change in the index in any given year is not influenced by whether it rose or fell in earlier years. Let X be the number of years among the next 6 years in which the index rises.

(a) What are n and p in the binomial distribution of X?

(b) Give the possible values that X can take and the probability of each value. Draw a probability histogram for the distribution of X.

(c) Find the mean of the number X of years in which the stock price index rises and mark the mean on your probability histogram.

(d) Find the standard deviation of X. What is the probability that X takes a value within one standard deviation of its mean?

5.13 Children inherit their blood type from their parents, with probabilities that reflect the parents' genetic makeup. Children of Juan and Maria each have probability 1/4 of having blood type A and inherit independently of each other. Juan and Maria plan to have 4 children; let X be the number who have blood type A.

(a) What are n and p in the binomial distribution of X?

(b) Find the probability of each possible value of X, and draw a probability histogram for this distribution.

(c) Find the mean number of children with type A blood, and mark the location of the mean on your probability histogram.

5.14 "What do you think is the ideal number of children for a family to have?" A Gallup poll asked this question of 1006 randomly chosen adults. Almost half (49%) thought two children was ideal.[3] Suppose that $p = 0.49$ is exactly true for the population of all adults. Gallup announced a margin of error of ± 3 percentage points for this poll. What is the probability that the sample proportion $\hat{p}$ for an SRS of size $n = 1006$ falls between 0.46 and 0.52? You see that it is likely, but not certain, that polls like this give results that are correct within their margin of error. We will say more about margins of error in Chapter 6.

5.15 A Gallup poll finds that 30% of adults visited a casino in the past 12 months, and that 6% bet on college sports.[4] These results come from a random sample of 1011 adults. For an SRS of size $n = 1011$:

(a) What is the probability that the sample proportion $\hat{p}$ is between 0.29 and 0.31 if the population proportion is $p = 0.30$?

(b) What is the probability that the sample proportion $\hat{p}$ is between 0.05 and 0.07 if the population proportion is $p = 0.06$?

(c) How does the probability that $\hat{p}$ falls within ±0.01 of the true p change as p gets closer to 0?

5.16 Return to the Gallup poll setting of Exercise 5.14. We are supposing that the proportion of all adults who think that two children is ideal is $p = 0.49$. What is the probability that a sample proportion $\hat{p}$ falls between 0.46 and 0.52 (that is, within ±3 percentage points of the true p) if the sample is an SRS of size $n = 250$? Of size $n = 4000$? Combine these results with your work in Exercise 5.14 to make a general statement about the effect of larger samples in a sample survey.

5.17 The Harvard College Alcohol Study finds that 67% of college students support efforts to "crack down on underage drinking." The study took a sample of almost 15,000 students, so the population proportion who support a crackdown is very close to $p = 0.67$.[5] The administration of your college surveys an SRS of 100 students and finds that 62 support a crackdown on underage drinking.

(a) What is the sample proportion who support a crackdown on underage drinking?

(b) If in fact the proportion of all students on your campus who support a crackdown is the same as the national 67%, what is the probability that the proportion in an SRS of 100 students is as small or smaller than the result of the administration's sample?

(c) A writer in the student paper says that support for a crackdown is lower on your campus than nationally. Write a short letter to the editor explaining why the survey does not support this conclusion.

5.18 The changing probabilities you found in Exercises 5.14 and 5.16 are due to the fact that the standard deviation of the sample proportion $\hat{p}$ gets smaller as the sample size n increases. If the population proportion is $p = 0.49$, how large a sample is needed to reduce the standard deviation of $\hat{p}$ to $\sigma_{\hat{p}} = 0.005$? (The 68–95–99.7 rule then says that about 95% of all samples will have $\hat{p}$ within 0.01 of the true p.)

5.19 In a test for ESP (extrasensory perception), the experimenter looks at cards that are hidden from the subject. Each card contains either a star, a circle, a wave, or a square. As the experimenter looks at each of 20 cards in turn, the subject names the shape on the card.

(a) If a subject simply guesses the shape on each card, what is the probability of a successful guess on a single card? Because the cards are independent, the count of successes in 20 cards has a binomial distribution.

(b) What is the probability that a subject correctly guesses at least 10 of the 20 shapes?

(c) In many repetitions of this experiment with a subject who is guessing, how many cards will the subject guess correctly on the average? What is the standard deviation of the number of correct guesses?

(d) A standard ESP deck actually contains 25 cards. There are five different shapes, each of which appears on 5 cards. The subject knows that the deck has this makeup. Is a binomial model still appropriate for the count of correct guesses in one pass through this deck? If so, what are n and p? If not, why not?

5.20   A selective college would like to have an entering class of 1200 students. Because not all students who are offered admission accept, the college admits more than 1200 students. Past experience shows that about 70% of the students admitted will accept. The college decides to admit 1500 students. Assuming that students make their decisions independently, the number who accept has the $B(1500, 0.7)$ distribution. If this number is less than 1200, the college will admit students from its waiting list.

(a) What are the mean and the standard deviation of the number X of students who accept?

(b) Use the normal approximation to find the probability that at least 1000 students accept.

(c) The college does not want more than 1200 students. What is the probability that more than 1200 will accept?

(d) If the college decides to increase the number of admission offers to 1700, what is the probability that more than 1200 will accept?

5.21   When the ESP study of Exercise 5.19 discovers a subject whose performance appears to be better than guessing, the study continues at greater length. The experimenter looks at many cards bearing one of five shapes (star, square, circle, wave, and cross) in an order determined by random numbers. The subject cannot see the experimenter as he looks at each card in turn, in order to avoid any possible nonverbal clues. The answers of a subject who does not have ESP should be independent observations, each with probability 1/5 of success. We record 1000 attempts.

(a) What are the mean and the standard deviation of the count of successes?

(b) What are the mean and standard deviation of the proportion of successes among the 1000 attempts?

(c) What is the probability that a subject without ESP will be successful in at least 24% of 1000 attempts?

(d) The researcher considers evidence of ESP to be a proportion of successes so large that there is only probability 0.01 that a subject could do this well or better by guessing. What proportion of successes must a subject have to meet this standard? (Example 1.30, page 79, shows how to do a normal calculation of the type required here.)

5.22   The mailing list of an agency that markets scuba-diving trips to the Florida Keys contains 70% males and 30% females. The agency calls 30 people chosen at random from its list.

(a) What is the probability that 20 of the 30 are men? (Use the binomial probability formula.)

(b) What is the probability that the first woman is reached on the fourth call? (That is, the first 4 calls give MMMF.)

5.23   One way of checking the effect of undercoverage, nonresponse, and other sources of error in a sample survey is to compare the sample with known demographic facts about the population. The 2000 census found that 23,772,494 of the 209,128,094 adults (age 18 and over) in the United States called themselves "Black or African American."

   (a) What is the population proportion p of blacks among American adults?

   (b) An opinion poll chooses 1500 adults at random. What is the mean number of blacks in such samples? (Explain the reasoning behind your calculation.)

   (c) Use a normal approximation to find the probability that such a sample will contain 170 or fewer blacks. Be sure to check that you can safely use the approximation.

5.24   Use the definition of binomial coefficients to show that each of the following facts is true. Then restate each fact in words in terms of the number of ways that k successes can be distributed among n observations.

   (a) $\binom{n}{n} = 1$ for any whole number $n \geq 1$.

   (b) $\binom{n}{n-1} = n$ for any whole number $n \geq 1$.

   (c) $\binom{n}{k} = \binom{n}{n-k}$ for any n and k with $k \leq n$.

5.25   Here is a simple probability model for multiple-choice tests. Suppose that each student has probability p of correctly answering a question chosen at random from a universe of possible questions. (A strong student has a higher p than a weak student.) The correctness of an answer to a question is independent of the correctness of answers to other questions. Jodi is a good student for whom $p = 0.75$.

   (a) Use the normal approximation to find the probability that Jodi scores 70% or lower on a 100-question test.

   (b) If the test contains 250 questions, what is the probability that Jodi will score 70% or lower?

   (c) How many questions must the test contain in order to reduce the standard deviation of Jodi's proportion of correct answers to half its value for a 100-item test?

   (d) Laura is a weaker student for whom $p = 0.6$. Does the answer you gave in (c) for the standard deviation of Jodi's score apply to Laura's standard deviation also?

5.26   You are tossing a balanced die that has probability 1/6 of coming up 1 on each toss. Tosses are independent. We are interested in how long we must wait to get the first 1.

   (a) The probability of a 1 on the first toss is 1/6. What is the probability that the first toss is not a 1 and the second toss is a 1?

   (b) What is the probability that the first two tosses are not 1s and the third toss is a 1? This is the probability that the first 1 occurs on the third toss.

(c) Now you see the pattern. What is the probability that the first 1 occurs on the fourth toss? On the fifth toss?

5.27 Generalize your work in Exercise 5.26. You have independent trials, each resulting in a success or a failure. The probability of a success is p on each trial. The binomial distribution describes the count of successes in a fixed number of trials. Now the number of trials is not fixed; instead, continue until you get a success. The random variable Y is the number of the trial on which the first success occurs. What are the possible values of Y? What is the probability $P(Y = k)$ for any of these values? (*Comment:* The distribution of the number **geometric** of trials to the first success is called a **geometric distribution.**)
distribution

5.2 The Sampling Distribution of a Sample Mean

Counts and proportions are discrete random variables that describe categorical data. The statistics most often used to describe quantitative data, on the other hand, are continuous random variables. The sample mean, percentiles, and standard deviation are examples of statistics based on quantitative data. Statistical theory describes the sampling distributions of these statistics. In this section we will concentrate on the sample mean. Because sample means are just averages of observations, they are among the most common statistics.

EXAMPLE 5.14 Figure 5.8 illustrates two striking facts about the sampling distribution of a sample mean. Figure 5.8(a) displays the distribution of customer service call lengths for a bank service center for a month. There are more than 30,000 calls in this population.[6] (We omitted a few extreme outliers, calls that lasted more than 20 minutes.) The distribution is extremely skewed to the right. The population mean is $\mu = 173.95$ seconds.

Table 1.1 (page 9) contains the lengths of a sample of 80 calls from this population. The mean of these 80 calls is $\bar{x} = 196.6$ seconds. If we take more samples, we will get different values of $\bar{x}$. To find the sampling distribution of $\bar{x}$, take many random samples of size 80 and calculate $\bar{x}$ for each sample. Figure 5.8(b) is the distribution of the values of $\bar{x}$ for 500 samples. The scales and choice of classes are exactly the same as in Figure 5.8(a), so that we can make a direct comparison.

The sample means are much less spread out than the individual call lengths. What is more, the distribution in Figure 5.8(b) is roughly symmetric rather than skewed. The normal quantile plot in Figure 5.9 confirms that the distribution is close to normal.

Example 5.14 illustrates two important facts that we will make precise in this section:

Averages are less variable than individual observations.

Averages are more normal than individual observations.

These two facts contribute to the popularity of sample means in statistical inference.

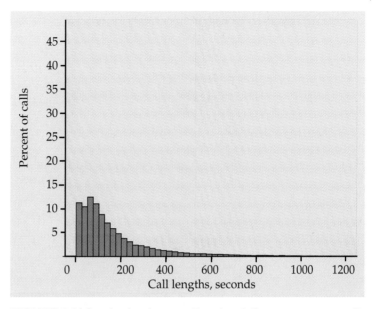

FIGURE 5.8(a)   The distribution of lengths of all customer service calls received by a bank in a month.

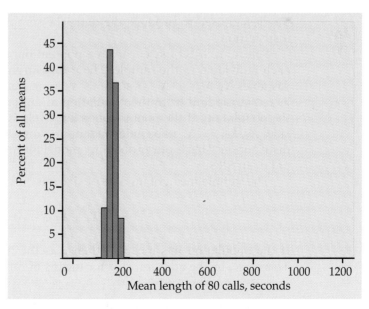

FIGURE 5.8(b)   The distribution of the sample means $\bar{x}$ for 500 random samples of size 80 from this population. The scales and histogram classes are exactly as in Figure 5.8(a).

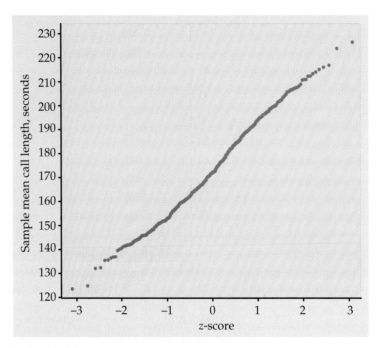

FIGURE 5.9 Normal quantile plot of the 500 sample means in Figure 5.8(b). The distribution is close to normal.

The mean and standard deviation of $\overline{x}$

The sample mean $\overline{x}$ from a sample or an experiment is an estimate of the mean μ of the underlying population, just as a sample proportion $\hat{p}$ is an estimate of a population proportion p. The sampling distribution of $\overline{x}$ is determined by the design used to produce the data, the sample size n, and the population distribution.

Select an SRS of size n from a population, and measure a variable X on each individual in the sample. The n measurements are values of n random variables $X_1, X_2, \ldots, X_n$. A single X_i is a measurement on one individual selected at random from the population and therefore has the distribution of the population. If the population is large relative to the sample, we can consider $X_1, X_2, \ldots, X_n$ to be independent random variables each having the same distribution. This is our probability model for measurements on each individual in an SRS.

The sample mean of an SRS of size n is

$$\overline{x} = \frac{1}{n}(X_1 + X_2 + \cdots + X_n)$$

If the population has mean μ, then μ is the mean of each observation X_i. Therefore, by the addition rule for means of random variables,

$$\mu_{\overline{x}} = \frac{1}{n}(\mu_{X_1} + \mu_{X_2} + \cdots + \mu_{X_n})$$

$$= \frac{1}{n}(\mu + \mu + \cdots + \mu) = \mu$$

That is, *the mean of $\bar{x}$ is the same as the mean of the population.* The sample mean $\bar{x}$ is therefore an unbiased estimator of the unknown population mean μ.

The observations are independent, so the addition rule for variances also applies:

$$\sigma_{\bar{x}}^2 = \left(\frac{1}{n}\right)^2 (\sigma_{X_1}^2 + \sigma_{X_2}^2 + \cdots + \sigma_{X_n}^2)$$

$$= \left(\frac{1}{n}\right)^2 (\sigma^2 + \sigma^2 + \cdots + \sigma^2)$$

$$= \frac{\sigma^2}{n}$$

Just as in the case of a sample proportion $\hat{p}$, the variability of the sampling distribution of a sample mean decreases as the sample size grows. Because the standard deviation of $\bar{x}$ is $\sigma/\sqrt{n}$, it is again true that the standard deviation of the statistic decreases in proportion to the square root of the sample size. Here is a summary of these facts.

MEAN AND STANDARD DEVIATION OF A SAMPLE MEAN

Let $\bar{x}$ be the mean of an SRS of size n from a population having mean μ and standard deviation σ. The mean and standard deviation of $\bar{x}$ are

$$\mu_{\bar{x}} = \mu$$

$$\sigma_{\bar{x}} = \frac{\sigma}{\sqrt{n}}$$

EXAMPLE 5.15 How accurately does a sample mean $\bar{x}$ estimate a population mean μ? Because the values of $\bar{x}$ vary from sample to sample, we must give an answer in terms of the sampling distribution. We know that $\bar{x}$ is an unbiased estimator of μ, so its values in repeated samples are not systematically too high or too low. Most samples will give an $\bar{x}$-value close to μ if the sampling distribution is concentrated close to its mean μ. So the accuracy of estimation depends on the spread of the sampling distribution.

The standard deviation of the population of service call lengths in Figure 5.8(a) is $\sigma = 184.81$ seconds. The length of a single call will often be far from the population mean. If we choose an SRS of 20 calls, the standard deviation of their mean length is

$$\sigma_{\bar{x}} = \frac{184.81}{\sqrt{20}} = 41.32 \text{ seconds}$$

Averaging over more calls reduces the variability and makes it more likely that $\bar{x}$ is close to μ. Our sample size of 80 calls is 4 times 20, so the standard deviation will be half as large:

$$\sigma_{\bar{x}} = \frac{184.81}{\sqrt{80}} = 20.66 \text{ seconds}$$

The central limit theorem

We have described the center and spread of the probability distribution of a sample mean $\bar{x}$, but not its shape. The shape of the distribution of $\bar{x}$ depends on the shape of the population distribution. Here is one important case: if the population distribution is normal, then so is the distribution of the sample mean.

SAMPLING DISTRIBUTION OF A SAMPLE MEAN

If a population has the $N(\mu, \sigma)$ distribution, then the sample mean $\bar{x}$ of n independent observations has the $N(\mu, \sigma/\sqrt{n})$ distribution.

This is a somewhat special result. Many population distributions are not normal. The service call lengths in Figure 5.8(a), for example, are strongly skewed. Yet Figures 5.8(b) and 5.9 show that means of samples of size 80 are close to normal. One of the most famous facts of probability theory says that as the sample size increases, the distribution of $\bar{x}$ gets closer to a normal distribution. This is true no matter what shape the population distribution has, as long as the population has a finite standard deviation σ. This is the *central limit theorem*. It is much more useful than the fact that the distribution of $\bar{x}$ is exactly normal if the population is exactly normal.

CENTRAL LIMIT THEOREM

Draw an SRS of size n from any population with mean μ and finite standard deviation σ. When n is large, the sampling distribution of the sample mean $\bar{x}$ is approximately normal:

$$\bar{x} \text{ is approximately } N\left(\mu, \frac{\sigma}{\sqrt{n}}\right)$$

EXAMPLE 5.16   With the normal distribution to work with, we can better describe how accurately a random sample of 80 calls estimates the mean length of all the calls in the population. The population standard deviation for the more than 30,000 calls in the population of Figure 5.8(a) is $\sigma = 184.81$ seconds. Example 5.15 calculates that the standard deviation of sample means is $\sigma_{\bar{x}} = 20.66$ seconds. By the 95 part of the 68–95–99.7 rule, 95% of all samples will have mean $\bar{x}$ within two standard deviations of μ, that is, within ± 41.32 seconds of μ.

That's not very accurate. The population is very spread out, so the sampling distribution of $\bar{x}$, though less spread out, is still quite variable. To divide the standard deviation of $\bar{x}$ by 4, we must take a sample 16 times as large, $n = 16 \times 80$, or 1280. Then

$$\sigma_{\bar{x}} = \frac{184.81}{\sqrt{1280}} = 5.165 \text{ seconds}$$

For samples of size 1280, 95% of the sample means will be within twice 5.165, or 10.33 seconds, of the population mean μ.

Example 5.16 reminds us that if the population is very spread out, the $\sqrt{n}$ in the standard deviation of $\bar{x}$ implies that even large samples will not estimate the population mean accurately. But the big point of the example is that the central limit theorem allows us to use normal probability calculations to answer questions about sample means even when the population distribution is not normal. How large a sample size n is needed for $\bar{x}$ to be close to normal depends on the population distribution. More observations are required if the shape of the population distribution is far from normal. Even for the very skewed call length population, however, samples of size 80 are large enough. Here is a more detailed example.

EXAMPLE 5.17   Figure 5.10 shows the central limit theorem in action for another very nonnormal population. Figure 5.10(a) displays the density curve of a single observation, that is, of the population. The distribution is strongly right-skewed, and the most probable outcomes are near 0. The mean μ of this distribution is 1, and

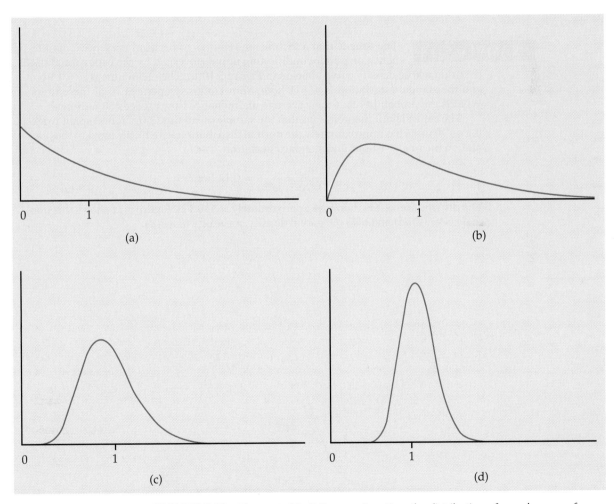

(a)

(b)

(c)

(d)

FIGURE 5.10   The central limit theorem in action: the distribution of sample means from a strongly nonnormal population becomes more normal as the sample size increases. (a) The distribution of 1 observation. (b) The distribution of $\bar{x}$ for 2 observations. (c) The distribution of $\bar{x}$ for 10 observations. (d) The distribution of $\bar{x}$ for 25 observations.

its standard deviation σ is also 1. This particular continuous distribution is called an **exponential distribution.** Exponential distributions are used as models for how long an electronic component will last and for the time required to serve a customer or repair a machine.

exponential distribution

Figures 5.10(b), (c), and (d) are the density curves of the sample means of 2, 10, and 25 observations from this population. As n increases, the shape becomes more normal. The mean remains at $\mu = 1$, but the standard deviation decreases, taking the value $1/\sqrt{n}$. The density curve for 10 observations is still somewhat skewed to the right but already resembles a normal curve having $\mu = 1$ and $\sigma = 1/\sqrt{10} = 0.32$. The density curve for $n = 25$ is yet more normal. The contrast between the shapes of the population distribution and of the distribution of the mean of 10 or 25 observations is striking.

The *Central Limit Theorem* applet on the text CD and Web site animates Figure 5.10. You can slide the sample size n from 1 to 100 and watch both the exact density curve of $\overline{x}$ and the normal approximation. As you increase n, the two curves move closer together.

EXAMPLE 5.18   The time X that a technician requires to perform preventive maintenance on an air-conditioning unit is governed by the exponential distribution whose density curve appears in Figure 5.10(a). The mean time is $\mu = 1$ hour and the standard deviation is $\sigma = 1$ hour. Your company operates 70 of these units. What is the probability that their average maintenance time exceeds 50 minutes?

The central limit theorem says that the sample mean time $\overline{x}$ (in hours) spent working on 70 units has approximately the normal distribution with mean equal to the population mean $\mu = 1$ hour and standard deviation

$$\frac{\sigma}{\sqrt{70}} = \frac{1}{\sqrt{70}} = 0.12 \text{ hour}$$

The distribution of $\overline{x}$ is therefore approximately $N(1, 0.12)$. Figure 5.11 shows this normal curve (*solid*) and also the actual density curve of $\overline{x}$ (*dashed*).

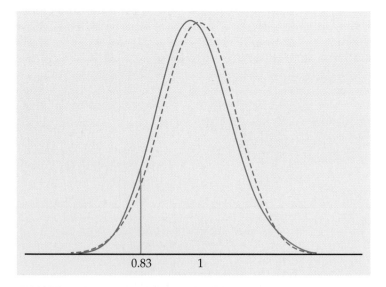

FIGURE 5.11   The exact distribution (*dashed*) and the normal approximation from the central limit theorem (*solid*) for the average time needed to maintain an air conditioner, for Example 5.18.

Because 50 minutes is 50/60 of an hour, or 0.83 hour, the probability we want is $P(\bar{x} > 0.83)$. A normal distribution calculation gives this probability as 0.9222. This is the area to the right of 0.83 under the solid normal curve in Figure 5.11. The exactly correct probability is the area under the dashed density curve in the figure. It is 0.9294. The central limit theorem normal approximation is off by only about 0.007.

Figure 5.12 summarizes the facts about the sampling distribution of $\bar{x}$ in a way that emphasizes the big idea of a sampling distribution. Keep taking random samples of size n from a population with mean μ. Find the sample mean $\bar{x}$ for each sample. Collect all the $\bar{x}$'s and display their distribution. That's the sampling distribution of $\bar{x}$. Sampling distributions are the key to understanding statistical inference. Keep this figure in mind as you go forward.

A few more facts

The central limit theorem is the big fact of this section. Here are three useful smaller facts related to our topic.

The normal approximation for sample proportions and counts is an example of the central limit theorem. This is true because a sample proportion can be thought of as a sample mean. Recall the idea that we used to find the mean and variance of a binomial random variable X. We wrote the count X as a sum

$$X = S_1 + S_2 + \cdots + S_n$$

of random variables S_i that take the value 1 if a success occurs on the ith trial and the value 0 otherwise. The variables S_i take only the values 0 and 1 and are far from normal. The proportion $\hat{p} = X/n$ is the sample mean of the S_i and, like all sample means, is approximately normal when n is large.

The fact that the sample mean of an SRS from a normal population has a normal distribution is a special case of a more general fact: **any linear combination of independent normal random variables is also normally distributed.** That is, if X and Y are independent normal random variables and a and b are any fixed numbers, $aX + bY$ is also normally distributed, and so it

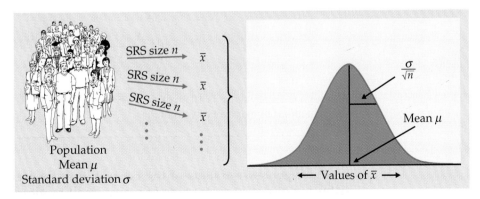

FIGURE 5.12   The sampling distribution of a sample mean $\bar{x}$ has mean μ and standard deviation $\sigma/\sqrt{n}$. The distribution is normal if the population distribution is normal; it is approximately normal for large samples in any case.

FIGURE 5.13 The normal probability calculation for Example 5.19.

is for any number of normal variables. In particular, the sum or difference of independent normal random variables has a normal distribution. The mean and standard deviation of $aX + bY$ are found as usual from the addition rules for means and variances. These facts are often used in statistical calculations.

EXAMPLE 5.19
Tom and George are playing in the club golf tournament. Their scores vary as they play the course repeatedly. Tom's score X has the $N(110, 10)$ distribution, and George's score Y varies from round to round according to the $N(100, 8)$ distribution. If they play independently, what is the probability that Tom will score lower than George and thus do better in the tournament? The difference $X - Y$ between their scores is normally distributed, with mean and variance

$$\mu_{X-Y} = \mu_X - \mu_Y = 110 - 100 = 10$$
$$\sigma_{X-Y}^2 = \sigma_X^2 + \sigma_Y^2 = 10^2 + 8^2 = 164$$

Because $\sqrt{164} = 12.8$, $X - Y$ has the $N(10, 12.8)$ distribution. Figure 5.13 illustrates the probability computation:

$$P(X < Y) = P(X - Y < 0)$$
$$= P\left(\frac{(X - Y) - 10}{12.8} < \frac{0 - 10}{12.8}\right)$$
$$= P(Z < -0.78) = 0.2177$$

Although George's score is 10 strokes lower on the average, Tom will have the lower score in about one of every five matches.

More general versions of the central limit theorem say that the distribution of a sum or average of many small random quantities is close to normal. This is true even if the quantities are not independent (as long as they are not too highly correlated) and even if they have different distributions (as long as no one random quantity is so large that it dominates the others).

The central limit theorem suggests why the normal distributions are common models for observed data. Any variable that is a sum of many small influences will have approximately a normal distribution.

BEYOND THE BASICS | Weibull distributions

Our discussion of sampling distributions has concentrated on the binomial model for count data and the normal model for quantitative variables. These models are important in statistical practice, but simplicity also contributes to their popularity. The parameters p in the binomial model and μ in the normal model are easy to understand. To estimate them from data we use statistics $\hat{p}$ and $\bar{x}$ that are also easy to understand and that have simple sampling distributions.

There are many other probability distributions that are used to model data in various circumstances. Estimation and sampling distributions are often more complex than for the binomial and normal distributions. Here is an example.

EXAMPLE 5.20

Weibull distributions

The normal distributions are not suitable models for all quantitative variables. The time that a product, such as a computer disk drive, lasts before failing rarely has a normal distribution, for example. Another class of continuous distributions, the **Weibull distributions,** is used to model time to failure. For engineers studying the reliability of products, Weibull distributions are more common than normal distributions.

Figure 5.14 shows the density curves of three members of the Weibull family. Each describes a different type of distribution for the time to failure of a product. The top curve in Figure 5.14 is a model for *infant mortality.* Many of these products fail immediately. If they do not fail at once, then most last a long time. The manufacturer tests these products and only ships the ones that do not fail immediately.

The middle curve in Figure 5.14 is a model for *early failure.* These products do not fail immediately, but many fail early in their lives after they are in the hands of customers. This is disastrous—the product or the process that makes it must be changed at once.

The bottom curve in Figure 5.14 is a model for *old-age wearout.* Very few of these products fail until they begin to wear out, and then many fail at about the same age.

A manufacturer certainly wants to know to which of these classes a new product belongs. To find out, engineers operate a random sample of products until they fail. From the failure time data we can estimate the parameter (called the "shape parameter") that distinguishes among the three Weibull distributions in Figure 5.14. The shape parameter has no simple definition like that of a population proportion or mean, and it cannot be estimated by a simple statistic such as $\hat{p}$ or $\bar{x}$.

Two things save the situation. First, statistical theory provides general approaches for finding good estimates of any parameter. These general methods tell us to use $\hat{p}$ and $\bar{x}$ in the binomial and normal settings, and they also tell us how to estimate the Weibull shape parameter. Second, modern software can calculate the estimate from data even though there is no algebraic formula that we can write for the estimate. Statistical practice often relies on

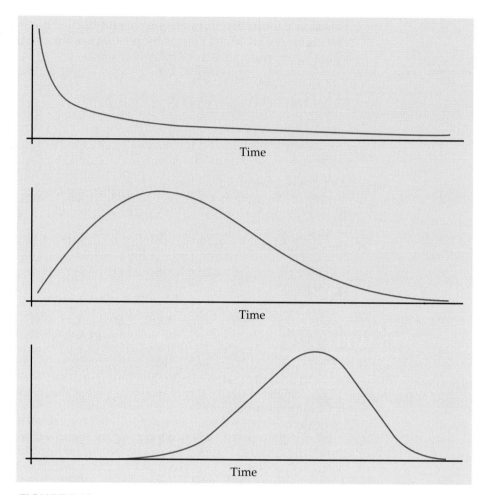

FIGURE 5.14 Density curves for three members of the Weibull family of distributions, for Example 5.20.

both mathematical theory and methods of computation more elaborate than the ones we will meet in this book. Fortunately, big ideas such as sampling distributions carry over to more complicated situations.[7]

SECTION 5.2 | Summary

The **sample mean** $\bar{x}$ of an SRS of size n drawn from a large population with mean μ and standard deviation σ has a sampling distribution with mean and standard deviation

$$\mu_{\bar{x}} = \mu$$

$$\sigma_{\bar{x}} = \frac{\sigma}{\sqrt{n}}$$

The sample mean $\bar{x}$ is therefore an unbiased estimator of the population mean μ and is less variable than a single observation.

Linear combinations of independent normal random variables have normal distributions. In particular, if the population has a normal distribution, so does $\bar{x}$.

The **central limit theorem** states that for large n the sampling distribution of $\bar{x}$ is approximately $N(\mu, \sigma/\sqrt{n})$ for any population with mean μ and finite standard deviation σ.

SECTION 5.2 | Exercises

5.28 Antonio measures the alcohol content of whiskey for his Chemistry 101 lab. He actually measures the mass of 5 milliliters of whiskey—a chemical calculation then finds the percent alcohol from the mass. The standard deviation of students' measurements of mass is $\sigma = 10$ milligrams (mg). Antonio repeats the measurement 3 times and records the mean $\bar{x}$ of his 3 measurements.

(a) What is the standard deviation $\sigma_{\bar{x}}$ of Antonio's mean result?

(b) How many times must Antonio repeat the measurement to reduce the standard deviation of $\bar{x}$ to 5 mg? Explain to someone who knows no statistics the advantage of reporting the average of several measurements rather than the result of a single measurement.

5.29 An automatic grinding machine in an auto parts plant prepares axles with a target diameter $\mu = 40.125$ millimeters (mm). The machine has some variability, so the standard deviation of the diameters is $\sigma = 0.002$ mm. A sample of 4 axles is inspected each hour for process control purposes, and records are kept of the sample mean diameter. If the process mean is exactly equal to the target value, what will be the mean and standard deviation of the numbers recorded?

5.30 Averages of several measurements are less variable than individual measurements. The true mass of the whiskey sample in Exercise 5.28 is 4.6 grams, or 4600 milligrams (mg). Antonio's measurements have the normal distribution with mean 4600 mg and standard deviation 10 mg. In this case, the mean of his 3 measurements also has a normal distribution.

(a) Sketch on the same graph the two normal curves, for individual measurements and for means of 3 measurements.

(b) What is the probability that Antonio misses the true mass by more than 10 mg in either direction if he makes one measurement?

(c) What is the probability that the mean of three independent measurements misses the true mass by more than 10 mg?

5.31 Averages are less variable than individual observations. Suppose that the axle diameters in Exercise 5.29 vary according to a normal distribution. In that case, the mean $\bar{x}$ of an SRS of axles also has a normal distribution.

(a) Make a sketch of the normal curve for a single axle. Add the normal curve for the mean of an SRS of 4 axles on the same sketch.

(b) What is the probability that the diameter of a single randomly chosen axle differs from the target value by 0.004 mm or more?

(c) What is the probability that the mean diameter of an SRS of 4 axles differs from the target value by 0.004 mm or more?

5.32   The number of lightning strikes on a square kilometer of open ground in a year has mean 6 and standard deviation 2.4. (These values are typical of much of the United States.) Counts of strikes on separate areas are independent. The National Lightning Detection Network uses automatic sensors to watch for lightning in an area of 10 square kilometers.

(a) What are the mean and standard deviation of the total number of lightning strikes observed?

(b) What are the mean and standard deviation of the mean number of strikes per square kilometer?

5.33   A study of the health of teenagers plans to measure the blood cholesterol level of an SRS of youth of ages 13 to 16 years. The researchers will report the mean $\bar{x}$ from their sample as an estimate of the mean cholesterol level μ in this population.

(a) Explain to someone who knows no statistics what it means to say that $\bar{x}$ is an "unbiased" estimator of μ.

(b) The sample result $\bar{x}$ is an unbiased estimator of the population truth μ no matter what size SRS the study chooses. Explain to someone who knows no statistics why a large sample gives more trustworthy results than a small sample.

5.34   The scores of high school seniors on the ACT college entrance examination in 2003 had mean $\mu = 20.8$ and standard deviation $\sigma = 4.8$. The distribution of scores is only roughly normal.

(a) What is the approximate probability that a single student randomly chosen from all those taking the test scores 23 or higher?

(b) Now take an SRS of 25 students who took the test. What are the mean and standard deviation of the sample mean score $\bar{x}$ of these 25 students?

(c) What is the approximate probability that the mean score $\bar{x}$ of these students is 23 or higher?

(d) Which of your two normal probability calculations in (a) and (c) is more accurate? Why?

5.35   The gypsy moth is a serious threat to oak and aspen trees. A state agriculture department places traps throughout the state to detect the moths. When traps are checked periodically, the mean number of moths trapped is only 0.5, but some traps have several moths. The distribution of moth counts is discrete and strongly skewed, with standard deviation 0.7.

(a) What are the mean and standard deviation of the average number of moths $\bar{x}$ in 50 traps?

(b) Use the central limit theorem to find the probability that the average number of moths in 50 traps is greater than 0.6.

5.36 North Carolina State University posts the grade distributions for its courses online.[8] You can find that the distribution of grades in Statistics 101 in the fall 2003 semester was

| | 4 | 3 | 2 | 1 | 0 |
Grade	A	B	C	D	F
Probability	0.21	0.43	0.30	0.05	0.01

(a) Using the common scale A = 4, B = 3, C = 2, D = 1, F = 0, take X to be the grade of a randomly chosen Statistics 101 student. Use the definitions of the mean (page 292) and standard deviation (page 300) for discrete random variables to find the mean μ and the standard deviation σ of grades in this course.

(b) Statistics 101 is a large course. We can take the grades of an SRS of 50 students to be independent of each other. If $\bar{x}$ is the average of these 50 grades, what are the mean and standard deviation of $\bar{x}$?

(c) What is the probability $P(X \geq 3)$ that a randomly chosen Statistics 101 student gets a B or better? What is the approximate probability $P(\bar{x} \geq 3)$ that the grade point average for 50 randomly chosen Statistics 101 students is B or better?

5.37 Sheila's doctor is concerned that she may suffer from gestational diabetes (high blood glucose levels during pregnancy). There is variation both in the actual glucose level and in the blood test that measures the level. A patient is classified as having gestational diabetes if the glucose level is above 140 milligrams per deciliter (mg/dl) one hour after a sugary drink is ingested. Sheila's measured glucose level one hour after ingesting the sugary drink varies according to the normal distribution with $\mu = 125$ mg/dl and $\sigma = 10$ mg/dl.

(a) If a single glucose measurement is made, what is the probability that Sheila is diagnosed as having gestational diabetes?

(b) If measurements are made instead on 4 separate days and the mean result is compared with the criterion 140 mg/dl, what is the probability that Sheila is diagnosed as having gestational diabetes?

5.38 A $1 bet in a state lottery's Pick 3 game pays $500 if the three-digit number you choose exactly matches the winning number, which is drawn at random. Here is the distribution of the payoff X:

Payoff X	$0	$500
Probability	0.999	0.001

Each day's drawing is independent of other drawings.

(a) What are the mean and standard deviation of X?

(b) Joe buys a Pick 3 ticket every day. What does the law of large numbers say about the average payoff Joe receives from his bets?

(c) What does the central limit theorem say about the distribution of Joe's average payoff after 365 bets in a year?

(d) Joe comes out ahead for the year if his average payoff is greater than $1 (the amount he spent each day on a ticket). What is the probability that Joe ends the year ahead?

5.39   In Exercise 5.37, Sheila's measured glucose level one hour after ingesting the sugary drink varies according to the normal distribution with $\mu = 125$ mg/dl and $\sigma = 10$ mg/dl. What is the level L such that there is probability only 0.05 that the mean glucose level of 4 test results falls above L for Sheila's glucose level distribution?

5.40   The number of flaws per square yard in a type of carpet material varies with mean 1.6 flaws per square yard and standard deviation 1.2 flaws per square yard. This population distribution cannot be normal, because a count takes only whole-number values. An inspector studies 200 square yards of the material, records the number of flaws found in each square yard, and calculates $\bar{x}$, the mean number of flaws per square yard inspected. Use the central limit theorem to find the approximate probability that the mean number of flaws exceeds 2 per square yard.

5.41   In response to the increasing weight of airline passengers, the Federal Aviation Administration in 2003 told airlines to assume that passengers average 190 pounds in the summer, including clothing and carry-on baggage. But passengers vary: the FAA gave a mean but not a standard deviation. A reasonable standard deviation is 35 pounds. Weights are not normally distributed, especially when the population includes both men and women, but they are not very nonnormal. A commuter plane carries 19 passengers. What is the approximate probability that the total weight of the passengers exceeds 4000 pounds? (*Hint:* To apply the central limit theorem, restate the problem in terms of the mean weight.)

5.42   The idea of insurance is that we all face risks that are unlikely but carry high cost. Think of a fire destroying your home. So we form a group to share the risk: we all pay a small amount, and the insurance policy pays a large amount to those few of us whose homes burn down. An insurance company looks at the records for millions of homeowners and sees that the mean loss from fire in a year is $\mu = \$250$ per house and that the standard deviation of the loss is $\sigma = \$1000$. (The distribution of losses is extremely right-skewed: most people have $0 loss, but a few have large losses.) The company plans to sell fire insurance for $250 plus enough to cover its costs and profit.

(a) Explain clearly why it would be unwise to sell only 12 policies. Then explain why selling many thousands of such policies is a safe business.

(b) If the company sells 10,000 policies, what is the approximate probability that the average loss in a year will be greater than $275?

5.43   The distribution of annual returns on common stocks is roughly symmetric, but extreme observations are more frequent than in a normal distribution. Because the distribution is not strongly nonnormal, the mean return over even a moderate number of years is close to normal. Annual real returns on the

Standard & Poor's 500-Stock Index over the period 1871 to 2004 have varied with mean 9.2% and standard deviation 20.6%. Andrew plans to retire in 45 years and is considering investing in stocks. What is the probability (assuming that the past pattern of variation continues) that the mean annual return on common stocks over the next 45 years will exceed 15%? What is the probability that the mean return will be less than 5%?

5.44   A hole in an engine block is 2.5 centimeters (cm) in diameter. Shafts manufactured to go through this hole must have 0.025 cm clearance for unforced fit. That is, shaft diameter cannot exceed 2.475 cm. The shafts vary in diameter according to the normal distribution with mean 2.45 cm and standard deviation 0.01 cm.

(a) What percent of shafts will fit into the hole?

(b) Redo the problem assuming that the hole diameter also varies, independently of the shaft diameter, following the normal distribution with mean 2.5 cm and standard deviation 0.01 cm. You must find the probability that the hole diameter exceeds the shaft diameter by at least 0.025 cm.

5.45   "Durable press" cotton fabrics are treated to improve their recovery from wrinkles after washing. Unfortunately, the treatment also reduces the strength of the fabric. The breaking strength of untreated fabric is normally distributed with mean 58 pounds and standard deviation 2.3 pounds. The same type of fabric after treatment has normally distributed breaking strength with mean 30 pounds and standard deviation 1.6 pounds.[9] A clothing manufacturer tests 5 specimens of each fabric. All 10 strength measurements are independent.

(a) What is the probability that the mean breaking strength of the 5 untreated specimens exceeds 50 pounds?

(b) What is the probability that the mean breaking strength of the 5 untreated specimens is at least 25 pounds greater than the mean strength of the 5 treated specimens?

5.46   Many companies place advertisements to improve the image of their brand rather than to promote specific products. In a randomized comparative experiment, business students read ads that cited either the *Wall Street Journal* or the *National Enquirer* for important facts about a fictitious company. The students then rated the trustworthiness of the source on a 7-point scale. Suppose that in the population of all students scores for the *Journal* have mean 4.8 and standard deviation 1.5, while scores for the *Enquirer* have mean 2.4 and standard deviation 1.6.[10]

(a) There are 30 students in each group. Although individual scores are discrete, the mean score for a group of 30 will be close to normal. Why?

(b) What are the means and standard deviations of the sample mean scores $\bar{y}$ for the *Journal* group and $\bar{x}$ for the *Enquirer* group?

(c) We can take all 60 scores to be independent because students are not told each other's scores. What is the distribution of the difference $\bar{y} - \bar{x}$ between the mean scores in the two groups?

(d) Find $P(\bar{y} - \bar{x} \geq 1)$.

5.47 The two previous exercises illustrate a common setting for statistical inference. This exercise gives the general form of the sampling distribution needed in this setting. We have a sample of n observations from a treatment group and an independent sample of m observations from a control group. Suppose that the response to the treatment has the $N(\mu_X, \sigma_X)$ distribution and that the response of control subjects has the $N(\mu_Y, \sigma_Y)$ distribution. Inference about the difference $\mu_Y - \mu_X$ between the population means is based on the difference $\bar{y} - \bar{x}$ between the sample means in the two groups.

(a) Under the assumptions given, what is the distribution of $\bar{y}$? Of $\bar{x}$?

(b) What is the distribution of $\bar{y} - \bar{x}$?

5.48 Linda invests her money in a portfolio that consists of 80% Fidelity 500 Index Fund and 20% Fidelity Diversified International Fund. Suppose that in the long run the annual real return X on the 500 Index Fund has mean 9% and standard deviation 20%, the annual real return Y on the Diversified International Fund has mean 10% and standard deviation 17%, and the correlation between X and Y is 0.6.

(a) The return on Linda's portfolio is $R = 0.8X + 0.2Y$. What are the mean and standard deviation of R?

(b) The distribution of returns is typically roughly symmetric but with more extreme high and low observations than a normal distribution. The average return over a number of years, however, is close to normal. If Linda holds her portfolio for 20 years, what is the approximate probability that her average return is less than 5%?

(c) The calculation you just made is not overly helpful, because Linda isn't really concerned about the mean return $\bar{R}$. To see why, suppose that her portfolio returns 12% this year and 6% next year. The mean return for the two years is 9%. If Linda starts with $1000, how much does she have at the end of the first year? At the end of the second year? How does this amount compare with what she would have if both years had the mean return, 9%? Over 20 years, there may be a large difference between the ordinary mean $\bar{R}$ and the *geometric mean*, which reflects the fact that returns in successive years multiply rather than add.

5.49 You are building a wall from precast concrete blocks. Standard "8 inch" blocks are $7\frac{5}{8}$ inches high to allow for a $\frac{3}{8}$ inch layer of mortar under each row of blocks. In practice, the height of a block-plus-mortar row varies according to a normal distribution with mean 8 inches and standard deviation 0.1 inch. Heights of successive rows are independent. Your wall has four rows of blocks. What is the distribution of the height of the wall? What is the probability that the height differs from the design height of 32 inches by more than half an inch?

5.50 The amount that households pay service providers for access to the Internet varies quite a bit, but the mean monthly fee is $28 and the standard deviation is $10. The distribution is not normal: many households pay about $10 for limited dial-up access or about $25 for unlimited dial-up access, but many pay more for broadband connections. A sample survey asks an SRS of 500 households with Internet access how much they pay. What is the probability that the average fee paid by the sample households exceeds $29?

5.51 A study of working couples measures the income X of the husband and the income Y of the wife in a large number of couples in which both partners are employed. Suppose that you knew the means μ_X and μ_Y and the variances σ_X^2 and σ_Y^2 of both variables in the population.

(a) Is it reasonable to take the mean of the total income $X + Y$ to be $\mu_X + \mu_Y$? Explain your answer.

(b) Is it reasonable to take the variance of the total income to be $\sigma_X^2 + \sigma_Y^2$? Explain your answer.

5.52 A particle moves along the line in a random walk. That is, the particle starts at the origin (position 0) and moves either right or left in independent steps of length 1. If the particle moves to the right with probability 3/4, its movement at the ith step is a random variable X_i with distribution

$$P(X_i = 1) = 0.75$$
$$P(X_i = -1) = 0.25$$

The position of the particle after k steps is the sum of these random movements,

$$Y = X_1 + X_2 + \cdots + X_k$$

Use the central limit theorem to find the approximate probability that the position of the particle after 500 steps is at least 200 to the right.

CHAPTER 5 | Exercises

5.53 The probability that a randomly chosen driver will be involved in an accident in the next year is about 0.2. This is based on the proportion of millions of drivers who have accidents. "Accident" includes things like crumpling a fender in your own driveway, not just highway accidents. Carlos, David, Jermaine, Ramon, Scott, and Sean are college students who live together in an off-campus apartment. Last year, 3 of the 6 had accidents. What is the probability that 3 or more of 6 randomly chosen drivers have an accident in the same year? Why does your calculation not apply to drivers like the 6 students?

5.54 Example 4.27 (page 303) notes that the total SAT scores of high school seniors in a recent year had mean $\mu = 1026$ and standard deviation $\sigma = 209$. The distribution of SAT scores is roughly normal.

(a) Ramon scored 1100. If scores have a normal distribution, what percentile of the distribution is this?

(b) Now consider the mean $\bar{x}$ of the scores of 70 randomly chosen students. If $\bar{x} = 1100$, what percentile of the sampling distribution of $\bar{x}$ is this?

(c) Which of your calculations, (a) or (b), is less accurate because SAT scores do not have an exactly normal distribution? Explain your answer.

5.55 Although cities encourage carpooling to reduce traffic congestion, most vehicles carry only one person. For example, 70% of vehicles on the roads in the Minneapolis–St. Paul metropolitan area are occupied by just the driver.

(a) If you choose 10 vehicles at random, what is the probability that more than half (that is, 6 or more) carry just one person?

(b) If you choose 100 vehicles at random, what is the probability that more than half (that is, 51 or more) carry just one person?

5.56 The Census Bureau says that the 10 most common names in the United States are (in order) Smith, Johnson, Williams, Jones, Brown, Davis, Miller, Wilson, Moore, and Taylor. These names account for 5.6% of all U.S. residents. Out of curiosity, you look at the authors of the textbooks for your current courses. There are 9 authors in all. Would you be surprised if none of the names of these authors were among the 10 most common? Give a probability to support your answer and explain the reasoning behind your calculation.

5.57 It is a striking fact that the first digits of numbers in legitimate records often follow a distribution known as Benford's law. Here it is:

First digit	1	2	3	4	5	6	7	8	9
Proportion	0.301	0.176	0.125	0.097	0.079	0.067	0.058	0.051	0.046

Fake records usually have fewer first digits 1, 2, and 3. What is the approximate probability, if Benford's law holds, that among 1200 randomly chosen invoices there are no more than 680 in amounts with first digit 1, 2, or 3?

5.58 According to genetic theory, the blossom color in the second generation of a certain cross of sweet peas should be red or white in a 3:1 ratio. That is, each plant has probability 3/4 of having red blossoms, and the blossom colors of separate plants are independent.

(a) What is the probability that exactly 6 out of 8 of these plants have red blossoms?

(b) What is the mean number of red-blossomed plants when 80 plants of this type are grown from seeds?

(c) What is the probability of obtaining at least 50 red-blossomed plants when 80 plants are grown from seeds?

5.59 The weight of the eggs produced by a certain breed of hen is normally distributed with mean 65 grams (g) and standard deviation 5 g. If cartons of such eggs can be considered to be SRSs of size 12 from the population of all eggs, what is the probability that the weight of a carton falls between 750 g and 825 g?

5.60 Suppose (as is roughly true) that 20% of all Internet users have posted photos online. A sample survey interviews an SRS of 1555 Internet users.

(a) What is the actual distribution of the number X in the sample who have posted photos online?

(b) What is the probability that 300 or fewer of the people in the sample have posted photos online? (Use either software or a suitable approximation.)

5.61 A study of rush hour traffic in San Francisco records the number of people in each car entering a freeway at a suburban interchange. Suppose that this number X has mean 1.5 and standard deviation 0.75 in the population of all cars that enter at this interchange during rush hours.

(a) Does the count X have a binomial distribution? Why or why not?

(b) Could the exact distribution of X be normal? Why or why not?

(c) Traffic engineers estimate that the capacity of the interchange is 700 cars per hour. According to the central limit theorem, what is the approximate distribution of the mean number of persons $\bar{x}$ in 700 randomly selected cars at this interchange?

(d) The count of people in 700 cars is $700\bar{x}$. Use your result from (c) to give an approximate distribution for the count. What is the probability that 700 cars will carry more than 1075 people?

5.62 Serving in a bomber crew in World War II was dangerous. The British estimated that the probability of an aircraft loss due to enemy action was 1/20 for each mission. A tour of duty for British airmen in Bomber Command was 30 missions. What is the probability that an airman would complete a tour of duty without being on an aircraft lost from enemy action?

5.63 A sample survey interviews an SRS of 267 college women. Suppose (as is roughly true) that 70% of all college women have been on a diet within the last 12 months. What is the probability that 75% or more of the women in the sample have been on a diet?

5.64 A machine fastens plastic screw-on caps onto containers of motor oil. If the machine applies more torque than the cap can withstand, the cap will break. Both the torque applied and the strength of the caps vary. The capping machine torque has the normal distribution with mean 7 inch-pounds and standard deviation 0.9 inch-pounds. The cap strength (the torque that would break the cap) has the normal distribution with mean 10 inch-pounds and standard deviation 1.2 inch-pounds.

(a) Explain why it is reasonable to assume that the cap strength and the torque applied by the machine are independent.

(b) What is the probability that a cap will break while being fastened by the capping machine?

5.65 The unique colors of the cashmere sweaters your firm makes result from heating undyed yarn in a kettle with a dye liquor. The pH (acidity) of the liquor is critical for regulating dye uptake and hence the final color. There are 5 kettles, all of which receive dye liquor from a common source. Past data show that pH varies according to a normal distribution with $\mu = 4.22$ and $\sigma = 0.127$. You use statistical process control to check the stability of the process. Twice each day, the pH of the liquor in each kettle is measured, each time giving a sample of size 5. The mean pH $\bar{x}$ is compared with "control limits" given by the 99.7 part of the 68–95–99.7 rule for normal distributions, namely, $\mu_{\bar{x}} \pm 3\sigma_{\bar{x}}$. What are the numerical values of these control limits for $\bar{x}$?

5.66 Does delaying oral practice hinder learning a foreign language? Researchers randomly assigned 23 beginning students of Russian to begin speaking practice immediately and another 23 to delay speaking for 4 weeks. At the end of the semester both groups took a standard test of comprehension of spoken Russian. Suppose that in the population of all beginning students, the test scores for early speaking vary according to the $N(32, 6)$ distribution and scores for delayed speaking have the $N(29, 5)$ distribution.

(a) What is the sampling distribution of the mean score $\bar{x}$ in the early-speaking group in many repetitions of the experiment? What is the sampling distribution of the mean score $\bar{y}$ in the delayed-speaking group?

(b) If the experiment were repeated many times, what would be the sampling distribution of the difference $\bar{y} - \bar{x}$ between the mean scores in the two groups?

(c) What is the probability that the experiment will find (misleadingly) that the mean score for delayed speaking is at least as large as that for early speaking?

5.67 Suppose (as is roughly true) that 88% of college men and 82% of college women were employed last summer. A sample survey interviews SRSs of 500 college men and 500 college women. The two samples are of course independent.

(a) What is the approximate distribution of the proportion $\hat{p}_F$ of women who worked last summer? What is the approximate distribution of the proportion $\hat{p}_M$ of men who worked?

(b) The survey wants to compare men and women. What is the approximate distribution of the difference in the proportions who worked, $\hat{p}_M - \hat{p}_F$? Explain the reasoning behind your answer.

(c) What is the probability that in the sample a higher proportion of women than men worked last summer?

EESEE CASE STUDIES

The Electronic Encyclopedia of Statistical Examples and Exercises (EESEE) is available on the text CD and Web site. These more elaborate stories, with data, provide settings for longer case studies.

5.68 Read the EESEE story "Psychic Probability." Answer Questions 1a, 2, and 4 for this case study.

5.69   Read the EESEE story "Habitat of the Spotted Owl." Write a report based on Questions 1, 2, 3, 4, 5, 6, 7a, and 7b in this case study. (Questions 1 to 6 involve data analysis, and Question 7 concerns the sampling distribution of the sample mean.)

5.70   Read the EESEE story "Rank and Yank: Is It a Good Idea?" Answer Question 3.

Testing Nuclear Weapons without Explosions

MARCEY ABATE *is Manager of the Stockpile Evaluation Department at Sandia National Laboratories.*

The Sandia National Laboratories in Albuquerque, New Mexico, use technology to solve challenging problems that affect the national security of the United States. Statisticians at the national laboratories participate in a full range of research and development projects that contribute to national security. The Sandia National Laboratories were born out of the World War II atomic bomb project, and many of their responsibilities still concern nuclear weapons. In particular, my group is responsible for guaranteeing that the nation's stockpile of nuclear weapons remains safe and reliable even though we no longer test weapons by detonating them. Statistics plays an important role in this work.

As a statistician and manager of the stockpile evaluation program, I oversee the development and operation of a statistical sampling plan for testing each weapon system in the nation's stockpile. On an annual basis, weapons are sampled from the stockpile, disassembled, inspected, denuclearized, and reassembled for testing either in a laboratory or in actual flights from military bases. We analyze the results of the laboratory and flight evaluations and investigate anything unusual disclosed by the data. The testing and investigation, in coordination with other assessment and certification efforts, help to ensure the health and performance of the overall stockpile. Other groups at Sandia are concerned with making certain that weapons cannot detonate accidentally and with improving our ability to detect weapons of mass destruction. Statistical simulations and data analyses play a role in these efforts as well.

As our stockpiles age and grow smaller due to nuclear weapons reduction agreements, and actual testing of weapons remains out of the question, evaluation will become yet more challenging. Because stockpile evaluation is based on statistical sampling, statistics will continue to play a key role in ensuring the continued safety, security, and reliability of the nation's stockpile.

WASSILY KANDINSKY, ANGULAR LINE, 1930 © ARS, NY (ALINARI/ART RESOURCE, NY)

Introduction to Inference

Introduction

The purpose of statistical inference is to draw conclusions from data. We have examined data and arrived at conclusions many times previously. Formal inference adds an emphasis on substantiating our conclusions by probability calculations.

Probability allows us to take chance variation into account and so to correct our judgment by calculation. Here is an example.

EXAMPLE 6.1 The Wade Tract in Thomas County, Georgia, is an old-growth forest of longleaf pine trees (*Pinus palustris*) that has survived in a relatively undisturbed state since before the settlement of the area by Europeans. Foresters who study these trees are interested in how the trees are distributed in the forest. Are the locations of the trees random, with no particular patterns? Or is there some sort of clustering, resulting in regions that have more trees than others? Figure 6.1 gives a plot of the locations of all 584 longleaf pine trees in a 200-meter by 200-meter region in the Wade Tract.[1] Do the locations appear to be random, or do there appear to be clusters of trees? One approach to the analysis of these data indicates that a pattern as clustered or more clustered than the one in Figure 6.1 would occur only 4% of the time if, in fact, the locations of longleaf pine trees in the Wade Tract are random.

Because this chance is fairly small, we conclude that there is some clustering of these trees. Our probability calculation helps us to distinguish

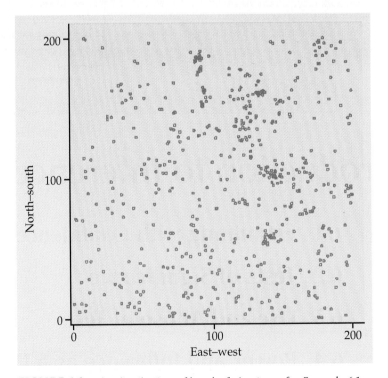

FIGURE 6.1 The distribution of longleaf pine trees, for Example 6.1.

between patterns that are consistent or inconsistent with the random location scenario.

Our unaided judgment can also err in the opposite direction, seeing a systematic effect when only chance variation is at work. Give a new drug and a placebo to 20 patients each; 12 of those taking the drug show improvement, but only 8 of the placebo patients improve. Is the drug more effective than the placebo? Perhaps, but a difference this large or larger between the results in the two groups would occur about one time in five simply because of chance variation. An effect that could so easily be just chance is not convincing.

In this chapter we introduce the two most prominent types of formal statistical inference. Section 6.1 concerns *confidence intervals* for estimating the value of a population parameter. Section 6.2 presents *tests of significance*, which assess the evidence for a claim. Both types of inference are based on the sampling distributions of statistics. That is, both report probabilities that state *what would happen if we used the inference method many times.* This kind of probability statement is characteristic of standard statistical inference. Users of statistics must understand the nature of the reasoning employed and the meaning of the probability statements that appear, for example, on computer output for statistical procedures.

Because the methods of formal inference are based on sampling distributions, they require a probability model for the data. Trustworthy probability models can arise in many ways, but the model is most secure and inference is most reliable when the data are produced by a properly randomized design. *When you use statistical inference, you are acting as if the data come from a random sample or a randomized experiment.* If this is not true, your conclusions may be open to challenge. Do not be overly impressed by the complex details of formal inference. This elaborate machinery cannot remedy basic flaws in producing the data such as voluntary response samples and confounded experiments. Use the common sense developed in your study of the first three chapters of this book, and proceed to detailed formal inference only when you are satisfied that the data deserve such analysis.

The primary purpose of this chapter is to describe the reasoning used in statistical inference. We will discuss only a few specific inference techniques, and these require an unrealistic assumption: that we know the standard deviation σ. Later chapters will present inference methods for use in most of the settings we met in learning to explore data. There are libraries—both of books and of computer software—full of more elaborate statistical techniques. Informed use of any of these methods requires an understanding of the underlying reasoning. A computer will do the arithmetic, but you must still exercise judgment based on understanding.

6.1 Estimating with Confidence

The SAT tests are widely used measures of readiness for college study. There are two parts, one for verbal reasoning ability (SATV) and one for mathematical reasoning ability (SATM). In April 1995, the scores were *recentered* so that the mean is approximately 500 in a large "standardization group." This scale is maintained from year to year so that scores have a constant interpretation. In 2003, 1,406,324 college-bound seniors took the SAT. Their mean SATV score was 507 with a standard deviation of 111. For the SATM the mean was 519 with a standard deviation of 115.

EXAMPLE 6.2   You want to estimate the mean SATM score for the more than 385,000 high school seniors in California. You know better than to trust data from the students who choose to take the SAT. Only about 45% of California students take the SAT. These self-selected students are planning to attend college and are not representative of all California seniors. At considerable effort and expense, you give the test to a simple random sample (SRS) of 500 California high school seniors. The mean score for your sample is $\bar{x} = 461$. What can you say about the mean score μ in the population of all 385,000 seniors?

The sample mean $\bar{x}$ is the natural estimator of the unknown population mean μ. We know that $\bar{x}$ is an unbiased estimator of μ. More important, the law of large numbers says that the sample mean must approach the population mean as the size of the sample grows. The value $\bar{x} = 461$ therefore appears to be a reasonable estimate of the mean score μ that all 385,000 students would achieve if they took the test. But how reliable is this estimate? A second sample would surely not give 461 again. Unbiasedness says only that there is no systematic tendency to underestimate or overestimate the truth. Could we plausibly get a sample mean of 410 or 510 on repeated samples? An estimate without an indication of its variability is of little value.

Statistical confidence

Just as unbiasedness of an estimator concerns the center of its sampling distribution, questions about variation are answered by looking at the spread. We know that if the entire population of SAT scores has mean μ and standard deviation σ, then in repeated samples of size 500 the sample mean $\bar{x}$ follows the $N(\mu, \sigma/\sqrt{500})$ distribution. Let us suppose that we know that the standard deviation σ of SATM scores in our California population is $\sigma = 100$. (This is not realistic. We will see in the next chapter how to proceed when σ is not known. For now, we are more interested in statistical reasoning than in details of realistic methods.) In repeated sampling the sample mean $\bar{x}$ has a normal distribution centered at the unknown population mean μ and having standard deviation

$$\sigma_{\bar{x}} = \frac{100}{\sqrt{500}} \doteq 4.5$$

Now we are in business. Consider this line of thought, which is illustrated by Figure 6.2:

- The 68–95–99.7 rule says that the probability is about 0.95 that $\bar{x}$ will be within 9 points (two standard deviations of $\bar{x}$) of the population mean score μ.

- To say that $\bar{x}$ lies within 9 points of μ is the same as saying that μ is within 9 points of $\bar{x}$.

- So 95% of all samples will capture the true μ in the interval from $\bar{x} - 9$ to $\bar{x} + 9$.

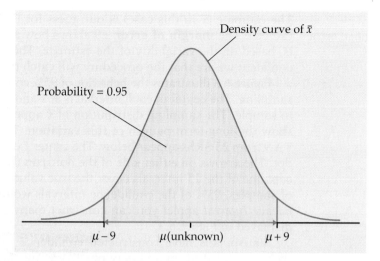

FIGURE 6.2 $\bar{x}$ lies within ± 9 of μ in 95% of all samples, so μ also lies within ± 9 of $\bar{x}$ in those samples.

We have simply restated a fact about the sampling distribution of $\bar{x}$. *The language of statistical inference uses this fact about what would happen in the long run to express our confidence in the results of any one sample.* Our sample gave $\bar{x} = 461$. We say that we are *95% confident* that the unknown mean score for all California seniors lies between

$$\bar{x} - 9 = 461 - 9 = 452$$

and

$$\bar{x} + 9 = 461 + 9 = 470$$

Be sure you understand the grounds for our confidence. There are only two possibilities:

1. The interval between 452 and 470 contains the true μ.

2. Our SRS was one of the few samples for which $\bar{x}$ is not within 9 points of the true μ. Only 5% of all samples give such inaccurate results.

We cannot know whether our sample is one of the 95% for which the interval $\bar{x} \pm 9$ catches μ or one of the unlucky 5%. The statement that we are 95% confident that the unknown μ lies between 552 and 570 is shorthand for saying, "We arrived at these numbers by a method that gives correct results 95% of the time."

Confidence intervals

The interval of numbers between the values $\bar{x} \pm 9$ is called a *95% confidence interval* for μ. Like most confidence intervals we will meet, this one has the form

$$\text{estimate} \pm \text{margin of error}$$

The estimate ($\bar{x}$ in this case) is our guess for the value of the unknown parameter. The **margin of error** ± 9 shows how accurate we believe our guess is, based on the variability of the estimate. The confidence level shows how confident we are that the procedure will catch the true population mean μ.

Figure 6.3 illustrates the behavior of 95% confidence intervals in repeated sampling. The center of each interval is at $\bar{x}$ and therefore varies from sample to sample. The sampling distribution of $\bar{x}$ appears at the top of the figure to show the long-term pattern of this variation. The 95% confidence intervals $\bar{x} \pm 9$ from 25 SRSs appear below. The center $\bar{x}$ of each interval is marked by a dot. The arrows on either side of the dot span the confidence interval. All except one of the 25 intervals cover the true value of μ. In a very large number of samples, 95% of the confidence intervals would contain μ. With the *Confidence Interval* applet you can construct many diagrams similar to the one displayed in Figure 6.3.

Statisticians have constructed confidence intervals for many different parameters based on a variety of designs for data collection. We will meet

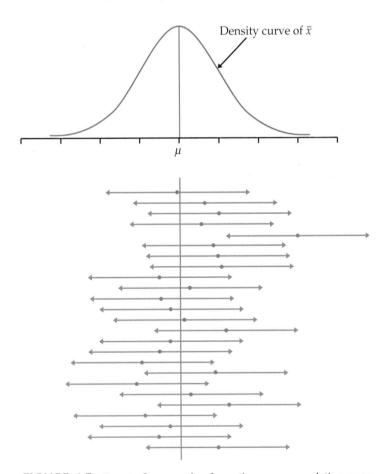

FIGURE 6.3 Twenty-five samples from the same population gave these 95% confidence intervals. In the long run, 95% of all samples give an interval that covers μ.

a number of these in later chapters. You need to know two important things about a confidence interval:

1. It is an interval of the form (a, b), where a and b are numbers computed from the data.

2. It has a property called a confidence level that gives the probability that the interval covers the parameter.

Users can choose the confidence level, but 95% is the standard for most situations. Occasionally, 90% or 99% is used. We will use C to stand for the confidence level in decimal form. For example, a 95% confidence level corresponds to $C = 0.95$.

> ## CONFIDENCE INTERVAL
>
> A level C **confidence interval** for a parameter is an interval computed from sample data by a method that has probability C of producing an interval containing the true value of the parameter.

Confidence interval for a population mean

We will now construct a level C confidence interval for the mean μ of a population when the data are an SRS of size n. The construction is based on the sampling distribution of the sample mean $\bar{x}$. This distribution is exactly $N(\mu, \sigma/\sqrt{n})$ when the population has the $N(\mu, \sigma)$ distribution. The central limit theorem says that this same sampling distribution is approximately correct for large samples whenever the population mean and standard deviation are μ and σ.

Our construction of a 95% confidence interval for the mean SAT score began by noting that any normal distribution has probability about 0.95 within ± 2 standard deviations of its mean. To construct a level C confidence interval we first catch the central C area under a normal curve. That is, we must find the number z^* such that any normal distribution has probability C within $\pm z^*$ standard deviations of its mean. Because all normal distributions have the same standardized form, we can obtain everything we need from the standard normal curve. Figure 6.4 shows how C and z^* are related. Values of z^* for many choices of C appear in the row labeled z^* at the bottom of Table D at the back of the book. Here are the most important entries from that row:

z^*	1.645	1.960	2.576
C	90%	95%	99%

Any normal curve has probability C between the point z^* standard deviations below the mean and the point z^* above the mean, as Figure 6.4 reminds us. The sample mean $\bar{x}$ has the normal distribution with mean μ and standard deviation $\sigma/\sqrt{n}$. So there is probability C that $\bar{x}$ lies between

$$\mu - z^* \frac{\sigma}{\sqrt{n}} \quad \text{and} \quad \mu + z^* \frac{\sigma}{\sqrt{n}}$$

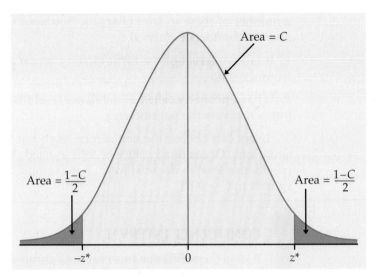

FIGURE 6.4 The area between $-z^*$ and z^* under the standard normal curve is C.

This is exactly the same as saying that the unknown population mean μ lies between

$$\bar{x} - z^*\frac{\sigma}{\sqrt{n}} \quad \text{and} \quad \bar{x} + z^*\frac{\sigma}{\sqrt{n}}$$

That is, there is probability C that the interval $\bar{x} \pm z^*\sigma/\sqrt{n}$ contains μ. That is our confidence interval. The estimate of the unknown μ is $\bar{x}$, and the margin of error is $z^*\sigma/\sqrt{n}$.

CONFIDENCE INTERVAL FOR A POPULATION MEAN

Choose an SRS of size n from a population having unknown mean μ and known standard deviation σ. The **margin of error** for a level C confidence interval for μ is

$$m = z^*\frac{\sigma}{\sqrt{n}}$$

Here z^* is the value on the standard normal curve with area C between the critical points $-z^*$ and z^*. The level C **confidence interval** for μ is

$$\bar{x} \pm m$$

This interval is exact when the population distribution is normal and is approximately correct for large n in other cases.

EXAMPLE 6.3 The National Student Loan Survey collects data to examine questions related to the amount of money that borrowers owe. The survey selected a sample of 1280 borrowers who began repayment of their loans between four and six months prior to the study.[2] The mean of the debt for undergraduate study was $18,900 and the standard deviation was about \$49,000. This distribution is clearly skewed but because our sample size is quite large, we can rely on the central limit theorem to assure us that the confidence interval based on the normal distribution will be a good approximation. Let's compute a 95% confidence interval for the true mean debt for all borrowers. Although the standard deviation is estimated from the data collected, we will treat it as a known quantity for our calculations here.

For 95% confidence, we see from Table D that $z^* = 1.960$. The margin of error for the 95% confidence interval for μ is therefore

$$m = z^* \frac{\sigma}{\sqrt{n}}$$

$$= 1.960 \frac{49,000}{\sqrt{1280}}$$

$$= 2684$$

We have computed the margin of error with more digits than we really need. Our mean is rounded to the nearest \$100, so we will do the same for the margin of error. Therefore, we will use $m = 2700$. The 95% confidence interval is

$$\bar{x} \pm m = 18,900 \pm 2700$$

$$= (16,200, \ 21,600)$$

We are 95% confident that the mean debt is between \$16,200 and \$21,600.

Note that we have rounded the results to the nearest \$100. Keeping additional digits would provide no additional useful information.

In this example we used a value for σ based on a large sample. Because of this, the confidence interval that we calculated will be the same as the one that we would compute using the methods discussed in the next chapter, where we no longer need to assume that σ is known.

Suppose the researchers who designed the National Student Loan Survey had used a different sample size. How would this affect the confidence interval? We can answer this question by changing the sample size in our calculations and assuming that the mean and standard deviation are the same.

EXAMPLE 6.4   Let's assume that the sample mean of the debt for undergraduate study is \$18,900 and the standard deviation is about \$49,000, as in Example 6.3. But suppose that the sample size is only 320. The margin of error for 95% confidence is

$$m = z^* \frac{\sigma}{\sqrt{n}}$$

$$= 1.960 \frac{49,000}{\sqrt{320}}$$

$$= 5400$$

and the 95% confidence interval is

$$\bar{x} \pm m = 18,900 \pm 5400$$

$$= (13,500, \ 24,300)$$

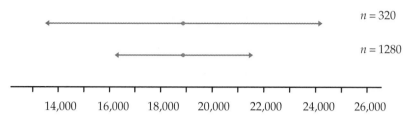

FIGURE 6.5   Confidence intervals for $n = 1280$ and $n = 320$, for Examples 6.3 and 6.4.

Notice that the margin of error for this example is twice as large as the margin of error that we computed in Example 6.3. The only change that we made was to assume that the sample size is 320 rather than 1280. The new sample size is exactly one-fourth of the original 1280. We double the margin of error when we reduce the sample size to one-fourth of the original value. Figure 6.5 illustrates the effect in terms of the intervals.

The argument leading to the form of confidence intervals for the population mean μ rested on the fact that the statistic $\bar{x}$ used to estimate μ has a normal distribution. Because many sample estimates have normal distributions (at least approximately), it is useful to notice that the confidence interval has the form

$$\text{estimate} \pm z^* \sigma_{\text{estimate}}$$

The estimate based on the sample is the center of the confidence interval. The margin of error is $z^* \sigma_{\text{estimate}}$. The desired confidence level determines z^* from Table D. The standard deviation of the estimate is found from a knowledge of the sampling distribution in a particular case. When the estimate is $\bar{x}$ from an SRS, the standard deviation of the estimate is $\sigma / \sqrt{n}$.

How confidence intervals behave

The margin of error $z^* \sigma / \sqrt{n}$ for the mean of a normal population illustrates several important properties that are shared by all confidence intervals in common use. The user chooses the confidence level, and the margin of error follows from this choice. High confidence is desirable and so is a small margin of error. High confidence says that our method almost always gives correct answers. A small margin of error says that we have pinned down the parameter quite precisely.

Suppose that you calculate a margin of error and decide that it is too large. Here are your choices to reduce it:

- Use a lower level of confidence (smaller C).

- Increase the sample size (larger n).

- Reduce σ.

For most problems you would choose a confidence level of 90%, 95%, or 99%. So z^* can be 1.645, 1.960, or 2.576. A look at Figure 6.4 will convince you that z^* will be smaller for lower confidence (smaller C). Table D shows that this is indeed the case. If n and σ are unchanged, a smaller z^* leads to a smaller margin of error. Similarly, increasing the sample size n reduces the

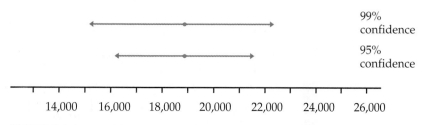

FIGURE 6.6   Confidence intervals for Examples 6.3 and 6.5.

margin of error for any fixed confidence level. The square root in the formula implies that we must multiply the number of observations by 4 in order to cut the margin of error in half. The standard deviation σ measures the variation in the population. You can think of the variation among individuals in the population as noise that obscures the average value μ. It is harder to pin down the mean μ of a highly variable population; that is why the margin of error of a confidence interval increases with σ. In practice, we can sometimes reduce σ by carefully controlling the measurement process or by restricting our attention to only part of a large population.

EXAMPLE 6.5   Suppose that for the student loan data in Example 6.3, we wanted 99% confidence rather than 95%. Table D tells us that for 99% confidence, $z^* = 2.576$. The margin of error for 99% confidence based on 1280 observations is

$$m = z^* \frac{\sigma}{\sqrt{n}}$$

$$= 2.576 \frac{49{,}000}{\sqrt{1280}}$$

$$= 3500$$

and the 99% confidence interval is

$$\bar{x} \pm m = 18{,}900 \pm 3500$$

$$= (15{,}400,\ 22{,}400)$$

Requiring 99%, rather than 95%, confidence has increased the margin of error from 2700 to 3500. Figure 6.6 compares the two intervals.

Choosing the sample size

A wise user of statistics never plans data collection without at the same time planning the inference. You can arrange to have both high confidence and a small margin of error. The margin of error of the confidence interval $\bar{x} \pm z^* \sigma / \sqrt{n}$ for a normal mean is

$$z^* \frac{\sigma}{\sqrt{n}}$$

To obtain a desired margin of error m, just set this expression equal to m, substitute the value of z^* for your desired confidence level, and solve for the sample size n. Here is the result.

> ### SAMPLE SIZE FOR DESIRED MARGIN OF ERROR
>
> The confidence interval for a population mean will have a specified margin of error m when the sample size is
>
> $$n = \left(\frac{z^*\sigma}{m} \right)^2$$

This formula is not the proverbial free lunch. In practice, taking observations costs time and money. The required sample size may be impossibly expensive. Do notice once again that it is the size of the *sample* that determines the margin of error. The size of the *population* (as long as the population is much larger than the sample) does not influence the sample size we need.

EXAMPLE 6.6   Suppose that we are planning a student loan survey similar to the one described in Example 6.3. If we want the margin of error to be $2000 with 95% confidence, how large a sample is required? For 95% confidence, Table D gives $z^* = 1.960$. For σ we will use the value from the previous study, $49,000. If the margin of error is $2000, we have

$$n = \left(\frac{z^*\sigma}{m} \right)^2 = \left(\frac{1.96 \times 49,000}{2000} \right)^2 = 2305.6$$

Because 2305 measurements will give a slightly wider interval than desired and 2306 measurements a slightly narrower interval, we could choose $n = 2306$. We need loan information from 2306 borrowers to determine an estimate of mean debt.

It is always safe to round *up* to the next higher whole number when finding n because this will give us a smaller margin of error. The purpose of this calculation is to determine a sample size that is sufficient to provide useful results, but the determination of what is useful is a matter of judgment. Would we need a much larger sample size to obtain a margin of error of $1500? Here is the calculation:

$$n = \left(\frac{z^*\sigma}{m} \right)^2 = \left(\frac{1.96 \times 49,000}{1500} \right)^2 = 4099$$

Because this number is so large, we decide to use 2300 as our target sample size.

Unfortunately, the actual number of usable observations is often less than what we plan at the beginning of a study. This is particularly true of data collected in surveys but is an important consideration in most studies. Careful study designers often assume a nonresponse rate or dropout rate that specifies what proportion of the originally planned sample will fail to provide data. We use this information to calculate the sample size to be used at the start of the study. For example, if in the survey above, we expect only 25% of those contacted to respond, we would need to start with a sample size of $4 \times 2300 = 9200$ to obtain usable information from 2300 borrowers.

Some cautions

We have already seen that small margins of error and high confidence can require large numbers of observations. You should also be keenly aware that *any formula for inference is correct only in specific circumstances*. If the government required statistical procedures to carry warning labels like those on drugs, most inference methods would have long labels indeed. Our handy formula $\bar{x} \pm z^*\sigma/\sqrt{n}$ for estimating a normal mean comes with the following list of warnings for the user:

- The data should be an SRS from the population. We are completely safe if we actually did a randomization and drew an SRS. We are not in great danger if the data can plausibly be thought of as independent observations from a population. That is the case in Examples 6.3 to 6.6, where we redefined our population to correspond to survey respondents.

- The formula is not correct for probability sampling designs more complex than an SRS. Correct methods for other designs are available. We will not discuss confidence intervals based on multistage or stratified samples. If you plan such samples, be sure that you (or your statistical consultant) know how to carry out the inference you desire.

- There is no correct method for inference from data haphazardly collected with bias of unknown size. Fancy formulas cannot rescue badly produced data.

- Because $\bar{x}$ is not resistant, outliers can have a large effect on the confidence interval. You should search for outliers and try to correct them or justify their removal before computing the interval. If the outliers cannot be removed, ask your statistical consultant about procedures that are not sensitive to outliers.

- If the sample size is small and the population is not normal, the true confidence level will be different from the value C used in computing the interval. Examine your data carefully for skewness and other signs of nonnormality. The interval relies only on the distribution of $\bar{x}$, which even for quite small sample sizes is much closer to normal than that of the individual observations. When $n \geq 15$, the confidence level is not greatly disturbed by nonnormal populations unless extreme outliers or quite strong skewness are present. Our debt data in Example 6.3 are clearly skewed but because of the large sample size, we are confident that the sample mean will be approximately normal. We will discuss this issue in more detail in the next chapter.

- You must know the standard deviation σ of the population. This unrealistic requirement renders the interval $\bar{x} \pm z^*\sigma/\sqrt{n}$ of little use in statistical practice. We will learn in the next chapter what to do when σ is unknown. If, however, the sample is large, the sample standard deviation s will be close to the unknown σ. The recipe $\bar{x} \pm z^*s/\sqrt{n}$ is then an approximate confidence interval for μ.

The most important caution concerning confidence intervals is a consequence of the first of these warnings. *The margin of error in a confidence interval covers only random sampling errors*. The margin of error is obtained from the sampling distribution and indicates how much error can be expected because of chance variation in randomized data production. *Practical difficulties such as undercoverage and nonresponse in a sample survey cause additional errors. These errors can be larger than the random sampling error.* This often happens when the sample size is large (so that $\sigma/\sqrt{n}$ is small). Remember this unpleasant fact when reading the results of an opinion poll or other sample survey. The practical conduct of the survey influences the trustworthiness of its results in ways that are not included in the announced margin of error.

Every inference procedure that we will meet has its own list of warnings. Because many of the warnings are similar to those above, we will not print the full warning label each time. It is easy to state (from the mathematics of probability) conditions under which a method of inference is exactly correct. These conditions are *never* fully met in practice. For example, no population is exactly normal. Deciding when a statistical procedure should be used in practice often requires judgment assisted by exploratory analysis of the data. Mathematical facts are therefore only a part of statistics. The difference between statistics and mathematics can be stated thus: mathematical theorems are true; statistical methods are often effective when used with skill.

Finally, you should understand what statistical confidence does not say. We are 95% confident that the mean SATM score for the California students in Example 6.2 lies between 452 and 470. This says that these numbers were calculated by a method that gives correct results in 95% of all possible samples. It does *not* say that the probability is 95% that the true mean falls between 452 and 470. No randomness remains after we draw a particular sample and get from it a particular interval. The true mean either is or is not between 452 and 470. Probability in its interpretation as long-run proportion makes no sense in this situation. The probability calculations of standard statistical inference describe how often the *method* gives correct answers.

BEYOND THE BASICS | The bootstrap

Confidence intervals are based on sampling distributions. In this section we have used the fact that the sampling distribution of $\bar{x}$ is $N(\mu, \sigma/\sqrt{n})$ when the data are an SRS from a $N(\mu, \sigma)$ population. If the data are not normal, the central limit theorem tells us that this sampling distribution is still a reasonable approximation as long as the distribution of the data is not strongly skewed and there are no outliers. Even a fair amount of skewness can be tolerated when the sample size is large. We used the sampling distribution of $\bar{x}$ to conclude that, for example, 95% of the time, $\bar{x}$ and μ will not differ by more than $1.96\sigma/\sqrt{n}$.

What if the population does not appear to be normal and we have only a small sample? Then we do not know what the sampling distribution of $\bar{x}$ looks like. The **bootstrap** is a procedure for approximating sampling distributions when theory cannot tell us their shape.[3]

The basic idea is to act as if our sample were the population. We take many samples from it. Each of these is called a **resample.** We calculate the mean $\bar{x}$ for each resample. We get different results from different resamples because we sample *with replacement.* An individual observation in the original sample can appear more than once in the resample.

For example, we have four measurements of a person's weight last month:

190.5   189.0   195.5   187.0

one resample with replacement might be

$$189.0 \quad 195.5 \quad 187.0 \quad 189.0$$

with $\bar{x} = 190.125$. Collect the $\bar{x}$'s from 1000 such resamples. Their distribution will be close to what we would get if we took 1000 samples from the entire population. We treat the distribution of $\bar{x}$'s from our 1000 resamples as if it were the sampling distribution. If we want a 95% confidence interval, for example, we use the middle 95% of this distribution.

The bootstrap is practical only when you can use a computer to take 1000 or more samples quickly. It is an example of how the use of fast and easy computing is changing the way we do statistics. More details about the bootstrap can be found in Chapter 14 (available in a separate booklet).

SECTION 6.1 | Summary

The purpose of a **confidence interval** is to estimate an unknown parameter with an indication of how accurate the estimate is and of how confident we are that the result is correct.

Any confidence interval has two parts: an interval computed from the data and a confidence level. The interval often has the form

$$\text{estimate} \pm \text{margin of error}$$

The **confidence level** states the probability that the method will give a correct answer. That is, if you use 95% confidence intervals often, in the long run 95% of your intervals will contain the true parameter value. When you apply the method once, you do not know if your interval gave a correct value (this happens 95% of the time) or not (this happens 5% of the time).

The **margin of error** for a level C confidence interval for the mean μ of a normal population with known standard deviation σ, based on an SRS of size n, is given by

$$m = z^* \frac{\sigma}{\sqrt{n}}$$

Here z^* is obtained from the row labeled z^* at the bottom of Table D. The probability is C that a standard normal random variable takes a value between $-z^*$ and z^*. The confidence interval is

$$\bar{x} \pm m$$

Other things being equal, the margin of error of a confidence interval decreases as

- the confidence level C decreases,
- the sample size n increases, and
- the population standard deviation σ decreases.

The sample size required to obtain a confidence interval of specified margin of error m for a normal mean is

$$n = \left(\frac{z^* \sigma}{m} \right)^2$$

where z^* is the critical point for the desired level of confidence.

A specific confidence interval recipe is correct only under specific conditions. The most important conditions concern the method used to produce the data. Other factors such as the form of the population distribution may also be important.

SECTION 6.1 | Exercises

6.1 A study based on a sample of size 25 reported a mean of 76 with a margin of error of 12 for 95% confidence. Give the 95% confidence interval.

6.2 Refer to the previous exercise. If you wanted 99% confidence for the same study, would your margin of error be greater than, equal to, or less than 12? Explain your answer.

6.3 Suppose that the sample mean is 50 and the standard deviation is assumed to be 5. Make a diagram similar to Figure 6.5 (page 390) that illustrates the effect of sample size on the width of a 95% interval. Use the following sample sizes: 10, 20, 40, and 100. Summarize what the diagram shows.

6.4 A study with 25 observations gave a mean of 70. Assume that the standard deviation is 15. Make a diagram similar to Figure 6.6 (page 391) that illustrates the effect of the confidence level on the width of the interval. Use 80%, 90%, 95%, and 99%. Summarize what the diagram shows.

6.5 For many important processes that occur in the body, direct measurement of characteristics of the process is not possible in live individuals. In many cases, however, we can measure a *biomarker*, a biochemical substance that is relatively easy to measure and is associated with the process of interest. Bone turnover is the net effect of two processes: the breaking down of old bone, called resorption, and the building of new bone, called formation. One biochemical measure of bone resorption is tartrate resistant acid phosphatase (TRAP), which can be measured in blood. In a study of bone turnover in young women, serum TRAP was measured in 31 subjects.[4] The units are units per liter (U/l). The mean was 13.2 U/l. Assume that the standard deviation is known to be 6.5 U/l. Give the margin of error and find a 95% confidence interval for the mean for young women represented by this sample.

6.6 Refer to the previous exercise. A biomarker for bone formation measured in the same study was osteocalcin (OC), measured in the blood. The units are nanograms per milliliter (ng/ml). For the 31 subjects in the study the mean was 33.4 ng/ml. Assume that the standard deviation is known to be 19.6 ng/ml. Report the 95% confidence interval.

6.7 Refer to Exercise 6.5. Repeat the calculations for a 99% confidence interval. How do the results compare with those in Exercise 6.5?

6.8 Refer to Exercise 6.6. Repeat the calculations for a 90% confidence interval. How do the results compare with those in Exercise 6.6?

6.9 Consider the following two scenarios. (A) Take a simple random sample of 100 students from an elementary school with children in grades kindergarten through fifth grade. (B) Take a simple random sample of 100 third-graders from the same school. For each of these samples you will measure the height of each child in the sample. Which sample should have the smaller margin of error for 95% confidence? Explain your answer.

6.10   You want to rent an unfurnished one-bedroom apartment for next semester. The mean monthly rent for a random sample of 10 apartments advertised in the local newspaper is $580. Assume that the standard deviation is $90. Find a 95% confidence interval for the mean monthly rent for unfurnished one-bedroom apartments available for rent in this community.

6.11   Refer to the previous exercise. Will the 95% confidence interval include approximately 95% of the rents of all unfurnished one-bedroom apartments in this area? Explain why or why not.

6.12   The mean OC for the 31 subjects in Exercise 6.6 was 33.4 ng/ml. In our calculations, we assumed that the standard deviation was known to be 19.6 ng/ml. Use the 68–95–99.7 rule from Chapter 1 (page 71) to find the approximate bounds on the values of OC that would include these percents of the population. If the assumed standard deviation is correct, it would appear that this distribution may be highly skewed. Why? (*Hint:* The measured values for a variable such as this are all positive.) Do you think that this skewness will invalidate the use of the normal confidence interval in this case? Explain your answer.

6.13   A questionnaire about study habits was given to a random sample of students taking a large introductory statistics class. The sample of 25 students reported that they spent an average of 80 minutes per week studying statistics. Assume that the standard deviation is 35 minutes.

(a) Give a 95% confidence interval for the mean time spent studying statistics by students in this class.

(b) Is it true that 95% of the students in the class have weekly study times that lie in the interval you found in part (a)? Explain your answer.

6.14   Refer to the previous exercise.

(a) Give the mean and standard deviation in hours.

(b) Calculate the 95% confidence interval in hours from your answer to part (a).

(c) Explain how you could have directly calculated this interval from the 95% interval that you calculated in the previous exercise.

6.15   The mean number of calories consumed by women in the United States who are 19 to 30 years of age is $\mu = 1791$ calories per day. The standard deviation is 31 calories.[5] You will study a sample of 200 women in this age range, and one of the variables you will collect is calories consumed per day.

(a) What is the standard deviation of the sample mean $\bar{x}$?

(b) The 68–95–99.7 rule says that the probability is about 0.95 that $\bar{x}$ is within _____ of the population mean μ. Fill in the blank.

(c) About 95% of all samples will capture the true mean of calories consumed per day in the interval $\bar{x}$ plus or minus _____. Fill in the blank.

6.16   Computers in some vehicles calculate various quantities related to performance. One of these is the fuel efficiency, or gas mileage, usually expressed as miles per gallon (MPG). For one vehicle equipped in this way, the MPG were recorded each time the gas tank was filled, and the computer was then reset.[6] Here are the MPG values for a random sample of 20 of these records:

15.8	13.6	15.6	19.1	22.4	15.6	22.5	17.2	19.4	22.6
19.4	18.0	14.6	18.7	21.0	14.8	22.6	21.5	14.3	20.9

Suppose that the standard deviation of the population is known to be $\sigma = 2.9$ MPG.

(a) What is $\sigma_{\bar{x}}$, the standard deviation of $\bar{x}$?

(b) Give a 95% confidence interval for μ, the mean MPG for this vehicle.

6.17   Refer to the previous exercise. Here are the values of the average speed in miles per hour (MPH) for the same sample:

21.0	19.0	18.7	39.2	45.8	19.8	48.4	21.0	29.1	35.7
31.6	49.0	16.0	34.6	36.3	19.0	43.3	37.5	16.5	34.5

Assume that the standard deviation is 10.3 MPH. Estimate the mean speed at which this vehicle was driven with a margin of error for 95% confidence.

6.18   In Exercise 6.17 you found an estimate with a margin of error for the average speed expressed in miles per hour. Convert your estimate and margin of error to the metric units kilometers per hour (KPH). To change MPH to KPH, multiply by 1.6.

6.19   Here are the Degree of Reading Power (DRP) scores for a sample of 44 third-grade students:

40	26	39	14	42	18	25	43	46	27	19
47	19	26	35	34	15	44	40	38	31	46
52	25	35	35	33	29	34	41	49	28	52
47	35	48	22	33	41	51	27	14	54	45

Suppose that the standard deviation of the population of DRP scores is known to be $\sigma = 11$. Give a 95% confidence interval for the population mean score.

6.20   The *Confidence Interval* applet lets you simulate large numbers of confidence intervals quickly. Select 95% confidence and then sample 50 intervals. Record the number of intervals that cover the true value (this appears in the "hit" box in the applet). Press the reset button and repeat 30 times. Make a stemplot of the results and find the mean. Describe the results. If you repeated this experiment very many times, what would you expect the average number of hits to be?

6.21   Repeat the previous exercise for 80% confidence.

6.22   You are planning a survey of starting salaries for recent liberal arts major graduates from your college. From a pilot study you estimate that the standard deviation is about $9000. What sample size do you need to have a margin of error equal to $400 with 95% confidence?

6.23   Suppose that in the setting of the previous exercise you are willing to settle for a margin of error of $800. Will the required sample size be larger or smaller? Verify your answer by performing the calculations.

6.24   How large a sample of one-bedroom apartments in Exercise 6.10 would be needed to estimate the mean μ within ±$20 with 90% confidence?

6.25   A new bone study is being planned that will measure the biomarker TRAP described in Exercise 6.5. Using the value of σ given there, 6.5 U/l, find the sample size required to provide an estimate of the mean TRAP with a margin of error of 2.0 U/l for 95% confidence.

6.26   Refer to the previous exercise. In similar previous studies, about 20% of the subjects drop out before the study is completed. Adjust your sample size requirement to have enough subjects at the end of the study to meet the margin of error criterion.

6.27   An advertisement in the student newspaper asks you to consider working for a telemarketing company. The ad states, "Earn between $500 and $1000 per week." Do you think that the ad is describing a confidence interval? Explain your answer.

6.28   To assess the accuracy of a laboratory scale, a standard weight known to weigh 10 grams is weighed repeatedly. The scale readings are normally distributed with unknown mean (this mean is 10 grams if the scale has no bias). The standard deviation of the scale readings is known to be 0.0002 gram.

  (a) The weight is weighed five times. The mean result is 10.0023 grams. Give a 98% confidence interval for the mean of repeated measurements of the weight.

  (b) How many measurements must be averaged to get a margin of error of ±0.0001 with 98% confidence?

6.29   As we prepare to take a sample and compute a 95% confidence interval, we know that the probability that the interval we compute will cover the parameter is 0.95. That's the meaning of 95% confidence. If we use several such intervals, however, our confidence that *all* give correct results is less than 95%. Suppose we take independent samples each month for four months and report a 95% confidence interval for each set of data.

  (a) What is the probability that all four intervals cover the true means? This probability (expressed as a percent) is our overall confidence level for the four simultaneous statements.

  (b) What is the probability that at least three of the four intervals cover the true means?

6.30 A newspaper invites readers to send email stating whether or not they are in favor of making full-day kindergarten available to all students in the state. A total of 320 responses are received and, of these, 80% are in favor of the new program. In an article describing the results, the authors state that the margin of error is 4% for 95% confidence. Assume that they have computed this number correctly.

(a) Use the sample proportion and the margin of error to compute the 95% confidence interval.

(b) Do you think that these results are trustworthy? Discuss your answer.

6.31 A recent Gallup poll conducted telephone interviews with a random sample of adults aged 18 and older. Data were obtained for 1011 people. Of these, 37% said that football is their favorite sport to watch on television.

(a) The poll announced a margin of error of ±3 percentage points for 95% confidence. Explain to someone who knows no statistics why we can't just say that 37% of Americans would say that football is their favorite sport to watch on television.

(b) Then explain clearly what "95% confidence" means.

(c) Give the 95% confidence interval.

(d) The poll was taken in December, an exciting part of the football season. Do you think that a similar poll conducted in June might produce different results? Explain why or why not.

6.2 Tests of Significance

Confidence intervals are one of the two most common types of formal statistical inference. They are appropriate when our goal is to estimate a population parameter. The second common type of inference is directed at a quite different goal: to assess the evidence provided by the data in favor of some claim about the population.

The reasoning of significance tests

A significance test is a formal procedure for comparing observed data with a hypothesis whose truth we want to assess. The hypothesis is a statement about the parameters in a population or model. The results of a test are expressed in terms of a probability that measures how well the data and the hypothesis agree. We use the following example to illustrate these ideas. We start by focusing on the concepts. Don't worry about the details; we will deal with them later.

EXAMPLE 6.7 One purpose of the National Student Loan Survey described in Example 6.3 is to compare the debt of different subgroups of students. For example, the 525 borrowers who last attended a private four-year college had a mean debt of $21,200, while those who last attended a public four-year college had a mean debt of $17,100. The difference of $4100 is fairly large, but we know that these numbers are estimates of the true means. If we took a different sample, we would get different estimates. Can we conclude from these data that the private four-year students have greater debt than public four-year borrowers?

One way to answer this question is to compute the probability of obtaining a difference as large or larger than the observed $4100 assuming that, in fact, there is no difference in the true means. This probability is 0.17. Because this probability is not particularly small, we conclude that observing a difference of $4100 is not very surprising when the true means are equal. The data do not provide evidence for us to conclude that the mean debts for private four-year borrowers and public four-year borrowers are different.

Here is an example with a different conclusion.

EXAMPLE 6.8   Another purpose of the National Student Loan Survey is to look for changes over time. For example, in 1997, the survey found that the mean debt for undergraduate study was $11,400. How does this compare with the value of $18,900 in the 2002 study? The difference is $7500. As we learned in the previous example, an observed difference in means is not necessarily sufficient for us to conclude that the true means are different. Do the data provide evidence that there is an increase in borrowing? Again, we answer this question with a probability calculated under the assumption that there is no difference in the true means. The probability is 0.00004 of observing an increase in mean debt that is $7500 or more. Because this probability is so small, we have sufficient evidence in the data to conclude that there has been a change in borrowing between 1997 and 2002.

What are the key steps in these examples? We first started each with a question about the difference between two mean debts. In Example 6.7, we compare private four-year borrowers with public four-year borrowers. In Example 6.8, we compare borrowers in 2002 with borrowers in 1997. We ask whether or not the data are compatible with no difference, that is, a difference of $0. Next we compared the data, $4100 in the first case and $7500 in the second, with the value that comes from the question, $0. The results of the comparisons are probabilities, 0.17 in the first case and 0.00004 in the second.

The 0.17 probability is not particularly small, so we have no evidence to question the possibility that the true difference is zero. In the second case, however, the probability is quite small. Something that happens with probability 0.00004 occurs only about 4 times out of 100,000. In this case we have two possible explanations:

1. we have observed something that is very unusual, or

2. the assumption that underlies the calculation, no difference in mean debt, is not true.

Because this probability is so small, we prefer the second conclusion: there has been a change in the mean debt between 1997 and 2002.

The probabilities in Examples 6.7 and 6.8 are measures of the compatibility of the data (a difference in means of $4100 and $7500) with the *null*

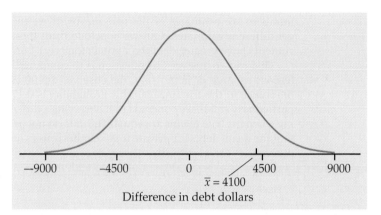

FIGURE 6.7 Comparison of the sample mean in Example 6.7 relative to the null hypothesized value 0.

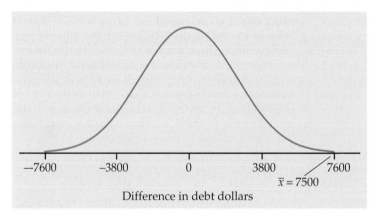

FIGURE 6.8 Comparison of the sample mean in Example 6.8 relative to the null hypothesized value 0.

hypothesis that there is no difference in the true means. Figures 6.7 and 6.8 compare the two results graphically. For each a normal curve centered at 0 is the sampling distribution. You can see that we are not particularly surprised to observe the difference $4100 in Figure 6.7, but the difference $7500 in Figure 6.8 is clearly an unusual observation. We will now consider some of the formal aspects of significance testing.

Stating hypotheses

In Example 6.7, we asked whether the difference in means observed is reasonable if, in fact, there is no difference in the true means. To answer this, we begin by supposing that the statement following the "if" in the previous sentence is true. In other words, we suppose that the true difference is $0. We then ask whether the data provide evidence against the supposition we have made. If so, we have evidence in favor of an effect (the means are different) we are seeking. The first step in a test of significance is to state a claim that we will try to find evidence *against*.

> ## NULL HYPOTHESIS
>
> The statement being tested in a test of significance is called the **null hypothesis.** The test of significance is designed to assess the strength of the evidence against the null hypothesis. Usually the null hypothesis is a statement of "no effect" or "no difference."

We abbreviate "null hypothesis" as H_0. A null hypothesis is a statement about a population, expressed in terms of some parameter or parameters. For example, our null hypothesis for Example 6.7 is

$$H_0: \text{there is no difference in the true means}$$

Note that the null hypothesis refers to the *true* means for all borrowers, including those for whom we do not have data.

alternative hypothesis   It is convenient also to give a name to the statement we hope or suspect is true instead of H_0. This is called the **alternative hypothesis** and is abbreviated as H_a. In Example 6.7, the alternative hypothesis states that the means are different. We write this as

$$H_a: \text{the true means are not the same}$$

Hypotheses always refer to some population or model, not to a particular outcome. For this reason, we must state H_0 and H_a in terms of population parameters.

Because H_a expresses the effect that we hope to find evidence *for*, we often begin with H_a and then set up H_0 as the statement that the hoped-for effect is not present. Stating H_a is often the more difficult task. It is not always clear, **one-sided or two-sided alternatives**   in particular, whether H_a should be **one-sided** or **two-sided,** which refers to whether a parameter differs from its null hypothesis value in a specific direction or in either direction.

The alternative hypothesis should express the hopes or suspicions we bring to the data. *It is cheating to first look at the data and then frame H_a to fit what the data show.* If you do not have a specific direction firmly in mind in advance, you must use a two-sided alternative. Moreover, some users of statistics argue that we should always use a two-sided alternative.

Test statistics

We will learn the form of significance tests in a number of common situations. Here are some principles that apply to most tests and that help in understanding these tests:

- The test is based on a statistic that estimates the parameter that appears in the hypotheses. Usually this is the same estimate we would use in a confidence interval for the parameter. When H_0 is true, we expect the estimate to take a value near the parameter value specified by H_0.

- Values of the estimate far from the parameter value specified by H_0 give evidence against H_0. The alternative hypothesis determines which directions count against H_0.

- To assess how far the estimate is from the parameter, standardize the estimate. In many common situations the test statistic has the form

$$z = \frac{\text{estimate} - \text{hypothesized value}}{\text{standard deviation of the estimate}}$$

Let's return to our student loan example.

EXAMPLE 6.9 In Example 6.7, the hypotheses are stated in terms of the difference in debt between borrowers who attended a private college and those who attended a public college:

H_0: there is no difference in the true means

H_a: there is a difference in the true means

Because H_a is two-sided, large values of both positive and negative differences count as evidence against the null hypothesis.

test statistic A **test statistic** measures compatibility between the null hypothesis and the data. We use it for the probability calculation that we need for our test of significance. It is a random variable with a distribution that we know.

EXAMPLE 6.10 In Example 6.7, we can state the null hypothesis as H_0: the true mean difference is 0. The estimate of the difference is $4100. Using methods that we will discuss in detail later, we can determine that the standard deviation of the estimate is $3000. For this problem the test statistic is

$$z = \frac{\text{estimate} - \text{hypothesized value}}{\text{standard deviation of the estimate}}$$

For our data,

$$z = \frac{4100 - 0}{3000} = 1.37$$

We have observed a sample estimate that is about one and a third standard deviations away from the hypothesized value of the parameter. Because the sample sizes are sufficiently large for us to conclude that the sample estimate is approximately normal, the standardized test statistic z will have approximately the $N(0, 1)$ distribution.

We will use facts about the normal distribution in what follows. Do you recall the 68–95–99.7 rule and the calculations that we did with the standard normal distribution (Table A)? If not, this would be a good time to refresh your memory by reviewing Section 1.3.

P-values

If all test statistics were normal, we could base our conclusions on the value of the z test statistic. In fact, the Supreme Court of the United States has said that "two or three standard deviations" ($z = 2$ or 3) is its criterion for rejecting

H_0 (see Exercise 8.77 on page 576), and this is the criterion used in most applications involving the law. Because not all test statistics are z's (random variables with the $N(0, 1)$ distribution), we translate the value of test statistics into a common language, the language of probability.

A test of significance finds the probability of getting an outcome *as extreme or more extreme than the actually observed outcome*. "Extreme" means "far from what we would expect if H_0 were true." The direction or directions that count as "far from what we would expect" are determined by H_a as well as H_0.

In Example 6.7 we want to know if the debt of private college borrowers is different from the debt of public college borrowers. The difference we calculated based on our sample is $4100, which corresponds to 1.37 standard deviations away from zero—that is, $z = 1.37$. Because we are using a two-sided alternative for this problem, the evidence against H_0 is measured by the probability

P(we observe a value of Z as extreme or more extreme than 1.37)

More formally, this probability is

$$P(Z \leq -1.37 \text{ or } Z \geq 1.37)$$

where Z has the standard normal distribution $N(0, 1)$.

P-VALUE

The probability, computed assuming that H_0 is true, that the test statistic would take a value as extreme or more extreme than that actually observed is called the **P-value** of the test. The smaller the P-value, the stronger the evidence against H_0 provided by the data.

The key to calculating the P-value is the sampling distribution of the test statistic. For the problems we consider in this chapter, we need only the standard normal distribution for the test statistic z.

EXAMPLE 6.11   In Example 6.10 we found that the test statistic for testing

H_0: the true mean difference is 0

versus

H_a: there is a difference in the true means

is

$$z = \frac{4100 - 0}{3000} = 1.37$$

If H_0 is true, then z is a single observation from the standard normal, $N(0, 1)$, distribution. Figure 6.9 illustrates this calculation. The P-value is the probability of observing a value of z at least as extreme as the one that we observed, $z = 1.37$. From Table A, our table of standard normal probabilities, we find

$$P(Z \geq 1.37) = 1 - 0.9147 = 0.0853$$

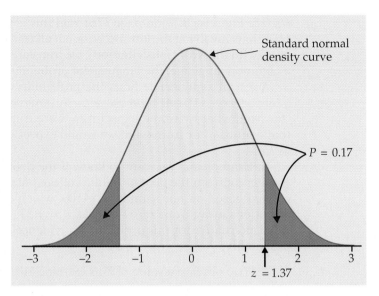

FIGURE 6.9 The *P*-value for Example 6.11. The *P*-value here is the probability (when H_0 is true) that $\bar{x}$ takes a value as extreme or more extreme than the actually observed value.

The probability for being extreme in the negative direction is the same:

$$P(Z \leq -1.37) = 0.0853$$

So the *P*-value is

$$P = 2P(Z \geq 1.37) = 2(0.0853) = 0.1706$$

This is the value that was reported on page 401. There is a 17% chance of observing a difference as extreme as the $4100 in our sample if the true population difference is zero. The *P*-value tells us that our outcome is not particularly extreme, so we conclude that the data do not provide evidence that would cause us to doubt the validity of the null hypothesis.

Statistical significance

We started our discussion of the reasoning of significance tests with the statement of null and alternative hypotheses. We then learned that a test statistic is the tool used to examine the compatibility of the observed data with the null hypothesis. Finally, we translated the test statistic into a *P*-value to quantify the evidence against H_0. One important final step is needed. What do we conclude?

We can compare the *P*-value we calculated with a fixed value that we regard as decisive. This amounts to announcing in advance how much evidence against H_0 we will require to reject H_0. The decisive value of *P* is called the **significance level**. It is denoted by α, the Greek letter alpha. If we choose $\alpha = 0.05$, we are requiring that the data give evidence against H_0 so strong that it would happen no more than 5% of the time (1 time in 20) when H_0 is true. If we choose $\alpha = 0.01$, we are insisting on stronger evidence against H_0,

significance level

evidence so strong that it would appear only 1% of the time (1 time in 100) if H_0 is in fact true. The Supreme Court's criterion ($z = 2$ or 3) translates into $\alpha = 0.0445$ or $\alpha = 0.0027$ for a two-sided test.

> ## STATISTICAL SIGNIFICANCE
>
> If the P-value is as small or smaller than α, we say that the data are **statistically significant at level α.**

"Significant" in the statistical sense does not mean "important." The original meaning of the word is "signifying something." In statistics the term is used to indicate only that the evidence against the null hypothesis reached the standard set by α. Significance at level 0.01 is often expressed by the statement "The results were significant ($P < 0.01$)." Here P stands for the P-value. The P-value is more informative than a statement of significance because we can then assess significance at any level we choose. For example, a result with $P = 0.03$ is significant at the $\alpha = 0.05$ level but is not significant at the $\alpha = 0.01$ level.

A test of significance is a recipe for assessing the significance of the evidence provided by data against a null hypothesis. The four steps common to all tests of significance are as follows:

1. State the *null hypothesis H_0* and the *alternative hypothesis H_a*. The test is designed to assess the strength of the evidence against H_0; H_a is the statement that we will accept if the evidence enables us to reject H_0.

2. Calculate the value of the *test statistic* on which the test will be based. This statistic usually measures how far the data are from H_0.

3. Find the *P-value* for the observed data. This is the probability, calculated assuming that H_0 is true, that the test statistic will weigh against H_0 at least as strongly as it does for these data.

4. State a conclusion. One way to do this is to choose a *significance level α*, how much evidence against H_0 you regard as decisive. If the P-value is less than or equal to α, you conclude that the alternative hypothesis is true; if it is greater than α, you conclude that the data do not provide sufficient evidence to reject the null hypothesis. Your conclusion is a sentence that summarizes what you have found by using a test of significance.

We will learn the details of many tests of significance in the following chapters. In most cases, Steps 2 and 3 of our outline are almost automatic once you have practiced a bit. The proper test statistic is determined by the hypotheses and the data collection design. We use computer software or a calculator to find its numerical value and the P-value. The computer will not formulate your hypotheses for you, however. Nor will it decide if significance testing is appropriate or help you to interpret the P-value that it presents to you. The most difficult and important step is the last one: stating a conclusion.

EXAMPLE 6.12   In Example 6.11 we found that the P-value is 0.1706. There is a 17% chance of observing a difference as extreme as the $4100 in our sample if the true population difference is zero. The P-value tells us that our outcome is not particularly extreme. We could report the result as "the data do not provide evidence that would cause us to conclude that there is a difference in student loan debt between private college borrowers and public college borrowers ($z = 1.37$, $P = 0.17$)."

If the P-value is small, we reject the null hypothesis. Here is an example.

EXAMPLE 6.13   In Example 6.8 we found that the average debt has risen by $7500 from 1997 to 2002. Since we would have a prior expectation that the debt would increase over this period because of rising costs of a college education, it is appropriate to use a one-sided alternative in this situation. So, our hypotheses are

$$H_0: \text{the true mean difference is } 0$$

versus

$$H_a: \text{the mean debt has increased between 1997 and 2002}$$

The standard deviation is $1900 (again we defer details regarding this calculation), and the test statistic is

$$z = \frac{\text{estimate} - \text{hypothesized value}}{\text{standard deviation of the estimate}}$$

$$z = \frac{7500 - 0}{1900}$$

$$= 3.95$$

Because only increases in debt, and not decreases, count against the null hypothesis, the one-sided alternative leads to the calculation of the P-value using the upper tail of the normal distribution. The P-value is

$$P = P(Z \geq 3.95)$$

$$= 0.00004$$

The calculation is illustrated in Figure 6.10. There is about a 4 in 100,000 chance of observing a difference as large or larger than the $7500 in our sample if the true pop-

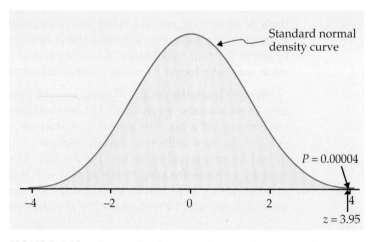

FIGURE 6.10   The P-value for Example 6.13. The P-value here is the probability (when H_0 is true) that $\bar{x}$ takes a value as large or larger than the actually observed value.

ulation difference is zero. This *P*-value tells us that our outcome is extremely rare. We conclude that the null hypothesis must be false. Here is one way to report the result: "The data clearly show that the mean debt for college loans has increased between 1997 and 2002 ($z = 3.95$, $P < 0.001$)."

Note that the calculated *P*-value for this example is 0.00004 but we reported the result as $P < 0.001$. The value 0.001, 1 in 1000, is sufficiently small to force a clear rejection of H_0. Standard practice is to report very small *P*-values as simply less than 0.001.

Tests for a population mean

Our discussion has focused on the reasoning of statistical tests, and we have outlined the key ideas for one type of procedure. Here is a summary. We want to test the hypothesis that a parameter has a specified value. This is the null hypothesis. For a test of a population mean, the null hypothesis is

$$H_0: \mu = \mu_0$$

where μ_0 is the specified value of μ that we would like to examine.

The test is based on data summarized as an estimate of the parameter. For a population mean this is the sample mean $\bar{x}$. Our test statistic measures the difference between the sample estimate and the hypothesized parameter in terms of standard deviations of the test statistic:

$$z = \frac{\text{estimate} - \text{hypothesized value}}{\text{standard deviation of the estimate}}$$

Recall from Chapter 5 (page 362) that the standard deviation of $\bar{x}$ is $\sigma/\sqrt{n}$. Therefore, the test statistic is

$$z = \frac{\bar{x} - \mu_0}{\sigma/\sqrt{n}}$$

Again recall from Chapter 5 that if the population is normal, then $\bar{x}$ will be normal (page 344) and z will have the standard normal distribution when H_0 is true. By the central limit theorem (page 362) both distributions will be approximately normal when the sample size is large even if the population is not normal.

Suppose we have calculated a test statistic $z = 1.7$. If the alternative is one-sided on the high side, then the *P*-value is the probability that a standard normal random variable Z takes a value as large or larger than the observed 1.7. That is,

$$P = P(Z \geq 1.7)$$

$$= 1 - P(Z < 1.7)$$

$$= 1 - 0.9554$$

$$= 0.0446$$

Similar reasoning applies when the alternative hypothesis states that the true μ lies below the hypothesized μ_0 (one-sided). When H_a states that μ is simply unequal to μ_0 (two-sided), values of z away from zero in either direction count against the null hypothesis. The *P*-value is the probability that a standard normal *Z* is at least as far from zero as the observed *z*. Again, if the test statistic is $z = 1.7$, the two-sided *P*-value is the probability that $Z \leq -1.7$ or $Z \geq 1.7$. Because the standard normal distribution is symmetric, we calculate this probability by finding $P(Z \geq 1.7)$ and *doubling* it:

$$P(Z \leq -1.7 \text{ or } Z \geq 1.7) = 2P(Z \geq 1.7)$$

$$= 2(1 - 0.9554) = 0.0892$$

We would make exactly the same calculation if we observed $z = -1.7$. It is the absolute value $|z|$ that matters, not whether *z* is positive or negative. Here is a statement of the test in general terms.

z TEST FOR A POPULATION MEAN

To test the hypothesis $H_0: \mu = \mu_0$ based on an SRS of size *n* from a population with unknown mean μ and known standard deviation σ, compute the test statistic

$$z = \frac{\bar{x} - \mu_0}{\sigma/\sqrt{n}}$$

In terms of a standard normal random variable *Z*, the *P*-value for a test of H_0 against

$H_a: \mu > \mu_0$ is $P(Z \geq z)$

$H_a: \mu < \mu_0$ is $P(Z \leq z)$

$H_a: \mu \neq \mu_0$ is $2P(Z \geq |z|)$

These *P*-values are exact if the population distribution is normal and are approximately correct for large *n* in other cases.

EXAMPLE 6.14   Do middle-aged male executives have different average blood pressure than the general population? The National Center for Health Statistics reports that the mean systolic blood pressure for males 35 to 44 years of age is 128 and the standard deviation in this population is 15. The medical director of a company looks at the medical records of 72 company executives in this age group and finds that the mean systolic blood pressure in this sample is $\bar{x} = 126.07$. Is this evidence that executive blood pressures differ from the national average?

The null hypothesis is "no difference" from the national mean $\mu_0 = 128$. The alternative is two-sided because the medical director did not have a particular direction in mind before examining the data. So the hypotheses about the unknown mean μ of the executive population are

$$H_0: \mu = 128$$

$$H_a: \mu \neq 128$$

As usual in this chapter, we make the unrealistic assumption that the population standard deviation is known, in this case that executives have the same $\sigma = 15$ as the general population of middle-aged males. The z test requires that the 72 executives in the sample are an SRS from the population of all middle-aged male executives in the company. We check this assumption by asking how the data were produced. If medical records are available only for executives with recent medical problems, for example, the data are of little value for our purpose. It turns out that all executives are given a free annual medical exam and that the medical director selected 72 exam results at random.

We can now proceed to compute the test statistic:

$$z = \frac{\bar{x} - \mu_0}{\sigma/\sqrt{n}} = \frac{126.07 - 128}{15/\sqrt{72}}$$

$$= -1.09$$

Figure 6.11 illustrates the P-value, which is the probability that a standard normal variable Z takes a value at least 1.09 away from zero. From Table A we find that this probability is

$$P = 2P(Z \geq |-1.09|) = 2P(Z \geq 1.09)$$

$$= 2(1 - 0.8621) = 0.2758$$

That is, more than 27% of the time an SRS of size 72 from the general male population would have a mean blood pressure at least as far from 128 as that of the executive sample. The observed $\bar{x} = 126.07$ is therefore not good evidence that executives differ from other men.

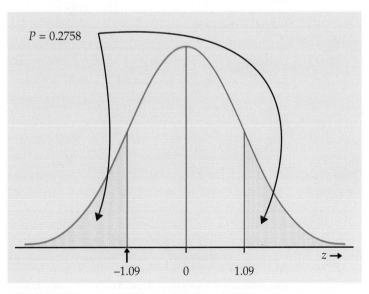

FIGURE 6.11   The P-value for the two-sided test in Example 6.14.

The data in Example 6.14 do *not* establish that the mean blood pressure μ for this company's middle-aged male executives is 128. We sought evidence that μ differed from 128 and failed to find convincing evidence. That is all we can say. No doubt the mean blood pressure of the entire executive population is not exactly equal to 128. A large enough sample would give evidence of the difference, even if it is very small. Tests of significance assess the evidence *against* H_0. If the evidence is strong, we can confidently reject H_0 in favor of the alternative. Failing to find evidence against H_0 means only that the data are consistent with H_0, not that we have clear evidence that H_0 is true.

EXAMPLE 6.15 In a discussion of SAT scores, someone comments: "Because only a minority of high school students take the test, the scores overestimate the ability of typical high school seniors. The mean SAT mathematics score is about 519, but I think that if all seniors took the test, the mean score would be no more than 450." You gave the test to an SRS of 500 seniors from California (Example 6.2). These students had a mean score of $\bar{x} = 461$. Is this good evidence against the claim that the mean for all California seniors is no more than 450? The hypotheses are

$$H_0: \mu = 450$$

$$H_a: \mu > 450$$

As we did in the discussion following Example 6.2, we assume that $\sigma = 100$. The z statistic is

$$z = \frac{\bar{x} - \mu_0}{\sigma/\sqrt{n}} = \frac{461 - 450}{100/\sqrt{500}}$$
$$= 2.46$$

Because H_a is one-sided on the high side, large values of z count against H_0. From Table A, we find that the P-value is

$$P = P(Z \geq 2.46) = 1 - 0.9931 = 0.0069$$

Figure 6.12 illustrates this P-value. A mean score as large as that observed would occur fewer than seven times in 1000 samples if the population mean were 450. This is convincing evidence that the mean SATM score for all California high school seniors is higher than 450.

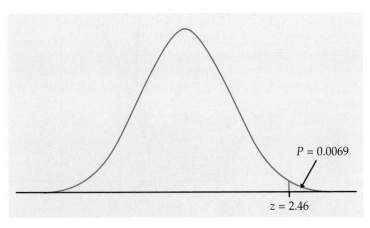

$P = 0.0069$

$z = 2.46$

FIGURE 6.12 The P-value for the one-sided test in Example 6.15.

Two-sided significance tests and confidence intervals

Recall the confidence intervals that we discussed in the first section of this chapter. The basic idea was that we constructed an interval that would include the true value of μ with a specified probability C. Suppose we use a 95% confidence interval ($C = 0.95$). Then the values of μ that are not in our interval would seem to be incompatible with the data. This sounds like a significance test with $\alpha = 0.05$ (or 5%) as our standard for drawing a conclusion. The following examples demonstrate that this is correct.

EXAMPLE 6.16   The Deely Laboratory analyzes specimens of a pharmaceutical product to determine the concentration of the active ingredient. Such chemical analyses are not perfectly precise. Repeated measurements on the same specimen will give slightly different results. The results of repeated measurements follow a normal distribution quite closely. The analysis procedure has no bias, so that the mean μ of the population of all measurements is the true concentration in the specimen. The standard deviation of this distribution is a property of the analytical procedure and is known to be $\sigma = 0.0068$ grams per liter. The laboratory analyzes each specimen three times and reports the mean result.

   The Deely Laboratory has been asked to evaluate the claim that the concentration of the active ingredient in a specimen is 0.86%. The true concentration is the mean μ of the population of repeated analyses. The hypotheses are

$$H_0: \mu = 0.86$$

$$H_a: \mu \neq 0.86$$

The lab chooses the 1% level of significance, $\alpha = 0.01$.
   Three analyses of one specimen give concentrations

$$0.8403 \quad 0.8363 \quad 0.8447$$

The sample mean of these readings is

$$\bar{x} = \frac{0.8403 + 0.8363 + 0.8447}{3} = 0.8404$$

The test statistic is

$$z = \frac{\bar{x} - \mu_0}{\sigma/\sqrt{n}} = \frac{0.8404 - 0.86}{0.0068/\sqrt{3}} = -4.99 \text{ standard deviations}$$

Because the alternative is two-sided, the P-value is

$$P = 2P(Z \geq |-4.99|) = 2P(Z \geq 4.99)$$

   We cannot find this probability in Table A. The largest value of z in that table is 3.49. All that we can say from Table A is that P is less than $2P(Z \geq 3.49) = 2(1 - 0.9998) = 0.0004$. If we use the bottom row of Table D, we find that the largest value of z^* is 3.291, corresponding to a P-value of $1 - 0.999 = 0.001$. Software could be used to give an accurate value of the P-value. However, because the P-value is clearly less than the company's standard of 1%, we reject H_0.

Suppose we compute a 99% confidence interval for the same data.

EXAMPLE 6.17

The 99% confidence interval for μ in Example 6.16 is

$$\bar{x} \pm z^* \frac{\sigma}{\sqrt{n}} = 0.8404 \pm 0.0101$$

$$= (0.8303, 0.8505)$$

The hypothesized value $\mu_0 = 0.86$ in Example 6.16 falls outside the confidence interval we computed in Example 6.17. We are therefore 99% confident that μ is *not* equal to 0.86, so we can reject

$$H_0: \mu = 0.86$$

at the 1% significance level. On the other hand, we cannot reject

$$H_0: \mu = 0.85$$

at the 1% level in favor of the two-sided alternative $H_a: \mu \neq 0.85$, because 0.85 lies inside the 99% confidence interval for μ. Figure 6.13 illustrates both cases.

The calculation in Example 6.16 for a 1% significance test is very similar to the calculation for a 99% confidence interval. In fact, a two-sided test at significance level α can be carried out directly from a confidence interval with confidence level $C = 1 - \alpha$.

TWO-SIDED SIGNIFICANCE TESTS AND CONFIDENCE INTERVALS

A level α two-sided significance test rejects a hypothesis $H_0: \mu = \mu_0$ exactly when the value μ_0 falls outside a level $1 - \alpha$ confidence interval for μ.

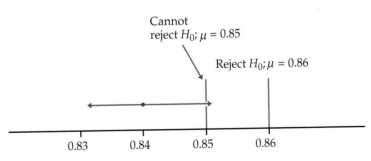

FIGURE 6.13 Values of μ falling outside a 99% confidence interval can be rejected at the 1% significance level; values falling inside the interval cannot be rejected.

P-values versus fixed α

The observed result in Example 6.16 was $z = -4.99$. The conclusion that this result is significant at the 1% level does not tell the whole story. The observed z is far beyond the z corresponding to 1%, and the evidence against H_0 is far stronger than 1% significance suggests. The *P*-value

$$2P(Z \geq 4.99) = 0.0000006$$

gives a better sense of how strong the evidence is. *The P-value is the smallest level α at which the data are significant.* Knowing the *P*-value allows us to assess significance at any level.

EXAMPLE 6.18   In Example 6.15, we tested the hypotheses

$$H_0: \mu = 450$$

$$H_a: \mu > 450$$

concerning the mean SAT mathematics score μ of California high school seniors. The test had the *P*-value $P = 0.0069$. This result is significant at the $\alpha = 0.01$ level because $0.0069 \leq 0.01$. It is not significant at the $\alpha = 0.005$ level, because the *P*-value is larger than 0.005. See Figure 6.14.

A *P*-value is more informative than a reject-or-not finding at a fixed significance level. But assessing significance at a fixed level α is easier, because no probability calculation is required. You need only look up a number in a table. A value z^* with a specified area to its right under the standard normal curve is called a **critical value** of the standard normal distribution. Because the practice of statistics almost always employs computer software that calculates *P*-values automatically, the use of tables of critical values is becoming outdated. We include the usual tables of critical values (such as Table D) at the end of the book for learning purposes and to rescue students without good computing facilities. The tables can be used directly to carry out fixed α tests. They also allow us to approximate *P*-values quickly without a probability calculation. The following example illustrates the use of Table D to find an approximate *P*-value.

critical value

EXAMPLE 6.19   In Example 6.10 we found the test statistic $z = 1.37$ for testing the null hypothesis that there was no difference in the mean debt between borrowers who attended a private college and those who attended a public college. The alternative was two-sided. Under the null hypothesis, z has a standard normal

FIGURE 6.14   An outcome with *P*-value *P* is significant at all levels α at or above *P* and is not significant at smaller levels α.

distribution, and from the last row in Table D we can see that there is a 95% chance that z is between ± 1.96. Therefore, we reject H_0 in favor of H_a whenever z is outside this range. Since our calculated value is 1.37, we are within the range and we do not reject the null hypothesis at the 5% level of significance.

SECTION 6.2 | Summary

A **test of significance** is intended to assess the evidence provided by data against a **null hypothesis H_0** in favor of an **alternative hypothesis H_a**.

The hypotheses are stated in terms of population parameters. Usually H_0 is a statement that no effect is present, and H_a says that a parameter differs from its null value, in a specific direction (**one-sided alternative**) or in either direction (**two-sided alternative**).

The test is based on a **test statistic**. The **P-value** is the probability, computed assuming that H_0 is true, that the test statistic will take a value at least as extreme as that actually observed. Small P-values indicate strong evidence against H_0. Calculating P-values requires knowledge of the sampling distribution of the test statistic when H_0 is true.

If the P-value is as small or smaller than a specified value α, the data are **statistically significant** at significance level α.

Significance tests for the hypothesis $H_0: \mu = \mu_0$ concerning the unknown mean μ of a population are based on the **z statistic**:

$$z = \frac{\bar{x} - \mu_0}{\sigma / \sqrt{n}}$$

The z test assumes an SRS of size n, known population standard deviation σ, and either a normal population or a large sample. P-values are computed from the normal distribution (Table A). Fixed α tests use the table of **standard normal critical values** (Table D).

SECTION 6.2 | Exercises

6.32 Here are several situations where there is an incorrect application of the ideas presented in this section. Write a short paragraph explaining what is wrong in each situation and why it is wrong.

(a) A climatologist wants to test the null hypothesis that it will rain tomorrow.

(b) A random sample of size 20 is taken from a population that is assumed to have a standard deviation of 15. The standard deviation of the sample mean is 15/20.

(c) A researcher tests the following null hypothesis: $H_0: \bar{x} = 10$.

6.33 Here are several situations where there is an incorrect application of the ideas presented in this section. Write a short paragraph explaining what is wrong in each situation and why it is wrong.

(a) A change is made that should improve student satisfaction with the way grades are processed at your college. The null hypothesis, that there is an improvement, is tested versus the alternative, that there is no change.

(b) A significance test rejected the null hypothesis that the sample mean is 25.

(c) A report on a study says that the results are statistically significant and the P-value is 0.95.

6.34 State the appropriate null hypothesis H_0 and alternative hypothesis H_a in each of the following cases.

(a) An experiment is designed to examine the effect of a diet high in soy products on the bone density of adult rats.

(b) The student newspaper at your college recently changed the format for their news stories. You take a random sample of students and select those who regularly read the newspaper. These are asked to indicate their opinions on the changes using a five-point scale: -2 if the new format is much worse than the old, -1 if the new format is somewhat worse than the old, 0 if the new format is the same as the old, $+1$ if the new format is somewhat better than the old, and $+2$ if the new format is much better than the old.

(c) The examinations in a large history class are scaled after grading so that the mean score is 75. A self-confident teaching assistant thinks that his students have a higher mean score than the class as a whole. His students this semester can be considered a sample from the population of all students he might teach, so he compares their mean score with 75.

6.35 State the null hypothesis H_0 and the alternative hypothesis H_a in each case.

(a) A national study reports that households spend an average of 30% of their food expenditures in restaurants. A restaurant association in your area wonders if the national finding applies locally. They interview a sample of 40 households and ask about their total food budget and the amount spent in restaurants.

(b) Experiments on learning in animals sometimes measure how long it takes a mouse to find its way through a maze. The mean time is 20 seconds for one particular maze. A researcher thinks that playing rap music will cause the mice to complete the maze faster. She measures how long each of 12 mice takes with the rap music as a stimulus.

(c) A dual X-ray absorptiometry (DXA) scanner is used to measure bone mineral density for people who may be at risk for osteoporosis. To be sure that the measurements are accurate, an object called a "phantom" that has known mineral density $\mu = 1.3$ grams per square centimeter is measured. This phantom is scanned 8 times.

6.36 In each of the following situations, state an appropriate null hypothesis H_0 and alternative hypothesis H_a. Be sure to identify the parameters that you use to state the hypotheses. (We have not yet learned how to test these hypotheses.)

(a) An education researcher randomly divides sixth-grade students into two groups for physical education class. He teaches both groups volleyball skills, using the same methods of instruction in both classes. He encourages Group A with compliments and other positive behavior but acts cool and neutral toward Group B. He hopes to show that positive teacher attitudes result in a higher mean score on a test of volleyball skills than do neutral attitudes.

(b) An education researcher believes that among college students there is a positive correlation between grade point average and self-esteem. To test this, she gathers grade point average and self-esteem data from a sample of students at your college.

(c) A sociologist asks a large sample of high school students which academic subject they like best. She suspects that a higher percent of females than of males will name English as their favorite subject.

6.37 Translate each of the following research questions into appropriate H_0 and H_a.

(a) Census Bureau data show that the mean household income in the area served by a shopping mall is \$72,500 per year. A market research firm questions shoppers at the mall to find out whether the mean household income of mall shoppers is higher than that of the general population.

(b) Last year, your company's service technicians took an average of 1.8 hours to respond to trouble calls from business customers who had purchased service contracts. Do this year's data show a different average response time?

6.38 A test statistic for a two-sided significance test for a population mean is $z = 2.3$. Sketch a standard normal curve and mark this value of z on it. Find the P-value and shade the appropriate areas under the curve to illustrate your calculations.

6.39 A test statistic for a two-sided significance test for a population mean is $z = -1.4$. Sketch a standard normal curve and mark this value of z on it. Find the P-value and shade the appropriate areas under the curve to illustrate your calculations.

6.40 The P-value for a significance test is 0.082.

(a) Do you reject the null hypothesis at level $\alpha = 0.05$?

(b) Do you reject the null hypothesis at level $\alpha = 0.01$?

(c) Explain your answers.

6.41 The P-value for a significance test is 0.032.

(a) Do you reject the null hypothesis at level $\alpha = 0.05$?

(b) Do you reject the null hypothesis at level $\alpha = 0.01$?

(c) Explain your answers.

6.42   A test of the null hypothesis $H_0: \mu = \mu_0$ gives test statistic $z = 1.6$.

(a) What is the P-value if the alternative is $H_a: \mu > \mu_0$?

(b) What is the P-value if the alternative is $H_a: \mu < \mu_0$?

(c) What is the P-value if the alternative is $H_a: \mu \neq \mu_0$?

6.43   A test of the null hypothesis $H_0: \mu = \mu_0$ gives test statistic $z = -1.6$.

(a) What is the P-value if the alternative is $H_a: \mu > \mu_0$?

(b) What is the P-value if the alternative is $H_a: \mu < \mu_0$?

(c) What is the P-value if the alternative is $H_a: \mu \neq \mu_0$?

6.44   The P-value for a two-sided test of the null hypothesis $H_0: \mu = 30$ is 0.09.

(a) Does the 95% confidence interval include the value 30? Why?

(b) Does the 90% confidence interval include the value 30? Why?

6.45   The P-value for a two-sided test of the null hypothesis $H_0: \mu = 30$ is 0.04.

(a) Does the 95% confidence interval include the value 30? Why?

(b) Does the 90% confidence interval include the value 30? Why?

6.46   A 95% confidence interval for a population mean is (57, 65).

(a) Can you reject the null hypothesis that $\mu = 68$ at the 5% significance level? Why?

(b) Can you reject the null hypothesis that $\mu = 62$ at the 5% significance level? Why?

6.47   A 90% confidence interval for a population mean is (12, 15).

(a) Can you reject the null hypothesis that $\mu = 13$ at the 10% significance level? Why?

(b) Can you reject the null hypothesis that $\mu = 10$ at the 10% significance level? Why?

6.48   A study found that the use of bed nets was associated with a lower prevalence of malarial infections in the Gambia.[7] A report of the study states that the significance is $P < 0.001$. Explain what this means in a way that could be understood by someone who has not studied statistics.

6.49   A study of alcohol awareness among college students reported a higher awareness for students enrolled in a health and safety class than for those enrolled in a statistics class.[8] The difference is described as being statistically significant. Explain what this means in simple terms and offer an explanation for why the health and safety students had a higher mean score.

6.50   A report based on the National Assessment of Educational Progress (NAEP)[9] states that the average score on their mathematics test for eighth-grade students in the District of Columbia was 243 in 2003. The report then says that this value is higher than the average in 2000, which was 235. A footnote states that comparisons (higher/lower/different) are determined by statistical tests with 0.05 as the level of significance. Explain what this means in language understandable to someone who knows no statistics. Do not use the word "significance" in your answer.

6.51 An NAEP report[10] similar to the one described in the previous exercise states that the average score on their mathematics test for eighth-grade students in Boston was 262. The report then says that this value was not significantly different from 287, the average score for eighth-grade students in U.S. public schools that are located in large central cities. A footnote states that comparisons (higher/lower/different) are determined by statistical tests with 0.05 as the level of significance. Explain what this means in language understandable to someone who knows no statistics. Do not use the word "significance" in your answer.

6.52 In Example 6.1 we looked at the distribution of longleaf pine trees in the Wade Tract. One way to formulate hypotheses about whether or not the trees are randomly distributed in the tract is to examine the average location in the north–south direction. The values range from 0 to 200, so if the trees are uniformly distributed in this direction, any difference from the middle value (100) should be due to chance variation. The sample mean for the 584 trees in the tract is 99.74. A theoretical calculation based on the assumption that the trees are uniformly distributed gives a standard deviation of 58. Carefully state the null and alternative hypotheses in terms of this variable. Note that this requires that you translate the research question about the random distribution of the trees into specific statements about the mean of a probability distribution. Test your hypotheses, report your results, and write a short summary of what you have found.

6.53 Answer the questions in the previous exercise for the east–west direction, where the sample mean is 113.8.

6.54 Statistics can help decide the authorship of literary works. Sonnets by a certain Elizabethan poet are known to contain an average of $\mu = 8.9$ new words (words not used in the poet's other works). The standard deviation of the number of new words is $\sigma = 2.5$. Now a manuscript with 6 new sonnets has come to light, and scholars are debating whether it is the poet's work. The new sonnets contain an average of $\bar{x} = 10.2$ words not used in the poet's known works. We expect poems by another author to contain more new words, so to see if we have evidence that the new sonnets are not by our poet we test

$$H_0: \mu = 8.9$$

$$H_a: \mu > 8.9$$

Give the z test statistic and its P-value. What do you conclude about the authorship of the new poems?

6.55 The Survey of Study Habits and Attitudes (SSHA) is a psychological test that measures the motivation, attitude toward school, and study habits of students. Scores range from 0 to 200. The mean score for U.S. college students is about 115, and the standard deviation is about 30. A teacher who suspects that older students have better attitudes toward school gives the SSHA to 25 students who are at least 30 years of age. Their mean score is $\bar{x} = 132.2$.

(a) Assuming that $\sigma = 30$ for the population of older students, carry out a test of

$$H_0: \mu = 115$$

$$H_a: \mu > 115$$

Report the P-value of your test, and state your conclusion clearly.

(b) Your test in (a) required two important assumptions in addition to the assumption that the value of σ is known. What are they? Which of these assumptions is most important to the validity of your conclusion in (a)?

6.56 The level of calcium in the blood in healthy young adults varies with mean about 9.5 milligrams per deciliter and standard deviation about $\sigma = 0.4$. A clinic in rural Guatemala measures the blood calcium level of 160 healthy pregnant women at their first visit for prenatal care. The mean is $\bar{x} = 9.57$. Is this an indication that the mean calcium level in the population from which these women come differs from 9.5?

(a) State H_0 and H_a.

(b) Carry out the test and give the P-value, assuming that $\sigma = 0.4$ in this population. Report your conclusion.

(c) Give a 95% confidence interval for the mean calcium level μ in this population. We are confident that μ lies quite close to 9.5. This illustrates the fact that a test based on a large sample ($n = 160$ here) will often declare even a small deviation from H_0 to be statistically significant.

6.57 Here are the Degree of Reading Power (DRP) scores for a sample of 44 third-grade students:

40	26	39	14	42	18	25	43	46	27	19
47	19	26	35	34	15	44	40	38	31	46
52	25	35	35	33	29	34	41	49	28	52
47	35	48	22	33	41	51	27	14	54	45

These students can be considered to be an SRS of the third-graders in a suburban school district. DRP scores are approximately normal. Suppose that the standard deviation of scores in this school district is known to be $\sigma = 11$. The researcher believes that the mean score μ of all third-graders in this district is higher than the national mean, which is 32.

(a) State the appropriate H_0 and H_a to test this suspicion.

(b) Carry out the test. Give the P-value, and then interpret the result in plain language.

6.58 The *Statistical Significance* applet illustrates statistical tests with a fixed level of significance for normally distributed data with known standard deviation. Open the applet and keep the default settings for the null ($\mu = 0$) and the alternative ($\mu > 0$) hypotheses, the sample size ($n = 10$), the standard deviation ($\sigma = 1$), and the significance level ($\alpha = 0.05$). In the "I have data, and the observed $\bar{x}$ is $\bar{x} =$" box enter the value 1. Is the difference between $\bar{x}$ and μ_0 significant at the 5% level? Repeat for $\bar{x}$ equal to 0.1, 0.2, 0.3, 0.4, 0.5, 0.6, 0.7, 0.8, 0.9. Make a table giving $\bar{x}$ and the results of the significance tests. What do you conclude?

6.59 Repeat the previous exercise with significance level $\alpha = 0.01$. How does the choice of α affect the values of $\bar{x}$ that are far enough away from μ_0 to be statistically significant?

6.60 The *P-Value of a Test of Significance* applet illustrates *P*-values of significance tests for normally distributed data with known standard deviation. Open the applet and keep the default settings for the null ($\mu = 0$) and the alternative ($\mu > 0$) hypotheses, the sample size ($n = 10$), the standard deviation ($\sigma = 1$), and the significance level ($\alpha = 0.05$). In the "I have data, and the observed $\bar{x}$ is $\bar{x} =$" box enter the value 1. What is the *P*-value? Repeat for $\bar{x}$ equal to 0.1, 0.2, 0.3, 0.4, 0.5, 0.6, 0.7, 0.8, 0.9. Make a table giving $\bar{x}$ and *P*-values. How does the *P*-value change as $\bar{x}$ moves farther away from μ_0?

6.61 To determine whether the mean nicotine content of a brand of cigarettes is greater than the advertised value of 1.4 milligrams, a health advocacy group tests

$$H_0: \mu = 1.4$$
$$H_a: \mu > 1.4$$

The calculated value of the test statistic is $z = 1.75$.

(a) Is the result significant at the 5% level?

(b) Is the result significant at the 1% level?

6.62 A computer has a random number generator designed to produce random numbers that are uniformly distributed on the interval from 0 to 1. If this is true, the numbers generated come from a population with $\mu = 0.5$ and $\sigma = 0.2887$. A command to generate 100 random numbers gives outcomes with mean $\bar{x} = 0.4365$. Assume that the population σ remains fixed. We want to test

$$H_0: \mu = 0.5$$
$$H_a: \mu \neq 0.5$$

(a) Calculate the value z of the z test statistic.

(b) Is the result significant at the 5% level ($\alpha = 0.05$)?

(c) Is the result significant at the 1% level ($\alpha = 0.01$)?

6.63 Explain in plain language why a significance test that is significant at the 1% level must always be significant at the 5% level.

6.64 You are told that a significance test is significant at the 5% level. From this information can you determine whether or not it is significant at the 1% level? Explain your answer.

6.65 Consider a significance test for a null hypothesis versus a two-sided alternative with a z test statistic. Give a value of z that will give a result significant at the 1% level but not at the 0.5% level.

6.66 You have performed a two-sided test of significance and obtained a value of $z = -4.3$. Use Table D to find the approximate P-value for this test.

6.67 You have performed a one-sided test of significance and obtained a value of $z = 0.22$. Use Table D to find the approximate P-value for this test.

6.68 You will perform a significance test of

$$H_0: \mu = 0$$

versus

$$H_a: \mu > 0$$

(a) What values of z would lead you to reject H_0 at the 5% level?

(b) If the alternative hypothesis was

$$H_a: \mu \neq 0$$

what values of z would lead you to reject H_0 at the 5% level?

(c) Explain why your answers to parts (a) and (b) are different.

6.69 Consider a significance test for a null hypothesis versus a two-sided alternative. Between what values from Table D does the P-value for an outcome $z = 1.34$ lie? Calculate the P-value using Table A, and verify that it lies between the values you found from Table D.

6.70 Refer to the previous exercise. Find the P-value for $z = -1.34$.

6.71 Radon is a colorless, odorless gas that is naturally released by rocks and soils and may concentrate in tightly closed houses. Because radon is slightly radioactive, there is some concern that it may be a health hazard. Radon detectors are sold to homeowners worried about this risk, but the detectors may be inaccurate. University researchers placed 12 detectors in a chamber where they were exposed to 105 picocuries per liter (pCi/l) of radon over 3 days. Here are the readings given by the detectors:[11]

91.9	97.8	111.4	122.3	105.4	95.0
103.8	99.6	96.6	119.3	104.8	101.7

Assume (unrealistically) that you know that the standard deviation of readings for all detectors of this type is $\sigma = 9$.

(a) Give a 95% confidence interval for the mean reading μ for this type of detector.

(b) Is there significant evidence at the 5% level that the mean reading differs from the true value 105? State hypotheses and base a test on your confidence interval from (a).

6.3   Use and Abuse of Tests

Carrying out a test of significance is often quite simple, especially if a fixed significance level α is used or if the P-value is given effortlessly by a computer. Using tests wisely is not so simple. Each test is valid only in certain circumstances, with properly produced data being particularly important. The z test, for example, should bear the same warning label that was attached in Section 6.1 to the corresponding confidence interval. Similar warnings accompany the other tests that we will learn. There are additional caveats that concern tests more than confidence intervals, enough to warrant this separate section. Some hesitation about the unthinking use of significance tests is a sign of statistical maturity.

The reasoning of significance tests has appealed to researchers in many fields, so that tests are widely used to report research results. In this setting H_a is a "research hypothesis" asserting that some effect or difference is present. The null hypothesis H_0 says that there is no effect or no difference. A low P-value represents good evidence that the research hypothesis is true. Here are some comments on the use of significance tests, with emphasis on their use in reporting scientific research.

Choosing a level of significance

The spirit of a test of significance is to give a clear statement of the degree of evidence provided by the sample against the null hypothesis. The P-value does this. It is common practice to report P-values and to describe results as statistically significant whenever $P \leq 0.05$. *However, there is no sharp border between "significant" and " not significant," only increasingly strong evidence as the P-value decreases.* Having both the P-value and the statement that we reject or fail to reject H_0 allows us to draw better conclusions from our data.

EXAMPLE 6.20   Suppose the test statistic for a two-sided significance test for a population mean is $z = 1.95$. From Table A we can calculate the P-value. It is

$$P = 2[1 - P(Z \leq 1.95)] = 2(1 - 0.9744) = 0.0512$$

We have failed to meet the standard of $\alpha = 0.05$. However, with the information provided by the P-value, we can see that the result just barely missed the standard. If the effect in question is interesting and potentially important, we might want to design another study with a larger sample to investigate it further.

Here is another example where the P-value provides useful information beyond that provided by the statement that we reject or fail to reject the null hypothesis.

EXAMPLE 6.21   We have a test statistic of $z = -4.66$ for a two-sided significance test on a population mean. Software tells us that the P-value is 0.000003. This means that there are 3 chances in 1,000,000 of observing a sample mean this far or farther away from the null hypothesized value of μ. This kind of event is virtually impossible if the null hypothesis is true. There is no ambiguity in the result; we can clearly reject the null hypothesis.

We frequently report small P-values such as that in the previous exercise as $P < 0.001$. This corresponds to a chance of 1 in 1000 and is sufficiently small to lead us to a clear rejection of the null hypothesis.

One reason for the common use of $\alpha = 0.05$ is the great influence of Sir R. A. Fisher, the inventor of formal statistical methods for analyzing experimental data. Here is his opinion on choosing a level of significance: "A scientific fact should be regarded as experimentally established only if a properly designed experiment *rarely fails* to give this level of significance."[12]

What statistical significance does not mean

When a null hypothesis ("no effect" or "no difference") can be rejected at the usual level $\alpha = 0.05$, there is good evidence that an effect is present. But that effect can be extremely small. *When large samples are available, even tiny deviations from the null hypothesis will be significant.*

EXAMPLE 6.22   Suppose that we are testing the hypothesis of no correlation between two variables. With 400 observations, an observed correlation of only $r = 0.1$ is significant evidence at the $\alpha = 0.05$ level that the correlation in the population is not zero. The low significance level does not mean there is a strong association, only that there is strong evidence of some association. The proportion of the variability in one of the variables explained by the other is $r^2 = 0.01$, or 1%.

For practical purposes, we might well decide to ignore this association. *Statistical significance is not the same as practical significance.*

The remedy for attaching too much importance to statistical significance is to pay attention to the actual experimental results as well as to the P-value. Plot your data and examine them carefully. Beware of outliers. *The foolish user of statistics who feeds the data to a computer without exploratory analysis will often be embarrassed.* It is usually wise to give a confidence interval for the parameter in which you are interested. Confidence intervals are not used as often as they should be, while tests of significance are perhaps overused.

Don't ignore lack of significance

There is a tendency to infer that there is no effect whenever a P-value fails to attain the usual 5% standard. A provocative editorial in the *British Medical Journal* entitled "Absence of Evidence Is Not Evidence of Absence" deals with this issue.[13] Here is one of the examples they cite.

EXAMPLE 6.23   A randomized trial of interventions for reducing transmission of HIV-1 reported an incident rate ratio of 1.00, meaning that the intervention group and the control group both had the same rate of HIV-1 infection. The 95% confidence interval was reported as 0.63 to 1.58.[14] The editorial notes that a summary of these results that says the intervention has no effect on HIV-1 infection is misleading. The confidence interval indicates that the intervention may be capable of achieving a 37% decrease in infection; it might also be harmful and produce a 58% increase in infection. Clearly, more data are needed to distinguish between these possibilities.

The situation can be worse. Research in some fields has rarely been published unless significance at the 0.05 level is attained.

EXAMPLE 6.24 A survey of four journals published by the American Psychological Association showed that of 294 articles using statistical tests, only 8 reported results that did not attain the 5% significance level.[15]

In some areas of research, small effects that are detectable only with large sample sizes can be of great practical significance. Data accumulated from a large number of patients taking a pharmaceutical product may be needed before we can conclude that there are life-threatening consequences for a small number of people. Here is a different example.

EXAMPLE 6.25 A sample of size 10 gave a correlation of $r = 0.5$ between two variables. The P-value is 0.102 for a two-sided significance test. In most situations, a correlation this large would be interesting and worthy of additional study.

Another important aspect of planning a study is to verify that the test you plan to use does have high probability of detecting an effect of the size you hope to find. This probability is the *power* of the test. Power calculations are discussed later in this chapter.

Statistical inference is not valid for all sets of data

We learned long ago that badly designed surveys or experiments often produce invalid results. *Formal statistical inference cannot correct basic flaws in the design.*

EXAMPLE 6.26 There is no doubt that there is a significant difference in English vocabulary scores between high school seniors who have studied a foreign language and those who have not. But because the effect of actually studying a language is confounded with the differences between students who choose language study and those who do not, this statistical significance is hard to interpret. The most plausible explanation is that students who were already good at English chose to study another language. A randomized comparative experiment would isolate the actual effect of language study and so make significance meaningful. However, such an experiment probably could not be done.

Tests of significance and confidence intervals are based on the laws of probability. Randomization in sampling or experimentation ensures that these laws apply. *But we must often analyze data that do not arise from randomized samples or experiments. To apply statistical inference to such data, we must have confidence in a probability model for the data.* The diameters of successive holes bored in auto engine blocks during production, for example, may behave like independent observations from a normal distribution. We can check this probability model by examining the data. If the model appears correct, we can apply the methods of this chapter to do inference about the process mean diameter μ.

Beware of searching for significance

Statistical significance is a commodity much sought after by researchers. It means (or ought to mean) that you have found an effect that you were looking for. *The reasoning behind statistical significance works well if you decide what effect you are seeking, design an experiment or sample to search for it, and use a test of significance to weigh the evidence you get.* But because a successful search for a new scientific phenomenon often ends with statistical significance, it is all too tempting to make significance itself the object of the search. There are several ways to do this, none of them acceptable in polite scientific society.

In genomics experiments on microarrays it is common to perform tens of thousands of significance tests. If each of these was examined separately and statistical significance declared for all that had *P*-values that pass the 0.05 standard, we would have quite a mess. In the absence of any real biological effects, we would expect that, by chance alone, approximately 5% of these tests will show statistical significance. Much research in genomics is directed toward appropriate ways to deal with this situation.[16]

We do not mean that searching data for suggestive patterns is not proper scientific work. It certainly is. Many important discoveries have been made by accident rather than by design. Exploratory analysis of data is an essential part of statistics. We do mean that the usual reasoning of statistical inference does not apply when the search for a pattern is successful. *You cannot legitimately test a hypothesis on the same data that first suggested that hypothesis.* The remedy is clear. Once you have a hypothesis, design a study to search specifically for the effect you now think is there. If the result of this study is statistically significant, you have real evidence at last.

SECTION 6.3 | Summary

P-values are more informative than the reject-or-not result of a fixed level α test. Beware of placing too much weight on traditional values of α, such as $\alpha = 0.05$.

Very small effects can be highly significant (small P), especially when a test is based on a large sample. A statistically significant effect need not be practically important. Plot the data to display the effect you are seeking, and use confidence intervals to estimate the actual values of parameters.

On the other hand, lack of significance does not imply that H_0 is true, especially when the test has low power.

Significance tests are not always valid. Faulty data collection, outliers in the data, and testing a hypothesis on the same data that suggested the hypothesis can invalidate a test. Many tests run at once will probably produce some significant results by chance alone, even if all the null hypotheses are true.

SECTION 6.3 | Exercises

6.72 You are the statistical expert on a team that is planning a study. After you have made a careful presentation of the mechanics of significance testing, one of the team members suggests using $\alpha = 0.50$ for the study because you would be more likely to obtain statistically significant results with this choice. Explain in simple terms why this would not be a good use of statistical methods.

6.73 A research report described two results that both achieved statistical significance at the 5% level. The P-value for the first is 0.049; for the second it is 0.00002. Do the P-values add any useful information beyond that conveyed by the statement that both results are statistically significant? Write a short paragraph explaining your views on this question.

6.74 You perform 1000 significance tests using $\alpha = 0.05$. Assuming that all null hypotheses are true, about how many of the test results would you expect to be statistically significant? Explain how you obtained your answer.

6.75 Assume that you are performing a large number of significance tests. Let n be the number of these tests. How large would n need to be for you to expect about one P-value to be 0.00001 or smaller? Use this information to write an explanation of how to interpret a result that has $P = 0.00001$ in this setting.

6.76 A study with 5000 subjects reported a result that was statistically significant at the 5% level. Explain why this result might not be particularly large or important.

6.77 A study with 12 subjects reported a result that failed to achieve statistical significance at the 5% level. The P-value was 0.052. Write a short summary of how you would interpret these findings.

6.78 A P-value of 0.95 is reported for a significance test for a population mean. Interpret this result.

6.79 Give an example of a situation where searching for significance would lead to misleading conclusions.

6.80   Find two journal articles that report results with statistical analyses. For each article, summarize how the results are reported and write a critique of the presentation. Be sure to include details regarding use of significance testing at a particular level of significance, *P*-values, and confidence intervals.

6.81   What distinguishes managerial trainees who eventually become executives from those who, after expensive training, don't succeed and leave the company? We have abundant data on past trainees—data on their personalities and goals, their college preparation and performance, even their family backgrounds and their hobbies. Statistical software makes it easy to perform dozens of significance tests on these dozens of variables to see which ones best predict later success. We find that future executives are significantly more likely than washouts to have an urban or suburban upbringing and an undergraduate degree in a technical field.

Explain clearly why using these "significant" variables to select future trainees is not wise. Then suggest a follow-up study using this year's trainees as subjects that should clarify the importance of the variables identified by the first study.

6.82   Every user of statistics should understand the distinction between statistical significance and practical importance. A sufficiently large sample will declare very small effects statistically significant. Let us suppose that SAT mathematics (SATM) scores in the absence of coaching vary normally with mean $\mu = 480$ and $\sigma = 100$. Suppose further that coaching may change μ but does not change σ. An increase in the SATM score from 480 to 483 is of no importance in seeking admission to college, but this unimportant change can be statistically very significant. To see this, calculate the *P*-value for the test of

$$H_0: \mu = 480$$

$$H_a: \mu > 480$$

in each of the following situations:

(a) A coaching service coaches 100 students; their SATM scores average $\bar{x} = 483$.

(b) By the next year, the service has coached 1000 students; their SATM scores average $\bar{x} = 483$.

(c) An advertising campaign brings the number of students coached to 10,000; their average score is still $\bar{x} = 483$.

6.83   Give a 99% confidence interval for the mean SATM score μ after coaching in each part of the previous exercise. For large samples, the confidence interval says, "Yes, the mean score is higher after coaching, but only by a small amount."

6.84   As in the previous exercises, suppose that SATM scores vary normally with $\sigma = 100$. One hundred students go through a rigorous training program designed to raise their SATM scores by improving their mathematics skills.

Carry out a test of

$$H_0: \mu = 480$$

$$H_a: \mu > 480$$

in each of the following situations:

(a) The students' average score is $\bar{x} = 496.4$. Is this result significant at the 5% level?

(b) The average score is $\bar{x} = 496.5$. Is this result significant at the 5% level?

The difference between the two outcomes in (a) and (b) is of no importance. Beware attempts to treat $\alpha = 0.05$ as sacred.

6.85 Give an example of a set of data for which statistical inference is not valid.

6.86 **Bonferroni procedure** One way to deal with the problem of misleading P-values when performing more than one significance test is to adjust the criterion you use for statistical significance. The **Bonferroni procedure** does this in a simple way. If you perform 2 tests and want to use the $\alpha = 5\%$ significance level, you would require a P-value of $0.05/2 = 0.025$ to declare either one of the null hypotheses significant. In general, if you perform k tests and want protection at level α, use α/k as your cutoff for statistical significance. You perform 7 tests and calculate P-values of 0.241, 0.003, 0.050, 0.004, 0.009, 0.010, 0.001. Which of these are statistically significant using the Bonferroni procedure with $\alpha = 0.05$?

6.87 Refer to the previous problem. A researcher has performed 10 tests of significance and wants to apply the Bonferroni procedure with $\alpha = 0.05$. The calculated P-values are 0.045, 0.888, 0.050, 0.004, 0.001, 0.041, 0.888, 0.010, 0.002, 0.223. Which of the null hypotheses are rejected with this procedure?

6.88 Table 1.5 (page 32) gives average property damage per year due to tornadoes for each of the states. Is it appropriate to use the statistical methods we discussed in this chapter for these data? Explain why or why not.

6.89 Give an example of an interesting set of data for which statistical inference is valid. Explain your answer.

6.4 Power and Inference as a Decision*

Although we prefer to use P-values rather than the reject-or-not view of the fixed α significance test, the latter view is very important for planning studies and for understanding statistical decision theory. We will discuss these two topics in this section.

Power

Fixed level α significance tests are closely related to confidence intervals—in fact, we saw that a two-sided test can be carried out directly from a confidence

---

*Although the topics in this section are important in planning and interpreting significance tests, they can be omitted without loss of continuity.

interval. The significance level, like the confidence level, says how reliable the method is in repeated use. If we use 5% significance tests repeatedly when H_0 is in fact true, we will be wrong (the test will reject H_0) 5% of the time and right (the test will fail to reject H_0) 95% of the time.

The ability of a test to detect that H_0 is false is measured by the probability that the test will reject H_0 when an alternative is true. The higher this probability is, the more sensitive the test is.

POWER

The probability that a fixed level α significance test will reject H_0 when a particular alternative value of the parameter is true is called the **power** of the test to detect that alternative.

EXAMPLE 6.28   Can a 6-month exercise program increase the total body bone mineral content (TBBMC) of young women? A team of researchers is planning a study to examine this question. Based on the results of a previous study, they are willing to assume that $\sigma = 2$ for the percent change in TBBMC over the 6-month period. A change in TBBMC of 1% would be considered important, and the researchers would like to have a reasonable chance of detecting a change this large or larger. Is 25 subjects a large enough sample for this project?

We will answer this question by calculating the power of the significance test that will be used to evaluate the data to be collected. The calculation consists of three steps:

1. State H_0, H_a, the particular alternative we want to detect, and the significance level α.

2. Find the values of $\bar{x}$ that will lead us to reject H_0.

3. Calculate the probability of observing these values of $\bar{x}$ when the alternative is true.

Step 1   The null hypothesis is that the exercise program has no effect on TBBMC. In other words, the mean percent change is zero. The alternative is that exercise is beneficial; that is, the mean change is positive. Formally, we have

$$H_0 : \mu = 0$$

$$H_a : \mu > 0$$

The alternative of interest is $\mu = 1\%$ increase in TBBMC. A 5% test of significance will be used.

Step 2   The z test rejects H_0 at the $\alpha = 0.05$ level whenever

$$z = \frac{\bar{x} - \mu_0}{\sigma/\sqrt{n}} = \frac{\bar{x} - 0}{2/\sqrt{25}} \geq 1.645$$

Be sure you understand why we use 1.645. Rewrite this in terms of $\bar{x}$:

$$\bar{x} \geq 1.645\frac{2}{\sqrt{25}}$$

$$\bar{x} \geq 0.658$$

Because the significance level is $\alpha = 0.05$, this event has probability 0.05 of occurring *when the population mean μ is 0.*

Step 3 The power against the alternative $\mu = 1\%$ is the probability that H_0 will be rejected *when in fact $\mu = 1\%$.* We calculate this probability by standardizing $\bar{x}$, using the value $\mu = 1$, the population standard deviation $\sigma = 2$, and the sample size $n = 25$. The power is

$$P(\bar{x} \geq 0.658 \quad \text{when} \quad \mu = 1) = P\left(\frac{\bar{x} - \mu}{\sigma/\sqrt{n}} \geq \frac{0.658 - 1}{2/\sqrt{25}}\right)$$

$$= P(Z \geq -0.855) = 0.80$$

Figure 6.15 illustrates the power with the sampling distribution of $\bar{x}$ when $\mu = 1$. This significance test rejects the null hypothesis that exercise has no

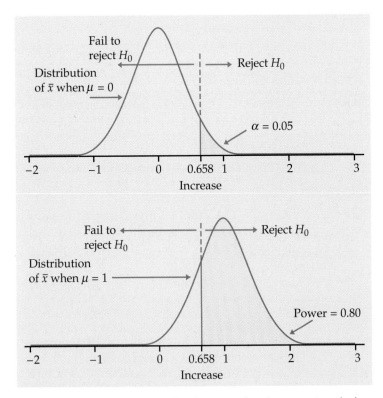

FIGURE 6.15 The sampling distributions of $\bar{x}$ when $\mu = 0$ and when $\mu = 1$ with the α and the power. Power is the probability that the test rejects H_0 when the alternative is true.

effect on TBBMC 80% of the time if the true effect of exercise is a 1% increase in TBBMC. If the true effect of exercise is a greater percent increase, the test will have greater power; it will reject with a higher probability.

High power is desirable. Along with 95% confidence intervals and 5% significance tests, 80% power is becoming a standard. Many U.S. government agencies that provide research funds require that the sample size for the funded studies be sufficient to detect important results 80% of the time using a 5% test of significance.

Increasing the power

Suppose you have performed a power calculation and found that the power is too small. What can you do to increase it? Here are four ways:

- Increase α. A 5% test of significance will have a greater chance of rejecting the alternative than a 1% test because the strength of evidence required for rejection is less.

- Consider a particular alternative that is farther away from μ_0. Values of μ that are in H_a but lie close to the hypothesized value μ_0 are harder to detect (lower power) than values of μ that are far from μ_0.

- Increase the sample size. More data will provide more information about $\bar{x}$ so we have a better chance of distinguishing values of μ.

- Decrease σ. This has the same effect as increasing the sample size: more information about μ. Improving the measurement process and restricting attention to a subpopulation are two common ways to decrease σ.

Power calculations are important in planning studies. Using a significance test with low power makes it unlikely that you will find a significant effect even if the truth is far from the null hypothesis. A null hypothesis that is in fact false can become widely believed if repeated attempts to find evidence against it fail because of low power. The following example illustrates this point.

EXAMPLE 6.29   The "efficient market hypothesis" for the time series of stock prices says that future stock prices (when adjusted for inflation) show only random variation. No information available now will help us predict stock prices in the future, because the efficient working of the market has already incorporated all available information in the present price. Many studies have tested the claim that one or another kind of information is helpful. In these studies, the efficient market hypothesis is H_0, and the claim that prediction is possible is H_a. Almost all the studies have failed to find good evidence against H_0. As a result, the efficient market hypothesis is quite popular. But an examination of the significance tests employed finds that the power is generally low. Failure to reject H_0 when using tests of low power is not evidence that H_0 is true. As one expert says, "The widespread impression that there is strong evidence for market efficiency may be due just to a lack of appreciation of the low power of many statistical tests."[17]

Here is another example of a power calculation, this time for a two-sided z test.

EXAMPLE 6.30 Example 6.16 (page 413) presented a test of

$$H_0: \mu = 0.86$$

$$H_a: \mu \neq 0.86$$

at the 1% level of significance. What is the power of this test against the specific alternative $\mu = 0.845$?

The test rejects H_0 when $|z| \geq 2.576$. The test statistic is

$$z = \frac{\bar{x} - 0.86}{0.0068/\sqrt{3}}$$

Some arithmetic shows that the test rejects when either of the following is true:

$$z \geq 2.576 \quad \text{(in other words, } \bar{x} \geq 0.870)$$

$$z \leq -2.576 \quad \text{(in other words, } \bar{x} \leq 0.850)$$

These are disjoint events, so the power is the sum of their probabilities, *computed assuming that the alternative $\mu = 0.845$ is true*. We find that

$$P(\bar{x} \geq 0.87) = P\left(\frac{\bar{x} - \mu}{\sigma/\sqrt{n}} \geq \frac{0.87 - 0.845}{0.0068/\sqrt{3}}\right)$$

$$= P(Z \geq 6.37) \doteq 0$$

$$P(\bar{x} \leq 0.85) = P\left(\frac{\bar{x} - \mu}{\sigma/\sqrt{n}} \leq \frac{0.85 - 0.845}{0.0068/\sqrt{3}}\right)$$

$$= P(Z \leq 1.27) = 0.8980$$

Figure 6.16 illustrates this calculation. Because the power is about 0.9, we are quite confident that the test will reject H_0 when this alternative is true.

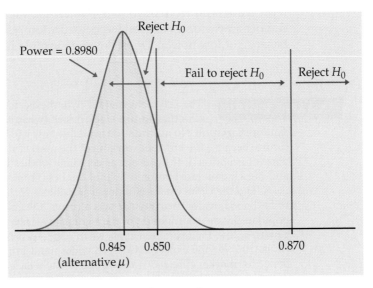

FIGURE 6.16 The power for Example 6.30.

Inference as decision*

We have presented tests of significance as methods for assessing the strength of evidence against the null hypothesis. This assessment is made by the *P*-value, which is a probability computed under the assumption that H_0 is true. The alternative hypothesis (the statement we seek evidence for) enters the test only to help us see what outcomes count against the null hypothesis.

There is another way to think about these issues. Sometimes we are really concerned about making a decision or choosing an action based on our evaluation of the data. **Acceptance sampling** is one such circumstance. A producer of bearings and the consumer of the bearings agree that each carload lot shall meet certain quality standards. When a carload arrives, the consumer chooses a sample of bearings to be inspected. On the basis of the sample outcome, the consumer will either accept or reject the carload. Let's examine how the idea of inference as a decision changes the reasoning used in tests of significance.

acceptance sampling

Two types of error

Tests of significance concentrate on H_0, the null hypothesis. If a decision is called for, however, there is no reason to single out H_0. There are simply two hypotheses, and we must accept one and reject the other. It is convenient to call the two hypotheses H_0 and H_a, but H_0 no longer has the special status (the statement we try to find evidence against) that it had in tests of significance. In the acceptance sampling problem, we must decide between

H_0: the lot of bearings meets standards

H_a: the lot does not meet standards

on the basis of a sample of bearings.

We hope that our decision will be correct, but sometimes it will be wrong. There are two types of incorrect decisions. We can accept a bad lot of bearings, or we can reject a good lot. Accepting a bad lot injures the consumer, while rejecting a good lot hurts the producer. To help distinguish these two types of error, we give them specific names.

> ### TYPE I AND TYPE II ERRORS
>
> If we reject H_0 (accept H_a) when in fact H_0 is true, this is a **Type I error.** If we accept H_0 (reject H_a) when in fact H_a is true, this is a **Type II error.**

The possibilities are summed up in Figure 6.17. If H_0 is true, our decision either is correct (if we accept H_0) or is a Type I error. If H_a is true, our decision either is correct or is a Type II error. Only one error is possible at one time. Figure 6.18 applies these ideas to the acceptance sampling example.

---

*The purpose of this discussion is to clarify the reasoning of significance tests by contrast with a related type of reasoning. It can be omitted without loss of continuity.

Truth about
the population

		H_0 true	H_a true
Decision based on sample	Reject H_0	Type I error	Correct decision
	Accept H_0	Correct decision	Type II error

FIGURE 6.17   The two types of error in testing hypotheses.

Truth about the lot

		Does meet standards	Does not meet standards
Decision based on sample	Reject the lot	Type I error	Correct decision
	Accept the lot	Correct decision	Type II error

FIGURE 6.18   The two types of error in the acceptance sampling setting.

Error probabilities

Any rule for making decisions is assessed in terms of the probabilities of the two types of error. This is in keeping with the idea that statistical inference is based on probability. We cannot (short of inspecting the whole lot) guarantee that good lots of bearings will never be rejected and bad lots never be accepted. But by random sampling and the laws of probability, we can say what the probabilities of both kinds of error are.

Significance tests with fixed level α give a rule for making decisions because the test either rejects H_0 or fails to reject it. If we adopt the decision-making way of thought, failing to reject H_0 means deciding that H_0 is true. We can then describe the performance of a test by the probabilities of Type I and Type II errors.

EXAMPLE 6.31   The mean diameter of a type of bearing is supposed to be 2.000 centimeters (cm). The bearing diameters vary normally with standard deviation $\sigma = 0.010$ cm. When a lot of the bearings arrives, the consumer takes an SRS of 5 bearings from the lot and measures their diameters. The consumer rejects the bearings if the sample mean diameter is significantly different from 2 at the 5% significance level.

This is a test of the hypotheses

$$H_0: \mu = 2$$

$$H_a: \mu \neq 2$$

To carry out the test, the consumer computes the z statistic:

$$z = \frac{\bar{x} - 2}{0.01/\sqrt{5}}$$

and rejects H_0 if

$$z < -1.96 \quad \text{or} \quad z > 1.96$$

A Type I error is to reject H_0 when in fact $\mu = 2$.

What about Type II errors? Because there are many values of μ in H_a, we will concentrate on one value. The producer and the consumer agree that a lot of bearings with mean 0.015 cm away from the desired mean 2.000 should be rejected. So a particular Type II error is to accept H_0 when in fact $\mu = 2.015$.

Figure 6.19 shows how the two probabilities of error are obtained from the two sampling distributions of $\bar{x}$, for $\mu = 2$ and for $\mu = 2.015$. When $\mu = 2$, H_0 is true and to reject H_0 is a Type I error. When $\mu = 2.015$, accepting H_0 is a Type II error. We will now calculate these error probabilities.

The probability of a Type I error is the probability of rejecting H_0 when it is really true. In Example 6.31, this is the probability that $|z| \geq 1.96$ when $\mu = 2$. But this is exactly the significance level of the test. The critical value 1.96 was chosen to make this probability 0.05, so we do not have to compute it again. The definition of "significant at level 0.05" is that sample outcomes this extreme will occur with probability 0.05 when H_0 is true.

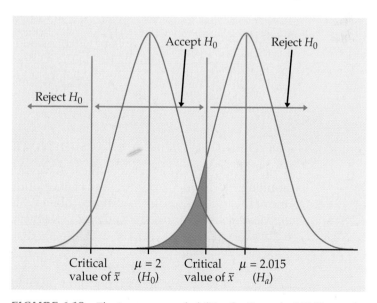

FIGURE 6.19 The two error probabilities for Example 6.31. The probability of a Type I error (tan area) is the probability of rejecting H_0: $\mu = 2$ when in fact $\mu = 2$. The probability of a Type II error (red area) is the probability of accepting H_0 when in fact $\mu = 2.015$.

SIGNIFICANCE AND TYPE I ERROR

The significance level α of any fixed level test is the probability of a Type I error. That is, α is the probability that the test will reject the null hypothesis H_0 when H_0 is in fact true.

The probability of a Type II error for the particular alternative $\mu = 2.015$ in Example 6.31 is the probability that the test will fail to reject H_0 when μ has this alternative value. The *power* of the test against the alternative $\mu = 2.015$ is just the probability that the test *does* reject H_0. By following the method of Example 6.30, we can calculate that the power is about 0.92. The probability of a Type II error is therefore $1 - 0.92$, or 0.08.

POWER AND TYPE II ERROR

The power of a fixed level test to detect a particular alternative is 1 minus the probability of a Type II error for that alternative.

The two types of error and their probabilities give another interpretation of the significance level and power of a test. The distinction between tests of significance and tests as rules for deciding between two hypotheses does not lie in the calculations but in the reasoning that motivates the calculations. In a test of significance we focus on a single hypothesis (H_0) and a single probability (the *P*-value). The goal is to measure the strength of the sample evidence against H_0. Calculations of power are done to check the sensitivity of the test. If we cannot reject H_0, we conclude only that there is not sufficient evidence against H_0, not that H_0 is actually true. If the same inference problem is thought of as a decision problem, we focus on two hypotheses and give a rule for deciding between them based on the sample evidence. We therefore must focus equally on two probabilities, the probabilities of the two types of error. We must choose one or the other hypothesis and cannot abstain on grounds of insufficient evidence.

The common practice of testing hypotheses

Such a clear distinction between the two ways of thinking is helpful for understanding. In practice, the two approaches often merge. We continued to call one of the hypotheses in a decision problem H_0. The common practice of *testing hypotheses* mixes the reasoning of significance tests and decision rules as follows:

1. State H_0 and H_a just as in a test of significance.

2. Think of the problem as a decision problem, so that the probabilities of Type I and Type II errors are relevant.

3. Because of Step 1, Type I errors are more serious. So choose an α (significance level) and consider only tests with probability of Type I error no greater than α.

4. Among these tests, select one that makes the probability of a Type II error as small as possible (that is, power as large as possible). If this probability is too large, you will have to take a larger sample to reduce the chance of an error.

Testing hypotheses may seem to be a hybrid approach. It was, historically, the effective beginning of decision-oriented ideas in statistics. An impressive mathematical theory of hypothesis testing was developed between 1928 and 1938 by Jerzy Neyman and Egon Pearson. The decision-making approach came later (1940s). Because decision theory in its pure form leaves you with two error probabilities and no simple rule on how to balance them, it has been used less often than either tests of significance or tests of hypotheses. Decision ideas have been applied in testing problems mainly by way of the Neyman-Pearson hypothesis-testing theory. That theory asks you first to choose α, and the influence of Fisher often has led users of hypothesis testing comfortably back to $\alpha = 0.05$ or $\alpha = 0.01$. Fisher, who was exceedingly argumentative, violently attacked the Neyman-Pearson decision-oriented ideas, and the argument still continues.

SECTION 6.4 | Summary

The **power** of a significance test measures its ability to detect an alternative hypothesis. The power to detect a specific alternative is calculated as the probability that the test will reject H_0 when that alternative is true. This calculation requires knowledge of the sampling distribution of the test statistic under the alternative hypothesis. Increasing the size of the sample increases the power when the significance level remains fixed.

An alternative to significance testing regards H_0 and H_a as two statements of equal status that we must decide between. This **decision theory** point of view regards statistical inference in general as giving rules for making decisions in the presence of uncertainty.

In the case of testing H_0 versus H_a, decision analysis chooses a decision rule on the basis of the probabilities of two types of error. A **Type I error** occurs if H_0 is rejected when it is in fact true. A **Type II error** occurs if H_0 is accepted when in fact H_a is true.

In a fixed level α significance test, the significance level α is the probability of a Type I error, and the power against a specific alternative is 1 minus the probability of a Type II error for that alternative.

SECTION 6.4 | Exercises

6.90   You are a reviewer for a research proposal that includes a section on sample size justification. A careful reading of this section indicates that the power is 20% for detecting an effect that most people would consider important. Write a short explanation of what this means and make a recommendation on whether or not this study should be run.

6.91   Two studies are identical in all respects except for the sample sizes. Consider the power versus a particular sample size. Will the study with the larger sample size have more power or less power than the one with the smaller sample size? Explain your answer in terms that could be understood by someone with very little knowledge of statistics.

6.92   The power for a two-sided test of the null hypothesis $\mu = 0$ versus the alternative $\mu = 10$ is 0.82. What is the power versus the alternative $\mu = -10$? Explain your answer.

6.93   A one-sided test of the null hypothesis $\mu = 50$ versus the alternative $\mu = 70$ has power equal to 0.5. Will the power for the alternative $\mu = 80$ be higher or lower than 0.5? Explain your answer.

6.94   In Exercise 6.52 (page 420) you performed a two-sided hypothesis test of the null hypothesis that the average north–south location of the longleaf pine trees sampled in the Wade Tract was $\mu = 100$. There were 584 trees in the sample and the standard deviation was assumed to be 58. The sample mean in that analysis was $\bar{x} = 99.74$. Use the *Power* applet to compute the power for the alternative $\mu = 99$ using a two-sided test at the 5% level of significance.

6.95   Refer to the previous exercise. Note that in the east–west direction, the average location was 113.8. Use the applet to find the power for the alternative $\mu = 110$.

6.96   You want to see if a redesign of the cover of a mail-order catalog will increase sales. A very large number of customers will receive the original catalog, and a random sample of customers will receive the one with the new cover. For planning purposes, you are willing to assume that the sales from the new catalog will be approximately normal with $\sigma = 60$ dollars and that the mean for the original catalog will be $\mu = 40$ dollars. You decide to use a sample size of $n = 1000$. You wish to test

$$H_0: \mu = 40$$

$$H_a: \mu > 40$$

You decide to reject H_0 if $\bar{x} > 43.12$ and to accept H_0 otherwise.

(a) Find the probability of a Type I error, that is, the probability that your test rejects H_0 when in fact $\mu = 40$ dollars.

(b) Find the probability of a Type II error when $\mu = 45$ dollars. This is the probability that your test accepts H_0 when in fact $\mu = 45$.

(c) Find the probability of a Type II error when $\mu = 50$.

(d) The distribution of sales is not normal, because many customers buy nothing. Why is it nonetheless reasonable in this circumstance to assume that the mean will be approximately normal?

6.97   Example 6.15 (page 412) gives a test of a hypothesis about the SAT scores of California high school students based on an SRS of 500 students. The hypotheses are

$$H_0: \mu = 450$$

$$H_a: \mu > 450$$

Assume that the population standard deviation is $\sigma = 100$. The test rejects H_0 at the 1% level of significance when $z \geq 2.326$, where

$$z = \frac{\bar{x} - 450}{100/\sqrt{500}}$$

Is this test sufficiently sensitive to usually detect an increase of 12 points in the population mean SAT score? Answer this question by calculating the power of the test against the alternative $\mu = 462$.

6.98 You must decide which of two discrete distributions a random variable X has. We will call the distributions p_0 and p_1. Here are the probabilities they assign to the values x of X:

x	0	1	2	3	4	5	6
p_0	0.1	0.1	0.1	0.1	0.2	0.1	0.3
p_1	0.3	0.2	0.1	0.1	0.1	0.1	0.1

You have a single observation on X and wish to test

$$H_0: p_0 \text{ is correct}$$

$$H_a: p_1 \text{ is correct}$$

One possible decision procedure is to reject H_0 only if $X = 0$ or $X = 1$.

(a) Find the probability of a Type I error, that is, the probability that you reject H_0 when p_0 is the correct distribution.

(b) Find the probability of a Type II error.

6.99 You are designing a computerized medical diagnostic program. The program will scan the results of routine medical tests (pulse rate, blood pressure, urinalysis, etc.) and either clear the patient or refer the case to a doctor. The program will be used as part of a preventive-medicine system to screen many thousands of persons who do not have specific medical complaints. The program makes a decision about each patient.

(a) What are the two hypotheses and the two types of error that the program can make? Describe the two types of error in terms of "false-positive" and "false-negative" test results.

(b) The program can be adjusted to decrease one error probability, at the cost of an increase in the other error probability. Which error probability would you choose to make smaller, and why? (This is a matter of judgment. There is no single correct answer.)

CHAPTER 6 | Exercises

6.100 A study of late adolescents and early adults reported average months of full-time employment for individuals aged 18 to 26.[18] The means are given in the

Age	18	19	20	21	22	23	24	25	26
Months employed	2.9	4.2	5.0	5.3	6.4	7.4	8.5	8.9	9.3

table. Assume that the standard deviation for each of these means is 4.5 months and that each sample size is 750.

(a) Calculate the 95% confidence interval for each mean.

(b) Plot the means versus age. Draw a vertical line through the first mean extending up to the upper confidence limit and down to the lower limit. At the ends of the line, draw a short dash. Do the same for each of the other means.

(c) Write a summary of what the data show. Note that in circumstances such as this, it is common practice not to make any adjustments for the fact that several confidence intervals are being reported. Be sure to include comments about this in your summary.

6.101 The Safety Climate Index (SCI) measures workers' perceptions about the safety of their work environment. A study of safe work practices of industrial workers reported mean SCI scores for workers classified by workplace size.[19] Here are the means:

Workplace size	Fewer than 50 workers	50 to 200 workers	More than 200 workers
Mean SCI	67.23	70.37	74.83

Assume that the standard deviation is 19 and the sample sizes are all 180. (We will discuss ways to compare three means such as these in Chapter 12.)

(a) Calculate the 95% confidence interval for each mean.

(b) Plot the means versus workplace size. Draw a vertical line through the first mean extending up to the upper confidence limit and down to the lower limit. At the ends of the line, draw a short dash. Do the same for each of the other means.

(c) One way to adjust for the fact that we are reporting three confidence intervals is a procedure that uses a larger value of z^* in the calculation of the margin of error. For this problem one recommendation would be to use $z^* = 2.40$. Repeat parts (a) and (b) making this adjustment.

(d) Summarize your results. Be sure to include comments on the effects of the adjustment on your results.

6.102 For this exercise you will use the *Confidence Interval* applet. Set the confidence level at 95% and click the "Sample" button 10 times to simulate 10 confidence intervals. Record the percent hit. Simulate another 10 intervals by clicking another 10 times (do not click the "Reset" button). Record the percent hit for your 20 intervals. Repeat the process of simulating 10 additional intervals and recording the results until you have a total of 200 intervals. Plot your results and write a summary of what you have found.

6.103 Refer to the previous exercise. Do the simulations and report the results for 90% confidence.

6.104 You are testing the null hypothesis that $\mu = 0$ versus the alternative $\mu > 0$ using $\alpha = 0.05$. Assume $\sigma = 17$. Suppose $\bar{x} = 5$ and $n = 10$. Calculate the test statistic and its P-value. Repeat assuming the same value of $\bar{x}$ but with $n = 20$. Do the same for sample sizes of 30, 40, and 50. Plot the values of the test statistic versus the sample size. Do the same for the P-values. Summarize what this demonstration shows about the effect of the sample size on significance testing.

6.105 Patients with chronic kidney failure may be treated by dialysis, using a machine that removes toxic wastes from the blood, a function normally performed by the kidneys. Kidney failure and dialysis can cause other changes, such as retention of phosphorus, that must be corrected by changes in diet. A study of the nutrition of dialysis patients measured the level of phosphorus in the blood of several patients on six occasions. Here are the data for one patient (in milligrams of phosphorus per deciliter of blood):[20]

$$5.4 \;\; 5.2 \;\; 4.5 \;\; 4.9 \;\; 5.7 \;\; 6.3$$

The measurements are separated in time and can be considered an SRS of the patient's blood phosphorus level. Assume that this level varies normally with $\sigma = 0.9$ mg/dl.

(a) Give a 95% confidence interval for the mean blood phosphorus level.

(b) The normal range of phosphorus in the blood is considered to be 2.6 to 4.8 mg/dl. Is there strong evidence that this patient has a mean phosphorus level that exceeds 4.8?

6.106 An agronomist examines the cellulose content of a variety of alfalfa hay. Suppose that the cellulose content in the population has standard deviation $\sigma = 8$ milligrams per gram (mg/g). A sample of 16 cuttings has mean cellulose content $\bar{x} = 140$ mg/g.

(a) Give a 95% confidence interval for the mean cellulose content in the population.

(b) A previous study claimed that the mean cellulose content was $\mu = 135$ mg/g, but the agronomist believes that the mean is higher than that figure. State H_0 and H_a and carry out a significance test to see if the new data support this belief.

(c) The statistical procedures used in (a) and (b) are valid when several assumptions are met. What are these assumptions?

6.107 Because sulfur compounds cause "off-odors" in wine, oenologists (wine experts) have determined the odor threshold, the lowest concentration of a compound that the human nose can detect. For example, the odor threshold for dimethyl sulfide (DMS) is given in the oenology literature as 25 micrograms per liter of wine (μg/l). Untrained noses may be less sensitive, however. Here are the DMS odor thresholds for 10 beginning students of oenology:

$$32 \quad 33 \quad 40 \quad 35 \quad 24 \quad 36 \quad 31 \quad 30 \quad 20 \quad 25$$

Assume (this is not realistic) that the standard deviation of the odor threshold for untrained noses is known to be $\sigma = 7$ μg/l.

(a) Make a stemplot to verify that the distribution is roughly symmetric with no outliers. (A normal quantile plot confirms that there are no systematic departures from normality.)

(b) Give a 95% confidence interval for the mean DMS odor threshold among all beginning oenology students.

(c) Are you convinced that the mean odor threshold for beginning students is higher than the published threshold, 25 μg/l? Carry out a significance test to justify your answer.

6.108 Consumers can purchase nonprescription medications at food stores, mass merchandise stores such as Kmart and Wal-Mart, or pharmacies. About 45% of consumers make such purchases at pharmacies. What accounts for the popularity of pharmacies, which often charge higher prices?

A study examined consumers' perceptions of overall performance of the three types of stores, using a long questionnaire that asked about such things as "neat and attractive store," "knowledgeable staff," and "assistance in choosing among various types of nonprescription medication." A performance score was based on 27 such questions. The subjects were 201 people chosen at random from the Indianapolis telephone directory. Here are the means and standard deviations of the performance scores for the sample:[21]

Store type	$\bar{x}$	s
Food stores	18.67	24.95
Mass merchandisers	32.38	33.37
Pharmacies	48.60	35.62

We do not know the population standard deviations, but a sample standard deviation s from so large a sample is usually close to σ. Use s in place of the unknown σ in this exercise.

(a) What population do you think the authors of the study want to draw conclusions about? What population are you certain they can draw conclusions about?

(b) Give 95% confidence intervals for the mean performance for each type of store.

(c) Based on these confidence intervals, are you convinced that consumers think that pharmacies offer higher performance than the other types of stores? (In Chapter 12, we will study a statistical method for comparing means of several groups.)

6.109 A study of the pay of corporate chief executive officers (CEOs) examined the increase in cash compensation of the CEOs of 104 companies, adjusted for inflation, in a recent year. The mean increase in real compensation was $\bar{x} = 6.8\%$, and the standard deviation of the increases was $s = 53\%$. Is this good evidence that the mean real compensation μ of all CEOs increased that year? The hypotheses are

$$H_0: \mu = 0 \quad \text{(no increase)}$$

$$H_a: \mu > 0 \quad \text{(an increase)}$$

Because the sample size is large, the sample s is close to the population σ, so take $\sigma = 53\%$.

(a) Sketch the normal curve for the sampling distribution of $\bar{x}$ when H_0 is true. Shade the area that represents the P-value for the observed outcome $\bar{x} = 6.8\%$.

(b) Calculate the P-value.

(c) Is the result significant at the $\alpha = 0.05$ level? Do you think the study gives strong evidence that the mean compensation of all CEOs went up?

6.110   When asked to explain the meaning of "statistically significant at the $\alpha = 0.05$ level," a student says, "This means there is only probability 0.05 that the null hypothesis is true." Is this an essentially correct explanation of statistical significance? Explain your answer.

6.111   Another student, when asked why statistical significance appears so often in research reports, says, "Because saying that results are significant tells us that they cannot easily be explained by chance variation alone." Do you think that this statement is essentially correct? Explain your answer.

6.112   Use a computer to generate $n = 12$ observations from a normal distribution with mean 20 and standard deviation 5: $N(20, 5)$. Find the 95% confidence interval for μ. Repeat this process 100 times and then count the number of times that the confidence interval includes the value $\mu = 20$. Explain your results.

6.113   Use a computer to generate $n = 12$ observations from a normal distribution with mean 20 and standard deviation 5: $N(20, 5)$. Test the null hypothesis that $\mu = 20$ using a two-sided significance test. Repeat this process 100 times and then count the number of times that you reject H_0. Explain your results.

6.114   Use the same procedure for generating data as in the previous exercise. Now test the null hypothesis that $\mu = 18$. Explain your results.

6.115   Search the Internet or some journals that report research in your field and find two reports that provide an estimate with a margin of error or a confidence interval. For each report:

(a) Describe the method used to collect the data.

(b) Describe the variable being studied.

(c) Give the estimate and the confidence interval.

(d) Describe any practical difficulties that may have led to errors in addition to the sampling errors quantified by the margin of error.

6.116 Persons aged 55 and over represented 21.3% of the U.S. population in the year 2000. This group is expected to increase to 30.5% by 2025. In terms of actual numbers of people, the increase is from 58.6 million to 101.4 million. Restauranteurs have found this market to be important and would like to make their businesses attractive to older customers. One study used a questionnaire to collect data from people aged 50 and over.[22] For one part of the analysis, individuals were classified into two age groups: 50 to 64 and 65 to 79. There were 267 people in the first group and 263 in the second. One set of items concerned ambience, menu design, and service. A series of statements were rated on a 1 to 5 scale with 1 representing "strongly disagree" and 5 representing "strongly agree." In some cases the wording has been shortened in the table below. Here are the means:

Statement	50–64	65–79
Ambience:		
Most restaurants are too dark	2.75	2.93
Most restaurants are too noisy	3.33	3.43
Background music is often too loud	3.27	3.55
Restaurants are too smoky	3.17	3.12
Tables are too small	3.00	3.19
Tables are too close together	3.79	3.81
Menu design:		
Print size is not large enough	3.68	3.77
Glare makes menus difficult to read	2.81	3.01
Colors of menus make them difficult to read	2.53	2.72
Service:		
It is difficult to hear the service staff	2.65	3.00
I would rather be served than serve myself	4.23	4.14
I would rather pay the server than a cashier	3.88	3.48
Service is too slow	3.13	3.10

First examine the means of the people who are 50 to 64. Order the statements according to the means and describe the results. Then do the same for the older group. For each statement compute the z statistic and the associated P-value for the comparison between the two groups. For these calculations you can assume that the standard deviation of the difference is 0.08, so z is simply the difference in the means divided by 0.08. Note that you are performing 13 significance tests in this exercise. Keep this in mind when you interpret your results. Write a report summarizing your work.

EESEE CASE STUDIES

The Electronic Encyclopedia of Statistical Examples and Exercises (EESEE) is available on the text CD and Web site. These more elaborate stories, with data, provide settings for longer case studies.

6.117   Read the EESEE story "Precision Time." Write a report that responds to all of the questions posed in this case study.

6.118   Read the EESEE material "Anecdotes of Significance Testing." Write a report answering all the questions for these mini–case studies.

From *P*-Values to President

IOGNÁID Ó MUIRCHEARTAIGH *is President of the National University of Ireland, Galway, Ireland.*

My work as a professional statistician has exposed me to many different areas of life and learning. Although the career path may not have been obvious, I was well prepared for the wide variety of activities involved in being president of a dynamic research-intensive university.

My doctoral research concerned statistical discrimination, with particular application to medical diagnosis. Essentially we matched the case histories of new and undiagnosed patients with a data base of previously diagnosed cases and symptoms. In other work over the years, I have been involved in estimating the forms of economic demand functions, using data on ocean whitecaps from orbiting satellites to infer wind speed, and a legal case where it was alleged that money went missing in bank branches more often when a particular bank official was present (statistical analysis resulted in the withdrawal of the accusation).

One of the more fun applications of statistics with which I was involved was the analysis of scoring on our local golf course. For handicap purposes, it is necessary to rank the holes on a golf course in order of difficulty, and every golf club does so. We considered the proposition that the difficulty of a hole would depend on wind speed and direction, developed a statistical model for this dependence, and applied it to data collected from our local club. This led to a reordering of the rankings for the course and an article in the *Journal of the Royal Statistical Society*.

This wide variety of applications and of interests helped to prepare me for the even wider spectrum of activities involved in being president of a university. Every day I deal with myriad issues and people, and certainly my quantitative background is a significant bonus in addressing many complex issues—although there are times when I think being an accountant might be more of a help than being a statistician!

Inference for Distributions

Introduction

We began our study of data analysis in Chapter 1 by learning graphical and numerical tools for describing the distribution of a single variable and for comparing several distributions. Our study of the practice of statistical inference begins in the same way, with inference about a single distribution and comparison of two distributions. Comparing more than two distributions requires more elaborate methods, which are presented in Chapters 12 and 13.

Two important aspects of any distribution are its center and spread. If the distribution is normal, we describe its center by the mean μ and its spread by the standard deviation σ. In this chapter, we will meet confidence intervals and significance tests for inference about a population mean μ and for comparing the means of two populations. The previous chapter emphasized the reasoning of tests and confidence intervals; now we emphasize statistical practice, so we no longer assume that population standard deviations are known. The t procedures for inference about means are among the most common statistical methods. Inference about the spread of a population, as we will see, poses some difficult practical problems.

7.1 Inference for the Mean of a Population

Both confidence intervals and tests of significance for the mean μ of a normal population are based on the sample mean $\bar{x}$, which estimates the unknown μ. The sampling distribution of $\bar{x}$ depends on σ. This fact causes no difficulty when σ is known. When σ is unknown, however, we must estimate σ even though we are primarily interested in μ. The sample standard deviation s is used to estimate the population standard deviation σ.

The t distributions

Suppose that we have a simple random sample (SRS) of size n from a normally distributed population with mean μ and standard deviation σ. The sample mean $\bar{x}$ then has the normal distribution with mean μ and standard deviation $\sigma/\sqrt{n}$. When σ is not known, we estimate it with the sample standard deviation s, and then we estimate the standard deviation of $\bar{x}$ by $s/\sqrt{n}$. This quantity is called the *standard error* of the sample mean $\bar{x}$ and we denote it by $\mathrm{SE}_{\bar{x}}$.

> ### STANDARD ERROR
>
> When the standard deviation of a statistic is estimated from the data, the result is called the **standard error** of the statistic. The standard error of the sample mean is
>
> $$\mathrm{SE}_{\bar{x}} = \frac{s}{\sqrt{n}}$$

The term "standard error" is sometimes used for the actual standard deviation of a statistic, $\sigma/\sqrt{n}$ in the case of $\bar{x}$. The estimated value $s/\sqrt{n}$ is then called the "estimated standard error." In this book we will use the term "standard error" only when the standard deviation of a statistic is estimated from the data. The term has this meaning in the output of many statistical computer packages and in reports of research in many fields that apply statistical methods.

The standardized sample mean, or one-sample z statistic,

$$z = \frac{\bar{x} - \mu}{\sigma/\sqrt{n}}$$

is the basis of the z procedures for inference about μ when σ is known. This statistic has the standard normal distribution $N(0, 1)$. When we substitute the standard error $s/\sqrt{n}$ for the standard deviation $\sigma/\sqrt{n}$ of $\bar{x}$, the statistic does *not* have a normal distribution. It has a distribution that is new to us, called a t *distribution*.

THE t DISTRIBUTIONS

Suppose that an SRS of size n is drawn from an $N(\mu, \sigma)$ population. Then the **one-sample t statistic**

$$t = \frac{\bar{x} - \mu}{s/\sqrt{n}}$$

has the **t distribution** with $n - 1$ **degrees of freedom.**

---

degrees of freedom

There is a different t distribution for each sample size. A particular t distribution is specified by giving the **degrees of freedom.** The degrees of freedom for this t statistic come from the sample standard deviation s in the denominator of t. We saw in Chapter 1 (page 51) that s has $n - 1$ degrees of freedom. There are other t statistics with different degrees of freedom, some of which we will meet later in this chapter.

We use $t(k)$ to stand for the t distribution with k degrees of freedom. The t distributions were discovered in 1908 by William S. Gosset. Gosset was a statistician employed by the Guinness brewing company, which required that he not publish his discoveries under his own name. He therefore wrote under the pen name "Student." The t distribution is often called "Student's t" in his honor. The density curves of the $t(k)$ distributions are similar in shape to the standard normal curve. That is, they are symmetric about 0 and are bell-shaped. The spread of the t distributions is a bit greater than that of the standard normal distribution. This is due to the extra variability caused by substituting the random variable s for the fixed parameter σ. As the degrees of freedom k increase, the $t(k)$ density curve approaches the $N(0, 1)$ curve ever more closely. This reflects the fact that s approaches σ as the sample size increases. Figure 7.1 compares the density curves of the standard normal distribution and the t distribution with 5 degrees of freedom. The similarity in shape is apparent, as is the fact that the t distribution has more probability in the tails and less in the center than does the standard normal distribution.

Table D in the back of the book gives critical values for the t distributions. For convenience, we have labeled the table entries by both the critical value p needed for significance tests and by the confidence level C (in percent) required for confidence intervals. The standard normal critical values

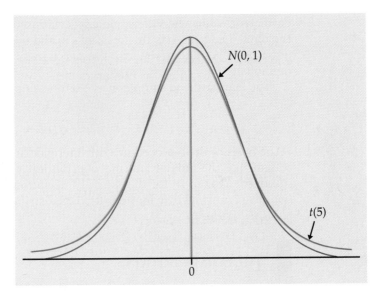

FIGURE 7.1   Density curves for the standard normal and $t(5)$ distributions. Both are symmetric with center 0. The t distributions have more probability in the tails than does the standard normal distribution.

in the bottom row of entries are labeled z^*. As in the case of the normal table (Table A), computer software often makes Table D unnecessary.

The one-sample t confidence interval

With the t distributions to help us, we can analyze samples from normal populations with unknown σ by replacing the standard deviation $\sigma/\sqrt{n}$ of $\bar{x}$ by its standard error $s/\sqrt{n}$ in the z procedures of Chapter 6. The z statistic then becomes the one-sample t statistic. We must now employ P-values or critical values from t in place of the corresponding normal values (z). Here are the details for the confidence interval.

THE ONE-SAMPLE t CONFIDENCE INTERVAL

Suppose that an SRS of size n is drawn from a population having unknown mean μ. A level C **confidence interval** for μ is

$$\bar{x} \pm t^* \frac{s}{\sqrt{n}}$$

where t^* is the value for the $t(n-1)$ density curve with area C between $-t^*$ and t^*. The quantity

$$t^* \frac{s}{\sqrt{n}}$$

is the **margin of error.** This interval is exact when the population distribution is normal and is approximately correct for large n in other cases.

The one-sample t confidence interval is similar in both reasoning and computational detail to the z confidence interval of Chapter 6. There, the margin of error for the population mean was $z^*\sigma/\sqrt{n}$. Here, we replace σ by its estimate s and z^* by t^*. So the margin of error for the population mean when we use the data to estimate σ is $t^*s/\sqrt{n}$.

EXAMPLE 7.1 In fiscal year 1996, the U.S. Agency for International Development provided 238,300 metric tons of corn soy blend (CSB) for development programs and emergency relief in countries throughout the world. CSB is a highly nutritious, low-cost fortified food that is partially precooked and can be incorporated into different food preparations by the recipients. As part of a study to evaluate appropriate vitamin C levels in this commodity, measurements were taken on samples of CSB produced in a factory.[1]

The following data are the amounts of vitamin C, measured in milligrams per 100 grams (mg/100 g) of blend (dry basis), for a random sample of size 8 from a production run:

$$26 \quad 31 \quad 23 \quad 22 \quad 11 \quad 22 \quad 14 \quad 31$$

We want to find a 95% confidence interval for μ, the mean vitamin C content of the CSB produced during this run.

The sample mean is $\bar{x} = 22.50$ and the standard deviation is $s = 7.19$ with degrees of freedom $n - 1 = 7$. The standard error is

$$SE_{\bar{x}} = s/\sqrt{n} = 7.19/\sqrt{8} = 2.54$$

From Table D we find $t^* = 2.365$. The 95% confidence interval is

$$\bar{x} \pm t^* \frac{s}{\sqrt{n}} = 22.50 \pm 2.365 \frac{7.19}{\sqrt{8}}$$

$$= 22.5 \pm (2.365)(2.54)$$

$$= 22.5 \pm 6.0$$

$$= (16.5, \ 28.5)$$

We are 95% confident that the mean vitamin C content of the CSB for this run is between 16.5 and 28.5 mg/100 g.

In this example we have given the actual interval $(16.5, \ 28.5)$ as our answer. Sometimes we prefer to report the mean and margin of error: the mean vitamin C content is 22.5 mg/100 g with a margin of error of 6.0 mg/100 g.

The use of the t confidence interval in Example 7.1 rests on assumptions that cannot easily be checked but are reasonable in this case. The samples from the production run were taken at regular time intervals, and experience has shown that this procedure gives data that can be treated as random samples from a normal population. The assumption that the population distribution is normal cannot be effectively checked with only 8 observations. On the other hand, when we make a stemplot, we can see clearly that there are no outliers:

```
3 | 1 1
2 | 2 2 3 6
1 | 1 4
```

The one-sample t test

Significance tests using the estimated standard error are very similar to the z test that we studied in the last chapter.

THE ONE-SAMPLE t TEST

Suppose that an SRS of size n is drawn from a population having unknown mean μ. To test the hypothesis $H_0: \mu = \mu_0$ based on an SRS of size n, compute the one-sample t statistic

$$t = \frac{\bar{x} - \mu_0}{s/\sqrt{n}}$$

In terms of a random variable T having the $t(n-1)$ distribution, the P-value for a test of H_0 against

$H_a: \mu > \mu_0$ is $P(T \geq t)$

$H_a: \mu < \mu_0$ is $P(T \leq t)$

$H_a: \mu \neq \mu_0$ is $2P(T \geq |t|)$

These P-values are exact if the population distribution is normal and are approximately correct for large n in other cases.

EXAMPLE 7.2 The specifications for the CSB described in Example 7.1 state that the mixture should contain 2 pounds of vitamin premix for every 2000 pounds of product. These specifications are designed to produce a mean (μ) vitamin C content in the final product of 40 mg/100 g. We can test a null hypothesis that the mean vitamin C content of the production run in Example 7.1 conforms to these specifications. Specifically, we test

$$H_0: \mu = 40$$
$$H_a: \mu \neq 40$$

Recall that $n = 8$, $\bar{x} = 22.50$, and $s = 7.19$. The t test statistic is

$$t = \frac{\bar{x} - \mu_0}{s/\sqrt{n}} = \frac{22.5 - 40}{7.2/\sqrt{8}}$$
$$= -6.88$$

This means that the sample mean $\bar{x} = 22.50$ is almost seven standard deviations

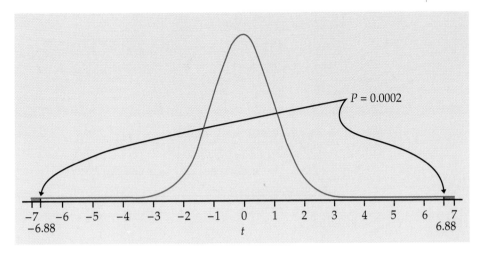

FIGURE 7.2   The *P*-value for Example 7.2.

df = 7		
p	0.001	0.0005
t*	4.785	5.408

away from the null hypothesized value $\mu = 40$. Because the degrees of freedom are $n - 1 = 7$, this t statistic has the $t(7)$ distribution. Figure 7.2 shows that the *P*-value is $2P(T \geq 6.88)$, where T has the $t(7)$ distribution. From Table D we see that $P(T \geq 5.408) = 0.0005$. Therefore, we conclude that the *P*-value is less than 2×0.0005. Software gives the exact value as $P = 0.0002$. Clearly, these data are incompatible with a process mean of $\mu = 40$ mg/100 g. We reject H_0 and conclude that the vitamin C content for this run is below the specifications.

In this example we tested the specifications $\mu = 40$ mg/100 g against the two-sided alternative $\mu \neq 40$ mg/100 g because we had no prior suspicion that the production run would produce CSB with too much or too little vitamin C. If we knew that the proper amount of vitamin C was added to the mixture but suspected that some of the vitamin might be lost or destroyed by the production process, we should use a one-sided test. *It is wrong, however, to examine the data first and then decide to do a one-sided test in the direction indicated by the data.* If in doubt, use a two-sided test. In the present circumstance, however, we could use our results from Example 7.2 to justify a one-sided test for *another* sample from the same or a similar process.

EXAMPLE 7.3   For the vitamin C problem described in the previous example, we want to test whether or not vitamin C is lost or destroyed in the production process. Here we test

$$H_0 : \mu = 40$$

versus

$$H_a : \mu < 40$$

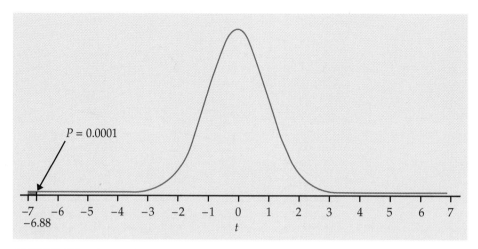

FIGURE 7.3   The *P*-value for Example 7.3.

The *t* test statistic does not change: $t = -6.88$. As Figure 7.3 illustrates, however, the *P*-value is now $P(T \le -6.88)$, half of the value in the previous exercise. From Table D we can determine that $P < 0.0005$; software gives the exact value as $P = 0.0001$. We conclude that the production process has lost or destroyed some of the vitamin C.

Note that the conclusion drawn in this example relies heavily on the assumption that the proper amount of vitamin C was added to the blend. If this assumption is not correct, it is still true that the CSB produced has a vitamin C content below the specification, but the explanation that this is caused by loss or destruction of the vitamin C in the production process is not necessarily true.

For small data sets, such as the one in Example 7.1, it is easy to perform the computations for confidence intervals and significance tests with an ordinary calculator. For larger data sets, however, we prefer to use software or a statistical calculator.

EXAMPLE 7.4   An investor with a stock portfolio worth several hundred thousand dollars sued his broker and brokerage firm because lack of diversification in his portfolio led to poor performance. Table 7.1 gives the rates of return for the 39 months that the account was managed by the broker.[2] Figure 7.4 gives a histogram for these data and Figure 7.5 gives the normal quantile plot. There are no outliers and

TABLE 7.1

Monthly rates of return on a portfolio (percent)

−8.36	1.63	−2.27	−2.93	−2.70	−2.93	−9.14	−2.64
6.82	−2.35	−3.58	6.13	7.00	−15.25	−8.66	−1.03
−9.16	−1.25	−1.22	−10.27	−5.11	−0.80	−1.44	1.28
−0.65	4.34	12.22	−7.21	−0.09	7.34	5.04	−7.24
−2.14	−1.01	−1.41	12.03	−2.56	4.33	2.35	

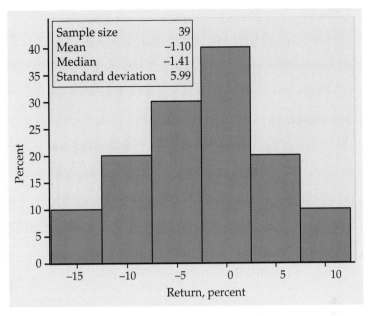

FIGURE 7.4   Histogram for Example 7.4.

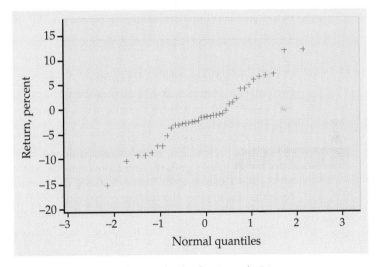

FIGURE 7.5   Normal quantile plot for Example 7.4.

the distribution shows no strong skewness. We are reasonably confident that the distribution of $\bar{x}$ is approximately normal, and we proceed with our inference based on normal theory.

The arbitration panel compared these returns with the average of the Standard & Poor's 500-Stock Index for the same period. Consider the 39 monthly returns as a random sample from the population of monthly returns the brokerage would generate if it managed the account forever. Are these returns compatible with a population mean of $\mu = 0.95\%$, the S&P 500 average? Our hypotheses are

$$H_0: \mu = 0.95$$

$$H_a: \mu \neq 0.95$$

Minitab

```
Test of mu = 0.950 vs mu not = 0.950

Variable    N      Mean      StDev     SE Mean          T          P
RETURN     39     -1.100     5.991       0.959       -2.14      0.039
```

SPSS

One-Sample Test						
	Test Value = 0.95					
					95% Confidence Interval of the Difference	
	t	df	Sig. (2-tailed)	Mean Difference	Lower	Upper
RETURN	−2.136	38	.039	−2.0467	−3.9868	−.1066

FIGURE 7.6 Minitab and SPSS output for Example 7.4.

Minitab and SPSS outputs appear in Figure 7.6. Output from other software will look similar.

Here is one way to report the conclusion: the mean monthly return on investment for this client's account was $\bar{x} = -1.1\%$. This differs significantly from the performance of the S&P 500-Stock Index for the same period ($t = -2.14$, df = 38, $P < 0.039$).

The hypothesis test in Example 7.4 leads us to conclude that the mean return on the client's account differs from that of the S&P 500-Stock Index. Now let's assess the return on the client's account with a confidence interval.

EXAMPLE 7.5 The mean monthly return on the client's portfolio was $\bar{x} = -1.1\%$ and the standard deviation was $s = 5.99\%$. Figure 7.7 gives the Minitab, SPSS, and Excel outputs for a 95% confidence interval for the population mean μ. Note that Excel gives the margin of error next to the label "Confidence Level(95.0%)" rather than the actual confidence interval. We see that the 95% confidence interval is $(-3.04, 0.84)$, or (from Excel) -1.0997 ± 1.9420.

Because the S&P 500 return, 0.95%, falls outside this interval, we know that μ differs significantly from 0.95% at the $\alpha = 0.05$ level. Example 7.4 gave the actual P-value as $P < 0.039$.

The confidence interval suggests that the broker's management of this account had a long-term mean somewhere between a loss of 3.04% and a gain of 0.84% per month. We are interested not in the actual mean but in the difference between the broker's process and the diversified S&P 500-Stock Index.

EXAMPLE 7.6 Following the analysis accepted by the arbitration panel, we are considering the S&P 500 monthly average return as a constant standard. (It is easy to envision scenarios where we would want to treat this type of quantity as random.) The difference between the mean of the investor's account and the S&P 500 is

Minitab

Variable	N	Mean	StDev	SE Mean	95.0% C.I.
RETURN	39	-1.100	5.991	0.959	(-3.042, 0.842)

SPSS

Descriptives

				Statistic	Std. Error
RETURN	MEAN			-1.0997	.9593
	95% Confidence Interval for Mean		Lower Bound	-3.0418	
			Upper Bound	.8423	

Excel

Mean	-1.09974359
Standard Error	0.95930991
Standard Deviation	5.990888471
Count	39
Confidence Level(95.0%)	1.942021452

FIGURE 7.7   Minitab, SPSS, and Excel output for Example 7.5.

$\bar{x} - \mu = -1.10 - 0.95 = -2.05\%$. In Example 7.5 we found that the 95% confidence interval for the investor's account was $(-3.04, \ 0.84)$. To obtain the corresponding interval for the difference, subtract 0.95 from each of the endpoints. The resulting interval is $(-3.04 - 0.95, \ 0.84 - 0.95)$, or $(-3.99, \ -0.11)$. We conclude with 95% confidence that the underperformance was between -3.99% and -0.11%. This estimate helps to set the compensation owed the investor.

Matched pairs t procedures

The CSB production problem of Example 7.1 concerns only a single population. We know that comparative studies are usually preferred to single-sample investigations because of the protection they offer against confounding. For that reason, inference about a parameter of a single distribution is less common than comparative inference. One common comparative design, however, **matched pairs** makes use of single-sample procedures. In a **matched pairs** study, subjects are matched in pairs and the outcomes are compared within each matched pair. The experimenter can toss a coin to assign two treatments to the two subjects in each pair. Matched pairs are also common when randomization is not possible. One situation calling for matched pairs is when observations are taken on the same subjects, under different conditions. See Chapter 3 (page 207) for additional discussion of matched pairs designs.

EXAMPLE 7.7   Many people believe that the moon influences the actions of some individuals. A study of dementia patients in nursing homes recorded various types of disruptive behaviors every day for 12 weeks. Days were classified as moon days

TABLE 7.2

Aggressive behaviors of dementia patients

Patient	Moon days	Other days	Difference	Patient	Moon days	Other days	Difference
1	3.33	0.27	3.06	9	6.00	1.59	4.41
2	3.67	0.59	3.08	10	4.33	0.60	3.73
3	2.67	0.32	2.35	11	3.33	0.65	2.68
4	3.33	0.19	3.14	12	0.67	0.69	−0.02
5	3.33	1.26	2.07	13	1.33	1.26	0.07
6	3.67	0.11	3.56	14	0.33	0.23	0.10
7	4.67	0.30	4.37	15	2.00	0.38	1.62
8	2.67	0.40	2.27				

if they were in a three-day period centered at the day of the full moon. For each patient the average number of disruptive behaviors was computed for moon days and for all other days. The data for the 15 subjects whose behaviors were classified as aggressive are presented in Table 7.2.[3] The patients in this study are not a random sample of dementia patients. However, we examine their data in the hope that what we find is not unique to this particular group of individuals and applies to other patients who have similar characteristics.

To analyze these data, we first subtract the disruptive behaviors for moon days from the disruptive behaviors for other days. These 15 differences form a single sample. They appear in the "Difference" columns in Table 7.2. The first patient, for example, averaged 3.33 aggressive behaviors on moon days but only 0.27 aggressive behaviors on other days. The difference $3.33 - 0.27 = 3.06$ is what we will use in our analysis. Examination of the data indicates that there are three patients with very small differences but there are no indications of extreme outliers or strong skewness. We proceed with our analysis using the normality-based methods of this section.

To assess whether there is a difference in aggressive behaviors on moon days versus other days, we test

$$H_0: \mu = 0$$

$$H_a: \mu \neq 0$$

Here μ is the mean difference in aggressive behaviors, moon versus other days, for patients of this type. The null hypothesis says that aggressive behaviors occur at the same frequency for both types of days, and H_a says that the behaviors on moon days are not the same as on other days.

The 15 differences have

$$\bar{x} = 2.433 \quad \text{and} \quad s = 1.460$$

The one-sample t statistic is therefore

$$t = \frac{\bar{x} - 0}{s/\sqrt{n}} = \frac{2.433}{1.460/\sqrt{15}}$$

$$= 6.45$$

df = 14		
p	0.001	0.0005
t^*	3.787	4.140

The P-value is found from the $t(14)$ distribution (remember that the degrees of freedom are 1 less than the sample size). Table D shows that 6.45 lies beyond the upper 0.0005 critical value of the $t(14)$ distribution. Since we are using a two-sided alternative, we know therefore that the P-value is less than two times this value, or 0.0010.

Software gives a value that is much smaller, $P = 0.000015$. In practice, there is little difference between these two P-values; the data provide clear evidence in favor of the alternative hypothesis. A difference this large is very unlikely to occur by chance if there is, in fact, no effect of the moon on aggressive behaviors. In scholarly publications, the details of routine statistical procedures are omitted; our test would be reported in the form: "There was more aggressive behavior on moon days than on other days ($t = 6.45$, df $= 14$, $P < 0.001$)."

Note that we could have justified a one-sided alternative in this example. Based on previous research, we expect more aggressive behaviors on moon days, and the alternative $H_a: \mu > 0$ is reasonable in this setting. The choice of the alternative here, however, has no effect on the conclusion: from Table D we determine that P is less than 0.0005; from software it is 0.000008. These are very small values and we would still report $P < 0.001$. *In most circumstances we cannot be absolutely certain about the outcome, and the safest strategy is to use the two-sided alternative.*

The results of the significance test allow us to conclude that dementia patients exhibit more aggressive behaviors in the days around a full moon. What are the implications of the study for the administrators who run the facilities where these patients live? For example, should they increase staff on these days? To make these kinds of decisions, an estimate of the magnitude of the problem, with a margin of error, would be helpful.

EXAMPLE 7.8   A 95% confidence interval for the mean difference in aggressive behaviors per day requires the critical value $t^* = 2.145$ from Table D. The margin of error is

$$t^* \frac{s}{\sqrt{n}} = 2.145 \frac{1.460}{\sqrt{15}}$$
$$= 0.81$$

and the confidence interval is

$$\bar{x} \pm t^* \frac{s}{\sqrt{n}} = 2.43 \pm 0.81$$
$$= (1.62, \ 3.24)$$

The estimated average difference is 2.43 aggressive behaviors per day, with margin of error 0.81 for 95% confidence. The increase needs to be interpreted in terms of the baseline values. The average number of aggressive behaviors per day on other days is 0.59; on moon days it is 1.50. This is approximately a 250% increase. If aggressive behaviors require a substantial amount of attention by staff, then the administrators should be aware of the increased level of these activities during the full-moon period. Additional staff may be needed.

The following are key points to remember concerning matched pairs:

1. A matched pairs analysis is needed when there are two measurements or observations on each individual and we want to examine the difference.

```
4 | 4 4
3 | 1 1 1 6 7
2 | 1 3 4 7
1 |
0 | 0 1 1
```

FIGURE 7.8 *Stemplot of differences in aggressive behaviors for Examples 7.7 and 7.8.*

2. For each individual, use the difference between the two measurements as the data for your analysis.

3. Use the one-sample confidence interval and significance-testing procedures that we learned in this section.

Use of the *t* procedures in Examples 7.7 and 7.8 faces several issues. First, no randomization is possible in a study like this. Our inference procedures assume that there is a process that generates these aggressive behaviors and that the process produces them at possibly different rates during the days near the full moon. Second, many of the patients in these nursing homes did not exhibit any disruptive behaviors. These were not included in our analysis. So our inference is restricted to patients who do exhibit disruptive behaviors.

A final difficulty is that the data show departures from normality. In a matched pairs analysis, we assume that the differences are normally distributed because the *t* procedures are applied to the differences. Figure 7.8 gives a stemplot of the differences. There are 3 patients with very small differences in aggressive behaviors while the other 12 have a large increase. We have a dilemma here. The data may not be normal, but our sample size is very small and we do not have very much data to support our view that moon phase influences behavior. We can try an alternative procedure that does not require the normality assumption—but there is a price to pay. The alternative procedures have less power to detect differences. For Example 7.7, the *P*-value is so small that we are very confident that we have found an effect of the moon phase on behavior.

Robustness of the *t* procedures

The results of one-sample *t* procedures are exactly correct only when the population is normal. Real populations are never exactly normal. The usefulness of the *t* procedures in practice therefore depends on how strongly they are affected by nonnormality. Procedures that are not strongly affected are called *robust*.

ROBUST PROCEDURES

A statistical inference procedure is called **robust** if the probability calculations required are insensitive to violations of the assumptions made.

The assumption that the population is normal rules out outliers, so the presence of outliers shows that this assumption is not valid. The t procedures are not robust against outliers, because $\bar{x}$ and s are not resistant to outliers.

There are three patients with fairly low values of the difference in Example 7.7. Whether or not these are outliers is a matter of judgment. If we rerun the analysis without these three patients, the t statistic would increase to 11.89 and the P-value would be much lower. Careful inspection of the records may reveal some characteristic of these patients which distinguishes them from the others in the study. Without such information, it is difficult to justify excluding them from the analysis. *In general, we should be very cautious about discarding suspected outliers, particularly when they make up a substantial proportion of the data, as they do in this example.*

Fortunately, the t procedures are quite robust against nonnormality of the population except in the case of outliers or strong skewness. Larger samples improve the accuracy of P-values and critical values from the t distributions when the population is not normal. This is true for two reasons. First, the sampling distribution of the sample mean $\bar{x}$ from a large sample is close to normal (that's the central limit theorem). We need be less concerned about the normality of the individual observations when the sample is large. Second, as the size n of a sample grows, the sample standard deviation s approaches the population standard deviation σ. This fact is closely related to the law of large numbers. In large samples, s will be an accurate estimate of σ whether or not the population has a normal distribution.

A normal quantile plot, stemplot, or boxplot to check for skewness and outliers is an important preliminary to the use of t procedures for small samples. For most purposes, the one-sample t procedures can be safely used when $n \geq 15$ unless an outlier or clearly marked skewness is present. *Except in the case of small samples, the assumption that the data are an SRS from the population of interest is more crucial than the assumption that the population distribution is normal.* Here are practical guidelines for inference on a single mean:[4]

- *Sample size less than 15:* Use t procedures if the data are close to normal. If the data are clearly nonnormal or if outliers are present, do not use t.

- *Sample size at least 15:* The t procedures can be used except in the presence of outliers or strong skewness.

- *Large samples:* The t procedures can be used even for clearly skewed distributions when the sample is large, roughly $n \geq 40$.

Consider, for example, some of the data we studied in Chapter 1. The breaking-strength data in Figure 1.32 (page 81) contain three outliers in a sample of size 23, which makes the use of t procedures risky. The guinea pig survival times in Figure 1.33 (page 82) are strongly skewed to the right with no outliers. Since there are 72 observations, we could use the t procedures here. On the other hand, many would prefer to use a transformation to make these data more nearly normal. (See the material on inference for nonnormal populations on page 465 and in Chapter 14.) Figure 1.34 (page 83) gives the normal quantile plot for 105 acidity measurements of rainwater. These data appear to be normal and we would apply the t procedures in this case.

The power of the t test*

The power of a statistical test measures its ability to detect deviations from the null hypothesis. In practice, we carry out the test in the hope of showing that the null hypothesis is false, so high power is important. The power of the one-sample t test against a specific alternative value of the population mean μ is the probability that the test will reject the null hypothesis when the alternative value of the mean is true. To calculate the power, we assume a fixed level of significance, often $\alpha = 0.05$.

Calculation of the exact power of the t test takes into account the estimation of σ by s and is a bit complex. But an approximate calculation that acts as if σ were known is almost always adequate for planning a study. This calculation is very much like that for the z test, presented in Section 6.4: write the event that the test rejects H_0 in terms of $\bar{x}$, and then find the probability of this event when the population mean has the alternative value.

Consider Example 7.7 where we examined the effect of the moon on the aggressive behavior of dementia patients in nursing homes. Suppose that we wanted to perform a similar study in a different setting. How many patients should we include in our new study? To answer this question, we do a power calculation. In Example 7.7 we found $\bar{x} = 2.433$ and $s = 1.460$. Let's use $s = 1.5$ for our calculations. *It is always better to use a value of the standard deviation that is a little larger than what we expect than one that is smaller.* This may give a sample size that is a little larger than we need. We want to avoid a situation where we fail to find the effect that we are looking for because we did not have enough data. Recall from Example 7.8 that the 95% confidence interval for the difference is (1.62, 3.24). Let's use $\mu = 1.0$ as the alternative of interest. We are very confident that the effect was larger than this in our previous study, and this amount of an increase in aggressive behavior would still be important to those who work in these facilities. Finally, based on the previous study, we can justify using a one-sided alternative; we expect the moon days to be associated with an increase in aggressive behavior.

EXAMPLE 7.9   Let's compute the power of the t test for

$$H_0\text{:}\ \mu = 0$$

$$H_a\text{:}\ \mu > 0$$

against the alternative $\mu = 1.0$. We will use a 5% level of significance. The t test with n observations rejects H_0 at the 5% significance level if the t statistic

$$t = \frac{\bar{x} - 0}{s/\sqrt{n}}$$

exceeds the upper 5% point of $t(19)$, which is 1.729. Taking $n = 20$ and $s = 1.5$, the event that the test rejects H_0 is therefore

$$t = \frac{\bar{x}}{1.5/\sqrt{20}} \geq 1.729$$

$$\bar{x} \geq 1.729\frac{1.5}{\sqrt{20}}$$

$$\bar{x} \geq 0.580$$

---

*This section can be omitted without loss of continuity.

The power is the probability that $\bar{x} \geq 0.580$ when $\mu = 1.0$. Taking $\sigma = 1.5$, this probability is found by standardizing $\bar{x}$:

$$P(\bar{x} \geq 0.580 \text{ when } \mu = 1.0) = P\left(\frac{\bar{x} - 1.0}{1.5/\sqrt{20}} \geq \frac{0.580 - 1.0}{1.5/\sqrt{20}}\right)$$
$$= P(Z \geq -1.25)$$
$$= 1 - 0.1056 = 0.89$$

The power is 89% that we will detect an increase of 1.0 aggressive behaviors per day during moon days. This is sufficient power for most situations. For many studies, 80% is considered the standard value for desirable power. We could repeat the calculations for some smaller values of n to determine the smallest value that would meet the 80% criterion.

Power calculations are used in planning studies to assure us that we have a reasonable chance of detecting effects of interest. They give us some guidance in selecting a sample size. In making these calculations, we need assumptions about the standard deviation and the alternative of interest. In our example we assumed that the standard deviation would be 1.5, but in practice we are hoping that the value will be somewhere around this value. Similarly, we have used a somewhat arbitrary alternative of 1.0. This is a guess based on the results of the previous study. *Beware of putting too much trust in fine details of the results of these calculations.* They serve as a guide, not a mandate.

Inference for nonnormal populations*

We have not discussed how to do inference about the mean of a clearly non-normal distribution based on a small sample. If you face this problem, you should consult an expert. Three general strategies are available:

- In some cases a distribution other than a normal distribution will describe the data well. There are many nonnormal models for data, and inference procedures for these models are available.

- Because skewness is the chief barrier to the use of t procedures on data without outliers, you can attempt to transform skewed data so that the distribution is symmetric and as close to normal as possible. Confidence levels and P-values from the t procedures applied to the transformed data will be quite accurate for even moderate sample sizes.

distribution-free procedures

- The third strategy is to use a **distribution-free** inference procedure. Such procedures do not assume that the population distribution has any specific form, such as normal. Distribution-free procedures are often called **nonparametric procedures.** Chapter 15 discusses several of these procedures.

nonparametric procedures

Each of these strategies can be effective, but each quickly carries us beyond the basic practice of statistics. We emphasize procedures based on normal distributions because they are the most common in practice, because their robustness makes them widely useful, and (most important) because

---

*This section can be omitted without loss of continuity.

we are first of all concerned with understanding the principles of inference. We will therefore not discuss procedures for nonnormal continuous distributions and will present here only one of the many distribution-free procedures that do not require the population to have any specific type of distribution. We will be content with illustrating by example the use of a transformation and of a simple distribution-free procedure.

Transforming data   When the distribution of a variable is skewed, it often happens that a simple transformation results in a variable whose distribution is symmetric and even close to normal. The most common transformation is the **logarithm,** or **log.** The logarithm tends to pull in the right tail of a distribution. For example, the data 2, 3, 4, 20 show an outlier in the right tail. Their logarithms 0.30, 0.48, 0.60, 1.30 are much less skewed. Taking logarithms is a possible remedy for right skewness. Instead of analyzing values of the original variable X, we first compute their logarithms and analyze the values of $\log X$. Here is an example of this approach.

(margin note: log transformation)

EXAMPLE 7.10   Table 7.3 presents data on the amounts (in grams per mile) of three pollutants in the exhaust of 46 vehicles of the same type, measured under standard conditions prescribed by the Environmental Protection Agency. We will

TABLE 7.3

Amounts of three pollutants emitted by light–duty engines (grams per mile)

EN	HC	CO	NOX	EN	HC	CO	NOX
1	0.50	5.01	1.28	2	0.65	14.67	0.72
3	0.46	8.60	1.17	4	0.41	4.42	1.31
5	0.41	4.95	1.16	6	0.39	7.24	1.45
7	0.44	7.51	1.08	8	0.55	12.30	1.22
9	0.72	14.59	0.60	10	0.64	7.98	1.32
11	0.83	11.53	1.32	12	0.38	4.10	1.47
13	0.38	5.21	1.24	14	0.50	12.10	1.44
15	0.60	9.62	0.71	16	0.73	14.97	0.51
17	0.83	15.13	0.49	18	0.57	5.04	1.49
19	0.34	3.95	1.38	20	0.41	3.38	1.33
21	0.37	4.12	1.20	22	1.02	23.53	0.86
23	0.87	19.00	0.78	24	1.10	22.92	0.57
25	0.65	11.20	0.95	26	0.43	3.81	1.79
27	0.48	3.45	2.20	28	0.41	1.85	2.27
29	0.51	4.10	1.78	30	0.41	2.26	1.87
31	0.47	4.74	1.83	32	0.52	4.29	2.94
33	0.56	5.36	1.26	34	0.70	14.83	1.16
35	0.51	5.69	1.73	36	0.52	6.35	1.45
37	0.57	6.02	1.31	38	0.51	5.79	1.51
39	0.36	2.03	1.80	40	0.48	4.62	1.47
41	0.52	6.78	1.15	42	0.61	8.43	1.06
43	0.58	6.02	0.97	44	0.46	3.99	2.01
45	0.47	5.22	1.12	46	0.55	7.47	1.39

FIGURE 7.9   Normal quantile plot for CO emissions, for Example 7.10. The distribution is skewed to the right.

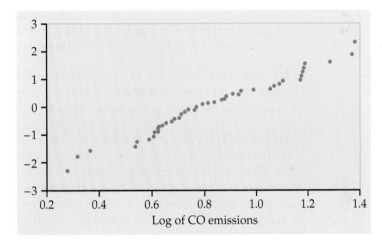

FIGURE 7.10   Normal quantile plot for the logarithms of the CO emissions, for Example 7.10. This distribution is close to normal.

concentrate on emissions of carbon monoxide (CO). We would like to give a confidence interval for the mean emissions μ for this vehicle type.[5]

A normal quantile plot of the CO data from Table 7.3 (Figure 7.9) shows that the distribution is skewed to the right. Because there are no extreme outliers, the sample mean of 46 observations will nonetheless have an approximately normal sampling distribution. The t procedures could be used for approximate inference. For more exact inference, we will seek to transform the data so that the distribution is more nearly normal. Figure 7.10 is a normal quantile plot of the logarithms of the CO measurements. The transformed data are very close to normal, so t procedures will give quite exact results.

The application of the t procedures to the transformed data is straightforward. Call the original CO values from Table 7.3 values of the variable X. The transformed data are values of $X_{new} = \log X$. In most software packages, it is an easy task to transform data in this way and then analyze the new variable.

EXAMPLE 7.11   Analysis of the logs of the CO values in Minitab produces the following output:

```
   N     MEAN    STDEV    SE MEAN    95.0 PERCENT C.I.
   46   0.8198   0.2654    0.0391    ( 0.7409, 0.8986)
```

For comparison, the 95% t confidence interval for the original mean μ is found from the original data as follows:

```
   N     MEAN    STDEV    SE MEAN    95.0 PERCENT C.I.
   46   7.960    5.261     0.776     (  6.398,  9.523)
```

The advantage of analyzing transformed data is that use of procedures based on the normal distributions is better justified and the results are more exact. The disadvantage is that a confidence interval for the mean μ in the original scale (in our example, emissions measured in grams per mile) cannot be recovered from the confidence interval for the mean of the logs.

The sign test   Perhaps the most straightforward way to cope with non-normal data is to use a *distribution-free*, or *nonparametric*, procedure. As the name indicates, these procedures do not require the population distribution to have any specific form, such as normal. Distribution-free significance tests are quite simple and are available in most statistical software systems. Distribution-free tests have two drawbacks. First, they are generally less powerful than tests designed for use with a specific distribution, such as the t test. Second, we must often modify the statement of the hypotheses in order to use a distribution-free test. A distribution-free test concerning the center of a distribution, for example, is usually stated in terms of the median rather than the mean. This is sensible when the distribution may be skewed. But the distribution-free test does not ask the same question (Has the mean changed?) that the t test does. The simplest distribution-free test, and one of the most

sign test   useful, is the **sign test.**

Let's examine again the aggressive-behavior data of Example 7.7 (page 459). In that example we concluded that there was more aggressive behavior on moon days than on other days. The stemplot given in Figure 7.8 was not very assuring concerning the assumption that the data are normal. There were 3 patients with low values that seemed to be somewhat different from the observations on the other 12 patients. How does the sign test deal with these data?

EXAMPLE 7.12   The sign test is based on the following simple observation: of the 15 patients in our sample, 14 had more aggressive behaviors on moon days than on other days. This sounds like convincing evidence in favor of a moon effect on behavior, but we need to do some calculations to confirm this.

Let p be the probability that a randomly chosen dementia patient will have more aggressive behaviors on moon days than on other days. The null hypothesis of "no moon effect" says that the moon days are no different from other days, so a patient is equally likely to have more aggressive behaviors on moon days as on other days. We therefore want to test

$$H_0 : p = 1/2$$

$$H_a : p > 1/2$$

There are 15 patients in the study, so the number who have more aggressive behaviors on moon days has the binomial distribution $B(15, 1/2)$ if H_0 is true. The P-value for the observed count 14 is therefore $P(X \geq 14)$, where X has the $B(15, 1/2)$ distribution. You can compute this probability with software or from the binomial probability formula (page 349):

$$P(X \geq 14) = P(X = 14) + P(X = 15)$$

$$= \binom{15}{14} \left(\frac{1}{2}\right)^{14} \left(\frac{1}{2}\right)^{1} + \binom{15}{15} \left(\frac{1}{2}\right)^{15} \left(\frac{1}{2}\right)^{0}$$

$$= (15) \left(\frac{1}{2}\right)^{15} + \left(\frac{1}{2}\right)^{15}$$

$$= 0.000488$$

Using Table C we would approximate this value as 0.0005. As in Example 7.7, there is very strong evidence in favor of an increase in aggressive behavior on moon days.

There are several varieties of sign test, all based on counts and the binomial distribution. The sign test for matched pairs (Example 7.12) is the most useful. The null hypothesis of "no effect" is then always $H_0 : p = 1/2$. The alternative can be one-sided in either direction or two-sided, depending on the type of change we are looking for. The test gets its name from the fact that we look only at the signs of the differences, not their actual values.

> ### THE SIGN TEST FOR MATCHED PAIRS
>
> Ignore pairs with difference 0; the number of trials n is the count of the remaining pairs. The test statistic is the count X of pairs with a positive difference. P-values for X are based on the binomial $B(n, 1/2)$ distribution.

The matched pairs t test in Example 7.7 tested the hypothesis that the mean of the distribution of differences (moon days minus other days) is 0. The sign test in Example 7.12 is in fact testing the hypothesis that the *median* of the differences is 0. If p is the probability that a difference is positive, then $p = 1/2$ when the median is 0. This is true because the median of the distribution is the point with probability 1/2 lying to its right. As Figure 7.11 illustrates, $p > 1/2$ when the median is greater than 0, again because the probability to the right of the median is always 1/2. The sign test of $H_0 : p = 1/2$ against $H_a : p > 1/2$ is a test of

$$H_0 : \text{population median} = 0$$

$$H_a : \text{population median} > 0$$

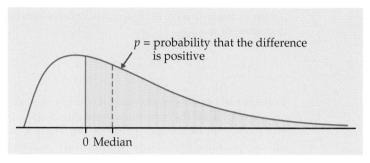

FIGURE 7.11 Why the sign test tests the median difference: when the median is greater than 0, the probability p of a positive difference is greater than 1/2, and vice versa.

The sign test in Example 7.12 makes no use of the actual differences—it just counts how many patients had more aggressive behaviors on moon days than on other days. Because the sign test uses so little of the available information, it is much less powerful than the t test when the population is close to normal. *It is better to use a test that is powerful when we believe our assumptions are approximately satisfied than a less powerful test with fewer assumptions.* There are other distribution-free tests that are more powerful than the sign test.[6]

SECTION 7.1 | Summary

Significance tests and confidence intervals for the mean μ of a normal population are based on the sample mean $\bar{x}$ of an SRS. Because of the central limit theorem, the resulting procedures are approximately correct for other population distributions when the sample is large.

The standardized sample mean, or **one-sample z statistic,**

$$z = \frac{\bar{x} - \mu}{\sigma / \sqrt{n}}$$

has the $N(0, 1)$ distribution. If the standard deviation $\sigma / \sqrt{n}$ of $\bar{x}$ is replaced by the **standard error** $s/\sqrt{n}$, the **one-sample t statistic**

$$t = \frac{\bar{x} - \mu}{s / \sqrt{n}}$$

has the **t distribution** with $n - 1$ degrees of freedom.

There is a t distribution for every positive **degrees of freedom k.** All are symmetric distributions similar in shape to normal distributions. The $t(k)$ distribution approaches the $N(0, 1)$ distribution as k increases.

A level C **confidence interval for the mean** μ of a normal population is

$$\bar{x} \pm t^* \frac{s}{\sqrt{n}}$$

where t^* is the value for the $t(n - 1)$ density curve with area C between $-t^*$ and t^*. The quantity

$$t^* \frac{s}{\sqrt{n}}$$

is the **margin of error.**

Significance tests for $H_0: \mu = \mu_0$ are based on the t statistic. P-values or fixed significance levels are computed from the $t(n - 1)$ distribution.

These one-sample procedures are used to analyze **matched pairs** data by first taking the differences within the matched pairs to produce a single sample.

The t procedures are relatively **robust** against nonnormal populations, especially for larger sample sizes. The t procedures are useful for nonnormal data when $n \geq 15$ unless the data show outliers or strong skewness.

The **power** of the t test is calculated like that of the z test, using an approximate value for both σ and s.

Small samples from skewed populations can sometimes be analyzed by first applying a **transformation** (such as the logarithm) to obtain an approximately normally distributed variable. The t procedures then apply to the transformed data.

The **sign test** is a **distribution-free test** because it uses probability calculations that are correct for a wide range of population distributions.

The sign test for "no treatment effect" in matched pairs counts the number of positive differences. The P-value is computed from the $B(n, 1/2)$ distribution, where n is the number of non-0 differences. The sign test is less powerful than the t test in cases where use of the t test is justified.

SECTION 7.1 | Exercises

7.1   A study of 584 longleaf pine trees in the Wade Tract in Thomas County, Georgia, is described in Example 6.1 (page 382). For each tree in the tract, the researchers measured the diameter at breast height (DBH). This is the diameter of the tree at 4.5 feet and the units are centimeters (cm). Only trees with DBH greater than 1.5 cm were sampled. Here are the diameters of a random sample of 40 of these trees:

10.5	13.3	26.0	18.3	52.2	9.2	26.1	17.6	40.5	31.8
47.2	11.4	2.7	69.3	44.4	16.9	35.7	5.4	44.2	2.2
4.3	7.8	38.1	2.2	11.4	51.5	4.9	39.7	32.6	51.8
43.6	2.3	44.6	31.5	40.3	22.3	43.3	37.5	29.1	27.9

(a) Use a histogram or stemplot and a boxplot to examine the distribution of DBHs. Include a normal quantile plot if you have the necessary software. Write a careful description of the distribution.

(b) Is it appropriate to use the methods of this section to find a 95% confidence interval for the mean DBH of all trees in the Wade Tract? Explain why or why not.

(c) Report the mean with the margin of error and the confidence interval. Write a short summary describing the meaning of the confidence interval.

(d) Do you think these results would apply to other similar trees in the same area? Give reasons for your answer.

7.2   C-reactive protein (CRP) is a substance that can be measured in the blood. Values increase substantially within 6 hours of an infection and reach a peak within 24 to 48 hours after. In adults, chronically high values have been linked to an increased risk of cardiovascular disease. In a study of apparently healthy children aged 6 to 60 months in Papua New Guinea, CRP was measured in 90 children.[7] The units are milligrams per liter (mg/l). Here are the data from a random sample of 40 of these children.

0.00	3.90	5.64	8.22	0.00	5.62	3.92	6.81	30.61	0.00
73.20	0.00	46.70	0.00	0.00	26.41	22.82	0.00	0.00	3.49
0.00	0.00	4.81	9.57	5.36	0.00	5.66	0.00	59.76	12.38
15.74	0.00	0.00	0.00	0.00	9.37	20.78	7.10	7.89	5.53

(a) Look carefully at the data above. Do you think that there are outliers or is this a skewed distribution? Now use a histogram or stemplot to examine the distribution. Write a short summary describing the distribution.

(b) Do you think that the mean is a good characterization of the center of this distribution? Explain why or why not.

(c) Find a 95% confidence interval for the mean CRP. Discuss the appropriateness of using this methodology for these data.

7.3   Refer to the previous exercise. With strongly skewed distributions such as this, we frequently reduce the skewness by taking a log transformation. We have a bit of a problem here, however, because some of the data are recorded as 0.00 and the logarithm of zero is not defined. For this variable, the value 0.00 is recorded whenever the amount of CRP in the blood is below the level that the measuring instrument is capable of detecting. The usual procedure in this circumstance is to add a small number to each observation before taking the logs. Transform these data by adding 1 to each observation and then taking the logarithm. Use the questions in the previous exercise as a guide to your analysis and prepare a summary contrasting this analysis with the one that you performed in the previous exercise.

7.4   In the Papua New Guinea study that provided the data for the previous two exercises, the researchers also measured serum retinol. A low value of this variable can be an indicator of vitamin A deficiency. Below are the data on the same sample of 40 children from this study. The units are micromoles per liter (μmol/l).

1.15	1.36	0.38	0.34	0.35	0.37	1.17	0.97	0.97	0.67
0.31	0.99	0.52	0.70	0.88	0.36	0.24	1.00	1.13	0.31
1.44	0.35	0.34	1.90	1.19	0.94	0.34	0.35	0.33	0.69
0.69	1.04	0.83	1.11	1.02	0.56	0.82	1.20	0.87	0.41

Analyze these data. Use the questions in the previous two exercises as a guide.

7.5   Children in a psychology study were asked to solve some puzzles and were then given feedback on their performance. Then they were asked to rate how luck played a role in determining their scores.[8] This variable was recorded on a 1 to 10 scale with 1 corresponding to very lucky and 10 corresponding to very unlucky. Here are the scores for 60 children:

1	10	1	10	1	1	10	5	1	1	8	1	10	2	1
9	5	2	1	8	10	5	9	10	10	9	6	10	1	5
1	9	2	1	7	10	9	5	10	10	10	1	8	1	6
10	1	6	10	10	8	10	3	10	8	1	8	10	4	2

(a) Use graphical methods to display the distribution. Describe any unusual characteristics. Do you think that these would lead you to hesitate before using the normality-based methods of this section?

(b) Give a 95% confidence interval for the mean luck score.

(c) The children in this study were volunteers whose parents agreed to have them participate in the study. To what extent do you think your results would apply to all similar children in this community?

7.6   Computers in some vehicles calculate various quantities related to the performance. One of these is the fuel efficiency, or gas mileage, usually expressed as miles per gallon (MPG). For one vehicle equipped in this way, the MPG was recorded each time the gas tank was filled and the computer was then reset.[9] Here are the MPG values for a random sample of 20 of these records:

15.8	13.6	15.6	19.1	22.4	15.6	22.5	17.2	19.4	22.6
19.4	18.0	14.6	18.7	21.0	14.8	22.6	21.5	14.3	20.9

(a) Describe the distribution using graphical methods and summarize the results.

(b) Is it appropriate to use methods based on normal distributions to analyze these data? Explain why or why not.

(c) Find the mean, the standard deviation, the standard error, and the margin of error for 95% confidence. Report the 95% confidence interval for μ, the mean MPG for this vehicle based on these data.

(d) Do you think that this interval would apply to other similar vehicles? Give reasons why and why not.

7.7   Refer to the previous exercise. Here are the values of the average speed in miles per hour (MPH) for the same sample:

21.0	19.0	18.7	39.2	45.8	19.8	48.4	21.0	29.1	35.7
31.6	49.0	16.0	34.6	36.3	19.0	43.3	37.5	16.5	34.5

Answer the questions given in the previous exercise.

7.8   In a study of children with attention deficit hyperactivity disorder (ADHD), parents were asked to rate their child on a variety of items related to how well their child performs different tasks.[10] One item was "Has difficulty organizing work," rated on a five-point scale of 0 to 4 with 0 corresponding to "not at all" and 4 corresponding to "very much." The mean rating for 282 boys with ADHD was reported as 2.22 with a standard deviation of 1.03.

(a) Do you think that these data are normally distributed? Explain why or why not.

(b) Is it appropriate to use the methods of this section to compute a 99% confidence interval? Explain why or why not.

(c) Find the 99% margin of error and the corresponding confidence interval. Write a sentence explaining the interval and the meaning of the 99% confidence level.

(d) The boys in this study were all evaluated at the Western Psychiatric Institute and Clinic at the University of Pittsburgh. To what extent do you think the results could be generalized to boys with ADHD in other locations?

7.9   Refer to the previous exercise. Compute the 90% and the 95% confidence intervals. Display the three intervals graphically and write a short explanation of the effect of the confidence level on the width of the interval using your display as an example.

7.10   If we increase our food intake, we generally gain weight. Nutrition scientists can calculate the amount of weight gain that would be associated with a given increase in calories. In one study, 16 nonobese adults, aged 25 to 36 years, were fed 1000 calories per day in excess of the calories needed to maintain a stable body weight. The subjects maintained this diet for 8 weeks, so they consumed a total of 56,000 extra calories.[11] According to theory, 3500 extra calories will translate into a weight gain of 1 pound. Therefore, we expect each of these subjects to gain 56,000/3500 = 16 pounds (lb). Here are the weights before and after the 8-week period expressed in kilograms (kg):

Subject	1	2	3	4	5	6	7	8
Weight before	55.7	54.9	59.6	62.3	74.2	75.6	70.7	53.3
Weight after	61.7	58.8	66.0	66.2	79.0	82.3	74.3	59.3

Subject	9	10	11	12	13	14	15	16
Weight before	73.3	63.4	68.1	73.7	91.7	55.9	61.7	57.8
Weight after	79.1	66.0	73.4	76.9	93.1	63.0	68.2	60.3

(a) For each subject, subtract the weight before from the weight after to determine the weight change.

(b) Find the mean and the standard deviation for the weight change.

(c) Calculate the standard error and the margin of error for 95% confidence. Report the 95% confidence interval in a sentence that explains the meaning of the 95%.

(d) Convert the mean weight gain in kilograms to mean weight gain in pounds. Because there are 2.2 kg per pound, multiply the value in kilograms by 2.2 to obtain pounds. Do the same for the standard deviation and the confidence interval.

(e) Test the null hypothesis that the mean weight gain is 16 lb. Be sure to specify the null and alternative hypotheses, the test statistic with degrees of freedom, and the P-value. What do you conclude?

(f) Write a short paragraph explaining your results.

7.11 Nonexercise activity thermogenesis (NEAT) provides a partial explanation for the results you found in the previous analysis. NEAT is energy burned by fidgeting, maintenance of posture, spontaneous muscle contraction, and other activities of daily living. In the study of the previous exercise, the 16 subjects increased their NEAT by 328 calories per day in response to the additional food intake. The standard deviation was 256.

(a) Test the null hypothesis that there was no change in NEAT versus the two-sided alternative. Summarize the results of the test and give your conclusion.

(b) Find a 95% confidence interval for the change in NEAT. Discuss the additional information provided by the confidence interval that is not evident from the results of the significance test.

7.12 You will have complete sales information for last month in a week, but right now you have data from a random sample of 40 stores. The mean change in sales in the sample is +3.8% and the standard deviation of the changes is 12%. Are average sales for all stores different from last month?

(a) State appropriate null and alternative hypotheses. Explain how you decided between the one- and two-sided alternatives.

(b) Find the t statistic and its P-value. State your conclusion.

(c) If the test gives strong evidence against the null hypothesis, would you conclude that sales are up in every one of your stores? Explain your answer.

7.13 For a sample of size 5, a test of a null hypothesis versus a two-sided alternative gives $t = 2.45$.

(a) Is the test result significant at the 5% level? Draw a sketch of the appropriate t distribution and illustrate your calculation with this sketch.

(b) Now assume that the same statistic was obtained for a sample size of $n = 10$. Assess the statistical significance of the result and illustrate the calculation with a sketch. How did the statistical significance change with the sample size? Explain your answer.

7.14   Assume a sample size of $n = 2000$. Draw a picture of the distribution of the t statistic under the null hypothesis. Use your picture to illustrate the values of the test statistic that would lead to rejection of the null hypothesis at the 1% level for a two-sided alternative.

7.15   Repeat the previous exercise for the two situations where the alternative is one-sided.

7.16   Computer software reports $\bar{x} = 12.3$ and $P = 0.08$ for a t test of $H_0: \mu = 0$ versus $H_a: \mu \neq 0$. Base on prior knowledge, you can justify testing the alternative $H_a: \mu > 0$. What is the P-value for your significance test?

7.17   Suppose that $\bar{x} = -12.3$ in the setting of the previous exercise. Would this change your answer? Use a sketch of the distribution of the test statistic under the null hypothesis to illustrate and explain your answer.

7.18   Use Table D to find the critical value t^* to be used for a confidence interval for the mean of the population in each of the following situations.

(a) A 95% confidence interval based on $n = 20$ observations.

(b) A 90% confidence interval from an SRS of 30 observations.

(c) An 80% confidence interval from a sample of size 50.

7.19   Use software to find the critical values t^* that you would use for 95% confidence intervals for sample sizes of 10, 20, 30, 40, 50, 100, 200, and 500. Plot the values of t^* versus the sample size and describe the relationship.

7.20   A sample of size $n = 15$ is used to perform a significance test for $H_0: \mu = 0$ versus $H_a: \mu > 0$. The test statistic is $t = 2.15$.

(a) What are the degrees of freedom for this statistic?

(b) Give the two critical values t^* from Table D that bracket t.

(c) What are the right-tail probabilities p for these two entries?

(d) Between what two values does the P-value of the test fall?

(e) Sketch the t distribution for this exercise and illustrate your answers to parts (c) and (d) with the sketch.

(f) Is the value $t = 2.15$ significant at the 5% level? Is it significant at the 1% level?

(g) If you have software available, find the exact P-value.

7.21   The hypotheses $H_0: \mu = 50$ and $H_a: \mu \neq 50$ are examined using a sample of size $n = 30$. The one-sample t statistic has the value $t = 1.35$.

(a) Give the degrees of freedom for the test statistic.

(b) Locate the two critical values t^* from Table D that bracket t. What are the right-tail probabilities p for these two values?

(c) How would you report the *P*-value for this test?

(d) Is the value $t = 1.35$ statistically significant at the 10% level? At the 5% level?

(e) Illustrate your answers to the previous parts of this exercise with a sketch of the *t* distribution.

(f) If you have software available, find the exact *P*-value.

7.22 The one-sample *t* statistic for a test of $H_0: \mu = 30$ versus $H_a: \mu < 30$ based on $n = 12$ observations has the value $t = -3.21$.

(a) Find the degrees of freedom for this statistic.

(b) Use Table D to find an approximate *P*-value. Use a sketch of the *t* distribution to illustrate your work.

(c) Find the exact *P*-value if you have software available.

7.23 Most software gives *P*-values for two-sided alternatives. Explain why you cannot always divide these *P*-values by 2 to obtain *P*-values for one-sided alternatives.

7.24 Here are estimates of the daily intakes of calcium (in milligrams) for 40 women between the ages of 18 and 24 years who participated in a study of women's bone health:

725	764	853	559	1225	456	555	1168	682	676
808	882	1062	970	909	802	574	576	784	997
651	716	438	1220	1475	948	1050	976	572	483
1256	685	1144	848	1213	3223	1270	1108	350	421

(a) Use a stemplot or other graphical summary to describe the distribution of intakes. If you have software available, make a normal quantile plot. Write a short paragraph describing the distribution. Be sure to refer to your graphics in your summary.

(b) Find the mean and the standard deviation. Sketch a normal curve with this mean and standard deviation.

(c) Write a sentence in which you give the 95% confidence interval for the mean and an explanation of how it should be interpreted.

(d) There is an outlier. Eliminate it and answer parts (a), (b), and (c) again. How do the results change?

(e) Take one side of the following issue and present reasons for your views. "These results do not apply to women aged 18 to 24 years from the same community who did not volunteer to participate in the study."

7.25 Refer to the previous exercise. For some nutrients, including calcium, recommendations are expressed as adequate intakes (AI). The AI for women in this age range is 1000 milligrams. Let's compare the intakes of the women in this sample with the AI using a significance test. Use the data for all 40 women.

(a) What null and alternative hypotheses would you use for this problem?

(b) Give the test statistic, the degrees of freedom, the *P*-value, and a sketch illustrating the *P*-value.

(c) Write a short paragraph giving the descriptive statistics and the significance test results for this problem.

(d) (Optional) Use the sign test for this problem. Compare the two approaches to the analysis of these data.

7.26   Many organizations are doing surveys to determine the satisfaction of their customers. Attitudes toward various aspects of campus life were the subject of one such study conducted at Purdue University. Each item was rated on a 1 to 5 scale, with 5 being the highest rating. The average response of 1406 first-year students to "Feeling welcomed at Purdue" was 3.9 with a standard deviation of 0.98. Assuming that the respondents are an SRS, give a 99% confidence interval for the mean of all first-year students.

7.27   Dual-energy X-ray absorptiometry (DXA) is a technique for measuring bone health. One of the most common measures is total body bone mineral content (TBBMC). A highly skilled operator is required to take the measurements. Recently a new DXA machine was purchased by a research lab and two operators were trained to take the measurements. TBBMC for eight subjects was measured by both operators.[12] The units are grams (g). A comparison of the means for the two operators provides a check on the training they received and allows us to determine if one of the operators is producing measurements that are consistently higher than the other. Here are the data:

Operator	Subject							
	1	2	3	4	5	6	7	8
1	1.328	1.342	1.075	1.228	0.939	1.004	1.178	1.286
2	1.323	1.322	1.073	1.233	0.934	1.019	1.184	1.304

(a) Take the difference between the TBBMC recorded for operator 1 and the TBBMC for operator 2. Describe the distribution of these differences.

(b) Use a significance test to examine the null hypothesis that the two operators have the same mean. Be sure to give the test statistic with its degrees of freedom, the *P*-value, and your conclusion.

(c) The sample here is rather small, so we may not have much power to detect differences of interest. Use a 95% confidence interval to provide a range of differences that are compatible with these data.

(d) The eight subjects used for this comparison were not a random sample. In fact, they were friends of the researchers whose ages and weights were similar to the types of people who would be measured with this DXA. Comment on the appropriateness of this procedure for selecting a sample, and discuss any consequences regarding the interpretation of the significance testing and confidence interval results.

7.28   Refer to the previous exercise. TBBMC measures the total amount of mineral in the bones. Another important variable is total body bone mineral density (TBBMD). This variable is calculated by dividing TBBMC by the area corresponding to bone in the DXA scan. The units are grams per squared centimeter (g/cm^2). Here are the TBBMD values for the same subjects:

Operator	Subject							
	1	2	3	4	5	6	7	8
1	4042	3703	2626	2673	1724	2136	2808	3322
2	4041	3697	2613	2628	1755	2140	2836	3287

Analyze these data using the questions in the previous exercise as a guide.

7.29   Do piano lessons improve the spatial-temporal reasoning of preschool children? Neurobiological arguments suggest that this may be true. A study designed to test this hypothesis measured the spatial-temporal reasoning of 34 preschool children before and after six months of piano lessons.[13] (The study also included children who took computer lessons and a control group; but we are not concerned with those here.) The changes in the reasoning scores are

2	5	7	−2	2	7	4	1	0	7	3	4	3	4	9	4	5
2	9	6	0	3	6	−1	3	4	6	7	−2	7	−3	3	4	4

(a) Display the data and summarize the distribution.

(b) Find the mean, the standard deviation, and the standard error of the mean.

(c) Give a 95% confidence interval for the mean improvement in reasoning scores.

7.30   Refer to the previous exercise. Test the null hypothesis that there is no improvement versus the alternative suggested by the neurobiological arguments. State the hypotheses, and give the test statistic with degrees of freedom and the P-value. What do you conclude? From your answer to part (c) of the previous exercise what can be concluded from this significance test?

7.31   The researchers studying vitamin C in CSB in Example 7.1 were also interested in a similar commodity called wheat soy blend (WSB). Both of these commodities are mixed with other ingredients and cooked. Loss of vitamin C as a result of this process was another concern of the researchers. One preparation used in Haiti called gruel (or "bouillie" in Creole) can be made from WSB, salt, sugar, milk, banana, and other optional items to improve the taste. Samples of gruel prepared in Haitian households were collected. The vitamin C content (in milligrams per 100 grams of blend, dry basis) was measured before and after cooking. Here are the results:

Sample	1	2	3	4	5
Before	73	79	86	88	78
After	20	27	29	36	17

Set up appropriate hypotheses and carry out a significance test for these data. (It is not possible for cooking to increase the amount of vitamin C.)

7.32 Refer to the previous exercise. The fact that vitamin C is destroyed by cooking is neither new nor surprising; so the significance test you performed in the previous exercise simply confirms that this fact is evident in the small sample of data collected. The real question here concerns how much of the vitamin C is lost.

(a) Give a 95% confidence interval for the amount of vitamin C lost by preparing and cooking gruel in Haiti.

(b) The units for your interval are mg/100 g of blend. To meaningfully interpret the amount lost, it is necessary to know something about the amount that we started with. For example, a loss of 50 mg/100 g would probably not be of much concern if we started with 5000 mg/100 g. The specifications call for the blend to contain 98 mg/100 g (dry basis). The difference between this specification and the "before" values above is due to sample variation in the manufacturing process and the handling of the product from the time it was manufactured until it was used to prepare gruel in these Haitian homes. Express the "after" data as percent of specification and give a 95% confidence interval for the mean percent.

7.33 A bank wonders whether omitting the annual credit card fee for customers who charge at least $3000 in a year would increase the amount charged on its credit card. The bank makes this offer to an SRS of 500 of its existing credit card customers. It then compares how much these customers charge this year with the amount that they charged last year. The mean increase is $565, and the standard deviation is $267.

(a) Is there significant evidence at the 1% level that the mean amount charged increases under the no-fee offer? State H_0 and H_a and carry out a t test.

(b) Give a 95% confidence interval for the mean amount of the increase.

(c) The distribution of the amount charged is skewed to the right, but outliers are prevented by the credit limit that the bank enforces on each card. Use of the t procedures is justified in this case even though the population distribution is not normal. Explain why.

(d) A critic points out that the customers would probably have charged more this year than last even without the new offer because the economy is more prosperous and interest rates are lower. Briefly describe the design of an experiment to study the effect of the no-fee offer that would avoid this criticism.

7.34   The level of various substances in the blood of kidney dialysis patients is of concern because kidney failure and dialysis can lead to nutritional problems. A researcher performed blood tests on several dialysis patients on 6 consecutive clinic visits. One variable measured was the level of phosphate in the blood. Phosphate levels for an individual tend to vary normally over time. The data on one patient, in milligrams of phosphate per deciliter (mg/dl) of blood, are given below:[14]

$$5.6 \quad 5.1 \quad 4.6 \quad 4.8 \quad 5.7 \quad 6.4$$

(a) Calculate the sample mean $\bar{x}$ and its standard error.

(b) Use the t procedures to give a 90% confidence interval for this patient's mean phosphate level.

7.35   The normal range of values for blood phosphate levels is 2.6 to 4.8 mg/dl. The sample mean for the patient in the previous exercise falls above this range. Is this good evidence that the patient's mean level in fact falls above 4.8? State H_0 and H_a and use the data in the previous exercise to carry out a t test. Between which levels from Table D does the P-value lie? Are you convinced that the patient's phosphate level is higher than normal?

7.36   In a randomized comparative experiment on the effect of dietary calcium on blood pressure, 54 healthy white males were divided at random into two groups. One group received calcium; the other, a placebo. At the beginning of the study, the researchers measured many variables on the subjects. The paper reporting the study gives $\bar{x} = 114.9$ and $s = 9.3$ for the seated systolic blood pressure of the 27 members of the placebo group.

(a) Give a 95% confidence interval for the mean blood pressure of the population from which the subjects were recruited.

(b) What assumptions about the population and the study design are required by the procedure you used in (a)? Which of these assumptions are important for the validity of the procedure in this case?

7.37   How accurate are radon detectors of a type sold to homeowners? To answer this question, university researchers placed 12 detectors in a chamber that exposed them to 105 picocuries per liter (pCi/l) of radon.[15] The detector readings were as follows:

91.9	97.8	111.4	122.3	105.4	95.0
103.8	99.6	96.6	119.3	104.8	101.7

(a) Make a stemplot of the data. The distribution is somewhat skewed to the right, but not strongly enough to forbid use of the t procedures.

(b) Is there convincing evidence that the mean reading of all detectors of this type differs from the true value of 105? Carry out a test in detail and write a brief conclusion.

7.38 Refer to the lengths of calls to a customer service center in Table 1.1 (page 9). Give graphical and numerical summaries for these data. Compute a 95% confidence interval for the mean call length. Comment on the validity of your interval.

7.39 The researchers studying vitamin C in CSB in Example 7.1 were also interested in a similar commodity called wheat soy blend (WSB). A major concern was the possibility that some of the vitamin C content would be destroyed as a result of storage and shipment of the commodity to its final destination. The researchers specially marked a collection of bags at the factory and took a sample from each of these to determine the vitamin C content. Five months later in Haiti they found the specially marked bags and took samples. The data consist of two vitamin C measures for each bag, one at the time of production in the factory and the other five months later in Haiti. The units are mg/100 g as in Example 7.1. Here are the data:

Factory	Haiti	Factory	Haiti	Factory	Haiti
44	40	45	38	39	43
50	37	32	40	52	38
48	39	47	35	45	38
44	35	40	38	37	38
42	35	38	34	38	41
47	41	41	35	44	40
49	37	43	37	43	35
50	37	40	34	39	38
39	34	37	40	44	36

(a) Describe the data graphically and numerically. Summarize your results.

(b) Set up hypotheses to examine the question of interest to these researchers.

(c) Perform the significance test and summarize your results.

(d) Find 95% confidence intervals for the mean at the factory, the mean five months later in Haiti, and for the change.

7.40 Table 1.5 (page 32) gives the average property damage per year due to tornadoes for each of the 50 states and Puerto Rico. It does not make sense to use the *t* procedures (or any other statistical procedures) to give a 95% confidence interval for the mean property damage per year due to tornadoes in the United States. Explain why not.

7.41 The National Endowment for the Humanities sponsors summer institutes to improve the skills of high school teachers of foreign languages. One such institute hosted 20 French teachers for 4 weeks. At the beginning of the period, the teachers were given the Modern Language Association's listening test of understanding of spoken French. After 4 weeks of immersion in French in and out of class, the listening test was given again. (The actual French spoken in the two tests was different, so that simply taking the first test should not improve the score on the second test.) The maximum possible score on the test is 36.[16] Here are the data:

Teacher	Pretest	Posttest	Gain	Teacher	Pretest	Posttest	Gain
1	32	34	2	11	30	36	6
2	31	31	0	12	20	26	6
3	29	35	6	13	24	27	3
4	10	16	6	14	24	24	0
5	30	33	3	15	31	32	1
6	33	36	3	16	30	31	1
7	22	24	2	17	15	15	0
8	25	28	3	18	32	34	2
9	32	26	−6	19	23	26	3
10	20	26	6	20	23	26	3

To analyze these data, we first subtract the pretest score from the posttest score to obtain the improvement for each teacher. These 20 differences form a single sample. They appear in the "Gain" columns. The first teacher, for example, improved from 32 to 34, so the gain is $34 - 32 = 2$.

(a) State appropriate null and alternative hypotheses for examining the question of whether or not the course improves French-spoken language skills.

(b) Describe the gain data. Use numerical and graphical summaries.

(c) Perform the significance test. Give the test statistic, the degrees of freedom, and the P-value. Summarize your conclusion.

(d) Give a 95% confidence interval for the mean improvement.

7.42   The table below gives the pretest and posttest scores on the MLA listening test in Spanish for 20 high school Spanish teachers who attended an intensive summer course in Spanish. The setting is identical to the one described in the previous exercise.

Teacher	Pretest	Posttest	Teacher	Pretest	Posttest
1	30	29	11	30	32
2	28	30	12	29	28
3	31	32	13	31	34
4	26	30	14	29	32
5	20	16	15	34	32
6	30	25	16	20	27
7	34	31	17	26	28
8	15	18	18	25	29
9	28	33	19	31	32
10	20	25	20	29	32

Summarize the data graphically and numerically. Then analyze the data using a significance test and a confidence interval. Write a short report summarizing your results.

7.43   Refer to the IQ test scores for fifth-grade students in Table 1.3 (page 14). Give numerical and graphical summaries of the data, and compute a 95% confidence interval. Comment on the validity of the interval.

The following exercises concern the optional material in the sections on the power of the t test and on nonnormal populations.

7.44   The differences in the TBBMC measures in Exercise 7.27 can also be analyzed using a sign test. Set up the appropriate null and alternative hypotheses, carry out the test, and summarize the results. How do these results compare with those that you obtained in Exercise 7.27?

7.45   TBBMD measures for the same subjects that you studied in the previous exercise are given in Exercise 7.28. Answer the questions given in the previous exercise for TBBMD.

7.46   Exercise 7.31 gives data on the amount of vitamin C in gruel made from wheat soy blend in 5 Haitian households before and after cooking. Is there evidence that the median amount of vitamin C is less after cooking? State hypotheses, carry out a sign test, and report your conclusion.

7.47   Apply the sign test to the data in Exercise 7.29 to assess the effects of piano lessons on spatial-temporal reasoning.

   (a) State the hypotheses two ways: in terms of a population median and in terms of the probability of an improvement in the test score.

   (b) Carry out the sign test. Find the approximate P-value using the normal approximation to the binomial distributions, and report your conclusion.

7.48   Use the sign test to assess whether the summer institute of Exercise 7.41 improves French listening skills. State the hypotheses, give the P-value using the binomial table (Table C), and report your conclusion.

7.49   Use the sign test to assess whether the summer institute of Exercise 7.42 improves Spanish listening skills. State the hypotheses, give the P-value using the binomial table (Table C), and report your conclusion.

7.50   A manufacturer of electric motors tests insulation at a high temperature (250° C) and records the number of hours until the insulation fails.[17]   The data for 5 specimens are

$$446 \quad 326 \quad 372 \quad 377 \quad 310$$

The small sample size makes judgment from the data difficult, but engineering experience suggests that the logarithm of the failure time will have a normal distribution. Take the logarithms of the 5 observations, and use t procedures to give a 90% confidence interval for the mean of the log failure time for insulation of this type.

7.51   Suppose that the bone researchers in Exercise 7.27 wanted to detect an alternative mean difference of 0.002. Find the power for this alternative for a sample size of 15. Use the standard deviation that you found in Exercise 7.27 for these calculations.

7.52   You are designing a study to test the null hypothesis that $\mu = 0$ versus the alternative that μ is positive. Assume that σ is 10. Suppose that it would be important to be able to detect the alternative that $\mu = 2$. Perform power calculations for a variety of sample sizes and determine how large a sample you would need to detect this alternative with power of at least 0.80.

7.2 Comparing Two Means

A medical researcher is interested in the effect on blood pressure of added calcium in our diet. She conducts a randomized comparative experiment in which one group of subjects receives a calcium supplement and a control group gets a placebo. A psychologist develops a test that measures social insight. He compares the social insight of male college students with that of female college students by giving the test to a large group of students of each sex. A bank wants to know which of two incentive plans will most increase the use of its credit cards. It offers each incentive to a random sample of credit card customers and compares the amount charged during the following six months. Two-sample problems such as these are among the most common situations encountered in statistical practice.

> **TWO-SAMPLE PROBLEMS**
>
> - The goal of inference is to compare the responses in two groups.
>
> - Each group is considered to be a sample from a distinct population.
>
> - The responses in each group are independent of those in the other group.

A two-sample problem can arise from a randomized comparative experiment that randomly divides the subjects into two groups and exposes each group to a different treatment. Comparing random samples separately selected from two populations is also a two-sample problem. Unlike the matched pairs designs studied earlier, there is no matching of the units in the two samples, and the two samples may be of different sizes. Inference procedures for two-sample data differ from those for matched pairs.

We can present two-sample data graphically by a back-to-back stemplot (for small samples) or by side-by-side boxplots (for larger samples). Now we will apply the ideas of formal inference in this setting. When both population distributions are symmetric, and especially when they are at least approximately normal, a comparison of the mean responses in the two populations is most often the goal of inference.

We have two independent samples, from two distinct populations (such as subjects given a treatment and those given a placebo). The same variable is measured for both samples. We will call the variable x_1 in the first population and x_2 in the second because the variable may have different distributions in the two populations. Here is the notation that we will use to describe the two populations:

Population	Variable	Mean	Standard deviation
1	x_1	μ_1	σ_1
2	x_2	μ_2	σ_2

We want to compare the two population means, either by giving a confidence interval for $\mu_1 - \mu_2$ or by testing the hypothesis of no difference, $H_0: \mu_1 = \mu_2$.

Inference is based on two independent SRSs, one from each population. Here is the notation that describes the samples:

Population	Sample size	Sample mean	Sample standard deviation
1	n_1	$\bar{x}_1$	s_1
2	n_2	$\bar{x}_2$	s_2

Throughout this section, the subscripts 1 and 2 show the population to which a parameter or a sample statistic refers.

The two-sample z statistic

The natural estimator of the difference $\mu_1 - \mu_2$ is the difference between the sample means, $\bar{x}_1 - \bar{x}_2$. If we are to base inference on this statistic, we must know its sampling distribution. Our knowledge of probability is equal to the task. First, the mean of the difference $\bar{x}_1 - \bar{x}_2$ is the difference of the means $\mu_1 - \mu_2$. This follows from the addition rule for means (page 298) and the fact that the mean of any $\bar{x}$ is the same as the mean of the population. Because the samples are independent, their sample means $\bar{x}_1$ and $\bar{x}_2$ are independent random variables. The addition rule for variances (page 302) says that the variance of the difference $\bar{x}_1 - \bar{x}_2$ is the sum of their variances, which is

$$\frac{\sigma_1^2}{n_1} + \frac{\sigma_2^2}{n_2}$$

We now know the mean and variance of the distribution of $\bar{x}_1 - \bar{x}_2$ in terms of the parameters of the two populations. If the two population distributions are both normal, then the distribution of $\bar{x}_1 - \bar{x}_2$ is also normal. This is true because each sample mean alone is normally distributed and because a difference of independent normal random variables is also normal.

EXAMPLE 7.13   A fourth-grade class has 10 girls and 7 boys. The children's heights are recorded on their tenth birthdays. What is the chance that the girls are taller than the boys? Of course, it is very unlikely that all of the girls are taller than all of the boys. We translate the question into the following: what is the probability that the mean height of the girls is greater than the mean height of the boys?

Based on information from the National Center for Health Statistics, we assume that the heights (in inches) of 10-year-old girls are $N(54.5, 2.7)$ and the heights of 10-year-old boys are $N(54.1, 2.4)$. The heights of the students in our class are assumed to be random samples from these populations. The two distributions are shown in Figure 7.12(a).

The difference $\bar{x}_1 - \bar{x}_2$ between the female and male mean heights varies in different random samples. The sampling distribution has mean

$$\mu_1 - \mu_2 = 54.5 - 54.1 = 0.4 \text{ inch}$$

and variance

$$\frac{\sigma_1^2}{n_1} + \frac{\sigma_2^2}{n_2} = \frac{2.7^2}{10} + \frac{2.4^2}{7}$$
$$= 1.55$$

The standard deviation of the difference in sample means is therefore $\sqrt{1.55} = 1.25$ inches.

If the heights vary normally, the difference in sample means is also normally distributed. The distribution of the difference in heights is shown in Figure 7.12(b). We

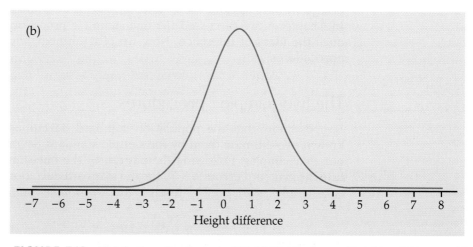

FIGURE 7.12 Distributions for Example 7.13. (a) Distributions of heights of 10-year-old boys and girls. (b) Distribution of the difference between means of 10 girls and 7 boys.

standardize $\bar{x}_1 - \bar{x}_2$ by subtracting its mean (0.4) and dividing by its standard deviation (1.25). Therefore, the probability that the girls are taller than the boys is

$$P(\bar{x}_1 - \bar{x}_2 > 0) = P\left(\frac{(\bar{x}_1 - \bar{x}_2) - 0.4}{1.25} > \frac{0 - 0.4}{1.25}\right)$$

$$= P(Z > -0.32) = 0.63$$

Even though the population mean height of 10-year-old girls is greater than the population mean height of 10-year-old boys, the probability that the sample mean of the girls is greater than the sample mean of the boys in our class is only 63%. *Large samples are needed to see the effects of small differences.*

As Example 7.13 reminds us, any normal random variable has the $N(0, 1)$ distribution when standardized. We have arrived at a new z statistic.

TWO-SAMPLE z STATISTIC

Suppose that $\bar{x}_1$ is the mean of an SRS of size n_1 drawn from an $N(\mu_1, \sigma_1)$ population and that $\bar{x}_2$ is the mean of an independent SRS of size n_2 drawn from an $N(\mu_2, \sigma_2)$ population. Then the **two-sample z statistic**

$$z = \frac{(\bar{x}_1 - \bar{x}_2) - (\mu_1 - \mu_2)}{\sqrt{\dfrac{\sigma_1^2}{n_1} + \dfrac{\sigma_2^2}{n_2}}}$$

has the standard normal $N(0, 1)$ sampling distribution.

In the unlikely event that both population standard deviations are known, the two-sample z statistic is the basis for inference about $\mu_1 - \mu_2$. Exact z procedures are seldom used, however, because σ_1 and σ_2 are rarely known. In Chapter 6, we discussed the one-sample z procedures in order to introduce the ideas of inference. Now we pass immediately to the more useful t procedures.

The two-sample t procedures

Suppose now that the population standard deviations σ_1 and σ_2 are not known. We estimate them by the sample standard deviations s_1 and s_2 from our two samples. Following the pattern of the one-sample case, we substitute the standard errors $s_i/\sqrt{n_i}$ for the standard deviations $\sigma_i/\sqrt{n_i}$ used in the two-sample z statistic. The result is the *two-sample t statistic*:

$$t = \frac{(\bar{x}_1 - \bar{x}_2) - (\mu_1 - \mu_2)}{\sqrt{\dfrac{s_1^2}{n_1} + \dfrac{s_2^2}{n_2}}}$$

Unfortunately, this statistic does *not* have a t distribution. A t distribution replaces an $N(0, 1)$ distribution only when a single standard deviation (σ) in a z statistic is replaced by a standard error (s). In this case, we replaced two standard deviations (σ_1 and σ_2) by their estimates (s_1 and s_2), which does not produce a statistic having a t distribution.

approximations for the degrees of freedom

Nonetheless, we can approximate the distribution of the two-sample t statistic by using the $t(k)$ distribution with an **approximation for the degrees of freedom k.** We use these approximations to find approximate values of t^* for confidence intervals and to find approximate P-values for significance tests.

Here are two approximations:

1. Use a value of k that is calculated from the data. In general it will not be a whole number.

2. Use k equal to the smaller of $n_1 - 1$ and $n_2 - 1$.

Most statistical software uses the first option to approximate the $t(k)$ distribution for two-sample problems unless the user requests another method. Use of this approximation without software is a bit complicated; we will give the details later in this section. If you are not using software, the second approximation is preferred. This approximation is appealing because it is conservative.[18] Margins of error for confidence intervals are a bit larger than they need to be, so the true confidence level is larger than C. For significance testing, the true P-values are a bit smaller than those we obtain from the approximation; for tests at a fixed significance level, we are a little less likely to reject H_0 when it is true. In practice, the choice of approximation almost never makes a difference in our conclusion.

The two-sample t significance test

THE TWO-SAMPLE t SIGNIFICANCE TEST

Suppose that an SRS of size n_1 is drawn from a normal population with unknown mean μ_1 and that an independent SRS of size n_2 is drawn from another normal population with unknown mean μ_2. To test the hypothesis $H_0: \mu_1 = \mu_2$, compute the **two-sample t statistic**

$$t = \frac{\bar{x}_1 - \bar{x}_2}{\sqrt{\dfrac{s_1^2}{n_1} + \dfrac{s_2^2}{n_2}}}$$

and use P-values or critical values for the $t(k)$ distribution, where the degrees of freedom k are either approximated by software or are the smaller of $n_1 - 1$ and $n_2 - 1$.

EXAMPLE 7.14   An educator believes that new directed reading activities in the classroom will help elementary school pupils improve some aspects of their reading ability. She arranges for a third-grade class of 21 students to take part in these activities for an eight-week period. A control classroom of 23 third-graders follows the

same curriculum without the activities. At the end of the eight weeks, all students are given a Degree of Reading Power (DRP) test, which measures the aspects of reading ability that the treatment is designed to improve. The data appear in Table 7.4.[19]
First examine the data:

```
            Control                              Treatment
              9 7 0 │ 1 │
              8 6 0 │ 2 │ 4
              7 7 3 │ 3 │ 3
      8 6 3 2 2 2 1 │ 4 │ 3 3 3 4 6 9 9
            5 5 4 3 │ 5 │ 2 3 4 6 7 7 8 9
                2 0 │ 6 │ 1 2 7
                    │ 7 │ 1
                  5 │ 8 │
```

A back-to-back stemplot suggests that there is a mild outlier in the control group but no deviation from normality serious enough to forbid use of t procedures. Separate normal quantile plots for both groups (Figure 7.13) confirm that both are approximately normal. The scores of the treatment group appear to be somewhat higher than those of the control group. The summary statistics are

Group	n	$\bar{x}$	s
Treatment	21	51.48	11.01
Control	23	41.52	17.15

Because we hope to show that the treatment (Group 1) is better than the control (Group 2), the hypotheses are

$$H_0: \mu_1 = \mu_2$$

$$H_a: \mu_1 > \mu_2$$

The two-sample t test statistic is

$$t = \frac{\bar{x}_1 - \bar{x}_2}{\sqrt{\dfrac{s_1^2}{n_1} + \dfrac{s_2^2}{n_2}}}$$

$$= \frac{51.48 - 41.52}{\sqrt{\dfrac{11.01^2}{21} + \dfrac{17.15^2}{23}}}$$

$$= 2.31$$

The P-value for the one-sided test is $P(T \geq 2.31)$. Software gives the approximate P-value as 0.0132 and uses 37.9 as the degrees of freedom. For the second approximation, the degrees of freedom k are equal to the smaller of

$$n_1 - 1 = 21 - 1 = 20 \quad \text{and} \quad n_2 - 1 = 23 - 1 = 22$$

TABLE 7.4

DRP scores for third-graders

Treatment group				Control group			
24	61	59	46	42	33	46	37
43	44	52	43	43	41	10	42
58	67	62	57	55	19	17	55
71	49	54		26	54	60	28
43	53	57		62	20	53	48
49	56	33		37	85	42	

FIGURE 7.13 Normal quantile plots for the DRP scores in Table 7.4.

df = 20		
p	0.02	0.01
t^*	2.197	2.528

Comparing 2.31 with the entries in Table D for 20 degrees of freedom, we see that P lies between 0.02 and 0.01. The data strongly suggest that directed reading activity improves the DRP score ($t = 2.31$, df = 20, $P < 0.02$).

Note that when we report a result such as this with $P < 0.02$, we imply that the result is *not* significant at the 0.01 level.

If your software gives only P-values for the two-sided alternative, $2P(T \geq |t|)$, you need to divide the reported value by 2 after checking that the means differ in the direction specified by the alternative hypothesis.

The two-sample t confidence interval

The same ideas that we used for the two-sample t significance tests also apply to give us *two-sample t confidence intervals*. We can use either software or the conservative approach with Table D to approximate the value of t^*.

THE TWO-SAMPLE t CONFIDENCE INTERVAL

Suppose that an SRS of size n_1 is drawn from a normal population with unknown mean μ_1 and that an independent SRS of size n_2 is drawn from another normal population with unknown mean μ_2. The **confidence interval for $\mu_1 - \mu_2$** given by

$$(\bar{x}_1 - \bar{x}_2) \pm t^* \sqrt{\frac{s_1^2}{n_1} + \frac{s_2^2}{n_2}}$$

has confidence level at least C no matter what the population standard deviations may be. Here, t^* is the value for the $t(k)$ density curve with area C between $-t^*$ and t^*. The value of the degrees of freedom k is approximated by software or we use the smaller of $n_1 - 1$ and $n_2 - 1$.

To complete the analysis of the DRP scores we examined in Example 7.14, we need to describe the size of the treatment effect. We do this with a confidence interval for the difference between the treatment group and the control group means.

EXAMPLE 7.15 We will find a 95% confidence interval for the mean improvement in the entire population of third-graders. The interval is

$$(\bar{x}_1 - \bar{x}_2) \pm t^* \sqrt{\frac{s_1^2}{n_1} + \frac{s_2^2}{n_2}} = (51.48 - 41.52) \pm t^* \sqrt{\frac{11.01^2}{21} + \frac{17.15^2}{23}}$$

$$= 9.96 \pm 4.31 t^*$$

Using software, the degrees of freedom are 37.9 and $t^* = 2.025$. This approximation gives

$$9.96 \pm (4.31 \times 2.025) = 9.96 \pm 8.72 = (1.2, \ 18.7)$$

The conservative approach uses the $t(20)$ distribution. Table D gives $t^* = 2.086$. With this approximation we have

$$9.96 \pm (4.31 \times 2.086) = 9.96 \pm 8.99 = (1.0, \ 18.9)$$

We can see that the conservative approach does, in fact, give a larger interval than the more accurate approximation used by software. However, the difference is rather small.

We estimate the mean improvement to be about 10 points, but with a margin of error of almost 9 points. Although we have good evidence of some improvement, the data do not allow a very precise estimate of the size of the average improvement.

The design of the study in Example 7.14 is not ideal. Random assignment of students was not possible in a school environment, so existing third-grade classes were used. The effect of the reading programs is therefore confounded with any other differences between the two classes. The classes were chosen to be as similar as possible—for example, in the social and economic status of the students. Extensive pretesting showed that the two classes were on the average quite similar in reading ability at the beginning of the experiment. To avoid the effect of two different teachers, the researcher herself taught reading in both classes during the eight-week period of the experiment. We can therefore be somewhat confident that the two-sample test is detecting the effect of the treatment and not some other difference between the classes. This example is typical of many situations in which an experiment is carried out but randomization is not possible.

Robustness of the two-sample procedures

The two-sample t procedures are more robust than the one-sample t methods. When the sizes of the two samples are equal and the distributions of the two populations being compared have similar shapes, probability values from the t table are quite accurate for a broad range of distributions when the sample sizes are as small as $n_1 = n_2 = 5$.[20] When the two population distributions have different shapes, larger samples are needed. The guidelines given on page 463 for the use of one-sample t procedures can be adapted to two-sample procedures by replacing "sample size" with the "sum of the sample sizes" $n_1 + n_2$. These guidelines are rather conservative, especially when the two samples are of equal size. *In planning a two-sample study, choose equal sample sizes if you can.* The two-sample t procedures are most robust against nonnormality in this case, and the conservative probability values are most accurate.

Here is an example with moderately large sample sizes that are not equal. Even if the distributions are not normal, we are confident that the sample means will be approximately normal. The two-sample t test is very robust in this case.

EXAMPLE 7.16   The U.S. Department of Agriculture (USDA) uses sample surveys to produce important economic estimates. One pilot study estimated wheat prices in July and in September using independent samples of wheat producers in the two months. Here are the summary statistics, in dollars per bushel:[21]

Month	n	$\bar{x}$	s
September	45	$3.61	$0.19
July	90	$2.95	$0.22

The September prices are higher on the average. But we have data from only a few producers each month. Can we conclude that national average prices in July and September are not the same? Or are these differences merely what we would expect to see due to random variation?

Because we did not specify a direction for the difference before looking at the data, we choose a two-sided alternative. The hypotheses are

$$H_0: \mu_1 = \mu_2$$

$$H_a: \mu_1 \neq \mu_2$$

Because the samples are moderately large, we can confidently use the t procedures even though we lack the detailed data and so cannot verify the normality condition.
The two-sample t statistic is

$$t = \frac{\bar{x}_1 - \bar{x}_2}{\sqrt{\dfrac{s_1^2}{n_1} + \dfrac{s_2^2}{n_2}}}$$

$$= \frac{3.61 - 2.95}{\sqrt{\dfrac{(0.19)^2}{45} + \dfrac{(0.22)^2}{90}}}$$

$$= 18.03$$

df = 40	
p	0.0005
t^*	3.551

The conservative approach finds the P-value by comparing 18.03 to critical values for the $t(44)$ distribution because the smaller sample has 45 observations. We must double the table tail area p because the alternative is two-sided. Table D does not have entries for 44 degrees of freedom. When this happens, we use the next smaller degrees of freedom. Our calculated value of t is larger than the $p = 0.0005$ entry in the table. Doubling 0.0005, we conclude that the P-value is less than 0.001. The data give conclusive evidence that the mean wheat prices were higher in September than they were in July ($t = 18.03$, df = 44, $P < 0.001$).

In this example the exact P-value is very small because $t = 18$ says that the observed difference in means is 18 standard errors above the hypothesized difference of zero ($\mu_1 = \mu_2$). This is so unlikely that the probability is zero for all practical purposes. The difference in mean prices is not only highly significant but large enough (66 cents per bushel) to be important to producers.

In this and other examples, we can choose which population to label 1 and which to label 2. After inspecting the data, we chose September as Population 1 because this choice makes the t statistic a positive number. This avoids any possible confusion from reporting a negative value for t. *Choosing the population labels is **not** the same as choosing a one-sided alternative after looking*

at the data. Choosing hypotheses after seeing a result in the data is a violation of sound statistical practice.

Inference for small samples

Small samples require special care. We do not have enough observations to examine the distribution shapes, and only extreme outliers stand out. The power of significance tests tends to be low, and the margins of error of confidence intervals tend to be large. Despite these difficulties, we can often draw important conclusions from studies with small sample sizes. If the size of an effect is as large as it was in the wheat price example, it should still be evident even if the n's are small.

EXAMPLE 7.17 In the setting of Example 7.16, a quick survey collects prices from only 5 producers each month. The data are

Month	Price of wheat ($/bushel)				
September	$3.5900	$3.6150	$3.5950	$3.5725	$3.5825
July	$2.9200	$2.9675	$2.9175	$2.9250	$2.9325

The prices are reported to the nearest quarter of a cent. First, examine the distributions with a back-to-back stemplot after rounding each price to the nearest cent.

```
September             July
           2.9 | 2 2 3 3 7
           3.0 |
           3.1 |
           3.2 |
           3.3 |
           3.4 |
    9 8 7  3.5 |
      2 0  3.6 |
```

The pattern is clear. There is little variation among prices within each month, and the distributions for the two months are far apart relative to the within-month variation.

A significance test can confirm that the difference between months is too large to easily arise just by chance. We test

$$H_0: \mu_1 = \mu_2$$

$$H_a: \mu_1 \neq \mu_2$$

The price is higher in September ($t = 56.99$, df $= 7.55$, $P < 0.0001$). The difference in sample means is 65.9 cents.

TABLE 7.6

Ratio of current assets to current liabilities

Healthy firms						Failed firms		
1.50	0.10	1.76	1.14	1.84	2.21	0.82	0.89	1.31
2.08	1.43	0.68	3.15	1.24	2.03	0.05	0.83	0.90
2.23	2.50	2.02	1.44	1.39	1.64	1.68	0.99	0.62
0.89	0.23	1.20	2.16	1.80	1.87	0.91	0.52	1.45
1.91	1.67	1.87	1.21	2.05	1.06	1.16	1.32	1.17
0.93	2.17	2.61	3.05	1.52	1.93	0.42	0.48	0.93
1.95	2.61	1.11	0.95	0.96	2.25	0.88	1.10	0.23
2.73	1.56	2.73	0.90	2.12	1.42	1.11	0.19	0.13
1.62	1.76	2.22	2.80	1.85	0.96	2.03	0.51	1.12
1.71	1.02	2.50	1.55	1.69	1.64	0.92	0.26	1.15
1.03	1.80	0.67	2.44	2.30	2.21	0.13	0.88	0.09
1.96	1.81							

7.80 In what ways are companies that fail different from those that continue to do business? A study compared various characteristics of 68 healthy and 33 failed firms. One of the variables was the ratio of current assets to current liabilities. Roughly speaking, this is the amount that the firm is worth divided by what it owes.[30] The data are given in Table 7.6.

(a) Display the data so that the two distributions can be compared. Describe the shapes of the distributions and any important characteristics.

(b) We expect that failed firms will have a lower ratio. Describe and test appropriate hypotheses for these data. What do you conclude?

(c) It is not possible to do a randomized experiment for this kind of question. Explain why.

7.81 A market research firm supplies manufacturers with estimates of the retail sales of their products from samples of retail stores. Marketing managers are prone to look at the estimate and ignore sampling error. Suppose that an SRS of 70 stores this month shows mean sales of 53 units of a small appliance, with standard deviation 15 units. During the same month last year, an SRS of 55 stores gave mean sales of 50 units, with standard deviation 18 units. An increase from 50 to 53 is a rise of 6%. The marketing manager is happy because sales are up 6%.

(a) Use the two-sample t procedure to give a 95% confidence interval for the difference in mean number of units sold at all retail stores.

(b) Explain in language that the manager can understand why he cannot be certain that sales rose by 6%, and that in fact sales may even have dropped.

7.82 Does cocaine use by pregnant women cause their babies to have low birth weight? To study this question, birth weights of babies of women who tested positive for cocaine/crack during a drug-screening test were compared with the birth weights for babies whose mothers either tested negative or were not

Excel

A1	▼	*fx* t-Test: Two-Sample Assuming Unequal Variances

	A	B	C
1	**t-Test: Two-Sample Assuming Unequal Variances**		
2			
3		*Variable 1*	*Variable 2*
4	Mean	2.9325	3.591
5	Variance	0.000415625	0.000251875
6	Observations	5	5
7	Hypothesized Mean Difference	0	
8	df	8	
9	t Stat	-56.9921639	
10	P(T<=t) one-tail	4.99E-12	
11	t Critical one-tail	1.85954832	
12	P(T<=t) two-tail	9.97E-12	
13	t Critical two-tail	2.306005626	

◄ ► ►│ \ **Sheet1** / Sheet2 / Sheet3 /

Ready

SPSS

Group Statistics

	MONTH	N	Mean	Std. Deviation	Std. Error Mean
PRICE	JULY	5	2.932500	2.03869E-02	9.11729E-03
	SEPT	5	3.591000	1.58706E-02	7.09753E-03

Independent Samples Test
t-test for
Equality of
Means

		t	df	sig (2-tailed)	Mean Difference	Std. Error Difference	95% Confidence Interval of the Difference	
							Lower	Upper
PRICE	Equal variances assumed	−56.99	8	.000	−.6585	1.155E-02	−.685	−.632
	Equal variances not assumed	−56.99	7.546	.000	−6585	1.155E-02	−.685	−.632

FIGURE 7.14 Excel, SPSS, Minitab, and SAS output for Example 7.17. *(continued)*

Minitab

Two sample T-Test and Confidnece Interval

```
Two sample T for Price

Month    N     Mean      StDev     SE MEAN
July     5     2.9325    0.204     0.0091
SEPT     5     3.5910    0.0159    0.0071


95% CI for mu (JULY) -mu (SEPT): (-0.6858, -0.6312
T-Test mu (JULY) =mu (SEPT) (VS not =): T=-56.99 P=
0.0000 DF=7
```

SAS

```
                        The TTEST Procedure

                           Statistics


                            Lower CL            Upper CL
Variable   month     N      Mean       Mean      Mean      Std Err

price      july      5      2.9072     2.9325    2.9578    0.0091
price      sept      5      3.5713     3.591     3.6107    0.0071
price      Diff (1-2)       -0.685     -0.659    -0.632    0.0116

                            T-Tests
Variable   Method          Variances     DF    t Value    Pr>  |t|
price      Pooled          Equal          8     -56.99     < .0001
price      Satterthwaite   Unequal      7.55    -56.99
```

FIGURE 7.14   *(Continued)* Excel, SPSS, Minitab, and SAS output for Example 7.17.

Figure 7.14 gives outputs for this analysis from several software systems. Although the formats differ, the basic information is the same. All report the sample sizes, the sample means and standard deviations (or variances), the t statistic, and its P-value. All agree that the P-value is very small, though some give more detail than others. Software often labels the groups in alphabetical order. In this example, July is then the first population and $t = -56.99$, the negative of our result. Always check the means first and report the statistic (you may need to change the sign) in an appropriate way. Be sure to also mention the size of the effect you observed, such as "The mean price for September was 65.9 cents higher than in July."

SPSS and SAS report the results of *two t* procedures: a special procedure that assumes that the two population variances are equal and the general two-sample procedure that we have just studied. We don't recommend the "equal-variances" procedures, but we describe them later, in the section on pooled two-sample t procedures.

Software approximation for the degrees of freedom*

We noted earlier that the two-sample t statistic does not have an exact t distribution. Moreover, the exact distribution changes as the unknown population standard deviations σ_1 and σ_2 change. However, the distribution can be approximated by a t distribution with degrees of freedom given by

$$df = \frac{\left(\dfrac{s_1^2}{n_1} + \dfrac{s_2^2}{n_2}\right)^2}{\dfrac{1}{n_1 - 1}\left(\dfrac{s_1^2}{n_1}\right)^2 + \dfrac{1}{n_2 - 1}\left(\dfrac{s_2^2}{n_2}\right)^2}$$

This is the approximation used by most statistical software. It is quite accurate when both sample sizes n_1 and n_2 are 5 or larger.

EXAMPLE 7.18   For the DRP study of Example 7.14, the following table summarizes the data:

Group	n	$\bar{x}$	s
1	21	51.48	11.01
2	23	41.52	17.15

For greatest accuracy, we will use critical points from the t distribution with degrees of freedom given by the equation above:

$$df = \frac{\left(\dfrac{11.01^2}{21} + \dfrac{17.15^2}{23}\right)^2}{\dfrac{1}{20}\left(\dfrac{11.01^2}{21}\right)^2 + \dfrac{1}{22}\left(\dfrac{17.15^2}{23}\right)^2}$$

$$= \frac{344.486}{9.099} = 37.86$$

This is the value that we reported in Examples 7.14 and 7.15 where we gave the results produced by software.

The number df given by the above approximation is always at least as large as the smaller of $n_1 - 1$ and $n_2 - 1$. On the other hand, df is never larger than the sum $n_1 + n_2 - 2$ of the two individual degrees of freedom. The number of degrees of freedom is generally not a whole number. There is a t distribution with any positive degrees of freedom, even though Table D contains entries only for whole-number degrees of freedom. When df is small and is not a whole number, interpolation between entries in Table D may be needed to obtain an accurate critical value or P-value. Because of this and the need to calculate df, we do not recommend regular use of this approximation if a computer is not doing the arithmetic. With a computer, however, the more accurate procedures are painless.

---

*This material can be omitted unless you are using statistical software and wish to understand what the software does.

The pooled two-sample t procedures*

There is one situation in which a t statistic for comparing two means has exactly a t distribution. Suppose that the two normal population distributions have the *same* standard deviation. In this case we need substitute only a single standard error in a z statistic, and the resulting t statistic has a t distribution. We will develop the z statistic first, as usual, and from it the t statistic.

Call the common—and still unknown—standard deviation of both populations σ. Both sample variances s_1^2 and s_2^2 estimate σ^2. The best way to combine these two estimates is to average them with weights equal to their degrees of freedom. This gives more weight to the information from the larger sample, which is reasonable. The resulting estimator of σ^2 is

$$s_p^2 = \frac{(n_1 - 1)s_1^2 + (n_2 - 1)s_2^2}{n_1 + n_2 - 2}$$

pooled estimator of σ^2   This is called the **pooled estimator of σ^2** because it combines the information in both samples.

When both populations have variance σ^2, the addition rule for variances says that $\bar{x}_1 - \bar{x}_2$ has variance equal to the *sum* of the individual variances, which is

$$\frac{\sigma^2}{n_1} + \frac{\sigma^2}{n_2} = \sigma^2 \left(\frac{1}{n_1} + \frac{1}{n_2} \right)$$

The standardized difference of means in this equal-variance case is therefore

$$z = \frac{(\bar{x}_1 - \bar{x}_2) - (\mu_1 - \mu_2)}{\sigma \sqrt{\dfrac{1}{n_1} + \dfrac{1}{n_2}}}$$

This is a special two-sample z statistic for the case in which the populations have the same σ. Replacing the unknown σ by the estimate s_p gives a t statistic. The degrees of freedom are $n_1 + n_2 - 2$, the sum of the degrees of freedom of the two sample variances. This statistic is the basis of the pooled two-sample t inference procedures.

THE POOLED TWO-SAMPLE t PROCEDURES

Suppose that an SRS of size n_1 is drawn from a normal population with unknown mean μ_1 and that an independent SRS of size n_2 is drawn from another normal population with unknown mean μ_2. Suppose also that the two populations have the same standard deviation. A level C confidence interval for $\mu_1 - \mu_2$ is

$$(\bar{x}_1 - \bar{x}_2) \pm t^* s_p \sqrt{\frac{1}{n_1} + \frac{1}{n_2}}$$

Here t^* is the value for the $t(n_1 + n_2 - 2)$ density curve with area C between $-t^*$ and t^*.

---

*This section can be omitted if desired, but it should be read if you plan to read Chapters 12 and 13.

To test the hypothesis $H_0: \mu_1 = \mu_2$, compute the pooled two-sample t statistic

$$ t = \frac{\bar{x}_1 - \bar{x}_2}{s_p\sqrt{\dfrac{1}{n_1} + \dfrac{1}{n_2}}} $$

In terms of a random variable T having the $t(n_1 + n_2 - 2)$ distribution, the P-value for a test of H_0 against

$$H_a: \mu_1 > \mu_2 \quad \text{is} \quad P(T \geq t)$$
$$H_a: \mu_1 < \mu_2 \quad \text{is} \quad P(T \leq t)$$
$$H_a: \mu_1 \neq \mu_2 \quad \text{is} \quad 2P(T \geq |t|)$$

EXAMPLE 7.19   Does increasing the amount of calcium in our diet reduce blood pressure? Examination of a large sample of people revealed a relationship between calcium intake and blood pressure, but such observational studies do not establish causation. Animal experiments then showed that calcium supplements do reduce blood pressure in rats, justifying an experiment with human subjects. A randomized comparative experiment gave one group of 10 black men a calcium supplement for 12 weeks. The control group of 11 black men received a placebo that appeared identical. (In fact, a block design with black and white men as the blocks was used. We will look only at the results for blacks, because the earlier survey suggested that calcium is more effective for blacks.) The experiment was double-blind. Table 7.5 gives the seated systolic (heart contracted) blood pressure for all subjects at the beginning and end of the 12-week period, in millimeters (mm) of mercury. Because the researchers were interested in decreasing blood pressure, Table 7.5 also shows the decrease for each subject. An increase appears as a negative entry.[22]

TABLE 7.5

Seated systolic blood pressure

Calcium group			Placebo group		
Begin	End	Decrease	Begin	End	Decrease
107	100	7	123	124	−1
110	114	−4	109	97	12
123	105	18	112	113	−1
129	112	17	102	105	−3
112	115	−3	98	95	3
111	116	−5	114	119	−5
107	106	1	119	114	5
112	102	10	114	112	2
136	125	11	110	121	−11
102	104	−2	117	118	−1
			130	133	−3

FIGURE 7.15 Normal quantile plots for the decrease in blood pressure from Table 7.5.

As usual, we first examine the data. To compare the effects of the two treatments, take the response variable to be the amount of the decrease in blood pressure. Inspection of the data reveals that there are no outliers. Normal quantile plots (Figure 7.15) give a more detailed picture. The calcium group has a somewhat short left tail, but there are no departures from normality that will prevent use of t procedures. To examine the question of the researchers who collected these data, we perform a significance test.

EXAMPLE 7.20 Take Group 1 to be the calcium group and Group 2 to be the placebo group. The evidence that calcium lowers blood pressure more than a placebo is assessed by testing

$$H_0: \mu_1 = \mu_2$$
$$H_a: \mu_1 > \mu_2$$

Here are the summary statistics for the decrease in blood pressure:

Group	Treatment	n	$\bar{x}$	s
1	Calcium	10	5.000	8.743
2	Placebo	11	−0.273	5.901

The calcium group shows a drop in blood pressure, and the placebo group has a small increase. The sample standard deviations do not rule out equal population standard deviations. A difference this large will often arise by chance in samples this small. We are willing to assume equal population standard deviations. The pooled sample variance is

$$s_p^2 = \frac{(n_1 - 1)s_1^2 + (n_2 - 1)s_2^2}{n_1 + n_2 - 2}$$

$$= \frac{(9)(8.743)^2 + (10)(5.901)^2}{10 + 11 - 2} = 54.536$$

so that

$$s_p = \sqrt{54.536} = 7.385$$

The pooled two-sample t statistic is

$$t = \frac{\bar{x}_1 - \bar{x}_2}{s_p\sqrt{\dfrac{1}{n_1} + \dfrac{1}{n_2}}}$$

$$= \frac{5.000 - (-0.273)}{7.385\sqrt{\dfrac{1}{10} + \dfrac{1}{11}}}$$

$$= \frac{5.273}{3.227} = 1.634$$

df = 19

p	0.10	0.05
t^*	1.328	1.729

The P-value is $P(T \geq 1.634)$, where T has the $t(19)$ distribution. From Table D we can see that P falls between the $\alpha = 0.10$ and $\alpha = 0.05$ levels. Statistical software gives the exact value $P = 0.059$. The experiment found evidence that calcium reduces blood pressure, but the evidence falls a bit short of the traditional 5% and 1% levels.

Sample size strongly influences the P-value of a test. An effect that fails to be significant at a specified level α in a small sample can be significant in a larger sample. In the light of the rather small samples in Example 7.20, the evidence for some effect of calcium on blood pressure is rather good. The published account of the study combined these results for blacks with the results for whites and adjusted for pretest differences among the subjects. Using this more detailed analysis, the researchers were able to report the P-value $P = 0.008$.

Of course, a P-value is almost never the last part of a statistical analysis. To make a judgment regarding the size of the effect of calcium on blood pressure, we need a confidence interval.

EXAMPLE 7.21   We estimate that the effect of calcium supplementation is the difference between the sample means of the calcium and the placebo groups, $\bar{x}_1 - \bar{x}_2 = 5.273$ mm. A 90% confidence interval for $\mu_1 - \mu_2$ uses the critical value $t^* = 1.729$ from the $t(19)$ distribution. The interval is

$$(\bar{x}_1 - \bar{x}_2) \pm t^* s_p \sqrt{\frac{1}{n_1} + \frac{1}{n_2}} = [5.000 - (-0.273)] \pm (1.729)(7.385)\sqrt{\frac{1}{10} + \frac{1}{11}}$$

$$= 5.273 \pm 5.579$$

We are 90% confident that the difference in means is in the interval $(-0.306, 10.852)$. The calcium treatment reduced blood pressure by about 5.3 mm more than a placebo on the average, but the margin of error for this estimate is 5.6 mm.

The pooled two-sample t procedures are anchored in statistical theory and so have long been the standard version of the two-sample t in textbooks. *But they require the assumption that the two unknown population standard deviations are equal.* As we shall see in Section 7.3, this assumption is hard to verify. The pooled t procedures are therefore a bit risky. They are reasonably robust against both nonnormality and unequal standard deviations when the sample sizes are nearly the same. When the samples are quite different in size, the pooled t procedures become sensitive to unequal standard deviations and should be used with caution unless the samples are large. Unequal standard deviations are quite common. In particular, it is not unusual for the spread of data to increase when the center moves up. Statistical software often calculates both the pooled and the unpooled t statistics, as in Figure 7.14.

SECTION 7.2 | Summary

Significance tests and confidence intervals for the difference of the means μ_1 and μ_2 of two normal populations are based on the difference $\bar{x}_1 - \bar{x}_2$ of the sample means from two independent SRSs. Because of the central limit theorem, the resulting procedures are approximately correct for other population distributions when the sample sizes are large.

When independent SRSs of sizes n_1 and n_2 are drawn from two normal populations with parameters μ_1, σ_1 and μ_2, σ_2 the **two-sample z statistic**

$$z = \frac{(\bar{x}_1 - \bar{x}_2) - (\mu_1 - \mu_2)}{\sqrt{\dfrac{\sigma_1^2}{n_1} + \dfrac{\sigma_2^2}{n_2}}}$$

has the $N(0, 1)$ distribution.

The **two-sample t statistic**

$$t = \frac{(\bar{x}_1 - \bar{x}_2) - (\mu_1 - \mu_2)}{\sqrt{\dfrac{s_1^2}{n_1} + \dfrac{s_2^2}{n_2}}}$$

does *not* have a t distribution. However, good approximations are available.

Conservative inference procedures for comparing μ_1 and μ_2 are obtained from the two-sample t statistic by using the $t(k)$ distribution with degrees of freedom k equal to the smaller of $n_1 - 1$ and $n_2 - 1$.

More accurate probability values can be obtained by estimating the degrees of freedom from the data. This is the usual procedure for statistical software.

An approximate level C **confidence interval** for $\mu_1 - \mu_2$ is given by

$$(\bar{x}_1 - \bar{x}_2) \pm t^* \sqrt{\frac{s_1^2}{n_1} + \frac{s_2^2}{n_2}}$$

Here, t^* is the value for the $t(k)$ density curve with area C between $-t^*$ and t^*, where k is computed from the data by software or is the smaller of $n_1 - 1$ and $n_2 - 1$. The quantity

$$t^* \sqrt{\frac{s_1^2}{n_1} + \frac{s_2^2}{n_2}}$$

is the **margin of error.**

Significance tests for $H_0: \mu_1 = \mu_2$ use the **two-sample t statistic**

$$t = \frac{\bar{x}_1 - \bar{x}_2}{\sqrt{\dfrac{s_1^2}{n_1} + \dfrac{s_2^2}{n_2}}}$$

The P-value is approximated using the $t(k)$ where k is estimated from the data using software or is the smaller of $n_1 - 1$ and $n_2 - 1$.

The guidelines for practical use of two-sample t procedures are similar to those for one-sample t procedures. Equal sample sizes are recommended.

If we can assume that the two populations have equal variances, **pooled two-sample t procedures** can be used. These are based on the **pooled estimator**

$$s_p^2 = \frac{(n_1 - 1)s_1^2 + (n_2 - 1)s_2^2}{n_1 + n_2 - 2}$$

of the unknown common variance and the $t(n_1 + n_2 - 2)$ distribution.

SECTION 7.2 | Exercises

In exercises that call for two-sample t procedures, you may use either of the two approximations for the degrees of freedom that we have discussed: the value given by your software or the smaller of $n_1 - 1$ and $n_2 - 1$. Be sure to state clearly which approximation you have used.

7.53   A multimedia program designed to improve dietary behavior among low-income women was evaluated by comparing women who were randomly assigned to intervention and control groups. The intervention was a 30-minute session in a computer kiosk in the Food Stamp office.[23] One of the outcomes was the score on a knowledge test taken about 2 months after the program.

Here is a summary of the data:

Group	n	$\bar{x}$	s
Intervention	165	5.08	1.15
Control	212	4.33	1.16

(a) The test had six multiple-choice items that were scored as correct or incorrect, so the total score was an integer between 0 and 6. Do you think that these data are normally distributed? Explain why or why not.

(b) Is it appropriate to use the two-sample t procedures that we studied in this section to analyze these data? Give reasons for your answer.

(c) Describe appropriate null and alternative hypotheses for evaluating the intervention. Some people would prefer a two-sided alternative in this situation while others would use a one-sided significance test. Give reasons for each point of view.

(d) Carry out the significance test. Report the test statistic with the degrees of freedom and the P-value. Write a short summary of your conclusion.

(e) Find a 95% confidence interval for the difference between the two means. Compare the information given by the interval with the information given by the significance test.

(f) The women in this study were all residents of Durham, North Carolina. To what extent do you think the results can be generalized to other populations?

7.54 Self-efficacy is a general concept that measures how well we think we can control different situations. In the study described in the previous exercise, the participants were asked, "How sure are you that you can eat foods low in fat over the next month?" The response was measured on a five-point scale with 1 corresponding to "not sure at all" and 5 corresponding to "very sure." Here is a summary of the self-efficacy scores obtained about 2 months after the intervention:

Group	n	$\bar{x}$	s
Intervention	165	4.10	1.19
Control	212	3.67	1.12

Analyze the data using the questions in the previous exercise as a guide.

7.55 Exposure to dust at work can lead to lung disease later in life. One study measured the workplace exposure of tunnel construction workers.[24] Part of the study compared 115 drill and blast workers with 220 outdoor concrete workers. Total dust exposure was measured in milligram years per cubic meter ($mg.y/m^3$). The mean exposure for the drill and blast workers was 18.0 $mg.y/m^3$ with a standard deviation of 7.8 $mg.y/m^3$. For the outdoor concrete workers, the corresponding values were 6.5 $mg.y/m^3$ and 3.4 $mg.y/m^3$.

(a) The sample included all workers for a tunnel construction company who received medical examinations as part of routine health checkups. Discuss the extent to which you think these results apply to other similar types of workers.

(b) Use a 95% confidence interval to describe the difference in the exposures. Write a sentence that gives the interval and provides the meaning of the 95% confidence.

(c) Test the null hypothesis that the exposures for these two types of workers are the same. Justify your choice of a one-sided or two-sided alternative. Report the test statistic, the degrees of freedom, and the P-value. Give a short summary of your conclusion.

(d) The authors of the article describing these results note that the distributions are somewhat skewed. Do you think that this fact makes your analysis invalid? Give reasons for your answer.

7.56 Not all dust particles that are in the air around us cause problems for our lungs. Some particles are too large and stick to other areas of our body before they can get to our lungs. Others are so small that we can breathe them in and out and they will not deposit on our lungs. The researchers in the study described in the previous exercise also measured respirable dust. This is dust that deposits in our lungs when we breathe it. For the drill and blast workers, the mean exposure to respirable dust was 6.3 mg.y/m^3 with a standard deviation of 2.8 mg.y/m^3. The corresponding values for the outdoor concrete workers were 1.4 mg.y/m^3 and 0.7 mg.y/m^3. Analyze these data using the questions in the previous exercise as a guide.

7.57 A recent study of food portion sizes reported that over a 17-year period, the average size of a soft drink consumed by Americans aged 2 years and older increased from 13.1 ounces (oz) to 19.9 oz. The authors state that the difference is statistically significant with $P < 0.01$.[25] Explain what additional information you would need to compute a confidence interval for the increase, and outline the procedure that you would use for the computations. Do you think that a confidence interval would provide useful additional information? Explain why or why not.

7.58 The results in the previous exercise were based on two national surveys with a very large number of individuals. Here is a study that also looked at beverage consumption but the sample sizes are much smaller. One part of this study compared 20 children who were 7 to 10 years old with 5 who were 11 to 13.[26] The younger children consumed an average of 8.2 oz of sweetened drinks per day while the older ones averaged 14.5 oz. The standard deviations were 10.7 oz and 8.2 oz respectively.

(a) Do you think that it is reasonable to assume that these data are normally distributed? Explain why or why not. (*Hint:* Think about the 68–95–99.7% rule.)

(b) Using the methods in this section, test the null hypothesis that the two groups of children consume equal amounts of sweetened drinks versus the two-sided alternative. Report all details of the significance-testing procedure with your conclusion.

(c) Give a 95% confidence interval for the difference in means.

(d) Do you think that the analyses performed in parts (b) and (c) are appropriate for these data? Explain why or why not.

(e) The children in this study were all participants in an intervention study at the Cornell Summer Day Camp at Cornell University. To what extent do you think that these results apply to other groups of children?

7.59 In each of the following situations explain what is wrong and why.

(a) A researcher wants to test $H_0: \bar{x}_1 = \bar{x}_2$ versus the two-sided alternative $H_1: \bar{x}_1 \neq \bar{x}_2$.

(b) A study recorded the scores of 20 children who were similar in age. The scores of the 10 boys in the study were compared with the scores of all 20 children using the two-sample methods of this section.

(c) A two-sample t statistic gave a P-value of 0.96. From this you can reject the null hypothesis with 95% confidence.

7.60 For each of the following, answer the question and give a short explanation of your reasoning.

(a) A 95% confidence interval for the difference between two means is reported as $(1.6, 2.3)$. What can you conclude about the results of a significance test of the null hypothesis that the population means are equal versus the two-sided alternative?

(b) Will larger samples generally give a larger or smaller margin of error for the difference between two sample means?

7.61 For each of the following, answer the question and give a short explanation of your reasoning.

(a) A significance test for comparing two means gave $t = -3.11$ with 23 degrees of freedom. Can you reject the null hypothesis that the μ's are equal versus the two-sided alternative at the 5% significance level?

(b) Answer part (a) for the one-sided alternative that the difference in means is positive.

7.62 Assume $\bar{x}_1 = 100$, $\bar{x}_2 = 120$, $s_1 = 10$, $s_2 = 12$, $n_1 = 50$, and $n_2 = 50$. Find a 95% confidence interval for the difference in the corresponding values of μ. Does this interval include more or fewer values than a 99% confidence interval? Explain your answer.

7.63 You want to compare the daily sales for two different designs of Web pages for your Internet business. You assign the next 60 days to either Design A or Design B, 30 days to each.

(a) Would you use a one-sided or two-sided significance test for this problem? Explain your choice.

(b) If you use Table D to find the critical value, what are the degrees of freedom?

(c) The t statistic for comparing the mean sales is 2.06. If you use Table D, what P-value would you report? What would you conclude?

7.64 If you perform the significance test in the previous exercise using level $\alpha = 0.05$, how large (positive or negative) must the t statistic be to reject the null hypothesis that the two designs give the same average sales?

7.65 Refer to the previous two exercises. You realize that day of the week may affect online sales. To compare two Web page designs, you choose two successive weeks in the middle of a month. You flip a coin to assign one Monday to the first design and the other Monday to the second. You repeat this for each of the seven days of the week. You now have 7 sales amounts for each design. It is *incorrect* to use the two-sample t test to see if the mean sales differ for the two designs. Carefully explain why.

7.66 The purchasing department has suggested that all new computer monitors for your company should be flat screens. You want data to assure you that employees will like the new screens. The next 20 employees needing a new computer are the subjects for an experiment.

(a) Label the employees 01 to 20. Randomly choose 10 to receive flat screens. The remaining 10 get standard monitors.

(b) After a month of use, employees express their satisfaction with their new monitors by responding to the statement "I like my new monitor" on a scale from 1 to 5, where 1 represents "strongly disagree," 2 is "disagree," 3 is "neutral," 4 is "agree," and 5 stands for "strongly agree." The employees with the flat screens have average satisfaction 4.8 with standard deviation 0.7. The employees with the standard monitors have average 3.0 with standard deviation 1.5. Give a 95% confidence interval for the difference in the mean satisfaction scores for all employees.

(c) Would you reject the null hypothesis that the mean satisfaction for the two types of monitors is the same versus the two-sided alternative at significance level 0.05? Use your confidence interval to answer this question. Explain why you do not need to calculate the test statistic.

7.67 Refer to the previous exercise. A coworker suggested that you give the flat screens to the next 10 employees who need new screens and the standard monitor to the following 10. Explain why your randomized design is better.

7.68 Does bread lose its vitamins when stored? Small loaves of bread were prepared with flour that was fortified with a fixed amount of vitamins. After baking, the vitamin C content of two loaves was measured. Another two loaves were baked at the same time, stored for three days, and then the vitamin C content was measured. The units are milligrams per hundred grams of flour (mg/100 g).[27] Here are the data:

Immediately after baking:	47.62,	49.79
Three days after baking:	21.25,	22.34

(a) When bread is stored, does it lose vitamin C? To answer this question, perform a two-sample t test for these data. Be sure to state your hypotheses, the test statistic with degrees of freedom, and the P-value.

(b) Give a 90% confidence interval for the amount of vitamin C lost.

7.69 Suppose that the researchers in the previous exercise could have measured the same two loaves of bread immediately after baking and again after three days. Assume that the data given had come from this study design. (Assume that the values given in the previous exercise are for first loaf and second loaf from left to right.)

(a) Explain carefully why your analysis in the previous exercise is *not correct* now, even though the data are the same.

(b) Redo the analysis for the design based on measuring the same loaves twice.

7.70 Refer to Exercise 7.68. The amount of vitamin E (in mg/100 g of flour) in the same loaves was also measured. Here are the data:

Immediately after baking:	94.6,	96.0
Three days after baking:	97.4,	94.3

(a) When bread is stored, does it lose vitamin E? To answer this question, perform a two-sample t test for these data. Be sure to state your hypotheses, the test statistic with degrees of freedom, and the P-value.

(b) Give a 90% confidence interval for the amount of vitamin E lost.

7.71 Refer to Exercises 7.68 and 7.70. Some people claim that significance tests with very small samples never lead to rejection of the null hypothesis. Discuss this claim using the results of these two exercises.

7.72 Corporate advertising tries to enhance the image of the corporation. A study compared two ads from two sources, the *Wall Street Journal* and the *National Enquirer*. Subjects were asked to pretend that their company was considering a major investment in Performax, the fictitious sportswear firm in the ads. Each subject was asked to respond to the question "How trustworthy was the source in the sportswear company ad for Performax?" on a 7-point scale. Higher values indicated more trustworthiness.[28] Here is a summary of the results:

Ad source	n	$\bar{x}$	s
Wall Street Journal	66	4.77	1.50
National Enquirer	61	2.43	1.64

(a) Compare the two sources of ads using a t test. Be sure to state your null and alternative hypotheses, the test statistic with degrees of freedom, the P-value, and your conclusion.

(b) Give a 95% confidence interval for the difference.

(c) Write a short paragraph summarizing the results of your analyses.

7.73 A friend has performed a significance test of the null hypothesis that two means are equal. His report states that the null hypothesis is rejected in favor of the alternative that the first mean is larger than the second. In a presentation on his work, he notes that the first sample mean was larger than the second mean and this is why he chose this particular one-sided alternative.

(a) Explain what is wrong with your friend's procedure and why.

(b) Suppose he reported $t = 1.70$ with a P-value of 0.06. What is the correct P-value that he should report?

7.74 You use statistical software to perform a significance test of the null hypothesis that two means are equal. The software reports P-values for the two-sided alternative. Your alternative is that the first mean is greater than the second mean.

(a) The software reports $t = 1.81$ with a P-value of 0.07. Would you reject H_0 with $\alpha = 0.05$? Explain your answer.

(b) The software reports $t = -1.81$ with a P-value of 0.07. Would you reject H_0 with $\alpha = 0.05$? Explain your answer.

7.75 The study of 584 longleaf pine trees in the Wade Tract in Thomas County, Georgia, had several purposes. Are trees in one part of the tract more or less like trees in any other part of the tract or are there differences? In Example 6.1 (page 382) we examined how the trees were distributed in the tract and found that the pattern was not random. In this exercise we will examine the sizes of the trees. In Exercise 7.1 we analyzed the sizes, measured as diameter at breast height (DBH), for a random sample of 40 trees. Here we divide the tract into northern and southern halves and take random samples of 30 trees from each half. Here are the diameters in centimeters (cm) of the sampled trees:

	27.8	14.5	39.1	3.2	58.8	55.5	25.0	5.4	19.0	30.6
North	15.1	3.6	28.4	15.0	2.2	14.2	44.2	25.7	11.2	46.8
	36.9	54.1	10.2	2.5	13.8	43.5	13.8	39.7	6.4	4.8

	44.4	26.1	50.4	23.3	39.5	51.0	48.1	47.2	40.3	37.4
South	36.8	21.7	35.7	32.0	40.4	12.8	5.6	44.3	52.9	38.0
	2.6	44.6	45.5	29.1	18.7	7.0	43.8	28.3	36.9	51.6

(a) Use a back-to-back stemplot and side-by-side boxplots to examine the data graphically. Describe the patterns in the data.

(b) Is it appropriate to use the methods of this section to compare the mean DBH of the trees in the north half of the tract with the mean DBH of trees in the south half? Give reasons for your answer.

(c) What are appropriate null and alternative hypotheses for comparing the two samples of tree DBHs? Give reasons for your choices.

(d) Perform the significance test. Report the test statistic, the degrees of freedom, and the *P*-value. Summarize your conclusion.

(e) Find a 95% confidence interval for the difference in mean DBHs. Explain how this interval provides additional information about this problem.

7.76 The Wade Tract can also be divided into eastern and western halves. Here are the DBHs of 30 randomly selected longleaf pine trees from each half:

East	23.5	43.5	6.6	11.5	17.2	38.7	2.3	31.5	10.5	23.7
	13.8	5.2	31.5	22.1	6.7	2.6	6.3	51.1	5.4	9.0
	43.0	8.7	22.8	2.9	22.3	43.8	48.1	46.5	39.8	10.9

West	17.2	44.6	44.1	35.5	51.0	21.6	44.1	11.2	36.0	42.1
	3.2	25.5	36.5	39.0	25.9	20.8	3.2	57.7	43.3	58.0
	21.7	35.6	30.9	40.6	30.7	35.6	18.2	2.9	20.4	11.4

Using the questions in the previous exercise, analyze these data.

7.77 Do piano lessons improve the spatial-temporal reasoning of preschool children? We examined this question in Exercises 7.29 and 7.30 (page 479) by analyzing the change in spatial-temporal reasoning of 34 preschool children after six months of piano lessons. Here we examine the same question by comparing the changes of those students with the changes of 44 children in a control group.[29] Here are the data for the children who took piano lessons:

2	5	7	−2	2	7	4	1	0	7	3	4	3	4	9	4	5
2	9	6	0	3	6	−1	3	4	6	7	−2	7	−3	3	4	4

The control group scores are

1	−1	0	1	−4	0	0	1	0	−1	0	1	1	−3	−2
4	−1	2	4	2	2	2	−3	−3	0	2	0	−1	3	−1
5	−1	7	0	4	0	2	1	−6	0	2	−1	0	−2	

(a) Display the data and summarize the distributions.

(b) Make a table with the sample size, the mean, the standard deviation, and the standard error of the mean for each of the two groups.

(c) Translate the question of interest into hypotheses, test them, and summarize your conclusions.

7.78 Refer to the previous exercise. Give a 95% confidence interval that describes the comparison between the children who took piano lessons and the controls.

7.79 Refer to Exercises 7.29 and 7.30 (page 479) and the previous two exercises. We have used four ways to address the question of interest. Discuss the relative merits of each approach.

tested, a group we call "other." Here are the summary statistics. The birth weights are measured in grams.[31]

Group	n	$\bar{x}$	s
Positive test	134	2733	599
Other	5974	3118	672

(a) Formulate appropriate hypotheses and carry out the test of significance for these data.

(b) Give a 95% confidence interval for the mean difference in birth weights.

(c) Discuss the limitations of the study design. What do you believe can be concluded from this study?

7.83 A study of iron deficiency among infants compared samples of infants following different feeding regimens. One group contained breast-fed infants, while the children in another group were fed a standard baby formula without any iron supplements. Here are summary results on blood hemoglobin levels at 12 months of age:[32]

Group	n	$\bar{x}$	s
Breast-fed	23	13.3	1.7
Formula	19	12.4	1.8

(a) Is there significant evidence that the mean hemoglobin level is higher among breast-fed babies? State H_0 and H_a and carry out a t test. Give the P-value. What is your conclusion?

(b) Give a 95% confidence interval for the mean difference in hemoglobin level between the two populations of infants.

(c) State the assumptions that your procedures in (a) and (b) require in order to be valid.

7.84 The Survey of Study Habits and Attitudes (SSHA) is a psychological test designed to measure the motivation, study habits, and attitudes toward learning of college students. These factors, along with ability, are important in explaining success in school. Scores on the SSHA range from 0 to 200. A selective private college gives the SSHA to an SRS of both male and female first-year students. The data for the women are as follows:

154	109	137	115	152	140	154	178	101
103	126	126	137	165	165	129	200	148

Here are the scores of the men:

108	140	114	91	180	115	126	92	169	146
109	132	75	88	113	151	70	115	187	104

(a) Examine each sample graphically, with special attention to outliers and skewness. Is use of a *t* procedure acceptable for these data?

(b) Most studies have found that the mean SSHA score for men is lower than the mean score in a comparable group of women. Test this supposition here. That is, state hypotheses, carry out the test and obtain a *P*-value, and give your conclusions.

(c) Give a 90% confidence interval for the mean difference between the SSHA scores of male and female first-year students at this college.

The following exercises concern optional material on the pooled two-sample t procedures and on the power of tests.

7.85   The knowledge test means for intervention and control groups were compared using the two-sample *t* procedures that do not assume equal standard deviations in Exercise 7.53 (page 504). Examine the standard deviations for the two groups and verify that it is appropriate to use the pooled procedures for these data. Compare the means using a significance test and find the 95% confidence interval for the difference using the pooled methods. How do the results compare with those you obtained in Exercise 7.53?

7.86   You used methods that do not require equal standard deviations when you analyzed the self-efficacy data in Exercise 7.54 (page 505). Can you justify using the pooled procedures for these data? Explain your answer. Analyze the data using these procedures and compare what you found in Exercise 7.54 with these results.

7.87   Refer to the Wade Tract DBH data in Exercise 7.75 (page 510), where we compared samples of trees in the northern half of the tract with those in the southern half. Because the standard deviations for the two samples are quite close, it is reasonable to analyze these data using the pooled procedures. Perform the significance test and find the 95% confidence interval for the difference in means using these methods. Summarize your results and compare them with what you found in Exercise 7.75.

7.88   Exercise 7.76 (page 511) compares the DBHs of samples of trees from the eastern and western halves of the Wade Tract. Is it reasonable to use the two-sample pooled procedures with these data? Analyze the data with these methods and compare your results with those you found in Exercise 7.76.

7.89   Use the Wade Tract data in Exercise 7.75 to calculate the software approximation to the degrees of freedom using the formula on page 498. Verify your calculation with software.

7.90   Use the Wade Tract data in Exercise 7.76 to calculate the software approximation to the degrees of freedom using the formula on page 498. Verify your calculation with software.

7.91   The data on occupational exposure to dust that we analyzed in Exercise 7.55 (page 505), come from two groups of workers that are quite different in size. This complicates the issue regarding pooling because the sample that is larger will dominate the calculations.

(a) Calculate the degrees of freedom approximation using the formula for the degrees of freedom given on page 498. Then verify your calculations with software.

(b) Find the pooled estimate of the standard deviation. Write a short summary comparing it with the estimates of the standard deviations that come from each group.

(c) Find the standard error of the difference in sample means that you would use for the method that does not assume equal variances. Do the same for the pooled approach. Compare these two estimates with each other.

(d) Perform the significance test and find the 95% confidence interval using the pooled methods. How do these results compare with those you found in Exercise 7.55?

(e) Exercise 7.56 has data for the same workers but for respirable dust. Here the standard deviations differ more than those in Example 7.55. Answer parts (a) through (d) for these data. Write a summary of what you have found in this exercise.

7.92 The analysis of the loss of vitamin C when bread is stored in Exercise 7.68 (page 508) is a rather unusual case involving very small sample sizes. There are only two observations per condition (immediately after baking and three days later). When the samples are so small, we have very little information to make a judgment about whether the population standard deviations are equal. The potential gain from pooling is large when the sample sizes are very small. Assume that we will perform a two-sided test using the 5% significance level.

(a) Find the critical value for the unpooled t test statistic that does not assume equal variances. Use the minimum of $n_1 - 1$ and $n_2 - 1$ for the degrees of freedom.

(b) Find the critical value for the pooled t test statistic.

(c) How does comparing these critical values show an advantage of the pooled test?

7.3   Optional Topics in Comparing Distributions*

In this section we discuss three topics that are related to the material that we have already covered in this chapter. If we can do inference for means, it is natural to ask if we can do something similar for spread. The answer is yes, but there are many cautions. We also discuss robustness and show how to find the power for the two-sample t test. If you plan to design studies, you should become familiar with this last topic.

---

*This section can be omitted without loss of continuity.

Inference for population spread

The two most basic descriptive features of a distribution are its center and spread. In a normal population, these aspects are measured by the mean and the standard deviation. We have described procedures for inference about population means for normal populations and found that these procedures are often useful for nonnormal populations as well. It is natural to turn next to inference about the standard deviations of normal populations. Our recommendation here is short and clear: Don't do it without expert advice.

We will describe the *F* test for comparing the spread of two normal populations. *Unlike the t procedures for means, the F test and other procedures for standard deviations are extremely sensitive to nonnormal distributions.* This lack of robustness does not improve in large samples. It is difficult in practice to tell whether a significant *F*-value is evidence of unequal population spreads or simply evidence that the populations are not normal. Consequently, we do not recommend use of inference about population standard deviations in basic statistical practice.[33]

It was once common to test equality of standard deviations as a preliminary to performing the pooled two-sample *t* test for equality of two population means. It is better practice to check the distributions graphically, with special attention to skewness and outliers, and to use the software-based two-sample *t* that does not require equal standard deviations. In the words of one distinguished statistician, "To make a preliminary test on variances is rather like putting to sea in a rowing boat to find out whether conditions are sufficiently calm for an ocean liner to leave port!"[34]

The *F* test for equality of spread

Because of the limited usefulness of procedures for inference about the standard deviations of normal distributions, we will present only one such procedure. Suppose that we have independent SRSs from two normal populations, a sample of size n_1 from $N(\mu_1, \sigma_1)$ and a sample of size n_2 from $N(\mu_2, \sigma_2)$. The population means and standard deviations are all unknown. The hypothesis of equal spread,

$$H_0: \sigma_1 = \sigma_2$$
$$H_a: \sigma_1 \neq \sigma_2$$

is tested by a simple statistic, the ratio of sample variances.

THE *F* STATISTIC AND *F* DISTRIBUTIONS

When s_1^2 and s_2^2 are sample variances from independent SRSs of sizes n_1 and n_2 drawn from normal populations, the *F* statistic

$$F = \frac{s_1^2}{s_2^2}$$

has the *F* distribution with $n_1 - 1$ and $n_2 - 1$ degrees of freedom when $H_0: \sigma_1 = \sigma_2$ is true.

F distributions   The **F distributions** are a family of distributions with two parameters: the degrees of freedom of the sample variances in the numerator and denominator of the F statistic. The F distributions are another of R. A. Fisher's contributions to statistics and are called F in his honor. Fisher introduced F statistics for comparing several means. We will meet these useful statistics in later chapters. The numerator degrees of freedom are always mentioned first. Interchanging the degrees of freedom changes the distribution, so the order is important. Our brief notation will be $F(j, k)$ for the F distribution with j degrees of freedom in the numerator and k in the denominator. The F distributions are not symmetric but are right-skewed. The density curve in Figure 7.16 illustrates the shape. Because sample variances cannot be negative, the F statistic takes only positive values and the F distribution has no probability below 0. The peak of the F density curve is near 1; values far from 1 in either direction provide evidence against the hypothesis of equal standard deviations.

Tables of F critical values are awkward because a separate table is needed for every pair of degrees of freedom j and k. Table E in the back of the book gives upper p critical values of the F distributions for p = 0.10, 0.05, 0.025, 0.01, and 0.001. For example, these critical values for the $F(9, 10)$ distribution shown in Figure 7.16 are

p	0.10	0.05	0.025	0.01	0.001
F^*	2.35	3.02	3.78	4.94	8.96

The skewness of the F distributions causes additional complications. In the symmetric normal and t distributions, the point with probability 0.05 below it is just the negative of the point with probability 0.05 above it. This is not true for F distributions. We therefore require either tables of both the upper and lower tails or means of eliminating the need for lower-tail critical values.

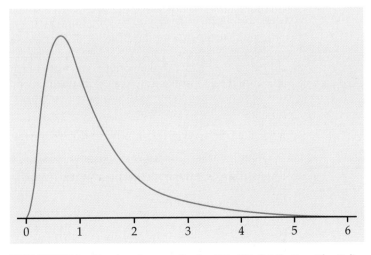

FIGURE 7.16   The density curve for the $F(9, 10)$ distribution. The F distributions are skewed to the right.

Statistical software that eliminates the need for tables is plainly very convenient. If you do not use statistical software, arrange the F test as follows:

1. Take the test statistic to be

$$F = \frac{\text{larger } s^2}{\text{smaller } s^2}$$

This amounts to naming the populations so that s_1^2 is the larger of the observed sample variances. The resulting F is always 1 or greater.

2. Compare the value of F with critical values from Table E. Then *double* the significance levels from the table to obtain the significance level for the two-sided F test.

The idea is that we calculate the probability in the upper tail and double to obtain the probability of all ratios on either side of 1 that are at least as improbable as that observed. Remember that the order of the degrees of freedom is important in using Table E.

EXAMPLE 7.22 Example 7.19 (page 500) recounts a medical experiment comparing the effects of calcium and a placebo on the blood pressure of black men. The analysis (Example 7.20) employed the pooled two-sample t procedures. Because these procedures require equal population standard deviations, it is tempting to first test

$$H_0: \sigma_1 = \sigma_2$$

$$H_a: \sigma_1 \neq \sigma_2$$

The larger of the two sample standard deviations is $s = 8.743$ from 10 observations. The other is $s = 5.901$ from 11 observations. The two-sided test statistic is therefore

$$F = \frac{\text{larger } s^2}{\text{smaller } s^2} = \frac{8.743^2}{5.901^2} = 2.195$$

We compare the calculated value $F = 2.20$ with critical points for the $F(9, 10)$ distribution. Table E shows that 2.20 is *less* than the 0.10 critical value of the $F(9, 10)$ distribution, which is $F^* = 2.35$. Doubling 0.10, we know that the observed F falls short of the 0.20 critical value. The results are not significant at the 20% level (or any lower level). Statistical software shows that the exact upper-tail probability is 0.118, and hence $P = 0.236$. *If* the populations were normal, the observed standard deviations would give little reason to suspect unequal population standard deviations. Because one of the populations shows some nonnormality, however, we cannot be fully confident of this conclusion.

Robustness of normal inference procedures

We have claimed that

- The t procedures for inference about means are quite robust against nonnormal population distributions. These procedures are particularly robust when the population distributions are symmetric and (for the two-sample case) when the two sample sizes are equal.

- The F test and other procedures for inference about variances are so lacking in robustness as to be of little use in practice.

Simulations with a large variety of nonnormal distributions support these claims. One set of simulations was carried out with samples of size 25 and used significance tests with fixed level $\alpha = 0.05$. The three types of tests studied were the one-sample and pooled two-sample t tests and the F test for comparing two variances.

The robustness of the one-sample and two-sample t procedures is remarkable. The true significance level remains between about 4% and 6% for a large range of populations. The t test and the corresponding confidence intervals are among the most reliable tools that statisticians use. Remember, however, that outliers can greatly disturb the t procedures. Also, two-sample procedures are less robust when the sample sizes are not similar.

The lack of robustness of the tests for variances is equally remarkable. The true significance levels depart rapidly from the target 5% as the population distribution departs from normality. The two-sided F test carried out with 5% critical values can have a true level of less than 1% or greater than 11% even in symmetric populations with no outliers. Results such as these are the basis for our recommendation that these procedures not be used.

The power of the two–sample t test

The two-sample t test is one of the most used statistical procedures. Unfortunately, because of inadequate planning, users frequently fail to find evidence for the effects that they believe to be true. Power calculations should be part of the planning of any statistical study. Information from a pilot study or previous research is needed.

In Section 7.1 (optional material), we learned how to find an approximation for the power of the one-sample t test. The basic concepts for the two-sample case are the same. Here, we give the exact method, which involves a new distribution, the **noncentral t distribution.** To perform the calculations, we simply need software to calculate probabilities for this distribution.

noncentral t distribution

We first present the method for the pooled two-sample t test, where the parameters are μ_1, μ_2, and the common standard deviation σ. Modifications to get approximate results when we do not pool are then described.

To find the power for the pooled two-sample t test, use the following steps. We consider only the case where the null hypothesis is $\mu_1 - \mu_2 = 0$.

1. Specify

    (a) an alternative that you consider important to detect; that is, a value for $\mu_1 - \mu_2$;

    (b) the sample sizes, n_1 and n_2;

    (c) the Type I error for a fixed significance level, α;

    (d) a guess at the standard deviation, σ.

2. Find the degrees of freedom df $= n_1 + n_2 - 2$ and the value of t^* that will lead to rejection of H_0.

noncentrality
parameter

3. Calculate the **noncentrality parameter**

$$\delta = \frac{|\mu_1 - \mu_2|}{\sigma\sqrt{\dfrac{1}{n_1} + \dfrac{1}{n_2}}}$$

4. Find the power as the probability that a noncentral t random variable with degrees of freedom df and noncentrality parameter δ will be greater than t^*. In SAS this is 1-PROBT(tstar, df, delta). If you do not have software that can perform this calculation, you can approximate the power as the probability that a standard normal random variable is greater than $t^* - \delta$, that is, $P(z > t^* - \delta)$, and use Table A.

Note that the denominator in the noncentrality parameter,

$$\sigma\sqrt{\frac{1}{n_1} + \frac{1}{n_2}}$$

is our guess at the standard error for the difference in the sample means. Therefore, if we wanted to assess a possible study in terms of the margin of error for the estimated difference, we would examine t^* times this quantity.

If we do not assume that the standard deviations are equal, we need to guess both standard deviations and then combine these for our guess at the standard error:

$$\sqrt{\frac{\sigma_1^2}{n_1} + \frac{\sigma_2^2}{n_2}}$$

This guess is then used in the denominator of the noncentrality parameter. For the degrees of freedom, the conservative approximation is appropriate.

EXAMPLE 7.23 In Example 7.20 we examined the effect of calcium on blood pressure by comparing the means of a treatment group and a placebo group using a two-sample t test. The P-value was 0.059, failing to achieve the usual standard of 0.05 for statistical significance. Suppose that we wanted to plan a new study that would provide convincing evidence, say at the 0.01 level, with high probability. Let's examine a study design with 45 subjects in each group ($n_1 = n_2 = 45$). Based on our previous results we choose $\mu_1 - \mu_2 = 5$ as an alternative that we would like to be able to detect with $\alpha = 0.01$. For σ we use 7.4, our pooled estimate from Example 7.20. The degrees of freedom are $n_1 + n_2 - 2 = 88$ and $t^* = 2.37$ for the significance test. The noncentrality parameter is

$$\delta = \frac{5}{7.4\sqrt{\dfrac{1}{45} + \dfrac{1}{45}}} = \frac{5}{1.56} = 3.21$$

Software gives the power as 0.7965, or 80%. The normal approximation gives 0.7983, a very accurate result. With this choice of sample sizes we would expect the margin of error for a 95% confidence interval ($t^* = 1.99$) for the difference in means to be

$$t^* \times 7.4\sqrt{\frac{1}{45} + \frac{1}{45}} = 1.99 \times 1.56 = 3.1$$

With software it is very easy to examine the effects of variations on a study design. In the above example, we might want to examine the power for $\alpha = 0.05$ and the effects of reducing the sample sizes.

SECTION 7.3 | Summary

Inference procedures for comparing the standard deviations of two normal populations are based on the **F statistic,** which is the ratio of sample variances:

$$F = \frac{s_1^2}{s_2^2}$$

If an SRS of size n_1 is drawn from the x_1 population and an independent SRS of size n_2 is drawn from the x_2 population, the F statistic has the **F distribution** $F(n_1 - 1, n_2 - 1)$ if the two population standard deviations σ_1 and σ_2 are in fact equal.

The **F test for equality of standard deviations** tests $H_0: \sigma_1 = \sigma_2$ versus $H_a: \sigma_1 \neq \sigma_2$ using the statistic

$$F = \frac{\text{larger } s^2}{\text{smaller } s^2}$$

and doubles the upper-tail probability to obtain the *P*-value.

The *t* procedures are quite **robust** when the distributions are not normal. The *F* tests and other procedures for inference about the spread of one or more normal distributions are so strongly affected by nonnormality that we do not recommend them for regular use.

The **power** of the pooled two-sample *t* test is found by first computing the critical value for the significance test, the degrees of freedom, and the **noncentrality parameter** for the alternative of interest. These are used to find the power from the **noncentral *t* distribution.** A normal approximation works quite well. Calculating margins of error for various study designs and assumptions is an alternative procedure for evaluating designs.

SECTION 7.3 | Exercises

In all exercises calling for use of the F test, assume that both population distributions are very close to normal. The actual data are not always sufficiently normal to justify use of the F test.

7.93   An *F* statistic will be used to compare two variances. How large does the ratio of the largest to the smallest variance need to be for the significance test to reject the null hypothesis that the population variances are the same in the following settings? Use the 5% level of significance.

(a) The two sample sizes are 5.

(b) The two sample sizes are 10.

(c) The two sample sizes are 26.

(d) What do you conclude?

7.94   Here are some summary statistics from two independent samples from normal distributions:

Sample	n	s^2
1	10	3.1
2	16	9.3

You want to test the null hypothesis that the two population standard deviations are equal versus the two-sided alternative at the 5% significance level.

(a) Calculate the test statistic.

(b) Find the appropriate value from Table E that you need to perform the significance test.

(c) What do you conclude?

7.95   Mean scores on a knowledge test are compared for two groups of women in Exercise 7.53 (page 504). Compare the standard deviations using an F test. What do you conclude? Comment on the normal assumption for these data. These standard deviations were so close that we are not particularly surprised at the result of the significance test. Assume that the sample standard deviation in the intervention is the value given in Exercise 7.53, 1.15. How large would the standard deviation in the control group need to be to reject the null hypothesis of equal standard deviations at the 5% level?

7.96   Compare the standard deviations of the self-efficacy scores in Exercise 7.54 (page 505). Give the test statistic, the degrees of freedom, and the P-value. Write a short summary of your analysis including comments on the assumptions for the test.

7.97   The two-sample problem in Exercise 7.55 (page 505) compares drill and blast workers with outdoor concrete workers with respect to the total dust that they are exposed to in the workplace. Here it may be useful to know whether or not the standard deviations differ in the two groups. Perform the F test and summarize the results. Are you concerned about the assumptions here? Explain why or why not.

7.98   Exercise 7.56 (page 506) is similar to Exercise 7.55 but the response variable here is exposure to dust particles that can enter and stay in the lungs. Compare the standard deviations with a significance test and summarize the results. Be sure to comment on the assumptions.

7.99   The diameters of trees in the Wade Tract for random samples selected from the north and south portions of the tract are compared in Exercise 7.75 (page 510). Are there statistically significant differences in the standard deviations for these two parts of the tract? Perform the significance test and summarize the results. Does the normal assumption appear reasonable for these data?

7.100 Tree diameters for the east and west halves of the Wade Tract are compared in Exercise 7.76 (page 511). Using the questions in the previous exercise as a guide, analyze these data.

7.101 Return to the SSHA data in Exercise 7.84 (page 513). SSHA scores are generally less variable among women than among men. We want to know whether this is true for this college.

(a) State H_0 and H_a. Note that H_a is one-sided in this case.

(b) Because Table E contains only upper critical values for F, a one-sided test requires that in calculating F the numerator s^2 belongs to the group that H_a claims to have the larger σ. Calculate this F.

(c) Compare F to the entries in Table E (no doubling of p) to obtain the P-value. Be sure the degrees of freedom are in the proper order. What do you conclude about the variation in SSHA scores?

7.102 We studied the loss of vitamin C when bread is stored in Exercise 7.68 (page 508). Recall that two loaves were measured immediately after baking and another two loaves were measured after three days of storage. These are very small sample sizes.

(a) Use Table E to find the value that the ratio of variances would have to exceed for us to reject the null hypothesis (at the 5% level) that the standard deviations are equal. What does this suggest about the power of the test?

(b) Perform the test and state your conclusion.

7.103 In Exercise 7.75 (page 510) DBH data for longleaf pine trees in two parts of the Wade Tract are compared. Suppose that you are planning a similar study where you will measure the diameters of longleaf pine trees. Based on Exercise 7.75, you are willing to assume that the standard deviation is 20 cm. Suppose that a difference in mean DBH of 10 cm or more would be important to detect. You will use a t statistic and a two-sided alternative for the comparison.

(a) Find the power if you randomly sample 20 trees from each area to be compared.

(b) Repeat the calculations for 60 trees in each sample.

(c) If you had to choose between the 20 or 60 trees per sample, which would be acceptable? Give reasons for your answer.

7.104 Refer to the previous exercise. Find the two standard deviations from Exercise 7.75. Do the same for the data in Exercise 7.76, which is a similar setting. These are somewhat smaller than the assumed value that you used in the previous exercise. Explain why it is generally a good idea to assume a standard deviation that is higher than you expect than one that is lower. Repeat the power calculations for some other reasonable values of σ and comment on the impact of this assumption for planning the new study.

7.105 In Exercise 7.82 (page 512) data on cocaine use and birth weight are summarized. The study has been criticized because of several design problems. Suppose that you are designing a new study. Based on the results in Exercise 7.82, you think that the true difference in mean birth weights may be about 350 grams (g); a difference this large is clinically important. For planning purposes assume that you will have 75 women in each group and that the common standard deviation is 650 g, a guess that is between the two standard deviations in Exercise 7.82. If you use a pooled two-sample t test with a Type I error of 0.05, what is the power of the test for this design?

7.106 Refer to the previous exercise. Repeat the power calculation for 20, 40, 60, 80, 100, and 120 women in each group. Plot the power versus the sample size and write a short summary of the results.

7.107 Refer to the previous two exercises. For each of the sample sizes considered, what is your guess at the margin of error for the 95% confidence interval for the difference in mean weights? Display these results with a graph or a sketch.

7.108 Refer to Exercise 7.72 (page 509), where we compared trustworthy ratings for ads from two different publications. Suppose that you are planning a similar study using two different publications that are not expected to show the differences seen when comparing the *Wall Street Journal* with the *National Enquirer*. You would like to detect a difference of 1.5 points using a two-sided significance test with a 5% level of significance. Based on Exercise 7.72, it is reasonable to use 1.6 as the value of the standard deviation for planning purposes.

(a) What is the power if you use sample sizes similar to those used in the previous study, for example, 65 for each publication?

(b) Repeat the calculations for 100 in each group.

(c) What sample size would you recommend for the new study?

CHAPTER 7 | Exercises

7.109 Individuals who consume large amounts of alcohol do not use the calories from this source as efficiently as calories from other sources. One study examined the effects of moderate alcohol consumption on body composition and the intake of other foods. Fourteen subjects participated in a crossover design where they either drank wine for the first 6 weeks and then abstained for the next 6 weeks or vice versa.[35] During the period when they drank wine, the subjects lost 0.4 kilograms (kg) of body weight; when they did not drink wine, they lost 1.1 kg. The standard deviation of the difference between the weight lost under these two conditions is 8.6 kg. During the wine period, they consumed an average of 2589 calories; with no wine, the mean consumption was 2575. The standard deviation of the difference was 210.

(a) Compute the differences in means and the standard errors for comparing body weight and caloric intake under the two experimental conditions.

(b) A report of the study indicated that there were no significant differences in these two outcome measures. Verify this result for each measure, giving the test statistic, degrees of freedom, and the *P*-value.

(c) One concern with studies such as this, with a small number of subjects, is that there may not be sufficient power to detect differences that are potentially important. Address this question by computing 95% confidence intervals for the two measures and discuss the information provided by the intervals.

(d) Here are some other characteristics of the study. The study periods lasted for 6 weeks. All subjects were males between the ages of 21 and 50 years who weighed between 68 and 91 kilograms (kg). They were all from the same city. During the wine period subjects were told to consume two 135 milliliter (ml) servings of red wine and no other alcohol. The entire six-week supply was given to each subject at the beginning of the period. During the other period, subjects were instructed to refrain from any use of alcohol. All subjects reported that they complied with these instructions except for three subjects, who said that they drank no more than 3 to 4 12-ounce bottles of beer during the no-alcohol period. Discuss how these factors could influence the interpretation of the results.

7.110   Healthy bones are continually being renewed by two processes. Through bone formation, new bone is built; through bone resorption, old bone is removed. If one or both of these processes is disturbed, by disease, aging, or space travel, for example, bone loss can be the result. Osteocalcin (OC) is a biochemical marker for bone formation: higher levels of bone formation are associated with higher levels of OC. A blood sample is used to measure OC, and it is much less expensive to obtain than direct measures of bone formation. The units are milligrams of OC per milliliter of blood (mg/ml). One study examined various biomarkers of bone turnover.[36] Here are the OC measurements on 31 healthy women aged 11 to 32 years who participated in this study:

68.9	56.3	54.6	31.2	36.4	31.4	52.8	38.4
35.7	76.5	44.4	40.2	77.9	54.6	9.9	20.6
20.0	17.2	24.2	20.9	17.9	19.7	15.9	20.8
8.1	19.3	16.9	10.1	47.7	30.2	17.2	

(a) Display the data with a stemplot or histogram and a boxplot. Describe the distribution.

(b) Find a 95% confidence interval for the mean OC. Comment on the suitability of using this procedure for these data.

7.111   Refer to the previous exercise. Tartrate resistant acid phosphatase (TRAP) is a biochemical marker for bone resorption that is also measured in blood. Here are the TRAP measurements, in units per liter (U/l), for the same 31 women:

19.4	25.5	19.0	9.0	19.1	14.6	25.2	14.6
28.8	14.9	10.7	5.9	23.7	19.0	6.9	8.1
9.5	6.3	10.1	10.5	9.0	8.8	8.2	10.3
3.3	10.1	9.5	8.1	18.6	14.4	9.6	

(a) Display the data with a stemplot or histogram and a boxplot. Describe the distribution.

(b) Find a 95% confidence interval for the mean TRAP. Comment on the suitability of using this procedure for these data.

7.112   Refer to Exercise 7.110 and the OC data for 31 women. Variables that measure concentrations such as this often have distributions that are skewed to the right. For this reason it is common to work with the logarithms of the measured values. Here are the OC values transformed with the (natural) log:

4.23	4.03	4.00	3.44	3.59	3.45	3.97	3.65
3.58	4.34	3.79	3.69	4.36	4.00	2.29	3.03
3.00	2.84	3.19	3.04	2.88	2.98	2.77	3.03
2.09	2.96	2.83	2.31	3.86	3.41	2.84	

(a) Display the data with a stemplot and a boxplot. Describe the distribution.

(b) Find a 95% confidence interval for the mean OC. Comment on the suitability of using this procedure for these data.

(c) Transform the mean and the endpoints of the confidence interval back to the original scale, mg/ml. Compare this interval with the one you computed in Exercise 7.110.

7.113   Refer to Exercise 7.111 and the TRAP data for 31 women. Variables that measure concentrations such as this often have distributions that are skewed to the right. For this reason it is common to work with the logarithms of the measured values. Here are the TRAP values transformed with the (natural) log:

2.97	3.24	2.94	2.20	2.95	2.68	3.23	2.68
3.36	2.70	2.37	1.77	3.17	2.94	1.93	2.09
2.25	1.84	2.31	2.35	2.20	2.17	2.10	2.33
1.19	2.31	2.25	2.09	2.92	2.67	2.26	

(a) Display the data with a stemplot and a boxplot. Describe the distribution.

(b) Find a 95% confidence interval for the mean TRAP. Comment on the suitability of using this procedure for these data.

(c) Transform the mean and the endpoints of the confidence interval back to the original scale, U/l. Compare this interval with the one you computed in Exercise 7.111.

7.114   The data used in Exercises 7.1 (page 471), 7.75, and 7.76 (page 511) were obtained by taking simple random samples from the 584 longleaf pine trees that were measured in the Wade Tract. The entire data set is given in the LONGLEAF data set. Find the 95% confidence interval for the mean DBH using the entire data set, and compare this interval with the one that you calculated in Exercise 7.1. Write a report about these data. Include comments on the effect of the sample size on the margin of error, the distribution of the data, the appropriateness of the normality-based methods for this problem, and the generalizability of the results to other similar stands of longleaf pine or other kinds of trees in this area of the United States and other areas.

7.115   Use the LONGLEAF data set to repeat the calculations that you performed in Exercises 7.75 and 7.76. Discuss the effect of the sample size on the results.

7.116   The DBH measures in the LONGLEAF data set do not appear to be normally distributed. Make a histogram of the data and a normal quantile plot if you have the software available. Mark the mean and the median on the histogram. Now, transform the data using a logarithm. Does this make the distribution appear to be normal? Use the same graphical summaries with the mean and the median marked on the histogram. Write a summary of your conclusions, paying particular attention to the use of data such as these for inference using the methods based on normal distributions.

7.117   A retailer entered into an exclusive agreement with a supplier who guaranteed to provide all products at competitive prices. The retailer eventually began to purchase supplies from other vendors who offered better prices. The original supplier filed a legal action claiming violation of the agreement. In defense, the retailer had an audit performed on a random sample of invoices. For each audited invoice, all purchases made from other suppliers were examined and the prices were compared with those offered by the original supplier. For each invoice, the percent of purchases for which the alternate supplier offered a lower price than the original supplier was recorded.[37] Here are the data:

0	100	0	100	33	34	100	48	78	100	77	100	38
68	100	79	100	100	100	100	100	100	89	100	100	

Report the average of the percents with a 95% margin of error. Do the sample invoices suggest that the original supplier's prices are not competitive on the average?

7.118   In a study of the effectiveness of weight-loss programs, 47 subjects who were at least 20% overweight took part in a group support program for 10 weeks. Private weighings determined each subject's weight at the beginning of the program and 6 months after the program's end. The matched pairs *t* test was used to assess the significance of the average weight loss. The paper reporting the study said, "The subjects lost a significant amount of weight over time, $t(46) = 4.68, p < 0.01$." It is common to report the results of statistical tests in this abbreviated style.[38]

(a) Why was the matched pairs statistic appropriate?

(b) Explain to someone who knows no statistics but is interested in weight-loss programs what the practical conclusion is.

(c) The paper follows the tradition of reporting significance only at fixed levels such as $\alpha = 0.01$. In fact, the results are more significant than "$p < 0.01$" suggests. What can you say about the *P*-value of the *t* test?

7.119   Some research suggests that women perform better than men in school but men score higher on standardized tests. Table 1.9 (page 39) presents data on a measure of school performance, grade point average (GPA), and a standardized test, IQ, for 78 seventh-grade students. Do these data lend further support to the previously found gender differences? Give graphical displays of the data and describe the distributions. Use significance tests and confidence intervals to examine this question, and prepare a short report summarizing your findings.

7.120   Refer to the previous exercise. Although self-concept in this study was measured on a scale with values in the data set ranging from 20 to 80, many prefer to think of this kind of variable as having only two possible values: low self-concept or high self-concept. Find the median of the self-concept scores in Table 1.9 and define those at or below the median to be low-self-concept students and those above the median to be high-self-concept students. Do high-self-concept students have grade point averages that are different from low-self-concept students? What about IQ? Prepare a report addressing these questions. Be sure to include graphical and numerical summaries and confidence intervals, and state clearly the details of significance tests.

7.121   On the morning of March 5, 1996, a train with 14 tankers of propane derailed near the center of the small Wisconsin town of Weyauwega. Six of the tankers were ruptured and burning when the 1700 residents were ordered to evacuate the town. Researchers study disasters like this so that effective relief efforts can be designed for future disasters. About half of the households with pets did not evacuate all of their pets. A study conducted after the derailment focused on problems associated with retrieval of the pets after the evacuation and characteristics of the pet owners. One of the scales measured "commitment to adult animals," and the people who evacuated all or some of their pets were compared with those who did not evacuate any of their pets. Higher scores indicate that the pet owner is more likely to take actions that benefit the pet.[39] Here are the data summaries:

Group	n	$\bar{x}$	s
Evacuated all or some pets	116	7.95	3.62
Did not evacuate any pets	125	6.26	3.56

Analyze the data and prepare a short report describing the results.

7.122   Do various occupational groups differ in their diets? A British study of this question compared 98 drivers and 83 conductors of London double-decker buses.[40] The conductors' jobs require more physical activity. The article reporting the study gives the data as "Mean daily consumption ($\pm$ se)." Some of the study results appear below:

	Drivers	Conductors
Total calories	2821 ± 44	2844 ± 48
Alcohol (grams)	0.24 ± 0.06	0.39 ± 0.11

(a) What does "se" stand for? Give $\bar{x}$ and s for each of the four sets of measurements.

(b) Is there significant evidence at the 5% level that conductors consume more calories per day than do drivers? Use the two-sample t method to give a P-value, and then assess significance.

(c) How significant is the observed difference in mean alcohol consumption? Use two-sample t methods to obtain the P-value.

(d) Give a 95% confidence interval for the mean daily alcohol consumption of London double-decker bus conductors.

(e) Give a 99% confidence interval for the difference in mean daily alcohol consumption between drivers and conductors.

7.123 (Optional) Use of the pooled two-sample t test is justified in part (b) of the previous exercise. Explain why. Find the P-value for the pooled t statistic, and compare with your result in the previous exercise.

7.124 The report cited in Exercise 7.122 says that the distribution of alcohol consumption among the individuals studied is "grossly skew."

(a) Do you think that this skewness prevents the use of the two-sample t test for equality of means? Explain your answer.

(b) (Optional) Do you think that the skewness of the distributions prevents the use of the F test for equality of standard deviations? Explain your answer.

7.125 Table 1.2 (page 11) gives literacy rates for men and women in 17 Islamic nations. Is it proper to apply the one-sample t method to these data to give a 95% confidence interval for the mean literacy rate of Islamic men? Explain your answer.

7.126 Polychlorinated biphenyls (PCBs) are a collection of compounds that are no longer produced in the United States but are still found in the environment. Evidence suggests that they can cause harmful health effects when consumed. Because PCBs can accumulate in fish, efforts have been made to identify areas where fish contain excessive amounts so that recommendations concerning consumption limits can be made. There are over 200 types of PCBs. Data from the Environmental Protection Agency National Study of Residues in Lake Fish are given in the data set PCB. More details about this data set can be found in the Data Appendix. Various lakes in the United States were sampled and the amounts of PCBs in fish were measured. The variable PCB is the sum of the amounts of all PCBs found in the fish. The units are parts per billion (ppb).

(a) Use graphical and numerical summaries to describe the distribution of this variable. Include a histogram with the location of the mean and the median clearly marked.

(b) Do you think it is appropriate to use methods based on normal distributions for these data? Explain why or why not.

(c) Find a 95% confidence interval for the mean. Will this interval contain approximately 95% of the observations in the data set? Explain your answer.

(d) Transform the PCB variable with a logarithm. Analyze the transformed data and summarize your results. Do you prefer to work with the raw data or with logs for this variable? Give reasons for your answer.

(e) Visit the Web site `http://epa.gov/waterscience/fishstudy/` to find details about how the data were collected. Write a summary describing these details and discuss how the results from this study can be generalized to other settings.

7.127   Refer to the previous exercise. Not all types of PCBs are equally harmful. A scale has been developed to convert the raw amount of each type of PCB to a toxic equivalent (TEQ). The PCB data set contains a variable TEQPCB that is the total TEQ from all PCBs found in each sample. Using the questions in the previous exercise, analyze these data and summarize the results.

7.128   In Exercise 7.2 (page 472) you analyzed the C-reactive protein (CRP) scores for a random sample of 40 children who participated in a study in Papua New Guinea. Serum retinol for the same children was analyzed in Exercise 7.4. Data for all 90 children who participated in the study are given in the data set PNG, described in the Data Appendix. Researchers who analyzed these data along with data from several other countries were interested in whether or not infections (as indicated by high CRP values) were associated with lower levels of serum retinol. A child with a value of CRP greater than 5.0 mg/l is classified as recently infected. Those whose CRP is less than or equal to 5.0 mg/l are not. Compare the serum retinol levels of the infected and noninfected children. Include graphical and numerical summaries, comments on all assumptions, and details of your analyses. Write a short report summarizing your results.

7.129   Refer to the previous exercise. The researchers in this study also measured α1-acid glycoprotein (AGP). This protein is similar to CRP in that it is an indicator of infection. However, it rises more slowly than CRP and reaches a maximum 2 to 3 days after an infection. The units for AGP are grams per liter (g/l), and any value greater than 1.0 g/l is an indication of infection. Analyze these data and write a report summarizing your results.

7.130   Table 7.3 (page 466) gives the levels of three pollutants in the exhaust of 46 randomly selected vehicles of the same type. You will investigate emissions of nitrogen oxides (NOX).

(a) Make a stemplot and, if your software allows, a normal quantile plot of the NOX levels. Do the plots suggest that the distribution of NOX emissions is approximately normal? Can you safely employ t procedures to analyze these data?

(b) Give a 95% confidence interval for the mean NOX level in vehicles of this type.

(c) Your supervisor hopes the average NOX level is less than 1 gram per mile. You will have to tell him that it's not so. Carry out a significance test to assess the strength of the evidence that the mean NOX level is greater than 1, and then write a short report to your supervisor based on your work in

(b) and (c). (Your supervisor has never heard of *P*-values, so you must use plain language.)

7.131 In the READING data set the response variable Post3 is to be compared for three methods of teaching reading. The Basal method is the standard, or control, method, and the two new methods are DRTA and Strat. We can use the methods of this chapter to compare Basal with DRTA and Basal with Strat. Note that to make comparisons among three treatments it is more appropriate to use the procedures that we will learn in Chapter 12.

(a) Is the mean reading score with the DRTA method higher than that for the Basal method? Perform an analysis to answer this question and summarize your results.

(b) Answer part (a) for the Strat method in place of DRTA.

7.132 Is there a difference between the average SAT scores of males and females? The CSDATA data set gives the math (SATM) and verbal (SATV) scores for a group of 224 computer science majors. The variable SEX indicates whether each individual is male or female.

(a) Compare the two distributions graphically, and then use the two-sample *t* test to compare the average SATM scores of males and females. Is it appropriate to use the pooled *t* test for this comparison? Write a brief summary of your results and conclusions that refers to both versions of the *t* test and to the *F* test for equality of standard deviations. Also give a 95% confidence interval for the difference in the means.

(b) Answer part (a) for the SATV scores.

(c) The students in the CSDATA data set were all computer science majors who began college during a particular year. To what extent do you think that your results would generalize to (*i*) computer science students entering in different years, (*ii*) computer science majors at other colleges and universities, and (*iii*) college students in general?

7.133 Refer to the previous exercise. Take a simple random sample of one-half of the data. Analyze the data for these 112 computer science majors. Compare your results with those you obtained in the previous exercise and comment on the effect of the sample size on these procedures.

7.134 (Optional) Example 7.13 (page 486) tells us that the mean height of 10-year-old girls is $N(54.5, 2.7)$ and for boys it is $N(54.1, 2.4)$. The null hypothesis that the mean heights of 10-year-old boys and girls are equal is clearly false. The difference in mean heights is $54.5 - 54.1 = 0.4$ inch. Small differences such as this can require large sample sizes to detect. To simplify our calculations, let's assume that the standard deviations are the same, say $\sigma = 2.5$, and that we will measure the heights of an equal number of girls and boys. How many would we need to measure to have a 90% chance of detecting the (true) alternative hypothesis?

7.135 (Optional) In Exercise 7.105 (page 524) you found the power for a study designed to compare birth weights of children born to cocaine users with those born to controls. Fix the sample size at 50 in each group and assume the standard deviation is 650 grams and the significance level is 0.05. Pick a set of alternatives that will give values of power ranging from fairly low values to fairly high values. Plot your results versus that alternative and give a short summary of what you have found.

7.136   As the degrees of freedom increase, the t distributions get closer and closer to the z ($N(0, 1)$) distribution. One way to see this is to look at how the value of t^* for a 95% confidence interval changes with the degrees of freedom. Make a plot with degrees of freedom from 2 to 100 on the x axis and t^* on the y axis. Draw a horizontal line on the plot corresponding to the value of $z^* = 1.96$. Summarize the main features of the plot.

7.137   Refer to the previous exercise. Make a similar plot and summarize its features for the value of t^* for a 90% confidence interval.

7.138   The margin of error for a confidence interval depends on the confidence level, the standard deviation, and the sample size. Fix the confidence level at 95% and the standard deviation at 1 to examine the effect of the sample size. Find the margin of error for sample sizes of 5 to 100 by 5's—that is, let $n = 5, 10, 15, \ldots, 100$. Plot the margins of error versus the sample size and summarize the relationship.

7.139   Refer to the previous exercise. Make a similar plot and summarize its features for a 99% confidence interval.

7.140   How much more would you expect to pay for a home that has four bedrooms than for a home that has three? Here are some data for West Lafayette, Indiana.[41] These are the selling prices (in dollars) that the owners of the homes are asking.

Four-bedroom homes:

| 121,900 | 139,900 | 157,000 | 159,900 | 176,900 | 224,900 | 235,000 | 245,000 | 294,000 |

Three-bedroom homes:

65,500	79,900	79,900	79,900	82,900	87,900	94,000	97,500	105,000
111,900	116,900	117,900	119,900	122,900	124,000	125,000	126,900	127,900
127,900	127,900	132,900	145,000	145,500	157,500	194,000	205,900	259,900
265,000								

(a) Plot the selling prices for the two sets of homes and describe the two distributions.

(b) Test the null hypothesis that the mean asking prices for the two sets of homes are equal versus the two-sided alternative. Give the test statistic with degrees of freedom, the P-value, and your conclusion.

(c) Would you consider using a one-sided alternative for this analysis? Explain why or why not.

(d) Give a 95% confidence interval for the difference in mean selling prices.

(e) These data are not SRSs from a population. Give a justification for use of the two-sample t procedures in this case.

7.141 Go to the Web site `www.realtor.com` and select two geographical areas of interest to you. You will compare the prices of similar types of homes in these two areas. State clearly how you define the areas and the type of homes. For example, you can use city names or ZIP codes to define the area and you can select single-family homes or condominiums. We view these homes as representative of the asking prices of homes for these areas at the time of your search. If the search gives a large number of homes, select a random sample. Be sure to explain exactly how you do this. Use the methods you have learned in this chapter to compare the selling prices. Be sure to include a graphical summary.

EESEE CASE STUDIES

The Electronic Encyclopedia of Statistical Examples and Exercises (EESEE) is available on the text CD and Web site. These more elaborate stories, with data, provide settings for longer case studies.

7.142 Read the EESEE story "Checkmating and Reading Skills." Write a report that answers Question 4 for this case study.

7.143 Read the EESEE material "Kicking a Helium Filled Football." Write a report answering all questions for this case study.

7.144 Read the EESEE material "Secondhand Smoking Bartenders." Write a report answering all questions for this case study.

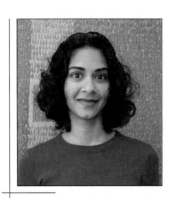

Management Consulting

AARTI SRIRAM *is a Statistical Consultant with Booz Allen Hamilton.*

Have you ever walked into a grocery store and seen a special display of canned goods or noticed the store selling two cartons of orange juice for the price of one? Have you watched a commercial on television advertising a set of tires as "buy one get one free"? How do companies choose these or other modes of advertising? What kind of advertising will give the best return on a company's marketing dollars, taking into account both the cost of the advertising and the response of consumers? When should special promotions be scheduled?

Companies that make consumer goods often hire management consulting firms to help answer questions such as these. As a statistical consultant with a major consulting firm, I work as part of a team that delves into these problems. I deal with large amounts of data on such things as overall product sales trends, seasonal effects, and the results of past advertising campaigns. I make sure the data are correct, gather more data when necessary, and build statistical models that answer the client's questions. Companies use the models we produce to schedule types and durations of promotions for the next calendar year.

Other companies seek help with managing their inventories. How large an inventory of tires should a tire manufacturer hold in order to supply its dealers and yet keep down the high cost of storing unsold tires? The answer depends on how well each tire model sells, which in turn depends on advertising, time of year, the choices made by tire dealerships, and many other factors. Using historical time-series data and statistical methods, I can guide the company toward a manufacturing and inventory policy that will ensure that dealers and consumers can get the tires they want while the manufacturer is protected from excessive costs. I enjoy the challenge of helping consumer-products companies optimize their business while delivering a high-quality product, with promotions that benefit you, the consumer.

Inference for Proportions

Introduction

Many statistical studies produce counts rather than measurements. An opinion poll asks a sample of adults whether they approve of the president's conduct of his office; the data are the counts of "Yes," "No," and "Don't know." An experiment compares the effectiveness of four treatments to prevent the common cold; the data are the number of subjects given each treatment who catch a cold during the next month. The proportions of men and women student athletes who gamble are compared using data from a survey. This chapter and the next present procedures for statistical inference in these settings.

The parameters we want to do inference about are population proportions. Just as in the case of inference about population means, we may be concerned with a single population or with comparing two populations. Inference about proportions in these one-sample and two-sample settings is very similar to inference about means, which we discussed in Chapter 7.

We begin in Section 8.1 with inference about a single population proportion. The statistical model for a count is then the binomial distribution, which we studied in Section 5.1. Section 8.2 concerns methods for comparing two proportions. Binomial distributions again play an important role.

8.1 Inference for a Single Proportion

We want to estimate the proportion p of some characteristic, such as approving the president's conduct of his office, among the members of a large population. We select a simple random sample (SRS) of size n from the population and record the count X of "successes" (such as "Yes" answers to a question about the president). We will use "success" as shorthand for whatever characteristic we are interested in. The sample proportion of successes $\hat{p} = X/n$ estimates the unknown population proportion p. If the population is much larger than the sample (say, at least 10 times as large), the individual responses are nearly independent and the count X has approximately the binomial distribution $B(n, p)$. In statistical terms, we are concerned with inference about the probability p of a success in the binomial setting.

If the sample size n is very small, we must base tests and confidence intervals for p on the binomial distributions. These are awkward to work with because of the discreteness of the binomial distributions.[1] But we know that when the sample is large, both the count X and the sample proportion $\hat{p}$ are approximately normal. We will consider only inference procedures based on the normal approximation. These procedures are similar to those for inference about the mean of a normal distribution.

Large-sample confidence interval for a single proportion

The unknown population proportion p is estimated by the sample proportion $\hat{p} = X/n$. We know (Chapter 5, page 344) that if the sample size n is sufficiently large, $\hat{p}$ has approximately the normal distribution, with mean $\mu_{\hat{p}} = p$ and standard deviation $\sigma_{\hat{p}} = \sqrt{p(1-p)/n}$. This means that approximately 95% of the time $\hat{p}$ will be within $2\sqrt{p(1-p)/n}$ of the unknown population proportion p.

Note that the standard deviation $\sigma_{\hat{p}}$ depends upon the unknown parameter p. To estimate this standard deviation using the data, we replace p in the formula by the sample proportion $\hat{p}$. As we did in Chapter 7, we use the term *standard error* for the standard deviation of a statistic that is estimated from data. Here is a summary of the procedure.

LARGE-SAMPLE CONFIDENCE INTERVAL FOR A POPULATION PROPORTION

Choose an SRS of size n from a large population with unknown proportion p of successes. The **sample proportion** is

$$\hat{p} = \frac{X}{n}$$

where X is the number of successes. The **standard error of $\hat{p}$** is

$$SE_{\hat{p}} = \sqrt{\frac{\hat{p}(1-\hat{p})}{n}}$$

and the **margin of error** for confidence level C is

$$m = z^{*}SE_{\hat{p}}$$

where z^{*} is the value for the standard normal density curve with area C between $-z^{*}$ and z^{*}. An **approximate level C confidence interval** for p is

$$\hat{p} \pm m$$

Use this interval for 90%, 95%, or 99% confidence when the number of successes and the number of failures are both at least 15.

EXAMPLE 8.1   Alcohol abuse has been described by college presidents as the number one problem on campus, and it is an important cause of death in young adults. How common is it? A survey of 17,096 students in U.S. four-year colleges collected information on drinking behavior and alcohol-related problems.[2] The researchers defined "frequent binge drinking" as having five or more drinks in a row three or more times in the past two weeks. According to this definition, 3314 students were classified as frequent binge drinkers. The proportion of drinkers is

$$\hat{p} = \frac{3314}{17,096} = 0.194$$

To find a 95% confidence interval, first compute the standard error:

$$SE_{\hat{p}} = \sqrt{\frac{\hat{p}(1-\hat{p})}{n}}$$

$$= \sqrt{\frac{(0.194)(1-0.194)}{17,096}}$$

$$= 0.00302$$

Approximately 95% of the time, $\hat{p}$ will be within two standard errors ($2 \times 0.00302 = 0.00604$) of the true p. From Table D we find the value of z^* to be 1.960. So the confidence interval is

$$\hat{p} \pm z^* \text{SE}_{\hat{p}} = 0.194 \pm (1.960)(0.00302)$$

$$= 0.194 \pm 0.006$$

$$= (0.188, \ 0.200)$$

We estimate with 95% confidence that between 18.8% and 20.0% of college students are frequent binge drinkers. Here is an alternative summary. We estimate that 19.4% of college students are frequent binge drinkers. For 95% confidence, the margin of error is 0.6%.

Because the calculations for statistical inference for a single proportion are relatively straightforward, many software packages do not include them. Figure 8.1 gives output from Minitab and Crunch It! for Example 8.1. As usual, the output reports more digits than are useful. *When you use software, be sure to think about how many digits are meaningful for your purposes.*

Remember that the margin of error in this confidence interval includes only random sampling error. There are other sources of error that are not taken into account. This survey used a design where the number of students sampled was proportional to the size of the college they attended; in this way we can treat the data as if we had an SRS. However, as is the case with many such surveys, we are forced to assume that the respondents provided accurate information. If the students did not answer the questions honestly, the results may be biased. Furthermore, we also have the typical problem of nonresponse. The response rate for this survey was 69%, an excellent rate for surveys of this type. Do the students who did not respond have different drinking habits than those who did? If so, this is another source of bias.

We recommend the large-sample confidence interval for 90%, 95%, and 99% confidence whenever the number of successes and the number of failures are both at least 15. For very small sample sizes, we recommend exact methods that use the binomial distribution. These are available as options in

Minitab

```
Sample     X        N      Sample p          95.0 % CI
1        3314    17096     0.193847     (0.187944, 0.199853)
```

Crunch It!

95% confidence interval results:
p = proportion of successes for population
Parameter: p

Proportion	Count	Total	Sample Prop.	Std. Err.	L. Limit	U. Limit
p	3314	17096	0.19384651	0.0030233653	0.18792082	0.1997722

FIGURE 8.1   Minitab and Crunch It! output for Example 8.1.

many statistical software packages and we do not cover them here. There is an intermediate case between large samples and very small samples where a slight modification of the large-sample approach works quite well.[3] We call this method the "plus four" procedure.

Plus four confidence interval for a single proportion

Computer studies reveal that confidence intervals based on the large-sample approach can be quite inaccurate. We will use a simple adjustment that works very well in practice. The adjustment is based on the following idea: act as if we have 4 additional observations, 2 of which are successes and 2 of which are failures. The new sample size is $n + 4$ and the count of successes is $X + 2$. The estimator of the population proportion based on this *plus four* rule is

$$\tilde{p} = \frac{X + 2}{n + 4}$$

plus four estimate   This estimate was first suggested by Edwin Bidwell Wilson in 1927 and we call it the **plus four estimate.** We base a confidence interval on the z statistic obtained by standardizing the plus four estimate $\tilde{p}$. Because $\tilde{p}$ is the sample proportion for our modified sample of size $n + 4$, it isn't surprising that the distribution of $\tilde{p}$ is close to the normal distribution with mean p and standard deviation $\sqrt{p(1 - p)/(n + 4)}$. To get a confidence interval, we estimate p by $\tilde{p}$ in this standard deviation to get the standard error of $\tilde{p}$. Here is the final result.

> ### PLUS FOUR CONFIDENCE INTERVAL FOR A SINGLE PROPORTION
>
> Choose an SRS of size n from a large population with unknown proportion p of successes. The **plus four estimate of the population proportion** is
>
> $$\tilde{p} = \frac{X + 2}{n + 4}$$
>
> where X is the number of successes. The **standard error of $\tilde{p}$** is
>
> $$\text{SE}_{\tilde{p}} = \sqrt{\frac{\tilde{p}(1 - \tilde{p})}{n + 4}}$$
>
> and the **margin of error** for confidence level C is
>
> $$m = z^* \text{SE}_{\tilde{p}}$$
>
> where z^* is the value for the standard normal density curve with area C between $-z^*$ and z^*. An **approximate level C confidence interval** for p is
>
> $$\tilde{p} \pm m$$
>
> Use this interval for 90%, 95%, or 99% confidence whenever the sample size is at least $n = 10$.

EXAMPLE 8.2   Research has shown that there are many health benefits associated with a diet that contains soy foods. Substances in soy called isoflavones are known to be responsible for these benefits. When soy foods are consumed, some subjects produce a chemical called equol, and it is thought that production of equol is a key factor in the health benefits of a soy diet. Unfortunately, not all people are equol producers; there appear to be two distinct subpopulations: equol producers and equol nonproducers.[4]

A nutrition researcher planning some bone health experiments would like to include some equol producers and some nonproducers among her subjects. A preliminary sample of 12 female subjects were measured, and 4 were found to be equol producers. We would like to estimate the proportion of equol producers in the population from which this researcher will draw her subjects.

The plus four estimate of the proportion of equol producers is

$$\tilde{p} = \frac{4+2}{12+4} = \frac{6}{16} = 0.375$$

For a 95% confidence interval, we will use Table D to find $z^* = 1.96$. We first compute the standard error

$$SE_{\tilde{p}} = \sqrt{\frac{\tilde{p}(1-\tilde{p})}{n+4}}$$

$$= \sqrt{\frac{(0.375)(1-0.375)}{16}}$$

$$= 0.12103$$

and then the margin of error

$$m = z^* SE_{\tilde{p}}$$

$$= 1.96(0.12103)$$

$$= 0.237$$

So the confidence interval is

$$\tilde{p} \pm m = 0.375 \pm 0.237$$

$$= (0.138, \ 0.612)$$

We estimate with 95% confidence that between 14% and 61% of women from this population are equol producers.

If the true proportion of equol users is near 14%, the lower limit of this interval, there may not be a sufficient number of equol producers in the study if subjects are tested only after they are enrolled in the experiment. It may be necessary to determine whether or not a potential subject is an equol producer. The study could then be designed to have the same number of equol producers and nonproducers.

Significance test for a single proportion

Recall that the sample proportion $\hat{p} = X/n$ is approximately normal, with mean $\mu_{\hat{p}} = p$ and standard deviation $\sigma_{\hat{p}} = \sqrt{p(1-p)/n}$. For confidence intervals, we substitute $\hat{p}$ for p in the last expression to obtain the standard error.

When performing a significance test, however, the null hypothesis specifies a value for p, and we assume that this is the true value when calculating the P-value. Therefore, when we test $H_0: p = p_0$, we substitute this p_0 into the expression for $\sigma_{\hat{p}}$ and then standardize $\hat{p}$. Here are the details.

LARGE-SAMPLE SIGNIFICANCE TEST FOR A POPULATION PROPORTION

Draw an SRS of size n from a large population with unknown proportion p of successes. To test the hypothesis $H_0: p = p_0$, compute the **z statistic**

$$z = \frac{\hat{p} - p_0}{\sqrt{\dfrac{p_0(1 - p_0)}{n}}}$$

In terms of a standard normal random variable Z, the approximate P-value for a test of H_0 against

$H_a: p > p_0$ is $P(Z \geq z)$

$H_a: p < p_0$ is $P(Z \leq z)$

$H_a: p \neq p_0$ is $2P(Z \geq |z|)$

We recommend the large-sample z significance test as long as the expected number of successes, np_0, and the expected number of failures, $n(1 - p_0)$, are both greater than 10. If this rule of thumb is not met, or if the population is less than 10 times as large as the sample, other procedures should be used. Here is an example.

EXAMPLE 8.3   According to the National Institute for Occupational Safety and Health,[5] job stress poses a major threat to the health of workers. A national survey of restaurant employees found that 75% said that work stress had a negative impact on their personal lives.[6] A sample of 100 employees of a restaurant chain finds that 68 answer "Yes" when asked, "Does work stress have a negative impact on your personal life?" Is this good reason to think that the proportion of all employees of this chain who would say "Yes" differs from the national proportion $p_0 = 0.75$?

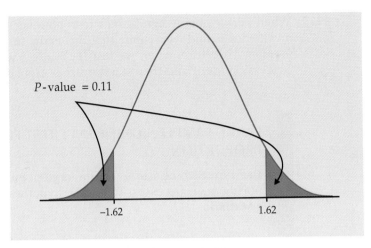

FIGURE 8.2   The *P*-value for Example 8.2.

To answer this question, we test

$$H_0 : p = 0.75$$

$$H_a : p \neq 0.75$$

The expected numbers of "Yes" and "No" responses are $100 \times 0.75 = 75$ and $100 \times 0.25 = 25$. Both are greater than 10, so we can use the z test. The test statistic is

$$z = \frac{\hat{p} - p_0}{\sqrt{\dfrac{p_0(1 - p_0)}{n}}}$$

$$= \frac{0.68 - 0.75}{\sqrt{\dfrac{(0.75)(0.25)}{100}}} = -1.62$$

From Table A we find $P(Z \leq -1.62) = 0.0526$. The *P*-value is the area in both tails, $P = 2 \times 0.0526 = 0.1052$. Figure 8.2 displays the *P*-value as an area under the standard normal curve. We conclude that the chain restaurant data are compatible with the survey results ($\hat{p} = 0.68$, $z = -1.62$, $P = 0.11$).

Figure 8.3 gives computer output from Minitab and Crunch It! for this example. *Note that for some entries software gives many more digits than we need.* You should decide how many digits are important for your analysis. In general, we will round proportions to two digits, for example, 0.68, and nonsignificant *P*-values to two digits, for example, $P = 0.11$.

In this example we have arbitrarily chosen to associate the response "Yes" that work stress has a negative impact on the respondent's personal life with success and "No" with failure. Suppose we reversed the choice. If we observed that 68 respondents said "Yes," then the other $100 - 68 = 32$ people said "No." Let's repeat the significance test with "No" as the success outcome. The national comparison value for the significance test is now 25%, the proportion in the national survey who responded "No."

Minitab

```
Test and Confidence Interval for One Proportion
Test of p = 0.75 vs p not = 0.75
Sample     X      N    Sample p       95.0 % CI        Z-Value   P-Value
1          68     100  0.680000  (0.588572, 0.771428)   -1.62     0.106
```

Crunch It!

Hypothesis test results:

p = proportion of successes for population

Parameter: p

H0 : Parameter = 0.75

HA : Parameter not = 0.75

Proportion	Count	Total	Sample Prop.	Std. Err.	Z-Stat	P-value
p	68	100	0.68	0.04330127	−1.6165807	0.106

FIGURE 8.3   Minitab and Crunch It! output for Example 8.3.

EXAMPLE 8.4   A sample of 100 restaurant workers were asked whether or not work stress had a negative impact on their personal lives and 32 of them responded "No." A large national survey reported that 25% of workers reported a negative impact. We test the null hypothesis

$$H_0: p = 0.25$$

$$H_a: p \neq 0.25$$

The test statistic is

$$z = \frac{\hat{p} - 0.25}{\sqrt{\dfrac{(0.25)(0.75)}{100}}} = \frac{0.32 - 0.25}{\sqrt{\dfrac{(0.25)(0.75)}{100}}} = 1.62$$

Using Table A, we find that $P = 0.11$.

When we interchanged "Yes" and "No" (or success and failure), we simply changed the sign of the test statistic z. The P-value remained the same. These facts are true in general. Our conclusion does not depend on an arbitrary choice of success and failure.

The results of our significance test have limited use in this example, as in many cases of inference about a single parameter. Of course, we do not expect the experience of the restaurant workers to be *exactly* the same as that of the workers in the national survey. *If the sample of restaurant workers is sufficiently large, we will have sufficient power to detect a very small difference. On the other hand, if our sample size is very small, we may be unable to detect differences that could be very important.* For these reasons we prefer to include a confidence interval as part of our analysis.

Confidence intervals provide additional information

To see what other values of p are compatible with the sample results, we will calculate a confidence interval.

EXAMPLE 8.5 The restaurant worker survey in Example 8.3 found that 68 of a sample of 100 employees agreed that work stress had a negative impact on their personal lives. That is, the sample size is $n = 100$ and the count of successes is $X = 68$. Because the number of successes and the number of failures are both at least 15, we will use the large-sample procedure to compute a 95% confidence interval. The sample proportion is

$$\hat{p} = \frac{X}{n}$$

$$= \frac{68}{100} = 0.68$$

The standard error is

$$SE_{\hat{p}} = \sqrt{\frac{\hat{p}(1 - \hat{p})}{n}}$$

$$= \sqrt{\frac{0.68(1 - 0.68)}{100}} = 0.0466$$

The z critical value for 95% confidence is $z^* = 1.96$, so the margin of error is

$$m = z^* SE_{\hat{p}}$$

$$= 1.96(0.0466)$$

$$= 0.91$$

The confidence interval is

$$\hat{p} \pm m = 0.68 \pm (1.96)(0.0466)$$

$$= 0.68 \pm 0.09$$

$$= (0.59,\ 0.77)$$

We are 95% confident that between 59% and 77% of the restaurant chain's employees feel that work stress is damaging their personal lives.

The confidence interval of Example 8.5 is much more informative than the significance test of Example 8.3. We have determined the values of p that are consistent with the observed results. Note that the standard error used for the confidence interval is estimated from the data, whereas the denominator in the test statistic z is based on the value assumed in the null hypothesis. A consequence of this fact is that the correspondence between the significance test result and the confidence interval is no longer exact. However, the correspondence is still very close. The confidence interval (0.59, 0.77) gives an approximate range of p_0's that would not be rejected by a test at the $\alpha = 0.05$ level of significance. We would not be surprised if the true proportion of restaurant workers who would say that work stress has a negative impact on their lives was as low as 60% or as high as 75%.

We do not often use significance tests for a single proportion, because it is uncommon to have a situation where there is a precise p_0 that we want to test. For physical experiments such as coin tossing or drawing cards from a well-shuffled deck, probability arguments lead to an ideal p_0. Even here, however, it can be argued, for example, that no real coin has a probability of heads *exactly*

CAUTION

equal to 0.5. Data from past large samples can sometimes provide a p_0 for the null hypothesis of a significance test. In some types of epidemiology research, for example, "historical controls" from past studies serve as the benchmark for evaluating new treatments. Medical researchers argue about the validity of these approaches, because the past never quite resembles the present. In general, we prefer comparative studies whenever possible.

Choosing a sample size

In Chapter 6, we showed how to choose the sample size n to obtain a confidence interval with specified margin of error m for a normal mean. Because we are using a normal approximation for inference about a population proportion, sample size selection proceeds in much the same way.

Recall that the margin of error for the large-sample confidence interval for a population proportion is

$$m = z^* \text{SE}_{\hat{p}} = z^* \sqrt{\frac{\hat{p}(1 - \hat{p})}{n}}$$

Choosing a confidence level C fixes the critical value z^*. The margin of error also depends on the value of $\hat{p}$ and the sample size n. Because we don't know the value of $\hat{p}$ until we gather the data, we must guess a value to use in the calculations. We will call the guessed value p^*. Here are two ways to get p^*:

- Use the sample estimate from a pilot study or from similar studies done earlier.

- Use $p^* = 0.5$. Because the margin of error is largest when $\hat{p} = 0.5$, this choice gives a sample size that is somewhat larger than we really need for the confidence level we choose. It is a safe choice no matter what the data later show.

Once we have chosen p^* and the margin of error m that we want, we can find the n we need to achieve this margin of error. Here is the result.

SAMPLE SIZE FOR DESIRED MARGIN OF ERROR

The level C confidence interval for a proportion p will have a margin of error approximately equal to a specified value m when the sample size satisfies

$$n = \left(\frac{z^*}{m}\right)^2 p^*(1 - p^*)$$

Here z^* is the critical value for confidence C, and p^* is a guessed value for the proportion of successes in the future sample.

The margin of error will be less than or equal to m if p^* is chosen to be 0.5. The sample size required is then given by

$$n = \left(\frac{z^*}{2m}\right)^2$$

The value of n obtained by this method is not particularly sensitive to the choice of p^* when p^* is fairly close to 0.5. However, if the value of p is smaller than about 0.3 or larger than about 0.7, use of $p^* = 0.5$ may select a sample size that is much larger than needed.

EXAMPLE 8.6 Your company has received complaints about its customer support service. You intend to hire a consulting company to carry out a sample survey of customers. Before contacting the consultant, you want some idea of the sample size you will have to pay for. One critical question is the degree of satisfaction with your customer service, measured on a five-point scale. You want to estimate the proportion p of your customers who are satisfied (that is, who choose either "satisfied" or "very satisfied," the two highest levels on the five-point scale).

You want to estimate p with 95% confidence and a margin of error less than or equal to 3%, or 0.03. For planning purposes, you are willing to use $p^* = 0.5$. To find the sample size required,

$$n = \left(\frac{z^*}{2m}\right)^2 = \left[\frac{1.96}{(2)(0.03)}\right]^2 = 1067.1$$

Round up to get $n = 1068$. (Always round up. Rounding down would give a margin of error slightly greater than 0.03.)

Similarly, for a 2.5% margin of error we have (after rounding up)

$$n = \left[\frac{1.96}{(2)(0.025)}\right]^2 = 1537$$

and for a 2% margin of error,

$$n = \left[\frac{1.96}{(2)(0.02)}\right]^2 = 2401$$

News reports frequently describe the results of surveys with sample sizes between 1000 and 1500 and a margin of error of about 3%. These surveys generally use sampling procedures more complicated than simple random sampling, so the calculation of confidence intervals is more involved than what we have studied in this section. The calculations in Example 8.6 nonetheless show in principle how such surveys are planned.

In practice, many factors influence the choice of a sample size. The following example illustrates one set of factors.

EXAMPLE 8.7 The Division of Recreational Sports (Rec Sports) at a major university is responsible for offering comprehensive recreational programs, services, and facilities to the students. Rec Sports is continually examining its programs to determine how well it is meeting the needs of the students. Rec Sports is considering adding some new programs and would like to know how much interest there would be in a new exercise program based on the Pilates method.[7] They will take a survey of undergraduate students. In the past, they emailed short surveys to all undergraduate students. The response rate obtained in this way was about 5%. This time they will send emails to a simple random sample of the students and will follow up with additional emails and eventually a phone call to get a higher response rate. Because of the work involved with the follow-up and limited staff, they would like to use a sample size of about 200. One of the questions they will ask is "Have you ever heard about the Pilates method of exercise?"

The primary purpose of the survey is to estimate various sample proportions for undergraduate students. Would the proposed sample size of $n = 200$ be adequate to provide Rec Sports with the needed information? To address this question, we calculate the margins of error of 95% confidence intervals for various values of $\hat{p}$.

EXAMPLE 8.8   In the Rec Sports survey, the margin of error of a 95% confidence interval for any value of $\hat{p}$ and $n = 200$ is

$$m = z^* SE_{\hat{p}}$$

$$= 1.96\sqrt{\frac{\hat{p}(1-\hat{p})}{200}}$$

$$= 0.139\sqrt{\hat{p}(1-\hat{p})}$$

The results for various values of $\hat{p}$ are

$\hat{p}$	m	$\hat{p}$	m
0.05	0.030	0.60	0.068
0.10	0.042	0.70	0.064
0.20	0.056	0.80	0.056
0.30	0.064	0.90	0.042
0.40	0.068	0.95	0.030
0.50	0.070		

Rec Sports judged these margins of error to be acceptable, and they used a sample size of 200 in their survey.

The table in Example 8.8 illustrates two points. First, the margins of error for $\hat{p} = 0.05$ and $\hat{p} = 0.95$ are the same. The margins of error will always be the same for $\hat{p}$ and $1 - \hat{p}$. This is a direct consequence of the form of the confidence interval. Second, the margin of error varies only between 0.064 and 0.070 as $\hat{p}$ varies from 0.3 to 0.7, and the margin of error is greatest when $\hat{p} = 0.5$, as we claimed earlier. It is true in general that the margin of error will vary relatively little for values of $\hat{p}$ between 0.3 and 0.7. Therefore, when planning a study, it is not necessary to have a very precise guess for p. If $p^* = 0.5$ is used and the observed $\hat{p}$ is between 0.3 and 0.7, the actual interval will be a little shorter than needed but the difference will be small.

Again, it is important to emphasize that these calculations consider only the effects of sampling variability that are quantified in the margin of error. Other sources of error, such as nonresponse and possible misinterpretation of questions, are not included in the table of margins of error for Example 8.8. Rec

Sports is trying to minimize these kinds of errors. They did a pilot study using a small group of current users of their facilities to check the wording of the questions, and they devised a careful plan to follow up with the students who did not respond to the initial email.

SECTION 8.1 | Summary

Inference about a population proportion p from an SRS of size n is based on the **sample proportion** $\hat{p} = X/n$. When n is large, $\hat{p}$ has approximately the normal distribution with mean p and standard deviation $\sqrt{p(1-p)/n}$.

For large samples, the **margin of error for confidence level C** is

$$m = z^* \mathrm{SE}_{\hat{p}}$$

where z^* is the value for the standard normal density curve with area C between $-z^*$ and z^*, and the **standard error of $\hat{p}$** is

$$\mathrm{SE}_{\hat{p}} = \sqrt{\frac{\hat{p}(1-\hat{p})}{n}}$$

The **level C large-sample confidence interval** is

$$\hat{p} \pm m$$

We recommend using this interval for 90%, 95%, and 99% confidence whenever the number of successes and the number of failures are both at least 15.

The **plus four estimate** of the population proportion is $\tilde{p} = (X+2)/(n+4)$, which is just the sample proportion when we add 2 successes and 2 failures to the data. For the plus four estimate, the **margin of error for confidence level C** is

$$m = z^* \mathrm{SE}_{\tilde{p}}$$

where z^* is the value for the standard normal density curve with area C between $-z^*$ and z^*, and the **standard error of $\tilde{p}$** is

$$\mathrm{SE}_{\tilde{p}} = \sqrt{\frac{\tilde{p}(1-\tilde{p})}{n+4}}$$

The **level C plus four confidence interval** is

$$\hat{p} \pm m$$

We recommend using this interval for 90%, 95%, and 99% confidence whenever the sample size is at least 10.

The **sample size** required to obtain a confidence interval of approximate margin of error m for a proportion is found from

$$n = \left(\frac{z^*}{m}\right)^2 p^*(1-p^*)$$

where p^* is a guessed value for the proportion, and z^* is the standard normal critical value for the desired level of confidence. To ensure that the margin of error of the interval is less than or equal to m no matter what $\hat{p}$ may be, use

$$n = \left(\frac{z^*}{2m}\right)^2$$

Tests of $H_0: p = p_0$ are based on the **z statistic**

$$z = \frac{\hat{p} - p_0}{\sqrt{\dfrac{p_0(1 - p_0)}{n}}}$$

with P-values calculated from the $N(0, 1)$ distribution. Use this procedure when the expected number of successes, np_0, and the expected number of failures, $n(1 - p_0)$, are both greater than 10.

SECTION 8.1 | Exercises

8.1   In each of the following circumstances state whether you would use the large-sample confidence interval, the plus four method, or neither for a 95% confidence interval.

(a) $n = 20, X = 15$.

(b) $n = 100, X = 15$.

(c) $n = 10, X = 2$.

(d) $n = 5, X = 2$.

(e) $n = 50, X = 20$.

8.2   In each of the following circumstances state whether you would use the large-sample confidence interval, the plus four method, or neither for a 95% confidence interval.

(a) $n = 8, X = 4$.

(b) $n = 1000, X = 12$.

(c) $n = 40, X = 18$.

(d) $n = 15, X = 2$.

(e) $n = 500, X = 225$.

8.3   Explain what is wrong with each of the following:

(a) An approximate 95% confidence interval for an unknown proportion p is $\hat{p}$ plus or minus its standard error.

(b) You can use a significance test to evaluate the hypothesis $H_0: \hat{p} = 0.3$ versus the two-sided alternative.

(c) The large-sample significance test for a population proportion is based on a t statistic.

8.4 Consider the binomial setting with $n = 60$ and $p = 0.4$.

(a) The sample proportion $\hat{p}$ will have a distribution that is approximately normal. Give the mean and the standard deviation of this normal distribution.

(b) Draw a sketch of this normal distribution. Mark the location of the mean.

(c) Find a value p^* for which the probability is 95% that $\hat{p}$ will be between $\pm p^*$. Mark these two values on your sketch.

8.5 Gambling is an issue of great concern to those involved in intercollegiate athletics. Because of this concern, the National Collegiate Athletic Association (NCAA) surveyed student-athletes concerning their gambling-related behaviors.[8] There were 5594 Division I male athletes in the survey. Of these, 3547 reported participation in some gambling behavior. This includes playing cards, betting on games of skill, buying lottery tickets, betting on sports, and similar activities.

(a) Find the sample proportion and the large-sample margin of error for 95% confidence. Explain in simple terms the meaning of the 95%.

(b) Because of the way that the study was designed to protect the anonymity of the student-athletes who responded, it was not possible to calculate the number of students who were asked to respond but did not. Does this fact affect the way that you interpret the results? Write a short paragraph explaining your answer.

8.6 In the study described in the previous exercise, 1447 out of a total of 3469 female student-athletes reported participation in some gambling activity.

(a) Use the large-sample methods to find an estimate of the true proportion with a 95% confidence interval.

(b) The margin of error for this sample is not the same as the margin of error calculated for the previous exercise. Explain why.

8.7 The National Survey of Student Engagement found that 87% of students report that their peers at least "sometimes" copy information from the Internet in their reports without reporting the source.[9] Assume that the sample size is 430,000.

(a) Find the margin of error for 99% confidence.

(b) Here are some items from the report that summarizes the survey. More than 430,000 students from 730 four-year colleges and universities participated. The average response rate was 43% and ranged from 15% to 89%. Institutions pay a participation fee of between $3000 and $7500 based on the size of their undergraduate enrollment. Discuss these as sources of error in this study. How do you think these errors would compare with the error that you calculated in part (a)?

8.8 Dogs are big and expensive. Rats are small and cheap. Can rats be trained to replace dogs in sniffing out illegal drugs? One study trained six male albino Sprague-Dawley rats to rear up on their hind legs in response to the smell of cocaine.[10] After training, each rat was tested 80 times. In the test a rat was presented with a large number of cups, one of which smelled like cocaine. A success was recorded if the rat correctly identified the cup containing cocaine by rearing up in front of it. The numbers of successes for the six rats were 80, 80, 73, 80, 74, and 80. You want to estimate the success rate in the future for each of the six rats. Compare the use of the large-sample estimates with the plus four estimates for this problem and make a recommendation concerning which is better. Write a short summary giving reasons for your recommendation.

8.9   The National Congregations Study collected data in a one-hour interview with a key informant—that is, a minister, priest, rabbi, or other staff person or leader.[11]   One question asked concerned the length of the typical sermon. For this question 390 out of 1191 congregations reported that the typical sermon lasted more than 30 minutes.

(a) Use the large-sample inference procedures to estimate the true proportion for this question with a 95% confidence interval.

(b) Compute the interval using the plus four method. Compare these results with those from part (a) and summarize what this example tells you about the two methods.

(c) There were 1236 congregations surveyed in this study. Calculate the non-response rate for this question. Does this influence how you interpret the results? Write a short discussion of this issue.

(d) The respondents to this question were not asked to use a stopwatch to record the lengths of a random sample of sermons at their congregations. They responded based on their impressions of the sermons. Do you think that ministers, priests, rabbis, or other staff persons or leaders might perceive sermon lengths differently from the people listening to the sermons? Discuss how your ideas would influence your interpretation of the results of this study.

8.10   The study described in the previous exercise also asked each respondent to classify his or her congregation according to theological orientation. For this question, 707 out of 1191 congregations were classified as "more conservative." Using the questions in the previous exercise as a guide, analyze and interpret these data. Compare your answers to parts (c) and (d) and discuss reasons why you think the answers should be similar or different.

8.11   A survey of 1280 student loan borrowers found that 448 had loans totaling more than $20,000 for their undergraduate education.[12]   Give a 95% confidence interval for the proportion of all student loan borrowers who have loans of $20,000 or more for their undergraduate education.

8.12   In the survey described in the previous exercise, there were 1050 borrowers whose total debt was $10,000 or more. Of these, 192 left school without completing a degree. Consider the population to be borrowers whose total debt was $10,000 or more. Find a 95% confidence interval for the proportion of borrowers who left school without completing a degree in this population.

8.13 Refer to Exercise 8.11. Would a 99% confidence interval be wider or narrower than the one that you found in that exercise? Verify your results by computing the interval.

8.14 Refer to Exercise 8.12. Would a 90% confidence interval be wider or narrower than the one that you found in that exercise? Verify your results by computing the interval.

8.15 For a study of unhealthy eating behaviors, 267 college women aged 18 to 25 years were surveyed.[13] Of these, 69% reported that they were on a diet sometime during the past year. Give a 95% confidence interval for the true proportion of college women aged 18 to 25 years in this population who dieted last year.

8.16 In the study described in the previous exercise, the researchers also surveyed 266 high school students who were 18 years old. In this sample 58.3% reported that they had dieted sometime in the past year. Give a 95% confidence interval for the true proportion of 18-year-old high school students in this population who were on a diet sometime during the past year.

8.17 In a study of the relationship between pet ownership and physical activity in older adults,[14] 594 subjects reported that they owned a pet, while 1939 reported that they did not. Give a 95% confidence interval for the proportion of older adults in this population who are pet owners.

8.18 In the study described in the previous exercise, 1434 subjects out of a total of 2533 reported that their annual income was $25,000 or more.

(a) Give a 95% confidence interval for the true proportion of subjects in this population with incomes above $25,000.

(b) Do you think that some respondents might not give truthful answers to a question about their income? Discuss the possible effects on your estimate and confidence interval.

(c) In the previous exercise, the question analyzed concerned pet ownership. Compare this question with the income question with respect to the possibility that the respondents were not truthful.

8.19 Yesterday, your top salesperson called on 10 customers and obtained orders for your new product from all 10. Suppose that it is reasonable to view these 10 customers as a random sample of all of her customers.

(a) Give the plus four estimate of the proportion of her customers who would buy the new product. Notice that we don't estimate that all customers will buy, even though all 10 in the sample did.

(b) Give the margin of error for 95% confidence. (You may see that the upper endpoint of the confidence interval is greater than 1. In that case, take the upper endpoint to be 1.)

(c) Do the results apply to all of your sales force? Explain why or why not.

8.20   In the United States approximately 900 people die in bicycle accidents each year. One study examined the records of 1711 bicyclists aged 15 or older who were fatally injured in bicycle accidents between 1987 and 1991 and were tested for alcohol. Of these, 542 tested positive for alcohol (blood alcohol concentration of 0.01% or higher).[15]

(a) Summarize the data with appropriate descriptive statistics.

(b) To do statistical inference for these data, we think in terms of a model where p is a parameter that represents the probability that a tested bicycle rider is positive for alcohol. Find a 99% confidence interval for p.

(c) Can you conclude from your statistical analysis of this study that alcohol causes fatal bicycle accidents?

(d) In this study 386 bicyclists had blood alcohol levels above 0.10%, a level defining legally drunk in many states. Give a 99% confidence interval for the proportion who were legally drunk according to this criterion.

8.21   In each of the following cases state whether or not the normal approximation to the binomial should be used for a significance test on the population proportion p.

(a) $n = 30$ and $H_0: p = 0.3$.

(b) $n = 30$ and $H_0: p = 0.6$.

(c) $n = 1000$ and $H_0: p = 0.5$.

(d) $n = 500$ and $H_0: p = 0.01$.

8.22   A matched pairs experiment compares the taste of instant versus fresh-brewed coffee. Each subject tastes two unmarked cups of coffee, one of each type, in random order and states which he or she prefers. Of the 40 subjects who participate in the study, 12 prefer the instant coffee. Let p be the probability that a randomly chosen subject prefers freshly brewed coffee to instant coffee. (In practical terms, p is the proportion of the population who prefer fresh-brewed coffee.)

(a) Test the claim that a majority of people prefer the taste of fresh-brewed coffee. Report the large-sample z statistic and its P-value.

(b) Draw a sketch of a standard normal curve and mark the location of your z statistic. Shade the appropriate area that corresponds to the P-value.

(c) Is your result significant at the 5% level? What is your practical conclusion?

(d) Find a 95% confidence interval for p using the large-sample method.

8.23   The English mathematician John Kerrich tossed a coin 10,000 times and obtained 5067 heads.

(a) Is this significant evidence at the 5% level that the probability that Kerrich's coin comes up heads is not 0.5? Use a sketch of the standard normal distribution to illustrate the P-value.

(b) Use a 95% confidence interval to find the range of probabilities of heads that would not be rejected at the 5% level.

8.24 One of your employees has suggested that your company develop a new product. You decide to take a random sample of your customers and ask whether or not there is interest in the new product. The response is on a 1 to 5 scale with 1 indicating "definitely would not purchase"; 2, "probably would not purchase"; 3, "not sure"; 4, "probably would purchase"; and 5, "definitely would purchase." For an initial analysis, you will record the responses 1, 2, and 3 as "No" and 4 and 5 as "Yes." What sample size would you use if you wanted the 95% margin of error to be 0.15 or less?

8.25 Refer to the previous exercise. Suppose that after reviewing the results of the previous survey, you proceeded with preliminary development of the product. Now you are at the stage where you need to decide whether or not to make a major investment to produce and market it. You will use another random sample of your customers but now you want the margin of error to be smaller. What sample size would you use if you wanted the 95% margin of error to be 0.075 or less?

8.26 You are planning an evaluation of an alcohol awareness program at your college that will take place six months after the program. Previous evaluations indicate that about 30% of the students who participate will respond "Yes" to the question "Do you think your behavior toward alcohol consumption has changed since the program?" How large a sample should you take if you want the margin of error for 95% confidence to be about 0.1?

 8.27 The evaluation in the previous exercise will also have questions that have not been asked before, so you do not have previous information about the possible value of p. Repeat the calculation above for the following values of p^*: 0.1, 0.2, 0.3, 0.4, 0.5, 0.6, 0.7, 0.8, and 0.9. Summarize the results in a table and graphically. What sample size will you use?

8.28 An automobile manufacturer would like to know what proportion of its customers are dissatisfied with the service received from their local dealer. The customer relations department will survey a random sample of customers and compute a 95% confidence interval for the proportion that are dissatisfied. From past studies, they believe that this proportion will be about 0.25. Find the sample size needed if the margin of error of the confidence interval is to be about 0.02. Suppose 15% of the sample say that they are dissatisfied. What is the margin of error of the 95% confidence interval?

 8.29 You have been asked to survey students at a large college to determine the proportion who favor an increase in student fees to support an expansion of the student newspaper. Each student will be asked whether he or she is in favor of the proposed increase. Using records provided by the registrar you can select a random sample of students from the college. After careful consideration of your resources, you decide that it is reasonable to conduct a study with a sample of 10 students. For this sample size, construct a table of the margins of error for 95% confidence intervals when $\hat{p}$ takes the values 0.1, 0.2, 0.3, 0.4, 0.5, 0.6, 0.7, 0.8, and 0.9.

8.30 A former editor of the student newspaper agrees to underwrite the study in the previous exercise because she believes the results will demonstrate that most students support an increase in fees. She is willing to provide funds for a sample of size 400. Write a short summary for your benefactor of why the increased sample size will provide better results.

8.2 Comparing Two Proportions

Because comparative studies are so common, we often want to compare the proportions of two groups (such as men and women) that have some characteristic. In the previous section we used data to estimate the proportion of college students who were frequent binge drinkers. Suppose we also wanted to compare the binge-drinking behaviors of men and women college students. Our problem now concerns the comparison of two proportions.

We call the two groups being compared Population 1 and Population 2, and the two population proportions of "successes" p_1 and p_2. The data consist of two independent SRSs, of size n_1 from Population 1 and size n_2 from Population 2. The proportion of successes in each sample estimates the corresponding population proportion. Here is the notation we will use in this section:

Population	Population proportion	Sample size	Count of successes	Sample proportion
1	p_1	n_1	X_1	$\hat{p}_1 = X_1/n_1$
2	p_2	n_2	X_2	$\hat{p}_2 = X_2/n_2$

To compare the two populations, we use the difference between the two sample proportions:

$$D = \hat{p}_1 - \hat{p}_2$$

When both sample sizes are sufficiently large, the sampling distribution of the difference D is approximately normal.

Inference procedures for comparing proportions are z procedures based on the normal approximation and on standardizing the difference D. The first step is to obtain the mean and standard deviation of D. By the addition rule for means, the mean of D is the difference of the means:

$$\mu_D = \mu_{\hat{p}_1} - \mu_{\hat{p}_2} = p_1 - p_2$$

That is, the difference $D = \hat{p}_1 - \hat{p}_2$ between the sample proportions is an unbiased estimator of the population difference $p_1 - p_2$. Similarly, the addition rule for variances tells us that the variance of D is the *sum* of the variances:

$$\sigma_D^2 = \sigma_{\hat{p}_1}^2 + \sigma_{\hat{p}_2}^2$$

$$= \frac{p_1(1 - p_1)}{n_1} + \frac{p_2(1 - p_2)}{n_2}$$

Therefore, when n_1 and n_2 are large, D is approximately normal with mean $\mu_D = p_1 - p_2$ and standard deviation

$$\sigma_D = \sqrt{\frac{p_1(1 - p_1)}{n_1} + \frac{p_2(1 - p_2)}{n_2}}$$

Large-sample confidence interval for a difference in proportions

To obtain a confidence interval for $p_1 - p_2$, we once again replace the unknown parameters in the standard deviation by estimates to obtain an estimated standard deviation, or standard error. Here is the confidence interval we want.

LARGE-SAMPLE CONFIDENCE INTERVAL FOR COMPARING TWO PROPORTIONS

Choose an SRS of size n_1 from a large population having proportion p_1 of successes and an independent SRS of size n_2 from another population having proportion p_2 of successes. The estimate of the difference in the population proportions is

$$D = \hat{p}_1 - \hat{p}_2$$

The **standard error of D** is

$$SE_D = \sqrt{\frac{\hat{p}_1(1 - \hat{p}_1)}{n_1} + \frac{\hat{p}_2(1 - \hat{p}_2)}{n_2}}$$

and the **margin of error** for confidence level C is

$$m = z^* SE_D$$

where z^* is the value for the standard normal density curve with area C between $-z^*$ and z^*. An **approximate level C confidence interval** for $p_1 - p_2$ is

$$D \pm m$$

Use this method for 90%, 95%, or 99% confidence when the number of successes and the number of failures in both samples are all at least 10.

In Example 8.1 (page 537) we estimated the proportion of college students who engage in frequent binge drinking. Are there characteristics of these students that relate to this behavior? For example, how similar is this behavior in men and women? This kind of follow-up question is natural in many studies such as this one. We will first use a confidence interval to examine it. In the binge-drinking study, data were also summarized by gender:

Population	n	X	$\hat{p} = X/n$
1 (men)	7,180	1,630	0.227
2 (women)	9,916	1,684	0.170
Total	17,096	3,314	0.194

In this table the $\hat{p}$ column gives the sample proportions of frequent binge drinkers. The last line gives the totals that we studied in Example 8.1.

EXAMPLE 8.9

Let's find a 95% confidence interval for the difference between the proportions of men and of women who are frequent binge drinkers. Output from Minitab and Crunch It! is given in Figure 8.4. To perform the computations using our formulas, we first find the difference in the proportions:

$$D = \hat{p}_1 - \hat{p}_2$$
$$= 0.227 - 0.170$$
$$= 0.057$$

Minitab

```
Test and Confidence Interval for Two Proportions

Sample      X        N    Sample p
1        1630     7180    0.227019
2        1684     9916    0.169827

Estimate for p(1) - p(2): 0.0571930
95% CI for p(1) - p(2): (0.450067, 0.0693792)
Test for p(1) - p(2) = 0 (vs not = 0):
                Z = 9.34 P-Value = 0.000
```

Crunch It!

95% confidence interval results:
p1 = proportion of successes for population 1
p2 = proportion of successes for population 2
Parameter: p1 – p2

Difference	Count1	Total1	Count2	Total2	Sample Diff.	Std. Err.	L. Limit	U. Limit
p1 – p2	1630	7180	1684	9916	0.057192955	0.0062175817	9.336581	0.06937919

FIGURE 8.4 Minitab and Crunch It! output for Example 8.9.

Then we calculate the standard error of D:

$$\text{SE}_D = \sqrt{\frac{\hat{p}_1(1 - \hat{p}_1)}{n_1} + \frac{\hat{p}_2(1 - \hat{p}_2)}{n_2}}$$

$$= \sqrt{\frac{(0.227)(0.773)}{7180} + \frac{(0.170)(0.830)}{9916}}$$

$$= 0.00622$$

For 95% confidence, we have $z^* = 1.96$, so the margin of error is

$$m = z^* \text{SE}_D$$

$$= (1.96)(0.00622)$$

$$= 0.012$$

The 95% confidence interval is

$$D \pm m = 0.057 \pm 0.012$$

$$= (0.045, 0.069)$$

With 95% confidence we can say that the difference in the proportions is between 0.045 and 0.069. Alternatively, we can report that the men are 5.7% more likely to be frequent binge drinkers, with a 95% margin of error of 1.2%.

In this example men and women were not sampled separately. The sample sizes are in fact random and reflect the gender distributions of the colleges that were randomly chosen. Two-sample significance tests and confidence intervals are still approximately correct in this situation. The authors of the report note that women are somewhat overrepresented partly because 6 of the 140 colleges in the study were all-women institutions.

In the example above we chose men to be the first population. Had we chosen women to be the first population, the estimate of the difference would be negative (-0.057). Because it is easier to discuss positive numbers, we generally choose the first population to be the one with the higher proportion.

Plus four confidence interval for a difference in proportions

Just as in the case of estimating a single proportion, a small modification of the sample proportions can greatly improve the accuracy of confidence intervals.[16] As before, add 2 successes and 2 failures to the actual data, and now divide them equally between the two samples. That is, *add 1 success and 1 failure to each sample.* We will again call the estimates produced by adding hypothetical observations **plus four estimates.** The plus four estimates of the two population proportions are

plus four estimates

$$\tilde{p}_1 = \frac{X_1 + 1}{n_1 + 2} \quad \text{and} \quad \tilde{p}_2 = \frac{X_2 + 1}{n_2 + 2}$$

The estimated difference between the populations is

$$\tilde{D} = \tilde{p}_1 - \tilde{p}_2$$

The standard deviation of $\tilde{D}$ is approximately

$$\sigma_{\tilde{D}} = \sqrt{\frac{p_1(1-p_1)}{n_1+2} + \frac{p_2(1-p_2)}{n_2+2}}$$

This is similar to the formula for σ_D, adjusted for the sizes of the modified samples.

To obtain a confidence interval for $p_1 - p_2$, we once again replace the unknown parameters in the standard deviation by estimates to obtain an estimated standard deviation, or standard error. Here is the confidence interval we want.

PLUS FOUR CONFIDENCE INTERVAL FOR COMPARING TWO PROPORTIONS

Choose an SRS of size n_1 from a large population having proportion p_1 of successes and an independent SRS of size n_2 from another population having proportion p_2 of successes. The **plus four estimate of the difference in proportions** is

$$\tilde{D} = \tilde{p}_1 - \tilde{p}_2$$

where

$$\tilde{p}_1 = \frac{X_1 + 1}{n_1 + 2} \qquad \tilde{p}_2 = \frac{X_2 + 1}{n_2 + 2}$$

The **standard error of $\tilde{D}$** is

$$SE_{\tilde{D}} = \sqrt{\frac{\tilde{p}_1(1-\tilde{p}_1)}{n_1+2} + \frac{\tilde{p}_2(1-\tilde{p}_2)}{n_2+2}}$$

and the **margin of error** for confidence level C is

$$m = z^* SE_{\tilde{D}}$$

where z^* is the value for the standard normal density curve with area C between $-z^*$ and z^*. An **approximate level C confidence interval** for $p_1 - p_2$ is

$$\tilde{D} \pm m$$

Use this method for 90%, 95%, or 99% confidence when both sample sizes are at least 5.

EXAMPLE 8.10   In studies that look for a difference between genders, a major concern is whether or not apparent differences are due to other variables that are associated with gender. Because boys mature more slowly than girls, a study of adolescents that compares boys and girls of the same age may confuse a gender effect with an effect of sexual maturity. The "Tanner score" is a commonly used measure of sexual maturity.[17] Subjects are asked to determine their score by placing a mark next to a rough drawing of an individual at their level of sexual maturity. There are five different drawings, so the score is an integer between 1 and 5.

A pilot study included 12 girls and 12 boys from a population that will be used for a large experiment. Four of the boys and three of the girls had Tanner scores of 4 or 5, a fairly high level of sexual maturity. Let's find a 95% confidence interval for the difference between the proportions of boys and of girls who have high (4 or 5) Tanner scores in this population. The numbers of successes and failures in both groups are not all at least 10, so the large-sample approach is not recommended. On the other hand, the sample sizes are both at least 5, so the plus four method is appropriate.

The plus four estimate of the population proportion for boys is

$$\tilde{p}_1 = \frac{X_1 + 1}{n_1 + 2} = \frac{4 + 1}{12 + 2} = 0.3571$$

For girls, the estimate is

$$\tilde{p}_2 = \frac{X_2 + 1}{n_2 + 2} = \frac{3 + 1}{12 + 2} = 0.2857$$

Therefore, the estimate of the difference is

$$\tilde{D} = \tilde{p}_1 - \tilde{p}_2 = 0.3571 - 0.2857 = 0.071$$

The standard error of $\tilde{D}$ is

$$SE_{\tilde{D}} = \sqrt{\frac{\tilde{p}_1(1 - \tilde{p}_1)}{n_1 + 2} + \frac{\tilde{p}_2(1 - \tilde{p}_2)}{n_2 + 2}}$$

$$= \sqrt{\frac{(0.3571)(1 - 0.3571)}{12 + 2} + \frac{(0.2857)(1 - 0.2857)}{12 + 2}}$$

$$= 0.1760$$

For 95% confidence, $z^* = 1.96$ and the margin of error is

$$m = z^* SE_{\tilde{D}} = (1.96)(0.1760) = 0.345$$

The confidence interval is

$$\tilde{D} \pm m = 0.071 \pm 0.345$$

$$= (-0.274, 0.416)$$

With 95% confidence we can say that the difference in the proportions is between −0.274 and 0.416. Alternatively, we can report that the difference in the proportions of boys and girls with high Tanner scores in this population is 7.1% with a 95% margin of error of 34.5%.

The very large margin of error in this example indicates that either boys or girls could be more sexually mature in this population and that the difference could be quite large. *Although the interval includes the possibility that there is no difference, corresponding to $p_1 = p_2$ or $p_1 - p_2 = 0$, we must be very cautious about concluding that there is **no** difference in the proportions.* With small sample sizes such as these, the data do not provide us with a lot of information for our inference. This fact is expressed quantitatively through the very large margin of error.

Significance test for a difference in proportions

Although we prefer to compare two proportions by giving a confidence interval for the difference between the two population proportions, it is sometimes useful to test the null hypothesis that the two population proportions are the same.

We standardize $D = \hat{p}_1 - \hat{p}_2$ by subtracting its mean $p_1 - p_2$ and then dividing by its standard deviation

$$\sigma_D = \sqrt{\frac{p_1(1 - p_1)}{n_1} + \frac{p_2(1 - p_2)}{n_2}}$$

If n_1 and n_2 are large, the standardized difference is approximately $N(0, 1)$. For the large-sample confidence interval we used sample estimates in place of the unknown population values in the expression for σ_D. Although this approach would lead to a valid significance test, we instead adopt the more common practice of replacing the unknown σ_D with an estimate that takes into account that our null hypothesis states that $p_1 = p_2$. If these two proportions are equal, then we can view all of the data as coming from a single population. Let p denote the common value of p_1 and p_2; then the standard deviation of $D = \hat{p}_1 - \hat{p}_2$ is

$$\sigma_D = \sqrt{\frac{p(1 - p)}{n_1} + \frac{p(1 - p)}{n_2}}$$

$$= \sqrt{p(1 - p)\left(\frac{1}{n_1} + \frac{1}{n_2}\right)}$$

We estimate the common value of p by the overall proportion of successes in the two samples:

$$\hat{p} = \frac{\text{number of successes in both samples}}{\text{number of observations in both samples}} = \frac{X_1 + X_2}{n_1 + n_2}$$

pooled estimate of p   This estimate of p is called the **pooled estimate** because it combines, or pools, the information from both samples.

To estimate σ_D under the null hypothesis, we substitute $\hat{p}$ for p in the expression for σ_D. The result is a standard error for D that assumes the truth of $H_0: p_1 = p_2$:

$$SE_{Dp} = \sqrt{\hat{p}(1 - \hat{p})\left(\frac{1}{n_1} + \frac{1}{n_2}\right)}$$

The subscript on SE_{Dp} reminds us that we pooled data from the two samples to construct the estimate.

SIGNIFICANCE TEST FOR COMPARING TWO PROPORTIONS

To test the hypothesis

$$H_0 : p_1 = p_2$$

compute the **z statistic**

$$z = \frac{\hat{p}_1 - \hat{p}_2}{\text{SE}_{Dp}}$$

where the **pooled standard error** is

$$\text{SE}_{Dp} = \sqrt{\hat{p}(1 - \hat{p}) \left(\frac{1}{n_1} + \frac{1}{n_2} \right)}$$

and where

$$\hat{p} = \frac{X_1 + X_2}{n_1 + n_2}$$

In terms of a standard normal random variable Z, the P-value for a test of H_0 against

$H_a : p_1 > p_2$ is $P(Z \geq z)$

$H_a : p_1 < p_2$ is $P(Z \leq z)$

$H_a : p_1 \neq p_2$ is $2P(Z \geq |z|)$

This z test is based on the normal approximation to the binomial distribution. As a general rule, we will use it when the number of successes and the number of failures in each of the samples is at least 5.

EXAMPLE 8.11 Are men and women college students equally likely to be frequent binge drinkers? We examine the survey data in Example 8.9 (page 557) to answer this question. Here is the data summary:

Population	n	X	$\hat{p} = X/n$
1 (men)	7,180	1,630	0.227
2 (women)	9,916	1,684	0.170
Total	17,096	3,314	0.194

The sample proportions are certainly quite different, but we will perform a significance test to see if the difference is large enough to lead us to believe that the population proportions are not equal. Formally, we test the hypotheses

$$H_0: p_1 = p_2$$

$$H_a: p_1 \neq p_2$$

The pooled estimate of the common value of p is

$$\hat{p} = \frac{1630 + 1684}{7180 + 9916} = \frac{3314}{17{,}096} = 0.194$$

Note that this is the estimate on the bottom line of the data summary above.

The test statistic is calculated as follows:

$$\text{SE}_{Dp} = \sqrt{(0.194)(0.806)\left(\frac{1}{7180} + \frac{1}{9916}\right)} = 0.006126$$

$$z = \frac{\hat{p}_1 - \hat{p}_2}{\text{SE}_{Dp}} = \frac{0.227 - 0.170}{0.006126}$$

$$= 9.34$$

The P-value is $2P(Z \geq 9.34)$. The largest value of z in Table A is 3.49, so from this table we can conclude $P < 2 \times 0.0002 = 0.0004$. Most software reports this result as simply 0 or a very small number. Output from Minitab and Crunch It! is given in Figure 8.5. Minitab reports the P-value as 0.000. This means that the calculated value is less than 0.0005; this is certainly a very small number. Crunch It! gives < 0.0001. The exact value is not particularly important; it is clear that we should reject the null hypothesis. For most situations 0.001 (1 chance in 1000) is sufficiently small. We report: among college students in the study, 22.7% of the men and 17.0% of the women were frequent binge drinkers; the difference is statistically significant ($z = 9.34$, $P < 0.001$).

We could have argued that we expect the proportion to be higher for men than for women in this example. This would justify using the one-sided alternative $H_a: p_1 > p_2$. The P-value would be half of the value obtained for the two-sided test. Because the z statistic is so large, this distinction is of no practical importance.

BEYOND THE BASICS | Relative risk

We summarized the comparison of the frequent binge-drinking proportions for men and for women by reporting a confidence interval for the difference in Example 8.9. Another way to summarize the comparison is to view each sample proportion as the **risk** that a college student of that gender is a

Minitab

Test and Confidence Interval for Two Proportions

```
Sample      X        N    Sample p
1        1630     7180    0.227019
2        1684     9916    0.169827

Estimate for p(1) - p(2): 0.0571930
95% CI for p(1) - p(2): (0.450067, 0.0693792)
Test for p(1) - p(2) = 0 (vs not = 0):
                    Z = 9.34 P-Value = 0.000
```

Crunch It!

Hypothesis test results:
p1 = proportion of successes for population 1
p2 = proportion of successes for population 2
Parameter: p1 – p2
H0 : Parameter = 0
HA : Parameter not = 0

Difference	Count1	Total1	Count2	Total2	Sample Diff.	Std. Err.	Z-Stat	P-value
p1 – p2	1630	7180	1684	9916	0.057192955	0.006125685	9.336581	<0.0001

FIGURE 8.5 Minitab and Crunch It! output for Example 8.11.

relative risk

frequent binge drinker. We then compare these two risks with the ratio of the two proportions, which is called the **relative risk** (RR) in many applications. Note that a relative risk of 1 means that the two proportions, $\hat{p}_1$ and $\hat{p}_2$, are equal. The procedure for calculating confidence intervals for relative risk is based on the same kind of principles that we have studied, but the details are somewhat more complicated. Fortunately, we can leave the details to software and concentrate on interpretation and communication of the results.

EXAMPLE 8.12

On page 557 we summarized the data on the proportions of men and women who are frequent binge drinkers with the following table:

Population	n	X	$\hat{p} = X/n$
1 (men)	7,180	1,630	0.227
2 (women)	9,916	1,684	0.170
Total	17,096	3,314	0.194

The relative risk is

$$\text{RR} = \frac{\hat{p}_1}{\hat{p}_2} = \frac{0.227}{0.170} = 1.34$$

Software gives the 95% confidence interval as 1.26 to 1.42. Men are 1.34 times as likely as women to be frequent binge drinkers; the 95% confidence interval is (1.26, 1.42).

In this example the confidence interval appears to be symmetric about the estimate: that is, 1.34 is the midpoint of 1.26 and 1.42. If we reported the results with more accuracy (RR = 1.337, 95% confidence interval = 1.258 to 1.421), we would see that the interval is *not* symmetric, and this is true in general.

SECTION 8.2 | Summary

The **large-sample estimate of the difference in two population proportions** is

$$D = \hat{p}_1 - \hat{p}_2$$

where $\hat{p}_1$ and $\hat{p}_2$ are the sample proportions

$$\hat{p}_1 = \frac{X_1}{n_1} \quad \text{and} \quad \hat{p}_2 = \frac{X_2}{n_2}$$

The **standard error of the difference D** is

$$\text{SE}_D = \sqrt{\frac{\hat{p}_1(1 - \hat{p}_1)}{n_1} + \frac{\hat{p}_2(1 - \hat{p}_2)}{n_2}}$$

The **margin of error for confidence level C** is

$$m = z^* \text{SE}_D$$

where z^* is the value for the standard normal density curve with area C between $-z^*$ and z^*. The **large-sample level C confidence interval** is

$$D \pm m$$

We recommend using this interval for 90%, 95%, or 99% confidence when the number of successes and the number of failures in both samples are all at least 10.

The **plus four estimate of the difference in two population proportions** is

$$\tilde{D} = \tilde{p}_1 - \tilde{p}_2$$

where $\tilde{p}_1$ and $\tilde{p}_2$ are the plus four estimates of the population proportions

$$\tilde{p}_1 = \frac{X_1 + 1}{n_1 + 2} \quad \text{and} \quad \tilde{p}_2 = \frac{X_2 + 1}{n_2 + 2}$$

The **standard error of the difference $\tilde{D}$** is

$$\text{SE}_{\tilde{D}} = \sqrt{\frac{\tilde{p}_1(1 - \tilde{p}_1)}{n_1 + 2} + \frac{\tilde{p}_2(1 - \tilde{p}_2)}{n_2 + 2}}$$

The **margin of error for confidence level C** is

$$m = z^* \text{SE}_{\tilde{D}}$$

where z^* is the value for the standard normal density curve with area C between $-z^*$ and z^*. The **plus four level C confidence interval** is

$$\tilde{D} \pm m$$

We recommend using this interval for 90%, 95%, or 99% confidence when both sample sizes are at least 5.

Significance tests of $H_0: p_1 = p_2$ use the **z statistic**

$$z = \frac{\hat{p}_1 - \hat{p}_2}{\text{SE}_{Dp}}$$

with P-values from the $N(0, 1)$ distribution. In this statistic,

$$\text{SE}_{Dp} = \sqrt{\hat{p}(1 - \hat{p})\left(\frac{1}{n_1} + \frac{1}{n_2}\right)}$$

and $\hat{p}$ is the **pooled estimate** of the common value of p_1 and p_2:

$$\hat{p} = \frac{X_1 + X_2}{n_1 + n_2}$$

Relative risk is the ratio of two sample proportions:

$$\text{RR} = \frac{\hat{p}_1}{\hat{p}_2}$$

Confidence intervals for relative risk are often used to summarize the comparison of two proportions.

SECTION 8.2 | Exercises

8.31 In each of the following circumstances state whether you would use the large-sample confidence interval, the plus four method, or neither for a 95% confidence interval.

(a) $n_1 = 25$, $n_2 = 25$, $X_1 = 10$, and $X_2 = 15$.

(b) $n_1 = 5$, $n_2 = 5$, $X_1 = 2$, and $X_2 = 5$.

(c) $n_1 = 25$, $n_2 = 25$, $X_1 = 8$, and $X_2 = 20$.

(d) $n_1 = 4$, $n_2 = 8$, $X_1 = 2$, and $X_2 = 7$.

(e) $n_1 = 100$, $n_2 = 10$, $X_1 = 40$, and $X_2 = 2$.

8.32 In each of the following circumstances state whether you would use the large-sample confidence interval, the plus four method, or neither for a 95% confidence interval.

(a) $n_1 = 4$, $n_2 = 100$, $X_1 = 1$, and $X_2 = 65$.

(b) $n_1 = 500$, $n_2 = 300$, $X_1 = 175$, and $X_2 = 208$.

(c) $n_1 = 6$, $n_2 = 10$, $X_1 = 4$, and $X_2 = 2$.

(d) $n_1 = 60$, $n_2 = 55$, $X_1 = 24$, and $X_2 = 37$.

(e) $n_1 = 200$, $n_2 = 100$, $X_1 = 128$, and $X_2 = 94$.

8.33   For each of the following, explain what is wrong and why.

(a) A z statistic is used to test the null hypothesis that $\hat{p}_1 = \hat{p}_2$.

(b) If two sample proportions are equal, then the sample counts are equal.

(c) A 95% confidence interval for the difference in two proportions includes errors due to nonresponse.

8.34   Suppose there are two binomial populations. For the first, the true proportion of successes is 0.4; for the second, it is 0.5. Consider taking independent samples from these populations, 50 from the first and 60 from the second.

(a) Find the mean and the standard deviation of the distribution of $\hat{p}_1 - \hat{p}_2$.

(b) This distribution is approximately normal. Sketch this normal distribution and mark the location of the mean.

(c) Find a value d for which the probability is 0.95 that the difference in sample proportions is within $\pm d$. Mark these values on your sketch.

8.35   Gambling behaviors of Division I intercollegiate student-athletes were analyzed in Exercises 8.5 and 8.6. Use the methods of this section to compare the males and females with a significance test and give an estimate of the differences in proportions of student-athletes who participate in any gambling activity with a 95% margin of error. In Exercise 8.5 it is noted that we do not have any information available to assess nonresponse. Consider the possibility that the response rates differ by gender and by whether or not the person participates in any gambling activity. Write a short summary of your opinions on this issue.

8.36   A study that evaluated the effects of a reduction in exposure to traffic-related air pollutants compared respiratory symptoms of 283 residents of an area with congested streets with 165 residents in a similar area where the congestion was removed because a bypass was constructed. The symptoms of the residents were evaluated at baseline and again a year after the bypass was completed.[18] For the residents of the congested streets, 17 reported that their symptoms of wheezing improved between baseline and one year later, while 35 of the residents of the bypass streets reported improvement.

(a) Find the two sample proportions.

(b) Report the difference in the proportions and the standard error of the difference.

(c) What are the appropriate null and alternative hypotheses for examining the question of interest? Be sure to explain your choice of the alternative hypothesis.

(d) Find the test statistic. Construct a sketch of the distribution of the test statistic under the assumption that the null hypothesis is true. Find the P-value and use your sketch to explain its meaning.

(e) Is no evidence of an effect the same as evidence that there is no effect? Use a 95% confidence interval to answer this question. Summarize your ideas in a way that could be understood by someone who has very little experience with statistics.

(f) The study was done in the United Kingdom. To what extent do you think that the results can be generalized to other circumstances?

8.37 In the Health ABC Study,[19] 595 subjects owned a pet and 1939 subjects did not. Among the pet owners, there were 285 women; 1024 of the non–pet owners were women. Find the proportion of pet owners who were women. Do the same for the non–pet owners. Give a 95% confidence interval for the difference in the two proportions. (Be sure to let Population 1 correspond to the group with the higher proportion of women so that the difference will be positive.)

8.38 Refer to the previous exercise. The 595 pet owners and 1939 non–pet owners were also classified according to whether or not they were married. For the pet owners, 53.3% were married, while for the non–pet owners, 57.7% were married. Find a 95% confidence interval for the difference. Write a short summary of your work.

8.39 Refer to Exercise 8.37. Redo the exercise in terms of the proportions of men in each classification. Explain how you could have obtained these results from the calculations you did in Exercise 8.37.

8.40 Refer to Exercise 8.38. Redo the exercise in terms of the proportions of unmarried people in each classification. Explain how you could have obtained these results from the calculations you did in Exercise 8.38.

8.41 A recent survey of Internet users reported that 18% downloaded music onto their computers. The filing of lawsuits by the recording industry may be a reason why this percent has decreased from the estimate of 29% from a survey taken one year before.[20] Assume that the sample sizes are both 1370. Using a significance test, evaluate whether or not there has been a change in the percent of Internet users who download music. Provide all details for the test and summarize your conclusion. Also report a 95% confidence interval for the difference in proportions and explain what information is provided in the interval that is not in the significance test results.

8.42 Refer to the previous exercise. Suppose we are not exactly sure about the sizes of the samples. Redo the calculations for the significance test and the confidence interval under the following assumptions: (*i*) both sample sizes are 1200, (*ii*) both sample sizes are 1600, (*iii*) the first sample size is 1200 and the second is 1600. Summarize the effects of the sample sizes on the results.

8.43 Different kinds of companies compensate their key employees in different ways. Established companies may pay higher salaries, while new companies may offer stock options that will be valuable if the company succeeds. Do high-tech companies tend to offer stock options more often than other companies? One study looked at a random sample of 200 companies. Of these, 91 were listed in the *Directory of Public High Technology Corporations* and

109 were not listed. Treat these two groups as SRSs of high-tech and non-high-tech companies. Seventy-three of the high-tech companies and 75 of the non-high-tech companies offered incentive stock options to key employees.[21] Give a 95% confidence interval for the difference in the proportions of the two types of companies that offer stock options.

8.44   Refer to the previous exercise. Compare the two groups of companies with a significance test. Summarize your analysis and conclusions.

8.45   A study was designed to find reasons why patients leave a health maintenance organization (HMO).[22] Patients were classified as to whether or not they had filed a complaint with the HMO. We want to compare the proportion of complainers who leave the HMO with the proportion of those who do not file complaints but who also leave the HMO. In the year of the study, 639 patients filed complaints, and 54 of these patients left the HMO voluntarily. For comparison, the HMO chose an SRS of 743 patients who had not filed complaints. Twenty-two of these patients left voluntarily. Give an estimate of the difference in the two proportions with a 95% confidence interval.

8.46   In the previous exercise you examined data from a study designed to find reasons why patients leave an HMO. There you compared the proportion of complainers who leave the HMO with the proportion of noncomplainers who leave. In the year of the study, 639 patients filed complaints and 54 of these patients left the HMO voluntarily. For comparison, the HMO chose an SRS of 743 patients who had not filed complaints. Twenty-two of those patients left voluntarily. We expect a higher proportion of complainers to leave. Do the data support this expectation? State hypotheses, find the test statistic and its P-value, and state your conclusion.

8.47   To what extent do syntax textbooks, which analyze the structure of sentences, illustrate gender bias? A study of this question sampled sentences from 10 texts.[23] One part of the study examined the use of the words "girl," "boy," "man," and "woman." We will call the first two words *juvenile* and the last two *adult*. Is the proportion of female references that are juvenile (girl) equal to the proportion of male references that are juvenile (boy)? Here are data from one of the texts:

Gender	n	X(juvenile)
Female	60	48
Male	132	52

(a) Find the proportion of juvenile references for females and its standard error. Do the same for the males.

(b) Give a 90% confidence interval for the difference and briefly summarize what the data show.

8.48   In Exercise 8.20 (page 553) we examined the percent of fatally injured bicyclists tested for alcohol who tested positive. Here we examine the same data with respect to gender.

Gender	n	X(tested positive)
Female	191	27
Male	1520	515

(a) Summarize the data by giving the estimates of the two population proportions and a 95% confidence interval for their difference.

(b) The standard error SE_D contains a contribution from each sample, $\hat{p}_1(1 - \hat{p}_1)/n_1$ and $\hat{p}_2(1 - \hat{p}_2)/n_2$. Which of these contributes the larger amount to the standard error of the difference? Explain why.

8.49   In Exercise 8.37 we compared the proportion of pet owners who were women with the proportion of non–pet owners who were women in the Health ABC study. Use a significance test to make the comparison and summarize the results of your analysis.

8.50   In Exercise 8.38 we compared the proportion of pet owners who were married with the proportion of non–pet owners who were married in the Health ABC study. Use a significance test to make the comparison and summarize the results of your analysis.

8.51   Exercise 8.47 addresses a question about gender bias with a confidence interval. Set up the problem as a significance test. Carry out the test and summarize the results.

8.52   The proportions of female and male fatally injured bicyclists were compared with a confidence interval in Exercise 8.48. Examine the same data with a test of significance.

8.53   A major court case on the health effects of drinking contaminated water took place in the town of Woburn, Massachusetts. A town well in Woburn was contaminated by industrial chemicals. During the period that residents drank water from this well, there were 16 birth defects among 414 births. In years when the contaminated well was shut off and water was supplied from other wells, there were 3 birth defects among 228 births. The plaintiffs suing the firm responsible for the contamination claimed that these data show that the rate of birth defects was higher when the contaminated well was in use.[24]   How statistically significant is the evidence? What assumptions does your analysis require? Do these assumptions seem reasonable in this case?

CHAPTER 8 | Exercises

8.54   The following quote is from a recent survey of Internet users.[25]   The sample size for the survey was 1371. Since 18% of those surveyed said they download music, the sample size for this subsample is 247.

>   Among current music downloaders, 38% say they are downloading less because of the RIAA suits. . . . About a third of current music downloaders

say they use peer-to-peer networks. . . . 24% of them say they swap files using email and instant messaging; 20% download files from music-related Web sites like those run by music magazines or musician homepages. And while online music services like iTunes are far from trumping the popularity of file-sharing networks, 17% of current music downloaders say they are using these paid services. Overall, 7% of Internet users say they have bought music at these new services at one time or another, including 3% who currently use paid services.

(a) For each percent quoted, give the margin of error. Note, you should express these in percents, as given in the quote.

(b) Rewrite the paragraph in a shorter form but include the margins of error.

(c) Pick either side (A) or side (B) and give arguments in favor of the view that you select. (A) The margins of error should be included because they are necessary for the reader to properly interpret the results. (B) The margins of error interfere with the flow of the important ideas. It would be better to just report one margin of error and say that all of the others are no greater than this number. If you choose view (B), be sure to give the value of the margin of error that you report.

8.55 A study of state lotteries included a random digit dialing (RDD) survey conducted by the National Opinion Research Center (NORC). The survey asked 2406 adults about their lottery spending.[26] A total of 248 individuals were classified as "heavy" players. Of these, 152 were male. The study notes that 48.5% of U.S. adults are male. For this analysis, assume that the 248 heavy lottery players are a random sample of all heavy lottery players and that the margin of error for the 48.5% estimate of the percent of males in the U.S. adult population is so small that it can be neglected. Use a significance test to compare the proportion of males among heavy lottery players with the proportion of males in the U.S. adult population. Construct a 95% confidence interval for the proportion. Write a summary of what you have found. Be sure to comment on the possibility that some people may be reluctant to provide information about their lottery spending and how this might affect the results.

8.56 Refer to Exercise 8.43. If you have not already done so, give a 95% confidence interval for the difference in the proportions of the two types of companies that offer stock options. Then compare the two types of companies using a significance test. Be sure to state your hypotheses, the test statistic, and the P-value. Write a short summary of your conclusions.

8.57 In Exercise 8.36 the effects of a reduction in air pollution on wheezing was examined by comparing the one-year change in symptoms in a group of residents who lived on congested streets with a group who lived in an area that had been congested but from which the congestion was removed when a bypass was built. The effect of the reduction in air pollution was assessed by comparing the proportions of residents in the two groups who reported that their wheezing symptoms improved. Here are some additional data from the same study:

Symptom	Bypass		Congested	
	n	Improved	n	Improved
Number of wheezing attacks	282	45	163	21
Wheezing disturbs sleep	282	45	164	12
Wheezing limits speech	282	12	164	4
Wheezing affects activities	281	26	165	13
Winter cough	261	15	156	14
Winter phlegm	253	12	144	10
Consulted doctor	247	29	140	18

The table gives the number of subjects in each group and the number reporting improvement. So, for example, the proportion who reported improvement in the number of wheezing attacks was 21/163 in the congested group.

(a) The reported sample sizes vary from symptom to symptom. Give possible reasons for this and discuss the possible impact on the results.

(b) Calculate the difference in the proportions for each symptom. Make a table of symptoms ordered from highest to lowest based on these differences. Include the estimates of the differences and the 95% confidence intervals in the table. Summarize your conclusions.

(c) Can you justify a one-sided alternative in this situation? Give reasons for your answer.

(d) Perform a significance test to compare the two groups for each of the symptoms. Summarize the results.

(e) Reanalyze the data using only the data from the bypass group. Give confidence intervals for the proportions that reported improved symptoms. Compare the conclusions that someone might make from these results with those you presented in part (b). Use your analyses of the data in this exercise to discuss the importance of a control group in studies such as this.

8.58 Following complaints about the working conditions in some apparel factories both in the United States and abroad, a joint government and industry commission recommended in 1998 that companies that monitor and enforce proper standards be allowed to display a "No Sweat" label on their products. Does the presence of these labels influence consumer behavior? A survey of U.S. residents aged 18 or older asked a series of questions about how likely they would be to purchase a garment under various conditions. For some conditions, it was stated that the garment had a "No Sweat" label; for others, there was no mention of such a label. On the basis of the responses, each person was classified as a "label user" or a "label nonuser."[27] There were 296 women surveyed. Of these, 63 were label users. On the other hand, 27 of 251 men were classified as users.

(a) Give a 95% confidence interval for the difference in the proportions.

(b) You would like to compare the women with the men. Set up appropriate hypotheses, and find the test statistic and the P-value. What do you conclude?

8.59 To devise effective marketing strategies it is helpful to know the characteristics of your customers. A study compared demographic characteristics of people who use the Internet for travel arrangements and of people who do not.[28] Of 1132 Internet users, 643 had completed college. Among the 852 nonusers, 349 had completed college.

(a) Do users and nonusers differ significantly in the proportion of college graduates?

(b) Give a 95% confidence interval for the difference in the proportions.

8.60 The study mentioned in the previous exercise also asked about income. Among Internet users, 493 reported income of less than $50,000 and 378 reported income of $50,000 or more. (Not everyone answered the income question.) The corresponding numbers for nonusers were 477 and 200. Perform a significance test to compare the incomes of users with nonusers and also give an estimate of the difference in proportions with a 95% margin of error.

8.61 Refer to the previous two exercises. Give the total number of users and the total number of nonusers for the analysis of education. Do the same for the analysis of income. The difference is due to respondents who chose "Rather not say" for the income question. Give the proportions of "Rather not say" individuals for users and nonusers. Perform a significance test to compare these and give a 95% confidence interval for the difference. People are often reluctant to provide information about their income. Do you think that this type of nonresponse for the income is a serious limitation for this study?

8.62 The Ping Company makes custom-built golf clubs and competes in the $4 billion golf equipment industry. To improve its business processes, Ping decided to seek ISO 9001 certification.[29] As part of this process, a study of the time it took to repair golf clubs sent to the company by mail determined that 16% of orders were sent back to the customers in 5 days or less. Ping examined the processing of repair orders and made changes. Following the changes, 90% of orders were completed within 5 days. Assume that each of the estimated percents is based on a random sample of 200 orders.

(a) How many orders were completed in 5 days or less before the changes? Give a 95% confidence interval for the proportion of orders completed in this time.

(b) Do the same for orders after the changes.

(c) Give a 95% confidence interval for the improvement. Express this both for a difference in proportions and for a difference in percents.

8.63 A Gallup poll used telephone interviews to survey a sample of 1006 U.S. residents over the age of 18 regarding their ideal family size.[30] The poll reported that 38% of Americans said that their ideal family would include three or more children. Assuming that this is an SRS of U.S. residents over the age of 18, give the 95% margin of error for this estimate.

8.64 According to literature on brand loyalty, consumers who are loyal to a brand are likely to consistently select the same product. This type of consistency could come from a positive childhood association. To examine brand loyalty among fans of the Chicago Cubs, 371 Cubs fans among patrons of a restaurant located in Wrigleyville were surveyed prior to a game at Wrigley Field, the Cubs' home field.[31] The respondents were classified as "die-hard fans" or "less loyal fans." Of the 134 die-hard fans, 90.3% reported that they had watched or listened to Cubs games when they were children. Among the 237 less loyal fans, 67.9% said that they had watched or listened as children.

(a) Find the numbers of die-hard Cubs fans who watched or listened to games when they were children. Do the same for the less loyal fans.

(b) Use a significance test to compare the die-hard fans with the less loyal fans with respect to their childhood experiences relative to the team.

(c) Express the results with a 95% confidence interval for the difference in proportions.

8.65 The study mentioned in the previous exercise found that two-thirds of the die-hard fans attended Cubs games at least once a month, but only 20% of the less loyal fans attended this often. Analyze these data using a significance test and a confidence interval. Write a short summary of your findings.

8.66 Refer to the study of gender bias and stereotyping described in Exercise 8.47 (page 569). Here are the counts of "girl," "woman," "boy," and "man" for all of the grammar texts studied. The one we analyzed in Exercise 8.47 was number 6.

	Text number									
	1	2	3	4	5	6	7	8	9	10
Girl	2	5	25	11	2	48	38	5	48	13
Woman	3	2	31	65	1	12	2	13	24	5
Boy	7	18	14	19	12	52	70	6	128	32
Man	27	45	51	138	31	80	2	27	48	95

For each text perform the significance test to compare the proportions of juvenile references for females and males. Summarize the results of the significance tests for the 10 texts studied. The researchers who conducted the study note that the authors of the last three texts are women, while the other seven texts were written by men. Do you see any pattern that suggests that the gender of the author is associated with the results?

8.67 Refer to the previous exercise. Let us now combine the categories "girl" with "woman" and "boy" with "man." For each text calculate the proportion of male references and test the hypothesis that male and female references are equally likely (that is, the proportion of male references is equal to 0.5). Summarize the results of your 10 tests. Is there a pattern that suggests a relation with the gender of the author?

8.68 Many new products introduced into the market are targeted toward children. The choice behavior of children with regard to new products is of particular interest to companies that design marketing strategies for these products. As part of one study, children in different age groups were compared on their ability to sort new products into the correct product category (milk or juice).[32] Here are some of the data:

Age group	n	Number who sorted correctly
4- to 5-year-olds	50	10
6- to 7-year-olds	53	28

Test the null hypothesis that the two age groups are equally skilled at sorting. Justify your choice of an alternative hypothesis. Also, give a 90% confidence interval for the difference. Summarize your results in a short paragraph.

8.69 The Gallup poll in Exercise 8.63 reported that in a similar poll in 1973, 43% of Americans said that their ideal family would include three or more children. Give the margin of error for this poll, assuming that the sample size was the same. Then compare the proportions with a significance test and give a 95% confidence interval for the difference. Write a summary of your results.

8.70 In a random sample of 875 students from a large public university, it was found that 411 of the students changed majors during their college years.

(a) Give a 95% confidence interval for the proportion of students at this university who change majors.

(b) Express your results from (a) in terms of the *percent* of students who change majors.

(c) University officials concerned with counseling students are interested in the number of students who change majors rather than the proportion. The university has 37,000 undergraduate students. Convert the confidence interval you found in (a) to a confidence interval for the *number* of students who change majors during their college years.

8.71 Go to the Gallup poll Web site `http://www.gallup.com/` and find a poll that has several questions of interest to you. Summarize the results of the poll giving margins of error and comparisons of interest. (For this exercise, you may assume that the data come from an SRS.)

8.72 Gastric freezing was once a recommended treatment for ulcers in the upper intestine. A randomized comparative experiment found that 28 of the 82 patients who were subjected to gastric freezing improved, while 30 of the 78 patients in the control group improved.

(a) State the appropriate null hypothesis and a two-sided alternative. Carry out a z test. What is the P-value?

(b) What do you conclude about the effectiveness of gastric freezing as a treatment for ulcers? (See Example 3.5 on page 197 for a discussion of gastric freezing.)

8.73 In this exercise we examine the effect of the sample size on the significance test for comparing two proportions. In each case suppose that $\hat{p}_1 = 0.6$ and $\hat{p}_2 = 0.5$, and take n to be the common value of n_1 and n_2. Use the z statistic to test $H_0: p_1 = p_2$ versus the alternative $H_a: p_1 \neq p_2$. Compute the statistic and the associated P-value for the following values of n: 10, 20, 40, 50, 80, 100, 500, and 1000. Summarize the results in a table. Explain what you observe about the effect of the sample size on statistical significance when the sample proportions $\hat{p}_1$ and $\hat{p}_2$ are unchanged.

8.74 In the first section of this chapter, we studied the effect of the sample size on the margin of error of the confidence interval for a single proportion. In this exercise we perform some calculations to observe this effect for the two-sample problem. Suppose that $\hat{p}_1 = 0.6$ and $\hat{p}_2 = 0.5$, and n represents the common value of n_1 and n_2. Compute the 95% margins of error for the difference in the two proportions for n = 10, 20, 40, 50, 80, 100, 500, and 1000. Be sure to use the plus four method where appropriate. Present the results in a table and with a graph. Write a short summary of your findings.

8.75 For a single proportion the margin of error of a confidence interval is largest for any given sample size n and confidence level C when $\hat{p} = 0.5$. This led us to use $p^* = 0.5$ for planning purposes. The same kind of result is true for the two-sample problem. The margin of error of the confidence interval for the difference between two proportions is largest when $\hat{p}_1 = \hat{p}_2 = 0.5$. Use these conservative values in the following calculations, and assume that the sample sizes n_1 and n_2 have the common value n. Calculate the margins of error of the 95% confidence intervals for the difference in two proportions for the following choices of n: 10, 20, 40, 50, 80, 100, 500, and 1000. Be sure to use the plus four method where appropriate. Present the results in a table and with a graph. Summarize your conclusions.

8.76 As the previous problem noted, using the guessed value 0.5 for both $\hat{p}_1$ and $\hat{p}_2$ gives a conservative margin of error in confidence intervals for the difference between two population proportions. You are planning a survey and will calculate a 95% confidence interval for the difference in two proportions when the data are collected. You would like the margin of error of the interval to be less than or equal to 0.10. You will use the same sample size n for both populations.

(a) How large a value of n is needed?

(b) Give a general formula for n in terms of the desired margin of error m and the critical value z^*.

8.77 *Castaneda v. Partida* is an important court case in which statistical methods were used as part of a legal argument.[33] When reviewing this case, the Supreme Court used the phrase "two or three standard deviations" as a criterion for statistical significance. This Supreme Court review has served as the basis for many subsequent applications of statistical methods in legal settings. (The two or three standard deviations referred to by the Court are values of the z statistic and correspond to P-values of approximately 0.05 and 0.0026.) In *Castaneda* the plaintiffs alleged that the method for selecting juries in a county in Texas was biased against Mexican Americans. For the period of time at issue, there were 181,535 persons eligible for jury duty, of whom

143,611 were Mexican Americans. Of the 870 people selected for jury duty, 339 were Mexican Americans.

(a) What proportion of eligible voters were Mexican Americans? Let this value be p_0.

(b) Let p be the probability that a randomly selected juror is a Mexican American. The null hypothesis to be tested is $H_0: p = p_0$. Find the value of $\hat{p}$ for this problem, compute the z statistic, and find the P-value. What do you conclude? (A finding of statistical significance in this circumstance does not constitute a proof of discrimination. It can be used, however, to establish a prima facie case. The burden of proof then shifts to the defense.)

(c) We can reformulate this exercise as a two-sample problem. Here we wish to compare the proportion of Mexican Americans among those selected as jurors with the proportion of Mexican Americans among those not selected as jurors. Let p_1 be the probability that a randomly selected juror is a Mexican American, and let p_2 be the probability that a randomly selected nonjuror is a Mexican American. Find the z statistic and its P-value. How do your answers compare with your results in (b)?

8.78 In many sports there is a home field or home court advantage. This means that the home team is more likely to win when playing at home than they are to win when playing at an opponent's field or court, all other things being equal. Go to the Web site of your favorite sports team and find the proportion of wins for home games and the proportion of wins for away games. Now consider these games to be a random sample of the process that generates wins and losses. A complete analysis of data like these requires methods that are beyond what we have studied, but the methods discussed in this chapter will give us a reasonable approximation. Examine the home court advantage for your team and write a summary of your results. Be sure to comment on the effect of the sample size.

8.79 The National Student Loan Survey asked the student loan borrowers in their sample about attitudes toward debt.[34] Here are some of the questions they asked, with the percent who responded in a particular way:

(a) "To what extent do you feel burdened by your student loan payments?" 55.5% said they felt burdened.

(b) "If you could begin again, taking into account your current experience, what would you borrow?" 54.4% said they would borrow less.

(c) "Since leaving school, my education loans have not caused me more financial hardship than I had anticipated at the time I took out the loans." 34.3% disagreed.

(d) "Making loan payments is unpleasant but I know that the benefits of education loans are worth it." 58.9% agreed.

(e) "I am satisfied that the education I invested in with my student loan(s) was worth the investment for career opportunities." 58.9% agreed.

(f) "I am satisfied that the education I invested in with my student loan(s) was worth the investment for personal growth." 71.5% agreed.

Assume that the sample size is 1280 for all of these questions. Compute 95% confidence intervals for each of the questions and write a short report about what student loan borrowers think about their debt.

EESEE CASE STUDIES

The Electronic Encyclopedia of Statistical Examples and Exercises (EESEE) is available on the text CD and Web site. These more elaborate stories, with data, provide settings for longer case studies.

8.80   Read the EESEE story "Leave Surgery After the Beep." Write a report that answers all questions for this case study.

8.81   Read the EESEE story "Drive-through Competition." Write a report that answers all questions for this case study.

Topics in Inference

Finding Genes in Plants and People

REBECCA DOERGE *is Professor of Statistics and Agronomy at Purdue University in Indiana.*

Genes lie behind much of the variation among people, animals, and plants. But where in the three billion "base pairs" of human DNA are the genes involved in a disease? I use statistics to track down genes and to link them to the functions they regulate.

My research aims to unravel the mystery of which genes are involved in diseases and traits that affect agricultural plants, trees, and humans. Not so surprisingly, many of the biological processes that lead to variation in humans lead to similar kinds of variation in animals, agricultural crops, and other organisms.

As a statistical geneticist, I try to identify the genes (segments of chromosomes made up of DNA) that are involved in human diseases such as multiple sclerosis and different types of cancer. Using statistics I am able to describe the variation of a particular disease in both sick and well individuals and then associate or correlate this variation with regions of chromosomes that contain genes influencing the disease. Once pointed to the correct chromosomes and regions of the chromosomes, researchers can study the genes in those areas so that doctors can understand how they influence a particular disease.

In agricultural plants a similar statistical analysis called quantitative trait locus mapping allows us to identify genes that are responsible for specific functions and to breed them into related species to carry over their beneficial properties. For example, there is a type of tomato that grows in South America that is naturally resistant to insects such as aphids and white flies. We can identify the genes that cause this natural resistance and breed them into the cultivated tomato that we eat. Consumers eat healthy foods at lower prices, and we protect the environment from harmful chemicals.

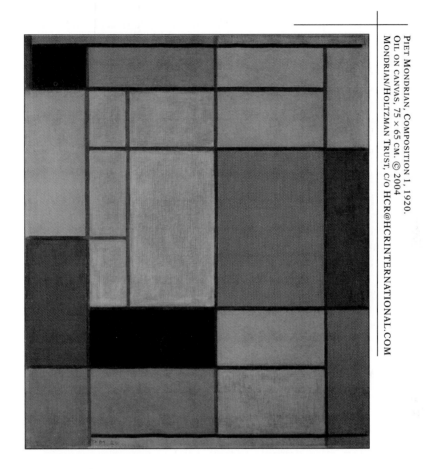

Analysis of Two-Way Tables

Introduction

We continue our study of methods for analyzing categorical data in this chapter. Inference about proportions in one-sample and two-sample settings was the focus of Chapter 8. We now study how to compare two or more populations when the response variable has two or more possible values and how to test whether two categorical variables are independent. A single statistical test handles both of these cases. The corresponding procedures for quantitative variables are more elaborate; they will appear in Chapters 12 and 13.

Of course, we need to do a careful job of describing the data before we apply our methods of statistical inference. Therefore, we begin our study with a discussion of data analysis methods for two-way tables. The second section of this chapter gives the basics of statistical inference that are appropriate in this setting. An optional third section provides some technical details. The goodness of fit test is presented in the last section.

9.1   Data Analysis for Two-Way Tables

When we studied inference for two proportions in Chapter 8, we started by summarizing the raw data by giving the number of observations in each group (n) and how many of these were classified as "successes" (X).

EXAMPLE 9.1   Here is a summary from Example 8.9 (page 557), where we compared frequent binge drinking of men and women by examining the proportions of each gender who engage in this activity.

Population	n	X
1 (men)	7,180	1,630
2 (women)	9,916	1,684
Total	17,096	3,314

The two-way table

In this chapter we use a different summary of the data. Rather than recording just the count of binge drinkers, we record counts for all outcomes in a **two-way table** **way table.**

two-way table

EXAMPLE 9.2   Here is the two-way table classifying students by gender and whether or not they are binge drinkers:

Two-way table for frequent binge drinking and gender

Frequent binge drinker	Men	Women	Total
	Gender		
Yes	1,630	1,684	3,314
No	5,550	8,232	13,782
Total	7,180	9,916	17,096

For this problem we could view gender as an explanatory variable and frequent binge drinking as a response variable. This is why we put gender in the columns (like the x axis in a regression) and frequent binge drinking in the rows (like the y axis in a regression).

Be sure that you understand how this table is obtained from the one in Example 9.1. *Most errors in the use of categorical-data methods come from a misunderstanding of how these tables are constructed.*

In Chapter 2 we studied relationships in which at least the response variable was quantitative. In this section we have a similar goal; but here both of the variables are categorical. Some variables—such as gender, race, and occupation—are inherently categorical. Other categorical variables are created by grouping values of a quantitative variable into classes. Published data are often reported in grouped form to save space. To describe categorical data, we use the *counts* (frequencies) or *percents* (relative frequencies) of individuals that fall into various categories.

row and column variables   The two-way table in Example 9.2 describes two categorical variables, gender and binge drinking. We call binge drinking the **row variable** because each horizontal row in the table describes the drinking behavior. Gender is the **column variable** because each vertical column describes one gender group. Each combination of values for these two variables is called a **cell.** For example, the cell corresponding to men who are frequent binge drinkers contains the number 1630. This table is called a 2×2 table because there are 2 rows and 2 columns.

Suppose we are interested in the men who are binge drinkers. The proportion of these is simply 1630 divided by 17,096, or 0.095. We would estimate that 9.5% of college students are male frequent binge drinkers. For each cell, a proportion is obtained by dividing the cell entry by the total sample size. The **joint distribution**   collection of these proportions is the **joint distribution** of the two categorical variables.

EXAMPLE 9.3   For the binge-drinking example, the joint distribution of binge drinking and gender is

Joint distribution of frequent binge drinking and gender		
	Gender	
Frequent binge drinker	Men	Women
Yes	0.095	0.099
No	0.325	0.482

Because this is a distribution, the sum of the proportions should be 1. For this example the sum is 1.001. The difference is due to roundoff error.

From the joint distribution we see that the proportions of men and women frequent binge drinkers are similar in the population of college students. For the men we have 9.5%; the women are slightly higher at 9.9%. Note, however, that the proportion of women who are not frequent binge drinkers is also higher than the proportion of men. One reason for this is that there are more women in the sample than men. To understand this set of data we will need to do some additional calculations. Let's look at the distribution of gender.

Marginal distributions

marginal
distribution

When we examine the distribution of a single variable in a two-way table, we are looking at a **marginal distribution.** There are two marginal distributions, one for each categorical variable in the two-way table. They are very easy to compute.

EXAMPLE 9.4   Look at the table in Example 9.2. The total numbers of men and women are given in the bottom row, labeled "Total." Our sample has 7180 men and 9916 women. To find the marginal distribution of gender we simply divide these numbers by the total sample size, 17,096. The marginal distribution of gender is

Marginal distribution of gender		
	Men	**Women**
Proportion	0.420	0.580

Note that the proportions sum to 1; there is no roundoff error.

Often we prefer to use percents rather than proportions. Here is the marginal distribution of gender described with percents:

Marginal distribution of gender		
	Men	**Women**
Proportion	42.0%	58.0%

Which form do you prefer?

The other marginal distribution for this example is the distribution of binge drinking.

EXAMPLE 9.5   In the table in Example 9.2, the numbers of frequent binge drinkers and non–frequent binge drinkers are given in the last column. This is the other margin of the table. There are 3314 frequent binge drinkers. The other 13,782 students are not frequent binge drinkers. If we convert these counts to percents, we obtain the marginal distribution of the frequent-binge-drinking variable (in terms of percents):

Marginal distribution of frequent binge drinking	
Yes	No
Percent 19.4%	80.6%

Each marginal distribution from a two-way table is a distribution for a single categorical variable. As we saw in Chapter 1 (page 7), we can use a bar graph or a pie chart to display such a distribution. For our two-way table, we will be content with numerical summaries: 58% of these college students are women, and 19.4% of the students are frequent binge drinkers. When we have more rows or columns, the graphical displays are particularly useful. We will illustrate these for other examples in Figures 9.3, 9.6, and 9.7 (pages 593 and 603).

Describing relations in two-way tables

The table in Example 9.2 contains much more information than the two marginal distributions of gender alone and frequent binge drinking alone. We need to do a little more work to examine the relationship. *Relationships among categorical variables are described by calculating appropriate percents from the counts given.* What percents do you think we should use to describe the relationship between gender and frequent binge drinking?

EXAMPLE 9.6   What percent of the women in our sample are frequent binge drinkers? This is the count of the women who are frequent binge drinkers as a percent of the number of women in the sample:

$$\frac{1684}{9916} = 0.170 = 17.0\%$$

A similar calculation gives the percent of men who are frequent binge drinkers:

$$\frac{1630}{7180} = 0.227 = 22.7\%$$

Recall that among all college students in the sample, 9.5% were male frequent binge drinkers and 9.9% were female frequent binge drinkers. These percents are based on the joint distribution. They are fairly similar because the counts for these two groups, 1630 and 1684, are close. The calculations that we just performed, however, give us a different view. When we look separately at women and men, we see that the proportions of frequent binge drinkers are somewhat different, 17.0% for women versus 22.7% for men.

Conditional distributions

In Example 9.6 we first looked at the women alone and then examined the distribution of the other categorical variable, frequent binge drinking. Another way to say this is that we conditioned on the value of gender being female. Similarly we conditioned on the value of gender being male. When we condition on the value of one variable and calculate the distribution of the other variable, we obtain a **conditional distribution.** Note that in Example 9.6 we calculated the percents for only frequent binge drinking. The complete conditional distribution gives the proportions or percents for all possible values.

conditional distribution

EXAMPLE 9.7   For women, the conditional distribution of the frequent-binge-drinking variable in terms of percents is

Conditional distribution of frequent binge drinking for women	
Yes	**No**
Percent 17.0%	83.0%

Note that we have included the percents for both of the possible values, Yes and No, of the frequent-binge-drinking variable. These percents sum to 100%. Similar calculations give the corresponding distribution for men:

Conditional distribution of frequent binge drinking for men	
Yes	**No**
Percent 22.7%	77.3%

Comparing the conditional distributions reveals the nature of the association between gender and frequent binge drinking. In this set of data the men are more likely to be frequent binge drinkers than the women.

Bar graphs can help us to see relationships between two categorical variables. We will use these in our example in Section 9.2. No single graph (such as a scatterplot) portrays the form of the relationship between categorical variables, and no single numerical measure (such as the correlation) summarizes the strength of an association. Bar graphs are flexible enough to be helpful, but you must think about what comparisons you want to display. For numerical measures, we must rely on well-chosen percents or on more advanced statistical methods.[1]

A two-way table contains a great deal of information in compact form. Making that information clear almost always requires finding percents. You must decide which percents you need. Of course, we prefer to use software to compute the joint, marginal, and conditional distributions.

EXAMPLE 9.8   Figure 9.1 gives computer output for the data in Example 9.2 using SPSS, Minitab, and SAS. There are minor variations among software packages, but these are typical of what is usually produced. Each cell in the 2 × 2 table has four entries. These are the count (the number of observations in the cell), the conditional distributions for rows and columns, and the joint distribution. Note

SPSS

BINGE * GENDER Crosstabulation

			GENDER Male	Female	Total
BINGE	Yes	Count	1630	1684	3314
		% within BINGE	49.2%	50.8%	100.0%
		% within GENDER	22.7%	17.0%	19.4%
		% of Total	9.5%	9.9%	19.4%
	No	Count	5550	8232	13782
		% within BINGE	40.3%	59.7%	100.0%
		% within GENDER	77.3%	83.0%	80.6%
		% of Total	32.5%	48.2%	80.6%
Total		Count	7180	9916	17096
		% within BINGE	42.0%	58.0%	100.0%
		% within GENDER	100.0%	100.0%	100.0%
		% of Total	42.0%	58.0%	100.0%

Minitab

	no	yes	All
men	5550	1630	7180
	77.30	22.70	100.00
	40.27	49.19	42.00
	32.46	9.53	42.00
women	8232	1684	9916
	83.02	16.98	100.00
	59.73	50.81	58.00
	48.15	9.85	58.00
All	13782	3314	17096
	80.62	19.38	100.00
	100.00	100.00	100.00
	80.62	19.38	100.00

Cell Contents --
               Count
               % of Row
               % of Col
               % of Tbl

FIGURE 9.1   Computer output for the binge-drinking study in Example 9.2. *(continued)*

SAS

```
Table of Binge by Gender

                Binge      Gender

                Frequency
                Percent
                Row Pct
                Col Pct    Men       Women     Total

                No             5550      8232    13782
                              32.46     48.15    80.62
                              40.27     59.73
                              77.30     83.02

                Yes            1630      1684     3314
                               9.53      9.85    19.38
                              49.19     50.81
                              22.70     16.98

                Total          7180      9916    17096
                              42.00     58.00   100.00
```

FIGURE 9.1 *(Continued)* Computer output for the binge-drinking study in Example 9.2.

that all of these are expressed as percents rather than proportions. Marginal totals and distributions are given in the rightmost column and the bottom row.

Most software packages order the row and column labels numerically or alphabetically. In general it is better to use words rather than numbers for the column labels. This sometimes involves some additional work but it avoids the kind of confusion that can result when you forget the real values associated with each numerical value. You should verify that the entries in Figure 9.1 correspond to the calculations that we performed in Examples 9.3 to 9.7. In addition, verify the calculations for the conditional distributions of gender for each value of the frequent-binge-drinking variable.

Simpson's paradox

As is the case with quantitative variables, the effects of lurking variables can strongly influence relationships between two categorical variables. Here is an example that demonstrates the surprises that can await the unsuspecting consumer of data.

EXAMPLE 9.9 Air travelers would like their flights to arrive on time. Airlines collect data on on-time arrivals and report them to the government. Following are one month's data for flights from two western cities for two airlines:

	Alaska Airlines	America West
On time	718	5534
Delayed	74	532
Total	792	6066

Alaska Airlines is delayed 9.3% (74/792) of the time, and America West is delayed only 8.8% (532/6066) of the time. It seems that you should choose America West to minimize delays.

Some cities are more likely to have delays than others, however. If we consider which city the flights left from in our analysis, we obtain a more complete picture of the two airlines' on-time flights.

EXAMPLE 9.10 Here are the data broken down by which city each flight left from.[2] Check that the entries in the original two-way table are just the sums of the city entries in this table.

	Los Angeles			Phoenix		
	On time	Delayed	Total	On time	Delayed	Total
Alaska Airlines	497	62	559	221	12	233
America West	694	117	811	4840	415	5255

Alaska Airlines beats America West for flights from Los Angeles: only 11.1% (62/559) delayed compared with 14.4% (117/811) for America West. Alaska Airlines wins again for flights from Phoenix, with 5.2% (12/233) delayed versus 7.9% (415/5255). So Alaska Airlines is the better choice for both Los Angeles and Phoenix.

The city of origin for each flight is a lurking variable when we compare the late-flight percents of the two airlines. When we ignore the lurking variable, America West seems better, even though Alaska Airlines does better for each city. How can Alaska Airlines do better for both cities, yet do worse overall? Let's look at the data. Seventy-one percent (559/792) of Alaska Airlines flights are from Los Angeles, where there are more delays for both airlines. On the other hand, 87% of the America West flights are from Phoenix, where there are few delays. *The original two-way table, which did not take account of the city of origin for each flight, was misleading.* This example illustrates *Simpson's paradox.*

> ## SIMPSON'S PARADOX
>
> An association or comparison that holds for all of several groups can reverse direction when the data are combined to form a single group. This reversal is called **Simpson's paradox.**

The lurking variables in Simpson's paradox are categorical. That is, they break the observations into groups, such as the city of origin for the airline flights. *Simpson's paradox is an extreme form of the fact that observed associations can be misleading when there are lurking variables.*

The perils of aggregation

three-way table The flight data in Example 9.10 is a **three-way table** that reports the frequencies of each combination of levels of three categorical variables: city, airline, and delayed or not. We present a three-way table as two or more two-way tables, one for each level of the third variable. In Example 9.10, there is a separate table of airline versus delayed or not for each city. We can obtain the original two-way table by adding the corresponding entries in these tables. **aggregation** This is called **aggregating** the data. Aggregation has the effect of ignoring the city variable, which then becomes a lurking variable. *Conclusions that seem obvious when we look only at aggregated data can become debatable when we examine lurking variables.* Here is another example.

EXAMPLE 9.11 A study by the National Science Foundation found that the median salary of newly graduated female engineers and scientists was only 73% of the median salary for males.[3] When the new graduates were broken down by field, however, the picture changed. Women's median salaries as a percent of the male median in the 16 fields studied were

94%	96%	98%	95%	85%	85%	84%	100%
103%	100%	107%	93%	104%	93%	106%	100%

How can women do nearly as well as men in every field, and better in some, yet fall far behind men when we look at all young engineers and scientists? Women are concentrated in the life and social sciences, where salaries are lower than in the physical sciences and engineering. When salaries from all fields are lumped together, the women place lower because the fields they favor pay less. Field of science is a lurking variable that influences the observed association between gender and salary.

SECTION 9.1 | Summary

A **two-way table** of counts organizes data about two categorical variables. Values of the **row variable** label the rows that run across the table, and values of the **column variable** label the columns that run down the table. Two-way tables are often used to summarize large amounts of data by grouping outcomes into categories.

The **joint distribution** of the row and column variables is found by dividing the count in each cell by the total number of observations.

The **row totals** and **column totals** in a two-way table give the **marginal distributions** of the two variables separately. It is clearer to present these distributions as percents of the table total. Marginal distributions do not give any information about the relationship between the variables.

To find the **conditional distribution** of the row variable for one specific value of the column variable, look only at that one column in the table. Find each entry in the column as a percent of the column total.

There is a conditional distribution of the row variable for each column in the table. Comparing these conditional distributions is one way to describe the association between the row and the column variables. It is particularly useful when the column variable is the explanatory variable. When the row variable is explanatory, find the conditional distribution of the column variable for each row and compare these distributions.

Bar graphs are a flexible means of presenting categorical data. There is no single best way to describe an association between two categorical variables.

We present data on three categorical variables in a **three-way table,** printed as separate two-way tables for each level of the third variable. A comparison between two variables that holds for each level of a third variable can be changed or even reversed when the data are **aggregated** by summing over all levels of the third variable. **Simpson's paradox** refers to the reversal of a comparison by aggregation. It is an example of the potential effect of lurking variables on an observed association.

9.2 Inference for Two-Way Tables

In the previous section we used a 2×2 table to illustrate the basic ideas about joint, marginal, and conditional distributions. There we learned that the key to examining the relationship between two categorical variables is to look at conditional distributions. In this section we discuss the significance test that we use to examine the relationship.

EXAMPLE 9.12 Many popular businesses are franchises—think of McDonald's. The owner of a local franchise benefits from the brand recognition, national advertising, and detailed guidelines provided by the franchise chain. In return, he or she pays fees to the franchise firm and agrees to follow its policies. The relationship between the local entrepreneur and the franchise firm is spelled out in a detailed contract.

One clause that the contract may or may not contain is the entrepreneur's right to an exclusive territory. This means that the new outlet will be the only representative of the franchise in a specified territory and will not have to compete with other outlets of the same chain. How does the presence of an exclusive-territory clause in the contract relate to the survival of the business? A study designed to address this question collected data from a sample of 170 new franchise firms.[4]

Two categorical variables were measured for each firm. First, the firm was classified as successful or not based on whether or not it was still franchising as of a certain date. Second, the contract each firm offered to franchisees was classified according to whether or not there was an exclusive-territory clause. Here are the data:

Observed numbers of firms

	Exclusive territory		
Success	Yes	No	Total
Yes	108	15	123
No	34	13	47
Total	142	28	170

The two categorical variables in this example are "Success," with values "Yes" and "No," and "Exclusive territory," with values "Yes" and "No." The objective in this example is to compare franchises that have exclusive territories with those that do not. We view "Exclusive territory" as an explanatory variable and therefore we make it the column variable. The row variable is a categorical response variable, "Success."

EXAMPLE 9.13   Figure 9.2 shows the output from Crunch It! for the data of Example 9.12. It is similar to Figure 9.1, which displayed the outputs from other statistical packages for Example 9.2, but the new figure includes some additional information needed for statistical inference which we deleted from those outputs. Check

	Yes	No	Total
Yes	108 87.8% 76.06% 63.53%	15 12.2% 53.57% 8.824%	123 100.00% 72.35% 72.35%
No	34 72.34% 23.94% 20%	13 27.66% 46.43% 7.647%	47 100.00% 27.65% 27.65%
Total	142 83.53% 100.00% 83.53%	28 16.47% 100.00% 16.47%	170 100.00% 100.00% 100.00%

Cell format:
Count
Row percent
Column percent
Total percent

Test for independence of Success and Exclusive Territory:

Statistic	DF	Value	P-value
Chi-square	1	5.9111857	0.015

FIGURE 9.2   Crunch It! computer output for Example 9.13.

Figure 9.2 carefully. Be sure that you can identify the joint distribution, the marginal distributions, and the conditional distributions.

In this example, we are interested in the effect of having an exclusive territory on the success of the firm.

EXAMPLE 9.14   To compare firms that have an exclusive territory with those that do not, we examine column percents. Here they are, rounded from the output for clarity:

Column percents for firms		
	Exclusive territory	
Success	**Yes**	**No**
Yes	76%	54%
No	24%	46%
Total	100%	100%

The "Total" row reminds us that 100% of each type of firm have been classified as successful or not. (The sums sometimes differ slightly from 100% because of roundoff error.) The bar graph in Figure 9.3 compares the percents. The data reveal a clear re-

FIGURE 9.3   Bar graph of the percents of firms that are successful, for Example 9.14.

lationship: 76% of firms with exclusive territories are successful, as opposed to only 54% of firms that were not offered exclusivity.

The difference between the percents of successes among the two types of firms is quite large. A statistical test will tell us whether or not these differences can be plausibly attributed to chance. Specifically, if there is no association between success and having an exclusive territory, how likely is it that a sample would show differences as large or larger than those displayed in Figure 9.3?

The hypothesis: no association

The null hypothesis H_0 of interest in a two-way table is: There is *no association* between the row variable and the column variable. In Example 9.12, this null hypothesis says that success and having an exclusive territory are not related. The alternative hypothesis H_a is that there is an association between these two variables. The alternative H_a does not specify any particular direction for the association. For two-way tables in general, the alternative includes many different possibilities. Because it includes all of the many kinds of association that are possible, we cannot describe H_a as either one-sided or two-sided.

In our example, the hypothesis H_0 that there is no association between success and having an exclusive territory is equivalent to the statement that the distributions of the success variable are the same among firms with and without exclusive territories. For other two-way tables, where the columns correspond to independent samples from distinct populations, there are c distributions for the row variable, one for each population. The null hypothesis then says that the c distributions of the row variable are identical. The alternative hypothesis is that the distributions are not all the same.

Expected cell counts

expected cell counts

To test the null hypothesis in $r \times c$ tables, we compare the observed cell counts with **expected cell counts** calculated under the assumption that the null hypothesis is true. A numerical summary of the comparison will be our test statistic.

EXAMPLE 9.15   The observed and expected counts for the successful-firms example appear in the Minitab computer output shown in Figure 9.4. The expected counts are given as the second entry in each cell and are labeled Exp Freq. For example, in the first cell the observed count is 108 and the expected count is 102.74.

How is this expected count obtained? Look at the percents in the right margin of the table in Figure 9.2. We see that 72.35% of all firms are successful. If the null hypothesis of no relation between success and exclusive territories is true, we expect this overall percent to apply to both firms with and firms without exclusive territories. In particular, we expect 72.35% of the firms with exclusive territories to be successful.

```
Rows: Success       Columns: Excl

            1_Yes       2_No        All

  1_Yes       108         15         123
            102.74      20.26      123.00

  2_No         34         13          47
             39.26       7.74       47.00

  All         142         28         170
            142.00      28.00      170.00

Chi-Square = 5.911, DF = 1, P    -Value = 0.015


   Cell Contents     --
                     Count
                     Exp Freq
```

FIGURE 9.4 Minitab computer output for Example 9.15.

Since there are 142 firms with exclusive territories, the expected count is 72.35% of 142, or 102.74. The other expected counts are calculated in the same way.

The reasoning of Example 9.15 leads to a simple formula for calculating expected cell counts. To compute the expected count for successful firms with exclusive territories, we multiplied the proportion of successful firms (123/170) by the number of firms with exclusive territories (142). From Figure 9.4 we see that the numbers 123 and 142 are the row and column totals for the cell of interest and that 170 is n, the total number of observations for the table. The expected cell count is therefore the product of the row and column totals divided by the table total.

$$\text{expected cell count} = \frac{\text{row total} \times \text{column total}}{n}$$

The chi-square test

To test the H_0 that there is no association between the row and column classifications, we use a statistic that compares the entire set of observed counts with the set of expected counts. First, take the difference between each observed count and its corresponding expected count, and square these values so that they are all 0 or positive. A large difference means less if it comes from a cell that we think will have a large count, so divide each squared difference by the expected count, a kind of standardization. Finally, sum over all cells. The result is called the *chi-square statistic* X^2. The chi-square statistic was invented by the English statistician Karl Pearson (1857–1936) in 1900, for purposes slightly different from ours. It is the oldest inference procedure still used in its original form. With the work of Pearson and his contemporaries at the beginning of this century, statistics first emerges as a separate discipline.

CHI-SQUARE STATISTIC

The **chi-square statistic** is a measure of how much the observed cell counts in a two-way table diverge from the expected cell counts. The formula for the statistic is

$$X^2 = \sum \frac{(\text{observed count} - \text{expected count})^2}{\text{expected count}}$$

where "observed" represents an observed sample count, "expected" represents the expected count for the same cell, and the sum is over all $r \times c$ cells in the table.

If the expected counts and the observed counts are very different, a large value of X^2 will result. Large values of X^2 provide evidence against the null hypothesis. To obtain a P-value for the test, we need the sampling distribution of X^2 under the assumption that H_0 (no association between the row and column variables) is true. We once again use an approximation, related to the normal approximation for binomial distributions. The result is a new distribution, the **chi-square distribution,** which we denote by χ^2 (χ is the lowercase Greek letter chi).

chi-square distribution χ^2

Like the t distributions, the χ^2 distributions form a family described by a single parameter, the degrees of freedom. We use $\chi^2(\text{df})$ to indicate a particular member of this family. Figure 9.5 displays the density curves of the $\chi^2(2)$ and $\chi^2(4)$ distributions. As the figure suggests, χ^2 distributions take only positive values and are skewed to the right. Table F in the back of the book gives upper critical values for the χ^2 distributions.

0

(a)

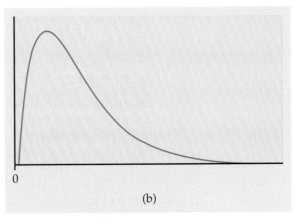
0

(b)

FIGURE 9.5   (a) The $\chi^2(2)$ density curve. (b) The $\chi^2(4)$ density curve.

CHI-SQUARE TEST FOR TWO-WAY TABLES

The null hypothesis H_0 is that there is no association between the row and column variables in a two-way table. The alternative is that these variables are related.

If H_0 is true, the chi-square statistic X^2 has approximately a χ^2 distribution with $(r-1)(c-1)$ degrees of freedom.

The P-value for the chi-square test is

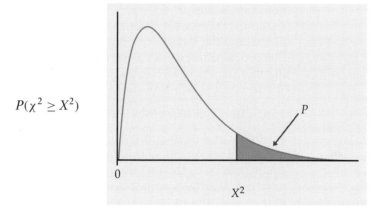

$P(\chi^2 \geq X^2)$

where χ^2 is a random variable having the $\chi^2(\text{df})$ distribution with df $= (r-1)(c-1)$.

The chi-square test always uses the upper tail of the χ^2 distribution, because any deviation from the null hypothesis makes the statistic larger. The approximation of the distribution of X^2 by χ^2 becomes more accurate as the cell counts increase. Moreover, it is more accurate for tables larger than 2×2 tables. For tables larger than 2×2, we will use this approximation whenever the average of the expected counts is 5 or more and the smallest expected count is 1 or more. For 2×2 tables, we require that all four expected cell counts be 5 or more.[5]

EXAMPLE 9.16   The results of the chi-square significance test for the successful-firms example appear in the computer outputs in Figures 9.2 and 9.4, labeled Chi-Square. Because all the expected cell counts are moderately large, the χ^2 distribution provides an accurate P-value. We see that $X^2 = 5.91$, df $= 1$, and the $P = 0.015$. As a check we verify that the degrees of freedom are correct for a 2×2 table:

$$\text{df} = (r-1)(c-1) = (2-1)(2-1) = 1$$

The chi-square test confirms that the data contain clear evidence against the null hypothesis that there is no relationship between success and exclusive territories. Under H_0, the chance of obtaining a value of X^2 greater than or equal to the calculated value of 5.91 is small—fewer than 15 times in 1000.

The test does not tell us what kind of relationship is present. It is up to us to see that the data show that firms with exclusive territories are more likely to be successful. You should always accompany a chi-square test by percents such as those in Example 9.14 and Figure 9.2 and by a description of the nature of the relationship.

The observational study of Example 9.12 cannot tell us whether offering exclusive territories is a *cause* of success in a franchise firm. The association may be explained by confounding of contract terms with other variables such as type of business. *Our data don't allow us to investigate possible confounding variables.* A randomized comparative experiment that assigns firms to have or not have exclusive territories would settle the issue of causation. As is often the case, however, an experiment isn't practical. The authors of this study supplemented the data we used with interviews of founders of franchises and a theoretical framework for the results. Still, they limit their conclusion to stating that the firms with exclusive territories are more likely to survive.

The chi-square test and the *z* test

A comparison of the proportions of "successes" in two populations leads to a 2 × 2 table. We can compare two population proportions either by the chi-square test or by the two-sample *z* test from Section 8.2. In fact, *these tests always give exactly the same result,* because the X^2 statistic is equal to the square of the *z* statistic, and $\chi^2(1)$ critical values are equal to the squares of the corresponding $N(0, 1)$ critical values. The advantage of the *z* test is that we can test either one-sided or two-sided alternatives. The chi-square test always tests the two-sided alternative. Of course, the chi-square test can compare more than two populations, whereas the *z* test compares only two.

BEYOND THE BASICS | Meta-analysis

meta-analysis

Policymakers wanting to make decisions based on research are sometimes faced with the problem of summarizing the results of many studies. These studies may show effects of different magnitudes, some highly significant and some not significant. What *overall conclusion* can we draw? **Meta-analysis** is a collection of statistical techniques designed to combine information from different but similar studies. Each individual study must be examined with care to ensure that its design and data quality are adequate. The basic idea is to compute a measure of the effect of interest for each study. These are then combined, usually by taking some sort of weighted average, to produce a summary measure for all of the studies. Of course, a confidence interval for the summary is included in the results. Here is an example.

EXAMPLE 9.17   Vitamin A is often given to young children in developing countries to prevent night blindness. It was observed that children receiving vitamin A appear to have reduced death rates. To investigate the possible relationship between vitamin A supplementation and death, a large field trial with over 25,000 children was undertaken in the Aceh province of Indonesia. About half of the children were given large doses of vitamin A, and the other half were controls. In 1986, the researchers reported a 34% reduction in mortality (deaths) for the treated children who were 1 to 6 years old compared with the controls. Several additional studies were then undertaken. Most of the results confirmed the association: treatment of young children in developing countries with vitamin A reduces the death rate; but the size of the effect varied quite a bit.

How can we use the results of these studies to guide policy decisions? To address this question, a meta-analysis was performed on data available from eight studies.[6] Although the designs varied, each study provided a two-way table of counts. Here is the table for Aceh. A total of $n = 25,200$ children were enrolled in the study. Approximately half received vitamin A supplements. One year after the start of the study, the number of children who died was determined.

	Vitamin A	Control
Dead	101	130
Alive	12,890	12,079
Total	12,991	12,209

relative risk

The summary measure chosen was the **relative risk:** the ratio formed by dividing the proportion of children who died in the vitamin A group by the proportion of children who died in the control group. For Aceh, the proportion who died in the vitamin A group was

$$\frac{101}{12,991} = 0.00777$$

or 7.7 per thousand; for the control group, the proportion who died was

$$\frac{130}{12,209} = 0.01065$$

or 10.6 per thousand. The relative risk is therefore

$$\frac{0.00777}{0.01065} = 0.73$$

Relative risk less than 1 means that the vitamin A group has the lower mortality rate.

The relative risks for the eight studies were

0.73	0.50	0.94	0.71	0.70	1.04	0.74	0.80

A meta-analysis combined these eight results to produce a relative risk estimate of 0.77 with a 95% confidence interval of $(0.68, 0.88)$. That is, vitamin A supplementation reduced the mortality rate to 77% of its value in an untreated group. In other words, there is a 23% reduction in the mortality rate. The confidence interval does not include 1, so the null hypothesis of no effect (a relative risk of 1) can be clearly rejected. The researchers examined many variations of this meta-analysis, such as using different weights and leaving out one study at a time. These variations had little effect on the final estimate.

After these findings were published, large-scale programs to distribute high-potency vitamin A supplements were started. These programs have saved hundreds of thousands of lives since the meta-analysis was conducted and the arguments and uncertainties were resolved.

SECTION 9.2 | Summary

The **null hypothesis** for $r \times c$ tables of count data is that there is no relationship between the row variable and the column variable.

Expected cell counts under the null hypothesis are computed using the formula

$$\text{expected count} = \frac{\text{row total} \times \text{column total}}{n}$$

The null hypothesis is tested by the **chi-square statistic,** which compares the observed counts with the expected counts:

$$X^2 = \sum \frac{(\text{observed} - \text{expected})^2}{\text{expected}}$$

Under the null hypothesis, X^2 has approximately the χ^2 distribution with $(r-1)(c-1)$ degrees of freedom. The P-value for the test is

$$P(\chi^2 \geq X^2)$$

where χ^2 is a random variable having the $\chi^2(\text{df})$ distribution with df $= (r-1)(c-1)$.

The chi-square approximation is adequate for practical use when the average expected cell count is 5 or greater and all individual expected counts are 1 or greater, except in the case of 2×2 tables. All four expected counts in a 2×2 table should be 5 or greater.

The section we just completed assumed that you have access to software or a statistical calculator. If you do not, you now need to study the material on computations in the following optional section. All exercises appear at the end of the chapter.

9.3 Formulas and Models for Two-Way Tables*

Computations

The calculations required to analyze a two-way table are straightforward but tedious. In practice, we recommend using software, but it is possible to do the work with a calculator, and some insight can be gained by examining the details. Here is an outline of the steps required.

---

*The analysis of two-way tables is based on computations that are a bit messy and on statistical models that require a fair amount of notation to describe. This section gives the details. By studying this material you will deepen your understanding of the methods described in this chapter, but this section is optional.

> ## COMPUTATIONS FOR TWO-WAY TABLES
>
> **1.** Calculate descriptive statistics that convey the important information in the table. Usually these will be column or row percents.
>
> **2.** Find the expected counts and use these to compute the X^2 statistic.
>
> **3.** Use chi-square critical values from Table F to find the approximate P-value.
>
> **4.** Draw a conclusion about the association between the row and column variables.

The following example illustrates these steps. The two-way table in this example does not compare several populations. Instead, it arises by classifying observations on a single population in two ways.

EXAMPLE 9.18   Market researchers know that background music can influence the mood and purchasing behavior of customers. One study in a supermarket in Northern Ireland compared three treatments: no music, French accordion music, and Italian string music. Under each condition, the researchers recorded the numbers of bottles of French, Italian, and other wine purchased.[7] Here is the two-way table that summarizes the data:

| | Music | | | |
Wine	None	French	Italian	Total
French	30	39	30	99
Italian	11	1	19	31
Other	43	35	35	113
Total	84	75	84	243

This is a 3 × 3 table, to which we have added the marginal totals obtained by summing across rows and columns. For example, the first-row total is 30 + 39 + 30 = 99. The grand total, the number of bottles of wine in the study, can be computed by summing the row totals, 99 + 31 + 113 = 243, or the column totals, 84 + 75 + 84 = 243. *It is easy to make an error in these calculations, so it is a good idea to do both as a check on your arithmetic.*

Computing conditional distributions

First, we summarize the observed relation between the music being played and the type of wine purchased. The researchers expected that music would influence sales, so music type is the explanatory variable and the type of wine purchased is the response variable. In general, the clearest way to describe this kind of relationship is to compare the conditional distributions of the response variable for each value of the explanatory variable. So we will compare the column percents that give the conditional distribution of purchases for each type of music played.

When no music was played, there were 84 bottles of wine sold. Of these, 30 were French wine. Therefore, the column proportion for this cell is

$$\frac{30}{84} = 0.357$$

That is, 35.7% of the wine sold was French when no music was played. Similarly, 11 bottles of Italian wine were sold under this condition, and this is 13.1% of the sales.

$$\frac{11}{84} = 0.131$$

In all, we calculate nine percents. Here are the results:

Column percents for wine and music

Wine	Music			Total
	None	French	Italian	
French	35.7	52.0	35.7	40.7
Italian	13.1	1.3	22.6	12.8
Other	51.9	46.7	41.7	46.5
Total	100.0	100.0	100.0	100.0

In addition to the conditional distributions of types of wine sold for each kind of music being played, the table also gives the marginal distribution of the types of wine sold. These percents appear in the rightmost column, labeled "Total."

The sum of the percents in each column should be 100, except for possible small roundoff errors. It is good practice to calculate each percent separately and then sum each column as a check. In this way we can find arithmetic errors that would not be uncovered if, for example, we calculated the column percent for the "Other" row by subtracting the sum of the percents for "French" and "Italian" from 100.

Figure 9.6 compares the distributions of types of wine sold for each of the three music conditions. There appears to be an association between the music played and the type of wine that customers buy. Sales of Italian wine are very low when French music is playing but are higher when Italian music or no music is playing. French wine is popular in this market, selling well under all music conditions but notably better when French music is playing.

Another way to look at these data is to examine the row percents. These fix a type of wine and compare its sales when different types of music are playing. Figure 9.7 displays these results. We see that more French wine is sold when French music is playing. Similarly for Italian wine. The negative effect of French music on sales of Italian wine is dramatic.

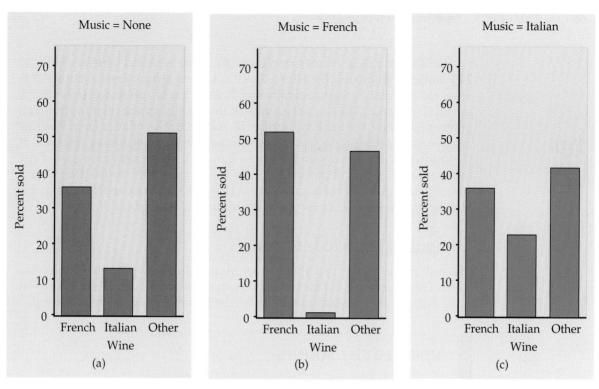

FIGURE 9.6 Comparison of the percents of different types of wine sold for different music conditions, for Example 9.19.

FIGURE 9.7 Comparison of the percents of different types of wine sold for different music conditions, for Example 9.19.

We observe a clear relationship between music type and wine sales for the 243 bottles sold during the study. The chi-square test assesses whether this observed association is statistically significant, that is, too strong to occur often just by chance. The test only confirms that there is some relationship. The percents we have compared describe the nature of the relationship. *The chi-square test does not in itself tell us what population our conclusion describes.* If the study was done in one market on a Saturday, the results may apply only to Saturday shoppers at this market. The researchers may invoke their understanding of consumer behavior to argue that their findings apply more generally, but that is beyond the scope of the statistical analysis.

Computing expected cell counts

The null hypothesis is that there is no relationship between music and wine sales. The alternative is that these two variables are related. Here is the formula for the expected cell counts under the hypothesis of "no relationship."

EXPECTED CELL COUNTS

$$\text{expected count} = \frac{\text{row total} \times \text{column total}}{n}$$

EXAMPLE 9.20   What is the expected count in the upper-left cell in the table of Example 9.18, bottles of French wine sold when no music is playing, under the null hypothesis that music and wine sales are independent?

The column total, the number of bottles of wine sold when no music is playing, is 84. The row total shows that 99 bottles of French wine were sold during the study. The total sales were 243. The expected cell count is therefore

$$\frac{(84)(99)}{243} = 34.222$$

Note that although any count of bottles sold must be a whole number, an expected count need not be. The expected count is the mean over many repetitions of the study, assuming no relationship.

Nine similar calculations produce this table of expected counts:

Expected counts for wine and music

Wine	None	French	Italian	Total
		Music		
French	34.222	30.556	34.222	99.000
Italian	10.716	9.568	10.716	31.000
Other	39.062	34.877	39.062	113.001
Total	84.000	75.001	84.000	243.001

We can check our work by adding the expected counts to obtain the row and column totals, as in the table. These should be the same as those in the table of observed counts, except for small roundoff errors, such as 113.001 rather than 113 for the total number of bottles of other wine sold.

The X^2 statistic and its P-value

The expected counts are all large, so we proceed with the chi-square test. We compare the table of observed counts with the table of expected counts using the X^2 statistic.[8] We must calculate the term for each cell, then sum over all nine cells. For French wine with no music, the observed count is 30 bottles and the expected count is 34.222. The contribution to the X^2 statistic for this cell is therefore

$$\frac{(30 - 34.222)^2}{34.222} = 0.5209$$

The X^2 statistic is the sum of nine such terms:

$$X^2 = \sum \frac{(\text{observed} - \text{expected})^2}{\text{expected}}$$

$$= \frac{(30 - 34.222)^2}{34.222} + \frac{(39 - 30.556)^2}{30.556} + \frac{(30 - 34.222)^2}{34.222}$$

$$+ \frac{(11 - 10.716)^2}{10.716} + \frac{(1 - 9.568)^2}{9.568} + \frac{(19 - 10.716)^2}{10.716}$$

$$+ \frac{(43 - 39.062)^2}{39.062} + \frac{(35 - 34.877)^2}{34.877} + \frac{(35 - 39.062)^2}{39.062}$$

$$= 0.5209 + 2.3337 + 0.5209 + 0.0075 + 7.6724$$

$$+ 6.4038 + 0.3971 + 0.0004 + 0.4223$$

$$= 18.28$$

Because there are $r = 3$ types of wine and $c = 3$ music conditions, the degrees of freedom for this statistic are

$$\text{df} = (r - 1)(c - 1) = (3 - 1)(3 - 1) = 4$$

df = 4		
p	0.0025	0.001
χ^2	16.42	18.47

Under the null hypothesis that music and wine sales are independent, the test statistic X^2 has a $\chi^2(4)$ distribution. To obtain the P-value, look at the df = 4 row in Table F. The calculated value $X^2 = 18.28$ lies between the critical points for probabilities 0.0025 and 0.001. The P-value is therefore between 0.0025 and 0.001. Because the expected cell counts are all large, the P-value from Table F will be quite accurate. There is strong evidence ($X^2 = 18.28$, df = 4, $P < 0.0025$) that the type of music being played has an effect on wine sales.

The size and nature of the relationship between music and wine sales are described by row and column percents. These are displayed in Figures 9.6 and 9.7. Here is another way to look at the data: we see that just two of the nine terms that make up the chi-square sum contribute about 14 of the total $X^2 = 18.28$. Comparing the observed and expected counts in these two cells, we see that sales of Italian wine are much below expectation when French music is playing and much above expectation when Italian music is playing. We are led to a specific conclusion: sales of Italian wine are strongly affected by Italian and French music. Figure 9.7(b) displays this effect.

Models for two-way tables

The chi-square test for the presence of a relationship between the two directions in a two-way table is valid for data produced from several different study designs. The precise statement of the null hypothesis "no relationship" in terms of population parameters is different for different designs. We now describe two of these settings in detail. *An essential requirement is that each experimental unit or subject is counted only once in the data table.* In the exclusive-territory study of Example 9.12, each firm is classified in two ways, having exclusive territories or not, and successful or not. In Example 9.18, the market researchers changed the background music and took three samples of wine sales in three distinct environments. Each column in the table represents one of these samples, and each row a type of wine.

Comparing several populations: the first model   Example 9.18 (wine sales in three environments) is an example of *separate and independent random samples* from each of c populations. The c columns of the two-way table represent the populations. There is a single categorical response variable, wine type. The r rows of the table correspond to the values of the response variable.

We know that the z test for comparing the two proportions of successes and the chi-square test for the 2×2 table are equivalent. The $r \times c$ table allows us to compare more than two populations or more than two categories of response, or both. In this setting, the null hypothesis "no relationship between column variable and row variable" becomes

H_0: The distribution of the response variable is the same in all c populations.

Because the response variable is categorical, its distribution just consists of the probabilities of its r possible values. The null hypothesis says that these probabilities (or population proportions) are the same in all c populations.

EXAMPLE 9.21   In the market research study of Example 9.18, we compare three populations:

Population 1: bottles of wine sold when no music is playing
Population 2: bottles of wine sold when French music is playing
Population 3: bottles of wine sold when Italian music is playing

We have three samples, of sizes 84, 75, and 84, a separate sample from each population. The null hypothesis for the chi-square test is

H_0: The proportions of each wine type sold are the same in all three populations.

The parameters of the model are the proportions of the three types of wine that would be sold in each of the three environments. There are three proportions (for French wine, Italian wine, and other wine) for each environment.

More generally, if we take an independent SRS from each of c populations and classify each outcome into one of r categories, we have an $r \times c$ table of population proportions. There are c different sets of proportions to be compared. There are c groups of subjects, and a single categorical variable with r possible values is measured for each individual.

> ## MODEL FOR COMPARING SEVERAL POPULATIONS USING TWO-WAY TABLES
>
> Select independent SRSs from each of c populations, of sizes $n_1, n_2, \ldots, n_c$. Classify each individual in a sample according to a categorical response variable with r possible values. There are c different probability distributions, one for each population.
>
> The null hypothesis is that the distributions of the response variable are the same in all c populations. The alternative hypothesis says that these c distributions are not all the same.

Testing independence: the second model A second model for which our analysis of $r \times c$ tables is valid is illustrated by the exclusive-territory study, Example 9.12. There, a *single* sample from a *single* population was classified according to two categorical variables.

EXAMPLE 9.22 The single population studied is recently established franchise firms. The researchers had a sample of 170 such firms. They measured two categorical variables for each firm: "Exclusive territory," with possible responses "Yes" and "No"; and "Firm success," with possible responses "Yes" and "No." The null hypothesis for the chi-square test is

H_0: "Exclusive territory" and "Firm success" are independent.

The parameters of the model are the probabilities for each of the four possible combinations of values of the row and column variables. If the null hypothesis is true, the multiplication rule for independent events (page 267) says that these can be found as the products of outcome probabilities for each variable alone.

More generally, take an SRS from a single population and record the values of two categorical variables, one with r possible values and the other with c possible values. The data are summarized by recording the numbers of individuals for each possible combination of outcomes for the two random variables. This gives an $r \times c$ table of counts. Each of these $r \times c$ possible outcomes **joint distribution** has its own probability. The probabilities give the **joint distribution** of the two categorical variables.

Each of the two categorical random variables has a distribution. These are **marginal** the **marginal distributions** because they are the sums of the population pro- **distributions** portions in the rows and columns.

The null hypothesis "no relationship" now states that the row and column variables are independent. The multiplication rule for independent events (page 267) tells us that the joint probabilities are the products of the marginal probabilities.

EXAMPLE 9.23 The joint probability distribution gives a probability for each of the four cells in our 2×2 table of "Exclusive territory" and "Firm success." The marginal distribution for "Exclusive territory" gives probabilities for each of the two possible exclusive-territory categories; the marginal distribution for "Firm success" gives probabilities for each of the two possible firm-success categories.

Independence between "Exclusive territory" and "Firm success" implies that the joint distribution can be obtained by multiplying the appropriate terms from the two marginal distributions. For example, the probability that a firm has an exclusive territory *and* is successful is the probability that the firm has an exclusive territory times the probability that the firm is successful. The hypothesis that "Exclusive territory" and "Firm success" are independent says that the multiplication rule applies to *all* outcomes.

> ### MODEL FOR EXAMINING INDEPENDENCE IN TWO-WAY TABLES
>
> Select an SRS of size n from a population. Measure two categorical variables for each individual.
>
> The null hypothesis is that the row and column variables are independent. The alternative hypothesis is that the row and column variables are dependent.

Concluding remarks

You can distinguish between the two models by examining the design of the study. In the independence model, there is a single sample. The column totals and row totals are random variables. The total sample size n is set by the researcher; the column and row sums are known only after the data are collected. For the comparison-of-populations model, on the other hand, there is a sample from each of two or more populations. The column sums are the sample sizes selected at the design phase of the research. The null hypothesis in both models says that there is no relationship between the column variable and the row variable. The precise statement of the hypothesis differs, depending on the sampling design. Fortunately, *the test of the hypothesis of "no relationship" is the same for both models;* it is the chi-square test. There are yet other statistical models for two-way tables that justify the chi-square test of the null hypothesis "no relation," made precise in ways suitable for these models. Statistical methods related to the chi-square test also allow the analysis of three-way and higher-way tables of count data. You can find a discussion of these topics in advanced texts on categorical data.[9]

SECTION 9.3 | Summary

For two-way tables we first compute percents or proportions that describe the relationship of interest. Then, we compute expected counts, the X^2 statistic, and the *P*-value.

Two different models for generating $r \times c$ tables lead to the chi-square test. In the first model, independent SRSs are drawn from each of c populations, and each observation is classified according to a categorical variable with r possible values. The null hypothesis is that the distributions of the row categorical variable are the same for all c populations. In the second model, a single SRS is drawn from a population, and observations are classified according to two categorical variables having r and c possible values. In this model, H_0 states that the row and column variables are independent.

9.4   Goodness of Fit*

In Example 9.2 (page 582) we introduced data on the binge drinking of men and women college students. Later, in Example 9.7 (page 586), we used conditional distributions to describe and compare the drinking behavior of these men and women. For men, 17.0% were frequent binge drinkers and 83.0% were not; for women, 22.7% were frequent binge drinkers and 77.3% were not. The chi-square test is the tool that we use to compare these two distributions. We now consider a slight variation on this scenario where one of the distributions is known and we want to compare it with a hypothesized distribution. Here is an example that illustrates the basic ideas.

EXAMPLE 9.24   Are you more likely to have a motor vehicle collision when using a cell phone? A study of 699 drivers who were using a cellular telephone when they were involved in a collision examined this question.[10] These drivers made 26,798 cellular-telephone calls during a 14-month study period. Each of the 699 collisions was classified in various ways. Here are the numbers for each day of the week:

Number of collisions by day of the week							
Day of the week							
Sun.	Mon.	Tue.	Wed.	Thu.	Fri.	Sat.	Total
20	133	126	159	136	113	12	699

We have a total of 699 accidents involving drivers who were using a cellular telephone at the time of their accident. Let's explore the relationship between these accidents and the day of the week. Are the accidents equally likely to occur on any day of the week?

We can think of this table of counts as a one-way table with seven cells, each with a count of the number of accidents that occurred on the particular day of the week. Our question is translated into a null hypothesis that motor vehicle accidents involving cellular phone use are equally likely to occur on each of the seven days of the week. The alternative is that the probabilities vary from day to day. Our analysis of these data is very similar to the analyses of two-way tables that we studied in Section 9.2. We first compute expected

---

*This section is optional.

counts. Since there are 699 accidents and 7 days, under the null hypothesis we expect one-seventh of the accidents to occur on each day. So the expected number of accidents for each day is $699/7 = 99.86$. Next, we construct a chi-square statistic that compares the actual numbers with the expected numbers:

$$X^2 = \sum \frac{(\text{observed count} - \text{expected count})^2}{\text{expected count}}$$

For Sunday, we have

$$\frac{(\text{observed count} - \text{expected count})^2}{\text{expected count}} = \frac{(20 - 99.86)^2}{99.86} = 63.86$$

and for Monday,

$$\frac{(\text{observed count} - \text{expected count})^2}{\text{expected count}} = \frac{(133 - 99.86)^2}{99.86} = 11.0$$

Performing the same calculations for the other days and summing the results over the seven cells gives

$$X^2 = 208.85$$

The degrees of freedom are 1 less than the number of cells, df $= 7 - 1 = 6$. We calculate the P-value using Table F or software. From Table F, we can determine $P < 0.0005$ and we conclude that these types of accidents are not equally likely to occur on each of the seven days of the week.

We have covered all of the basics for chi-square goodness of fit tests in our example. Now we will summarize the details for the general case.

THE CHI-SQUARE GOODNESS OF FIT TEST

Data for n observations on a categorical variable with k possible outcomes are summarized as observed counts, $n_1, n_2, \ldots, n_k$ in k cells. A null hypothesis specifies probabilities $p_1, p_2, \ldots, p_k$ for the possible outcomes.

For each cell, multiply the total number of observations n by the specified probability to determine the expected counts:

$$\text{expected count} = np_i$$

The **chi-square statistic** measures how much the observed cell counts differ from the expected cell counts. The formula for the statistic is

$$X^2 = \sum \frac{(\text{observed count} - \text{expected count})^2}{\text{expected count}}$$

The degrees of freedom are $k - 1$, and P-values are computed from the chi-square distribution.

In our example about cellular telephones and motor vehicle accidents we examined the accident rates for all seven days of the week. A traffic expert would very likely tell us that accidents are less likely on weekends. Indeed, the two days contributing the largest amounts to the X^2 statistic are Saturday and Sunday. Let's examine the pattern of accidents for weekdays. *Note, this is not the same as making up hypotheses after looking at the data.* After some reflection about the data and the context of the problem, we have determined that it is reasonable to ask this question. Such decisions require careful judgment and the rationale must be clearly explained. More advanced methods of inference allow the examination of this kind of question within the framework of the overall analysis.[11]

EXAMPLE 9.25   Here are the data from Example 9.24 for weekdays only:

Number of collisions by day of the week					
Day of the week					
Mon.	Tue.	Wed.	Thu.	Fri.	Total
133	126	159	136	113	667

We now have a total of 667 accidents because we are not considering the $32 = 20 + 12$ accidents that occurred on weekends. For five equally likely outcomes we have $p_i = 1/5$ for each. The expected count for each of the five days is $667(1/5) = 133.4$, and the contribution to the chi-square statistic for Monday is

$$\frac{\text{(observed count} - \text{expected count)}^2}{\text{expected count}} = \frac{(133 - 133.4)^2}{133.4} = 0.0012$$

Summing the results for the five weekdays gives $X^2 = 8.49$ with $5 - 1 = 4$ degrees of freedom. The *P*-value is 0.08. For weekdays, the data do not provide evidence in support of differences due to the day of the week in the accident rates for motor vehicle collisions where the driver was using a cellular phone.

Many software packages do not provide routines for computing the chi-square goodness of fit test. However, there is a very simple trick that can be used to produce the results from software that can analyze two-way tables. Make a two-way table where the first column contains k cells with the observed counts. Make a second column with counts that correspond *exactly* to the probabilities specified by the null hypothesis, with a very large number of observations. For the problem in Exercise 9.25, we will have a second column with 10,000 accidents on each of the five days. When analyzed as a two-way table, software gives $X^2 = 8.37$ with $P = 0.08$, which agrees reasonably well with our exact calculations.

CHAPTER 9 | Exercises

9.1   The Census Bureau provides estimates of numbers of people in the United States classified in various ways.[12] Let's look at college students. The following table gives us data to examine the relation between age and full-time or part-time student status. The numbers in the table are expressed as thousands of U.S. college students.

	U.S. college students by age and status	
	Status	
Age	**Full-time**	**Part-time**
15–19	3388	389
20–24	5238	1164
25–34	1703	1699
35 and over	762	2045

(a) What is the U.S. Census Bureau estimate of the number of full-time college students aged 15 to 19?

(b) Give the joint distribution of age and status for this table.

(c) What is the marginal distribution of age? Display the results graphically.

(d) What is the marginal distribution of status? Display the results graphically.

9.2   Refer to the previous exercise. Find the conditional distribution of status for each of the four age categories. Display the distributions graphically and summarize their differences and similarities.

9.3   Refer to the previous two exercises. Compute the conditional distribution of age for each of the two status categories. Summarize the distributions graphically and write a short paragraph describing the distributions and how they differ.

9.4   The table below gives some census data concerning the enrollment status of recent high school graduates aged 16 to 24 years.[13] The table entries are in thousands of students.

	Enrollment and gender	
Status	**Men**	**Women**
Two-year college, full-time	890	969
Two-year college, part-time	340	403
Four-year college, full-time	2897	3321
Four-year college, part-time	249	383
Graduate school	306	366
Vocational school	160	137

(a) How many male recent high school graduates aged 16 to 24 years were enrolled full-time in two-year colleges?

(b) Give the marginal distribution of gender for these students. Display the results graphically.

(c) What is the marginal distribution of status for these students? Display the results graphically.

9.5   Refer to the previous exercise. Find the conditional distribution of gender for each status. Describe the distributions graphically and write a short summary comparing the major features of these distributions.

9.6   Refer to the previous two exercises. Find the conditional distribution of status for each gender. Describe the distributions graphically and write a short summary comparing the major features of these distributions.

9.7   Here are the row and column totals for a two-way table with two rows and two columns:

$$
\begin{array}{cc|c}
a & b & 200 \\
c & d & 200 \\
\hline
200 & 200 & 400
\end{array}
$$

Find *two different* sets of counts a, b, c, and d for the body of the table that give these same totals. This shows that the relationship between two variables cannot be obtained from the two individual distributions of the variables.

9.8   Construct a 3×3 table of counts where there is no apparent association between the row and column variables.

9.9   A marketing research firm conducted a survey of companies in its state. They mailed a questionnaire to 300 small companies, 300 medium-sized companies, and 300 large companies. The rate of nonresponse is important in deciding how reliable survey results are. Here are the data on response to this survey:

	Response		
Size of company	Yes	No	Total
Small	175	125	300
Medium	145	155	300
Large	120	180	300

(a) What was the overall percent of nonresponse?

(b) Describe how nonresponse is related to the size of the business. (Use percents to make your statements precise.)

(c) Draw a bar graph to compare the nonresponse percents for the three size categories.

(d) Using the total number of responses as a base, compute the percent of responses that come from each of small, medium, and large businesses.

(e) The sampling plan was designed to obtain equal numbers of responses from small, medium, and large companies. In preparing an analysis of the survey results, do you think it would be reasonable to proceed as if the responses represented companies of each size equally?

9.10 A study of the career plans of young women and men sent questionnaires to all 722 members of the senior class in the College of Business Administration at the University of Illinois. One question asked which major within the business program the student had chosen.[14] Here are the data from the students who responded:

	Gender	
Major	Female	Male
Accounting	68	56
Administration	91	40
Economics	5	6
Finance	61	59

(a) Describe the differences between the distributions of majors for women and men with percents, with a graph, and in words.

(b) What percent of the students did not respond to the questionnaire? The nonresponse weakens conclusions drawn from these data.

9.11 Cocaine addiction is hard to break. Addicts need cocaine to feel any pleasure, so perhaps giving them an antidepressant drug will help. A 3-year study with 72 chronic cocaine users compared an antidepressant drug called desipramine with lithium and a placebo. (Lithium is a standard drug to treat cocaine addiction. A placebo is a dummy drug, used so that the effect of being in the study but not taking any drug can be seen.) One-third of the subjects, chosen at random, received each treatment.[15] Here are the results:

	Cocaine relapse?	
Treatment	Yes	No
Desipramine	10	14
Lithium	18	6
Placebo	20	4

Compare the effectiveness of the three treatments in preventing relapse. Use percents and draw a bar graph. Write a brief summary of your conclusions.

9.12 Asia has become a major competitor of the United States and Western Europe in education as well as economics. Here are counts of first university degrees in science and engineering in the three regions:[16]

	Region		
Field	United States	Western Europe	Asia
Engineering	61,941	158,931	280,772
Natural science	111,158	140,126	242,879
Social science	182,166	116,353	236,018

Direct comparison of counts of degrees would require us to take into account Asia's much larger population. We can, however, compare the distribution of degrees by field of study in the three regions. Do this using calculations and graphs, and write a brief summary of your findings.

9.13 A study examined patterns and characteristics of volunteer service for young people from high school through early adulthood.[17] Here are some data that can be used to compare males and females on participation in unpaid volunteer service or community service and motivation for participation:

	Participants			
	Motivation			
Gender	Strictly voluntary	Court-ordered	Other	Non-participants
Men	31.9%	2.1%	6.3%	59.7%
Women	43.7%	1.1%	6.5%	48.7%

Note that the percents in each row sum to 100%.

(a) Graphically compare the volunteer-service profiles for men and women. Describe any differences that are striking.

(b) Find the proportion of men who volunteer. Do the same for women. Refer to the section on relative risk in Chapter 8 (page 563) and the discussion in this chapter on page 599. Compute the relative risk of being a volunteer for females versus males. Write a clear sentence contrasting females and males using relative risk as your numerical summary.

9.14 Refer to the previous exercise. Recompute the table for volunteers only. To do this take the entries for each motivation and divide by the percent of volunteers. Do this separately for each gender. Verify that the percents sum to 100% for each gender. Give a graphical summary to compare the motivation of men and women who are volunteers. Compare this with your summary in part (a) of the previous exercise and write a short paragraph describing similarities and differences in these two views of the data.

9.15 Mountain View University has professional schools in business and law. Here is a three-way table of applicants to these professional schools, categorized by gender, school, and admission decision.[18]

	Business				Law		
		Admit				Admit	
Gender	Yes	No		Gender	Yes	No	
Male	400	200		Male	90	110	
Female	200	100		Female	200	200	

(a) Make a two-way table of gender by admission decision for the combined professional schools by summing entries in the three-way table.

(b) From your two-way table, compute separately the percents of male and female applicants admitted. Male applicants are admitted to Mountain View's professional schools at a higher rate than female applicants.

(c) Now compute separately the percents of male and female applicants admitted by the business school and by the law school.

(d) Explain carefully, as if speaking to a skeptical reporter, how it can happen that Mountain View appears to favor males when this is not true within each of the professional schools.

9.16 Refer to the previous exercise. Make up a similar table for a hypothetical university having four different schools that illustrates the same point. Carefully summarize your table with the appropriate percents.

9.17 A university classifies its classes as either "small" (fewer than 40 students) or "large." A dean sees that 62% of Department A's classes are small, while Department B has only 40% small classes. She wonders if she should cut Department A's budget and insist on larger classes. Department A responds to the dean by pointing out that classes for third- and fourth-year students tend to be smaller than classes for first- and second-year students. The three-way table below gives the counts of classes by department, size, and student audience. Write a short report for the dean that summarizes these data. Start by computing the percents of small classes in the two departments, and include other numerical and graphical comparisons as needed. (Do not perform any statistical significance tests.) Here are the numbers of classes to be analyzed:

	Department A			Department B		
Year	Large	Small	Total	Large	Small	Total
First	2	0	2	18	2	20
Second	9	1	10	40	10	50
Third	5	15	20	4	16	20
Fourth	4	16	20	2	14	16

9.18 For each of the following situations give the degrees of freedom and an appropriate bound on the *P*-value (give the exact value if you have software available) for the X^2 statistic for testing the null hypothesis of no association between the row and column variables.

(a) A 4 by 5 table with $X^2 = 26.23$.

(b) A 4 by 3 table with $X^2 = 26.23$.

(c) A 5 by 4 table with $X^2 = 26.23$.

(d) A 7 by 6 table with $X^2 = 26.23$.

9.19 For each of the following situations give the degrees of freedom and an appropriate bound on the *P*-value (give the exact value if you have software available) for the X^2 statistic for testing the null hypothesis of no association between the row and column variables.

(a) A 2 by 2 table with $X^2 = 1.31$.

(b) A 4 by 4 table with $X^2 = 17.54$.

(c) A 2 by 8 table with $X^2 = 22.10$.

(d) A 5 by 3 table with $X^2 = 12.61$.

9.20 In what ways do advertisers in magazines use sexual imagery to appeal to youth? One study classified each of 1509 full-page or larger ads as "not sexual" or "sexual," according to the amount and style of the dress of the male or female model in the ad. The ads were also classified according to the target readership of the magazine.[19] Here is the two-way table of counts:

Model dress	Women	Men	General interest	Total
Not sexual	351	514	248	1113
Sexual	225	105	66	396
Total	576	619	314	1509

(a) Summarize the data numerically and graphically.

(b) Perform the significance test that compares the model dress for the three categories of magazine readership. Summarize the results of your test and give your conclusion.

(c) All of the ads were taken from the March, July, and November issues of six magazines in one year. Discuss this fact from the viewpoint of the validity of the significance test and the interpretation of the results.

9.21 The ads in the study described in the previous exercise were also classified according to the age group of the intended readership. A summary of the data is given in the table.

	Magazine readership age group	
Model dress	Young adult	Mature adult
Not sexual (percent)	72.3%	76.1%
Sexual (percent)	27.7%	23.9%
Number of ads	1006	503

Using parts (a) and (b) in the previous exercise as a guide, analyze these data and write a report summarizing your work.

9.22 A study of identity theft looked at how well consumers protect themselves from this increasingly prevalent crime. The behaviors of 61 college students were compared with the behaviors of 59 nonstudents.[20] One of the questions was "When asked to create a password, I have used either my mother's maiden name, or my pet's name, or my birth date, or the last four digits of my social security number, or a series of consecutive numbers." For the students, 22 agreed with this statement while 30 of the nonstudents agreed.

(a) Display the data in a two-way table and perform the chi-square test. Summarize the results.

(b) Reanalyze the data using the methods for comparing two proportions that we studied in the previous chapter. Compare the results and verify that the chi-square statistic is the square of the z statistic.

(c) The students in this study were junior and senior college students from two sections of a course in Internet marketing at a large northeastern university. The nonstudents were a group of individuals who were recruited to attend commercial focus groups on the West Coast conducted by a lifestyle marketing organization. Discuss how the method of selecting the subjects in this study relates to the conclusions that can be drawn from it.

9.23 A survey of student athletes that asked questions about gambling behavior classified students according to the National Collegiate Athletic Association (NCAA) division.[21] For male student athletes, the percent who reported wagering on collegiate sports are given here along with the numbers of respondents in each division.

Division	I	II	III
Percent	17.2%	21.0%	24.4%
Number	5619	2957	4089

(a) Use a significance test to compare the percents for the three NCAA divisions. Give details and a short summary of your conclusion.

(b) The percents in the table above are given in the NCAA report but the numbers of male student athletes in each division who responded to the survey question are estimated based on other information in the report. To what extent do you think this has an effect on the results? (*Hint:* Rerun your analysis a few times, with slightly different numbers of students but the same percents.)

(c) Some student athletes may be reluctant to provide this kind of information, even in a survey where there is no possibility that they can be identified. Discuss how this fact may affect your conclusions.

(d) The chi-square test for this set of data assumes that the responses of the student athletes are independent. However, some of the students are at the same school and even on the same team. Discuss how you think this might affect the results.

9.24 One possible effect of air pollution is genetic damage. A study designed to examine this problem exposed one group of mice to air near a steel mill and another group to air in a rural area and compared the numbers of mutations in each group.[22] Here are the data for a mutation at the *Hm-2* gene locus:

	Location	
Mutation	Steel mill air	Rural air
Yes	30	23
No		
Total	96	150

(a) Fill in the missing entries in the table.

(b) Summarize the data numerically and graphically.

(c) Is there evidence to conclude that the location is related to the occurrence of mutations? Perform the significance test and summarize the results.

9.25 *E. jugularis* is a type of hummingbird that lives in the forest preserves of the Caribbean island of Saint Lucia. The males and the females of this species have bills that are shaped somewhat differently. Researchers who study these birds thought that the bill shape might be related to the shape of the flowers that they visit for food. The researchers observed 49 females and 21 males. Of the females, 20 visited the flowers of *H. bihai*, while none of the males visited these flowers.[23] Display the data in a two-way table and perform the chi-square test. Summarize the results and give a brief statement of your conclusion. Your two-way table has a count of zero in one cell. Does this invalidate your significance test? Explain why or why not.

9.26 The World Wide Web (WWW) has led to an enormous increase in the amount of information that is easily available to anyone with Internet access. References to Internet pages are becoming quite common in the scientific literature. One study examined Internet references in articles in three prominent journals: the *New England Journal of Medicine (NEJM)*, the *Journal of the American Medical Association (JAMA)*, and *Science*.[24] In one part of the study, Internet references were classified according to the top-level domain. Here are the data:

	Journal		
Top-level domain	*NEJM*	*JAMA*	*Science*
.gov	41	103	111
.org	37	46	162
.com	6	17	14
.edu	4	8	47
Other	9	15	52

Analyze the data. Include numerical and graphical summaries as well as a significance test. Summarize your results and conclusions.

9.27 The Health, Aging, and Body Composition (Health ABC) study is a 10-year study of older adults. A research project based on this study examined the relationship between physical activity and pet ownership.[25] The data collected included information concerning pet owner characteristics and the type of pet owned. Here is a table of counts of subjects classified by pet ownership status and education level:

	Pet ownership status		
Education level	Non–pet owners	Dog owners	Cat owners
Less than high school	421	93	28
High school graduate	666	100	40
Postsecondary	845	135	99

Note that "Dog owners" and "Cat owners" designate individuals who own a dog only or a cat only, respectively. Individuals who own a dog and a cat are not included in this table. Analyze the data. Include numerical and graphical summaries as well as a significance test. Summarize your results and conclusions.

9.28 Refer to the previous exercise. Here are similar data giving the relationship between pet ownership status and gender:

	Pet ownership status		
Gender	Non–pet owners	Dog owners	Cat owners
Female	1024	157	85
Male	915	171	82

Analyze the data. Include numerical and graphical summaries as well as a significance test. Summarize your results and conclusions.

9.29 A task force set up to examine retention of students in the majors that they chose when starting college examined data on transfers to other majors.[26] Here are some data giving counts of students classified by initial major and the area that they transferred to:

	Area transferred to				
Initial major	Engineering	Management	Liberal arts	Other	Total
Biology	13	25	158		398
Chemistry	16	15	19		114
Mathematics	3	11	20		72
Physics	9	5	14		61

Complete the table by computing the values for the "Other" column. Write a short paragraph explaining what conclusions you can draw about the relationship between initial major and area transferred to. Be sure to include numerical and graphical summaries as well as the details of your significance test.

9.30 To be competitive in global markets, many U.S. corporations are undertaking major reorganizations. Often these involve "downsizing," sometimes called a "reduction in force," (RIF), where substantial numbers of employees are terminated. Federal and various state laws require that employees be treated equally regardless of their age. In particular, employees over the age of 40 years are in a "protected" class, and many allegations of discrimination focus on comparing employees over 40 with their younger coworkers. Here are the data for a recent RIF:

	Over 40	
Terminated	No	Yes
Yes	16	82
No	585	771

(a) Make a table that includes the following information for each group (over 40 or not): total number of employees, the proportion of employees who were terminated, and the standard error for the proportion.

(b) Perform the chi-square test for this two-way table. Give the test statistic, the degrees of freedom, the P-value, and your conclusion.

9.31 A major issue that arises in the kind of case described in the previous exercise concerns the extent to which the employees are similar. This often involves examining other variables. For example, suppose that the employees over 40 did not do as good a job as the younger workers. Let's examine the last performance appraisal to explore this idea. The possible values are as follows: "partially meets expectations," "fully meets expectations," "usually exceeds expectations," and "continually exceeds expectations." Because there were very few employees who partially exceeded expectations, we combine the first two categories. Here are the data:

	Over 40	
Performance appraisal	No	Yes
Partially or fully meets expectations	82	237
Usually exceeds expectations	357	492
Continually exceeds expectations	63	32

Analyze the data. Do the older employees appear to have lower performance evaluations?

9.32 Shopping at secondhand stores is becoming more popular and has even attracted the attention of business schools. A study of customers' attitudes toward secondhand stores interviewed samples of shoppers at two secondhand stores of the same chain in two cities. The breakdown of the respondents by gender is as follows:[27]

Gender	City 1	City 2
Men	38	68
Women	203	150
Total	241	218

Is there a significant difference between the proportions of women customers in the two cities?

(a) State the null hypothesis, find the sample proportions of women in both cities, do a two-sided z test, and give a P-value using Table A.

(b) Calculate the X^2 statistic and show that it is the square of the z statistic. Show that the P-value from Table F agrees (up to the accuracy of the table) with your result from (a).

(c) Give a 95% confidence interval for the difference between the proportions of women customers in the two cities.

9.33 The study of shoppers in secondhand stores cited in the previous exercise also compared the income distributions of shoppers in the two stores. Here is the two-way table of counts:

Income	City 1	City 2
Under $10,000	70	62
$10,000 to $19,999	52	63
$20,000 to $24,999	69	50
$25,000 to $34,999	22	19
$35,000 or more	28	24

Verify that the X^2 statistic for this table is $X^2 = 3.955$. Give the degrees of freedom and the P-value. Is there good evidence that customers at the two stores have different income distributions?

9.34 Many furniture pieces are built with veneer, a thin layer of fine wood that is fastened to less expensive wood products underneath. Face checks are cracks that sometimes develop in the veneer. When face checks appear, the furniture needs to be reconstructed. Because this is a fairly expensive process, researchers seek ways to minimize the occurrence of face checks by controlling the manufacturing process. In one study, the type of adhesive used was one of the factors examined.[28] Because of the way that the veneer is cut, it has two different sides, called loose and tight, either of which can face out. Here are tables giving the numbers of veneer panels with and without face checks for two different adhesives, PVA and UF. Separate tables are given for the loose side and the tight side.

	Loose side		Tight side	
	Face checks		Face checks	
Adhesive	No	Yes	No	Yes
PVA	10	54	44	20
UF	21	43	37	27

Analyze the data. Write a summary of your results concerning the relationship between the adhesive and the occurrence of face checks. Be sure to include numerical and graphical summaries as well as the details of your significance tests.

9.35 Euthanasia of healthy but unwanted pets by animal shelters is believed to be the leading cause of death for cats and dogs. A study designed to find factors associated with bringing a cat to an animal shelter compared data on cats that were brought to an animal shelter with controls from the same county that were not brought in.[29] One of the factors examined was the source of the cat: the categories were private owner or breeder, pet store, and other (includes born in home, stray, and obtained from a shelter). This kind of study is called
case-control study   a **case-control study** by epidemiologists.

Here are the data:

| | Source | | |
Group	Private	Pet store	Other
Cases	124	16	76
Controls	219	24	203

The same researchers did a similar study for dogs.[30] The data are given in the following table:

| | Source | | |
Group	Private	Pet store	Other
Cases	188	7	90
Controls	518	68	142

(a) Analyze the data for the dogs and the cats separately. Be sure to include graphical and numerical summaries. Is there evidence to conclude that the source of the animal is related to whether or not the pet is brought to an animal shelter?

(b) Write a discussion comparing the results for the cats with those for the dogs.

(c) These data were collected using a telephone interview with pet owners in Mishawaka, Indiana. The animal shelter was run by the Humane Society of Saint Joseph County. The control group data were obtained by a random digit dialing telephone survey. Discuss how these facts relate to your interpretation of the results.

9.36 Can you increase the response rate for a mail survey by contacting the respondents before they receive the survey? A study designed to address this question compared three groups of subjects.[31] The first group received a preliminary letter about the survey, the second group was phoned, and the third received no preliminary contact. A positive response was defined as returning the survey within two weeks. Here are the counts:

| | Intervention | | |
Response	Letter	Phone call	None
Yes	171	146	118
No	220	68	455
Total	391	214	573

Another study also attempted to evaluate the effect of a letter sent before the survey on the response rate. In this study subjects who received a prenotification letter were compared with subjects who received no letter.[32] Here are the data:

Response	Letter	No letter
Yes	2570	2645
No	2448	2384
Total	5018	5029

(a) Summarize the results of each study graphically and numerically.

(b) Perform the appropriate significance tests, and report the results and the conclusions.

(c) Based on these two studies, write a report about methods to improve responses for surveys. The subjects in the first study were students from three Houston universities. The subjects in the second study were physicians in the United States. Be sure to include comments on the extent to which you think the results can be generalized to other populations of people from other places.

9.37 Gastric freezing was once a recommended treatment for ulcers in the upper intestine. A randomized comparative experiment found that 28 of the 82 patients who were subjected to gastric freezing improved, while 30 of the 78 patients in the control group improved. The hypothesis of "no difference" between the two groups can be tested in two ways: using a z statistic or using the X^2 statistic.

(a) State the appropriate hypothesis and a two-sided alternative, and carry out a z test. What is the P-value?

(b) Present the data in a 2×2 table. State the appropriate hypotheses and carry out the chi-square test. What is the P-value? Verify that the X^2 statistic is the square of the z statistic.

(c) What do you conclude about the effectiveness of gastric freezing as a treatment for ulcers? (See Example 3.5 on page 200 for a discussion of gastric freezing.)

9.38 A recent study of 865 college students found that 42.5% had student loans.[33] The students were randomly selected from the approximately 30,000 undergraduates enrolled in a large public university. The overall purpose of the study was to examine the effects of student loan burdens on the choice of a career. A student with a large debt may be more likely to choose a field where starting salaries are high so that the loan can more easily be repaid. The following table classifies the students by field of study and whether or not they have a loan:

Field of study	Student loan	
	Yes	No
Agriculture	32	35
Child development and family studies	37	50
Engineering	98	137
Liberal arts and education	89	124
Management	24	51
Science	31	29
Technology	57	71

Carry out a complete analysis of the association between having a loan and field of study, including a description of the association and an assessment of its statistical significance.

9.39   In one part of the study described in the previous exercise, students were asked to respond to some questions regarding their interests and attitudes. Some of these questions form a scale called PEOPLE that measures altruism, or an interest in the welfare of others. Each student was classified as low, medium, or high on this scale. Is there an association between PEOPLE score and field of study? Here are the data:

Field of study	PEOPLE score		
	Low	Medium	High
Agriculture	5	27	35
Child development and family studies	1	32	54
Engineering	12	129	94
Liberal arts and education	7	77	129
Management	3	44	28
Science	7	29	24
Technology	2	62	64

Analyze the data and summarize your results. Are there some fields of study that have very large or very small proportions of students in the high-PEOPLE category?

9.40   Following complaints about the working conditions in some apparel factories both in the United States and abroad, a joint government and industry commission recommended in 1998 that companies that monitor and enforce proper standards be allowed to display a "No Sweat" label on their products. Does the presence of these labels influence consumer behavior? A survey of U.S. residents aged 18 or older asked a series of questions about how likely they would be to purchase a garment under various conditions. For some conditions, it was stated that the garment had a "No Sweat" label; for others, there was no mention of such a label. On the basis of the responses, each person was classified as a "label user" or a "label nonuser."[34]   There were 296 women surveyed. Of these, 63 were label users. On the other hand, 27 of 251 men were classified as users.

(a) Construct the 2 × 2 table of counts for this problem. Include the marginal totals for your table.

(b) Use a X^2 statistic to examine the question of whether or not there is a relationship between gender and use of No Sweat labels. Give the test statistic, degrees of freedom, the P-value, and your conclusion.

(c) You examined this question using the methods of the previous chapter in Exercise 8.58 (page 572). Verify that if you square the z statistic you calculated for that exercise, you obtain the X^2 statistic that you calculated for this exercise.

9.41 Refer to Exercise 8.77 (page 576) concerning *Castaneda v. Partida*, the case where the Supreme Court review used the phrase "two or three standard deviations" as a criterion for statistical significance. Recall that there were 181,535 persons eligible for jury duty, of whom 143,611 were Mexican Americans. Of the 870 people selected for jury duty, 339 were Mexican Americans. We are interested in finding out if there is an association between being a Mexican American and being selected as a juror. Formulate this problem using a two-way table of counts. Construct the 2 × 2 table using the variables Mexican American or not and juror or not. Find the X^2 statistic and its P-value. Square the z statistic that you obtained in Exercise 8.77 and verify that the result is equal to the X^2 statistic.

9.42 Refer to Exercise 9.35 (page 623) concerning the case-control study of factors associated with bringing a cat to an animal shelter and the similar study for dogs. The last category for the source of the pet was given as "Other" and includes born in home, stray, and obtained from a shelter. The following two-way table lists these categories separately for cats:

			Source		
Group	**Private**	**Pet store**	**Home**	**Stray**	**Shelter**
Cases	124	16	20	38	18
Controls	219	24	38	116	49

Here is the same breakdown for dogs:

			Source		
Group	**Private**	**Pet store**	**Home**	**Stray**	**Shelter**
Cases	188	7	11	23	56
Controls	518	68	20	55	67

Analyze these 2 × 5 tables and compare the results that you obtained for the 2 × 3 tables in Exercise 9.35. With a large number of cells, the chi-square test sometimes does not have very much power.

9.43   A study designed to evaluate the effects of the herbal remedy *Echinacea purpurea* randomly assigned healthy children who were 2 to 11 years old to receive either echinacea or a placebo.[35] Each time a child had an upper respiratory infection (URI) treatment with echinacea or the placebo was given for the duration of the URI. The dose for the echinacea was based on the age of the child according to the recommendation of the manufacturer. The echinacea children had 337 URIs, while the placebo children had 370 URIs. For each URI many variables were measured. One of these was the parental assessment of the illness severity. Here are the data:

	Group	
Parental assessment	Echinacea	Placebo
Mild	153	170
Moderate	128	157
Severe	48	40

They also recorded the presence or absence of various types of *adverse events*. Here is a summary:

	Group	
Adverse event	Echinacea	Placebo
Itchiness	13	7
Rash	24	10
"Hyper" behavior	30	23
Diarrhea	38	34
Vomiting	22	21
Headache	33	24
Stomachache	52	41
Drowsiness	63	48
Other	63	48
Any adverse event	152	146

(a) Analyze the parental assessment data. Write a summary of your analysis and conclusion. Be sure to include graphical and numerical summaries.

(b) Analyze each adverse event. Display the results graphically in a single graph. Make a table of the relevant descriptive statistics.

(c) Use a statistical significance test to compare the echinacea URIs with those of the placebo URIs for each type of adverse event. Summarize the results in a table and write a short report giving your conclusions about the effect of echinacea on URIs in healthy children who are 2 to 11 years old. Explain why you need to analyze each type of adverse event separately rather than performing a chi-square test on the 10 × 2 table above.

(d) One concern about analyzing several outcome variables in situations like this is that we may be able to find statistical significance by chance if we look at a sufficiently large number of outcome variables. Explain why this is a concern in general but it is not a concern that is important for the interpretation of the results of your analysis here.

(e) The authors of the paper describing these results note that the unit of analysis for their computations is the URI and not the child. They state that similar results were found using more sophisticated statistical methods. Based on the descriptive statistics that you have computed, are you inclined to agree or to disagree with this statement of the authors? Explain your answer.

(f) This study was published in the *Journal of the American Medical Association (JAMA)* and was criticized in an article that appeared in *Alternative & Complementary Therapies.*[36] Three herbalists gave responses to the original article. Among their criticisms were *(i)* the dose of echinacea was too low, *(ii)* the treatment should have been given before the URI, not at the onset of symptoms, *(iii)* we should be skeptical of any positive trials on pharmaceuticals or negative trials on natural remedies that are published in *JAMA*, and *(iv) Echinacea angustifolia* (not *E. purpurea*) should have been used in combination with other herbs. Discuss these criticisms and write a summary of your opinions regarding echinacea.

 9.44 Go to the Web site for the National Congregations Study,

http://saint-denis.library.arizona.edu/natcong/,

and select data for one or more two-way tables on a topic that interests you. Analyze the data and write a report describing your conclusions.

9.45 Refer to Exercises 9.11, 9.23, 9.24, and 9.27. For each, state whether you are comparing two or more populations (the first model for two-way tables) or testing independence between two categorical variables (the second model).

The following exercises concern the optional material on goodness of fit discussed in Section 9.4.

9.46 In Example 6.1 (page 382) we examined data concerning the longleaf pine trees in the Wade Tract and concluded that the distribution of trees in the tract was not random. Here is another way to examine the same question. First, we divide the tract into four equal parts, or quadrants, in the east–west direction. Call the four parts, Q_1 to Q_4. Then we take a random sample of 100 trees and count the number of trees in each quadrant. Here are the data:

Quadrant	Q_1	Q_2	Q_3	Q_4
Count	18	22	39	21

(a) If the trees are randomly distributed, we expect to find 25 trees in each quadrant. Why? Explain your answer.

(b) We do not really expect to get *exactly* 25 trees in each quadrant. Why? Explain your answer.

(c) Perform the goodness of fit significance test for these data. Write a short report giving the details of your analysis and your conclusion.

9.47 At your college 29% of the students are in their first year, 27% in their second, 25% in their third, and 19% in their fourth year. You have taken a survey of students, and when you classify them by year of study, you have 54, 66, 56, and 30 students in the first, second, third, and fourth years, respectively. Use a goodness of fit test to examine how well your sample reflects the population of your college.

9.48 Computer software generated 500 random numbers that should look like they are from the standard normal distribution. They are categorized into five groups: (1) less than or equal to −0.8, (2) greater than −0.8 and less than or equal to −0.2, (3) greater than −0.2 and less than or equal to 0.2, (4) greater than 0.2 and less than or equal to 0.8, and (5) greater than 0.8. The counts in the five groups are 98, 112, 79, 111, and 100, respectively. Find the probabilities for these five intervals using Table A. Then compute the expected number for each interval for a sample of 500. Finally, perform the goodness of fit test and summarize your results.

9.49 Refer to the previous exercise. Use software to generate your own sample of 500 standard normal random variables, and perform the goodness of fit test. Choose a different set of intervals from the ones used in the previous exercise.

9.50 Computer software generated 500 random numbers that should look like they are from the uniform distribution on the interval 0 to 1 (see page 282). They are categorized into five groups: (1) less than or equal to 0.2, (2) greater than 0.2 and less than or equal to 0.4, (3) greater than 0.4 and less than or equal to 0.6, (4) greater than 0.6 and less than or equal to 0.8, and (5) greater than 0.8. The counts in the five groups are 113, 95, 108, 99, and 85, respectively. The probabilities for these five intervals are all the same. What is this probability? Compute the expected number for each interval for a sample of 500. Finally, perform the goodness of fit test and summarize your results.

9.51 Refer to the previous exercise. Use software to generate your own sample of 500 uniform random variables on the interval from 0 to 1, and perform the goodness of fit test. Choose a different set of intervals from the ones used in the previous exercise.

9.52 An instructor who assigned an exercise similar to the one described in the previous exercise received homework from a student who reported a *P*-value of 0.999. The instructor suspected that the student did not use the computer for the assignment but just made up some numbers for the homework. Why was the instructor suspicious? How would this scenario change if there were 1000 students in the class?

EESEE CASE STUDIES

The Electronic Encyclopedia of Statistical Examples and Exercises (EESEE) is available on the text CD and Web site. These more elaborate stories, with data, provide settings for longer case studies.

9.53 Read the EESEE story "Surgery in a Blanket." Write a report that answers all questions for this case study.

9.54 Read the EESEE story "Trilobite Bites." Write a report that answers all questions for this case study.

9.55 Read the EESEE story "Hip Protectors for the Elderly." Write a report that answers all questions for this case study.

Making Sure Medical Devices Are Safe and Effective

GREGORY CAMPBELL *is the Director of Biostatistics at the Center for Devices and Radiological Health, U.S. Food and Drug Administration.*

Modern medicine makes use of thousands of different kinds of devices, with new and revolutionary technologies being invented every year. New medical products, such as artificial hips, defibrillators, lasers for eye surgery, and coronary stents, must be approved by the Food and Drug Administration (FDA) before they can be sold in the United States. Companies must show that their new medical devices are both safe and effective. Often companies recruit volunteer patients on whom to test the new devices in statistically designed experiments called clinical trials. Companies submit the results of these trials as evidence that their devices are safe and work as claimed. I direct a group of 30 statisticians who review the statistical designs and data analyses that the companies submit. These reviews are team efforts by statisticians, clinicians, and engineers. I become personally involved in a review when the approval decision is unusually complicated or controversial. My training as a statistician is essential to my work both as a manager and as a scientist.

As a scientist, I am interested in developing new statistical methods to help the FDA make rapid and accurate decisions. For example, there is often information available about a condition or therapy before an experiment starts. We can then apply methods called Bayesian statistics that combine existing information with experimental results. Another of my personal scientific interests is in improving the evaluation of new devices used to diagnose medical conditions rather than to treat them. At the FDA we have used new statistical methods to evaluate, for example, microarrays for gene expression and digital mammography.

Statistics plays a very important role in the FDA, helping us make scientifically based decisions that improve public health and protect everyone's safety. I am stimulated by both the scientific challenges and the responsibility to the public in this important work.

Inference for Regression

Introduction

In this chapter we describe methods for inference when there is a single response variable and one quantitative explanatory variable. The descriptive tools we learned in Chapter 2—scatterplots, least-squares regression, and correlation—are essential preliminaries to inference and also provide a foundation for confidence intervals and significance tests.

We first met the sample mean $\bar{x}$ in Chapter 1 as a measure of the center of a collection of observations. Later we learned that when the data are a random sample from a population, the sample mean is an estimate of the population mean μ. In Chapters 6 and 7, we used $\bar{x}$ as the basis for confidence intervals and significance tests for inference about μ.

Now we will follow the same program for the problem of fitting straight lines to data. In Chapter 2 we met the least-squares regression line $\hat{y} = a + bx$ as a description of a straight-line relationship between a response variable y and an explanatory variable x. At that point we did not distinguish between sample and population. Now we will think of the least-squares line computed from a sample as an estimate of a *true* regression line for the population, just as $\bar{x}$ is an estimate of the true population mean μ. To emphasize the change in thinking, we introduce new notation. Following the common practice of using Greek letters for population parameters, we will write the population line as $\beta_0 + \beta_1 x$. The least-squares line fitted to sample data is now $b_0 + b_1 x$. This notation reminds us that the intercept b_0 of the fitted line estimates the intercept β_0 of the population line, and the slope b_1 estimates the slope β_1.

10.1 Simple Linear Regression

Statistical model for linear regression

Simple linear regression studies the relationship between a response variable y and a single explanatory variable x. We expect that different values of x will produce different mean responses. We encountered a similar but simpler situation in Chapter 7 when we discussed methods for comparing two population means. Figure 10.1 illustrates the statistical model for a comparison of

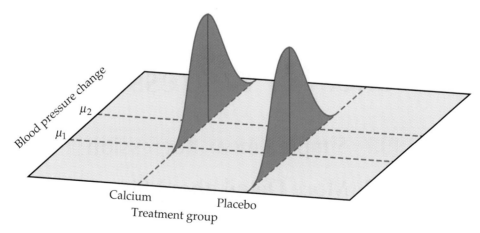

FIGURE 10.1 The statistical model for comparing responses to two treatments; the mean response varies with the treatment.

the blood pressure in two groups of experimental subjects, one group taking a calcium supplement and the other a placebo. In each population the observed change in blood pressure varies according to a normal distribution. The mean change will generally be different for the two populations. In the figure, μ_1 and μ_2 represent the two population means. The two normal curves in Figure 10.1 have the same spread, indicating that the population standard deviations are assumed to be equal. We can think of the treatment (placebo or calcium) as the explanatory variable in this example.

In linear regression the explanatory variable x can have many different values. Imagine, for example, giving different amounts x of calcium to different groups of subjects. We can think of the values of x as defining different **subpopulations,** one for each possible value of x. Each subpopulation consists of all individuals in the population having the same value of x. If we conducted an experiment with five different amounts of calcium, we could view these values as defining five different subpopulations.

subpopulations

The statistical model for simple linear regression assumes that for each value of x the observed values of the response variable y are normally distributed about a mean that depends on x. We use μ_y to represent these means. Rather than just two means μ_1 and μ_2, we are interested in how the many means μ_y change as x changes. In general the means μ_y can change according to any sort of pattern as x changes. In **simple linear regression** we assume that they all lie on a line when plotted against x. The equation of the line is

simple linear regression

$$\mu_y = \beta_0 + \beta_1 x$$

with intercept β_0 and slope β_1. This is the **population regression line;** it describes how the mean response changes with x. Actually observed y's will vary about these means. The model assumes that this variation, measured by the standard deviation σ, is the same for all values of x. Figure 10.2 displays this statistical model. The line is the population regression line, and the three normal curves show how the response y will vary for three different values of the explanatory variable x.

population regression line

Data for simple linear regression

The data for a linear regression are observed values of y and x. The model takes each x to be a fixed known quantity. In practice, x may not be exactly known. If the error in measuring x is large, more advanced inference methods

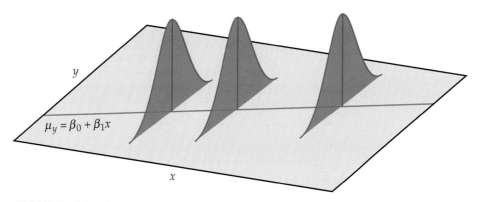

FIGURE 10.2   The statistical model for linear regression; the mean response is a straight-line function of the explanatory variable.

are needed. The response y to a given x is a random variable that will take different values if we have several observations with the same x-value. The model describes the mean and standard deviation of the random variable y. These unknown parameters must be estimated from the data.

We will use the following example to explain the fundamentals of simple linear regression. Because regression calculations in practice are always done by statistical software, we will rely on computer output for the arithmetic in this example. In the next section, we give an example that illustrates how to do the work with a calculator if software is unavailable.

EXAMPLE 10.1   Computers in some vehicles calculate various quantities related to the performance. One of these is the fuel efficiency, or gas mileage, expressed as miles per gallon (MPG). Another is the average speed in miles per hour (MPH). For one vehicle equipped in this way, MPG and MPH were recorded each time the gas tank was filled and the computer was then reset.[1] How does the speed at which the vehicle is driven affect the fuel efficiency? There are 234 observations available. We will work with a simple random sample of size 60.

Before starting our analysis, it is appropriate to consider the extent to which our results can reasonably be generalized. Because we have a simple random sample from a population of size 234, we are on firm ground in making inferences about this particular vehicle. However, as a practical matter, no one really cares about this particular vehicle. Our results are only interesting if they can be applied to other similar vehicles that are driven under similar conditions. Our statistical modeling for this data set is concerned about the process by which speed affects the fuel efficiency. Although we would not expect the parameters that describe the relationship between speed and fuel efficiency to be *exactly* the same for similar vehicles, we would expect to find qualitatively similar results.

In the statistical model for predicting fuel efficiency from speed, subpopulations are defined by the explanatory variable, speed. For a particular value of speed, say 30 MPH, we can think about operating this vehicle repeatedly at this average speed. Variation in driving conditions and the behavior of the driver would be sources of variation that would give different values of MPG for this subpopulation.

EXAMPLE 10.2   We start our analysis with a graphical display of the data. Figure 10.3 is a plot of fuel efficiency versus speed for our sample of 60 observations. The least-squares regression line and a smooth function are also shown in the plot. Although there is a positive association between MPG and MPH, the fit is not linear. The smooth function shows us that the relationship levels off somewhat with increasing speed.

Always start with a graphical display of the data. There is no point in trying to do statistical inference if our data do not, at least approximately, meet the assumptions that are the foundation for our inference. At this point we need to make a choice. One possibility would be to confine our interest to speeds

FIGURE 10.3 Scatterplot of MPG versus MPH with a smooth function and the least-squares line, for Example 10.2.

that are 30 MPH or less, a region where it appears that a line would be a good fit to the data. Another possibility is to make some sort of transformation that will make the relationship approximately linear for the entire set of data. We will choose the second option.

EXAMPLE 10.3 One type of function that looks similar to the smooth-function fit in Figure 10.3 is a logarithm. Therefore, we will examine the effect of transforming speed by taking the logarithm. The result is shown in Figure 10.4. In this plot the smooth function and the line are quite close. We are satisfied that the

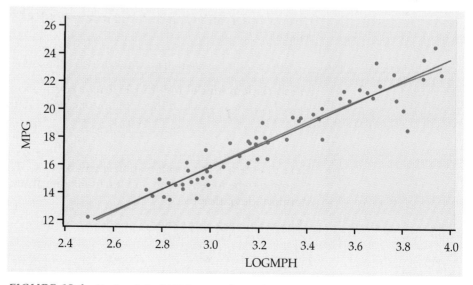

FIGURE 10.4 Scatterplot of MPG versus logarithm of MPH with a smooth function and the least-squares line, for Example 10.3.

relationship between the log of speed and fuel efficiency is approximately linear for this set of data.

Consider a particular value of x: for example, $x = \log(30) = 3.4$. We can think about the subpopulation as repeated driving at an average speed of 30 MPH. The fuel efficiency will, of course, vary within this subpopulation because of differing driving conditions and driver behavior. Our statistical model assumes that these fuel efficiencies are normally distributed with a mean μ_y that depends upon x in a linear way. Specifically,

$$\mu_y = \beta_0 + \beta_1 x$$

This population regression line gives the mean fuel efficiency for all values of x. We cannot observe this line, because the observed responses y vary about their means. The statistical model for linear regression consists of the population regression line and a description of the variation of y about the line. Figure 10.2 displays the line and the variation in a picture, and the following equation expresses the idea in an equation:

$$\text{DATA} = \text{FIT} + \text{RESIDUAL}$$

The FIT part of the model consists of the subpopulation means, given by the expression $\beta_0 + \beta_1 x$. The RESIDUAL part represents deviations of the data from the line of population means. We assume that these deviations are normally distributed with standard deviation σ. We use ϵ (the Greek letter epsilon) to stand for the RESIDUAL part of the statistical model. A response y is the sum of its mean and a chance deviation ϵ from the mean. The deviations ϵ represent "noise," that is, variation in y due to other causes that prevent the observed (x, y)-values from forming a perfectly straight line on the scatterplot.

SIMPLE LINEAR REGRESSION MODEL

Given n observations on the explanatory variable x and the response variable y,

$$(x_1, y_1), \quad (x_2, y_2), \quad \ldots, \quad (x_n, y_n)$$

the **statistical model for simple linear regression** states that the observed response y_i when the explanatory variable takes the value x_i is

$$y_i = \beta_0 + \beta_1 x_i + \epsilon_i$$

Here $\beta_0 + \beta_1 x_i$ is the mean response when $x = x_i$. The deviations ϵ_i are assumed to be independent and normally distributed with mean 0 and standard deviation σ.

The parameters of the model are β_0, β_1, and σ.

Because the means μ_y lie on the line $\mu_y = \beta_0 + \beta_1 x$, they are all determined by β_0 and β_1. Once we have estimates of β_0 and β_1, the linear relationship determines the estimates of μ_y for all values of x. Linear regression allows us to do inference not only for subpopulations for which we have data but also for those corresponding to x's not present in the data. We will learn how to do inference about

- the slope β_1 and the intercept β_0 of the population regression line,
- the mean response μ_y for a given value of x, and
- an individual future response y for a given value of x.

Estimating the regression parameters

The method of least squares presented in Chapter 2 fits a line to summarize a relationship between the observed values of an explanatory variable and a response variable. Now we want to use the least-squares line as a basis for inference about a population from which our observations are a sample. We can do this only when the statistical model just presented holds. In that setting, the slope b_1 and intercept b_0 of the least-squares line

$$\hat{y} = b_0 + b_1 x$$

estimate the slope β_1 and the intercept β_0 of the population regression line.

Using the formulas from Chapter 2 (page 137) and our new notation, the slope of the least-squares line is

$$b_1 = r\frac{s_y}{s_x}$$

and the intercept is

$$b_0 = \bar{y} - b_1\bar{x}$$

Here, r is the correlation between y and x, s_y is the standard deviation of y, and s_x is the standard deviation of x. Some algebra based on the rules for means of random variables (Section 4.4) shows that b_0 and b_1 are unbiased estimators of β_0 and β_1. Furthermore, b_0 and b_1 are normally distributed with means β_0 and β_1 and standard deviations that can be estimated from the data. Normality of these sampling distributions is a consequence of the assumption that the ϵ_i are distributed normally. A general form of the central limit theorem tells us that the distributions of b_0 and b_1 will still be approximately normal even if the ϵ_i are not. On the other hand, outliers and influential observations can invalidate the results of inference for regression.

The predicted value of y for a given value x^* of x is the point on the least-squares line $\hat{y} = b_0 + b_1 x^*$. This is an unbiased estimator of the mean response **residuals** μ_y when $x = x^*$. The **residuals** are

$$e_i = \text{observed response} - \text{predicted response}$$
$$= y_i - \hat{y}_i$$
$$= y_i - b_0 - b_1 x_i$$

The residuals e_i correspond to the model deviations ϵ_i. The e_i sum to 0, and the ϵ_i come from a population with mean 0.

The remaining parameter to be estimated is σ, which measures the variation of y about the population regression line. Because this parameter is the standard deviation of the model deviations, it should come as no surprise that we use the residuals to estimate it. As usual, we work first with the variance and take the square root to obtain the standard deviation. For simple linear regression the estimate of σ^2 is the average squared residual

$$s^2 = \frac{\sum e_i^2}{n-2}$$

$$= \frac{\sum (y_i - \hat{y}_i)^2}{n-2}$$

We average by dividing the sum by $n-2$ in order to make s^2 an unbiased estimate of σ^2. The sample variance of n observations uses the divisor $n-1$ for this same reason. The quantity $n-2$ is called the **degrees of freedom** for s^2. The estimate of σ is given by

degrees of freedom

$$s = \sqrt{s^2}$$

We will use statistical software to calculate the regression for predicting fuel efficiency with the log of speed for Example 10.1. In entering the data, we chose the names LOGMPH for the log of speed and MPG for fuel efficiency. *It is good practice to use names, rather than just x and y, to remind yourself which data the output describes.*

EXAMPLE 10.4   Figure 10.5 gives the outputs for four commonly used statistical software packages and Excel. Other software will give similar information. The SPSS output reports estimates of our three parameters as $b_0 = -7.796$, $b_2 = 7.874$,

SPSS

Model Summary

Model	R	R Square	Std. Error of the Estimate
1	.946	.895	.9995

a Predictors: (Constant), LOGMPH

Model		Coefficients B	Std. Error	t	Sig.	95% Confidence Interval for B Lower Bound	Upper Bound
1	(Constant)	-7.796	1.155	-6.750	.000	-10.108	-5.484
	LOGMPH	7.874	.354	22.237	.000	7.165	8.583

a Dependent Variable: MPG

Minitab

```
The regression equation is
MPG = - 7.80 + 7.87 logmph

Predictor        Coef        StDev           T           P
Constant       -7.796        1.155        -6.75       0.000
logmph          7.8742       0.3541       22.24       0.000

S = 0.9995      R-Sq = 89.5%      R-Sq(adj) = 89.3%
```

FIGURE 10.5   Regression output from SPSS, Minitab, Crunch It!, Excel, and SAS for the fuel efficiency example. *(continued)*

Crunch It!

Simple linear regression results:
Dependent Variable: MPG
Independent Variable: logmph
MPG = 7.7962503 + 7.874219 logmph
Sample size: 60
R (correlation coefficient) = 0.9461
R–sq = 0.8950163
Estimate of error standard deviation: 0.99951637

Parameter estimates:

Parameter	Estimate	Std. Err.	DF	T-Stat	P-Value
Intercept	-7.7962503	1.1549443	58	-6.7503257	<0.0001
Slope	7.874219	0.3541106	58	22.236609	<0.0001

Excel

	A	B	C	D	E	F	G
1	SUMMARY OUTPUT						
2							
3	Regression Statistics						
4	Multiple R	0.946053015					
5	R Square	0.895016308					
6	Adjusted R Square	0.893206244					
7	Standard Error	0.999516364					
8	Observations	60					
9							
10	ANOVA						
11		df	SS	MS	F	Significance F	
12	Regression	1	493.9885883	493.9886	494.4668	4.50949E-30	
13	Residual	58	57.94391174	0.999033			
14	Total	59	551.9325				
15							
16		Coefficients	Standard Error	tStet	P-value	Lower 95%	Upper 95%
17	intercept	-7.796250129	1.154944262	-6.75033	7.69E-09	-10.10812052	-5.48437974
18	logmph	7.874219013	0.354110611	22.23661	4.51E-30	7.165390143	8583047883

SAS

```
           Root MSE              0.99952    R-Square    0.8950
           Dependent Mean       17.72500    Adj R-Sq    0.8932
           Coeff Var             5.63902

                    Parameter     Standard
Variable   DF        Estimate        Error   t Value   Pr > |t|    95% Confidence Limits
Intercept   1        -7.79625      1.15494     -6.75    <.0001    -10.10812     -5.48438
logmph      1         7.87422      0.35411     22.24    <.0001      7.16539      8.58305
```

FIGURE 10.5 *(Continued)* Regression output from SPSS, Minitab, Crunch It!, Excel, and SAS for the fuel efficiency example.

and $s = 0.9995$. Be sure that you can find these entries in this output and the corresponding values in the other outputs.

The least-squares regression line is the straight line that is plotted in Figure 10.4. We would report it as

$$\widehat{\text{MPG}} = -7.80 + 7.87\text{LOGMPH}$$

with a model standard deviation of $s = 1.00$. Note that the number of digits provided varies with the software used and we have rounded off the values to three significant digits. *It is important to avoid cluttering up your report of the results of a statistical analysis with many digits that are not relevant.* Software often reports many more digits than are meaningful or useful.

The outputs contain much other information that we ignore for now. Computer outputs often give more information than we want or need. *The experienced user of statistical software learns to ignore the parts of the output that are not needed for the current problem.* On the other hand, it is very frustrating to find that a software package does not print out the particular statistics that we want for an analysis.

Now that we have fitted a line, we should examine the residuals. We usually plot the residuals both against the case number (especially if this reflects the order in which the observations were taken) and against the explanatory variable. For this example, in place of case number, we prefer another variable that is similar but is recorded in a more useful scale. It is the total number of miles that the vehicle has been driven. Figure 10.6 gives a plot of the residuals versus miles driven with a smooth-function fit. The smooth function suggests that the residuals increase slightly up to about 50,000 miles and then tend to decrease somewhat. With the data that we have for this example, it is difficult to decide if this effect is real or due to chance variation. It is not unreasonable to think that the vehicle performance decreases with age. Since the effect does

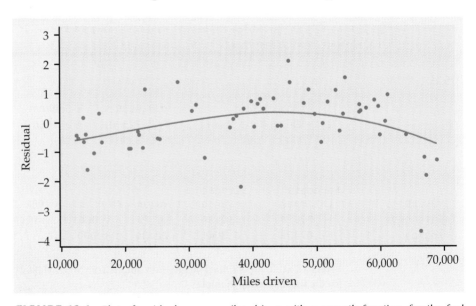

FIGURE 10.6 Plot of residuals versus miles driven with a smooth function, for the fuel efficiency example.

not appear to be particularly large, we will ignore it for the present analysis. With more data, however, it may be an interesting phenomenon to study.

The residuals are plotted versus the explanatory variable, log of MPH, in Figure 10.7. No clear pattern is evident. There is one residual that is somewhat low, and we have seen it in all of our plots. Inspection of Figure 10.4 reveals that this observation does not appear to distort our least-squares regression line.

Finally, Figure 10.8 is a normal quantile plot of the residuals. Because the plot looks fairly straight, we are confident that we do not have a serious

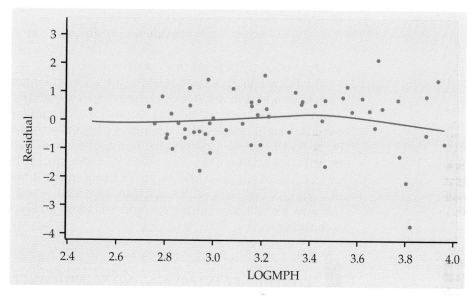

FIGURE 10.7   Plot of residuals versus log of MPH with a smooth function, for the fuel efficiency example.

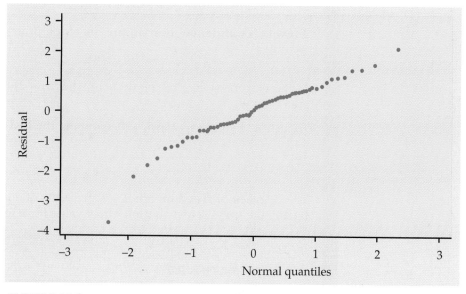

FIGURE 10.8   Normal quantile plot of the residuals for the fuel efficiency example.

violation of our assumption that the residuals are normally distributed. Observe that the low outlier is also visible in this plot.

Confidence intervals and significance tests

Chapter 7 presented confidence intervals and significance tests for means and differences in means. In each case, inference rested on the standard errors of estimates and on t distributions. Inference for the intercept and slope in a linear regression is similar in principle, although the recipes are more complicated. All of the confidence intervals, for example, have the form

$$\text{estimate} \pm t^*\text{SE}_{\text{estimate}}$$

where t^* is a critical point of a t distribution.

Confidence intervals and tests for the slope and intercept are based on the normal sampling distributions of the estimates b_1 and b_0. Standardizing these estimates gives standard normal z statistics. The standard deviations of these estimates are multiples of σ, the model parameter that describes the variability about the true regression line. Because we do not know σ, we estimate it by s, the variability of the data about the least-squares line. When we do this, we get t distributions with degrees of freedom $n - 2$, the degrees of freedom of s. We give formulas for the standard errors SE_{b_1} and SE_{b_0} in Section 10.2. For now we will concentrate on the basic ideas and let the computer do the computations.

CONFIDENCE INTERVALS AND SIGNIFICANCE TESTS FOR REGRESSION SLOPE AND INTERCEPT

A **level C confidence interval for the intercept** β_0 is

$$b_0 \pm t^*\text{SE}_{b_0}$$

A **level C confidence interval for the slope** β_1 is

$$b_1 \pm t^*\text{SE}_{b_1}$$

In these expressions t^* is the value for the $t(n - 2)$ density curve with area C between $-t^*$ and t^*.

To test the hypothesis $H_0: \beta_1 = 0$, compute the **test statistic**

$$t = \frac{b_1}{\text{SE}_{b_1}}$$

The **degrees of freedom** are $n - 2$. In terms of a random variable T having the $t(n - 2)$ distribution, the P-value for a test of H_0 against

$H_a: \beta_1 > 0$ is $P(T \geq t)$

$$H_a: \beta_1 < 0 \text{ is } P(T \leq t)$$

$$H_a: \beta_1 \neq 0 \text{ is } 2P(T \geq |t|)$$

There are similar formulas for significance tests about the intercept β_0, using SE_{b_0} and the $t(n - 2)$ distribution. Although computer outputs often include a test of $H_0: \beta_0 = 0$, this information usually has little practical value. From the equation for the population regression line, $\mu_y = \beta_0 + \beta_1 x$, we see that β_0 is the mean response corresponding to $x = 0$. In many practical situations, this subpopulation does not exist or is not interesting.

On the other hand, the test of $H_0: \beta_1 = 0$ is quite useful. When we substitute $\beta_1 = 0$ in the model, the x term drops out and we are left with

$$\mu_y = \beta_0$$

This model says that the mean of y does not vary with x. All of the y's come from a single population with mean β_0, which we would estimate by $\bar{y}$. The hypothesis $H_0: \beta_1 = 0$ therefore says that there is no straight-line relationship between y and x and that linear regression of y on x is of no value for predicting y.

EXAMPLE 10.5   The computer outputs in Figure 10.5 for the fuel efficiency problem contain the information needed for inference about the regression slope and intercept. Let's look at the SPSS output. The column labeled Std. Error gives the standard errors of the estimates. The value of SE_{b_1} appears on the line labeled with the variable name for the explanatory variable, LOGMPH. It is given as .354. In a summary we would report that the regression coefficient for the log of speed is 7.87 with a standard error of 0.35.

The t statistic and P-value for the test of $H_0: \beta_1 = 0$ against the two-sided alternative $H_a: \beta_1 \neq 0$ appear in the columns labeled t and Sig. We can verify the t calculation from the formula for the standardized estimate:

$$t = \frac{b_1}{SE_{b_1}} = \frac{7.874}{0.354} = 22.24$$

The P-value is given as .000. This is a rounded number and from that information we can conclude that $P < 0.0005$. The other outputs in Figure 10.5 also indicate that the P-value is very small. We will report the result as $P < 0.001$ because 1 chance in 1000 is sufficiently small for us to decisively reject the null hypothesis.

We have found a statistically significant linear relationship between fuel efficiency and log speed. The estimated slope is more than 22 standard deviations away from zero. Because this is extremely unlikely to happen if the true slope is zero, we have strong evidence for our claim. Note, however, that this is not the same as concluding that we have found a strong relationship between the response and explanatory variables in this example. *A very small P-value for the significance test for a zero slope does not necessarily imply that we have found a strong relationship.* A confidence interval will provide additional information about the relationship.

EXAMPLE 10.6   A confidence interval for β_1 requires a critical value t^* from the $t(n-2) = t(58)$ distribution. In Table D there are entries for 50 and 60 degrees of freedom. The values for these rows are very similar. To be conservative, we will use the larger critical value, for 50 degrees of freedom. Find the confidence level values at the bottom of the table. In the 95% confidence column the entry for 50 degrees of freedom is $t^* = 2.009$.

To compute the 95% confidence interval for β_1 we combine the estimate of the slope with the margin of error:

$$b_1 \pm t^* SE_{b_1} = 7.874 \pm (2.009)(0.354)$$

$$= 7.874 \pm 0.711$$

The interval is $(7.16, 8.58)$. This agrees with the value given by the software outputs that provide this information in Figure 10.5. We estimate that an increase of 1 in the logarithm of speed is associated with an increase of between 7.16 and 8.58 MPG.

To interpret the interval, it is useful to translate the statement back to the original MPH scale. From Figure 10.4 we can see that the values for LOGMPH range from about 2.5 to 3.9. Let's translate the increase of 1 unit in LOGMPH to the MPH scale by considering a change from 2.8 to 3.8. Since $\log(16.4) = 2.8$ and $\log(44.7) = 3.8$, the change corresponds to an increase in speed from 16.4 MPH to 44.7 MPH. An increase in average speed from 16.4 to 44.7 MPH is associated with an increase of 7.8 ± 0.7 in MPG.

Note that the intercept in this example is not of practical interest. It estimates MPG when the logarithm MPH (that's x) is 0, a value that cannot occur. For this reason, we do not compute a confidence interval for β_0.

Confidence intervals for mean response

For any specific value of x, say x^*, the mean of the response y in this subpopulation is given by

$$\mu_y = \beta_0 + \beta_1 x^*$$

To estimate this mean from the sample, we substitute the estimates b_0 and b_1 for β_0 and β_1:

$$\hat{\mu}_y = b_0 + b_1 x^*$$

A confidence interval for μ_y adds to this estimate a margin of error based on the standard error $SE_{\hat{\mu}}$. (The formula for the standard error is given in Section 10.2.)

> ## CONFIDENCE INTERVAL FOR A MEAN RESPONSE
>
> A **level C confidence interval for the mean response** μ_y when x takes the value x^* is
>
> $$\hat{\mu}_y \pm t^* \mathrm{SE}_{\hat{\mu}}$$
>
> where t^* is the value for the $t(n-2)$ density curve with area C between $-t^*$ and t^*.

Many computer programs calculate confidence intervals for the mean response corresponding to each of the x-values in the data. Some can calculate an interval for any value x^* of the explanatory variable. We will use a plot to illustrate these intervals.

EXAMPLE 10.7 Figure 10.9 shows the upper and lower confidence limits on a graph with the data and the least-squares line. The 95% confidence limits appear as curved black lines. For any x^*, the confidence interval for the mean response extends from the lower black line to the upper black line. The intervals are narrowest for values of x^* near the mean of the observed x's and increase as x^* moves away from $\bar{x}$.

Some software will do these calculations directly if you input a value for the explanatory variable. Others will calculate the intervals for each value of x in the data set. Creating a new data set with an additional observation with x equal to the value of interest and y missing will often work.

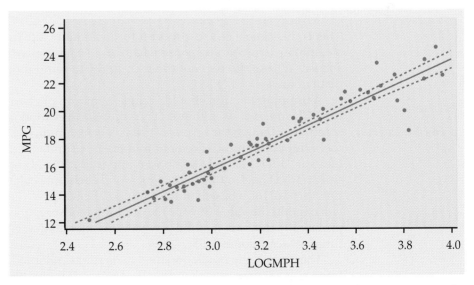

FIGURE 10.9   The 95% confidence limits (blue dashed curves) for the mean response for the fuel efficiency example.

EXAMPLE 10.8   Let's find the confidence interval for mean response corresponding to 30 MPH. We use $x = \log(30) = 3.4$ as the value for the explanatory variable. Our predicted fuel efficiency is

$$\widehat{MPG} = -7.80 + 7.87LOGMPH$$
$$= -7.80 + (7.87)(3.4)$$
$$= 19.0$$

Software tells us that the 95% confidence interval for the mean response is 18.7 MPG to 19.3 MPG.

If we operated this vehicle a large number of times under similar conditions at an average speed of 30 MPH, we would expect the fuel efficiency to be between 18.7 and 19.3 MPG. Note that many of the observations in Figure 10.9 lie outside the confidence bands. *These confidence intervals do not tell us what mileage to expect for a single observation at a particular average speed such as 30 MPH.* We need a different kind of interval for this purpose.

Prediction intervals

In the last example, we predicted the mean fuel efficiency when the average speed is 30 MPH. Suppose we now want to predict a future observation of fuel efficiency when the vehicle is driven at 30 MPH under similar conditions. Our best guess at the fuel efficiency is what we obtained before using the regression equation, that is, 19.0 MPG. The margin of error, on the other hand, is larger because it is harder to predict an individual value than to predict the mean.

The predicted response y for an individual case with a specific value x^* of the explanatory variable x is

$$\hat{y} = b_0 + b_1 x^*$$

This is the same as the expression for $\hat{\mu}_y$. That is, the fitted line is used both to estimate the mean response when $x = x^*$ and to predict a single future response. We use the two notations $\hat{\mu}_y$ and $\hat{y}$ to remind ourselves of these two distinct uses.

A useful prediction should include a margin of error to indicate its accuracy. The interval used to predict a future observation is called a **prediction interval.** Although the response y that is being predicted is a random variable, the interpretation of a prediction interval is similar to that for a confidence interval. Draw a sample of n observations (x_i, y_i) and then one additional observation (x^*, y). Do this many times, and each time calculate the 95% prediction interval for y based on the sample. Then 95% of the prediction intervals will contain the value of y for the additional observation. In other words, the probability that this method produces an interval that contains the value of a future observation is 0.95.

The form of the prediction interval is very similar to that of the confidence interval for the mean response. The difference is that the standard error $SE_{\hat{y}}$ used in the prediction interval includes both the variability due to the fact that the least-squares line is not exactly equal to the true regression line *and* the

prediction interval

variability of the future response variable y around the subpopulation mean. (The formula for $SE_{\hat{y}}$ appears in Section 10.2.)

PREDICTION INTERVAL FOR A FUTURE OBSERVATION

A **level *C* prediction interval for a future observation** on the response variable y from the subpopulation corresponding to x^* is

$$\hat{y} \pm t^* SE_{\hat{y}}$$

where t^* is the value for the $t(n-2)$ density curve with area C between $-t^*$ and t^*.

Again, we use a graph to illustrate the results.

EXAMPLE 10.9   Figure 10.10 shows the upper and lower prediction limits, along with the data and the least-squares line. The 95% prediction limits are indicated by the black curves.

Compare this figure with Figure 10.9, which shows the 95% confidence limits drawn to the same scale. The upper and lower limits of the prediction intervals are farther from the least-squares line than are the confidence limits. *The figures remind us that the interval for a single future observation must be larger than an interval for the mean of its subpopulation.*

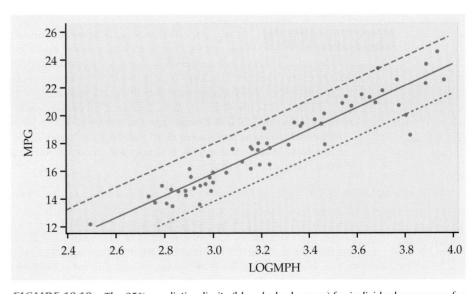

FIGURE 10.10   The 95% prediction limits (blue dashed curves) for individual responses for the fuel efficiency example. Compare with Figure 10.9.

EXAMPLE 10.10   Let's find the prediction interval for a future observation of fuel efficiency when the vehicle is driven at 30 MPH. The predicted value is the same as our estimate of the mean response that we calculated in Example 10.8, that is, 19.0 MPG. Software tells us that the 95% prediction interval is 17.0 MPG to 21.0 MPG.

If we operated this vehicle a single time under similar conditions at an average speed of 30 MPH, we would expect the fuel efficiency to be between 17.0 and 21.0 MPG. Note that most, but not all, of the observations in Figure 10.10 lie within the prediction bands.

BEYOND THE BASICS | Nonlinear regression

The regression model that we have studied assumes that the relationship between the response variable and the explanatory variable can be summarized with a straight line. When the relationship is not linear, we can sometimes make it linear by a transformation. In other circumstances, we use models that allow for various types of curved relationships. These models are called **nonlinear models.**

nonlinear models

Technical details are much more complicated for nonlinear models. In general we cannot write down simple formulas for the parameter estimates; we use a computer to solve systems of equations to find the estimates. However, the basic principles are those that we have already learned. For example,

$$DATA = FIT + RESIDUAL$$

still applies. The FIT is a nonlinear (curved) function and the residuals are assumed to be an SRS from the $N(0, \sigma)$ distribution. The nonlinear function contains parameters that must be estimated from the data. Approximate standard errors for these estimates are part of the standard output provided by software. Here is an example.

EXAMPLE 10.11   As we age, our bones become weaker and are more likely to break. Osteoporosis (or weak bones) is the major cause of bone fractures in older women. Some researchers have studied this problem by looking at how and when bone mass is accumulated by young women. Understanding the relationship between age and bone mass is an important part of this approach to the problem.

Figure 10.11 displays data for a measure of bone strength, called "total body bone mineral density" (TBBMD), and age for a sample of 256 young women.[2] TBBMD is measured in grams per square centimeter (g/cm^2), and age is recorded in years. The solid curve is the nonlinear fit, and the dashed curves are 95% prediction limits. The fitted nonlinear equation is

$$\hat{y} = 1.162 \frac{e^{-1.162+0.28x}}{1 + e^{-1.162+0.28x}}$$

In this equation, $\hat{y}$ is the predicted value of TBBMD, the response variable; and x is age, the explanatory variable. A straight line would not do a very good job of summarizing the relationship between TBBMD and age. At first, TBBMD increases with age, but then it levels off as age increases. The value of the function where it is level is called "peak bone mass"; it is a parameter in the nonlinear model. The estimate is 1.162 and the standard error is 0.008. Software gives the 95% confidence interval as (1.146, 1.178).

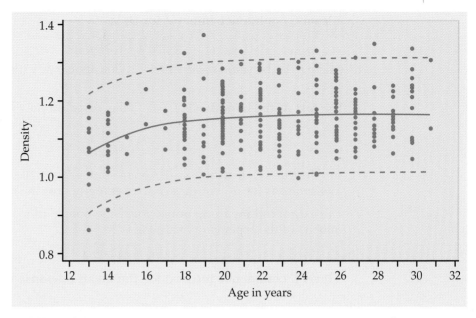

FIGURE 10.11   Plot of total body bone mineral density versus age.

The long-range goals of the researchers who conducted this study include developing intervention programs (exercise and increasing calcium intake have been shown to be effective) for young women that will increase their TBBMD. What age groups should be the target of these interventions? The fitted nonlinear model can be used to obtain estimates of the age (with a standard error) at which any given percent of the peak bone mass is attained. We estimate that the age at which the population reaches 95% of peak bone mass is 16.2 years (SE = 1.1 years). For 99% of peak bone mass, the age is 22.1 years (SE = 2.5 years). Intervention programs should be directed toward high school and college-aged women.

SECTION 10.1 | Summary

The statistical model for **simple linear regression** is

$$y_i = \beta_0 + \beta_1 x_i + \epsilon_i$$

where $i = 1, 2, \ldots, n$. The ϵ_i are assumed to be independent and normally distributed with mean 0 and standard deviation σ. The **parameters** of the model are β_0, β_1, and σ.

The intercept and slope β_0 and β_1 are estimated by the intercept and slope of the **least-squares regression line,** b_0 and b_1. The parameter σ is estimated by

$$s = \sqrt{\frac{\sum e_i^2}{n - 2}}$$

where the e_i are the **residuals**

$$e_i = y_i - \hat{y}_i$$

A **level *C* confidence interval for β_1** is

$$b_1 \pm t^* \text{SE}_{b_1}$$

where t^* is the value for the $t(n-2)$ density curve with area C between $-t^*$ and t^*.

The **test of the hypothesis** $H_0: \beta_1 = 0$ is based on the ***t* statistic**

$$t = \frac{b_1}{\text{SE}_{b_1}}$$

and the $t(n-2)$ distribution. There are similar formulas for confidence intervals and tests for β_0, but these are meaningful only in special cases.

The **estimated mean response** for the subpopulation corresponding to the value x^* of the explanatory variable is

$$\hat{\mu}_y = b_0 + b_1 x^*$$

A **level *C* confidence interval for the mean response** is

$$\hat{\mu}_y \pm t^* \text{SE}_{\hat{\mu}}$$

where t^* is the value for the $t(n-2)$ density curve with area C between $-t^*$ and t^*.

The **estimated value of the response variable** y for a future observation from the subpopulation corresponding to the value x^* of the explanatory variable is

$$\hat{y} = b_0 + b_1 x^*$$

A **level *C* prediction interval** for the estimated response is

$$\hat{y} \pm t^* \text{SE}_{\hat{y}}$$

where t^* is the value for the $t(n-2)$ density curve with area C between $-t^*$ and t^*.

The following section contains material that you should study if you plan to read Chapter 11 on multiple regression. In addition, the section we just completed assumes that you have access to software or a statistical calculator. If you do not, you now need to study the material on computations in the following optional section. The exercises are given at the end of the chapter.

10.2 More Detail about Simple Linear Regression*

In this section we study three optional topics. The first is analysis of variance for regression. If you plan to study the next chapter on multiple regression, you should study this material. The second topic concerns computations for regression inference. Here we present and illustrate the use of formulas for the inference procedures that we have just studied. Finally, we discuss inference for correlation.

---

*This material is optional.

Analysis of variance for regression

The usual computer output for regression includes additional calculations called **analysis of variance.** Analysis of variance, often abbreviated ANOVA, is essential for multiple regression (Chapter 11) and for comparing several means (Chapters 12 and 13). Analysis of variance summarizes information about the sources of variation in the data. It is based on the

analysis of variance

$$\text{DATA} = \text{FIT} + \text{RESIDUAL}$$

framework.

The total variation in the response y is expressed by the deviations $y_i - \bar{y}$. If these deviations were all 0, all observations would be equal and there would be no variation in the response. There are two reasons why the individual observations y_i are not all equal to their mean $\bar{y}$. First, the responses y_i correspond to different values of the explanatory variable x and will differ because of that. The fitted value $\hat{y}_i$ estimates the mean response for the specific x_i. The differences $\hat{y}_i - \bar{y}$ reflect the variation in mean response due to differences in the x_i. This variation is accounted for by the regression line, because the $\hat{y}$'s lie exactly on the line. Second, individual observations will vary about their mean because of variation within the subpopulation of responses to a fixed x_i. This variation is represented by the residuals $y_i - \hat{y}_i$ that record the scatter of the actual observations about the fitted line. The overall deviation of any y observation from the mean of the y's is the sum of these two deviations:

$$(y_i - \bar{y}) = (\hat{y}_i - \bar{y}) + (y_i - \hat{y}_i)$$

For deviations, this equation expresses the idea that

$$\text{DATA} = \text{FIT} + \text{RESIDUAL}$$

We have several times measured variation by an average of squared deviations. If we square each of the three deviations above and then sum over all n observations, it is an algebraic fact that the sums of squares add:

$$\sum(y_i - \bar{y})^2 = \sum(\hat{y}_i - \bar{y})^2 + \sum(y_i - \hat{y}_i)^2$$

We rewrite this equation as

$$\text{SST} = \text{SSM} + \text{SSE}$$

where

$$\text{SST} = \sum(y_i - \bar{y})^2$$
$$\text{SSM} = \sum(\hat{y}_i - \bar{y})^2$$
$$\text{SSE} = \sum(y_i - \hat{y}_i)^2$$

sum of squares

The SS in each abbreviation stands for **sum of squares,** and the T, M, and E stand for total, model, and error, respectively. ("Error" here stands for deviations from the line, which might better be called "residual" or "unexplained variation.") The total variation, as expressed by SST, is the sum of the variation due to the straight-line model (SSM) and the variation due to deviations from this model (SSE). This partition of the variation in the data between two sources is the heart of analysis of variance.

If $H_0: \beta_1 = 0$ were true, there would be no subpopulations and all of the y's should be viewed as coming from a single population with mean μ_y. The variation of the y's would then be described by the sample variance

$$s_y^2 = \frac{\sum(y_i - \bar{y})^2}{n - 1}$$

degrees of freedom

The numerator in this expression is SST. The denominator is the total **degrees of freedom,** or simply DFT.

Just as the total sum of squares SST is the sum of SSM and SSE, the total degrees of freedom DFT is the sum of DFM and DFE, the degrees of freedom for the model and for the error:

$$\text{DFT} = \text{DFM} + \text{DFE}$$

mean square

The model has one explanatory variable x, so the degrees of freedom for this source are DFM $= 1$. Because DFT $= n - 1$, this leaves DFE $= n - 2$ as the degrees of freedom for error. For each source, the ratio of the sum of squares to the degrees of freedom is called the **mean square,** or simply MS. The general formula for a mean square is

$$\text{MS} = \frac{\text{sum of squares}}{\text{degrees of freedom}}$$

Each mean square is an average squared deviation. MST is just s_y^2, the sample variance that we would calculate if all of the data came from a single population. MSE is familiar to us:

$$\text{MSE} = s^2 = \frac{\sum(y_i - \hat{y}_i)^2}{n - 2}$$

It is our estimate of σ^2, the variance about the population regression line.

SUMS OF SQUARES, DEGREES OF FREEDOM, AND MEAN SQUARES

Sums of squares represent variation present in the responses. They are calculated by summing squared deviations. **Analysis of variance** partitions the total variation between two sources.

The sums of squares are related by the formula

$$\text{SST} = \text{SSM} + \text{SSE}$$

That is, the total variation is partitioned into two parts, one due to the model and one due to deviations from the model.

Degrees of freedom are associated with each sum of squares. They are related in the same way:

$$\text{DFT} = \text{DFM} + \text{DFE}$$

To calculate **mean squares,** use the formula

$$\text{MS} = \frac{\text{sum of squares}}{\text{degrees of freedom}}$$

interpretation of r^2   In Section 2.3 we noted that r^2 is the fraction of variation in the values of y that is explained by the least-squares regression of y on x. The sums of squares make this interpretation precise. Recall that SST = SSM + SSE. It is an algebraic fact that

$$r^2 = \frac{\text{SSM}}{\text{SST}} = \frac{\sum(\hat{y}_i - \bar{y})^2}{\sum(y_i - \bar{y})^2}$$

Because SST is the total variation in y and SSM is the variation due to the regression of y on x, this equation is the precise statement of the fact that r^2 is the fraction of variation in y explained by x.

The ANOVA F test

F statistic   The null hypothesis $H_0: \beta_1 = 0$ that y is not linearly related to x can be tested by comparing MSM with MSE. The ANOVA test statistic is an **F statistic,**

$$F = \frac{\text{MSM}}{\text{MSE}}$$

When H_0 is true, this statistic has an F distribution with 1 degree of freedom in the numerator and $n - 2$ degrees of freedom in the denominator. These degrees of freedom are those of MSM and MSE. The F distributions were introduced in Section 7.3. Figure 7.16 (page 517) shows a typical F density curve. Just as there are many t statistics, there are many F statistics. The ANOVA F statistic is not the same as the F statistic of Section 7.3.

When $\beta_1 \neq 0$, MSM tends to be large relative to MSE. So large values of F are evidence against H_0 in favor of the two-sided alternative.

ANALYSIS OF VARIANCE F TEST

In the simple linear regression model, the hypotheses

$$H_0: \beta_1 = 0$$
$$H_a: \beta_1 \neq 0$$

are tested by the **F statistic**

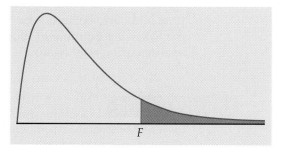

$$F = \frac{\text{MSM}}{\text{MSE}}$$

The P-value is the probability that a random variable having the $F(1, n - 2)$ distribution is greater than or equal to the calculated value of the F statistic.

The F statistic tests the same null hypothesis as one of the t statistics that we encountered earlier in this chapter, so it is not surprising that the two are related. It is an algebraic fact that $t^2 = F$ in this case. For linear regression with one explanatory variable, we prefer the t form of the test because it more easily allows us to test one-sided alternatives and is closely related to the confidence interval for β_1.

ANOVA table
The ANOVA calculations are displayed in an *analysis of variance table,* often abbreviated **ANOVA table.** Here is the format of the table for simple linear regression:

Source	Degrees of freedom	Sum of squares	Mean square	F
Model	1	$\sum(\hat{y}_i - \bar{y})^2$	SSM/DFM	MSM/MSE
Error	$n - 2$	$\sum(y_i - \hat{y}_i)^2$	SSE/DFE	
Total	$n - 1$	$\sum(y_i - \bar{y})^2$	SST/DFT	

EXAMPLE 10.12 The entire output generated by SPSS for the fuel efficiency data in Example 10.3 is given in Figure 10.12. Note that SPSS uses the labels Regression, Residual, and Total for the three sources of variation. We have called these Model, Error, and Total. Other statistical software packages may use slightly different labels. We round the calculated value of the F statistic to 494.47; the P-value is given as 0.000. This is a rounded value and we can conclude that $P < 0.0005$. (The actual value is much less than this.) There is strong evidence against the null hypothesis that there is no relationship between fuel efficiency and the logarithm of speed. Now look at the output for the regression coefficients. The t statistic for LOGMPH is given as 22.237, which we round to 22.24. If we square this number we obtain the F statistic (accurate up to roundoff error). The value of r^2 is also given in the output. Log of speed explains 89.5% of the variability in fuel efficiency.

Calculations for regression inference

We recommend using statistical software for regression calculations. With time and care, however, the work is feasible with a calculator. We will use the following example to illustrate how to perform inference for regression analysis using a calculator.

EXAMPLE 10.13 Nitrogen balance studies are used to determine protein requirements for people. Each subject is fed three different controlled diets during three separate experimental periods. The three diets are similar with regard to all nutrients except protein.

Nitrogen balance is the difference between the amount of nitrogen consumed and the amount lost in feces and urine and by other means. Since virtually all of the nitrogen in a diet comes from protein, nitrogen balance is an indicator of the amount of protein retained by the body. The protein requirement for an individual is the intake corresponding to a balance of zero.

ANOVA[b]

Model		Sum of Squares	df	Mean Square	F	Sig.
1	Regression	493.989	1	493.989	494.467	.000[a]
	Residual	57.944	58	.999		
	Total	551.932	59			

a. Predictors: (Constant), LOGMPH
b. Dependent Variable: MPG

Model Summary

Model	R	R Square	Adjusted R Square	Std. Error of the Estimate
1	.946[a]	.895	.893	.9995

a. Predictors: (Constant), LOGMPH

Coefficients[a]

Model		Unstandardized Coefficients		Standardized Coefficients	t	Sig.
		B	Std. Error	Beta		
1	(Constant)	-7.796	1.155		-6.750	.000
	LOGMPH	7.874	.354	.946	22.237	.000

a. Dependent Variable: MPG

FIGURE 10.12   Regression output with ANOVA table for Example 10.12.

Linear regression is used to model the relationship between nitrogen balance, measured in milligrams of nitrogen per kilogram of body weight per day (mg/kg/d), and protein intake, measured in grams of protein per kilogram of body weight per day (g/kg/d). Here are the data for one subject:[3]

Protein intake (x)	0.543	0.797	1.030
Nitrogen balance (y)	−23.4	17.8	67.3

The data and the least-squares line are plotted in Figure 10.13. The strong straight-line pattern suggests that we can use linear regression to model the relationship between nitrogen balance and protein intake.

FIGURE 10.13   Data and regression line for Example 10.13.

We begin our regression calculations by fitting the least-squares line. Fitting the line gives estimates b_1 and b_0 of the model parameters β_1 and β_0. Next we examine the residuals from the fitted line and obtain an estimate s of the remaining parameter σ. These calculations are preliminary to inference. Finally, we use s to obtain the standard errors needed for the various interval estimates and significance tests. *Roundoff errors that accumulate during these calculations can ruin the final results. Be sure to carry many significant digits and check your work carefully.*

Preliminary calculations   After examining the scatterplot (Figure 10.13) to verify that the data show a straight-line pattern, we begin our calculations.

EXAMPLE 10.14   We start by making a table with the mean and standard deviation for each of the variables, the correlation, and the sample size. These calculations should be familiar from Chapters 1 and 2. Here is the summary:

Variable	Mean	Standard deviation	Correlation	Sample size
Intake	$\bar{x} = 0.79000$	$s_x = 0.24357545$	$r = 0.99698478$	$n = 3$
N balance	$\bar{y} = 20.56667$	$s_y = 45.4132506$		

These quantities are the building blocks for our calculations.

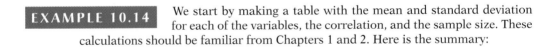

We will need one additional quantity for the calculations to follow. It is the expression $\sum(x_i - \bar{x})^2$. We obtain this quantity as an intermediate step when we calculate s_x. You could also find it using the fact that $\sum(x_i - \bar{x})^2 = (n-1)s_x^2$. You should verify that the value for our example is

$$\sum(x_i - \bar{x})^2 = 0.118658$$

Our first task is to find the least-squares line. This is easy with the building blocks that we have assembled.

EXAMPLE 10.15   The slope of the least-squares line is

$$b_1 = r\frac{s_y}{s_x}$$

$$= 0.99698478\frac{45.4132506}{0.24357545}$$

$$= 185.882$$

The intercept is

$$b_0 = \bar{y} - b_1\bar{x}$$

$$= 20.56667 - (185.882)(0.79000)$$

$$= -126.280$$

The equation of the least-squares regression line is therefore

$$\hat{y} = -126 + 186x$$

This is the line shown in Figure 10.13.

We now have estimates of the first two parameters, β_0 and β_1, of our linear regression model. We now find the estimate of the third parameter, σ: the standard deviation s about the fitted line. To do this we need to find the predicted values and then the residuals.

The first observation is an intake of $x = 0.543$. The corresponding predicted value of balance is

$$\hat{y}_1 = b_0 + b_1 x_1$$

$$= -126.280 + (185.882)(0.543)$$

$$= -25.346$$

and the residual is

$$e_1 = y_1 - \hat{y}_1$$

$$= -23.4 - (-25.346)$$

$$= 1.946$$

The residuals for the other intakes are calculated in the same way. You should verify that they are -4.068 and 2.122. Also check that the three residuals sum to zero.

EXAMPLE 10.16 The estimate of σ^2 is s^2, the sum of the squares of the residuals divided by $n - 2$. The estimated standard deviation about the line is the square root of this quantity.

$$s^2 = \frac{\sum e_i^2}{n - 2}$$

$$= \frac{(1.946)^2 + (-4.068)^2 + (2.122)^2}{1}$$

$$= 24.84$$

So the estimate of the standard deviation about the line is

$$s = \sqrt{24.84} = 4.984$$

Now that we have estimates of the three parameters of our model, we can proceed to the more detailed calculations needed for regression inference.

Inference for slope and intercept Confidence intervals and significance tests for the slope β_1 and intercept β_0 of the population regression line make use of the estimates b_1 and b_0 and their standard errors.

Some algebra using the rules for variances (Section 4.4) establishes that the standard deviation of b_1 is

$$\sigma_{b_1} = \frac{\sigma}{\sqrt{\sum (x_i - \bar{x})^2}}$$

Similarly, the standard deviation of b_0 is

$$\sigma_{b_0} = \sigma \sqrt{\frac{1}{n} + \frac{\bar{x}^2}{\sum (x_i - \bar{x})^2}}$$

To estimate these standard deviations, we need only replace σ by its estimate s.

STANDARD ERRORS FOR ESTIMATED REGRESSION COEFFICIENTS

The standard error of the slope b_1 of the least-squares regression line is

$$\text{SE}_{b_1} = \frac{s}{\sqrt{\sum (x_i - \bar{x})^2}}$$

The standard error of the intercept b_0 is

$$\text{SE}_{b_0} = s \sqrt{\frac{1}{n} + \frac{\bar{x}^2}{\sum (x_i - \bar{x})^2}}$$

The plot of the regression line with the data in Figure 10.13 *appears* to show a very strong relationship, but our sample size is very small. We assess the situation with a significance test for the slope.

EXAMPLE 10.17 First we need the standard error of the estimated slope:

$$SE_{b_1} = \frac{s}{\sqrt{\sum(x_i - \bar{x})^2}}$$

$$= \frac{4.984}{\sqrt{0.118658}}$$

$$= 14.47$$

To test

$$H_0: \beta_1 = 0$$
$$H_a: \beta_1 \neq 0$$

calculate the t statistic:

$$t = \frac{b_1}{SE_{b_1}}$$

$$= \frac{185.882}{14.47} = 12.85$$

Using Table D with $n - 2 = 1$ degrees of freedom, we conclude $P < 0.05$. (The exact value obtained from software is 0.0494.) The data provide evidence in favor of a relationship between nitrogen balance and protein intake ($t = 12.85$, df $= 1$, $P < 0.05$).

Three things are important to note about this example. First, the sample size is very small. Even though the estimated slope is more than 12 standard deviations away from zero, we have only barely attained the 0.05 standard for statistical significance. *It is important to remember that we need to have a very large effect if we expect to detect it with a small sample size.* Second, we would, of course, prefer to have more than three observations for this analysis. However, for each diet, data are collected for about a month. Because the requirement is assumed to be a distribution rather than a single number for everyone, it is important to measure several subjects. Because of the enormous expense involved, researchers typically use only three levels of intake. Third, because we expect balance to increase with increasing intake, a one-sided significance test is justified in this setting.

The significance test tells us that the data provide sufficient information to conclude that intake and balance are related. We use the estimate b_1 and its confidence interval to further describe the relationship.

EXAMPLE 10.18 For the protein requirement problem, let's find a 95% confidence interval for the slope β_1. The degrees of freedom are $n - 2 = 1$, so t^* from Table D is 12.71. We compute

$$185.88214.47b_1 \pm t^*SE_{b_1} = 185.882 \pm (12.71)(14.47)$$

$$= 186 \pm 184$$

The interval is (2, 370).

Note the effect of the small sample size on the critical value t^*. With one additional observation, it would decrease to 4.303.

In this example, the intercept β_0 does not have a meaningful interpretation. A protein intake of zero is theoretically possible, but we would not expect our linear model to be reasonable when extended to such an extreme value. For problems where inference for β_0 is appropriate, the calculations are performed in the same way as those for β_1. Note that there is a different formula for the standard error, however.

Confidence intervals for the mean response and prediction intervals for a future observation   When we substitute a particular value x^* of the explanatory variable into the regression equation and obtain a value of $\hat{y}$, we can view the result in two ways. We have estimated the mean response μ_y, and we have also predicted a future value of the response y. The margins of error for these two uses are quite different. Prediction intervals for an individual response are wider than confidence intervals for estimating a mean response. We now proceed with the details of these calculations. Once again, standard errors are the essential quantities. And once again, these standard errors are multiples of s, our basic measure of the variability of the responses about the fitted line.

STANDARD ERRORS FOR $\hat{\mu}$ AND $\hat{y}$

The standard error of $\hat{\mu}$ is

$$SE_{\hat{\mu}} = s\sqrt{\frac{1}{n} + \frac{(x^* - \overline{x})^2}{\sum(x_i - \overline{x})^2}}$$

The standard error for predicting an individual response $\hat{y}$ is[4]

$$SE_{\hat{y}} = s\sqrt{1 + \frac{1}{n} + \frac{(x^* - \overline{x})^2}{\sum(x_i - \overline{x})^2}}$$

Note that the only difference between the formulas for these two standard errors is the extra 1 under the square root sign in the standard error for prediction. This standard error is larger due to the additional variation of individual responses about the mean response. It produces prediction intervals that are wider than the confidence intervals for the mean response.

For the nitrogen balance example, we can think about the mean balance that would result if a particular protein intake was consumed many times. The confidence interval for the mean response would provide an interval estimate of this population value. On the other hand, we might want to predict a future observation under conditions similar to those used in the study, that is, for a one-month period, at a particular intake level. A prediction interval attempts to capture this future observation.

EXAMPLE 10.19   Let's find a 95% confidence interval for the mean balance corresponding to an intake of 0.7 g/kg/d. The estimated mean balance is

$$\hat{\mu} = b_0 + b_1 x_1$$

$$= -126.280 + (185.882)(0.7)$$

$$= 3.837$$

The standard error is

$$\text{SE}_{\hat{\mu}} = s\sqrt{\frac{1}{n} + \frac{(x^* - \bar{x})^2}{\sum(x_i - \bar{x})^2}}$$

$$= 4.984\sqrt{\frac{1}{3} + \frac{(0.79 - 0.70)^2}{0.118658}}$$

$$= 3.136$$

To find the 95% confidence interval we compute

$$\hat{y} \pm t^* \text{SE}_{\hat{\mu}} = 3.837 \pm (12.71)(3.136)$$

$$= 3.837 \pm 39.859$$

$$= 4 \pm 40$$

The interval is -36 to 44 mg/kg/d of nitrogen.

Calculations for the prediction intervals are similar. The only difference is the use of the formula for $\text{SE}_{\hat{y}}$ in place of $\text{SE}_{\hat{\mu}}$.

Since the confidence interval for mean response includes the value 0, the corresponding intake 0.7 g/kg/d should be considered as a possible value for the intake requirement for this individual. Other intakes would also produce confidence intervals that would include the value of 0 for mean balance. Here is one method that is commonly used to determine a single value of the requirement for an individual.

EXAMPLE 10.20   We define the estimated requirement for an individual to be the intake corresponding to zero balance using the fitted regression equation. To do this, we set the equation

$$\hat{\mu} = b_0 + b_1 x$$

equal to 0 and solve for the intake x. So,

$$x = -b_0/b_1$$

$$= -(-126.280)/185.882$$

$$= 0.68$$

The estimated protein requirement for this individual is 0.68 g/kg/d.

If we repeat these calculations using data collected on a large number of individuals, we can estimate the requirement distribution for a population. There are many interesting statistical issues related to this problem.[5]

Inference for correlation

The correlation coefficient is a measure of the strength and direction of the linear association between two variables. Correlation does not require an explanatory-response relationship between the variables. We can consider the sample correlation r as an estimate of the correlation in the population and base inference about the population correlation on r.

The correlation between the variables x and y when they are measured for every member of a population is the **population correlation.** As usual, we use Greek letters to represent population parameters. In this case ρ (the Greek letter rho) is the population correlation. When $\rho = 0$, there is no linear association in the population. In the important case where the two variables x and y are both normally distributed, the condition $\rho = 0$ is equivalent to the statement that x and y are independent. That is, there is no association of any kind between x and y. (Technically, the condition required is that x and y be **jointly normal.** This means that the distribution of x is normal and also that the conditional distribution of y given any fixed value of x is normal.) We therefore may wish to test the null hypothesis that a population correlation is 0.

population correlation ρ

jointly normal variables

TEST FOR A ZERO POPULATION CORRELATION

To test the hypothesis $H_0: \rho = 0$, compute the t statistic:

$$t = \frac{r\sqrt{n-2}}{\sqrt{1-r^2}}$$

where n is the sample size and r is the sample correlation.

In terms of a random variable T having the $t(n-2)$ distribution, the P-value for a test of H_0 against

$H_a: \rho > 0$ is $P(T \geq t)$

$H_a: \rho < 0$ is $P(T \leq t)$

$H_a: \rho \neq 0$ is $2P(T \geq |t|)$

Most computer packages have routines for calculating correlations and some will provide the significance test for the null hypothesis that ρ is zero.

Correlations

		LOGMPH	MPG
LOGMPH	Pearson Correlation	1	.946**
	Sig. (2-tailed)	.	.000
	N	60	60
MPG	Pearson Correlation	.946**	1
	Sig. (2-tailed)	.000	.
	N	60	60

**. Correlation is significant at the 0.01 level (2-tailed).

FIGURE 10.14   Correlation output for Example 10.21.

EXAMPLE 10.21 For the fuel efficiency example, the SPSS output appears in Figure 10.14. The sample correlation between fuel efficiency and the logarithm of speed is $r = 0.946$. SPSS calls this a Pearson correlation to distinguish it from other kinds of correlations that it can calculate. The P-value for a two-sided test of $H_0: \rho = 0$ is given as 0.000. This means that the actual P-value is less than 0.0005. We conclude that there is a nonzero correlation between MPG and LOGMPH.

If we wanted to test the one-sided alternative that the population correlation is negative, we divide the P-value in the output by 2, after checking that the sample coefficient is in fact negative.

If your software does not give the significance test, you can do the computations easily with a calculator.

EXAMPLE 10.22 The correlation between MPG and LOGMPH is $r = 0.946$. Recall that $n = 60$. The t statistic for testing the null hypothesis that the population correlation is zero is

$$t = \frac{r\sqrt{n-2}}{\sqrt{1-r^2}}$$

$$= \frac{0.946\sqrt{60-2}}{\sqrt{1-(0.946)^2}}$$

$$= 22.2$$

The degrees of freedom are $n - 2 = 58$. From Table D we conclude $P < 0.0001$. This agrees with the SPSS output in Figure 10.14, where the P-value is given as 0.000. The data provide clear evidence that fuel efficiency and the log of speed are related.

There is a close connection between the significance test for a correlation and the test for the slope in a linear regression. Recall that

$$b_1 = r\frac{s_y}{s_x}$$

From this fact we see that if the slope is 0, so is the correlation, and vice versa. It should come as no surprise to learn that the procedures for testing $H_0: \beta_1 = 0$ and $H_0: \rho = 0$ are also closely related. In fact, the t statistics for testing these hypotheses are numerically equal. That is,

$$\frac{b_1}{s_{b_1}} = \frac{r\sqrt{n-2}}{\sqrt{1-r^2}}$$

Check that this holds in both of our examples.

In our examples, the conclusion that there is a statistically significant correlation between the two variables would not come as a surprise to anyone familiar with the meaning of these variables. The significance test simply tells us whether or not there is evidence in the data to conclude that the population correlation is different from 0. The actual size of the correlation is of considerably more interest. We would therefore like to give a confidence interval for the population correlation. Unfortunately, most software packages do not perform this calculation. Because hand calculation of the confidence interval is very tedious, we do not give the method here.[6]

SECTION 10.2 | Summary

The **ANOVA table** for a linear regression gives the degrees of freedom, sum of squares, and mean squares for the model, error, and total sources of variation. The **ANOVA F statistic** is the ratio MSM/MSE. Under $H_0: \beta_1 = 0$, this statistic has an $F(1, n-2)$ distribution and is used to test H_0 versus the two-sided alternative.

The **square of the sample correlation** can be expressed as

$$r^2 = \frac{\text{SSM}}{\text{SST}}$$

and is interpreted as the proportion of the variability in the response variable y that is explained by the explanatory variable x in the linear regression.

The **standard errors** for b_0 and b_1 are

$$\text{SE}_{b_0} = s\sqrt{\frac{1}{n} + \frac{\bar{x}^2}{\sum(x_i - \bar{x})^2}}$$

$$\text{SE}_{b_1} = \frac{s}{\sqrt{\sum(x_i - \bar{x})^2}}$$

The **standard error** that we use for a confidence interval for the estimated mean response for the subpopulation corresponding to the value x^* of the explanatory variable is

$$\text{SE}_{\hat{\mu}} = s\sqrt{\frac{1}{n} + \frac{(x^* - \bar{x})^2}{\sum(x_i - \bar{x})^2}}$$

The **standard error** that we use for a prediction interval for a future observation from the subpopulation corresponding to the value x^* of the explanatory

variable is

$$SE_{\hat{y}} = s\sqrt{1 + \frac{1}{n} + \frac{(x^* - \bar{x})^2}{\sum (x_i - \bar{x})^2}}$$

When the variables y and x are jointly normal, the sample correlation is an estimate of the **population correlation** ρ. The test of $H_0: \rho = 0$ is based on the **t statistic**

$$t = \frac{r\sqrt{n - 2}}{\sqrt{1 - r^2}}$$

which has a $t(n - 2)$ distribution under H_0. This test statistic is numerically identical to the t statistic used to test $H_0: \beta_1 = 0$.

CHAPTER 10 | Exercises

10.1   Find a 95% confidence interval for the slope in each of the following settings:

(a) $n = 25$, $\hat{y} = 12.3 + 16.10x$, and $SE_{b_1} = 8.05$.

(b) $n = 25$, $\hat{y} = 1.3 + 6.10x$, and $SE_{b_1} = 8.05$.

(c) $n = 125$, $\hat{y} = 12.3 + 16.10x$, and $SE_{b_1} = 8.05$.

10.2   For each of the settings in the previous exercise, test the null hypothesis that the slope is zero versus the two-sided alternative.

10.3   For each of the following, explain what is wrong and why.

(a) The parameters of the simple linear regression model are b_0, b_1, and s.

(b) To test $H_0: b_1 = 0$, use a t test.

(c) For a particular value of the explanatory variable, the confidence interval for the mean response will be wider than the prediction interval for a future observation.

10.4   The index of biotic integrity (IBI) is a measure of the water quality in streams. IBI and land-use measures for a collection of streams in the Ozark Highland ecoregion of Arkansas were collected as part of a study.[7] Table 10.1 gives the data for IBI and the area of the watershed in square kilometers for streams in the original sample with area less than or equal to 70 km^2.

(a) Use numerical and graphical methods to describe the variable IBI. Do the same for area. Summarize your results.

(b) Plot the data and describe the relationship. Are there any outliers or unusual patterns?

(c) Give the statistical model for simple linear regression for this problem.

(d) State the null and alternative hypotheses for examining the relationship between IBI and area.

TABLE 10.1

Watershed area and index of biotic integrity

Area	IBI	Area	IBI	Area	IBI	Area	IBI	Area	IBI
21	47	29	61	31	39	32	59	34	72
34	76	49	85	52	89	2	74	70	89
6	33	28	46	21	32	59	80	69	80
47	78	8	53	8	43	58	88	54	84
10	62	57	55	18	29	19	29	39	54
49	78	9	71	5	55	14	58	9	71
23	33	31	59	18	81	16	71	21	75
32	64	10	41	26	82	9	60	54	84
12	83	21	82	27	82	23	86	26	79
16	67	26	56	26	85	28	91		

(e) Run the simple linear regression and summarize the results.

(f) Obtain the residuals and plot them versus area. Is there anything unusual in the plot?

(g) Do the residuals appear to be approximately normal? Give reasons for your answer.

(h) Do the assumptions for the analysis of these data using the model you gave in part (c) appear to be reasonable? Explain your answer.

10.5 The researchers who conducted the study described in the previous exercise also recorded the percent of the watershed area that was forest for each of the streams. The data are given in Table 10.2. Analyze these data using the questions in the previous exercise as a guide.

TABLE 10.2

Percent forest and index of biotic integrity

Forest	IBI	Forest	IBI	Forest	IBI	Forest	IBI	Forest	IBI
0	47	0	61	0	39	0	59	0	72
0	76	3	85	3	89	7	74	8	89
9	33	10	46	10	32	11	80	14	80
17	78	17	53	18	43	21	88	22	84
25	62	31	55	32	29	33	29	33	54
33	78	39	71	41	55	43	58	43	71
47	33	49	59	49	81	52	71	52	75
59	64	63	41	68	82	75	60	79	84
79	83	80	82	86	82	89	86	90	79
95	67	95	56	100	85	100	91		

10.6 Refer to Exercise 10.4.

  (a) Find a 95% confidence interval for mean response corresponding to an area of 30 km².

  (b) Find a 95% prediction interval for a future response.

  (c) Write a short paragraph interpreting the meaning of the intervals in terms of Ozark Highland streams.

  (d) Do you think that these results can be applied to other streams in Arkansas or in other states? Explain why or why not.

10.7 Exercise 2.100 (page 182) gives the modulus of elasticity (MOE) and the modulus of rupture (MOR) for 32 plywood specimens. Because measuring MOR involves breaking the wood but measuring MOE does not, we would like to predict the destructive test result, MOR, using the nondestructive test result, MOE.

  (a) Describe the distribution of MOR using graphical and numerical summaries. Do the same for MOE.

  (b) Make a plot of the two variables. Which should be plotted on the x axis? Give a reason for your answer.

  (c) Give the statistical model for this analysis, run the analysis, summarize the results, and write a short summary of your conclusions.

  (d) Examine the assumptions needed for the analysis. Are you satisfied that there are no serious violations that would cause you to question the validity of your conclusions?

10.8 Refer to the previous exercise. Consider an MOE of 2,000,000.

  (a) Interpret the confidence interval for mean response and the prediction interval for a future observation for this value of MOE.

  (b) Which interval will include more values? Give a reason for your answer.

  (c) (Optional) Calculate the two intervals.

10.9 In Exercise 7.2 (page 472) we examined the distribution of C-reactive protein (CRP) in a sample of 40 children from Papua New Guinea. Serum retinol values for the same children were studied in Exercise 7.4. One important question that can be addressed with these data is whether or not infections, as indicated by CRP, cause a decrease in the measured values of retinol, low values of which indicate a vitamin A deficiency. The data are given in Table 10.3.

  (a) Examine the distributions of CRP and serum retinol. Use graphical and numerical methods.

  (b) Forty percent of the CRP values are zero. Does this violate any assumption that we need to do a regression analysis using CRP to predict serum retinol? Explain your answer.

  (c) Run the regression, summarize the results, and write a short paragraph explaining your conclusions.

TABLE 10.3

C-reactive protein and serum retinol

CRP	Retinol	CRP	Retinol	CRP	Retinol	CRP	Retinol	CRP	Retinol
0.00	1.15	30.61	0.97	22.82	0.24	5.36	1.19	0.00	0.83
3.90	1.36	0.00	0.67	0.00	1.00	0.00	0.94	0.00	1.11
5.64	0.38	73.20	0.31	0.00	1.13	5.66	0.34	0.00	1.02
8.22	0.34	0.00	0.99	3.49	0.31	0.00	0.35	9.37	0.56
0.00	0.35	46.70	0.52	0.00	1.44	59.76	0.33	20.78	0.82
5.62	0.37	0.00	0.70	0.00	0.35	12.38	0.69	7.10	1.20
3.92	1.17	0.00	0.88	4.81	0.34	15.74	0.69	7.89	0.87
6.81	0.97	26.41	0.36	9.57	1.90	0.00	1.04	5.53	0.41

(d) Explain the assumptions needed for your results to be valid. Examine the data with respect to these assumptions and report your results.

10.10 In Exercise 7.110 (page 525) we looked at the distribution of osteocalcin (OC), a biomarker for bone formation, in a sample of 31 healthy women aged 11 to 32 years. This biomarker is relatively inexpensive to measure, requiring only a single sample of blood. Measuring bone formation (VO+), on the other hand, is very expensive. Oral and intravenous administration of stable isotopes of calcium are needed, 25 blood samples over a period of two weeks are drawn, and the collection of all urine and fecal samples for two weeks is required. If a biomarker can reliably predict bone formation, then we could avoid the cost of the expensive VO+ measures. Studies designed to assess the effects of interventions intended to increase bone formation could include many more subjects if only the biomarker measurement is needed. The measured values of VO+ and OC for the 31 women in this study are given in Table 10.4.

(a) Use numerical and graphical summaries to describe the distributions of VO+ and OC.

(b) Plot the data. Give a reason for your choice of variables for the x and y axes. Describe the pattern and note any unusual observations. Do the

TABLE 10.4

VO+ and osteocalcin

VO+	OC	VO+	OC	VO+	OC	VO+	OC
476	8.1	1032	40.2	624	17.2	285	9.9
694	10.1	445	20.6	479	15.9	403	19.7
753	17.9	896	31.2	572	16.9	391	20.0
687	17.2	968	19.3	512	24.2	513	20.8
628	20.9	985	44.4	838	30.2	878	31.4
1100	38.4	1251	76.5	870	47.7	2221	54.6
1303	54.6	2545	36.4	1606	68.9	1126	77.9
1682	52.8	2240	56.3	1557	35.7		

assumptions needed for regression analysis appear to be approximately satisfied?

(c) Run the regression using OC to predict VO+. Summarize the results.

10.11 In Exercise 7.111 (page 525) we looked at the distribution of tartrate resistant acid phosphatase (TRAP), a biomarker for bone resorption formation. Table 10.5 gives values for this biomarker and a measure of bone resorption VO−. Analyze these data using the questions in the previous exercise as a guide.

TABLE 10.5

VO− and TRAP

VO−	TRAP	VO−	TRAP	VO−	TRAP	VO−	TRAP
407	3.3	874	5.9	445	6.3	351	6.9
980	8.1	493	8.1	572	8.2	634	8.8
1028	9.0	1116	9.0	857	9.5	536	9.5
701	9.6	934	10.1	477	10.1	254	10.3
766	10.5	496	10.7	924	14.4	954	14.6
918	14.6	1065	14.9	722	18.6	1486	19.0
1018	19.0	2236	19.1	903	19.4	960	23.7
1251	25.2	1761	25.5	1446	28.8		

10.12 Refer to the OC and VO+ data in Exercise 10.10. For variables such as these, it is common to work with the logarithms of the measured values. Reanalyze these data using the logs of both OC and VO+. Summarize your results and compare them with those you obtained in Exercise 10.10.

10.13 Refer to the TRAP and VO− data in Exercise 10.11. Reanalyze these data using the logs of both TRAP and VO−. Summarize your results and compare them with those you obtained in Exercise 10.11.

10.14 In Exercise 2.3 (page 113) you examined the relationship between a reading test score and an IQ score for a sample of 60 fifth-grade children.

(a) Run the regression and summarize the results of the significance tests.

(b) Rerun the analysis with the four possible outliers removed. Summarize your findings, paying particular attention to the effects of removing the outliers.

10.15 Exercise 2.99 (page 182) gives data on neuron responses to pure tones and to monkey calls.

(a) Describe each variable graphically and numerically.

(b) Plot the data with the pure tone response on the x axis and the monkey call response on the y axis. Describe the relationship and mark the point with the largest residual and the point with the extreme value of the tone response.

(c) Analyze the entire set of 37 observations and summarize the results.

(d) Perform additional analyses to assess the effects of the two marked points on the results. Summarize your findings.

10.16 Suppose that there is a linear relationship between the number of students x in an elementary school and the annual budget y. Write a population regression model to describe this relationship.

(a) Which parameter in your model is the fixed cost in the budget (for example, the salary of the principal and some administrative costs) that does not change as x increases?

(b) Which parameter in your model shows how total cost changes when there are more students in the school? Do you expect this number to be greater than 0 or less than 0?

(c) Actual data from schools will not fit a straight line exactly. What term in your model allows variation among schools of the same size x?

10.17 How is the flow of investors' money into stock mutual funds related to the flow of money into bond mutual funds? Here are data on the net new money flowing into stock and bond mutual funds in the years 1985 to 2000, in billions of dollars.[8] "Net" means that funds flowing out are subtracted from those flowing in. If more money leaves than arrives, the net flow will be negative. To eliminate the effect of inflation, all dollar amounts are in "real dollars" with constant buying power equal to that of a dollar in the year 2000.

Year	1985	1986	1987	1988	1989	1990	1991	1992
Stocks	12.8	34.6	28.8	−23.3	8.3	17.1	50.6	97.0
Bonds	100.8	161.8	10.6	−5.8	−1.4	9.2	74.6	87.1

Year	1993	1994	1995	1996	1997	1998	1999	2000
Stocks	151.3	133.6	140.1	238.2	243.5	165.9	194.3	309.0
Bonds	84.6	−72.0	−6.8	3.3	30.0	79.2	−6.2	−48.0

(a) Make a scatterplot with cash flow into stock funds as the explanatory variable. Find the least-squares line for predicting net bond investments from net stock investments. What do the data suggest?

(b) Is there statistically significant evidence that there is some straight-line relationship between the flows of cash into bond funds and stock funds? (State hypotheses, give a test statistic and its P-value, and state your conclusion.)

(c) What fact about the scatterplot explains why the relationship described by the least-squares line is not significant?

10.18 Can a pretest on mathematics skills predict success in a statistics course? The 82 students in an introductory statistics class took a pretest at the beginning of the semester. The least-squares regression line for predicting the score y on the final exam from the pretest score x was $\hat{y} = 9.7 + 0.76x$. The standard error of b_1 was 0.44.

(a) Test the null hypothesis that there is no linear relationship between the pretest score and the score on the final exam against the two-sided alternative.

(b) Would you reject this null hypothesis versus the one-sided alternative that the slope is positive? Explain your answer.

10.19   Metatarsus adductus (call it MA) is a turning in of the front part of the foot that is common in adolescents and usually corrects itself. Hallux abducto valgus (call it HAV) is a deformation of the big toe that is not common in youth and often requires surgery. Perhaps the severity of MA can help predict the severity of HAV. Table 2.2 (page 118) gives data on 38 consecutive patients who came to a medical center for HAV surgery.[9] Using X-rays, doctors measured the angle of deformity for both MA and HAV. They speculated that there is a positive association—more serious MA is associated with more serious HAV.

(a) Make a scatterplot of the data in Table 2.2. (Which is the explanatory variable?)

(b) Describe the form, direction, and strength of the relationship between MA angle and HAV angle. Are there any clear outliers in your graph?

(c) Give a statistical model that provides a framework for asking the question of interest for this problem.

(d) Translate the question of interest into null and alternative hypotheses.

(e) Test these hypotheses and write a short description of the results. Be sure to include the value of the test statistic, the degrees of freedom, the P-value, and a clear statement of what you conclude.

10.20   Refer to the previous exercise. Give a 95% confidence interval for the slope. Explain how this interval can tell you what to conclude from a significance test for this parameter.

10.21   We assume that our wages will increase as we gain experience and become more valuable to our employers. Wages also increase because of inflation. By examining a sample of employees at a given point in time, we can look at part of the picture. How does length of service (LOS) relate to wages? Table 10.6 gives data on the LOS in months and wages for 60 women who work in Indiana banks. Wages are yearly total income divided by the number of weeks worked. We have multiplied wages by a constant for reasons of confidentiality.[10]

(a) Plot wages versus LOS. Describe the relationship. There is one woman with relatively high wages for her length of service. Circle this point and do not use it in the rest of this exercise.

(b) Find the least-squares line. Summarize the significance test for the slope. What do you conclude?

TABLE 10.6

Bank wages and length of service

Wages	LOS	Size	Wages	LOS	Size	Wages	LOS	Size
48.3355	94	Large	64.1026	24	Large	41.2088	97	Small
49.0279	48	Small	54.9451	222	Small	67.9096	228	Small
40.8817	102	Small	43.8095	58	Large	43.0942	27	Large
36.5854	20	Small	43.3455	41	Small	40.7000	48	Small
46.7596	60	Large	61.9893	153	Large	40.5748	7	Large
59.5238	78	Small	40.0183	16	Small	39.6825	74	Small
39.1304	45	Large	50.7143	43	Small	50.1742	204	Large
39.2465	39	Large	48.8400	96	Large	54.9451	24	Large
40.2037	20	Large	34.3407	98	Large	32.3822	13	Small
38.1563	65	Small	80.5861	150	Large	51.7130	30	Large
50.0905	76	Large	33.7163	124	Small	55.8379	95	Large
46.9043	48	Small	60.3792	60	Large	54.9451	104	Large
43.1894	61	Small	48.8400	7	Large	70.2786	34	Large
60.5637	30	Large	38.5579	22	Small	57.2344	184	Small
97.6801	70	Large	39.2760	57	Large	54.1126	156	Small
48.5795	108	Large	47.6564	78	Large	39.8687	25	Large
67.1551	61	Large	44.6864	36	Large	27.4725	43	Small
38.7847	10	Small	45.7875	83	Small	67.9584	36	Large
51.8926	68	Large	65.6288	66	Large	44.9317	60	Small
51.8326	54	Large	33.5775	47	Small	51.5612	102	Large

(c) State carefully what the slope tells you about the relationship between wages and length of service.

(d) Give a 95% confidence interval for the slope.

10.22 Refer to the previous exercise. Analyze the data with the outlier included.

(a) How does this change the estimates of the parameters β_0, β_1, and σ?

(b) What effect does the outlier have on the results of the significance test for the slope?

(c) How has the width of the 95% confidence interval changed?

10.23 The Leaning Tower of Pisa is an architectural wonder. Engineers concerned about the tower's stability have done extensive studies of its increasing tilt. Measurements of the lean of the tower over time provide much useful information. The following table gives measurements for the years 1975 to 1987. The variable "lean" represents the difference between where a point on the tower would be if the tower were straight and where it actually is. The data are coded as tenths of a millimeter in excess of 2.9 meters, so that the 1975 lean, which was 2.9642 meters, appears in the table as 642. Only the last two digits of the year were entered into the computer.[11]

Year	75	76	77	78	79	80	81	82	83	84	85	86	87
Lean	642	644	656	667	673	688	696	698	713	717	725	742	757

(a) Plot the data. Does the trend in lean over time appear to be linear?

(b) What is the equation of the least-squares line? What percent of the variation in lean is explained by this line?

(c) Give a 99% confidence interval for the average rate of change (tenths of a millimeter per year) of the lean.

10.24 Refer to the previous exercise.

(a) In 1918 the lean was 2.9071 meters. (The coded value is 71.) Using the least-squares equation for the years 1975 to 1987, calculate a predicted value for the lean in 1918. (Note that you must use the coded value 18 for year.)

(b) Although the least-squares line gives an excellent fit to the data for 1975 to 1987, this pattern did not extend back to 1918. Write a short statement explaining why this conclusion follows from the information available. Use numerical and graphical summaries to support your explanation.

10.25 Refer to the previous two exercises.

(a) How would you code the explanatory variable for the year 2007?

(b) The engineers working on the Leaning Tower of Pisa were most interested in how much the tower would lean if no corrective action is taken. Use the least-squares equation to predict the tower's lean in the year 2007.

(c) To give a margin of error for the lean in 2007, would you use a confidence interval for a mean response or a prediction interval? Explain your choice.

10.26 The capacity (bits) of the largest DRAM (dynamic random access memory) chips commonly available at retail has increased as follows:[12]

Year	1971	1980	1987	1993	1999	2000
Bits	1,024	64,000	1,024,000	16,384,000	256,000,000	512,000,000

(a) Make a scatterplot of the data. Growth is much faster than linear.

(b) Plot the logarithm of DRAM capacity against year. These points are close to a straight line.

(c) Regress the logarithm of DRAM capacity on year. Give a 95% confidence interval for the slope of the population regression line.

10.27 Is there a nonzero correlation between net flow of money into stock mutual funds and into bond funds? Use the regression analysis you did in Exercise 10.17 (page 672) to answer this question with no additional calculations.

10.28 Chinese students from public schools in Hong Kong were the subjects of a study designed to investigate the relationship between various measures of parental behavior and other variables. The sample size was 713. The data were obtained from questionnaires filled in by the students. One of the variables examined was parental control, an indication of the amount of control that the parents exercised over the behavior of the students. Another was the self-esteem of the students.[13]

(a) The correlation between parental control and self-esteem was $r = -0.19$. Calculate the t statistic for testing the null hypothesis that the population correlation is 0.

(b) Find an approximate P-value for testing H_0 versus the two-sided alternative and report your conclusion.

10.29 How are returns on common stocks in overseas markets related to returns in U.S. markets? Measure U.S. returns by the annual rate of return on the Standard & Poor's 500-Stock Index and overseas returns by the annual rate of return on the Morgan Stanley EAFE (Europe, Australasia, Far East) index. Both are recorded in percents. Regress the EAFE returns on the S&P 500 returns for the 30 years 1971 to 2000. Here is part of the Minitab output for this regression:

```
The regression equation is
EAFE = 4.76 + 0.663 S&P


Analysis of Variance

Source           DF        SS         MS        F       P
Regression        1      3445.9     3445.9     9.50    0.005
Residual Error
Total            29     13598.3
```

Complete the analysis of variance table by filling in the "Residual Error" row.

10.30 Refer to the previous exercise. What are the values of the regression standard error s and the squared correlation r^2?

10.31 Refer to the previous two exercises. The standard deviation of the S&P 500 returns for these years is 16.45%. From this and your work in the previous exercise, find the standard error for the least-squares slope b_1. Give a 95% confidence interval for the slope β_1 of the population regression line.

10.32 Concern about the quality of life for chronically ill patients is becoming as important as treating their physical symptoms. The SF-36, a questionnaire for measuring the health quality of life, was given to 50 patients with chronic obstructive lung disease.[14] A correlation of 0.68 was reported between the component of the questionnaire called general health perceptions (GHP) and a measure of lung function called forced vital capacity (FVC), expressed as a percent of normal. The mean and standard deviation of GHP are 43.5 and 20.3, and for FVC the values are 80.9 and 17.2.

(a) Find the equation of the least-squares line for predicting GHP from FVC.

(b) Give the results of the significance test for the null hypothesis that the slope is 0. (*Hint:* What is the relation between this test and the test for a zero correlation?)

10.33 A study reported a correlation $r = 0.5$ based on a sample size of $n = 20$; another reported the same correlation based on a sample size of $n = 10$. For each, perform the test of the null hypothesis that $\rho = 0$. Describe the results and explain why the conclusions are different.

10.34 Refer to the bank wages data given in Table 10.6 and described in Exercise 10.21 (page 673). The data also include a variable "Size," which classifies the bank as large or small. Obtain the residuals from the regression used to predict wages from LOS, and plot them versus LOS using different symbols for the large and small banks. Include on your plot a horizontal line at 0 (the mean of the residuals). Describe the important features of this plot. Explain what they indicate about wages in this set of data.

10.35 The SAT and the ACT are the two major standardized tests that colleges use to evaluate candidates. Most students take just one of these tests. However, some students take both. Table 10.7 gives the scores of 60 students who did this. How can we relate the two tests?

   (a) Plot the data with SAT on the x axis and ACT on the y axis. Describe the overall pattern and any unusual observations.

   (b) Find the least-squares regression line and draw it on your plot. Give the results of the significance test for the slope.

   (c) What is the correlation between the two tests?

TABLE 10.7							
SAT and ACT scores							
SAT	ACT	SAT	ACT	SAT	ACT	SAT	ACT
1000	24	870	21	1090	25	800	21
1010	24	880	21	860	19	1040	24
920	17	850	22	740	16	840	17
840	19	780	22	500	10	1060	25
830	19	830	20	780	12	870	21
1440	32	1190	30	1120	27	1120	25
490	7	800	16	590	12	800	18
1050	23	830	16	990	24	960	27
870	18	890	23	700	16	880	21
970	21	880	24	930	22	1020	24
920	22	980	27	860	23	790	14
810	19	1030	23	420	21	620	18
1080	23	1220	30	800	20	1150	28
1000	19	1080	22	1140	24	970	20
1030	25	970	20	920	21	1060	24

10.36 Refer to the previous exercise. Find the predicted value of ACT for each observation in the data set.

   (a) What is the mean of these predicted values? Compare it with the mean of the ACT scores.

(b) Compare the standard deviation of the predicted values with the standard deviation of the actual ACT scores. If least-squares regression is used to predict ACT scores for a large number of students such as these, the average predicted value will be accurate but the variability of the predicted scores will be too small.

(c) Find the SAT score for a student who is one standard deviation above the mean ($z = (x - \bar{x})/s = 1$). Find the predicted ACT score and standardize this score. (Use the means and standard deviations from this set of data for these calculations.)

(d) Repeat part (c) for a student whose SAT score is one standard deviation below the mean ($z = -1$).

(e) What do you conclude from parts (c) and (d)? Perform additional calculations for different z's if needed.

10.37 Refer to the previous two exercises. An alternative to the least-squares method is based on matching standardized scores. Specifically, we set

$$\frac{(y - \bar{y})}{s_y} = \frac{(x - \bar{x})}{s_x}$$

and solve for y. Let's use the notation $y = a_0 + a_1 x$ for this line. The slope is $a_1 = s_y/s_x$ and the intercept is $a_0 = \bar{y} - a_1\bar{x}$. Compare these expressions with the formulas for the least-squares slope and intercept (page 137).

(a) Using the data in Table 10.7, find the values of a_0 and a_1.

(b) Plot the data with the least-squares line and the new prediction line.

(c) Use the new line to find predicted ACT scores. Find the mean and the standard deviation of these scores. How do they compare with the mean and standard deviation of the ACT scores?

10.38 Here are data for 12 perch caught in a lake in Finland:[15]

Weight (grams)	Length (cm)	Width (cm)	Weight (grams)	Length (cm)	Width (cm)
5.9	8.8	1.4	300.0	28.7	5.1
100.0	19.2	3.3	300.0	30.1	4.6
110.0	22.5	3.6	685.0	39.0	6.9
120.0	23.5	3.5	650.0	41.4	6.0
150.0	24.0	3.6	820.0	42.5	6.6
145.0	25.5	3.8	1000.0	46.6	7.6

In this exercise we will examine different models for predicting weight.

(a) Run the regression using length to predict weight. Do the same using width as the explanatory variable. Summarize the results. Be sure to include the value of r^2.

(b) Plot weight versus length and weight versus width. Include the least-squares lines in these plots. Do these relationships appear to be linear? Explain your answer.

10.39 Refer to the previous exercise.

(a) Try to find a better model using a transformation of length. One possibility is to use the square. Make a plot and perform the regression analysis. Summarize the results.

(b) Do the same for width.

10.40 Refer to the previous two exercises.

(a) Create a new variable that is the product of length and width. Do a plot and run the regression using this new variable. Summarize the results.

(b) Write a short report summarizing and comparing the different regression analyses that you performed in this exercise and the previous two exercises.

10.41 Refer to the data on the index of biotic integrity and area in Exercise 10.4 (page 667) and the additional data on percent watershed area that was forest in Exercise 10.5. Find the correlations among these three variables, perform the test of statistical significance, and summarize the results. Which of these test results could have been obtained from the analyses that you performed in Exercises 10.4 and 10.5?

10.42 Food neophobia is a personality trait associated with avoiding unfamiliar foods. In one study of 564 children who were 2 to 6 years of age, food neophobia and the frequency of consumption of different types of food were measured.[16] Here is a summary of the correlations:

Type of food	Correlation
Vegetables	−0.27
Fruit	−0.16
Meat	−0.15
Eggs	−0.08
Sweet/fatty snacks	0.04
Starchy staples	−0.02

Perform the significance test for each correlation and write a summary about food neophobia and the consumption of different types of food.

10.43 A study reported correlations between several personality traits and scores on the Graduate Record Examinations (GRE) for a sample of 342 test takers.[17]

Here is a table of the correlations:

Personality trait	GRE score		
	Analytical	Quantitative	Verbal
Conscientiousness	−0.17	−0.14	−0.12
Rationality	−0.06	−0.03	−0.08
Ingenuity	−0.06	−0.08	−0.02
Quickness	0.21	0.15	0.26
Creativity	0.24	0.26	0.29
Depth	0.06	0.08	0.15

For each correlation, test the null hypothesis that the corresponding true correlation is zero. Reproduce the table and mark the correlations that have $P < 0.001$ with ***, those for which $P < 0.01$ with **, and those for which $P < 0.05$ with *. Some critics of standardized tests have suggested that the tests penalize students who are "deep thinkers" and those who are very creative. Others have suggested that students who work quickly do better on these tests. Write a summary of the results of your significance tests taking into account these comments.

10.44 Metabolic rate, the rate at which the body consumes energy, is important in studies of weight gain, dieting, and exercise. The table below gives data on the lean body mass and resting metabolic rate for 12 women and 7 men who are subjects in a study of dieting. Lean body mass, given in kilograms, is a person's weight leaving out all fat. Metabolic rate is measured in calories burned per 24 hours, the same calories used to describe the energy content of foods. The researchers believe that lean body mass is an important influence on metabolic rate.

Subject	Sex	Mass	Rate	Subject	Sex	Mass	Rate
1	M	62.0	1792	11	F	40.3	1189
2	M	62.9	1666	12	F	33.1	913
3	F	36.1	995	13	M	51.9	1460
4	F	54.6	1425	14	F	42.4	1124
5	F	48.5	1396	15	F	34.5	1052
6	F	42.0	1418	16	F	51.1	1347
7	M	47.4	1362	17	F	41.2	1204
8	F	50.6	1502	18	M	51.9	1867
9	F	42.0	1256	19	M	46.9	1439
10	M	48.7	1614				

(a) Make a scatterplot of the data, using different symbols or colors for men and women. Summarize what you see in the plot.

(b) Run the regression to predict metabolic rate from lean body mass for the women in the sample and summarize the results. Do the same for the men.

10.45 Refer to the previous exercise. It is tempting to conclude that there is a strong linear relationship for the women but no relationship for the men. Let's look at this issue a little more carefully.

(a) Find the confidence interval for the slope in the regression equation that you ran for the females. Do the same for the males. What do these suggest about the possibility that these two slopes are the same? (The formal method for making this comparison is a bit complicated and is beyond the scope of this chapter.)

(b) Examine the formula for the standard error of the regression slope given on page 660. The term in the denominator is $\sqrt{\Sigma(x_i - \bar{x})^2}$. Find this quantity for the females; do the same for the males. How do these calculations help to explain the results of the significance tests?

(c) Suppose you were able to collect additional data for males. How would you use lean body mass in deciding which subjects to choose?

10.46 Think about what would happen if you analyzed a subset of a set of data by analyzing only data for a restricted range of values of the explanatory variable. What results would you expect to change? Examine your ideas by analyzing the fuel efficiency data studied in Example 10.4 (page 640) for only those cases with speed less than or equal to 30 MPH. Note that this corresponds to 3.4 in the log scale. Now, let's do the same analysis with a restriction on the response variable. Run the analysis with only those cases with fuel efficiency less than or equal to 20 MPG. Write a summary comparing the effects of these two restrictions with each other and with the results in Example 10.4.

EESEE CASE STUDIES

The Electronic Encyclopedia of Statistical Examples and Exercises (EESEE) is available on the text CD and Web site. These more elaborate stories, with data, provide settings for longer case studies.

10.47 Read the EESEE story "Blood Alcohol Content." Write a report that answers all questions for this case study.

10.48 Read the EESEE story "Weighing Trucks in Motion." Write a report that answers all questions for this case study.

10.49 Read the EESEE story "What's Driving Car Sales?" Write a report that answers all questions for this case study.

STATISTICS IN PRACTICE

Statistics for Science, and More

CHRISTOPHER PRESLEY *is Advanced Statistician in the Corporate Communications function of Eastman Chemical Company.*

While growing up, I had imagined myself as a scientist devising experiments for the next great discovery. This childhood dream led me to enter college as a physics major. As is the case with most students, I changed my major while working on my undergraduate degree. I switched from physics to mathematics, then added statistics as another major. My interest in statistics began when I took a course in design of experiments during my junior year. I saw that statistics provided an opportunity to combine the skills learned from numerous math classes with my early love for science.

After graduate school, I accepted a position with Eastman Chemical Company as an industrial statistician. Initially, my job responsibilities involved consulting with engineers and scientists on experiments and analyzing the resulting data. However, over the past several years my job responsibilities have grown, and I have been involved with many different aspects of the company. I have applied statistics to problems in manufacturing, analytical (testing) services, customer service, research and development, human resources, marketing, engineering, communications, employee research, and customer satisfaction. For example, I currently manage the Employee Voice, a survey of our employees that allows management and employees themselves to track employee sentiment as our business environment changes. I divide employees into 12 similar groups, survey one group each month, and generate control charts and comparisons with 43 other large companies. Learning statistics has made me very versatile within the company and has opened many possible career paths.

I believe that, regardless of your field of study, having a good foundation of statistical knowledge will provide you with a tremendous advantage in your career. The skills gained from statistics can be applied to a wide range of industries and functions. You will also be able to have a significant impact on decision making by providing data that are accurate, useful, and comprehensible to people at various levels.

WASSILY KANDINSKY, COMPOSITION WITH CHESSBOARD—ORANGE 1923
© ARS, NY (THE NEW YORK PUBLIC LIBRARY/ART RESOURCE, NY)

Multiple Regression

Introduction

In Chapter 10 we presented methods for inference in the setting of a linear relationship between a response variable *y* and a *single* explanatory variable *x*. With multiple linear regression we use *more than one* explanatory variable to explain or predict a single response variable. The descriptive tools we learned in Chapter 2—scatterplots, least-squares regression, and correlation—are essential preliminaries to inference and also provide a foundation for confidence intervals and significance tests.

Many of the ideas that we encountered in our study of simple linear regression carry over to this more complicated case. However, the introduction of several explanatory variables leads to many additional considerations. In this short chapter we cannot explore all of these issues. Rather, we will outline some basic facts about inference in the multiple regression setting and then illustrate multiple regression analysis with a case study.

11.1   Inference for Multiple Regression

Population multiple regression equation

The simple linear regression model assumes that the mean of the response variable *y* depends on the explanatory variable *x* according to a linear equation

$$\mu_y = \beta_0 + \beta_1 x$$

For any fixed value of *x*, the response *y* varies normally around this mean and has a standard deviation σ that is the same for all values of *x*.

In the multiple regression setting, the response variable *y* depends on not one but *p* explanatory variables. We will denote these explanatory variables by $x_1, x_2, \ldots, x_p$. The mean response is a linear function of the explanatory variables:

$$\mu_y = \beta_0 + \beta_1 x_1 + \beta_2 x_2 + \cdots + \beta_p x_p$$

population regression equation

This expression is the **population regression equation.** We do not observe the mean response because the observed values of *y* vary about their means. We can think of subpopulations of responses, each corresponding to a particular set of values for *all* of the explanatory variables $x_1, x_2, \ldots, x_p$. In each subpopulation, *y* varies normally with a mean given by the population regression equation. The regression model assumes that the standard deviation σ of the responses is the same in all subpopulations.

EXAMPLE 11.1 Our case study uses data collected at a large university on all first-year computer science majors in a particular year.[1] The purpose of the study was to attempt to predict success in the early university years. One measure of success was the cumulative grade point average (GPA) after three semesters. Among the explanatory variables recorded at the time the students enrolled in the university were average high school grades in mathematics (HSM), science (HSS), and English (HSE).

We will use high school grades to predict the response variable GPA. There are $p = 3$ explanatory variables: $x_1 =$ HSM, $x_2 =$ HSS, and $x_3 =$ HSE. The high school grades are coded on a scale from 1 to 10, with 10 corresponding to A, 9 to A−, 8 to B+, and so on. These grades define the subpopulations. For example, the straight-C students are the subpopulation defined by HSM = 4, HSS = 4, and HSE = 4.

One possible multiple regression model for the subpopulation mean GPAs is

$$\mu_{\text{GPA}} = \beta_0 + \beta_1 \text{HSM} + \beta_2 \text{HSS} + \beta_3 \text{HSE}$$

For the straight-C subpopulation of students, the model gives the subpopulation mean as

$$\mu_{\text{GPA}} = \beta_0 + 4\beta_1 + 4\beta_2 + 4\beta_3$$

Data for multiple regression

The data for a simple linear regression problem consist of observations (x_i, y_i) on the two variables. Because there are several explanatory variables in multiple regression, the notation needed to describe the data is more elaborate. In our case study, each observation or case consists of a value for the response variable and for each of the explanatory variables for one student. Call x_{ij} the value of the jth explanatory variable for the ith student. The data are then

$$\text{Student 1:} \quad (x_{11}, x_{12}, \ldots, x_{1p}, y_1)$$
$$\text{Student 2:} \quad (x_{21}, x_{22}, \ldots, x_{2p}, y_2)$$
$$\vdots$$
$$\text{Student } n: \quad (x_{n1}, x_{n2}, \ldots, x_{np}, y_n)$$

Here, n is the number of cases and p is the number of explanatory variables. Data are often entered into computer regression programs in this format. Each row is a case and each column corresponds to a different variable. The data for Example 11.1, with several additional explanatory variables, appear in this format in the CSDATA data set described in the Data Appendix.

Multiple linear regression model

We combine the population regression equation and assumptions about variation to construct the multiple linear regression model. The subpopulation means describe the FIT part of our statistical model. The RESIDUAL part represents the variation of observations about the means. We will use the same notation for the residual that we used in the simple linear regression model. The symbol ϵ represents the deviation of an individual observation from its subpopulation mean. We assume that these deviations are normally distributed with mean 0 and an unknown standard deviation σ that does not depend on the values of the x variables. *This is an assumption that we can check by examining the residuals in the same way that we did for simple linear regression.*

> ## MULTIPLE LINEAR REGRESSION MODEL
>
> The **statistical model for multiple linear regression** is
>
> $$y_i = \beta_0 + \beta_1 x_{i1} + \beta_2 x_{i2} + \cdots + \beta_p x_{ip} + \epsilon_i$$
>
> for $i = 1, 2, \ldots, n$.
>
> The **mean response** μ_y is a linear function of the explanatory variables:
>
> $$\mu_y = \beta_0 + \beta_1 x_1 + \beta_2 x_2 + \cdots + \beta_p x_p$$
>
> The **deviations** ϵ_i are independent and normally distributed with mean 0 and standard deviation σ. In other words, they are an SRS from the $N(0, \sigma)$ distribution.
>
> The parameters of the model are $\beta_0, \beta_1, \beta_2, \ldots, \beta_p$, and σ.

The assumption that the subpopulation means are related to the regression coefficients β by the equation

$$\mu_y = \beta_0 + \beta_1 x_1 + \beta_2 x_2 + \cdots + \beta_p x_p$$

implies that we can estimate all subpopulation means from estimates of the β's. To the extent that this equation is accurate, we have a useful tool for describing how the mean of y varies with the x's.

Estimation of the multiple regression parameters

For simple linear regression we used the principle of least squares first described in Section 2.3 to obtain estimators of the intercept and slope of the regression line. For multiple regression the principle is the same but the details are more complicated. Let

$$b_0, \ b_1, \ b_2, \ \ldots, \ b_p$$

denote the estimators of the parameters

$$\beta_0, \ \beta_1, \ \beta_2, \ \ldots, \ \beta_p$$

For the ith observation the predicted response is

$$\hat{y}_i = b_0 + b_1 x_{i1} + b_2 x_{i2} + \cdots + b_p x_{ip}$$

residual The ith **residual,** the difference between the observed and predicted response, is therefore

$$
\begin{aligned}
e_i &= \text{observed response} - \text{predicted response} \\
&= y_i - \hat{y}_i \\
&= y_i - b_0 - b_1 x_{i1} - b_2 x_{i2} - \cdots - b_p x_{ip}
\end{aligned}
$$

method of least squares   The **method of least squares** chooses the values of the b's that make the sum of the squares of the residuals as small as possible. In other words, the parameter estimates $b_0, b_1, b_2, \ldots, b_p$ minimize the quantity

$$\sum(y_i - b_0 - b_1 x_{i1} - b_2 x_{i2} - \cdots - b_p x_{ip})^2$$

The formula for the least-squares estimates is complicated. We will be content to understand the principle on which they are based and to let software do the computations.

The parameter σ^2 measures the variability of the responses about the population regression equation. As in the case of simple linear regression, we estimate σ^2 by an average of the squared residuals. The estimator is

$$s^2 = \frac{\sum e_i^2}{n - p - 1}$$

$$= \frac{\sum(y_i - \hat{y}_i)^2}{n - p - 1}$$

degrees of freedom   The quantity $n - p - 1$ is the **degrees of freedom** associated with s^2. The degrees of freedom equal the sample size n minus $p + 1$, the number of β's we must estimate to fit the model. In the simple linear regression case there is just one explanatory variable, so $p = 1$ and the degrees of freedom are $n - 2$. To estimate σ we use

$$s = \sqrt{s^2}$$

Confidence intervals and significance tests for regression coefficients

We can obtain confidence intervals and significance tests for each of the regression coefficients β_j as we did in simple linear regression. The standard errors of the b's have more complicated formulas, but all are multiples of s. We again rely on statistical software for the calculations.

> **CONFIDENCE INTERVALS AND SIGNIFICANCE TESTS FOR β_j**
>
> A **level C confidence interval** for β_j is
>
> $$b_j \pm t^* \text{SE}_{b_j}$$
>
> where SE_{b_j} is the standard error of b_j and t^* is the value for the $t(n - p - 1)$ density curve with area C between $-t^*$ and t^*.
>
> To test the hypothesis $H_0: \beta_j = 0$, compute the **t statistic**
>
> $$t = \frac{b_j}{\text{SE}_{b_j}}$$
>
> In terms of a random variable T having the $t(n - p - 1)$ distribution, the P-value for a test of H_0 against

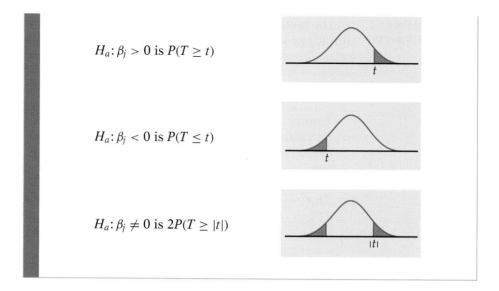

$$H_a: \beta_j > 0 \text{ is } P(T \geq t)$$

$$H_a: \beta_j < 0 \text{ is } P(T \leq t)$$

$$H_a: \beta_j \neq 0 \text{ is } 2P(T \geq |t|)$$

confidence intervals for mean response

prediction intervals

Because regression is often used for prediction, we may wish to construct **confidence intervals for a mean response** and **prediction intervals** for a future observation from multiple regression models. The basic ideas are the same as in the simple linear regression case. In most software systems, the same commands that give confidence and prediction intervals for simple linear regression work for multiple regression. The only difference is that we specify a list of explanatory variables rather than a single variable. Modern software allows us to perform these rather complex calculations without an intimate knowledge of all of the computational details. This frees us to concentrate on the meaning and appropriate use of the results.

ANOVA table for multiple regression

ANOVA *F* test

In simple linear regression the F test from the ANOVA table is equivalent to the two-sided t test of the hypothesis that the slope of the regression line is 0. For multiple regression there is a corresponding **ANOVA *F* test,** but it tests the hypothesis that *all* of the regression coefficients (with the exception of the intercept) are 0. Here is the general form of the ANOVA table for multiple regression:

Source	Degrees of freedom	Sum of squares	Mean square	F
Model	p	$\sum(\hat{y}_i - \bar{y})^2$	SSM/DFM	MSM/MSE
Error	$n - p - 1$	$\sum(y_i - \hat{y}_i)^2$	SSE/DFE	
Total	$n - 1$	$\sum(y_i - \bar{y})^2$	SST/DFT	

The ANOVA table is similar to that for simple linear regression. The degrees of freedom for the model increase from 1 to p to reflect the fact that we now have p explanatory variables rather than just one. As a consequence, the degrees of freedom for error decrease by the same amount. *It is always a good idea to cal-*

culate the degrees of freedom by hand and then check that your software agrees with your calculations. In this way you can verify that your software is using the number of cases and number of explanatory variables that you intended.

The sums of squares represent sources of variation. Once again, both the sums of squares and their degrees of freedom add:

$$SST = SSM + SSE$$

$$DFT = DFM + DFE$$

The estimate of the variance σ^2 for our model is again given by the MSE in the ANOVA table. That is, $s^2 = MSE$.

F statistic   The ratio MSM/MSE is an ***F* statistic** for testing the null hypothesis

$$H_0: \beta_1 = \beta_2 = \cdots = \beta_p = 0$$

against the alternative hypothesis

$$H_a: \text{at least one of the } \beta_j \text{ is not } 0$$

The null hypothesis says that none of the explanatory variables are predictors of the response variable when used in the form expressed by the multiple regression equation. The alternative states that at least one of them is linearly related to the response. As in simple linear regression, large values of F give evidence against H_0. When H_0 is true, F has the $F(p, n - p - 1)$ distribution. The degrees of freedom for the F distribution are those associated with the model and error in the ANOVA table. *A common error in the use of multiple regression is to assume that all of the regression coefficients are statistically different from zero whenever the F statistic has a small P-value. Be sure that you understand the difference between the F test and the t tests for individual coefficients.*

ANALYSIS OF VARIANCE *F* TEST

In the multiple regression model, the hypothesis

$$H_0: \beta_1 = \beta_2 = \cdots = \beta_p = 0$$

is tested by the analysis of variance F statistic

$$F = \frac{MSM}{MSE}$$

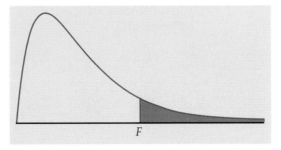

The *P*-value is the probability that a random variable having the $F(p, n - p - 1)$ distribution is greater than or equal to the calculated value of the F statistic.

Squared multiple correlation R^2

For simple linear regression we noted that the square of the sample correlation could be written as the ratio of SSM to SST and could be interpreted as the proportion of variation in y explained by x. A similar statistic is routinely calculated for multiple regression.

> **THE SQUARED MULTIPLE CORRELATION**
>
> The statistic
>
> $$R^2 = \frac{\text{SSM}}{\text{SST}} = \frac{\sum (\hat{y}_i - \bar{y})^2}{\sum (y_i - \bar{y})^2}$$
>
> is the proportion of the variation of the response variable y that is explained by the explanatory variables $x_1, x_2, \ldots, x_p$ in a multiple linear regression.

multiple correlation coefficient

Often, R^2 is multiplied by 100 and expressed as a percent. The square root of R^2, called the **multiple correlation coefficient,** is the correlation between the observations y_i and the predicted values $\hat{y}_i$.

11.2   A Case Study

Preliminary analysis

In this section we illustrate multiple regression by analyzing the data from the study described in Example 11.1. The response variable is the cumulative GPA after three semesters for a group of computer science majors at a large university. The explanatory variables previously mentioned are average high school grades, represented by HSM, HSS, and HSE. We also examine the SAT mathematics and SAT verbal scores as explanatory variables. We have data for $n = 224$ students in the study. We use SAS to illustrate the outputs that are given by most software.

The first step in the analysis is to carefully examine each of the variables. Means, standard deviations, and minimum and maximum values appear in Figure 11.1.

Variable	N	Mean	Std Dev	Minimum	Maximum
GPA	224	2.6352232	0.7793949	0.1200000	4.0000000
SATM	224	595.2857143	86.4014437	300.0000000	800.0000000
SATV	224	504.5491071	92.6104591	285.0000000	760.0000000
HSM	224	8.3214286	1.6387367	2.0000000	10.0000000
HSS	224	8.0892857	1.6996627	3.0000000	10.0000000
HSE	224	8.0937500	1.5078736	3.0000000	10.0000000

FIGURE 11.1   Descriptive statistics for the computer science student case study.

The minimum value for the SAT mathematics (SATM) variable appears to be rather extreme; it is $(595 - 300)/86 = 3.43$ standard deviations below the mean. We do not discard this case at this time but will take care in our subsequent analyses to see if it has an excessive influence on our results. The mean for the SATM score is higher than the mean for the verbal score (SATV), as we might expect for a group of computer science majors. The two standard deviations are about the same. The means of the three high school grade variables are similar, with the mathematics grades being a bit higher. The standard deviations for the high school grade variables are very close to each other. The mean GPA is 2.635 on a 4-point scale, with standard deviation 0.779.

Because the variables GPA, SATM, and SATV have many possible values, we could use stemplots or histograms to examine the shapes of their distributions. Normal quantile plots indicate whether or not the distributions look normal. It is important to note that the multiple regression model *does not* require any of these distributions to be normal. Only the deviations of the responses y from their means are assumed to be normal. The purpose of examining these plots is to understand something about each variable alone before attempting to use it in a complicated model. Extreme values of any variable should be noted and checked for accuracy. If found to be correct, the cases with these values should be carefully examined to see if they are truly exceptional and perhaps do not belong in the same analysis with the other cases. When these data are examined in this way, no obvious problems are evident.

The high school grade variables HSM, HSS, and HSE have relatively few values and are best summarized by giving the relative frequencies for each possible value. The output in Figure 11.2 provides these summaries. The distributions are all skewed, with a large proportion of high grades (10 = A and 9 = A−). Again we emphasize that these distributions need not be normal.

Relationships between pairs of variables

The second step in our analysis is to examine the relationships between all pairs of variables. Scatterplots and correlations are our tools for studying two-variable relationships. The correlations appear in Figure 11.3 (page 693). The output includes the P-value for the test of the null hypothesis that the population correlation is 0 versus the two-sided alternative for each pair. Thus, we see that the correlation between GPA and HSM is 0.44, with a P-value of 0.0001, whereas the correlation between GPA and SATV is 0.11, with a P-value of 0.087. The first is statistically significant by any reasonable standard, and the second is rather marginal.

The high school grades all have higher correlations with GPA than do the SAT scores. As we might expect, math grades have the highest correlation ($r = 0.44$), followed by science grades (0.33) and then English grades (0.29). The two SAT scores have a rather high correlation with each other (0.46), and the high school grades also correlate well with each other (0.45 to 0.58). The SAT mathematics score correlates well with HSM (0.45), less well with HSS (0.24), and rather poorly with HSE (0.11). The correlations of the SAT verbal score with the three high school grades are about equal, ranging from 0.22 to 0.26.

It is important to keep in mind that by examining pairs of variables we are seeking a better understanding of the data. The fact that the correlation of a particular explanatory variable with the response variable does not achieve

HSM	Frequency	Percent	Cumulative Frequency	Cumulative Percent
2	1	0.4	1	0.4
3	1	0.4	2	0.9
4	4	1.8	6	2.7
5	6	2.7	12	5.4
6	23	10.3	35	15.6
7	28	12.5	63	28.1
8	36	16.1	99	44.2
9	59	26.3	158	70.5
10	66	29.5	224	100.0

HSM	Frequency	Percent	Cumulative Frequency	Cumulative Percent
3	1	0.4	1	0.4
4	7	3.1	8	3.6
5	9	4.0	17	7.6
6	24	10.7	41	18.3
7	42	18.8	83	37.1
8	31	13.8	114	50.9
9	50	22.3	164	73.2
10	60	26.8	224	100.0

HSM	Frequency	Percent	Cumulative Frequency	Cumulative Percent
3	1	0.4	1	0.4
4	4	1.8	5	2.2
5	5	2.2	10	4.5
6	23	10.3	33	14.7
7	43	19.2	76	33.9
8	49	21.9	125	55.8
9	42	23.2	177	79.0
10	47	21.0	224	100.0

FIGURE 11.2   The distributions of the high school grade variables.

statistical significance does not *necessarily* imply that it will not be a useful (and significant) predictor in a multiple regression.

Numerical summaries such as correlations are useful, but plots are generally more informative when seeking to understand data. Plots tell us whether the numerical summary gives a fair representation of the data. For a multiple regression, each pair of variables should be plotted. For the six variables in our case study, this means that we should examine 15 plots. In general there are $p + 1$ variables in a multiple regression analysis with p explanatory variables, so that $p(p + 1)/2$ plots are required. Multiple regression is a compli-

```
Pearson  Correlation  Coefficients   / Prob > |R|  under  Ho:  Rho=0   / N = 224


              GPA         SATM        SATV         HSM         HSS         HSE

GPA        1.00000     0.25171     0.11449     0.43650     0.32943     0.28900
           0.0         0.0001      0.0873      0.0001      0.0001      0.0001


SATM       0.25171     1.00000     0.46394     0.45351     0.24048     0.10828
           0.0001      0.0         0.0001      0.0001      0.0003      0.1060


SATV       0.11449     0.46394     1.00000     0.22112     0.26170     0.24371
           0.0873      0.0001      0.0         0.0009      0.0001      0.0002


HSM        0.43650     0.45351     0.22112     1.00000     0.57569     0.44689
           0.0001      0.0001      0.0009      0.0         0.0001      0.0001


HSS        0.32943     0.24048     0.26170     0.57569     1.00000     0.57937
           0.0001      0.0003      0.0001      0.0001      0.0         0.0001


HSE        0.28900     0.10828     0.24371     0.44689     0.57937     1.00000
           0.0001      0.1060      0.0002      0.0001      0.0001      0.0
```

FIGURE 11.3   Correlations among the case study variables.

cated procedure. If we do not do the necessary preliminary work, we are in serious danger of producing useless or misleading results. We leave the task of making these plots as an exercise.

Regression on high school grades

To explore the relationship between the explanatory variables and our response variable GPA, we run several multiple regressions. The explanatory variables fall into two classes. High school grades are represented by HSM, HSS, and HSE, and standardized tests are represented by the two SAT scores. We begin our analysis by using the high school grades to predict GPA. Figure 11.4 gives the multiple regression output.

ANOVA table   The output contains an **ANOVA table,** some additional descriptive statistics, and information about the parameter estimates. When examining any ANOVA table, it is a good idea to first verify the degrees of freedom. This ensures that we have not made some serious error in specifying the model for the program or in entering the data. Because there are $n = 224$ cases, we have DFT $= n - 1 = 223$. The three explanatory variables give DFM $= p = 3$ and DFE $= n - p - 1 = 223 - 3 = 220$.

F statistic   The ANOVA **F statistic** is 18.86, with a P-value of 0.0001. Under the null hypothesis

$$H_0: \beta_1 = \beta_2 = \beta_3 = 0$$

the F statistic has an $F(3, 220)$ distribution. According to this distribution, the chance of obtaining an F statistic of 18.86 or larger is 0.0001. We therefore

```
Dependent  Variable:   GPA

                          Analysis  of Variance

                        Sum of              Mean
Source           DF     Squares            Square      F Value       Prob>F

Model             3      27.71233          9.23744       18.861       0.0001
Error           220     107.75046          0.48977
C Total         223     135.46279

       Root  MSE          0.69984      R-Square         0.2046
       Dep  Mean          2.63522      Adj  R-sq        0.1937
       C.V.              26.55711

                          Parameter  Estimates

                     Parameter        Standard      T for H0:
Variable    DF       Estimate            Error   Parameter=0    Prob  >  |T|

INTERCEP     1       0.589877        0.29424324        2.005       0.0462
HSM          1       0.168567        0.03549214        4.749       0.0001
HSS          1       0.034316        0.03755888        0.914       0.3619
HSE          1       0.045102        0.03869585        1.166       0.2451
```

FIGURE 11.4   Multiple regression output for regression using high school grades to predict GPA.

conclude that at least one of the three regression coefficients for the high school grades is different from 0 in the population regression equation.

In the descriptive statistics that follow the ANOVA table we find that Root MSE is 0.6998. This value is the square root of the MSE given in the ANOVA table and is s, the **estimate of the parameter σ** of our model. The value of R^2 is 0.20. That is, 20% of the observed variation in the GPA scores is explained by linear regression on high school grades. Although the P-value is very small, the model does not explain very much of the variation in GPA. Remember, a small P-value does not necessarily tell us that we have a large effect, particularly when the sample size is large.

estimate of σ

R^2

fitted regression equation

From the Parameter Estimates section we obtain the **fitted regression equation**

$$\widehat{GPA} = 0.590 + 0.169HSM + 0.034HSS + 0.045HSE$$

Let's find the predicted GPA for a student with an A− average in HSM, B+ in HSS, and B in HSE. The explanatory variables are HSM = 9, HSS = 8, and HSE = 7. The predicted GPA is

$$\widehat{GPA} = 0.590 + 0.169(9) + 0.034(8) + 0.045(7)$$
$$= 2.7$$

t statistics   Recall that the **t statistics** for testing the regression coefficients are obtained by dividing the estimates by their standard errors. Thus, for the coefficient of HSM we obtain the *t*-value given in the output by calculating

$$t = \frac{b}{\text{SE}_b} = \frac{0.168567}{0.03549214} = 4.749$$

P-values   The **P-values** appear in the last column. Note that these *P*-values are for the two-sided alternatives. HSM has a *P*-value of 0.0001, and we conclude that the regression coefficient for this explanatory variable is significantly different from 0. The *P*-values for the other explanatory variables (0.36 for HSS and 0.25 for HSE) do not achieve statistical significance.

Interpretation of results

The significance tests for the individual regression coefficients seem to contradict the impression obtained by examining the correlations in Figure 11.3. In that display we see that the correlation between GPA and HSS is 0.33 and the correlation between GPA and HSE is 0.29. The *P*-values for both of these correlations are 0.0001. In other words, if we used HSS alone in a regression to predict GPA, or if we used HSE alone, we would obtain statistically significant regression coefficients.

This phenomenon is not unusual in multiple regression analysis. Part of the explanation lies in the correlations between HSM and the other two explanatory variables. These are rather high (at least compared with the other correlations in Figure 11.3). The correlation between HSM and HSS is 0.58, and that between HSM and HSE is 0.45. Thus, when we have a regression model that contains all three high school grades as explanatory variables, there is considerable overlap of the predictive information contained in these variables. *The significance tests for individual regression coefficients assess the significance of each predictor variable assuming that all other predictors are included in the regression equation.* Given that we use a model with HSM and HSS as predictors, the coefficient of HSE is not statistically significant. Similarly, given that we have HSM and HSE in the model, HSS does not have a significant regression coefficient. HSM, however, adds significantly to our ability to predict GPA even after HSS and HSE are already in the model.

Unfortunately, we cannot conclude from this analysis that the *pair* of explanatory variables HSS and HSE contribute nothing significant to our model for predicting GPA once HSM is in the model. The impact of relations among the several explanatory variables on fitting models for the response is the most important new phenomenon encountered in moving from simple linear regression to multiple regression. We can only hint at the many complicated problems that arise.

Residuals

As in simple linear regression, we should always examine the residuals as an aid to determining whether the multiple regression model is appropriate for the data. Because there are several explanatory variables, we must examine several residual plots. It is usual to plot the residuals versus the predicted values $\hat{y}$ and also versus each of the explanatory variables. Look for outliers, influential observations, evidence of a curved (rather than linear) relation, and

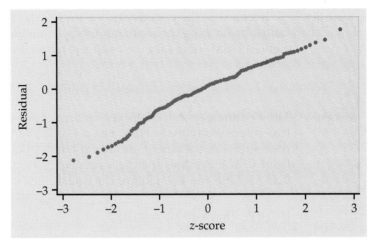

FIGURE 11.5 Normal quantile plot of the residuals from the high school grade model. There are no important deviations from normality.

anything else unusual. Again, we leave the task of making these plots as an exercise. The plots all appear to show more or less random noise around the center value of 0.

If the deviations ϵ in the model are normally distributed, the residuals should be normally distributed. Figure 11.5 presents a normal quantile plot of the residuals. The distribution appears to be approximately normal. There are many other specialized plots that help detect departures from the multiple regression model. Discussion of these, however, is more than we can undertake in this chapter.

Refining the model

Because the variable HSS has the largest *P*-value of the three explanatory variables (see Figure 11.4) and therefore appears to contribute the least to our explanation of GPA, we rerun the regression using only HSM and HSE as explanatory variables. The SAS output appears in Figure 11.6. The *F* statistic indicates that we can reject the null hypothesis that the regression coefficients for the two explanatory variables are both 0. The *P*-value is still 0.0001. The value of R^2 has dropped very slightly compared with our previous run, from 0.2046 to 0.2016. Dropping HSS from the model resulted in the loss of very little explanatory power. The measure *s* of variation about the fitted equation (Root MSE in the printout) is nearly identical for the two regressions, another indication that dropping HSS loses very little. The *t* statistics for the individual regression coefficients indicate that HSM is still clearly significant ($P = 0.0001$), while the coefficient of HSE is larger than before (1.747 versus 1.166) and approaches the traditional 0.05 level of significance ($P = 0.082$).

Comparison of the fitted equations for the two multiple regression analyses tells us something more about the intricacies of this procedure. For the first run we have

$$\widehat{\text{GPA}} = 0.590 + 0.169\text{HSM} + 0.034\text{HSS} + 0.045\text{HSE}$$

```
Dependent Variable: GPA

                          Analysis of Variance

                        Sum of          Mean
Source            DF    Squares         Square      F Value      Prob>F

Model              2    27.30349       13.65175      27.894      0.0001
Error            221   108.15930        0.48941
C Total          223   135.46279


        Root MSE        0.69958     R-Square       0.2016
        Dep Mean        2.63522     Adj R-sq       0.1943
        C.V.           26.54718


                          Parameter Estimates

                     Parameter      Standard    T for H0:
Variable   DF        Estimate          Error    Parameter=0    Prob > |T|

INTERCEP    1        0.624228     0.29172204        2.140        0.0335
HSM         1        0.182654     0.03195581        5.716        0.0001
HSE         1        0.060670     0.03472914        1.747        0.0820
```

FIGURE 11.6 Multiple regression output for regression using HSM and HSE to predict GPA.

whereas the second gives us

$$\widehat{\text{GPA}} = 0.624 + 0.183\text{HSM} + 0.061\text{HSE}$$

Eliminating HSS from the model changes the regression coefficients for all of the remaining variables and the intercept. This phenomenon occurs quite generally in multiple regression. *Individual regression coefficients, their standard errors, and significance tests are meaningful only when interpreted in the context of the other explanatory variables in the model.*

Regression on SAT scores

We now turn to the problem of predicting GPA using the two SAT scores. Figure 11.7 gives the output. The fitted model is

$$\widehat{\text{GPA}} = 1.289 + 0.002283\text{SATM} - 0.000025\text{SATV}$$

The degrees of freedom are as expected: 2, 221, and 223. The F statistic is 7.476, with a P-value of 0.0007. We conclude that the regression coefficients for SATM and SATV are not both 0. Recall that we obtained the P-value 0.0001 when we used high school grades to predict GPA. Both multiple regression equations are highly significant, but this obscures the fact that the two models have quite different explanatory power. For the SAT regression, $R^2 = 0.0634$, whereas for the high school grades model even with only HSM and HSE

```
Dependent Variable: GPA

                          Analysis of Variance

                         Sum of          Mean
  Source          DF     Squares        Square      F Value      Prob>F

  Model            2     8.58384        4.29192       7.476       0.0007
  Error          221   126.87895        0.57411
  C Total        223   135.46279

        Root MSE         0.75770     R-Square        0.0634
        Dep Mean         2.63522     Adj R-sq        0.0549
        C.V.            28.75287

                          Parameter Estimates

                      Parameter      Standard     T for H0:
  Variable   DF        Estimate         Error    Parameter=0     Prob > |T|

  INTERCEP    1        1.288677    0.37603684        3.427         0.0007
  SATM        1        0.002283    0.00066291        3.444         0.0007
  SATV        1       -0.000024562 0.00061847       -0.040         0.9684
```

FIGURE 11.7 Multiple regression output for regression using SAT scores to predict GPA.

(Figure 11.6), we have $R^2 = 0.2016$, a value more than three times as large. Stating that we have a statistically significant result is quite different from saying that an effect is large or important.

Further examination of the output in Figure 11.7 reveals that the coefficient of SATM is significant ($t = 3.44$, $P = 0.0007$), and that for SATV is not ($t = -0.04$, $P = 0.9684$). For a complete analysis we should carefully examine the residuals. Also, we might want to run the analysis with SATM as the only explanatory variable.

Regression using all variables

We have seen that either the high school grades or the SAT scores give a highly significant regression equation. The mathematics component of each of these groups of explanatory variables appears to be the key predictor. Comparing the values of R^2 for the two models indicates that high school grades are better predictors than SAT scores. Can we get a better prediction equation using all of the explanatory variables together in one multiple regression?

To address this question we run the regression with all five explanatory variables. The output appears in Figure 11.8. The F statistic is 11.69, with a P-value of 0.0001, so at least one of our explanatory variables has a nonzero regression coefficient. This result is not surprising given that we have already seen that HSM and SATM are strong predictors of GPA. The value of R^2 is 0.2115, not much higher than the value of 0.2046 that we found for the high school grades regression.

```
Dependent Variable: GPA

                           Analysis of Variance

                           Sum of        Mean
Source              DF     Squares       Square      F Value      Prob>F

Model                5     28.64364      5.72873      11.691      0.0001
Error              218    106.81914      0.49000
C Total            223    135.46279

       Root MSE         0.70000     R-Square     0.2115
       Dep Mean         2.63522     Adj R-sq     0.1934
       C.V.            26.56311


                           Parameter Estimates

                    Parameter      Standard     T for H0:
   Variable   DF    Estimate        Error      Parameter=0    Prob > |T|

   INTERCEP    1     0.326719     0.39999643      0.817         0.4149
   SATM        1     0.000944     0.00068566      1.376         0.1702
   SATV        1    -0.000408     0.00059189     -0.689         0.4915
   HSM         1     0.145961     0.03926097      3.718         0.0003
   HSS         1     0.035905     0.03779841      0.950         0.3432
   HSE         1     0.055293     0.03956869      1.397         0.1637


Test: SAT      Numerator:        0.4657   DF:     2   F value:    0.9503
               Denominator:   0.489996    DF:   218   Prob>F:     0.3882


Test: HS       Numerator:        6.6866   DF:     3   F value:   13.6462
               Denominator:   0.489996    DF:   218   Prob>F:     0.0001
```

FIGURE 11.8 Multiple regression output for regression using all variables to predict GPA.

Examination of the t statistics and the associated P-values for the individual regression coefficients reveals that HSM is the only one that is significant ($P = 0.0003$). That is, only HSM makes a significant contribution when it is added to a model that already has the other four explanatory variables. Once again it is important to understand that this result does not necessarily mean that the regression coefficients for the four other explanatory variables are *all* 0.

Figure 11.9 gives the Excel and Minitab multiple regression outputs for this problem. Although the format and organization of outputs differ among software packages, the basic results that we need are easy to find.

Many statistical software packages provide the capability for testing whether a collection of regression coefficients in a multiple regression model are *all* 0. We use this approach to address two interesting questions about this set of data. We did not discuss such tests in the outline that opened this section, but the basic idea is quite simple.

Excel

	A	B	C	D	E	F	G
1	SUMMARY OUTPUT						
2							
3	*Regression Statistics*						
4	Multiple R	0.459837234					
5	R Square	0.211450282					
6	Adjusted R Square	0.193364279					
7	Standard Error	0.699997195					
8	Observations	224					
9							
10	ANOVA						
11		*df*	*SS*	*MS*	*F*	*Signifcance F*	
12	Regression	5	28.64364489	5.728729	11.69138	5.06E-10	
13	Residual	218	106.8191439	0.489996			
14	Total	223	135.4627888				
15							
16		*Coefficients*	*Standard Error*	*t Stat*	*P-value*	*Lower 95%*	*upper 95%*
17	intercept	0.326718739	0.399996431	0.816804	0.414932	-0.461636967	1.115074446
18	HSM	0.14596108	0.039260974	3.717714	0.000256	0.068581358	0.223340801
19	HSS	0.03590532	0.037798412	0.949916	0.343207	-0.03859183	0.11040247
20	HSE	0.055292581	0.039568691	1.397382	0.163719	-0.022693622	0.133278785
21	SATM	0.000843593	0.000685657	1.376187	0.170176	-0.000407774	0.002294959
22	SATV	0.00040785	0.000591893	-0.68906	0.491518	-0.001574415	0.00075816

Sheet1 \ **Sheet2** / e data /

Ready   Sum=0   CARS

Minitab

```
The regression equation is
GPA = 0.327 + 0.146 HSM + 0.0359 HSS + 0.0553 HSE +0.000944 SATM -
0.000408 SATV

Predictor        Coef        StDev           T          P
Constant       0.3267       0.4000        0.82       0.415
HSM            0.14596      0.03926       3.72       0.000
HSS            0.03591      0.03780       0.95       0.343
HSE            0.05529      0.03957       1.40       0.164
SATM        0.0009436    0.0006857       1.38       0.170
SATV       -0.0004078    0.0005919      -0.69       0.492


S = 0.7000      R-Sq = 21.1%     R-Sq(adj) = 19.3%

Analysis of Variance

Source         DF          SS           MS          F          P
Regression      5       28.6436      5.7287      11.69      0.000
Error         218      106.8191      0.4900
Total         223      135.4628
```

FIGURE 11.9   Excel and Minitab multiple regression outputs for regression using all variables to predict GPA.

Test for a collection of regression coefficients

In the context of the multiple regression model with all five predictors, we ask first whether or not the coefficients for the two SAT scores are both 0. In other words, do the SAT scores add any significant predictive information to that already contained in the high school grades? To be fair, we also ask the complementary question—do the high school grades add any significant predictive information to that already contained in the SAT scores?

The answers are given in the last two parts of the output in Figure 11.8. For the first test we see that $F = 0.9503$. Under the null hypothesis that the two SAT coefficients are 0, this statistic has an $F(2, 218)$ distribution and the P-value is 0.39. We conclude that the SAT scores are not significant predictors of GPA in a regression that already contains the high school scores as predictor variables. Recall that the model with just SAT scores has a highly significant F statistic. We now see that whatever predictive information is in the SAT scores can also be found in the high school grades. In this sense, the SAT scores are superfluous.

The test statistic for the three high school grade variables is $F = 13.6462$. Under the null hypothesis that these three regression coefficients are 0, the statistic has an $F(3, 218)$ distribution and the P-value is 0.0001. We conclude that high school grades contain useful information for predicting GPA that is not contained in SAT scores.

Of course, our statistical analysis of these data does not imply that SAT scores are less useful than high school grades for predicting college grades for all groups of students. We have studied a select group of students—computer science majors—from a specific university. Generalizations to other situations are beyond the scope of inference based on these data alone.

BEYOND THE BASICS | Multiple logistic regression

Many studies have yes/no or success/failure response variables. A surgery patient lives or dies; a consumer does or does not purchase a product after viewing an advertisement. Because the response variable in a multiple regression is assumed to have a normal distribution, this methodology is not suitable for predicting such responses. However, there are models that apply the ideas of regression to response variables with only two possible outcomes.

logistic regression

One type of model that can be used is called **logistic regression.** We think in terms of a binomial model for the two possible values of the response variable and use one or more explanatory variables to explain the probability of success. Details are more complicated than those for multiple regression and are given in the supplemental Chapter 16 on this topic. However, the fundamental ideas are very much the same. Here is an example.

EXAMPLE 11.2 Marketers sometimes use sexual imagery in advertisements targeted at teens and young adults. One study designed to examine this issue analyzed how models were dressed in 1509 ads in magazines read by young and mature adults.[2] The dress of the models in the ads was classified as not sexual or sexual. Logistic regression was used to model the probability that the model's dress was sexual as a function of four explanatory variables. Here, model dress with values 1 for sexual dress and 0 for not sexual dress is the response variable.

The explanatory variables were x_1, a variable having the value 1 if the median age of the readers of the magazine is 20 to 29, and 0 if the median age of the readers of the magazine is 40 to 49; x_2, the gender of the model, coded as 1 for female and 0 for male; x_3, a code to indicate men's magazines with values 1 for a men's magazine and 0 otherwise; and x_4, a code to indicate women's magazines with values 1 for a women's magazine and 0 otherwise. Note that general-interest magazines are coded as 0 for both x_3 and x_4.

Similar to the F test in multiple regression, there is a chi-square test for multiple logistic regression that tests the null hypothesis that *all* coefficients of the explanatory variables are zero. The value is $\chi^2 = 168.2$, and the degrees of freedom are the number of explanatory variables, 4 in this case. (Recall that we studied the chi-square distribution in Chapter 9, where we analyzed two-way tables of counts.) The P-value is reported as $P = 0.001$. (You can verify that it is less than 0.0005 using Table F.) We conclude that not all of the explanatory variables have zero coefficients.

Interpretation of the coefficients is a little more difficult in multiple logistic regression because of the form of the model. For our example, the fitted model is

$$\log\left(\frac{p}{1-p}\right) = -2.32 + 0.50x_1 + 1.31x_2 - 0.05x_3 + 0.45x_4$$

The expression $p/(1-p)$ is the odds that the model is sexually dressed. Logistic regression models the "log odds" as a linear combination of the explanatory variables. Positive coefficients are associated with a higher probability that the model is dressed sexually. We see that ads in magazines with younger readers, female models, and women's magazines are more likely to have sexual dress for their models.

In place of the t tests for individual coefficients in multiple regression, chi-square tests, each with 1 degree of freedom, are used to test whether individual coefficients are zero. For reader age and model gender, $P < 0.01$, while for the indicator for women's magazines, $P < 0.05$. The indicator for men's magazines is not statistically significant.

CHAPTER 11 | Summary

The statistical model for **multiple linear regression** with response variable y and p explanatory variables $x_1, x_2, \ldots, x_p$ is

$$y_i = \beta_0 + \beta_1 x_{i1} + \beta_2 x_{i2} + \cdots + \beta_p x_{ip} + \epsilon_i$$

where $i = 1, 2, \ldots, n$. The ϵ_i are assumed to be independent and normally distributed with mean 0 and standard deviation σ. The **parameters** of the model are $\beta_0, \beta_1, \beta_2, \ldots, \beta_p$, and σ.

The β's are estimated by $b_0, b_1, b_2, \ldots, b_p$, which are obtained by the **method of least squares**. The parameter σ is estimated by

$$s = \sqrt{\text{MSE}} = \sqrt{\frac{\sum e_i^2}{n - p - 1}}$$

where the e_i are the **residuals,**

$$e_i = y_i - \hat{y}_i$$

A **level C confidence interval for** β_j is

$$b_j \pm t^* \text{SE}_{b_j}$$

where t^* is the value for the $t(n - p - 1)$ density curve with area C between $-t^*$ and t^*.

The test of the hypothesis $H_0: \beta_j = 0$ is based on the t **statistic**

$$t = \frac{b_j}{\text{SE}_{b_j}}$$

and the $t(n - p - 1)$ distribution.

The estimate b_j of β_j and the test and confidence interval for β_j are all based on a specific multiple linear regression model. The results of all of these procedures change if other explanatory variables are added to or deleted from the model.

The **ANOVA table** for a multiple linear regression gives the degrees of freedom, sum of squares, and mean squares for the model, error, and total sources of variation. The **ANOVA F statistic** is the ratio MSM/MSE and is used to test the null hypothesis

$$H_0: \beta_1 = \beta_2 = \cdots = \beta_p = 0$$

If H_0 is true, this statistic has an $F(p, \ n - p - 1)$ distribution.

The **squared multiple correlation** is given by the expression

$$R^2 = \frac{\text{SSM}}{\text{SST}}$$

and is interpreted as the proportion of the variability in the response variable y that is explained by the explanatory variables $x_1, x_2, \dots, x_p$ in the multiple linear regression.

CHAPTER 11 | Exercises

11.1 In each of the following settings, give a 95% confidence interval for the coefficients of x_1 and x_2.

   (a) $n = 25, \hat{y} = 10.6 + 12.1x_1 + 17.3x_2$, $\text{SE}_{b_1} = 7.2$, and $\text{SE}_{b_2} = 4.1$.

   (b) $n = 103, \hat{y} = 15.6 + 12.1x_1 + 7.3x_2$, $\text{SE}_{b_1} = 7.2$, and $\text{SE}_{b_2} = 4.1$.

11.2 For each of the settings in the previous exercise, test the null hypotheses that each of the two coefficients is zero versus the two-sided alternative.

11.3 In each of the following situations, explain what is wrong and why.

   (a) The multiple correlation gives the proportion of the variation in the response variable that is explained by the explanatory variables.

(b) In a multiple regression with a sample size of 50 and four explanatory variables, the test statistic for the null hypothesis $H_0: b_2 = 0$ is a t statistic that follows the $t(45)$ distribution when the null hypothesis is true.

(c) One of the assumptions for multiple regression is that the distribution of each explanatory variable should be normal.

11.4   The fitted regression equation for a multiple regression is

$$\hat{y} = -2.6 + 4.1x_1 - 3.2x_2$$

(a) If $x_1 = 5$ and $x_2 = 3$, what is the predicted value of y?

(b) For the answer to part (a) to be valid, is it necessary that the values $x_1 = 5$ and $x_2 = 3$ correspond to a case in the data set? Explain why or why not.

(c) If you hold x_1 at a fixed value, what is the effect of an increase of one unit in x_2 on the predicted value of y?

11.5   What factors predict substance abuse among high school students? One study designed to answer this question collected data from 89 high school seniors in a suburban Florida high school.[3] One of the response variables was marijuana use, which was rated on a four-point scale. A multiple regression analysis used grade point average (GPA), popularity, and a depression score to predict marijuana use. The results were reported in a table similar to this:

	b	t	P
GPA	−0.597	4.55	< 0.001
Popularity	0.340	2.69	< 0.01
Depression	0.030	2.69	< 0.01

A footnote to the table gives $R^2 = 0.34$, $F(3, 85) = 14.83$, and $P < 0.001$.

(a) State the null and alternative hypotheses that are tested by each of the t statistics. Give the results of these significance tests.

(b) Interpret the sign of each of the regression coefficients given in the table.

(c) In the expression $F(3, 85)$, what do the numbers 3 and 85 represent?

(d) State the null and alternative hypotheses that are tested by the F statistic. What is the conclusion?

(e) Each of the variables in this analysis was measured by having the students complete a questionnaire. Discuss how this might affect the results.

(f) How well do you think that these results can be applied to other populations of high school students?

11.6   Refer to the previous exercise. The researchers also studied cigarette use, alcohol use, and cocaine use. Here is a summary of the results for the individual regression coefficients:

		b	**t**	**P**
Cigarette	GPA	−0.340	2.16	< 0.05
	Popularity	0.338	2.24	< 0.05
	Depression	0.034	2.60	< 0.05
Alcohol	GPA	−0.321	3.83	< 0.001
	Popularity	0.185	2.29	< 0.05
	Depression	0.015	2.19	< 0.05
Cocaine	GPA	−0.583	5.99	< 0.001
	Popularity	−1.90	2.25	< 0.05
	Depression	0.002	0.27	Not sig.

And here are other relevant results:

Response variable	R^2	F	P
Cigarette	0.18	6.38	< 0.001
Alcohol	0.27	10.37	< 0.001
Cocaine	0.38	12.21	< 0.001

Using the questions given in the previous exercise, summarize the results for each of these response variables. Then write a short essay comparing the results for the four different response variables.

11.7   A study designed to determine how willing consumers are to pay a premium for non-biotech breakfast cereals included both U.S. and U.K. subjects.[4] The response variable was a measure of how much extra they would be willing to pay, and the explanatory variables included items related to perceived risks and benefits, demographic variables, and country. Country was coded as 0 for the U.K. subjects and 1 for the U.S. subjects. The parameter estimate for country was reported as −0.2304 with $t = −4.196$. The total number of subjects was 1810.

(a) Interpret the regression coefficient. Are subjects in the U.S. more willing to pay extra for non-biotech breakfast cereals than U.K. subjects, or are they less willing?

(b) Use the t statistic to find a bound on the P-value. Explain the hypothesis tested by this statistic and summarize the result of the significance test.

(c) The U.S. data were collected using questionnaires that were sent to a nationally representative sample of 5200 households maintained by the National Dairy Panel Group (NDP). The response rate was 58%. The same questionnaire was used for an online survey of the 9000 U.K. customers enrolled in another NDP group. The response rate was 28.5%. Several of the items used in the analysis included "Don't Know" as a possible response. Respondents choosing this option were excluded from the analysis. Discuss the implications of these considerations on the results.

11.8   Although the benefits of physical exercise are well known, most people do not exercise and many who start exercise programs drop out after a short time. A study designed to determine factors associated with exercise enjoyment collected data from 282 female volunteers who were participants in not-for-credit aerobic dance classes at two university centers.[5] Exercise enjoyment was the response variable, with a possible range of 18 to 136. Three explanatory variables were analyzed: satisfaction with the music used (range 4 to 28), satisfaction with the instructor (range 6 to 42), and identity, a variable that measured the extent to which the subject viewed herself as an exerciser. A table of correlations among the four variables was given, and the text noted that all were significant with $P < 0.01$. The coefficients for music (1.02), instructor (0.96), and identity (0.30) were given in another table, where it was noted that $R^2 = 0.33$.

(a) Can you give the fitted regression equation? If your answer is Yes, write the equation; if No, explain what additional information you would need.

(b) Does the fact that all of the correlations between the four variables are significant at $P < 0.01$ tell us that the regression coefficients for each of the three explanatory variables will be statistically significant? Explain your answer.

(c) The statistic for testing the null hypothesis that the population regression coefficients for the three explanatory variables are all zero is $F = 45.64$. Give the degrees of freedom for this statistic, and carry out the significance test. What do you conclude?

(d) What proportion of the variation in exercise enjoyment is explained by music, instructor, and identity?

(e) The authors of the study note that males were not included because there were too few of them in these classes. Do you think that these results would apply to males? Explain why or why not.

11.9   Six explanatory variables are used to predict a response variable using a multiple regression. There are 200 observations.

(a) Write the statistical model that is the foundation for this analysis. Be sure to include a description of all assumptions.

(b) Outline the analysis of variance table giving the sources of variation and numerical values for the degrees of freedom.

11.10   A multiple regression analysis of 54 cases was performed with 3 explanatory variables. Suppose that SSM = 18 and SSE = 100.

(a) Find the value of the F statistic for testing the null hypothesis that the coefficients of all of the explanatory variables are zero.

(b) What are the degrees of freedom for this statistic?

(c) Find bounds on the P-value using Table E. Give the two entries from the table that you use to determine the bounds.

11.11   Refer to the previous exercise. What proportion of the variation in the response variable is explained by the explanatory variables?

11.12 Using a new software package, you ran a multiple regression. The output reported an F statistic with $P < 0.001$, but none of the t tests for the individual coefficients were significant. Does this mean that there is something wrong with the software? Explain your answer.

11.13 Nutrition facts labels provide consumers with information about the nutritional value of food products that they buy. A study of these labels collected data from 152 consumers who were sent information about a frozen chicken dinner. Each subject was asked to give an overall product nutrition score and also evaluated each of 10 nutrients on a 9-point scale, with higher values indicating that the product has a healthy value for the given nutrient. Composite scores for favorable nutrients (such as protein and fiber) and unfavorable nutrients (such as fat and sodium) were used in a multiple regression to predict the overall product nutrition score.[6] The following was reported in a table:

Explanatory variables	b	se	t	Model F	R^2
				33.7**	.31
Unfavorable nutrients	0.82	.12	6.8**		
Favorable nutrients	0.57	.10	5.5**		
Constant	3.33	.13	26.1**		

$**p < .01.$

(a) What is the equation of the least-squares line?

(b) Give the null and alternative hypotheses associated with the entry labeled "Model F" and interpret this result.

(c) The column labeled "t" contains three entries. Explain what each of these means.

(d) What are the degrees of freedom associated with the t statistics that you explained in part (c)?

11.14 The product used in the previous exercise was described by the researchers as a poor-nutrition product. The label information for this product had high values for unfavorable nutrients such as fat and low values for favorable nutrients such as fiber. The researchers who conducted this study collected a parallel set of data from subjects who were provided label information for a good-nutrition product. This label had low values for the unfavorable nutrients and high values for the favorable ones. The same type of multiple regression model was run for the 162 consumers who participated in this part of the study. Here are the regression results:

Explanatory variables	b	se	t	Model F	R^2
				44.0**	.36
Unfavorable nutrients	0.86	.12	6.9**		
Favorable nutrients	0.66	.10	6.9**		
Constant	3.96	.12	32.8**		

$**p < .01.$

For this analysis, answer the questions in parts (a) to (d) of the previous exercise.

11.15   Refer to the previous two exercises. When the researchers planned these studies, they expected both unfavorable nutrients and favorable nutrients to be positively associated with the overall product nutrition score. They also expected the unfavorable nutrients to have a stronger effect. Examine the regression coefficients and the associated t statistics for the two regression models. Then, use this information to discuss how well the researchers' expectations were fulfilled.

The following five exercises use the data given in the next exercise.

11.16   Online stock trading through the Internet has increased dramatically during the past several years. An article discussing this new method of investing provided data on the major Internet stock brokerages who provide this service.[7] Below are some data for the top 10 Internet brokerages. The variables are Mshare, the market share of the firm; Accts, the number of Internet accounts in thousands; and Assets, the total assets in billions of dollars. These firms are not a random sample from any population but we will use multiple regression methods to develop statistical models that relate assets to the other two variables.

ID	Broker	Mshare	Accts	Assets
1	Charles Schwab	27.5	2500	219.0
2	E*Trade	12.9	909	21.1
3	TD Waterhouse	11.6	615	38.8
4	Datek	10.0	205	5.5
5	Fidelity	9.3	2300	160.0
6	Ameritrade	8.4	428	19.5
7	DLJ Direct	3.6	590	11.2
8	Discover	2.8	134	5.9
9	Suretrade	2.2	130	1.3
10	National Discount Brokers	1.3	125	6.8

(a) Plot assets versus accounts and describe the relationship.

(b) Perform a simple linear regression to predict assets from the number of accounts. Give the least-squares line and the results of the hypothesis test for the slope.

(c) Obtain the residuals from part (b) and plot them versus accounts. Describe the plot. What do you conclude?

(d) Construct a new variable that is the square of the number of accounts. Rerun the regression analysis with accounts and the square as explanatory variables. Summarize the results.

11.17   In the multiple regression you performed in the previous exercise, the P-value for the number of accounts was 0.8531, while the P-value for the square was 0.0070. Unless we have a strong theoretical reason for considering a model with a quadratic term and no linear term, we prefer not to do this. One problem with these two explanatory variables is that they are highly correlated. Here is a way to construct a version of the quadratic term that is less correlated with the linear term. We first find the mean for accounts, and then we subtract this value from accounts before squaring. The mean is 793.6, so the new quadratic explanatory variable will be $(\text{Accts} - 793.6)^2$. Run the multiple regression to predict assets using accounts and the new quadratic term. Compare these results with what you found in the previous exercise.

11.18   To one person, the plot of assets versus the number of accounts indicates that the relationship is curved. Another person might see this as a linear relationship with two outliers. Identify the two outliers and rerun the linear regression and the multiple regression with the linear and quadratic terms. Summarize your results.

11.19   Sometimes we attempt to model curved relationships by transforming variables. Take the logarithm of assets and the logarithm of the number of accounts. Does the relationship between the logs appear to be approximately linear? Analyze the data and provide a summary of your results. Be sure to include plots along with the results of your statistical inference.

11.20   Recall that the relationship between an explanatory variable and a response variable can depend on what other explanatory variables are included in the model.

(a) Use a simple linear regression to predict assets using the number of accounts. Give the regression equation and the results of the significance test for the regression coefficient.

(b) Do the same using market share to predict assets.

(c) Run a multiple regression using both the number of accounts and market share to predict assets. Give the multiple regression equation and the results of the significance tests for the two regression coefficients.

(d) Compare the results in parts (a), (b), and (c). If you had to choose one of these three models, which one do you prefer? Give an explanation for your answer.

11.21   Banks charge different interest rates for different loans. A random sample of 2229 loans made for the purchase of new automobiles was studied to identify variables that explain the interest rate charged. All of these loans were made directly by the bank. A multiple regression was run with interest rate as the response variable and 13 explanatory variables.[8]

(a) The F statistic reported is 71.34. State the null and alternative hypotheses for this statistic. Give the degrees of freedom and the P-value for this test. What do you conclude?

(b) The value of R^2 is 0.297. What percent of the variation in interest rates is explained by the 13 explanatory variables?

(c) The researchers report a t statistic for each of the regression coefficients. State the null and alternative hypotheses tested by each of these statistics. What are the degrees of freedom for these t statistics? What values of t will lead to rejection of the null hypothesis at the 5% level?

(d) The following table gives the explanatory variables and the t statistics (these are given without the sign, assuming that all tests are two-sided) for the regression coefficients. Which of the explanatory variables are significantly different from zero in this model?

Variable	b	t
Intercept	15.47	
Loan size (in dollars)	−0.0015	10.30
Length of loan (in months)	−0.906	4.20
Percent down payment	−0.522	8.35
Cosigner (0 = no, 1 = yes)	−0.009	3.02
Unsecured loan (0 = no, 1 = yes)	0.034	2.19
Total payments (borrower's monthly installment debt)	0.100	1.37
Total income (borrower's total monthly income)	−0.170	2.37
Bad credit report (0 = no, 1 = yes)	0.012	1.99
Young borrower (0 = older than 25, 1 = 25 or younger)	0.027	2.85
Male borrower (0 = female, 1 = male)	−0.001	0.89
Married (0 = no, 1 = yes)	−0.023	1.91
Own home (0 = no, 1 = yes)	−0.011	2.73
Years at current address	−0.124	4.21

(e) The signs of many of these coefficients are what we might expect before looking at the data. For example, the negative coefficient of loan size means that larger loans get a smaller interest rate. This is very reasonable. Examine the signs of each of the statistically significant coefficients and give a short explanation of what they mean.

11.22 Refer to the previous exercise. The researchers also looked at loans made indirectly, that is, through an auto dealer. They studied 5664 indirect loans. Multiple regression was used to predict the interest rate using the same set of explanatory variables.

(a) The F statistic reported is 27.97. State the null and alternative hypotheses for this statistic. Give the degrees of freedom and the P-value for this test. What do you conclude?

(b) The value of R^2 is 0.141. What percent of the variation in interest rates is explained by the 13 explanatory variables? Compare this value with the percent explained for direct loans in the previous exercise.

(c) The researchers report a t statistic for each of the regression coefficients. State the null and alternative hypotheses tested by each of these statistics. What are the degrees of freedom for these t statistics? What values of t will lead to rejection of the null hypothesis at the 5% level?

(d) The following table gives the explanatory variables and the t statistics (these are given without the sign, assuming that all tests are two-sided) for the regression coefficients. Which of the explanatory variables are significantly different from zero in this model?

Variable	b	t
Intercept	15.89	
Loan size (in dollars)	−0.0029	17.40
Length of loan (in months)	−1.098	5.63
Percent down payment	−0.308	4.92
Cosigner (0 = no, 1 = yes)	−0.001	1.41
Unsecured loan (0 = no, 1 = yes)	0.028	2.83
Total payments (borrower's monthly installment debt)	−0.513	1.37
Total income (borrower's total monthly income)	0.078	0.75
Bad credit report (0 = no, 1 = yes)	0.039	1.76
Young borrower (0 = older than 25, 1 = 25 or younger)	−0.036	1.33
Male borrower (0 = female, 1 = male)	−0.179	1.03
Married (0 = no, 1 = yes)	−0.043	1.61
Own home (0 = no, 1 = yes)	−0.047	1.59
Years at current address	−0.086	1.73

(e) The signs of many of these coefficients are what we might expect before looking at the data. For example, the negative coefficient of loan size means that larger loans get a smaller interest rate. This is very reasonable. Examine the signs of each of the statistically significant coefficients and give a short explanation of what they mean.

11.23 Refer to the previous two exercises. The authors conclude that banks take higher risks with indirect loans because they do not take into account borrower characteristics when setting the loan rate. Explain how the results of the multiple regressions lead to this conclusion.

11.24 Refer to the educational data for 78 seventh-grade students given in Table 1.9 (page 39). We view GPA as the response variable. IQ, gender, and self-concept are the explanatory variables.

(a) Find the correlation between GPA and each of the explanatory variables. What percent of the total variation in student GPAs can be explained by the straight-line relationship for each of the explanatory variables?

(b) The importance of IQ in explaining GPA is not surprising. The purpose of the study is to assess the influence of self-concept on GPA. So we will include IQ in the regression model and ask, "How much does self-concept contribute to explaining GPA after the effect of IQ on GPA is taken into account?" Give a model that can be used to answer this question.

(c) Run the model and report the fitted regression equation. What percent of the variation in GPA is explained by the explanatory variables in your model?

(d) Translate the question of interest into appropriate null and alternative hypotheses about the model parameters. Give the value of the test statistic and its P-value. Write a short summary of your analysis with an emphasis on your conclusion.

11.25 Multiple regressions are sometimes used in litigation. In the case of *Cargill, Inc. v. Hardin*, the prosecution charged that the cash price of wheat was manipulated in violation of the Commodity Exchange Act. In a statistical study conducted for this case, a multiple regression model was constructed to predict the price of wheat using three supply-and-demand explanatory variables. Data for 14 years were used to construct the regression equation, and a prediction for the suspect period was computed from this equation. The value of R^2 was 0.989. The predicted value was reported as $2.136 with a standard error of $0.013. Express the prediction with a 95% interval. (The degrees of freedom were large for this analysis, so use 100 as the df to determine t^*.) The actual price for the period in question was $2.13. The judge in this case decided that the analysis provided evidence that the price was not artificially depressed, and the opinion was sustained by the court of appeals. Write a short summary of the results of the analysis that relate to the decision and explain why you agree or disagree with it.

The following six exercises use the BIOMARKERS data set described in the Data Appendix.

11.26   Consider the following four variables: VO+, a measure of bone formation; VO−, a measure of bone resorption; OC, a biomarker of bone formation; and TRAP, a biomarker of bone resorption.

(a) Using numerical and graphical summaries, describe the distribution of each of these variables.

(b) Using numerical and graphical summaries, describe the relationship between each pair of variables in this set.

11.27   Let's use regression methods to predict VO+, the measure of bone formation.

(a) Since OC is a biomarker of bone formation, we start with a simple linear regression using OC as the explanatory variable. Run the regression and summarize the results. Be sure to include an analysis of the residuals.

(b) Because the processes of bone formation and bone resorption are highly related, it is possible that there is some information in the bone resorption variables that can tell us something about bone formation. Use a model with both OC and TRAP, the biomarker of bone resorption, to predict VO+. Summarize the results. In the context of this model, it appears that TRAP is a better predictor of bone formation, VO+, than the biomarker of bone formation, OC. Is this view consistent with the pattern of relationships that you described in the previous exercise? One possible explanation is that while all of these variables are highly related, TRAP is measured with more precision than OC.

11.28   Now consider a regression model for predicting VO+ using OC, TRAP, and VO−.

(a) Write out the statistical model for this analysis including all assumptions.

(b) Run the multiple regression to predict VO+ using OC, TRAP, and VO−. Summarize the results.

(c) Make a table giving the estimated regression coefficients, standard errors, and t statistics with P-values. Include this analysis and the two that you ran in the previous exercise. Describe how the coefficients and the *P*-values differ for the three analyses.

(d) Give the percent of variation in VO+ explained by each of the three models and the estimate of σ. Give a short summary.

(e) The results you found in part (b) suggest another model. Run that model, summarize the results, and compare them with the results in (b).

 11.29 Because the distributions of VO+, VO−, OC, and TRAP tend to be skewed, it is common to work with logarithms rather than the measured values. Using the questions in the previous three exercises as a guide, analyze the log data.

 11.30 Refer to Exercises 11.26 to 11.28. Answer these questions with the roles of VO+ and VO− reversed; that is, run models to predict VO−, with VO+ as a possible explanatory variable.

 11.31 Refer to the previous exercise. Rerun using logs.

The following eleven exercises use the PCB data set described in the Data Appendix.

11.32 Production of polychlorinated biphenyls (PCBs) was banned in the United States in 1977, but because of their widespread use, these compounds are found in many species of fish. As a result, 38 states have issued advisories about limiting consumption of certain species caught in some areas. Specific advisories are targeted toward pregnant and lactating women and all women who are of childbearing age. Unfortunately, there are 209 different varieties, called congeners, of PCBs. Measurement of all of these congeners in a fish specimen is an expensive and time-consuming process. If the total amount of PCBs in a specimen can be estimated with data collected on a few congeners, costs can be reduced and more specimens can be measured. Consider the following variables: PCB (the total amount of PCB), and four congeners: PCB52, PCB118, PCB138, and PCB180.

(a) Using numerical and graphical summaries, describe the distribution of each of these variables.

(b) Using numerical and graphical summaries, describe the relationship between each pair of variables in this set.

11.33 Use the four congeners, PCB52, PCB118, PCB138, and PCB180, in a multiple regression to predict PCB.

(a) Write the statistical model for this analysis. Include all assumptions.

(b) Run the regression and summarize the results.

(c) Examine the residuals. Do they appear to be approximately normal? When you plot them versus each of the explanatory variables, are any patterns evident?

11.34 The examination of the residuals in part (c) of the previous exercise suggests that there may be two outliers, one with a high residual and one with a low residual.

(a) Because of safety issues, we are more concerned about underestimating PCB in a specimen than about overestimating. Give the specimen number for each of the two suspected outliers. Which one corresponds to an overestimate of PCB?

(b) Rerun the analysis with the two suspected outliers deleted, summarize these results, and compare them with those you obtained in the previous exercise.

11.35 Run a regression to predict PCB using the variables PCB52, PCB118, and PCB138. Note that this is similar to the analysis that you did in Exercise 11.33 with the change that PCB180 is not included as an explanatory variable.

(a) Summarize the results.

(b) In this analysis, the regression coefficient for PCB118 is not statistically significant. Give the estimate of the coefficient and the associated P-value.

(c) Find the estimate of the coefficient for PCB118 and the associated P-value for the model analyzed in Exercise 11.33.

(d) Using the results in parts (b) and (c), write a short paragraph explaining how the inclusion of other variables in a multiple regression can have an effect on the estimate of a particular coefficient and the results of the associated significance test.

11.36 Dioxins and furans are other classes of chemicals that can cause undesirable health effects similar to those caused by PCBs. The three types of chemicals are combined using toxic equivalent scores (TEQs), which attempt to measure the health effects on a common scale. The PCB data set contains TEQs for PCBs, dioxins, and furans. The variables are called TEQPCB, TEQDIOXIN, and TEQFURAN. The data set also includes the total TEQ, defined to be the sum of these three variables.

(a) Consider using a multiple regression to predict TEQ using the three components TEQPCB, TEQDIOXIN, and TEQFURAN as explanatory variables. Write the multiple regression model in the form

$$TEQ = \beta_0 + \beta_1 TEQPCB + \beta_2 TEQDIOXIN + \beta_3 TEQFURAN + \epsilon$$

Give numerical values for the parameters β_0, β_1, β_2, and β_3.

(b) The multiple regression model assumes that the ϵ's are normal with mean zero and standard deviation σ. What is the numerical value of σ?

(c) Use software to run this regression and summarize the results.

11.37 The information summarized in TEQ is used to assess and manage risks from these chemicals. For example, the World Health Organization (WHO) has established the tolerable daily intake (TDI) as 1 to 4 TEQs per kilogram of body weight per day. Therefore, it would be very useful to have a procedure for estimating TEQ using just a few variables that can be measured cheaply. Use the four PCB congeners, PCB52, PCB118, PCB138, and PCB180, in a multiple regression to predict TEQ. Give a description of the model and assumptions, summarize the results, examine the residuals, and write a summary of what you have found.

11.38 Because distributions of variables such as PCB, the PCB congeners, and TEQ tend to be skewed, researchers frequently analyze the logarithms of the measured variables. Create a data set that has the logs of each of the variables in the PCB data set. Note that zero is a possible value for PCB126; most software packages will eliminate these cases when you request a log transformation.

(a) If you do not do anything about the 16 zero values of PCB126, what does your software do with these cases? Is there an error message of some kind?

(b) If you attempt to run a regression to predict the log of PCB using the log of PCB126 and the log of PCB52, are the cases with the zero values of PCB126 eliminated? Do you think that is a good way to handle this situation?

(c) The smallest nonzero value of PCB126 is 0.0052. One common practice when taking logarithms of measured values is to replace the zeros by one-half of the smallest observed value. Create a logarithm data set using this procedure; that is, replace the 16 zero values of PCB126 by 0.0026 before taking logarithms. Use numerical and graphical summaries to describe the distributions of the log variables.

11.39 Refer to the previous exercise.

(a) Use numerical and graphical summaries to describe the relationships between each pair of log variables.

(b) Compare these summaries with the summaries that you produced in Exercise 11.32 for the measured variables.

11.40 Use the log data set that you created in Exercise 11.38 to find a good multiple regression model for predicting the log of PCB. Use only log PCB variables for this analysis. Write a report summarizing your results.

11.41 Use the log data set that you created in Exercise 11.38 to find a good multiple regression model for predicting the log of TEQ. Use only log PCB variables for this analysis. Write a report summarizing your results and comparing them with the results that you obtained in the previous exercise.

11.42 Use the results of your analysis of the log PCB data in Exercise 11.40 to write an explanation of how regression coefficients, standard errors of regression coefficients, and tests of significance for explanatory variables can change depending on what other explanatory variables are included in the multiple regression analysis.

The following nine exercises use the CHEESE data set described in the Data Appendix.

11.43 For each of the four variables in the CHEESE data set, find the mean, median, standard deviation, and interquartile range. Display each distribution by means of a stemplot and use a normal quantile plot to assess normality of the data. Summarize your findings. Note that when doing regressions with these data, we do not assume that these distributions are normal. Only the residuals from our model need be (approximately) normal. The careful study of each variable to be analyzed is nonetheless an important first step in any statistical analysis.

11.44 Make a scatterplot for each pair of variables in the CHEESE data set (you will have six plots). Describe the relationships. Calculate the correlation for each pair of variables and report the P-value for the test of zero population correlation in each case.

11.45 Perform a simple linear regression analysis using Taste as the response variable and Acetic as the explanatory variable. Be sure to examine the residuals carefully. Summarize your results. Include a plot of the data with the least-squares regression line. Plot the residuals versus each of the other two chemicals. Are any patterns evident? (The concentrations of the other chemicals are lurking variables for the simple linear regression.)

11.46 Repeat the analysis of Exercise 11.45 using Taste as the response variable and H2S as the explanatory variable.

11.47 Repeat the analysis of Exercise 11.45 using Taste as the response variable and Lactic as the explanatory variable.

11.48 Compare the results of the regressions performed in the three previous exercises. Construct a table with values of the F statistic, its P-value, R^2, and the estimate s of the standard deviation for each model. Report the three regression equations. Why are the intercepts in these three equations different?

11.49 Carry out a multiple regression using Acetic and H2S to predict Taste. Summarize the results of your analysis. Compare the statistical significance of Acetic in this model with its significance in the model with Acetic alone as a predictor (Exercise 11.45). Which model do you prefer? Give a simple explanation for the fact that Acetic alone appears to be a good predictor of Taste but with H2S in the model, it is not.

11.50 Carry out a multiple regression using H2S and Lactic to predict Taste. Comparing the results of this analysis with the simple linear regressions using each of these explanatory variables alone, it is evident that a better result is obtained by using both predictors in a model. Support this statement with explicit information obtained from your analysis.

11.51 Use the three explanatory variables Acetic, H2S, and Lactic in a multiple regression to predict Taste. Write a short summary of your results, including an examination of the residuals. Based on all of the regression analyses you have carried out on these data, which model do you prefer and why?

11.52 Search the Internet to find an example of the use of multiple regression. Give the setting of the example, describe the data, give the model, and summarize the results. Explain why the use of multiple regression in this setting was appropriate or inappropriate.

EESEE CASE STUDIES

The Electronic Encyclopedia of Statistical Examples and Exercises (EESEE) is available on the text CD and Web site. These more elaborate stories, with data, provide settings for longer case studies.

11.53 Read the EESEE story "Is Old Faithful Faithful?" Write a report that answers all questions for this case study.

11.54 Read the EESEE story "Seat Belt Safety." Write a report that answers all questions for this case study.

11.55 Read the EESEE story "Diamonds: How Precious Are They?" Write a report that answers all questions for this case study.

STATISTICS IN PRACTICE

Improving the Quality of a Bank's Services

DEXTER WILSON *is a Six Sigma Quality Analyst with HSBC Corporate, Investment Banking and Markets.*

In a competitive marketplace, the quality of a company's products or services is a key to satisfying customers. Using data is a key to improving quality. That's what I do: as a Quality Analyst, I use statistical tools to analyze data, whether they describe the time to process a new hire or the results of a customer satisfaction survey.

My work is organized by a rigorous and disciplined methodology called "Six Sigma." Yes, "Sigma" is the standard deviation. The Six Sigma method emphasizes improving quality through measuring, analyzing, and eliminating any weaknesses that the data reveal. I have been trained as a "black belt," an expert in this methodology. Six Sigma originated in the manufacturing industry but is now applied by leaders in service industries, such as the large bank that I work for.

I use an assortment of tools to analyze processes. Many are standard statistical methods such as analysis of variance (ANOVA). Using tools like ANOVA, I can tell if the responses of subgroups in a customer satisfaction survey differ. Once I identify a group's responses as unusual (whether good or bad), we investigate to find causes and make changes that will improve our service to our customers. When looking at the complex process of hiring an employee, our team used simulation to model the process. We entered historical data into a detailed map of the process to simulate a year's worth of new employee hires and predict where bottlenecks would occur and what level of budget and staffing would make the process run most smoothly. Management then used the model results in planning for the next year.

I enjoy the variety and complexity in my work—I solve problems from many areas of business, I cooperate with different groups of colleagues, and I use a variety of statistical tools, all with the overall goal of improving what we offer our customers. The most important aspect of my job is that everything is backed by factual data. Unlike in the past, when mere opinion usually dictated how and when improvements were made, now I can be much more certain of my conclusions and recommendations.

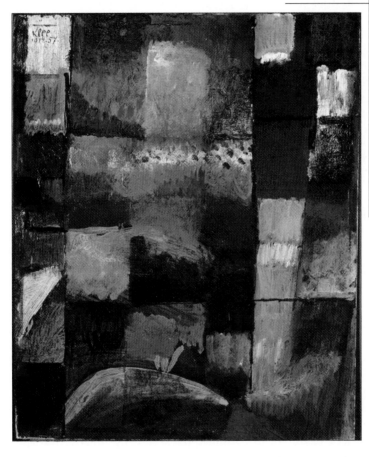

One–Way Analysis of Variance

Introduction

Many of the most effective statistical studies are comparative, whether we compare the salaries of women and men in a sample of firms or the responses to two treatments in a clinical trial. We display comparisons with back-to-back stemplots or side-by-side boxplots, and we measure them with five-number summaries or with means and standard deviations. Now we ask: "Is the difference between groups statistically significant?"

When only two groups are compared, Chapter 7 provides the tools we need. Two-sample t procedures compare the means of two normal populations, and we saw that these procedures, unlike comparisons of spread, are sufficiently robust to be widely useful. Now we will compare any number of means by techniques that generalize the two-sample t and share its robustness and usefulness.

12.1 Inference for One-Way Analysis of Variance

Which of four advertising offers mailed to sample households produces the highest sales in dollars? Which of ten brands of automobile tires wears longest? How long do cancer patients live under each of three therapies for their cancer? In each of these settings we wish to compare several treatments. In each case the data are subject to sampling variability—if we mailed the advertising offers to another set of households, we would get different data. We therefore pose the question for inference in terms of the *mean* response. We compare, for example, the mean tread lifetime of different brands of tires. In Chapter 7 we met procedures for comparing the means of two populations. We are now ready to extend those methods to problems involving more than two populations. The statistical methodology for comparing several means is called **analysis of variance,** or simply **ANOVA.**

We will consider two ANOVA techniques. When there is only one way to classify the populations of interest, we use **one-way ANOVA** to analyze the data. For example, to compare the tread lifetimes of ten specific brands of tires we use one-way ANOVA. This chapter presents the details for one-way ANOVA.

In many practical situations there is more than one way to classify the populations. A mail-order firm might want to compare mailings that offer different discounts and also have different layouts. Will a lower price offered in a plain format draw more sales on the average than a higher price offered in a fancy brochure? Analyzing the effect of price and layout together requires **two-way ANOVA,** and adding yet more factors necessitates higher-way ANOVA techniques. Most of the new ideas in ANOVA with more than one factor already appear in two-way ANOVA, which we discuss in Chapter 13.

ANOVA

one-way ANOVA

two-way ANOVA

Data for one-way ANOVA

Do two population means differ? If we have random samples from the two populations, we compute a two-sample t statistic and its P-value to assess the statistical significance of the difference in the sample means. We compare several means in much the same way. Instead of a t statistic, ANOVA uses an

F statistic and its *P*-value to evaluate the null hypothesis that all of several population means are equal.

In the sections that follow, we will examine the basic ideas and assumptions that are needed for the analysis of variance. Although the details differ, many of the concepts are similar to those discussed in the two-sample case.

One-way analysis of variance is a statistical method for comparing several population means. We draw a simple random sample (SRS) from each population and use the data to test the null hypothesis that the population means are all equal.

EXAMPLE 12.1   A magazine publisher wants to compare three different styles of covers for a magazine that will be offered for sale at supermarket checkout lines. She assigns 60 stores at random to the three styles of covers and records the number of magazines that are sold in a one-week period.

EXAMPLE 12.2   How do five bookstores in the same city differ in the demographics of their customers? A market researcher asks 50 customers of each store to respond to a questionnaire. One variable of interest is the customer's age.

These two examples have several similarities. In both there is a single quantitative response variable measured on several units; the units are stores in the first example and customers in the second. We expect the data to be approximately normal and will consider a transformation if they are not. We wish to compare several populations: stores displaying three cover styles in the first example and customers of five bookstores in the second. There is also an important difference between the examples. Example 12.1 describes an **experiment** in which stores are randomly assigned to cover styles. Example 12.2 is an **observational study** in which customers are selected during a particular time period and not all agree to provide data. We will treat our samples of customers as random samples even though this is only approximately true. In both examples, we will use ANOVA to compare the mean responses. The same ANOVA methods apply to data from random samples and to data from randomized experiments. *It is important to keep the data-production method in mind when interpreting the results. A strong case for causation is best made by a randomized experiment.* We will often use the term *groups* for the populations to be compared in a one-way ANOVA.

experiment
observational study

Comparing means

The question we ask in ANOVA is "Do all groups have the same population mean?" To answer this question we compare the sample means. Figure 12.1 displays the sample means for Example 12.1. Design 2 had the highest average sales. But is the observed difference in sample means among the groups just the result of chance variation? We do not expect sample means to be equal even if the population means are all identical. The purpose of ANOVA is to assess whether the observed differences among sample means are *statistically significant. Could a variation this large be plausibly due to chance, or is it good*

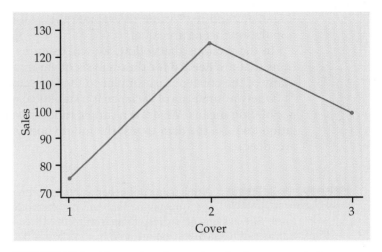

FIGURE 12.1 Mean sales of magazines for three cover designs.

evidence for a difference among the population means? This question can't be answered from the sample means alone. Because the standard deviation of a sample mean $\bar{x}$ is the population standard deviation σ divided by $\sqrt{n}$, the answer depends upon both the variation within the groups of observations and the sizes of the samples.

Side-by-side boxplots help us see the within-group variation. Compare Figures 12.2(a) and 12.2(b). The sample medians are the same in both figures, but the large variation within the groups in Figure 12.2(a) suggests that the differences among the sample medians could be due simply to chance variation. The data in Figure 12.2(b) are much more convincing evidence that the populations differ. Even the boxplots omit essential information, however. To assess the observed differences, we must also know how large the samples

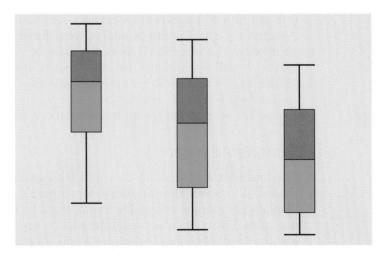

FIGURE 12.2(a) Side-by-side boxplots for three groups with large within-group variation. The differences among centers may be just chance variation.

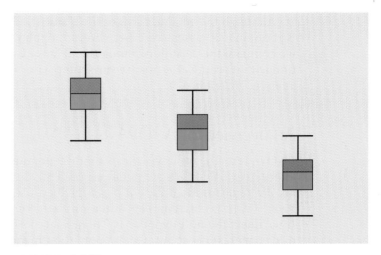

FIGURE 12.2(b)   Side-by-side boxplots for three groups with the same centers as in Figure 12.2(a) but with small within-group variation. The differences among centers are more likely to be significant.

are. Nonetheless, boxplots are a good preliminary display of ANOVA data. (ANOVA compares means, and boxplots display medians. If the distributions are nearly symmetric, these two measures of center will be close together.)

The two-sample t statistic

Two-sample t statistics compare the means of two populations. If the two populations are assumed to have equal but unknown standard deviations and the sample sizes are both equal to n, the t statistic is (page 500)

$$t = \frac{\bar{x} - \bar{y}}{s_p \sqrt{\dfrac{1}{n} + \dfrac{1}{n}}} = \frac{\sqrt{\dfrac{n}{2}}(\bar{x} - \bar{y})}{s_p}$$

The square of this t statistic is

$$t^2 = \frac{\dfrac{n}{2}(\bar{x} - \bar{y})^2}{s_p^2}$$

If we use ANOVA to compare two populations, the ANOVA F statistic is exactly equal to this t^2. We can therefore learn something about how ANOVA works by looking carefully at the statistic in this form.

between-group variation   The numerator in the t^2 statistic measures the variation **between** the groups in terms of the difference between their sample means $\bar{x}$ and $\bar{y}$. It includes a factor for the common sample size n. The numerator can be large because of a large difference between the sample means or because the sample **within-group variation** sizes are large. The denominator measures the variation **within** groups by s_p^2, the pooled estimator of the common variance. If the within-group variation is small, the same variation between the groups produces a larger statistic and a more significant result.

Although the general form of the F statistic is more complicated, the idea is the same. To assess whether several populations all have the same mean, we compare the variation *among* the means of several groups with the variation *within* groups. Because we are comparing variation, the method is called *analysis of variance*.

An overview of ANOVA

ANOVA tests the null hypothesis that the population means are *all* equal. The alternative is that they are not all equal. This alternative could be true because all of the means are different or simply because one of them differs from the rest. This is a more complex situation than comparing just two populations. If we reject the null hypothesis, we need to perform some further analysis to draw conclusions about which population means differ from which others.

The computations needed for an ANOVA are more lengthy than those for the t test. For this reason we generally use computer programs to perform the calculations. Automating the calculations frees us from the burden of arithmetic and allows us to concentrate on interpretation. *Complicated computations do not guarantee a valid statistical analysis. We should always start our ANOVA with a careful examination of the data using graphical and numerical summaries.*

EXAMPLE 12.3   In a study of workplace safety, workers were asked to rate the safety of their work environment, and a composite score called the Safety Climate Index (SCI) was calculated.[1] The index is the sum of the responses to 10 different questions about safety. The response for each of these questions is an integer ranging from 0 to 10, so the SCI has values from 0 to 100. The workers were classified

FIGURE 12.3   Histograms and descriptive statistics for the worker safety example.

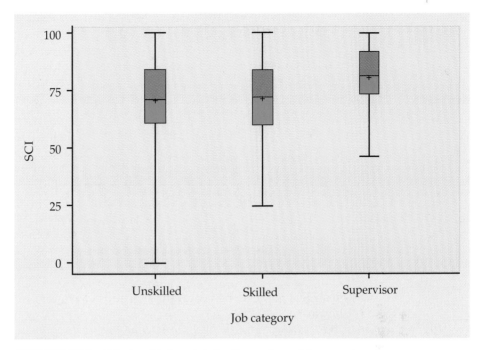

FIGURE 12.4   Side-by-side boxplots for the worker safety example.

according to their job category as unskilled, skilled, and supervisor. Here is a summary of the data:

Job category	n	$\bar{x}$	s
Unskilled workers	448	70.42	18.27
Skilled workers	91	71.21	18.83
Supervisors	51	80.51	14.58

Histograms and descriptive statistics for the three groups of workers are given in Figure 12.3. Note that the heights of the bars in the histograms are percents rather than counts. If we had used counts with the same scale on the y axis, then the bars for the skilled workers and the supervisors would be very small because of the smaller sample sizes in these groups. Figure 12.4 gives side-by-side boxplots for these data. We see that the largest and the smallest possible values are present in the data. The distributions are somewhat skewed toward lower values. Our sample sizes, however, are sufficiently large that we are confident that the sample means are approximately normal.

The three sample means are plotted in Figure 12.5. It appears that the means for the unskilled workers and the skilled workers are similar, while the supervisors have a higher mean. To apply ANOVA in this setting, we view the three samples that we have as three independent random samples from three distinct populations. Each of these populations has a mean and our inference asks questions about these means. We first ask whether or not there is sufficient evidence in the data to conclude that the corresponding population means are not all equal. Our null hypothesis here states that the population

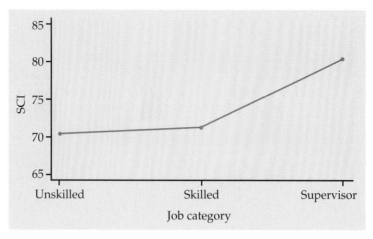

FIGURE 12.5   SCI means for the worker safety example.

mean SCI is the same for all three groups of workers. The alternative is that they are not all the same.

Rejecting the null hypothesis that the means are not all the same using ANOVA is not the same as concluding that all of the means are different from one another. Our inspection of the data for our example suggests that the means for the skilled workers and the unskilled workers may be the same while the mean for the supervisors is higher. The ANOVA null hypothesis can be false in many different ways. Additional analysis is required to distinguish among these possibilities.

contrasts

When there are particular versions of the alternative hypothesis that are of interest, we use **contrasts** to examine them. In our example, we might want to compare the supervisors with all of the other workers. *Note that to use contrasts, it is necessary that the questions of interest be formulated before examining the data. It is cheating to make up these questions after analyzing the data.*

multiple comparisons

If we have no specific relations among the means in mind before looking at the data, we instead use a **multiple-comparisons** procedure to determine which pairs of population means differ significantly. In later sections we will explore both contrasts and multiple comparisons in detail.

The ANOVA model

When analyzing data, the following equation reminds us that we look for an overall pattern and deviations from it:

$$DATA = FIT + RESIDUAL$$

In the regression model of Chapter 10, the FIT was the population regression line, and the RESIDUAL represented the deviations of the data from this line. We now apply this framework to describe the statistical models used in ANOVA. These models provide a convenient way to summarize the assumptions that are the foundation for our analysis. They also give us the necessary notation to describe the calculations needed.

First, recall the statistical model for a random sample of observations from a single normal population with mean μ and standard deviation σ. If the observations are

$$x_1, x_2, \ldots, x_n$$

we can describe this model by saying that the x_j are an SRS from the $N(\mu, \sigma)$ distribution. Another way to describe the same model is to think of the x's varying about their population mean. To do this, write each observation x_j as

$$x_j = \mu + \epsilon_j$$

The ϵ_j are then an SRS from the $N(0, \sigma)$ distribution. Because μ is unknown, the ϵ's cannot actually be observed. This form more closely corresponds to our

$$\text{DATA} = \text{FIT} + \text{RESIDUAL}$$

way of thinking. The FIT part of the model is represented by μ. It is the systematic part of the model, like the line in a regression. The RESIDUAL part is represented by ϵ_j. It represents the deviations of the data from the fit and is due to random, or chance, variation.

There are two unknown parameters in this statistical model: μ and σ. We estimate μ by $\bar{x}$, the sample mean, and σ by s, the sample standard deviation. The differences $e_j = x_j - \bar{x}$ are the sample residuals and correspond to the ϵ_j in the statistical model.

The model for one-way ANOVA is very similar. We take random samples from each of I different populations. The sample size is n_i for the ith population. Let x_{ij} represent the jth observation from the ith population. The I population means are the FIT part of the model and are represented by μ_i. The random variation, or RESIDUAL, part of the model is represented by the deviations ϵ_{ij} of the observations from the means.

THE ONE-WAY ANOVA MODEL

The **one-way ANOVA model** is

$$x_{ij} = \mu_i + \epsilon_{ij}$$

for $i = 1, \ldots, I$ and $j = 1, \ldots, n_i$. The ϵ_{ij} are assumed to be from an $N(0, \sigma)$ distribution. The **parameters of the model** are the population means $\mu_1, \mu_2, \ldots, \mu_I$ and the common standard deviation σ.

Note that the sample sizes n_i may differ, but the standard deviation σ is assumed to be the same in all of the populations. Figure 12.6 pictures this model for $I = 3$. The three population means μ_i are different, but the shapes of the three normal distributions are the same, reflecting the assumption that all three populations have the same standard deviation.

EXAMPLE 12.4   In our worker safety example there are three groups of workers that we want to compare, so $I = 3$. The population means μ_1, μ_2, and μ_3 are the mean SCI values for unskilled workers, for skilled workers, and for supervisors, respectively. The sample sizes n_i are 448, 91, and 51.

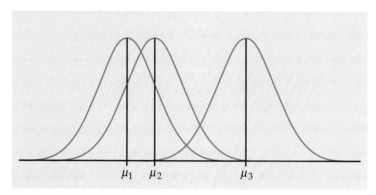

FIGURE 12.6 Model for one-way ANOVA with three groups. The three populations have normal distributions with the same standard deviation.

The observation $x_{1,1}$ is the SCI score for the first unskilled worker. The data for the other unskilled workers are denoted by $x_{1,2}, x_{1,3}, \ldots, x_{1,448}$. Similarly, the data for the other groups have a first subscript indicating the group and a second subscript indicating the worker in the group.

According to our model, the SCI for the first worker is $x_{1,1} = \mu_1 + \epsilon_{1,1}$, where μ_1 is the average for *all* unskilled workers and $\epsilon_{1,1}$ is the chance variation due to this particular worker. The ANOVA model assumes that the ϵ_{ij} are independent and normally distributed (at least approximately) with mean 0 and standard deviation σ. We have clear evidence that the data are not normal in our example. The values are numbers ranging from 0 to 100, and we saw some skewness for all three groups in Figures 12.3 and 12.4. However, because our inference is based on the sample means, which will be approximately normal, we are not overly concerned about these violations of our assumptions.

Formulating a clear definition of the populations being compared with ANOVA can be difficult, as in our example. Often some expert judgment is required, and different consumers of the results may have differing opinions. The workers in this study all worked in the same industry in a particular region. They certainly do represent some larger population of similar workers. We are more confident in generalizing our conclusions to similar populations when the results are clearly significant than when the level of significance just barely passes the standard of $P = 0.05$.

It is common to use numerical subscripts to distinguish the different means, and some software requires that levels of factors in ANOVA be specified as numerical values. An alternative is to use subscripts that suggest the actual groups. In our example, we could replace μ_1, μ_2, and μ_3 by μ_{UN}, μ_{SK}, and μ_{SU}.

Estimates of population parameters

The unknown parameters in the statistical model for ANOVA are the I population means μ_i and the common population standard deviation σ. To estimate

μ_i we use the sample mean for the ith group:

$$\bar{x}_i = \frac{1}{n_i} \sum_{j=1}^{n_i} x_{ij}$$

The residuals $e_{ij} = x_{ij} - \bar{x}_i$ reflect the variation about the sample means that we see in the data.

The ANOVA model assumes that the population standard deviations are all equal. If we have unequal standard deviations, we generally try to transform the data so that they are approximately equal. We might, for example, work with $\sqrt{x_{ij}}$ or $\log x_{ij}$. Fortunately, we can often find a transformation that *both* makes the group standard deviations more nearly equal and also makes the distributions of observations in each group more nearly normal. If the standard deviations are markedly different and cannot be made similar by a transformation, inference requires different methods.

Unfortunately, formal tests for the equality of standard deviations in several groups share the lack of robustness against nonnormality that we noted in Chapter 7 for the case of two groups. Because ANOVA procedures are not extremely sensitive to unequal standard deviations, we do *not* recommend a formal test of equality of standard deviations as a preliminary to the ANOVA. Instead, we will use the following rule as a guideline.

RULE FOR EXAMINING STANDARD DEVIATIONS IN ANOVA

If the largest standard deviation is less than twice the smallest standard deviation, we can use methods based on the assumption of equal standard deviations, and our results will still be approximately correct.[2]

When we assume that the population standard deviations are equal, each sample standard deviation is an estimate of σ. To combine these into a single estimate, we use a generalization of the pooling method introduced in Chapter 7.

POOLED ESTIMATOR OF σ

Suppose we have sample variances $s_1^2, s_2^2, \ldots, s_I^2$ from I independent SRSs of sizes $n_1, n_2, \ldots, n_I$ from populations with common variance σ^2. The **pooled sample variance**

$$s_p^2 = \frac{(n_1 - 1)s_1^2 + (n_2 - 1)s_2^2 + \cdots + (n_I - 1)s_I^2}{(n_1 - 1) + (n_2 - 1) + \cdots + (n_I - 1)}$$

is an unbiased estimator of σ^2. The **pooled standard deviation**

$$s_p = \sqrt{s_p^2}$$

is the estimate of σ.

Pooling gives more weight to groups with larger sample sizes. If the sample sizes are equal, s_p^2 is just the average of the I sample variances. Note that s_p is *not* the average of the I sample standard deviations.

EXAMPLE 12.5 In the worker safety study there are $I = 3$ groups and the sample sizes are $n_1 = 448$, $n_2 = 91$, and $n_3 = 51$. The sample standard deviations are $s_1 = 18.27$, $s_2 = 18.83$, and $s_3 = 14.58$.

Because the largest standard deviation (18.83) is less than twice the smallest ($2 \times 14.58 = 29.16$), our rule indicates that we can use the assumption of equal population standard deviations.

The pooled variance estimate is

$$s_p^2 = \frac{(n_1 - 1)s_1^2 + (n_2 - 1)s_2^2 + (n_3 - 1)s_3^2}{(n_1 - 1) + (n_2 - 1) + (n_3 - 1)}$$

$$= \frac{(447)(18.27)^2 + (90)(18.83)^2 + (50)(14.58)^2}{447 + 90 + 50}$$

$$= \frac{191,745}{587} = 326.7$$

The pooled standard deviation is

$$s_p = \sqrt{326.7} = 18.07$$

This is our estimate of the common standard deviation σ of the SCI scores in the three populations of workers.

Testing hypotheses in one-way ANOVA

Comparison of several means is accomplished by using an F statistic to compare the variation among groups with the variation within groups. We now show how the F statistic expresses this comparison. Calculations are organized in an **ANOVA table,** which contains numerical measures of the variation among groups and within groups.

First we must specify our hypotheses for one-way ANOVA. As usual, I represents the number of populations to be compared.

> **HYPOTHESES FOR ONE-WAY ANOVA**
>
> The **null and alternative hypotheses** for one-way ANOVA are
>
> H_0: $\mu_1 = \mu_2 = \cdots = \mu_I$
>
> H_a: not all of the μ_i are equal

We will now use our worker safety example to illustrate how to do a one-way ANOVA. Because the calculations are generally performed using statistical software, we focus on interpretation of the output.

Descriptives

SCI

	N	Mean	Std. Deviation	Std. Error	95% Confidence Interval for Mean		Minimum	Maximum
					Lower Bound	Upper Bound		
1	448	70.42	18.269	.863	68.73	72.12	0	100
2	91	71.21	18.830	1.974	67.29	75.13	25	100
3	51	80.51	14.577	2.041	76.41	84.61	46	100
Total	590	71.42	18.260	.752	69.94	72.89	0	100

FIGURE 12.7   Software output with descriptive statistics for the worker safety example.

EXAMPLE 12.6   Figure 12.7 gives descriptive statistics generated by SPSS for the ANOVA of the worker safety example. Summaries for each group are given on the first three lines. In addition to the sample size, the mean, and the standard deviation, this output also gives the standard error of the mean and the 95% confidence interval for the mean of each group. Refer to the material in Chapter 7 (pages 450–452) if you need to refresh your memory about these quantities. The three sample means $\bar{x}_i$ given in the output are estimates of the three unknown population means μ_i.

The output gives the estimates of the standard deviations for each group, the s_i, but does not provide s_p, the pooled estimate of the model standard deviation, σ. We could perform the calculation using a calculator as we did in Example 12.5. We will see an easier way to obtain this quantity from the ANOVA table in Figure 12.8. Some software packages report s_p as part of the standard ANOVA output. Sometimes you are not sure whether or not a quantity given by software is what you think it is. *A good way to resolve this dilemma is to do a sample calculation with a simple example to check the numerical results.*

Note that s_p is not the standard deviation given in the Total row of Figure 12.7. This quantity is the standard deviation that we would obtain if we viewed the data as a single sample of 590 workers and ignored the possibility that the group means could be different. As we have mentioned many times before, it is important to use care when reading and interpreting software output.

ANOVA

SCI

	Sum of Squares	df	Mean Square	F	Sig.
Between Groups	4662.233	2	2331.116	7.137	.001
Within Groups	191729.2	587	326.626		
Total	196391.4	589			

FIGURE 12.8   Software output giving the ANOVA table for the worker safety example.

EXAMPLE 12.7   Additional output generated by SPSS for the ANOVA of the worker safety example is given in Figure 12.8. We will discuss some details in the next section. For now, we observe that the results of our significance test are given in the last two columns of the output. The null hypothesis that the three population means are the same is tested by the statistic $F = 7.137$, and the associated P-value is reported as $P = 0.001$. The data provide clear evidence to support the claim that these three groups of workers have different mean SCI values.

The ANOVA table

ANOVA table   The information in an analysis of variance is organized in an **ANOVA table.** To understand the table, it is helpful to think in terms of our

$$DATA = FIT + RESIDUAL$$

view of statistical models. For one-way ANOVA, this corresponds to

$$x_{ij} = \mu_i + \epsilon_{ij}$$

We can think of these three terms as sources of variation. The ANOVA table separates the variation in the data into two parts: that due to the fit and the remainder, which we call residual. Different software packages use different terms for these sources of variation but the basic concept is common to all.

EXAMPLE 12.8   The SPSS output in Figure 12.8 gives the sources of variation in the first column. Here, FIT is called Between Groups, RESIDUAL is called Within Groups, and DATA is the last entry, Total.

In place of Between Groups, some software packages use Model or the name of the factor. Similarly, terms like Error or Residual are frequently used in place of Within Groups.

The Between Groups row in the table corresponds to the FIT term in our DATA = FIT + RESIDUAL way of thinking. It gives information related to the variation **among** group means. In writing ANOVA tables we will use the generic label "groups" or some other term that describes the factor being studied for this row.

variation among groups

The Within Groups row in the table corresponds to the RESIDUAL term in our DATA = FIT + RESIDUAL model. It gives information related to the variation **within** groups. We noted that the term "error" is frequently used for this source of variation, particularly for more general statistical models. This label is most appropriate for experiments in the physical sciences where the observations within a group differ because of measurement error. In business and the biological and social sciences, on the other hand, the within-group variation is often due to the fact that not all firms or plants or people are the same. This sort of variation is not due to errors and is better described as "residual" or "within-group" variation. We will use the generic label "error" for this source of variation in writing ANOVA tables.

variation within groups

Finally, the Total row in the ANOVA table corresponds to the DATA term in our DATA = FIT + RESIDUAL framework. So, for analysis of variance,

$$DATA = FIT + RESIDUAL$$

translates into

$$Total = Between\ Groups + Within\ Groups$$

sum of squares   The second column in the software output given in Figure 12.8 is labeled Sum of Squares. As you might expect, each **sum of squares** is a sum of squared deviations. We use SSG, SSE, and SST for the entries in this column, corresponding to groups, error, and total. Each sum of squares measures a different type of variation. SST measures variation of the data around the overall mean, $x_{ij} - \bar{x}$. Variation of the group means around the overall mean $\bar{x}_i - \bar{x}$ is measured by SSG. Finally, SSE measures variation of each observation around its group mean, $x_{ij} - \bar{x}_i$.

EXAMPLE 12.9   The Sum of Squares column in Figure 12.8 gives the values for the three sums of squares. Verify that SST = SSG + SSE for this example.

This fact is true in general. The total variation is always equal to the among-group variation plus the within-group variation. Note that software output frequently gives many more digits than we need, as in this case. In this example it appears that most of the variation is coming from within groups.

degrees of freedom   Associated with each sum of squares is a quantity called the **degrees of freedom.** Because SST measures the variation of all N observations around the overall mean, its degrees of freedom are DFT = $N - 1$. This is the same as the degrees of freedom for the ordinary sample variance. Similarly, because SSG measures the variation of the I sample means around the overall mean, its degrees of freedom are DFG = $I - 1$. Finally, SSE is the sum of squares of the deviations $x_{ij} - \bar{x}_i$. Here we have N observations being compared with I sample means and DFE = $N - I$.

EXAMPLE 12.10   In our worker safety example, we have $I = 3$ and $N = 590$. Therefore,

$$DFT = N - 1 = 590 - 1 = 589$$
$$DFG = I - 1 = 3 - 1 = 2$$
$$DFE = N - I = 590 - 3 = 587$$

These are the entries in the df column of Figure 12.8.

Note that the degrees of freedom add in the same way that the sums of squares add. That is, DFT = DFG + DFE.

mean square   For each source of variation, the **mean square** is the sum of squares divided by the degrees of freedom. You can verify this by doing the divisions for the values given on the output in Figure 12.8.

SUMS OF SQUARES, DEGREES OF FREEDOM, AND MEAN SQUARES

Sums of squares represent variation present in the data. They are calculated by summing squared deviations. In one-way ANOVA there are three **sources of variation:** groups, error, and total. The sums of squares are related by the formula

$$SST = SSG + SSE$$

Thus, the total variation is composed of two parts, one due to groups and one due to error.

Degrees of freedom are related to the deviations that are used in the sums of squares. The degrees of freedom are related in the same way as the sums of squares:

$$DFT = DFG + DFE$$

To calculate each **mean square,** divide the corresponding sum of squares by its degrees of freedom.

We can use the error mean square to find s_p, the pooled estimate of the **estimate of σ**   parameter σ of our model. It is true in general that

$$s_p^2 = MSE = \frac{SSE}{DFE}$$

In other words, the error mean square is an estimate of the within-group variance, σ^2. The estimate of σ is therefore the square root of this quantity. So,

$$s_p = \sqrt{MSE}$$

EXAMPLE 12.11   From the SPSS output in Figure 12.8 we see that the MSE is reported as 326.626. The pooled estimate of σ is therefore

$$s_p = \sqrt{MSE} = \sqrt{326.626} = 18.07$$

The F test

If H_0 is true, there are no differences among the group means. The ratio MSG/MSE is a statistic that is approximately 1 if H_0 is true and tends to be larger if H_a is true. This is the ANOVA F statistic. In our example,

MSG = 2331.116 and MSE = 326.626, so the ANOVA F statistic is

$$F = \frac{\text{MSG}}{\text{MSE}} = \frac{2331.116}{326.626} = 7.137$$

When H_0 is true, the F statistic has an F distribution that depends upon two numbers: the *degrees of freedom for the numerator* and the *degrees of freedom for the denominator*. These degrees of freedom are those associated with the mean squares in the numerator and denominator of the F statistic. For one-way ANOVA, the degrees of freedom for the numerator are DFG = $I - 1$, and the degrees of freedom for the denominator are DFE = $N - I$. We use the notation $F(I - 1, N - I)$ for this distribution.

The *One-Way ANOVA* applet available on the Web site www.whfreeman.com/ips is an excellent way to see how the value of the F statistic and the P-value depend upon the variability of the data within the groups and the differences between the means. See Exercises 12.56 and 12.57 for use of this applet.

THE ANOVA F TEST

To test the null hypothesis in a one-way ANOVA, calculate the **F statistic**

$$F = \frac{\text{MSG}}{\text{MSE}}$$

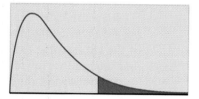

When H_0 is true, the F statistic has the $F(I - 1, N - I)$ distribution. When H_a is true, the F statistic tends to be large. We reject H_0 in favor of H_a if the F statistic is sufficiently large.

The **P-value** of the F test is the probability that a random variable having the $F(I - 1, N - I)$ distribution is greater than or equal to the calculated value of the F statistic.

Tables of F critical values are available for use when software does not give the P-value. Table E in the back of the book contains the F critical values for probabilities $p = 0.100, 0.050, 0.025, 0.010,$ and 0.001. For one-way ANOVA we use critical values from the table corresponding to $I - 1$ degrees of freedom in the numerator and $N - I$ degrees of freedom in the denominator.

EXAMPLE 12.12 In the study of worker safety, we found $F = 7.14$. (Note that it is standard practice to round F statistics to two places after the decimal point.) There were three populations, so the degrees of freedom in the numerator are DFG = $I - 1 = 2$. For this example the degrees of freedom in the denominator are DFE = $N - I = 590 - 3 = 587$. In Table E we first find the column corresponding to 2 degrees of freedom in the numerator. For the degrees of freedom in the denominator, we see that there are entries for 200 and 1000. These entries are very close. To be conservative we use critical values corresponding to 200 degrees of freedom in the denominator since these are slightly larger.

p	Critical value
0.100	2.33
0.050	3.04
0.025	3.76
0.010	4.71
0.001	7.15

We have $F = 7.14$. This is very close to the critical value for $P = 0.001$. Using the table, however, we can conclude only that $P < 0.010$ because our calculated F does not exceed 7.15. (Note that the more accurate calculations performed by software indicated that, in fact, $P < 0.001$.) For this example, we reject H_0 and conclude that the population means are not all the same.

Remember that the F test is always one-sided because any differences among the group means tend to make F large. The ANOVA F test shares the robustness of the two-sample t test. It is relatively insensitive to moderate nonnormality and unequal variances, especially when the sample sizes are similar.

The following display shows the general form of a one-way ANOVA table with the F statistic. The formulas in the sum of squares column can be used for calculations in small problems. There are other formulas that are more efficient for hand or calculator use, but ANOVA calculations are usually done by computer software.

Source	Degrees of freedom	Sum of squares	Mean square	F
Groups	$I - 1$	$\sum_{\text{groups}} n_i (\bar{x}_i - \bar{x})^2$	SSG/DFG	MSG/MSE
Error	$N - I$	$\sum_{\text{groups}} (n_i - 1)s_i^2$	SSE/DFE	
Total	$N - 1$	$\sum_{\text{obs}} (x_{ij} - \bar{x})^2$	SST/DFT	

One other item given by some software for ANOVA is worth noting. For an analysis of variance, we define the **coefficient of determination** as

coefficient of determination

$$R^2 = \frac{\text{SSG}}{\text{SST}}$$

The coefficient of determination plays the same role as the squared multiple correlation R^2 in a multiple regression. We can easily calculate the value from the ANOVA table entries.

EXAMPLE 12.13   The software-generated ANOVA table for the worker safety study is given in Figure 12.8. From that display, we see that SSG = 4662.233 and SST = 196,391.4. The coefficient of determination is

$$R^2 = \frac{\text{SSG}}{\text{SST}} = \frac{4662.233}{196,391.4} = 0.02$$

About 2% of the variation in SCI scores is explained by membership in the groups of workers, unskilled workers, skilled workers, and supervisors. The other 98% of the variation is due to worker-to-worker variation within each of the three groups. We can see this in the histograms of Figure 12.3. Each of the groups has a large amount of variation and there is a substantial amount of overlap in the distributions. *The fact that we have strong evidence (P < 0.001) against the null hypothesis that the three population means are not all the same does not tell us that the distributions of values are far apart.*

12.2   Comparing the Means

Contrasts

The ANOVA *F* test gives a general answer to a general question: Are the differences among observed group means significant? Unfortunately, a small *P*-value simply tells us that the group means are not all the same. It does not tell us specifically which means differ from each other. Plotting and inspecting the means give us some indication of where the differences lie, but we would like to supplement inspection with formal inference.

In the ideal situation, specific questions regarding comparisons among the means are posed before the data are collected. We can answer specific questions of this kind and attach a level of confidence to the answers we give. We now explore these ideas through our worker safety example.

EXAMPLE 12.14   In the worker safety study we compared the SCI scores for three groups of workers: unskilled workers, skilled workers, and supervisors. Let's use $\bar{x}_{\text{UN}}$, $\bar{x}_{\text{SK}}$, and $\bar{x}_{\text{SU}}$ to represent the three sample means and a similar notation for the population means. From Figure 12.7 (page 731) we see that the three sample means are

$$\bar{x}_{\text{UN}} = 70.42, \quad \bar{x}_{\text{SK}} = 71.21, \quad \text{and} \quad \bar{x}_{\text{SU}} = 80.51$$

The null hypothesis we tested was

$$H_0: \mu_{\text{UN}} = \mu_{\text{SK}} = \mu_{\text{SU}}$$

versus the alternative that the three population means are not all the same. We would report these results as $F(2, 587) = 7.14$ with $P < 0.001$. (Note that we have given the degrees of freedom for the *F* statistic in parentheses.) Because the *P*-value is very small, we conclude that the data provide clear evidence that the three population means are not all the same.

Having evidence that the three population means are not the same does not really tell us anything useful. We would really like our analysis to provide us with more specific information. The alternative hypothesis is true if

$$\mu_{UN} \neq \mu_{SK}$$

or if

$$\mu_{UN} \neq \mu_{SU}$$

or if

$$\mu_{SK} \neq \mu_{SU}$$

or if any combination of these statements is true. *When you reject the ANOVA null hypothesis, additional analyses are required to obtain useful results.*

Experts on safety in workplaces would suggest that supervisors face a very different safety environment than the other types of workers. Therefore, a reasonable question to ask is whether or not the supervisors are different from the others. We can take this question and translate it into a testable hypothesis.

EXAMPLE 12.15

To compare the supervisors with the other two groups of workers we construct the following null hypothesis:

$$H_{01}: \frac{1}{2}(\mu_{UN} + \mu_{SK}) = \mu_{SU}$$

We could use the two-sided alternative

$$H_{a1}: \frac{1}{2}(\mu_{UN} + \mu_{SK}) \neq \mu_{SU}$$

but we could also argue that the one-sided alternative

$$H_{a1}: \frac{1}{2}(\mu_{UN} + \mu_{SK}) < \mu_{SU}$$

is appropriate for this problem because we expect the unskilled workers and the skilled workers to have a work environment that is less safe than the supervisors.

In the example above we used H_{01} and H_{a1} to designate the null and alternative hypotheses. The reason for this is that there is a natural additional set of hypotheses that we should examine for this example. We use H_{02} and H_{a2} for these hypotheses.

EXAMPLE 12.16

Do the data provide any evidence to support a conclusion that the unskilled workers and the skilled workers have different mean SCI scores? We translate this question into the following null and alternative hypotheses:

$$H_{02}: \mu_{UN} = \mu_{SK}$$

$$H_{a2}: \mu_{UN} \neq \mu_{SK}$$

Each of H_{01} and H_{02} says that a combination of population means is 0. These combinations of means are called **contrasts.** We use ψ, the Greek letter psi, for contrasts among population means. For comparing the supervisors with the other two groups of workers, we have

contrast ψ

$$\psi_1 = -\frac{1}{2}(\mu_{\mathrm{UN}} + \mu_{\mathrm{SK}}) + \mu_{\mathrm{SU}}$$

$$= (-0.5)\mu_{\mathrm{UN}} + (-0.5)\mu_{\mathrm{SK}} + (1)\mu_{\mathrm{SU}}$$

and for comparing the unskilled workers with the skilled workers

$$\psi_2 = \mu_{\mathrm{UN}} - \mu_{\mathrm{SK}}$$

In each case, the value of the contrast is 0 when H_0 is true. *Note that we have chosen to define the contrasts so that they will be positive when the alternative of interest (what we expect) is true. Whenever possible, this is a good idea because it makes some computations easier.*

sample contrast
A contrast expresses an effect in the population as a combination of population means. To estimate the contrast, form the corresponding **sample contrast** by using sample means in place of population means. Under the ANOVA assumptions, a sample contrast is a linear combination of independent normal variables and therefore has a normal distribution. We can obtain the standard error of a contrast by using the rules for variances given in Section 4.4. Inference is based on t statistics. Here are the details.

CONTRASTS

A **contrast** is a combination of population means of the form

$$\psi = \sum a_i \mu_i$$

where the coefficients a_i have sum 0. The corresponding **sample contrast** is

$$c = \sum a_i \bar{x}_i$$

The **standard error of c** is

$$\mathrm{SE}_c = s_p \sqrt{\sum \frac{a_i^2}{n_i}}$$

To test the null hypothesis

$$H_0: \psi = 0$$

use the **t statistic**

$$t = \frac{c}{\mathrm{SE}_c}$$

with degrees of freedom DFE that are associated with s_p. The alternative hypothesis can be one-sided or two-sided.

> **A level C confidence interval for ψ is**
>
> $$c \pm t^* \mathrm{SE}_c$$
>
> where t^* is the value for the $t(\mathrm{DFE})$ density curve with area C between $-t^*$ and t^*.

Because each $\bar{x}_i$ estimates the corresponding μ_i, the addition rule for means tells us that the mean μ_c of the sample contrast c is ψ. In other words, c is an unbiased estimator of ψ. Testing the hypothesis that a contrast is 0 assesses the significance of the effect measured by the contrast. It is often more informative to estimate the size of the effect using a confidence interval for the population contrast.

EXAMPLE 12.17   In our example the coefficients in the contrasts are $a_1 = -0.5$, $a_2 = -0.5$, $a_3 = 1$ for ψ_1 and $a_1 = 0$, $a_2 = 1$, $a_3 = -1$ for ψ_2, where the subscripts 1, 2, and 3 correspond to UN, SK, and SU. In each case the sum of the a_i is 0. We look at inference for each of these contrasts in turn.

EXAMPLE 12.18   The sample contrast that estimates ψ_1 is

$$c_1 = \bar{x}_{\mathrm{UN}} + \frac{1}{2}(\bar{x}_{\mathrm{SK}} + \bar{x}_{\mathrm{SU}})$$

$$= -(0.5)70.42 - (0.5)71.21 + 80.51 = 9.69$$

with standard error

$$\mathrm{SE}_{c_1} = 18.07\sqrt{\frac{(0.5)^2}{448} + \frac{(-0.5)^2}{91} + \frac{(1)^2}{51}}$$

$$= 2.53$$

The t statistic for testing $H_{01}: \psi_1 = 0$ versus $H_{a1}: \psi_1 > 0$ is

$$t = \frac{c_1}{\mathrm{SE}_{c_1}}$$

$$= \frac{9.69}{2.53} = 3.83$$

Because s_p has 587 degrees of freedom, software using the $t(587)$ distribution gives the one-sided P-value as < 0.0001. If we used Table D, we would conclude that $P < 0.0005$. The P-value is small, so there is strong evidence against H_{01}.

We have evidence to conclude that the mean SCI score for supervisors is higher than the average of the means for the unskilled workers and the skilled workers. The size of the difference can be described with a confidence interval.

EXAMPLE 12.19   To find the 95% confidence interval for ψ_1, we combine the estimate with its margin of error:

$$c_1 \pm t^* \mathrm{SE}_{c_1} = 9.69 \pm (1.984)(2.53)$$
$$= 9.69 \pm 5.02$$

The interval is (4.67, 14.71). We are 95% confident that the difference is between 4.67 and 14.71.

We use the same method for the second contrast.

EXAMPLE 12.20   The second sample contrast, which compares the unskilled workers with the skilled workers, is

$$c_2 = 70.42 - 71.21$$
$$= -0.79$$

with standard error

$$\mathrm{SE}_{c_2} = 18.07 \sqrt{\frac{(1)^2}{448} + \frac{(-1)^2}{91}}$$
$$= 2.08$$

The t statistic for assessing the significance of this contrast is

$$t = \frac{-0.79}{2.08} = 0.36$$

The P-value for the two-sided alternative is 0.706. The data do not provide us with evidence in favor of a difference in population mean SCI scores between unskilled workers and skilled workers.

Note that we have not concluded that there is *no* difference between the population means. A confidence interval will tell us what values of the population difference are compatible with the data.

EXAMPLE 12.21   To find the 95% confidence interval for ψ_2, we combine the estimate with its margin of error:

$$c_2 \pm t^* \mathrm{SE}_{c_2} = -0.79 \pm (1.984)(2.08)$$
$$= -0.79 \pm 4.13$$

The interval is (−4.92, 3.34). With 95% confidence we state that the difference between the population means for these two groups of workers is between −4.92 and 3.34.

Some statistical software packages report the test statistics associated with contrasts as F statistics rather than t statistics. These F statistics are the squares of the t statistics described above.

Contrast Coefficients			
	JOBC		
Contrast	1	2	3
1	-.5	-.5	1
2	1	-1	0

Contrast Tests

	Contrast	Value of Contrast	Std. Error	t	df	Sig. (2-tailed)
SCI	1	9.69	2.736	3.543	587	.000
	2	-.78	2.078	-.378	587	.706

FIGURE 12.9 Software output giving the contrast analysis for the worker safety example.

SPSS output for contrasts is given in Figure 12.9. The results agree with the calculations that we performed in Examples 12.18 and 12.20 except for minor differences due to roundoff error in our calculations. Note that the output does not give the confidence intervals that we calculated in Examples 12.19 and 12.21. These are easily computed, however, from the contrast estimates and standard errors provided in the output.

As with much statistical software output, P-values for significance tests are reported for the two-sided alternative. *If the software you are using gives P-values for the two-sided alternative, and you are using a one-sided alternative, divide the reported P-value by 2.* In our example, we argued that a one-sided alternative was appropriate for the first contrast. The software reported the P-value as 0.000, so we can conclude $P < 0.0005$. Dividing this value by 2 has no effect on the conclusion.

Questions about population means are expressed as hypotheses about contrasts. A contrast should express a specific question that we have in mind when designing the study. When contrasts are formulated before seeing the data, *inference about contrasts is valid whether or not the ANOVA H_0 of equality of means is rejected.* Because the F test answers a very general question, it is less powerful than tests for contrasts designed to answer specific questions. Specifying the important questions before the analysis is undertaken enables us to use this powerful statistical technique.

Multiple comparisons

multiple comparisons

In many studies, specific questions cannot be formulated in advance of the analysis. If H_0 is not rejected, we conclude that the population means are indistinguishable on the basis of the data given. On the other hand, if H_0 is rejected, we would like to know which pairs of means differ. **Multiple-comparisons** methods address this issue. It is important to keep in mind that multiple-comparisons methods are used only *after rejecting* the ANOVA H_0.

EXAMPLE 12.22 Return once more to the worker safety data with three groups of workers. We can make three comparisons between pairs of means: unskilled workers versus skilled workers, unskilled workers versus supervisors, and skilled workers versus supervisors. We can write a t statistic for each of these pairs.

For example, the statistic

$$t_{12} = \frac{\bar{x}_1 - \bar{x}_2}{s_p \sqrt{\dfrac{1}{n_1} + \dfrac{1}{n_2}}}$$

$$= \frac{70.42 - 71.21}{18.07 \sqrt{\dfrac{1}{448} + \dfrac{1}{91}}}$$

$$= -0.38$$

compares populations 1 and 2. The subscripts on t specify which groups are compared. The t statistics for the other two pairs are

$$t_{13} = \frac{\bar{x}_1 - \bar{x}_3}{s_p \sqrt{\dfrac{1}{n_1} + \dfrac{1}{n_3}}}$$

$$= \frac{70.42 - 80.51}{18.07 \sqrt{\dfrac{1}{448} + \dfrac{1}{51}}}$$

$$= -3.78$$

and

$$t_{23} = \frac{\bar{x}_2 - \bar{x}_3}{s_p \sqrt{\dfrac{1}{n_2} + \dfrac{1}{n_3}}}$$

$$= \frac{71.21 - 80.51}{18.07 \sqrt{\dfrac{1}{91} + \dfrac{1}{51}}}$$

$$= -2.94$$

We performed the first calculation when we analyzed the contrast ψ_2 in the previous section, because that contrast was $\mu_1 - \mu_2$. These t statistics are very similar to the pooled two-sample t statistic for comparing two population means, which we described in Chapter 7 (page 499). The difference is that we now have more than two populations, so each statistic uses the pooled estimator s_p from all groups rather than the pooled estimator from just the two groups being compared. This additional information about the common σ increases the power of the tests. The degrees of freedom for all of these statistics are DFE $= 587$, those associated with s_p.

Because we do not have any specific ordering of the means in mind as an alternative to equality, we must use a two-sided approach to the problem of deciding which pairs of means are significantly different.

MULTIPLE COMPARISONS

To perform a **multiple-comparisons procedure,** compute *t* **statistics** for all pairs of means using the formula

$$t_{ij} = \frac{\overline{x}_i - \overline{x}_j}{s_p\sqrt{\dfrac{1}{n_i} + \dfrac{1}{n_j}}}$$

If

$$|t_{ij}| \geq t^{**}$$

we declare that the population means μ_i and μ_j are different. Otherwise, we conclude that the data do not distinguish between them. The value of t^{**} depends upon which multiple comparisons procedure we choose.

One obvious choice for t^{**} is the upper $\alpha/2$ critical value for the t(DFE) distribution. This choice simply carries out as many separate significance tests of fixed level α as there are pairs of means to be compared. The procedure based

LSD method

on this choice is called the **least-significant differences method,** or simply LSD. *LSD has some undesirable properties, particularly if the number of means being compared is large.* Suppose, for example, that there are $I = 20$ groups and we use LSD with $\alpha = 0.05$. There are 190 different pairs of means. If we perform 190 t tests, each with an error rate of 5%, our overall error rate will be unacceptably large. We expect about 5% of the 190 to be significant even if the corresponding population means are the same. Since 5% of 190 is 9.5, we expect 9 or 10 false rejections.

The LSD procedure fixes the probability of a false rejection for each single pair of means being compared. It does not control the overall probability of *some* false rejection among all pairs. Other choices of t^{**} control possible errors in other ways. The choice of t^{**} is therefore a complex problem, and a detailed discussion of it is beyond the scope of this text. Many choices for t^{**} are used in practice. One major statistical package allows selection from a list of over a dozen choices.

Bonferroni method

We will discuss only one of these, called the **Bonferroni method.** Use of this procedure with $\alpha = 0.05$, for example, guarantees that the probability of *any* false rejection among all comparisons made is no greater than 0.05. This is much stronger protection than controlling the probability of a false rejection at 0.05 for *each separate* comparison.

EXAMPLE 12.23   We apply the Bonferroni multiple-comparisons procedure with $\alpha = 0.05$ to the data from the worker safety study. The value of t^{**} for this procedure (from software or special tables) is 2.13. Of the statistics $t_{12} = -0.38$, $t_{13} = -3.78$, and $t_{23} = -2.94$ calculated in the beginning of this section, only t_{13} and t_{23} are significant.

Of course, we prefer to use software for the calculations.

Multiple Comparisons

Dependent Variable: SCI

Bonferroni

(I) JOBC	(J) JOBC	Mean Difference (I-J)	Std. Error	Sig.	95% Confidence Interval	
					Lower Bound	Upper Bound
1	2	-.78	2.078	1.000	-5.77	4.20
	3	-10.09*	2.671	.001	-16.50	-3.67
2	1	.78	2.078	1.000	-4.20	5.77
	3	-9.30*	3.161	.010	-16.89	-1.71
3	1	10.09*	2.671	.001	3.67	16.50
	2	9.30*	3.161	.010	1.71	16.89

*. The mean difference is significant at the .05 level.

FIGURE 12.10   Software output giving the multiple comparison analysis for the worker safety example.

EXAMPLE 12.24   The output generated by SPSS for Bonferroni comparisons appears in Figure 12.10. The software uses an asterisk to indicate that the difference in a pair of means is statistically significant. These results agree with the calculations that we performed in Examples 12.22 and 12.23. Note that each comparison is given twice in the output.

The data in the worker safety study provided a clear result: the supervisors have the highest mean SCI score, and we are unable to see a difference between the unskilled workers and the skilled workers. Unfortunately, this type of clarity does not always emerge from a multiple-comparisons analysis. For example, with three groups, we can (*a*) fail to detect a difference between Groups 1 and 2, (*b*) fail to detect a difference between Groups 2 and 3, and (*c*) conclude that Groups 1 and 3 are not the same. *This kind of apparent contradiction points out dramatically the nature of the conclusions of statistical tests of significance.* The conclusion appears to be illogical. If μ_1 is the same as μ_2 and μ_2 is the same as μ_3, doesn't it follow that μ_1 is the same as μ_3? Logically, the answer must be yes.

Some of the difficulty can be resolved by noting the choice of words used above. In describing the inferences, we talk about failing to detect a difference or concluding that two groups are different. In making logical statements, we say things like "is the same as." There is a big difference between the two modes of thought. Statistical tests ask, "Do we have adequate evidence to distinguish two means?" It is not illogical to conclude that we have sufficient evidence to distinguish μ_1 from μ_3, but not μ_1 from μ_2 or μ_2 from μ_3.

One way to deal with these difficulties of interpretation is to give confidence intervals for the differences. The intervals remind us that the differences are not known exactly. We want to give **simultaneous confidence intervals,** that is, intervals for *all* differences among the population means

simultaneous
confidence
intervals

at once. Again, we must face the problem that there are many competing procedures—in this case, many methods of obtaining simultaneous intervals.

> ## SIMULTANEOUS CONFIDENCE INTERVALS FOR DIFFERENCES BETWEEN MEANS
>
> **Simultaneous confidence intervals** for all differences $\mu_i - \mu_j$ between population means have the form
>
> $$(\bar{x}_i - \bar{x}_j) \pm t^{**} s_p \sqrt{\frac{1}{n_i} + \frac{1}{n_j}}$$
>
> The critical values t^{**} are the same as those used for the multiple-comparisons procedure chosen.

The confidence intervals generated by a particular choice of t^{**} are closely related to the multiple comparisons results for that same method. If one of the confidence intervals includes the value 0, then that pair of means will not be declared significantly different, and vice versa.

EXAMPLE 12.25 The SPSS output for the Bonferroni multiple-comparisons procedure given in Figure 12.10 includes the simultaneous 95% confidence intervals. We can see, for example, that the interval for $\mu_1 - \mu_2$ is −5.77 to 4.20. The fact that the interval includes 0 is consistent with the fact that we failed to detect a difference between these two means using this procedure. Note that the interval for $\mu_2 - \mu_1$ is also provided. This is not really a new piece of information, because it can be obtained from the other interval by reversing the signs and reversing the order, that is, −4.20 to 5.77. So, in fact, we really have only three intervals. Use of the Bonferroni procedure provides us with 95% confidence that *all three* intervals simultaneously contain the true values of the population mean differences.

Software

We have used SPSS to illustrate the analysis of the worker safety data. Other statistical software gives similar output and you should be able to read it without any difficulty.

EXAMPLE 12.26 Research suggests that customers think that a product is of high quality if it is heavily advertised. An experiment designed to explore this idea collected quality ratings (on a 1 to 7 scale) of a new line of take-home refrigerated entrées based on reading a magazine ad. Three groups were compared. The first group's ad included information that would undermine (U) the expected positive association between quality and advertising; the second group's ad contained information that would affirm (A) the association; and the third group was a control (C).[3] The data are given in Table 12.1. Outputs from SAS, Excel, Minitab, and the TI-83 are given in Figure 12.11.

TABLE 12.1

Quality ratings in three groups

Group	Quality ratings																			
Undermine ($n = 55$)	6	5	5	5	4	5	4	6	5	5	5	5	3	3	5	4	5	5	5	4
	5	4	4	5	4	4	4	5	5	5	4	5	5	5	4	4	5	5	4	5
	5	4	4	5	4	3	4	5	5	5	3	4	4	4	4					
Affirm ($n = 36$)	4	6	4	6	5	5	5	6	4	5	5	5	4	6	6	5	5	7	4	6
	6	4	5	4	5	5	6	4	5	5	4	6	4	6	5	5				
Control ($n = 36$)	5	4	5	6	5	7	5	6	7	5	7	5	4	5	4	4	6	6	5	6
	5	5	4	5	5	6	6	6	5	6	6	7	6	6	5	5				

SAS

```
General Linear Models Procedure

Dependent Variable: QUALITY
                                    Sum of            Mean
Source                   DF        Squares          Square   F Value    Pr > F
Model                     2      18.828255        9.414127    15,28    0.0001
Error                   124      76.384343        0.616003
Corrected Total         126      95.212598

              R-Square            C.V.        Root MSE         QUALITY Mean
              0.197750         15.94832         0.7849              4.9213

Level of          -----------QUALITY-----------
        GROUP        N         Mean                  SD

        affirm      36      5.05555556          0.82615960
        control     36      5.41666667          0.87423436
        under       55      4.50909091          0.69048365
```

Excel

	A	B	C	D	E	F	G
1	Anova Single Factor						
2							
3	SUMMARY						
4	*Groups*	*Count*	*Sum*	*Average*	*Variance*		
5	Column 1	55	248	4.509091	0.476768		
6	Column 2	36	182	5.055556	0.68254		
7	Column 3	36	195	5.416667	0.764286		
8							
9							
10	ANOVA						
11	*Source of Variation*	*SS*	*df*	*MS*	*F*	*P-value*	*F crit*
12	Between Groups	18.82825	2	9.414127	15.28261	1.17E-06	3.069289
13	Within Groups	76.38434	124	0.616003			
14							
15	Total	95.2126	126				

Sheet17 **Sheet18** Sheet 1 Sheet 2 Sheet 3

Ready   Sum=0

FIGURE 12.11   SAS, Excel, Minitab, and TI-83 output for the advertising study. *(continued)*

Minitab

```
One-Way Analysis of Variance

Analysis of Variance
Source        DF          SS         MS         F          P
Factor         2      18.828      9.414     15.28     0.0000
Error        124      76.384      0.616
Total        126      95.123
                                    Individual 95% CIs For Mean
                                    Based on Pooled StDev
Level          N       Mean      StDev    ---+---------+---------+---------+---
Undermin      55     4.5091     0.6905    (-----*----)
Affirm        36     5.0556     0.8262               (-----*------)
Control       36     5.4167     0.8742                            (-----*------)
                                          ---+---------+---------+---------+---
Pooled StDev =       0.7849               4.40      4.80      5.20      5.60
```

TI-83

```
One-way ANOVA
 F=15.28260579
 p=1.1673896E-6
 Factor
  df=2
  SS=18.828255
  MS=9.4141275
 Error
  df=124
  SS=76.3843434
  MS=.61600277
 Sxp=.784858439
```

FIGURE 12.11 *(Continued)* SAS, Excel, Minitab, and TI-83 output for the advertising study.

Power*

Recall that the power of a test is the probability of rejecting H_0 when H_a is in fact true. Power measures how likely a test is to detect a specific alternative. When planning a study in which ANOVA will be used for the analysis, it is important to perform power calculations to check that the sample sizes are adequate to detect differences among means that are judged to be important. Power calculations also help evaluate and interpret the results of studies in which H_0 was not rejected. We sometimes find that the power of the test was

---

*This section is optional.

so low against reasonable alternatives that there was little chance of obtaining a significant F.

In Chapter 7 we found the power for the two-sample t test. One-way ANOVA is a generalization of the two-sample t test, so it is not surprising that the procedure for calculating power is quite similar. Here are the steps that are needed:

1. Specify

   (a) an alternative (H_a) that you consider important; that is, values for the true population means $\mu_1, \mu_2, \ldots, \mu_I$;

   (b) sample sizes $n_1, n_2, \ldots, n_I$; usually these will all be equal to the common value n;

   (c) a level of significance α, usually equal to 0.05; and

   (d) a guess at the standard deviation σ.

2. Find the degrees of freedom DFG $= I - 1$ and DFE $= N - I$ and the critical value that will lead to rejection of H_0. This value, which we denote by F^*, is the upper α critical value for the $F(\text{DFG, DFE})$ distribution.

noncentrality parameter

3. Calculate the **noncentrality parameter**[4]

$$\lambda = \frac{\sum n_i (\mu_i - \overline{\mu})^2}{\sigma^2}$$

where $\overline{\mu}$ is a weighted average of the group means,

$$\overline{\mu} = \sum w_i \mu_i$$

and the weights are proportional to the sample sizes,

$$w_i = \frac{n_i}{\sum n_i} = \frac{n_i}{N}$$

4. Find the power, which is the probability of rejecting H_0 when the alternative hypothesis is true, that is, the probability that the observed F is greater than F^*. Under H_a, the F statistic has a distribution known as the **noncentral F distribution.** Using the SAS function for this distribution, we find the power as

noncentral F distribution

$$\text{Power} = 1 - \text{PROBF}(F^*, \text{DFG}, \text{DFE}, \lambda)$$

Note that if the n_i are all equal to the common value n, then $\overline{\mu}$ is the ordinary average of the μ_i and

$$\lambda = \frac{n \sum (\mu_i - \overline{\mu})^2}{\sigma^2}$$

If the means are all equal (the ANOVA H_0), then $\lambda = 0$. The noncentrality parameter measures how unequal the given set of means is. Large λ points to an alternative far from H_0, and we expect the ANOVA F test to have high power. Software makes calculation of the power quite easy, but tables and charts are also available.

EXAMPLE 12.27 Suppose that a study on reading comprehension for three different teaching methods has 10 students in each group. How likely is this study to detect differences in the mean responses that would be viewed as important? A previous study performed in a different setting found sample means of 41, 47, and 44, and the pooled standard deviation was 7. Based on these results, we will use $\mu_1 = 41$, $\mu_2 = 47$, $\mu_3 = 44$, and $\sigma = 7$ in a calculation of power. The n_i are equal, so $\bar{\mu}$ is simply the average of the μ_i:

$$\bar{\mu} = \frac{41 + 47 + 44}{3} = 44$$

The noncentrality parameter is therefore

$$\lambda = \frac{n \sum (\mu_i - \bar{\mu})^2}{\sigma^2}$$

$$= \frac{(10)[(41 - 44)^2 + (47 - 44)^2 + (44 - 44)^2]}{49}$$

$$= \frac{(10)(18)}{49} = 3.67$$

Because there are three groups with 10 observations per group, DFG = 2 and DFE = 27. The critical value for $\alpha = 0.05$ is $F^* = 3.35$. The power is therefore

$$1 - \text{PROBF}(3.35, 2, 27, 3.67) = 0.3486$$

The chance that we reject the ANOVA H_0 at the 5% significance level is only about 35%.

If the assumed values of the μ_i in this example describe differences among the groups that the experimenter wants to detect, then we would want to use more than 10 subjects per group. Although H_0 is assumed to be false, the chance of rejecting it is only about 35%. This chance can be increased to acceptable levels by increasing the sample sizes.

EXAMPLE 12.28 To decide on an appropriate sample size for the experiment described in the previous example, we repeat the power calculation for different values of n, the number of subjects in each group. Here are the results:

n	DFG	DFE	F^*	λ	Power
20	2	57	3.16	7.35	0.65
30	2	87	3.10	11.02	0.84
40	2	117	3.07	14.69	0.93
50	2	147	3.06	18.37	0.97
100	2	297	3.03	36.73	≈ 1

With $n = 40$ the experimenters have a 93% chance of rejecting H_0 with $\alpha = 0.05$ and thereby demonstrating that the groups have different means. In the long run, 93 out of every 100 such experiments would reject H_0 at the $\alpha = 0.05$ level of significance. Using 50 subjects per group increases the chance of finding significance to 97%. With 100 subjects per group, the experimenters are virtually certain to reject H_0. The exact power for $n = 100$ is 0.99989. In most real-life situations the additional cost of increasing the sample size from 50 to 100 subjects per group would not be justified by the relatively small increase in the chance of obtaining statistically significant results.

CHAPTER 12 | Summary

One-way analysis of variance (ANOVA) is used to compare several population means based on independent SRSs from each population. The populations are assumed to be normal with possibly different means and the same standard deviation.

To do an analysis of variance, first compute sample means and standard deviations for all groups. Side-by-side boxplots give an overview of the data. Examine normal quantile plots (either for each group separately or for the residuals) to detect outliers or extreme deviations from normality. Compute the ratio of the largest to the smallest sample standard deviation. If this ratio is less than 2 and the normal quantile plots are satisfactory, ANOVA can be performed.

The **null hypothesis** is that the population means are *all* equal. The **alternative hypothesis** is true if there are *any* differences among the population means.

ANOVA is based on separating the total variation observed in the data into two parts: variation **among group means** and variation **within groups.** If the variation among groups is large relative to the variation within groups, we have evidence against the null hypothesis.

An **analysis of variance table** organizes the ANOVA calculations. **Degrees of freedom, sums of squares, and mean squares** appear in the table. The **F statistic** and its **P-value** are used to test the null hypothesis.

Specific questions formulated before examination of the data can be expressed as **contrasts.** Tests and confidence intervals for contrasts provide answers to these questions.

If no specific questions are formulated before examination of the data and the null hypothesis of equality of population means is rejected, **multiple-comparisons** methods are used to assess the statistical significance of the differences between pairs of means.

The **power** of the F test depends upon the sample sizes, the variation among population means, and the within-group standard deviation.

CHAPTER 12 | Exercises

12.1   For each of the following, explain what is wrong and why.

(a) Use one-way ANOVA when the response variable has only two possible values.

(b) You cannot use one-way ANOVA when there are more than three means to be compared.

(c) The pooled estimate s_p is a parameter of the ANOVA model.

12.2 For each of the following, explain what is wrong and why.

(a) The ANOVA F statistic tests the null hypothesis that the three sample means are equal.

(b) The mean squares in an ANOVA table will add, that is, MST = MSG + MSE.

(c) Within-group variation is the variation in the data due to the differences in the sample means.

12.3 A study compared 4 groups with 11 observations per group. An F statistic of 3.52 was reported.

(a) Give the degrees of freedom for this statistic and the entries from Table E that correspond to this distribution.

(b) Sketch a picture of this F distribution with the information from the table included.

(c) Based on the table information, how would you report the P-value?

(d) Can you conclude that all pairs of means are different? Explain your answer.

12.4 For each of the following situations, state how large the F statistic needs to be for rejection of the null hypothesis at the 0.05 level.

(a) Compare 6 groups with 2 observations per group.

(b) Compare 6 groups with 4 observations per group.

(c) Compare 6 groups with 11 observations per group.

(d) Summarize what you have learned about F distributions from this exercise.

12.5 For each of the following situations, find the F statistic and the degrees of freedom. Then draw a sketch of the distribution under the null hypothesis and shade in the portion corresponding to the P-value. State how you would report the P-value.

(a) Compare 5 groups with 8 observations per group, MSE = 50, and MSG = 57.

(b) Compare 3 groups with 7 observations per group, SSG = 40, and SSE = 90.

12.6 For each of the following situations, draw a picture of the ANOVA model similar to Figure 12.6 (page 728). Use numerical values for the μ_i. To sketch the normal curves, you may want to review the 68–95–99.7 rule on page 71.

(a) $\mu_1 = 10$, $\mu_2 = 25$, $\mu_3 = 24$, and $\sigma = 5$.

(b) $\mu_1 = 10$, $\mu_2 = 20$, $\mu_3 = 30$, $\mu_4 = 30.1$, and $\sigma = 5$.

(c) $\mu_1 = 10$, $\mu_2 = 25$, $\mu_3 = 24$, and $\sigma = 2$.

12.7 For each of the following situations, identify the response variable and the populations to be compared, and give I, the n_i, and N.

(a) Last semester, an alcohol awareness program was conducted for three groups of students at an eastern university. Follow-up questionnaires were sent to the participants two months after each presentation. There were 220 responses from students in an elementary statistics course, 145 from a health and safety course, and 76 from a cooperative housing unit. One of the questions was "Did you discuss the presentation with any of your friends?" The answers were rated on a five-point scale with 1 corresponding to "not at all" and 5 corresponding to "a great deal."

(b) A study examined the effects of consuming three different varieties of onions on blood cholesterol levels. Five cross-bred (Large White by Landrace) pigs were randomly assigned to each treatment.

(c) A researcher for a video game developer wanted to evaluate a prototype of a new game. Free copies were given to 25 eighth-graders, 25 third-year high school students, and 25 second-year college students who were regular video game players. Two weeks later they were asked to tell how much they liked the game on a 1 to 10 scale.

12.8 For each of the following situations, identify the response variable and the populations to be compared, and give I, the n_i, and N.

(a) The effects of six different treatments designed to make fabrics stronger were compared. A batch of 120 samples of cloth was available for the experiment, and an equal number were randomly assigned to each of the treatments. The breaking strength of each cloth sample was measured.

(b) A waiter designed a study to see the effects of his behaviors on the amount of tips that he received. For some customers, he would tell a joke; for others, he would describe two of the food items as being particularly good that night; and for others, he would behave normally. Using a table of random numbers, he assigned equal numbers of his next 30 customers to his different behaviors.

(c) A supermarket wants to compare the effects of providing free samples of cheddar cheese on sales. An experiment will be conducted from 5:00 p.m. to 6:00 p.m. for the next 20 weekdays. On each day, customers will be offered one of the following: a small cube of cheese pierced by a toothpick, a small slice of cheese on a cracker, a cracker with no cheese, or nothing.

12.9 Refer to Exercise 12.7. For each situation, give the following:

(a) Degrees of freedom for the model, for error, and for the total.

(b) Null and alternative hypotheses.

(c) Numerator and denominator degrees of freedom for the F statistic.

12.10 Refer to Exercise 12.8. For each situation, give the following:

(a) Degrees of freedom for the model, for error, and for the total.

(b) Null and alternative hypotheses.

(c) Numerator and denominator degrees of freedom for the F statistic.

12.11 Refer to Exercise 12.7. For each situation, discuss the method of obtaining the data and how this would affect the extent to which the results could be generalized.

12.12 Refer to Exercise 12.8. For each situation, discuss the method of obtaining the data and how this would affect the extent to which the results could be generalized.

12.13 An experiment was run to compare three groups. The sample sizes were 20, 18, and 15, and the corresponding estimated standard deviations were 220, 190, and 200.

(a) Is it reasonable to use the assumption of equal standard deviations when we analyze these data? Give a reason for your answer.

(b) Give the values of the variances for the three groups.

(c) Find the pooled variance.

(d) What is the value of the pooled standard deviation?

12.14 An experiment was run to compare four groups. The sample sizes were 30, 22, 180, and 25, and the corresponding estimated standard deviations were 20, 21, 10, and 19.

(a) Is it reasonable to use the assumption of equal standard deviations when we analyze these data? Give a reason for your answer.

(b) Give the values of the variances for the four groups.

(c) Find the pooled variance.

(d) What is the value of the pooled standard deviation?

(e) Explain why your answer in part (c) is much closer to the standard deviation for the third group than to any of the other standard deviations.

12.15 Sleep deprivation experienced by physicians during residency training and the possible negative consequences are of concern to many in the health care community. One study of 33 resident anesthesiologists compared their changes from baseline in reaction times on four tasks.[5] Under baseline conditions, the physicians reported getting an average of 7.04 hours of sleep. While on duty, however, the average was 1.66 hours. For each of the tasks the researchers reported a statistically significant increase in the reaction time when the residents were working in a state of sleep deprivation.

(a) If each task is analyzed separately as the researchers did in their report, what is the appropriate statistical method to use? Explain your answer.

(b) Is it appropriate to use a one-way ANOVA with $I = 4$ to analyze these data? Explain why or why not.

12.16 The Americans with Disabilities Act (ADA) requires that students with learning disabilities (LD) and/or attention deficit disorder (ADD) be given certain accommodations when taking examinations. One study designed to assess the effects of these accommodations examined the relationship between end-of-term grades and the number of accommodations given.[6] The researchers reported the mean grades with sample sizes and standard deviations versus the number of accommodations in a table similar to this:

Accommodations	Mean grade	n	s
0	2.7894	160	0.85035
1	2.8605	38	0.83068
2	2.5757	37	0.82745
3	2.6286	7	1.03072
4	2.4667	3	1.66233
Total	2.7596	245	0.85701

(a) Plot the means versus the number of accommodations. Is there a pattern evident?

(b) A large number of digits are reported for the means and the standard deviations. Do you think that all of these are necessary? Give reasons for your answer and describe how you would report these results.

(c) Should we pool to obtain an estimate of an assumed standard deviation for these data? Explain your answer and give the pooled estimate if your answer is yes.

(d) The small numbers of observations with 3 or 4 accommodations lead to estimates that are highly variable in these groups compared with the other groups. Inclusion of groups with relatively few observations in an ANOVA can also lead to low power. We could eliminate these two levels from the analysis or we could combine them with the 37 observations in the group above to form a new group with 2 or more accommodations. Which of these options would you prefer? Give reasons for your answer.

(e) The 245 grades reported in the table were from a sample of 61 students who completed three, four, or five courses during a spring term at one college and who were qualified to receive accommodations. Students in the sample were self-identified, in the sense that they had to request qualification. Even when qualified, some students choose not to request accommodations for some or all of their courses. Based on these facts, would you advise that ANOVA methods be used for these data? Explain your answer. (The authors did not present the results of an ANOVA in their publication.)

(f) To what extent do you think the results of this study can be generalized to other settings? Give reasons for your answer.

(g) Most reasonable approaches to the analysis of these data would conclude that the data fail to provide evidence that the number of accommodations is related to the mean grades. Does this imply that the accommodations are not needed or does it suggest that they are effective? Discuss your answer.

12.17 Do people from different cultures experience emotions differently? One study designed to examine this question collected data from 416 college students from five different cultures.[7] The participants were asked to record, on a 1 (never) to 7 (always) scale, how much of the time they typically felt eight specific emotions. These were averaged to produce the global emotion score for each participant. Here is a summary of this measure:

Culture	n	Mean (s)
European American	46	4.39 (1.03)
Asian American	33	4.35 (1.18)
Japanese	91	4.72 (1.13)
Indian	160	4.34 (1.26)
Hispanic American	80	5.04 (1.16)

(a) From the information given do you think that we need to be concerned that a possible lack of normality in the data will invalidate the conclusions that we might draw using ANOVA to analyze the data? Give reasons for your answer.

(b) Is it reasonable to used a pooled standard deviation for these data? Why or why not?

(c) The ANOVA F statistic was reported as 5.69. Give the degrees of freedom and either an approximate (from a table) or an exact (from software) P-value. Sketch a picture of the F distribution that illustrates the P-value. What do you conclude?

(d) Without doing any additional formal analysis, describe the pattern in the means that appears to be responsible for your conclusion in part (c). Are there pairs of means that are quite similar?

12.18 Refer to the previous exercise. The experimenters also measured emotions in some different ways. For a period of a week, each participant carried a device that sounded an alarm at random times during a 3-hour interval 5 times a day. When the alarm sounded, participants recorded several mood ratings indicating their emotions for the time immediately preceding the alarm. These responses were combined to form two variables: frequency, the number of emotions recorded, expressed as a percent; and intensity, an average of the intensity scores measured on a scale of 0 to 6. At the end of the 1-week experimental period, the subjects were asked to recall the percent of time that they experienced different emotions. This variable was called "recall." Here is a summary of the results:

Culture	n	Frequency Mean (s)	Intensity Mean (s)	Recall Mean (s)
European American	46	82.87 (18.26)	2.79 (0.72)	49.12 (22.33)
Asian American	33	72.68 (25.15)	2.37 (0.60)	39.77 (23.24)
Japanese	91	73.36 (22.78)	2.53 (0.64)	43.98 (22.02)
Indian	160	82.71 (17.97)	2.87 (0.74)	49.86 (21.60)
Hispanic American	80	92.25 (8.85)	3.21 (0.64)	59.99 (24.64)
F statistic		11.89	13.10	7.06

Note that the convention of giving the standard deviations in parentheses after the means saves a great deal of space in a table such as this.

(a) For each response variable state whether or not it is reasonable to use a pooled standard deviation to analyze these data. Give reasons for your answer.

(b) Give the degrees of freedom for the F statistics and find the associated P-values. Summarize what you can conclude from these ANOVA analyses.

(c) Summarize the means paying particular attention to similarities and differences across cultures and across variables. Include the means from the previous exercise in your summary.

(d) The European American and Asian American subjects were from the University of Illinois, the Japanese subjects were from two universities in Tokyo, the Indian subjects were from eight universities in or near Calcutta, and the Hispanic American subjects were from California State University at Fresno. Participants were paid \$25 or an equivalent monetary incentive for the Japanese and Indians. Ads were posted on or near the campuses to recruit volunteers for the study. Discuss how these facts influence your conclusions and the extent to which you would generalize the results.

(e) The percent of female students in the samples were: European American, 83%; Asian American, 67%; Japanese, 63%; Indian, 64%; and Hispanic American, 79%. Use a chi-square test to compare these proportions (see Section 9.2) and discuss how this information influences your interpretation of the results that you have found in this exercise.

12.19 Does bread lose its vitamins when stored? Small loaves of bread were prepared with flour that was fortified with a fixed amount of vitamins. After baking, the vitamin C content of two loaves was measured. Another two loaves were baked at the same time, stored for one day, and then the vitamin C content was measured. In a similar manner, two loaves were stored for three, five, and seven days before measurements were taken. The units are milligrams of vitamin C per hundred grams of flour (mg/100 g).[8] Here are the data:

Condition	Vitamin C (mg/100 g)	
Immediately after baking	47.62	49.79
One day after baking	40.45	43.46
Three days after baking	21.25	22.34
Five days after baking	13.18	11.65
Seven days after baking	8.51	8.13

(a) Give a table with sample size, mean, standard deviation, and standard error for each condition.

(b) Perform a one-way ANOVA for these data. Be sure to state your hypotheses, the test statistic with degrees of freedom, and the P-value.

(c) Summarize the data and the means with a plot. Use the plot and the ANOVA results to write a short summary of your conclusions.

12.20 Refer to the previous exercise. Use the Bonferroni or another multiple comparisons procedure to compare the group means. Summarize the results.

12.21 Refer to Exercise 12.19. Measurements of the amounts of vitamin A (β-carotene) and vitamin E in each loaf are given below. Use an analysis of variance to study the data for each of these vitamins.

Condition	Vitamin A (mg/100 g)		Vitamin E (mg/100 g)	
Immediately after baking	3.36	3.34	94.6	96.0
One day after baking	3.28	3.20	95.7	93.2
Three days after baking	3.26	3.16	97.4	94.3
Five days after baking	3.25	3.36	95.0	97.7
Seven days after baking	3.01	2.92	92.3	95.1

12.22 Refer to the previous exercise.

(a) Explain why it is inappropriate to perform a multiple-comparisons analysis for the vitamin E data.

(b) Perform the Bonferroni or another multiple-comparisons procedure for the vitamin A data and summarize the results.

12.23 Refer to Exercises 12.19 to 12.22. Write a report summarizing what happens to vitamins A, C, and E after bread is baked. Include appropriate statistical inference results and graphs.

12.24 Refer to the corporate advertising data in Exercise 7.72 (page 509). Find the two-sample pooled t statistic. Then formulate the problem as an ANOVA and report the results of this analysis. Verify that $F = t^2$.

12.25 Any pooled two-sample t problem can be run as a one-way ANOVA with $I = 2$. Explain why it is inappropriate to analyze the data using contrasts or multiple-comparisons procedures in this setting.

12.26 The air in poultry-processing plants often contains fungus spores. If the ventilation is inadequate, this can affect the health of the workers. To measure the presence of spores, air samples are pumped to an agar plate, and "colony-forming units (CFUs)" are counted after an incubation period. Here are data from the "kill room" of a plant that slaughters 37,000 turkeys per day, taken at four seasons of the year. The units are CFUs per cubic meter of air.[9]

Fall	Winter	Spring	Summer
1231	384	2105	3175
1254	104	701	2526
1088	97	842	1090

(a) Examine the data using graphs and descriptive measures. How do airborne fungus spores vary with the seasons?

(b) Is the effect of season statistically significant?

12.27 Different varieties of the tropical flower *Heliconia* are fertilized by different species of hummingbirds. Over time, the lengths of the flowers and the form of the hummingbirds' beaks have evolved to match each other. Here are data on the lengths in millimeters of three varieties of these flowers on the island of Dominica:[10]

H. bihai							
47.12	46.75	46.81	47.12	46.67	47.43	46.44	46.64
48.07	48.34	48.15	50.26	50.12	46.34	46.94	48.36

H. caribaea red							
41.90	42.01	41.93	43.09	41.47	41.69	39.78	40.57
39.63	42.18	40.66	37.87	39.16	37.40	38.20	38.07
38.10	37.97	38.79	38.23	38.87	37.78	38.01	

H. caribaea yellow							
36.78	37.02	36.52	36.11	36.03	35.45	38.13	37.1
35.17	36.82	36.66	35.68	36.03	34.57	34.63	

Do a complete analysis that includes description of the data and a significance test to compare the mean lengths of the flowers for the three species.

12.28 Refer to Exercise 12.26. There is not sufficient information to examine the distributions in detail, but it is not unreasonable to expect count data such as these to be skewed. Reanalyze the data after taking logs of the CFU counts.

Summarize your work and compare the results you have found here with what you obtained in Exercise 12.26.

12.29 The distributions of the flower lengths in Exercise 12.27 are somewhat skewed. Take logs of the lengths and reanalyze the data. Write a summary of your results and include a comparison with the results you found in Exercise 12.27.

12.30 Do poets die young? According to William Butler Yeats, "She is the Gaelic muse, for she gives inspiration to those she persecutes. The Gaelic poets die young, for she is restless, and will not let them remain long on earth." One study designed to investigate this question examined the age at death for writers from different cultures and genders.[11] Three categories of writers examined were novelists, poets, and nonfiction writers. The ages at death for female writers in these categories from North America are given in Table 12.2. Most of the writers are from the United States, but Canadian and Mexican writers are also included.

(a) Use graphical and numerical methods to describe the data.

(b) Examine the assumptions necessary for ANOVA. Summarize your findings.

(c) Run the ANOVA and report the results.

(d) Use a contrast to compare the poets with the two other types of writers. Do you think that the quote from Yeats justifies the use of a one-sided alternative for examining this contrast? Explain your answer.

(e) Use another contrast to compare the novelists with the nonfiction writers. Explain your choice for an alternative hypothesis for this contrast.

(f) Use a multiple-comparisons procedure to compare the three means. How do the conclusions from this approach compare with those using the contrasts?

TABLE 12.2

Age at death for women writers

Type	Age at death														
Novels (n = 67)	57	90	67	56	90	72	56	90	80	74	73	86	53	72	86
	82	74	60	79	80	79	77	64	72	88	75	79	74	85	71
	78	57	54	50	59	72	60	77	50	49	73	39	73	61	90
	77	57	72	82	54	62	74	65	83	86	73	79	63	72	85
	91	77	66	75	90	35	86								
Poems (n = 32)	88	69	78	68	72	60	50	47	74	36	87	55	68	75	78
	85	69	38	58	51	72	58	84	30	79	90	66	45	70	48
	31	43													
Nonfiction (n = 24)	74	86	87	68	76	73	63	78	83	86	40	75	90	47	91
	94	61	83	75	89	77	86	66	97						

12.31 Kudzu is a plant that was imported to the United States from Japan and now covers over seven million acres in the South. The plant contains chemicals called isoflavones that have been shown to have beneficial effects on bones. One study used three groups of rats to compare a control group with rats that were fed either a low dose or a high dose of isoflavones from kudzu.[12] One of the outcomes examined was the bone mineral density in the femur (in grams per square centimeter). Here are the data:

Treatment	Bone mineral density (g/cm^2)							
Control	0.228	0.207	0.234	0.220	0.217	0.228	0.209	0.221
	0.204	0.220	0.203	0.219	0.218	0.245	0.210	
Low dose	0.211	0.220	0.211	0.233	0.219	0.233	0.226	0.228
	0.216	0.225	0.200	0.208	0.198	0.208	0.203	
High dose	0.250	0.237	0.217	0.206	0.247	0.228	0.245	0.232
	0.267	0.261	0.221	0.219	0.232	0.209	0.255	

(a) Use graphical and numerical methods to describe the data.

(b) Examine the assumptions necessary for ANOVA. Summarize your findings.

(c) Run the ANOVA and report the results.

(d) Use a multiple-comparisons method to compare the three groups.

(e) Write a short report explaining the effect of kudzu isoflavones on the femur of the rat.

12.32 If a supermarket product is offered at a reduced price frequently, do customers expect the price of the product to be lower in the future? This question was examined by researchers in a study conducted on students enrolled in an introductory management course at a large midwestern university. For 10 weeks 160 subjects received information about the products. The treatment conditions corresponded to the number of promotions (1, 3, 5, or 7) that were described during this 10-week period. Students were randomly assigned to four groups.[13] Table 12.3 (page 762) gives the data.

(a) Make a normal quantile plot for the data in each of the four treatment groups. Summarize the information in the plots and draw a conclusion regarding the normality of these data.

(b) Summarize the data with a table containing the sample size, mean, standard deviation, and standard error for each group.

(c) Is the assumption of equal standard deviations reasonable here? Explain why or why not.

(d) Run the one-way ANOVA. Give the hypotheses tested, the test statistic with degrees of freedom, and the P-value. Summarize your conclusion.

12.33 Refer to the previous exercise. Use the Bonferroni or another multiple-comparisons procedure to compare the group means. Summarize the results and support your conclusions with a graph of the means.

TABLE 12.3

Price promotion data

Number of promotions	Expected price (dollars)									
1	3.78	3.82	4.18	4.46	4.31	4.56	4.36	4.54	3.89	4.13
	3.97	4.38	3.98	3.91	4.34	4.24	4.22	4.32	3.96	4.73
	3.62	4.27	4.79	4.58	4.46	4.18	4.40	4.36	4.37	4.23
	4.06	3.86	4.26	4.33	4.10	3.94	3.97	4.60	4.50	4.00
3	4.12	3.91	3.96	4.22	3.88	4.14	4.17	4.07	4.16	4.12
	3.84	4.01	4.42	4.01	3.84	3.95	4.26	3.95	4.30	4.33
	4.17	3.97	4.32	3.87	3.91	4.21	3.86	4.14	3.93	4.08
	4.07	4.08	3.95	3.92	4.36	4.05	3.96	4.29	3.60	4.11
5	3.32	3.86	4.15	3.65	3.71	3.78	3.93	3.73	3.71	4.10
	3.69	3.83	3.58	4.08	3.99	3.72	4.41	4.12	3.73	3.56
	3.25	3.76	3.56	3.48	3.47	3.58	3.76	3.57	3.87	3.92
	3.39	3.54	3.86	3.77	4.37	3.77	3.81	3.71	3.58	3.69
7	3.45	3.64	3.37	3.27	3.58	4.01	3.67	3.74	3.50	3.60
	3.97	3.57	3.50	3.81	3.55	3.08	3.78	3.86	3.29	3.77
	3.25	3.07	3.21	3.55	3.23	2.97	3.86	3.14	3.43	3.84
	3.65	3.45	3.73	3.12	3.82	3.70	3.46	3.73	3.79	3.94

12.34 Do piano lessons improve the spatial-temporal reasoning of preschool children? In Exercise 7.77 (page 511) we examined this question by comparing the change scores (after treatment minus before treatment) of 34 children who took piano lessons with the scores of 44 children who did not. The latter group actually contained three groups of children: 10 were given singing lessons, 20 had some computer instruction, and 14 received no extra lessons. The data appear in Table 12.4.

TABLE 12.4

Piano lesson data

Lessons	Scores									
Piano	2	5	7	−2	2	7	4	1	0	7
	3	4	3	4	9	4	5	2	9	6
	0	3	6	−1	3	4	6	7	−2	7
	−3	3	4	4						
Singing	1	−1	0	1	−4	0	0	1	0	−1
Computer	0	1	1	−3	−2	4	−1	2	4	2
	2	2	−3	−3	0	2	0	−1	3	−1
None	5	−1	7	0	4	0	2	1	−6	0
	2	−1	0	−2						

(a) Make a table giving the sample size, the mean, the standard deviation, and the standard error for each group.

(b) Analyze the data using one-way analysis of variance. State the null and alternative hypotheses, the test statistic with degrees of freedom, the P-value, and your conclusion.

12.35   Refer to the previous exercise. Use the Bonferroni or another multiple-comparisons procedure to compare the group means. Summarize the results and support your conclusions with a graph of the means.

12.36   The researchers in Exercise 12.34 based their research on a biological argument for a causal link between music and spatial-temporal reasoning. Therefore, it is natural to test the contrast that compares the mean of the piano lesson group with the average of the three other means. Construct this contrast, perform the significance test, and summarize the results. Note that this is not the test that we performed in Exercise 7.77 (page 511), where we did not differentiate among the three groups who did not receive piano instruction.

12.37   Recommendations regarding how long infants in developing countries should be breast-fed are controversial. If the nutritional quality of the breast milk is inadequate because the mothers are malnourished, then there is risk of inadequate nutrition for the infant. On the other hand, the introduction of other foods carries the risk of infection from contamination. Further complicating the situation is the fact that companies that produce infant formulas and other foods benefit when these foods are consumed by large numbers of customers. One question related to this controversy concerns the amount of energy intake for infants who have other foods introduced into the diet at different ages. Part of one study compared the energy intakes, measured in kilocalories per day (kcal/d), for infants who were breast-fed exclusively for 4, 5, or 6 months.[14] Here are the data:

Breast-fed for:	Energy intake (kcal/d)									
4 months	499	620	469	485	660	588	675	517	649	209
	404	738	628	609	617	704	558	653	548	
5 months	490	395	402	177	475	617	616	587	528	518
	370	431	518	639	368	538	519	506		
6 months	585	647	477	445	485	703	528	465		

(a) Make a table giving the sample size, mean, and standard deviation for each group of infants. Is it reasonable to pool the variances?

(b) Run the analysis of variance. Report the F statistic with its degrees of freedom and P-value. What do you conclude?

12.38   Refer to the previous exercise.

(a) Examine the residuals. Is the normality assumption reasonable for these data?

(b) Explain why you do not need to use a multiple-comparisons procedure for these data.

12.39 Many studies have suggested that there is a link between exercise and healthy bones. Exercise stresses the bones and this causes them to get stronger. One study examined the effect of jumping on the bone density of growing rats.[15] There were three treatments: a control with no jumping, a low-jump condition (the jump height was 30 centimeters), and a high-jump condition (60 centimeters). After 8 weeks of 10 jumps per day, 5 days per week, the bone density of the rats (expressed in mg/cm^3) was measured. Here are the data:

Group	Bone density (mg/cm^3)									
Control	611	621	614	593	593	653	600	554	603	569
Low jump	635	605	638	594	599	632	631	588	607	596
High jump	650	622	626	626	631	622	643	674	643	650

(a) Make a table giving the sample size, mean, and standard deviation for each group of rats. Is it reasonable to pool the variances?

(b) Run the analysis of variance. Report the F statistic with its degrees of freedom and P-value. What do you conclude?

12.40 Refer to the previous exercise.

(a) Examine the residuals. Is the normality assumption reasonable for these data?

(b) Use the Bonferroni or another multiple-comparisons procedure to determine which pairs of means differ significantly. Summarize your results in a short report. Be sure to include a graph.

12.41 Iron-deficiency anemia is the most common form of malnutrition in developing countries, affecting about 50% of children and women and 25% of men. Iron pots for cooking foods had traditionally been used in many of these countries, but they have been largely replaced by aluminum pots, which are cheaper and lighter. Some research has suggested that food cooked in iron pots will contain more iron than food cooked in other types of pots. One study designed to investigate this issue compared the iron content of some Ethiopian foods cooked in aluminum, clay, and iron pots.[16] One of the foods was *yesiga wet'*, beef cut into small pieces and prepared with several Ethiopian spices. The iron content of four samples of *yesiga wet'* cooked in each of the three types of pots is given below. The units are milligrams of iron per 100 grams of cooked food.

Type of pot	Iron (mg/100 g food)			
Aluminum	1.77	2.36	1.96	2.14
Clay	2.27	1.28	2.48	2.68
Iron	5.27	5.17	4.06	4.22

(a) Make a table giving the sample size, mean, and standard deviation for each type of pot. Is it reasonable to pool the variances? Note that with the small sample sizes in this experiment, we expect a large amount of variability in the sample standard deviations.

(b) Run the analysis of variance. Report the F statistic with its degrees of freedom and P-value. What do you conclude?

12.42   Refer to the previous exercise.

(a) Examine the residuals. Is the normality assumption reasonable for these data?

(b) Use the Bonferroni or another multiple-comparisons procedure to determine which pairs of means differ significantly. Summarize your results in a short report. Be sure to include a graph.

12.43   One way to repair serious wounds is to insert some material as a scaffold for the body's repair cells to use as a template for new tissue. Scaffolds made from extracellular material (ECM) are particularly promising for this purpose. Because they are made from biological material, they serve as an effective scaffold and are then resorbed. Unlike biological material that includes cells, however, they do not trigger tissue rejection reactions in the body. One study compared 6 types of scaffold material.[17] Three of these were ECMs and the other three were made of inert materials. There were three mice used per scaffold type. The response measure was the percent of glucose phosphated isomerase (Gpi) cells in the region of the wound. A large value is good, indicating that there are many bone marrow cells sent by the body to repair the tissue.

Material	Gpi (%)		
ECM1	55	70	70
ECM2	60	65	65
ECM3	75	70	75
MAT1	20	25	25
MAT2	5	10	5
MAT3	10	15	10

(a) Make a table giving the sample size, mean, and standard deviation for each of the six types of material. Is it reasonable to pool the variances? Note that the sample sizes are small and the data are rounded.

(b) Run the analysis of variance. Report the F statistic with its degrees of freedom and P-value. What do you conclude?

12.44   Refer to the previous exercise.

(a) Examine the residuals. Is the normality assumption reasonable for these data?

(b) Use the Bonferroni or another multiple-comparisons procedure to determine which pairs of means differ significantly. Summarize your results in a short report. Be sure to include a graph.

12.45 Refer to the previous two exercises. Use a contrast to compare the three ECM materials with the three other materials. Summarize your conclusions. How do these results compare with those that you obtained from the multiple-comparisons procedure in the previous exercise?

12.46 Analysis of variance methods are often used in clinical trials where the goal is to assess the effectiveness of one or more treatments for a particular medical condition. One such study compared three treatments for dandruff and a placebo. The treatments were 1% pyrithione zinc shampoo (PyrI), the same shampoo but with instructions to shampoo two times (PyrII), 2% ketoconazole shampoo (Keto), and a placebo shampoo (Placebo). After six weeks of treatment, eight sections of the scalp were examined and given a score that measured the amount of scalp flaking on a 0 to 10 scale. The response variable was the sum of these eight scores. An analysis of the baseline flaking measure indicated that randomization of patients to treatments was successful in that no differences were found between the groups. At baseline there were 112 subjects in each of the three treatment groups and 28 subjects in the Placebo group. During the clinical trial, 3 dropped out from the PyrII group and 6 from the Keto group. No patients dropped out of the other two groups. The data are given in the DANDRUFF data set described in the Data Appendix.

   (a) Find the mean, standard deviation, and standard error for the subjects in each group. Summarize these, along with the sample sizes, in a table and make a graph of the means.

   (b) Run the analysis of variance on these data. Write a short summary of the results and your conclusion. Be sure to include the hypotheses tested, the test statistic with degrees of freedom, and the *P*-value.

12.47 Refer to the previous exercise.

   (a) Plot the residuals versus case number (the first variable in the data set). Describe the plot. Is there any pattern that would cause you to question the assumption that the data are independent?

   (b) Examine the standard deviations for the four treatment groups. Is there a problem with the assumption of equal standard deviations for ANOVA in this data set? Explain your answer.

   (c) Prepare normal quantile plots for each treatment group. What do you conclude from these plots?

   (d) Obtain the residuals from the analysis of variance and prepare a normal quantile plot of these. What do you conclude?

12.48 Refer to Exercise 12.46. Use the Bonferroni or another multiple-comparisons procedure that your software provides to compare the individual group means in the dandruff study. Write a short summary of your conclusions.

12.49 Refer to Exercise 12.46. There are several natural contrasts in this experiment that describe comparisons of interest to the experimenters. They are (1) Placebo versus the average of the three treatments; (2) Keto versus the average of the two Pyr treatments; and (3) PyrI versus PyrII.

(a) Express each of these three contrasts in terms of the means (μ's) of the treatments.

(b) Give estimates with standard errors for each of the contrasts.

(c) Perform the significance tests for the contrasts. Summarize the results of your tests and your conclusions.

12.50 Refer to Exercise 12.19 (page 757), where we studied the effects of storage on the vitamin C content of bread. In this experiment 64 mg of vitamin C per 100 g of flour was added to the flour that was used to make each loaf.

(a) Convert the vitamin C amounts (mg/100 g) to percents of the amounts originally in the loaves by dividing the amounts in Exercise 12.19 by 64 and multiplying by 100. Calculate the transformed means, standard deviations, and standard errors and summarize them with the sample sizes in a table.

(b) Explain how you could have calculated the table entries directly from the table you gave in part (a) of Exercise 12.19.

(c) Analyze the percents using analysis of variance. Compare the test statistic, degrees of freedom, P-value, and conclusion you obtain here with the corresponding values that you found in Exercise 12.19.

12.51 Refer to the previous exercise and Exercise 12.21 (page 758). The flour used to make the loaves contained 5 mg of vitamin A per 100 g of flour and 100 mg of vitamin E per 100 g of flour. Summarize the effects of transforming the data to percents for the three vitamins.

12.52 Refer to the previous exercise. Can you suggest a general conclusion regarding what happens to the test statistic, degrees of freedom, P-value, and conclusion when you perform analysis of variance on data that have been transformed by multiplying the raw data by a constant and then adding another constant? (That is, if y is the original data, we analyze y^*, where $y^* = a + by$ and a and $b \neq 0$ are constants.)

12.53 A study of reading comprehension in children compared three methods of instruction.[18] The three methods of instruction are called Basal, DRTA, and Strategies. As is common in such studies, several pretest variables were measured before any instruction was given. One purpose of the pretest was to see if the three groups of children were similar in their comprehension skills. The READING data set described in the Data Appendix gives two pretest measures that were used in this study. Use one-way ANOVA to analyze these data and write a summary of your results.

12.54 In the study described in the previous exercise, Basal is the traditional method of teaching, while DRTA and Strategies are two innovative methods based on similar theoretical considerations. The READING data set includes three response variables that the new methods were designed to improve. Analyze these variables using ANOVA methods. Be sure to include multiple comparisons or contrasts as needed. Write a report summarizing your findings.

12.55 Refer to the price promotion study that we examined in Exercise 12.32 (page 761). The explanatory variable in this study is the number of price promotions in a 10-week period, with possible values of 1, 3, 5, and 7. When using analysis of variance we treat the explanatory variable as categorical. An alternative analysis is to use simple linear regression. Perform this analysis and summarize the results. Plot the residuals from the regression model versus the number of promotions. What do you conclude?

12.56 The *One-Way ANOVA* applet lets you see how the F statistic and the P-value depend on the variability of the data within groups and the differences among the means.

(a) The black dots are at the means of the three groups. Move these up and down until you get a configuration that gives a P-value of about 0.01. What is the value of the F statistic?

(b) Now increase the variation within the groups by dragging the mark on the pooled standard error scale to the right. Describe what happens to the F statistic and the P-value. Explain why this happens.

12.57 Set the pooled standard error for the *One-Way ANOVA* applet at a middle value. Drag the black dots so that they are approximately equal.

(a) What is the F statistic? Give its P-value.

(b) Drag the mean of the second group up and the mean of the third group down. Describe the effect on the F statistic and its P-value. Explain why they change in this way.

12.58 The last line of the summary table given in Exercise 12.16 (page 755) gives the mean and the standard deviation for all of the data combined. Compare this standard deviation with the pooled standard deviation that you would use as an estimate of the model standard deviation. Explain why you would expect this standard deviation to be larger than the pooled standard deviation.

12.59 Search the Internet or your library to find a study that is interesting to you and that used one-way ANOVA to analyze the data. First describe the question or questions of interest and then give the details of how ANOVA was used to provide answers. Be sure to include how the study authors examined the assumptions for the analysis. Evaluate how well the authors used ANOVA in this study. If your evaluation finds the analysis deficient, make suggestions for how it could be improved.

12.60 In Example 12.27 (page 750) the power calculation indicated that there was a fairly small chance of detecting the alternative given. Redo the calculations for the alternative $\mu_1 = 40$, $\mu_2 = 48$, and $\mu_3 = 44$. Do you think that the choice of 10 students per treatment is adequate for this alternative?

12.61 Scores on an emotional scale were compared for five different cultures in Exercise 12.17 (page 756). Suppose that you are planning a new study using the same outcome variable. Your study will use European American, Asian American, and Hispanic American students from a large university.

(a) Explain how you would select the students to participate in your study.

(b) Use the data from Exercise 12.17 to perform power calculations to determine sample sizes for your study.

(c) Write a report that could be understood by someone with limited background in statistics and that describes your proposed study and why you think it is likely that you will obtain interesting results.

12.62 Exercise 12.31 (page 761) gave data for a bone health study that examined the effect of isoflavones on rat bone mineral density. In this study there were three groups. Controls received a placebo, and the other two groups received either a low or a high dose of isoflavones from kudzu. You are planning a similar study of a new kind of isoflavone. Use the results of the study described in Exercise 12.31 to plan your study. Write a proposal explaining why your study should be funded.

EESEE CASE STUDIES

The Electronic Encyclopedia of Statistical Examples and Exercises (EESEE) is available on the text CD and Web site. These more elaborate stories, with data, provide settings for longer case studies.

12.63 Read the EESEE story "Blinded Knee Doctors." Write a report that answers all questions for this case study.

12.64 Read the EESEE story "Stress among Pets and Friends." Write a report that answers all questions for this case study.

12.65 Read the EESEE story "Nutrition and Breakfast Cereals." Write a report that answers all questions for this case study.

STATISTICS IN PRACTICE

Developing New Drugs

STEVE SUN *is Manager of Statistics at Johnson & Johnson Pharmaceutical Research & Development.*

I first became interested in statistics for drug development when I was a student intern at Eli Lilly. It was then that I realized how prominent a role statistics plays in the pharmaceutical industry.

Before the Food and Drug Administration (FDA) will approve marketing of a new drug, the drug is tested through three stages of clinical trials. Phase I trials enroll healthy volunteers to characterize the safety and tolerability of the drug and to study its absorption and elimination by the body. Phase II trials are small-scale studies of the safety and effectiveness of the drug in patients for whom it is designed. Phase III trials are typically conducted on a large number of patients to evaluate the effectiveness and safety of the drug in detail. Statistical analysis of the results of these clinical trials provides the evidence on which the FDA bases its final decision on whether or not to approve the new drug application.

As a clinical statistician, I am responsible for collaborating with research scientists and physicians. My responsibilities include designing clinical trials and determining how many subjects will be required for the study. I also plan the statistical analysis and interpret the results. In addition, I carry out methodological research to find new ways to handle the many challenging statistical issues that arise in clinical trials.

Because of the diverse and complex nature of clinical trials, each individual trial raises many unique and specific issues that require creative solutions. For example, how should we deal with missing data due to disease progression in cancer patients? When should we stop a trial if safety concerns appear to outweigh the benefits, or if the drug works so well that we should give it to all patients? These are the kinds of challenges that deeply attract me to this profession. I experience great satisfaction from knowing that my work benefits patients by contributing to high scientific standards in clinical trials.

Two-Way Analysis of Variance

Introduction

The t procedures of Chapter 7 compare the means of two populations. We generalized these in Chapter 12 so that we could compare the means of several populations.

Two-way ANOVA compares the means of populations that are classified in two ways or the mean responses in two-factor experiments. Many of the key concepts are similar to those of one-way ANOVA, but the presence of more than one factor also introduces some new ideas. We once more assume that the data are approximately normal and that groups may have different means but do have the same standard deviation; we again pool to estimate the variance; and we again use F statistics for significance tests. The major difference between one-way and two-way ANOVA is in the FIT part of the model. We will carefully study this term, and we will find much that is both new and useful.

13.1   The Two-Way ANOVA Model

We begin with a discussion of the advantages of the two-way ANOVA design and illustrate these with some examples. Then we discuss the model and the assumptions.

Advantages of two-way ANOVA

In one-way ANOVA, we classify populations according to one categorical variable, or factor. In the two-way ANOVA model, there are two factors, each with its own number of levels. When we are interested in the effects of two factors, a two-way design offers great advantages over several single-factor studies. Several examples will illustrate these advantages.

EXAMPLE 13.1   You are a Web page designer and one of your clients sells spring break vacation packages to college students. The client has three different logos for the home page and would like to perform an experiment to determine if the logo used has an effect on sales. Data will be collected from 18 one-hour time periods, from 6:00 a.m. to midnight, for a week. Each day, the three logos will be randomly assigned to the 18 periods. So, each day there will be 6 time periods for each logo, and the total for the week will be 42 periods for each. A second experiment using the same format is planned for the following week to compare three different colors for the background of the home page. Total sales during each period is the outcome variable.

Here is a picture of the design of the first experiment with the sample sizes:

Logo	n
1	42
2	42
3	42
Total	126

And this represents the second experiment:

Color	n
Red	42
Green	42
Blue	42
Total	126

Each experiment has 42 periods of data collection for each level of the factor and a total of 126 periods. The total amount of time for the two experiments is two weeks with 252 periods. Each experiment will be analyzed using one-way ANOVA. The factor in the first experiment is logo with three levels, and the factor in the second experiment is color with three levels. Let's consider combining the two experiments into one.

EXAMPLE 13.2   Suppose we use a two-way approach for the Web page design problem. There are two factors, logo and color. Each has three levels so this is a 3×3 design. This gives a total of 9 possible combinations of logo and color. For each day, we have 18 periods, so we could assign each combination of logo and color to 2 periods. Let's consider running this two-way design for one week. With 2 periods per day, this gives 14 observation periods for each combination.

Here is a picture of the two-way design with the sample sizes:

Logo	Color Red	Green	Blue	Total
1	14	14	14	42
2	14	14	14	42
3	14	14	14	42
Total	42	42	42	126

With the two-way design for logo, notice that we have 42 periods of data collection for each logo, the same as what we had for the one-way experiment for logo alone. Similarly, there are 42 periods of data collection for each color. Therefore, the two-way design gives us the same amount of information for estimating the sales for each level of each factor as we had with the two one-way designs. The difference is that we can collect all of the information in only one week. By combining the two factors into one experiment, we have increased our efficiency by reducing the amount of data to be collected by half.

cell   Each combination of the factors in a two-way design corresponds to a **cell.** The 3×3 ANOVA for the Web page experiment has nine cells, each corresponding to a particular combination of logo and color.

EXAMPLE 13.3 Malaria is a serious health problem causing an estimated 2.7 million deaths per year, mostly in Africa.[1] Some research suggests that vitamin A can reduce episodes of malaria in young children. Red palm oil is a good source of vitamin A and is readily available in Nigeria, a country where malaria accounts for about 30% of the deaths of young children. Can an increase in the consumption of red palm oil reduce the occurrence and severity of malaria in this region?[2]

To design a study to answer this question we first need to determine an appropriate target group. Since malaria is a serious problem for young children, we will concentrate on children who are 2 to 5 years of age. A supplement will be prepared that contains either a placebo, a low dose of red palm oil, or a high dose of red palm oil. Because boys and girls may differ in exposure to malaria and the response to the red palm oil supplement, our design should also take gender into account. Let's consider a two-way ANOVA for this study.

EXAMPLE 13.4 The factors for our two-way ANOVA are red palm oil with three levels and gender with two levels. There are $3 \times 2 = 6$ cells in our study. Suppose we recruit 75 boys and 75 girls. We will then randomly assign 25 of each gender to each of the red palm oil levels. The outcome variable will be the amount of an acute-phase protein in the blood, which measures the severity of infection.

Here is a table that summarizes the design:

	Gender		
Red palm oil	Girls	Boys	Total
Placebo	25	25	50
Low dose	25	25	50
High dose	25	25	50
Total	75	75	150

This example illustrates another advantage of two-way designs. Although we are primarily interested in the possible benefit of red palm oil, we included gender in the design because we thought that there might be differences between the boys and the girls. Consider an alternative one-way design where we assign 150 children to the three levels of red palm oil and ignore gender. With this design we will have the same number of children at each of the red palm oil levels, so in this way it is similar to our two-way design. However, suppose that there are, in fact, differences between boys and girls. In this case, the one-way ANOVA would assign this variation to the RESIDUAL (within groups) part of the model. In the two-way ANOVA, gender is included as a factor, and therefore this variation is included in the FIT part of the model. Whenever we can move variation from RESIDUAL to FIT, we reduce the σ of our model and increase the power of our tests.

EXAMPLE 13.5   Osteoporosis is a disease primarily of the elderly. People with osteoporosis have low bone mass and an increased risk of bone fractures. Over 10 million people in the United States, 1.4 million Canadians, and many millions throughout the world have this disease. Adequate calcium in the diet is necessary for strong bones, but vitamin D is also needed for the body to efficiently utilize calcium. High doses of calcium in the diet will not prevent osteoporosis unless there is adequate vitamin D. Exposure of the skin to the ultraviolet rays in sunlight enables our bodies to make vitamin D. However, elderly people often avoid sunlight and in northern areas such as Canada, there is not sufficient ultraviolet light to make vitamin D, particularly in the winter months.

Suppose we wanted to see if calcium supplements will increase bone mass (or prevent a decrease in bone mass) in an elderly Canadian population. Because of the vitamin D complication, we will make this a factor in our design.

EXAMPLE 13.6   We will use a 2 × 2 design for our osteoporosis study. The two factors are calcium and vitamin D. The levels of each factor will be zero (placebo) and an amount that is expected to be adequate, 800 mg/day for calcium and 300 international units per day (IU/day) for vitamin D. Women between the ages of 70 and 80 will be recruited as subjects. Bone mineral density (BMD) will be measured at the beginning of the study, and supplements will be taken for one year. The change in BMD over the one-year period is the outcome variable. We expect a dropout rate of 25% and we would like to have about 20 subjects providing data in each group at the end of the study. We will therefore recruit 100 subjects and randomly assign 25 to each treatment combination.

Here is a table that summarizes the design with the sample sizes at baseline:

	Vitamin D		
Calcium	Placebo	300 IU/day	Total
Placebo	25	25	50
800 mg/day	25	25	50
Total	50	50	100

interaction

main effects

This example illustrates a third reason for using two-way designs. The effectiveness of the calcium supplement on BMD depends on having adequate vitamin D. We call this an **interaction.** In contrast, the average values for the calcium effect and the vitamin D effect are represented as **main effects.** The two-way model represents FIT as the sum of a main effect for each of the two factors *and* an interaction. One-way designs that vary a single factor and hold other factors fixed cannot discover interactions. We will discuss interactions more fully in a later section.

These examples illustrate several reasons why two-way designs are preferable to one-way designs.

> **ADVANTAGES OF TWO-WAY ANOVA**
>
> **1.** It is more efficient to study two factors simultaneously rather than separately.
>
> **2.** We can reduce the residual variation in a model by including a second factor thought to influence the response.
>
> **3.** We can investigate interactions between factors.

These considerations also apply to study designs with more than two factors. We will be content to explore only the two-way case. The choice of sampling or experimental design is fundamental to any statistical study. *Factors and levels must be carefully selected by an individual or team who understands both the statistical models and the issues that the study will address.*

The two-way ANOVA model

When discussing two-way models in general, we will use the labels A and B for the two factors. For particular examples and when using statistical software, it is better to use names for these categorical variables that suggest their meaning. Thus, in Example 13.2 we would say that the factors are logo and color. The numbers of levels of the factors are often used to describe the model. Again using Example 13.2 as an illustration, we would call this a 3×3 ANOVA. Similarly, Example 13.4 illustrates a 3×2 ANOVA. In general, factor A will have I levels and factor B will have J levels. Therefore, we call the general two-way problem an $I \times J$ ANOVA.

In a two-way design every level of A appears in combination with every level of B, so that $I \times J$ groups are compared. The sample size for level i of factor A and level j of factor B is n_{ij}.[3]

The total number of observations is

$$N = \sum n_{ij}$$

> **ASSUMPTIONS FOR TWO-WAY ANOVA**
>
> We have independent SRSs of size n_{ij} from each of $I \times J$ normal populations. The population means μ_{ij} may differ, but all populations have the same standard deviation σ. The μ_{ij} and σ are unknown parameters.
>
> Let x_{ijk} represent the kth observation from the population having factor A at level i and factor B at level j. The statistical model is
>
> $$x_{ijk} = \mu_{ij} + \epsilon_{ijk}$$
>
> for $i = 1, \ldots, I$ and $j = 1, \ldots, J$ and $k = 1, \ldots, n_{ij}$, where the deviations ϵ_{ijk} are from an $N(0, \sigma)$ distribution.

The FIT part of the model is the means μ_{ij}, and the RESIDUAL part is the deviations ϵ_{ijk} of the individual observations from their group means. To estimate a population mean μ_{ij} we use the sample mean of the observations in the samples from this group:

$$\bar{x}_{ij} = \frac{1}{n_{ij}} \sum_{k} x_{ijk}$$

The k below the $\sum$ means that we sum the n_{ij} observations that belong to the (i, j)th sample.

The RESIDUAL part of the model contains the unknown σ. We calculate the sample variances for each SRS and pool these to estimate σ^2:

$$s_p^2 = \frac{\sum (n_{ij} - 1) s_{ij}^2}{\sum (n_{ij} - 1)}$$

Just as in one-way ANOVA, the numerator in this fraction is SSE and the denominator is DFE. Also as in the one-way analysis, DFE is the total number of observations minus the number of groups. That is, $\text{DFE} = N - IJ$. The estimator of σ is s_p.

Main effects and interactions

In this section we will explore in detail the FIT part of the two-way ANOVA, which is represented in the model by the population means μ_{ij}. The two-way design gives some structure to the set of means μ_{ij}.

Because we have independent samples from each of $I \times J$ groups, we can think of the problem initially as a one-way ANOVA with IJ groups. Each population mean μ_{ij} is estimated by the corresponding sample mean $\bar{x}_{ij}$. We can calculate sums of squares and degrees of freedom as in one-way ANOVA. In accordance with the conventions used by many computer software packages, we use the term *model* when discussing the sums of squares and degrees of freedom calculated as in one-way ANOVA with IJ groups. Thus, SSM is a model sum of squares constructed from deviations of the form $\bar{x}_{ij} - \bar{x}$, where $\bar{x}$ is the average of all of the observations and $\bar{x}_{ij}$ is the mean of the (i, j)th group. Similarly, DFM is simply $IJ - 1$.

In two-way ANOVA, the terms SSM and DFM are broken down into terms corresponding to a main effect for A, a main effect for B, and an AB interaction. Each of SSM and DFM is then a sum of terms:

$$\text{SSM} = \text{SSA} + \text{SSB} + \text{SSAB}$$

and

$$\text{DFM} = \text{DFA} + \text{DFB} + \text{DFAB}$$

The term SSA represents variation among the means for the different levels of the factor A. Because there are I such means, $\text{DFA} = I - 1$ degrees of freedom. Similarly, SSB represents variation among the means for the different levels of the factor B, with $\text{DFB} = J - 1$.

Interactions are a bit more involved. We can see that SSAB, which is SSM − SSA − SSB, represents the variation in the model that is not accounted for by the main effects. By subtraction we see that its degrees of freedom are

$$\text{DFAB} = (IJ - 1) - (I - 1) - (J - 1) = (I - 1)(J - 1)$$

There are many kinds of interactions. The easiest way to study them is through examples.

EXAMPLE 13.7   There is a general consensus that food portions have been increasing, but there is little scientific evidence that documents this change. One study used data from three nationally representative surveys to examine this issue. More than 63,380 individuals provided data for these three surveys. Three time points were examined: 1978, 1991, and 1996.[4] Here are the means for the number of calories per portion in soft drinks consumed at home and in restaurants:

	Year			
Location	1978	1991	1996	Mean
Home	130	133	158	140
Restaurant	125	126	155	135
Mean	127	129	156	137

The table includes averages of the means in the rows and columns. For example, in 1978 the mean of calories in soft drinks consumed at home and in restaurants is

$$\frac{130 + 125}{2} = 127.5$$

which is rounded to 127 in the table. Similarly, the corresponding value for 1996 is

$$\frac{158 + 155}{2} = 156.5$$

marginal means   which is rounded to 156 in the table. These averages are called **marginal means** (because of their location at the *margins* of such tabulations). The grand mean can be obtained by averaging either set of marginal means. *It is always a good idea to do both as a check on your arithmetic.*

Figure 13.1 is a plot of the group means. From the plot we see that the soft drinks consumed at home have about 3 to 7 more calories per serving than soft drinks consumed in restaurants for the three years. In statistical language, there is a main effect for location. We also see that the means for 1978 and 1991 are similar but there is a large increase by 1996. This is the main

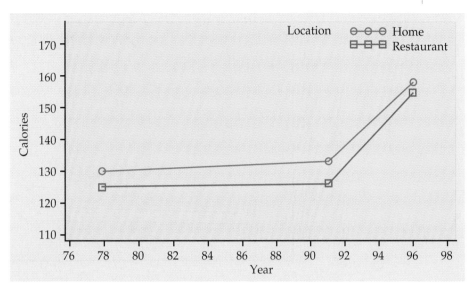

FIGURE 13.1   Plot of the mean calories of soft drinks consumed at home and in restaurants in 1978, 1991, and 1996, for Example 13.7.

effect of year. These main effects can be described by differences between the marginal means. For example, the mean for 1978 is 127 calories, it increases by 2 calories to 129 calories in 1991, and then it jumps by 27 calories to 156 calories in 1996.

To examine two-way ANOVA data for a possible interaction, always construct a plot similar to Figure 13.1. In this plot, we see that the patterns of means over the years are similar for home and restaurant, with the restaurant means being about 5 calories lower than the home means. The two profiles are roughly parallel. This is another way of saying that there is no clear interaction between location and year evident in these data.

When no interaction is present, the marginal means provide a reasonable description of the two-way table of means. *On the other hand, if there is an interaction, then the marginal means do not tell the whole story.* Here is an example that illustrates this point.

EXAMPLE 13.8   The surveys described in the previous example also obtained data on soft drinks consumed in fast-food restaurants. Here are the data from the previous example with the fast-food means added:

	Year			
Location	1978	1991	1996	Mean
Home	130	133	158	140
Restaurant	125	126	155	135
Fast food	131	143	191	155
Mean	127	134	168	143

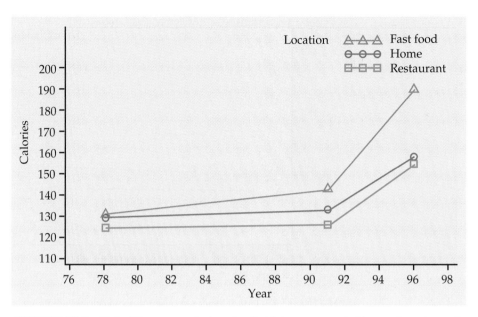

FIGURE 13.2 Plot of the mean calories of soft drinks consumed at home, in restaurants, and in fast-food restaurants in 1978, 1991, and 1996, for Example 13.8.

Including the additional fast-food restaurants changes the marginal means for years and the overall mean. Figure 13.2 is a plot of the group means.

In this figure we see that home and the fast-food restaurants were quite similar in 1978. The increase from 1978 to 1991 was somewhat similar, with fast-food increasing a little more than home. The most noticeable feature of the plot is the very large jump for the fast-food restaurants by 1996. The mean calories in soft drinks increased from 131 in 1978 to 191 in 1996, an increase of about 46%. This change is thought to be closely related to the trend toward "supersizing" food portions at fast-food restaurants.

The three patterns of means in Figure 13.2 are clearly not all parallel. The change over time differs for the fast-food restaurants. We have an interaction between location and year. *However, the presence of an interaction does not necessarily mean that the main effects are uninformative.* The calories in soft drinks have increased from 1978 to 1991 and from 1991 to 1996. Similarly, there is a tendency for the means to be lowest in restaurants, a little higher at home, and highest at fast-food restaurants.

Interactions come in many shapes and forms. *When we find an interaction, a careful examination of the means is needed to properly interpret the data.* Simply stating that interactions are significant tells us little. Plots of the group means are essential.

EXAMPLE 13.9   Some research has shown that people eat more when they eat in groups. One possible mechanism for this phenomenon is that they may spend more time eating when in a larger group. A study designed to examine this idea measured the length of time spent (in minutes) eating lunch in different settings.[5]

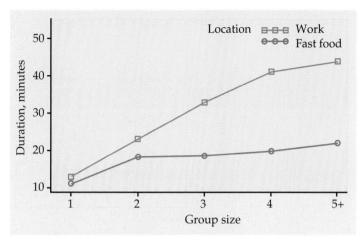

FIGURE 13.3 Plot of the mean meal duration versus lunch setting and group size, for Example 13.9.

Here are some data from this study:

| | Number of people eating | | | | | |
Lunch setting	1	2	3	4	5 or more	Mean
Workplace	12.6	23.0	33.0	41.1	44.0	30.7
Fast-food restaurant	10.7	18.2	18.4	19.7	21.9	17.8
Mean	11.6	20.6	25.7	30.4	32.9	24.2

Figure 13.3 gives the plot of the means for this example. The patterns are not parallel, so it appears that we have an interaction. Meals take longer when more people are present, but this phenomenon is much greater for the meals consumed at work. For fast-food eating, the meal durations are fairly similar when more than one person is present.

A different kind of interaction is present in the next example. Here, we must be very cautious in our interpretation of the main effects since one of them can lead to a distorted conclusion.

EXAMPLE 13.10   When we hear music that is familiar to us, we can quickly pick up the beat and our mind synchronizes with the music. However, if the music is unfamiliar, it takes us longer to synchronize. In a study that investigated the theoretical framework for this phenomenon, French and Tunisian nationals listened to French and Tunisian music.[6] Each subject was asked to tap in time with the music being played. A synchronization score, recorded in milliseconds, measured how well the subjects synchronized with the music. A higher score indicates better synchronization. Six songs of each music type were used. Here are the means:

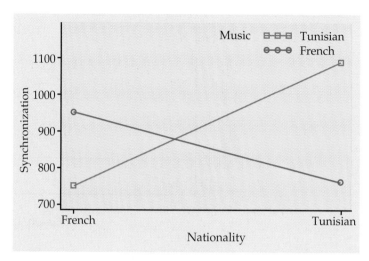

FIGURE 13.4 Plot of mean synchronization score versus type of music for French and Tunisian nationals, for Example 13.10.

	Music		
Nationality	French	Tunisian	Mean
French	950	750	850
Tunisian	760	1090	925
Mean	855	920	887

The means are plotted in Figure 13.4. In the study the researchers were not interested in main effects. Their theory predicted the interaction that we see in the figure. Subjects synchronize better with music from their own culture.

The interaction in Figure 13.4 is very different from those that we saw in Figures 13.2 and 13.3. These examples illustrate the point that it is necessary to plot the means and carefully describe the patterns when interpreting an interaction.

The design of the study in Example 13.10 allows us to examine two main effects and an interaction. However, this setting does not meet all of the assumptions needed for statistical inference using the two-way ANOVA framework of this chapter. *As with one-way ANOVA, we require that observations be independent.* In this study, we have a design that has each subject contributing data for two types of music, so these two scores will be dependent. The framework is similar to the matched pairs setting that we studied in Chapter 7 (page 459). The design is called a **repeated-measures design.** More advanced texts on statistical methods cover this important design.

repeated-measures design

13.2   Inference for Two-Way ANOVA

Inference for two-way ANOVA involves F statistics for each of the two main effects and an additional F statistic for the interaction. As with one-way ANOVA, the calculations are organized in an ANOVA table.

The ANOVA table for two-way ANOVA

Two-way ANOVA is the statistical analysis for a two-way design with a quantitative response variable. The results of a two-way ANOVA are summarized in an ANOVA table based on splitting the total variation SST and the total degrees of freedom DFT among the two main effects and the interaction. Both the sums of squares (which measure variation) and the degrees of freedom add:

$$SST = SSA + SSB + SSAB + SSE$$

$$DFT = DFA + DFB + DFAB + DFE$$

The sums of squares are always calculated in practice by statistical software. When the n_{ij} are not all equal, some methods of analysis can give sums of squares that do not add. From each sum of squares and its degrees of freedom we find the mean square in the usual way:

$$\text{mean square} = \frac{\text{sum of squares}}{\text{degrees of freedom}}$$

The significance of each of the main effects and the interaction is assessed by an F statistic that compares the variation due to the effect of interest with the within-group variation. Each F statistic is the mean square for the source of interest divided by MSE. Here is the general form of the two-way ANOVA table:

Source	Degrees of freedom	Sum of squares	Mean square	F
A	$I - 1$	SSA	SSA/DFA	MSA/MSE
B	$J - 1$	SSB	SSB/DFB	MSB/MSE
AB	$(I - 1)(J - 1)$	SSAB	SSAB/DFAB	MSAB/MSE
Error	$N - IJ$	SSE	SSE/DFE	
Total	$N - 1$	SST	SST/DFT	

There are three null hypotheses in two-way ANOVA, with an F test for each. We can test for significance of the main effect of A, the main effect of B, and the AB interaction. It is generally good practice to examine the test for interaction first, since the presence of a strong interaction may influence the interpretation of the main effects. Be sure to plot the means as an aid to interpreting the results of the significance tests.

SIGNIFICANCE TESTS IN TWO-WAY ANOVA

To test the main effect of A, use the F statistic

$$F_A = \frac{MSA}{MSE}$$

To test the main effect of B, use the F statistic

$$F_B = \frac{MSB}{MSE}$$

To test the interaction of A and B, use the F statistic

$$F_{AB} = \frac{MSAB}{MSE}$$

If the effect being tested is zero, the calculated F statistic has an F distribution with numerator degrees of freedom corresponding to the effect and denominator degrees of freedom equal to DFE. Large values of the F statistic lead to rejection of the null hypothesis. The P-value is the probability that a random variable having the corresponding F distribution is greater than or equal to the calculated value.

The following example illustrates how to do a two-way ANOVA. As with the one-way ANOVA, we focus our attention on interpretation of the computer output.

EXAMPLE 13.11 A study of cardiovascular risk factors compared runners who averaged at least 15 miles per week with a control group described as "generally sedentary." Both men and women were included in the study.[7] The design is a 2 × 2 ANOVA with the factors group and gender. There were 200 subjects in each of the four combinations. One of the variables measured was the heart rate after 6 minutes of exercise on a treadmill. SAS computer analysis produced the outputs in Figure 13.5 and Figure 13.6.

We begin with the usual preliminary examination. From Figure 13.5 we see that the ratio of the largest to the smallest standard deviation is less than 2. Therefore, we are not concerned about violating the assumption of equal population standard deviations. Normal quantile plots (not shown) do not reveal any outliers, and the data appear to be reasonably normal.

The ANOVA table at the top of the output in Figure 13.6 is in effect a one-way ANOVA with four groups: female control, female runner, male control, and male runner. In this analysis Model has 3 degrees of freedom, and Error has 796 degrees of freedom. The F test and its associated P-value for this analysis refer to the hypothesis that all four groups have the same population mean. We are interested in the main effects and interaction, so we ignore this test.

```
Analysis Variable : HR

  GROUP=Control GENDER=Female
   N            Mean         Std Dev      Std Error
  ---------------------------------------------------
  200           148.00        16.27           1.15
  ---------------------------------------------------


  GROUP=Control GENDER=Male
   N            Mean         Std Dev      Std Error
  ---------------------------------------------------
  200           130.00        17.10           1.21
  ---------------------------------------------------


  GROUP=Runners GENDER=Female
   N            Mean         Std Dev      Std Error
  ---------------------------------------------------
  200           115.99        15.97           1.13
  ---------------------------------------------------


  GROUP=Runners GENDER=Male
   N            Mean         Std Dev      Std Error
  ---------------------------------------------------
  200           103.98        12.50           0.88
  ---------------------------------------------------
```

FIGURE 13.5   Summary statistics for heart rates in the four groups of a 2×2 ANOVA, for Example 13.11.

```
General Linear Models Procedure
Dependent Variable: HR
                                 Sum of          Mean
Source                  DF       Squares        Square   F Value   Pr > F
Model                    3     215256.09      71752.03    296.35   0.0001
Error                  796     192729.83        242.12
Corrected Total        799     407985.92

              R-square          C.V.      Root MSE              HR Mean
              0.527607      12.49924        15.560               124.49


Source                  DF      Type I SS  Mean Square   F Value   Pr > F
GROUP                    1     168432.08     168432.08    695.65   0.0001
GENDER                   1      45030.00      45030.00    185.98   0.0001
GROUP*GENDER             1       1794.01       1794.01      7.41   0.0066
```

FIGURE 13.6   Two-way ANOVA output for heart rates, for Example 13.11.

The sums of squares for the group and gender main effects and the group-by-gender interaction appear at the bottom of Figure 13.6 under the heading Type I SS. These sum to the sum of squares for Model. Similarly, the degrees of freedom for these sums of squares sum to the degrees of freedom for Model. Two-way ANOVA splits the variation among the means (expressed by the Model sum of squares) into three parts that reflect the two-way layout.

Because the degrees of freedom are all 1 for the main effects and the interaction, the mean squares are the same as the sums of squares. The F statistics for the three effects appear in the column labeled F Value, and the P-values are under the heading Pr > F. For the group main effect, we verify the calculation of F as follows:

$$F = \frac{\text{MSG}}{\text{MSE}} = \frac{168{,}432}{242.12} = 695.65$$

All three effects are statistically significant. The group effect has the largest F, followed by the gender effect and then the group-by-gender interaction. To interpret these results, we examine the plot of means with bars indicating one standard error in Figure 13.7. Note that the standard errors are quite small due to the large sample sizes. The significance of the main effect for group is due to the fact that the controls have higher average heart rates than the runners for both genders. This is the largest effect evident in the plot.

The significance of the main effect for gender is due to the fact that the females have higher heart rates than the men in both groups. The differences are not as large as those for the group effect, and this is reflected in the smaller value of the F statistic.

The analysis indicates that a complete description of the average heart rates requires consideration of the interaction in addition to the main effects. The two lines in the plot are not parallel. This interaction can be described in two ways. The female-male difference in average heart rates is greater for the controls than for the runners. Alternatively, the difference in average heart rates between controls and runners is greater for women than for men. As the plot suggests, the interaction is not large. It is statistically significant because there were 800 subjects in the study.

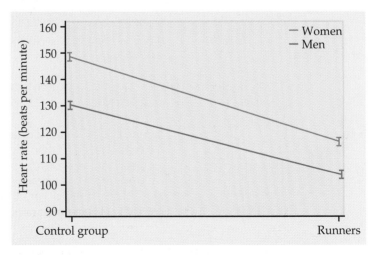

FIGURE 13.7   Plot of the group means with standard errors for heart rates in the 2 × 2 ANOVA, for Example 13.11.

Excel

	Control	Runners	Total				
Anova: Two Factor With Replication							
SUMMARY	Control	Runners	Total				
Female							
Count	200	200	400				
Sum	29600	23197	52797				
Average	148	115.985	131.9925				
Variance	264.7437	255.0902	516.1478				
Male							
Count	200	200	400				
Sum	26000	20795	46795				
Average	130	103.975	116.9875				
Variance	292.4221	156.2356.2	393.5161				
Total							
Count	400	400					
Sum	55600	43992					
Average	139	109.98					
Variance	359.0877	241.2978					

ANOVA						
Source of Variation	SS	df	MS	F	P-value	F crit
Sample	45030.01	1	45030.01	185.9799	3.29E-38	3.85316
Columns	168432.1	1	168432.1	695.647	1.10E-110	3.85316
Interaction	1794.005	1	1794.005	7.409481	0.00663	3.85316
Within	192729.8	796	242.1229			
Total	407985.9	799				

Minitab

```
Analysis of Variance for HR
Source         DF         SS          MS
Group           1     168432      168432
Gender          1      45030       45030
Interaction     1       1794        1794
Error         796     192730         242
Total         799     407986
```

FIGURE 13.8   Excel and Minitab two-way ANOVA output for the heart rate study, for Example 13.11.

Two-way ANOVA output for other software is similar to that given by SAS. Figure 13.8 gives the analysis of the heart rate data using Excel and Minitab.

CHAPTER 13 | Summary

Two-way analysis of variance is used to compare population means when populations are classified according to two factors.

ANOVA assumes that the populations are normal with possibly different means and the same standard deviation and that independent SRSs are drawn from each population.

As with one-way ANOVA, preliminary analysis includes examination of means, standard deviations, and normal quantile plots. **Marginal means** are calculated by taking averages of the cell means across rows and columns. Pooling is used to estimate the within-group variance.

ANOVA separates the total variation into parts for the **model** and **error.** The model variation is separated into parts for each of the **main effects** and the **interaction.**

The calculations are organized into an **ANOVA table.** F statistics and P-values are used to test hypotheses about the main effects and the interaction.

Careful inspection of the means is necessary to interpret significant main effects and interactions. Plots are a useful aid.

CHAPTER 13 | Exercises

13.1 For each of the following, explain what is wrong and why.

(a) The FIT part of the model in a two-way ANOVA represents the variation that is sometimes called error or residual.

(b) You should reject the null hypothesis that there is no interaction in a two-way ANOVA when the test statistic is small.

(c) Sums of squares are equal to mean squares divided by degrees of freedom.

13.2 For each of the following, explain what is wrong and why.

(a) The significance tests for the main effects in a two-way ANOVA have a chi-square distribution when the null hypothesis is true.

(b) You can perform a two-way ANOVA only when the sample sizes are the same in all cells.

(c) The cell means $\bar{x}_{ij}$ are parameters of the two-way ANOVA model.

13.3 A 2×4 ANOVA was run with 5 observations per cell.

(a) Give the degrees of freedom for the F statistic that is used to test for interaction in this analysis and the entries from Table E that correspond to this distribution.

(b) Sketch a picture of this distribution with the information from the table included.

(c) The calculated value of the F statistic is 2.59. How would you report the P-value?

(d) Would you expect a plot of the means to look parallel? Explain your answer.

13.4 For each of the following situations, state how large the F statistic needs to be for rejection of the null hypothesis at the 5% level. Sketch each distribution and indicate the region where you would reject.

(a) The main effect for the first factor in a 3×3 ANOVA with 4 observations per cell.

(b) The interaction in a 3×3 ANOVA with 4 observations per cell.

(c) The interaction in a 2×2 ANOVA with 251 observations per cell.

13.5 The National Crime Victimization Survey estimates that there were over 400,000 violent crimes committed against women by their intimate partner that resulted in physical injury. An intervention study designed to increase safety behaviors of abused women compared the effectiveness of six telephone intervention sessions with a control group of abused women who received standard care. Fifteen different safety behaviors were examined.[8] One of the variables analyzed was the total number of behaviors (out of 15) that each woman performed. Here is a summary of the means of this variable at baseline (just before the first telephone call) and at follow-up 3 and 6 months later:

		Time	
Group	Baseline	3 months	6 months
Intervention	10.4	12.5	11.9
Control	9.6	9.9	10.4

(a) Find the marginal means. Are they useful for understanding the results of this study?

(b) Plot the means. Do you think there is an interaction? Describe the meaning of an interaction for this study.

(Note: This exercise is from a repeated-measures design, and the data are not particularly normal because they are counts with values from 1 to 10. Although we cannot use the methods in this chapter for statistical inference in this setting, the example does illustrate ideas about interactions.)

13.6 Refer to the previous exercise. Table 13.1 gives the percents of women who responded that they performed each of the 15 safety behaviors studied.

TABLE 13.1

Safety behaviors of abused women

Behavior	Intervention group (%)			Control group (%)		
	Baseline	3 months	6 months	Baseline	3 months	6 months
Hide money	68.0	60.0	62.7	60.0	37.8	35.1
Hide extra keys	52.7	76.0	68.9	53.3	33.8	39.2
Abuse code to alert family	30.7	74.7	60.0	22.7	27.0	43.2
Hide extra clothing	37.3	73.6	52.7	42.7	32.9	27.0
Asked neighbors to call police	49.3	73.0	66.2	32.0	45.9	40.5
Know social security number	93.2	93.2	100.0	89.3	93.2	98.6
Keep rent, utility receipts	75.3	95.5	89.4	70.3	84.7	80.9
Keep birth certificates	84.0	90.7	93.3	77.3	90.4	93.2
Keep driver's license	93.3	93.3	97.3	94.7	95.9	98.6
Keep telephone numbers	96.0	98.7	100.0	90.7	97.3	100.0
Removed weapons	50.0	70.6	38.5	40.7	23.8	5.9
Keep bank account numbers	81.0	94.3	96.2	76.2	85.5	94.4
Keep insurance policy number	70.9	90.4	89.7	68.3	84.2	94.8
Keep marriage license	71.1	92.3	84.6	63.3	73.2	80.0
Hide valuable jewelry	78.7	84.5	83.9	74.0	75.0	80.3

(a) Summarize these data graphically. Do you think that your graphical display is more effective than Table 13.1 for describing the results of this study? Explain why or why not.

(b) Note any particular patterns in the data that would be important to someone who would use these results to design future intervention programs for abused women.

(c) The study was conducted "at a family violence unit of a large urban District Attorney's Office that serves an ethnically diverse population of three million citizens." To what extent do you think that this fact limits the conclusions that can be drawn?

13.7 Analysis of data for a 3×2 ANOVA with 5 observations per cell gave the F statistics in the following table:

Effect	F
A	1.21
B	3.63
AB	2.04

What can you conclude from the information given?

13.8 A study reported the following results for data analyzed using the methods that we studied in this chapter:

Effect	F	P-value
A	7.50	0.006
B	18.66	0.001
AB	6.14	0.011

(a) What can you conclude from the information given?

(b) What additional information would you need to write a summary of the results for this study?

13.9 Does repetition of an advertising message increase its effectiveness? One theory suggests that there are two phases in the process. In the first phase, called "wearin," negative or unfamiliar views are transformed into positive views. In the second phase, called "wearout," the effectiveness of the ad is decreased because of boredom or other factors. One study designed to investigate this theory examined two factors. The first was familiarity of the ad, with two factors, familiar and unfamiliar; the second was repetition, with three levels, 1, 2, and 3.[9] One of the response variables collected was attitude toward the ad. This variable was the average of four items, each measured on a seven-point scale, anchored by bad-good, low quality-high quality, unappealing-appealing, and unpleasant-pleasant. Here are the means for attitude:

Familiarity	Repetition 1	2	3
Familiar	4.56	4.73	5.24
Unfamiliar	4.14	5.26	4.41

(a) Make a plot of the means and describe the patterns that you see.

(b) Does the plot suggest that there is an interaction between familiarity and repetition? If your answer is Yes, describe the interaction.

13.10 Refer to the previous exercise. In settings such as this, researchers collect data for several response variables. For this study, they also constructed variables that were called attitude toward the brand, total thoughts, support arguments, and counterarguments. Here are the means:

Familiarity	Attitude to brand Repetition 1	2	3	Total Repetition 1	2	3	Support Repetition 1	2	3	Counter Repetition 1	2	3
Familiar	4.67	4.65	5.06	1.33	1.93	2.55	0.63	0.67	0.98	0.54	0.70	0.49
Unfamiliar	3.94	4.79	4.26	1.52	3.06	3.17	0.76	1.40	0.64	0.52	0.75	1.14

For each of the four response variables, give a graphical summary of the means. Use this summary to discuss any interactions that are evident. Write a short report summarizing the effect of repetition on the response variables measured, using the data in this exercise and the previous one.

13.11 Refer to the previous exercise. Here are the standard deviations for attitude toward brand:

Familiarity	Repetition 1	2	3
Familiar	1.16	1.46	1.16
Unfamiliar	1.39	1.22	1.42

Find the pooled estimate of the standard deviation for these data. Use the rule for examining standard deviations in ANOVA from Chapter 12 (page 729) to determine if it is reasonable to use a pooled standard deviation for the analysis of these data.

13.12 Refer to Exercise 13.10. Here are the standard deviations for total thoughts:

Familiarity	Repetition 1	2	3
Familiar	1.63	1.42	1.52
Unfamiliar	1.64	2.16	1.59

Find the pooled estimate of the standard deviation for these data. Use the rule for examining standard deviations in ANOVA from Chapter 12 (page 729) to determine if it is reasonable to use a pooled standard deviation for the analysis of these data.

13.13 Refer to Exercises 13.9 and 13.10. The subjects were 94 adult staff members at a West Coast university. They watched a half-hour local news show from a different state that included the ads. The selected ads were judged to be "good" by some experts and had been shown in regions other than where the study was conducted. The real names of the products were replaced by either familiar or unfamiliar brand names by a professional video editor. The ads were pretested and no one in the pretest sample suggested that the ads were not real. Discuss each of these facts in terms of how you would interpret the results of this study.

13.14 Refer to Exercises 13.9 and 13.10. The ratings for this study were each measured on a seven-point scale, anchored by bad-good, low quality-high quality, unappealing-appealing, and unpleasant-pleasant. The results presented were averaged over three ads for different products: a bank, women's clothing, and a health care plan. Write a short report summarizing the normality assumption for two-way ANOVA and the extent to which it is reasonable for the analysis of these data.

13.15 The amount of chromium in the diet has an effect on the way the body processes insulin. In an experiment designed to study this phenomenon, four diets were fed to male rats. There were two factors. Chromium had two levels: low (L) and normal (N). The rats were allowed to eat as much as they wanted (M) or the total amount that they could eat was restricted (R). We call the second factor Eat. One of the variables measured was the amount of an enzyme called GITH.[10] The means for this response variable appear in the following table:

| | Eat | |
Chromium	M	R
L	4.545	5.175
N	4.425	5.317

(a) Make a plot of the mean GITH for these diets, with the factor Chromium on the x axis and GITH on the y axis. For each Eat group connect the points for the two Chromium means.

(b) Describe the patterns you see. Does the amount of chromium in the diet appear to affect the GITH mean? Does restricting the diet rather than letting the rats eat as much as they want appear to have an effect? Is there an interaction?

(c) Compute the marginal means. Compute the differences between the M and R diets for each level of Chromium. Use this information to summarize numerically the patterns in the plot.

13.16   A study of undergraduate computer science students examined changes in major after the first year.[11] The study examined the fates of 256 students who enrolled as first-year computer science students in the same fall semester. The students were classified according to gender and their declared major at the beginning of the second year. For convenience we use the labels CS for computer science majors, EO for engineering and other science majors, and O for other majors. The explanatory variables included several high school grade summaries coded as $10 = A, 9 = A-$, etc. Here are the mean high school mathematics grades for these students:

| | Major | | |
Gender	CS	EO	O
Males	8.68	8.35	7.65
Females	9.11	9.36	8.04

Describe the main effects and interaction using appropriate graphs and calculations.

13.17   The mean SAT mathematics scores for the students in the previous exercise are summarized in the following table:

| | Major | | |
Gender	CS	EO	O
Males	628	618	589
Females	582	631	543

Summarize the results of this study using appropriate plots and calculations to describe the main effects and interaction.

13.18   The students studied in the previous two exercises were enrolled at a large midwestern university several years ago. Discuss how you would conduct a similar study at a college or university of your choice today. Include a description of all variables that you would collect for your study.

13.19   One way to repair serious wounds is to insert some material as a scaffold for the body's repair cells to use as a template for new tissue. Scaffolds made from extracellular material (ECM) are particularly promising for this purpose. Because they are made from biological material, they serve as an effective scaffold and are then resorbed. Unlike biological material that includes cells, however, they do not trigger tissue rejection reactions in the body. One study compared 6 types of scaffold material.[12] Three of these were ECMs and the other three were made of inert materials. There were three mice used per scaffold type. The response measure was the percent of glucose phosphated isomerase (Gpi) cells in the region of the wound. A large value is good, indicating that there are many bone marrow cells sent by the body to repair the tissue. In Exercise 12.43 (page 765) we analyzed the data for rats whose tissues were measured 4 weeks after the repair. The experiment included additional groups of rats who received the same types of scaffold but were measured at different times. Here are the data for 4 weeks and 8 weeks after the repair:

Material	Gpi (%)					
	4 weeks			8 weeks		
ECM1	55	70	70	60	65	65
ECM2	60	65	65	60	70	60
ECM3	75	70	75	70	80	70
MAT1	20	25	25	15	25	25
MAT2	5	10	5	10	5	5
MAT3	10	15	10	5	15	10

(a) Make a table giving the sample size, mean, and standard deviation for each of the material-by-time combinations. Is it reasonable to pool the variances? Because the sample sizes in this experiment are very small, we expect a large amount of variability in the sample standard deviations. Although they vary more than we would prefer, we will proceed with the ANOVA.

(b) Make a plot of the means. Describe the main features of the plot.

(c) Run the analysis of variance. Report the F statistics with degrees of freedom and P-values for each of the main effects and the interaction. What do you conclude? Write a short paragraph summarizing the results of your analysis.

13.20 Refer to the previous exercise. Here are the data that were collected at 2 weeks, 4 weeks, and 8 weeks:

Material	Gpi (%)								
	2 weeks			4 weeks			8 weeks		
ECM1	70	75	65	55	70	70	60	65	65
ECM2	60	65	70	60	65	65	60	70	60
ECM3	80	60	75	75	70	75	70	80	70
MAT1	50	45	50	20	25	25	15	25	25
MAT2	5	10	15	5	10	5	10	5	5
MAT3	30	25	25	10	15	10	5	15	10

Rerun the analyses that you performed in the previous exercise. How does the addition of the data for 2 weeks change the conclusions? Write a summary comparing these results with those in the previous exercise.

13.21 Refer to the previous exercise. Analyze the data for each time period separately using a one-way ANOVA. Use a multiple-comparisons procedure where needed. Summarize the results.

13.22 Iron-deficiency anemia is the most common form of malnutrition in developing countries, affecting about 50% of children and women and 25% of men. Iron pots for cooking foods had traditionally been used in many of these countries, but they have been largely replaced by aluminum pots, which are cheaper and lighter. Some research has suggested that food cooked in iron pots will contain more iron than food cooked in other types of pots. One study designed to investigate this issue compared the iron content of some Ethiopian foods cooked in aluminum, clay, and iron pots.[13] In Exercise 12.41 (page 764), we analyzed the iron content of *yesiga wet'*, beef cut into small pieces and prepared with several Ethiopian spices. The researchers who conducted this study also examined the iron content of *shiro wet'*, a legume-based mixture of chickpea flour and Ethiopian spiced pepper, and *ye-atkilt allych'a*, a lightly spiced vegetable casserole. In the table below, these three foods are labeled meat, legumes, and vegetables. Four samples of each food were cooked in each type of pot. The iron in the food is measured in milligrams of iron per 100 grams of cooked food. Here are the data:

| Type of pot | Iron content | | | | | | | | | | | |
	Meat				Legumes				Vegetables			
Aluminum	1.77	2.36	1.96	2.14	2.40	2.17	2.41	2.34	1.03	1.53	1.07	1.30
Clay	2.27	1.28	2.48	2.68	2.41	2.43	2.57	2.48	1.55	0.79	1.68	1.82
Iron	5.27	5.17	4.06	4.22	3.69	3.43	3.84	3.72	2.45	2.99	2.80	2.92

(a) Make a table giving the sample size, mean, and standard deviation for each type of pot. Is it reasonable to pool the variances? Although the standard deviations vary more than we would like, this is partially due to the small sample sizes and we will proceed with the analysis of variance.

(b) Plot the means. Give a short summary of how the iron content of foods depends upon the cooking pot.

(c) Run the analysis of variance. Give the ANOVA table, the *F* statistics with degrees of freedom and *P*-values, and your conclusions regarding the hypotheses about main effects and interactions.

13.23 Refer to the previous exercise. Although there is a statistically significant interaction, do you think that these data support the conclusion that foods cooked in iron pots contain more iron than foods cooked in aluminum or clay pots? Discuss.

13.24 Refer to Exercise 13.22. Rerun the analysis as a one-way ANOVA with 9 groups and 4 observations per group. Report the results of the *F* test. Examine differences in means using a multiple-comparisons procedure. Summarize your results and compare them with those you obtained in Exercise 13.22.

13.25 One step in the manufacture of large engines requires that holes of very precise dimensions be drilled. The tools that do the drilling are regularly examined and are adjusted to ensure that the holes meet the required specifications. Part of the examination involves measurement of the diameter of the drilling tool. A team studying the variation in the sizes of the drilled holes selected this

TABLE 13.2				
Tool diameter data				
Tool	Time		Diameter	
1	1	25.030	25.030	25.032
1	2	25.028	25.028	25.028
1	3	25.026	25.026	25.026
2	1	25.016	25.018	25.016
2	2	25.022	25.020	25.018
2	3	25.016	25.016	25.016
3	1	25.005	25.008	25.006
3	2	25.012	25.012	25.014
3	3	25.010	25.010	25.008
4	1	25.012	25.012	25.012
4	2	25.018	25.020	25.020
4	3	25.010	25.014	25.018
5	1	24.996	24.998	24.998
5	2	25.006	25.006	25.006
5	3	25.000	25.002	24.999

measurement procedure as a possible cause of variation in the drilled holes. They decided to use a designed experiment as one part of this examination. Some of the data are given in Table 13.2. The diameters in millimeters (mm) of five tools were measured by the same operator at three times (8:00 a.m., 11:00 a.m., and 3:00 p.m.). Three measurements were taken on each tool at each time. The person taking the measurements could not tell which tool was being measured, and the measurements were taken in random order.[14]

(a) Make a table of means and standard deviations for each of the 5 × 3 combinations of the two factors.

(b) Plot the means and describe how the means vary with tool and time. Note that we expect the tools to have slightly different diameters. These will be adjusted as needed. It is the process of measuring the diameters that is important.

(c) Use a two-way ANOVA to analyze these data. Report the test statistics, degrees of freedom, and P-values for the significance tests.

(d) Write a short report summarizing your results.

13.26   Refer to the previous exercise. Multiply each measurement by 0.04 to convert from millimeters to inches. Redo the plots and rerun the ANOVA using the transformed measurements. Summarize what parts of the analysis have changed and what parts have remained the same.

13.27   How does the frequency that a supermarket product is promoted at a discount affect the price that customers expect to pay for the product? Does the per-

TABLE 13.3

Expected price data

Number of promotions	Percent discount	Expected price ($)									
1	40	4.10	4.50	4.47	4.42	4.56	4.69	4.42	4.17	4.31	4.59
1	30	3.57	3.77	3.90	4.49	4.00	4.66	4.48	4.64	4.31	4.43
1	20	4.94	4.59	4.58	4.48	4.55	4.53	4.59	4.66	4.73	5.24
1	10	5.19	4.88	4.78	4.89	4.69	4.96	5.00	4.93	5.10	4.78
3	40	4.07	4.13	4.25	4.23	4.57	4.33	4.17	4.47	4.60	4.02
3	30	4.20	3.94	4.20	3.88	4.35	3.99	4.01	4.22	3.70	4.48
3	20	4.88	4.80	4.46	4.73	3.96	4.42	4.30	4.68	4.45	4.56
3	10	4.90	5.15	4.68	4.98	4.66	4.46	4.70	4.37	4.69	4.97
5	40	3.89	4.18	3.82	4.09	3.94	4.41	4.14	4.15	4.06	3.90
5	30	3.90	3.77	3.86	4.10	4.10	3.81	3.97	3.67	4.05	3.67
5	20	4.11	4.35	4.17	4.11	4.02	4.41	4.48	3.76	4.66	4.44
5	10	4.31	4.36	4.75	4.62	3.74	4.34	4.52	4.37	4.40	4.52
7	40	3.56	3.91	4.05	3.91	4.11	3.61	3.72	3.69	3.79	3.45
7	30	3.45	4.06	3.35	3.67	3.74	3.80	3.90	4.08	3.52	4.03
7	20	3.89	4.45	3.80	4.15	4.41	3.75	3.98	4.07	4.21	4.23
7	10	4.04	4.22	4.39	3.89	4.26	4.41	4.39	4.52	3.87	4.70

cent reduction also affect this expectation? These questions were examined by researchers in a study conducted on students enrolled in an introductory management course at a large midwestern university. For 10 weeks 160 subjects received information about the products. The treatment conditions corresponded to the number of promotions (1, 3, 5, or 7) that were described during this 10-week period and the percent that the product was discounted (10%, 20%, 30%, and 40%). Ten students were randomly assigned to each of the $4 \times 4 = 16$ treatments.[15] Table 13.3 gives the data.

(a) Summarize the means and standard deviations in a table and plot the means. Summarize the main features of the plot.

(b) Analyze the data with a two-way ANOVA. Report the results of this analysis.

(c) Using your plot and the ANOVA results, prepare a short report explaining how the expected price depends on the number of promotions and the percent of the discount.

13.28   Refer to the previous exercise. Rerun the analysis as a one-way ANOVA with $4 \times 4 = 16$ treatments. Summarize the results of this analysis. Use a multiple-comparisons procedure to describe combinations of number of promotions and percent discounts that are similar or different.

13.29   Do left-handed people live shorter lives than right-handed people? A study of this question examined a sample of 949 death records and contacted next of kin to determine handedness.[16] Note that there are many possible definitions of "left-handed." The researchers examined the effects of different definitions on the results of their analysis and found that their conclusions were not sensitive to the exact definition used. For the results presented here, people were defined to be right-handed if they wrote, drew, and threw a ball with the right hand. All others were defined to be left-handed. People were classified by gender (female or male) and handedness (left or right), and a 2×2 ANOVA was run with the age at death as the response variable. The F statistics were 22.36 (handedness), 37.44 (gender), and 2.10 (interaction). The following marginal mean ages at death (in years) were reported: 77.39 (females), 71.32 (males), 75.00 (right-handed), and 66.03 (left-handed).

(a) For each of the F statistics given above find the degrees of freedom and an approximate P-value. Summarize the results of these tests.

(b) Using the information given, write a short summary of the results of the study.

13.30 Scientists believe that exposure to the radioactive gas radon is associated with some types of cancers in the respiratory system. Radon from natural sources is present in many homes in the United States. A group of researchers decided to study the problem in dogs because dogs get similar types of cancers and are exposed to environments similar to those of their owners. Radon detectors are available for home monitoring but the researchers wanted to obtain actual measures of the exposure of a sample of dogs. To do this they placed the detectors in holders and attached them to the collars of the dogs. One problem was that the holders might in some way affect the radon readings. The researchers therefore devised a laboratory experiment to study the effects of the holders. Detectors from four series of production were available, so they used a two-way ANOVA design (series with 4 levels and holder with 2, representing the presence or absence of a holder). All detectors were exposed to the same radon source and the radon measure in picocuries per liter was recorded.[17] The F statistic for the effect of series is 7.02, for holder it is 1.96, and for the interaction it is 1.24.

(a) Using Table E or statistical software find approximate P-values for the three test statistics. Summarize the results of these three significance tests.

(b) The mean radon readings for the four series were 330, 303, 302, and 295. The results of the significance test for series were of great concern to the researchers. Explain why.

13.31 The PLANTS1 data set in the Data Appendix gives the percent of nitrogen in four different species of plants grown in a laboratory. The species are *Leucaena leucocephala*, *Acacia saligna*, *Prosopis juliflora*, and *Eucalyptus citriodora*. The researchers who collected these data were interested in commercially growing these plants in parts of the country of Jordan where there is very little rainfall. To examine the effect of water, they varied the amount per day from 50 millimeters (mm) to 650 mm in 100 mm increments. There were nine plants per species-by-water combination. Because the plants are to be used primarily for animal food, with some parts that can be consumed by people, a high nitrogen content is very desirable.

(a) Find the means for each species-by-water combination. Plot these means versus water for the four species, connecting the means for each species by lines. Describe the overall pattern.

(b) Find the standard deviations for each species-by-water combination. Is it reasonable to pool the standard deviations for this problem? Note that with sample sizes of size 9, we expect these standard deviations to be quite variable.

(c) Run the two-way analysis of variance. Give the results of the hypothesis tests for the main effects and the interaction.

13.32   Refer to the previous exercise. Examine the residuals. Are there any unusual patterns or outliers? If you think that there are one or more points that are somewhat extreme, rerun the two-way analysis without these observations. Does this change the results in any substantial way?

13.33   Refer to Exercise 13.31. Run a separate one-way analysis of variance for each water level. If there is evidence that the species are not all the same, use a multiple-comparisons procedure to determine which pairs of species are significantly different. In what way, if any, do the differences appear to vary by water level? Write a short summary of your conclusions.

13.34   Refer to Exercise 13.31. Run a separate one-way analysis of variance for each species and summarize the results. Since the amount of water is a quantitative factor, we can also analyze these data using regression. Run simple linear regressions separately for each species to predict nitrogen percent from water. Use plots to determine whether or not a line is a good way to approximate this relationship. Summarize the regression results and compare them with the one-way ANOVA results.

13.35   Refer to Exercise 13.31. Additional data collected by the same researchers according to a similar design are given in the PLANTS2 data set in the Data Appendix. Here, there are two response variables. They are fresh biomass and dry biomass. High values for both of these variables are desirable. The same four species and seven levels of water are used for this experiment. Here, however, there are four plants per species-by-water combination. Analyze each of the response variables in the PLANTS2 data set using the outline from Exercise 13.31.

13.36   Perform the tasks described in Exercise 13.32 for the two response variables in the PLANTS2 data set.

13.37   Perform the tasks described in Exercise 13.33 for the two response variables in the PLANTS2 data set.

13.38   Perform the tasks described in Exercise 13.34 for the two response variables in the PLANTS2 data set.

13.39   Refer to the data given for the change-of-majors study in the data set MAJORS described in the Data Appendix. Analyze the data for SATV, the SAT verbal score. Your analysis should include a table of sample sizes, means, and standard deviations; normal quantile plots; a plot of the means; and a two-way ANOVA using sex and major as the factors. Write a short summary of your conclusions.

13.40 Refer to the data given for the change-of-majors study in the data set MAJORS described in the Data Appendix. Analyze the data for HSS, the high school science grades. Your analysis should include a table of sample sizes, means, and standard deviations; normal quantile plots; a plot of the means; and a two-way ANOVA using sex and major as the factors. Write a short summary of your conclusions.

13.41 Refer to the data given for the change-of-majors study in the data set MAJORS described in the Data Appendix. Analyze the data for HSE, the high school English grades. Your analysis should include a table of sample sizes, means, and standard deviations; normal quantile plots; a plot of the means; and a two-way ANOVA using sex and major as the factors. Write a short summary of your conclusions.

13.42 Refer to the data given for the change-of-majors study in the data set MAJORS described in the Data Appendix. Analyze the data for GPA, the college grade point average. Your analysis should include a table of sample sizes, means, and standard deviations; normal quantile plots; a plot of the means; and a two-way ANOVA using sex and major as the factors. Write a short summary of your conclusions.

13.43 Search the Internet or your library to find a study that is interesting to you and uses a two-way ANOVA to analyze the data. First describe the question or questions of interest and then give the details of how ANOVA was used to provide answers. Be sure to include how the study authors examined the assumptions for the analysis. Evaluate how well the authors used ANOVA in this study. If your evaluation finds the analysis deficient, make suggestions for how it could be improved.

Some of the computer exercises in the text refer to thirteen relatively large data sets that are on the CD that accompanies this text. The CD also contains data from many other exercises and examples.

Background information for each of the thirteen data sets is presented below. For each, the first five cases are given here.

1 | BIOMARKERS

Text Reference:
Exercises
7.110–7.113,
10.10–10.13,
and 11.26–11.31

Healthy bones are continually being renewed by two processes. Through bone formation, new bone is built; through bone resorption, old bone is removed. If one or both of these processes is disturbed, by disease, aging, or space travel, for example, bone loss can be the result. The variables VOPLUS and VOMINUS measure bone formation and bone resorption, respectively. Osteocalcin (OC) is a biochemical marker for bone formation: higher levels of bone formation are associated with higher levels of OC. A blood sample is used to measure OC, and it is much less expensive to obtain than direct measures of bone formation. The units are milligrams of OC per milliliter of blood (mg/ml). Similarly, tartrate resistant acid phosphatase (TRAP) is a biochemical marker for bone resorption that is also measured in blood. It is measured in units per liter (U/l). These variables were measured in a study of 31 healthy women aged 11 to 32 years. The results were published in C. M. Weaver et al., "Quantification of biochemical markers of bone turnover by kinetic measures of bone formation and resorption in young healthy females," *Journal of Bone and Mineral Research*, 12 (1997), pp. 1714–1720. Variables with the first letter "L" are the logarithms of the measured variables. The data were provided by Linda McCabe.

VOPLUS	VOMINUS	OC	LOC	TRAP	LTRAP	LVOPLUS	LVOMINUS
1606	903	68.9	4.233	19.4	2.965	7.382	6.806
2240	1761	56.3	4.031	25.5	3.239	7.714	7.474
2221	1486	54.6	4.000	19.0	2.944	7.706	7.304
896	1116	31.2	3.440	9.0	2.197	6.798	7.018
2545	2236	36.4	3.595	19.1	2.950	7.842	7.712

2 | CHEESE

Text Reference:
Exercises
11.43–11.51

As cheddar cheese matures, a variety of chemical processes take place. The taste of matured cheese is related to the concentration of several chemicals in the final product. In a study of cheddar cheese from the LaTrobe Valley of Victoria, Australia, samples of cheese were analyzed for their chemical composition and were subjected to taste tests.

Data for one type of cheese-manufacturing process appear below. The variable "Case" is used to number the observations from 1 to 30. "Taste" is the response variable of interest. The taste scores were obtained by combining the scores from several tasters.

Three of the chemicals whose concentrations were measured were acetic acid, hydrogen sulfide, and lactic acid. For acetic acid and hydrogen sulfide (natural) log transformations were taken. Thus the explanatory variables are the transformed concentrations of acetic acid ("Acetic") and hydrogen sulfide ("H2S") and the untransformed concentration of lactic acid ("Lactic"). These data are based on experiments performed by G. T. Lloyd and E. H. Ramshaw of the CSIRO Division of Food Research, Victoria, Australia. Some results of the statistical analyses of these data are given in G. P. McCabe, L. McCabe, and A. Miller, "Analysis of taste and chemical composition of cheddar cheese, 1982–83 experiments," CSIRO Division of Mathematics and Statistics Consulting Report VT85/6; and in I. Barlow et al., "Correlations and changes in flavour and chemical parameters of cheddar cheeses during maturation," *Australian Journal of Dairy Technology*, 44 (1989), pp. 7–18. The table below gives the data for the first five cases.

Case	Taste	Acetic	H2S	Lactic
01	12.3	4.543	3.135	0.86
02	20.9	5.159	5.043	1.53
03	39.0	5.366	5.438	1.57
04	47.9	5.759	7.496	1.81
05	5.6	4.663	3.807	0.99

3 | CSDATA

Text Reference: Exercises 1.147, 2.113, 3.77, 3.78, 7.132, and 7.133

The computer science department of a large university was interested in understanding why a large proportion of their first-year students failed to graduate as computer science majors. An examination of records from the registrar indicated that most of the attrition occurred during the first three semesters. Therefore, they decided to study all first-year students entering their program in a particular year and to follow their progress for the first three semesters.

The variables studied included the grade point average after three semesters and a collection of variables that would be available as students entered their program. These included scores on standardized tests such as the SATs and high school grades in various subjects. The individuals who conducted the study were also interested in examining differences between men and women in this program. Therefore, sex was included as a variable.

Data on 224 students who began study as computer science majors in a particular year were analyzed. A few exceptional cases were excluded, such as students who did not have complete data available on the variables of interest (a few students were admitted who did not take the SATs). Data for the first five students appear below. There are eight variables for each student. OBS is a variable used to identify the student. The data files kept by the registrar identified students by social security number, but for this study they were simply given a number from 1 to 224. The grade point average after three semesters is the variable GPA. This university uses a four-point scale, with A corresponding to 4, B to 3, C to 2, etc. A straight-A student has a 4.00 GPA.

The high school grades included in the data set are the variables HSM, HSS, and HSE. These correspond to average high school grades in math, science, and English. High schools use different grading systems (some high schools have a grade higher than A for honors courses), so the university's task in constructing these variables is not easy. The researchers were willing to accept the university's judgment and used its values. High school grades were recorded on a scale from 1 to 10, with 10 corresponding to A, 9 to A−, 8 to B+, etc.

The SAT scores are SATM and SATV, corresponding to the mathematics and verbal parts of the SAT. Gender was recorded as 1 for men and 2 for women. This is an arbitrary code. For software packages that can use alphanumeric variables (values do not have to be numbers), it is more convenient to use M and F or Men and Women as values for the sex variable. With this kind of user-friendly capability, you do not have to remember who are the 1s and who are the 2s.

Results of the study are reported in P. F. Campbell and G. P. McCabe, "Predicting the success of freshmen in a computer science major," *Communications of the ACM*, 27 (1984), pp. 1108–1113. The table below gives data for the first five students.

OBS	GPA	HSM	HSS	HSE	SATM	SATV	SEX
001	3.32	10	10	10	670	600	1
002	2.26	6	8	5	700	640	1
003	2.35	8	6	8	640	530	1
004	2.08	9	10	7	670	600	1
005	3.38	8	9	8	540	580	1

4 | DANDRUFF

Text Reference:
Exercises
12.46–12.49

The DANDRUFF data set is based on W. L. Billhimer et al., "Results of a clinical trial comparing 1% pyrithione zinc and 2% ketoconazole shampoos," *Cosmetic Dermatology*, 9 (1996), pp. 34–39. The study reported in this paper is a clinical trial that compared three treatments for dandruff and a placebo. The treatments were 1% pyrithione zinc shampoo (PyrI), the same shampoo but with instructions to shampoo two times (PyrII), 2% ketoconazole shampoo (Keto), and a placebo shampoo (Placebo). After six weeks of treatment, eight sections of the scalp were examined and given a score that measured the amount of scalp flaking on a 0 to 10 scale. The response variable was the sum of these eight scores. An analysis of the baseline flaking measure indicated that randomization of patients to treatments was successful in that no differences were found between the groups. At baseline there were 112 subjects in each of the three treatment groups and 28 subjects in the Placebo group. During the clinical trial 3 dropped out from the PyrII group and 6 from the Keto group. No patients dropped out of the other two groups. Summary statistics given in the paper were used to generate random data that give the same conclusions. Here are the first five cases:

OBS	Treatment	Flaking
001	PyrI	17
002	PyrI	16
003	PyrI	18
004	PyrI	17
005	PyrI	18

5 | LONGLEAF

Text Reference:
Example 6.1 and
Exercises 7.75, 7.76,
7.87–7.90, 7.99,
7.100, 7.114–7.116

The Wade Tract in Thomas County, Georgia, is an old-growth forest of longleaf pine trees (*Pinus palustris*) that has survived in a relatively undisturbed state since before the settlement of the area by Europeans. This data set includes observations on 584 longleaf pine trees in a 200-meter by 200-meter region in the Wade Tract and is described in Noel Cressie, *Statistics for Spatial Data*, Wiley, 1993. The variable NS is the location within the region in the north–south direction, EW is the location within the region in the east–west direction, and DBH is the diameter of the tree at breast height, measured in centimeters. Here are the first five cases:

NS	EW	DBH
200.0	8.8	32.9
199.3	10.0	53.5
193.6	22.4	68.0
167.7	35.6	17.7
183.9	45.4	36.9

6 | MAJORS

Text Reference:
Exercises
13.39–13.42 See the description of the CSDATA data set for background information on the study for this data set. In this data file, the variables described for CSDATA are given with an additional variable "Maj" that specifies the student's major field of study at the end of three semesters. The codes 1, 2, and 3 correspond to Computer Science, Engineering and Other Sciences, and Other. All available data were used in the analyses performed, which resulted in sample sizes that were unequal in the six sex-by-major groups.

For a one-way ANOVA this causes no particular problems. However, for a two-way ANOVA several complications arise when the sample sizes are unequal. A detailed discussion of these complications is beyond the scope of this text. To avoid these difficulties and still use these interesting data, simulated data based on the results of this study are given on the data disk. ANOVA based on these simulated data gives the same qualitative conclusions as those obtained with the original data. Here are the first five cases:

OBS	SEX	Maj	SATM	SATV	HSM	HSS	HSE	GPA
001	1	1	640	530	8	6	8	2.35
002	1	1	670	600	9	10	7	2.08
003	1	1	600	400	8	8	7	3.21
004	1	1	570	480	7	7	6	2.34
005	1	1	510	530	6	8	8	1.40

7 | PCB

Text Reference:
Exercises 7.126,
7.127, 11.32–11.42 Polychlorinated biphenyls (PCBs) are a collection of compounds that are no longer produced in the United States but are still found in the environment. Evidence suggests that they can cause harmful health effects when consumed. Because PCBs can accumulate in fish, efforts have been made to identify areas where fish contain excessive amounts so that recommendations concerning consumption limits can be made. There are over 200 types of PCBs. Data from the Environmental Protection Agency National Study of Residues in Lake Fish are given in the data set PCB. Various lakes in the United States were sampled and the amounts of PCBs in fish were measured. The variable PCB is the sum of the amounts of all PCBs found in the fish, while the other variables with the prefix PCB are particular types of PCBs. The units are parts per billion (ppb). Not all types of PCBs are equally harmful. A scale has been developed to convert the raw amount of each type of PCB to a toxic equivalent (TEQ). TEQPCB is the total TEQ from all PCBs found in each sample. Dioxins and furans are also contaminants that were measured in these samples. The variables TEQDIOXIN and TEQFURAN give the TEQ of these, while TEQ is the total of TEQPCB, TEQDIOXIN, and TEQFURAN. More information is available online at http://epa.gov/waterscience/fishstudy/. This data set was provided by Joanne Lasrado of the Purdue University Department of Foods and Nutrition. Here are the first five cases:

PCB138	PCB153	PCB180	PCB28	PCB52	PCB126	PCB118
1.46	1.59	0.738	0.421	0.532	0.0000	0.720
0.64	0.92	0.664	0.025	0.030	0.0000	0.236
3.29	3.90	1.150	0.076	0.134	0.0000	1.540
3.94	5.44	1.330	0.152	0.466	0.0055	1.940
3.18	3.65	2.140	0.116	0.243	0.0059	1.470

PCB	TEQ	TEQPCB	TEQDIOXIN	TEQFURAN
19.9959	0.93840	0.18892	0.60948	0.1400
6.0996	0.96881	0.06837	0.90044	0.0000
24.9655	0.97992	0.32992	0.62000	0.0300
37.4436	0.99850	0.92350	0.00500	0.0700
30.1830	1.01654	0.92654	0.00000	0.0900

8 | PLANTS1

Text Reference:
Exercises
13.31–13.34

These data were collected by Maher Tadros from the Purdue University Department of Forestry and Natural Resources under the direction of his major professor, Andrew Gillespie. Maher is from Jordan, a Middle Eastern country where there is very little rainfall in many areas. His research concerns four species of plants that may be suitable for commercial development in his country. Products produced by these species can be used as feed for animals and in some cases for humans. The four species of plants are *Leucaena leucocephala*, *Acacia saligna*, *Prosopis juliflora*, and *Eucalyptus citriodora*. A major research question concerns how well these species can tolerate drought stress.

PLANTS1 gives data for a laboratory experiment performed by Maher at Purdue University. Seven different amounts of water were given daily to plants of each species. For each of the species-by-water combinations, there were nine plants. The response variable is the percent of the plant that consists of nitrogen. A high nitrogen content is desirable for plant products that are used for food.

The actual experiment performed to collect these data had an additional factor that is not given in the data set. The 4 × 7, species-by-water combinations were actually run with three plants per combination. This design was then repeated three times. The repeat factor is often called a replicate, or rep, and is a standard part of most well-designed experiments of this type. For our purposes we ignore this additional factor and analyze the design as a 4 × 7 two-way ANOVA with 9 observations per treatment combination. The first five cases are listed below. The four species, *Leucaena leucocephala*, *Acacia saligna*, *Prosopis juliflora*, and *Eucalyptus citriodora*, are coded 1 to 4. The water levels, 50 mm, 150 mm, 250 mm, 350 mm, 450 mm, 550 mm, and 650 mm, are coded 1 to 7.

OBS	Species	Water	pctnit
001	1	1	3.644
002	1	1	3.500
003	1	1	3.509
004	1	1	3.137
005	1	1	3.100

9 | PLANTS2

Text Reference:
Exercises
13.35–13.38

PLANTS2 gives data for a second experiment conducted by Maher Tadros in a lab at Purdue University. As in PLANTS1, there are the same four species of plants and the same seven levels of water. Here, however, there are four plants per species-by-water combination. The two response variables in the data set are fresh biomass and dry biomass. High values for these response variables indicate that the plants of the given species are resistant to drought at the given water level. Here are the first five cases:

OBS	Species	Water	fbiomass	dbiomass
001	1	1	105.13	37.65
002	1	1	138.95	48.85
003	1	1	90.05	38.85
004	1	1	102.25	36.91
005	1	1	207.90	74.35

10 | PNG

Text Reference:
Exercises 7.2–7.4,
7.128, 7.129,
and 10.9

A randomized double-blind placebo-controlled trial to assess the benefits of giving high-dose vitamin A supplements to young children was performed in the North Wosera District of East Sepik Province in Papua New Guinea. The results of the trial are reported in F. Rosales et al., "Relation of serum retinol to acute phase proteins and malarial morbidity in Papua New Guinea children," *American Journal of Clinical Nutrition*, 71 (2000), pp. 1580–1588. The data here were collected at the start of the study and were provided by Francisco Rosales of the Department of Nutritional Sciences, Pennsylvania State University. AGEY is age in years, RETINOL is serum retinol, CRP is C-reactive protein, AGP is α1-acid glycoprotein, and ACT is α1-antichymotrypsin. Serum retinol is a measure of vitamin A status, and CRP, AGP, and ACT are acute-phase proteins that have high values when there is an infection. Here are the first five cases:

ID	AGEY	RETINOL	CRP	AGP	ACT
1	4.33	0.31	3.49	0.61	0.28
2	3.33	0.26	8.64	0.82	0.34
3	2.50	0.24	33.59	0.73	0.36
4	3.33	0.67	3.47	1.05	0.34
5	4.83	0.59	0.00	0.55	0.08

11 | READING

Text Reference:
Exercises 7.131,
12.53, and 12.54

Jim Baumann and Leah Jones of the Purdue University School of Education conducted a study to compare three methods of teaching reading comprehension. The 66 students who participated in the study were randomly assigned to the methods (22 to each). The standard practice of comparing new methods with a traditional one was used in this study. The traditional method is called Basal and the two innovative methods are called DRTA and Strat.

In the data set the variable Subject is used to identify the individual students. The values are 1 to 66. The method of instruction is indicated by the variable Group, with values B, D, and S, corresponding to Basal, DRTA, and Strat. Two pretests and three posttests were given to all students. These are the variables Pre1, Pre2, Post1, Post2, and Post3. Data for the first five subjects are given below.

Subject	Group	Pre1	Pre2	Post1	Post2	Post3
01	B	4	3	5	4	41
02	B	6	5	9	5	41
03	B	9	4	5	3	43
04	B	12	6	8	5	46
05	B	16	5	10	9	46

12 | RUNNERS

Text Reference:
Exercise 1.61 and
Example 13.11

A study of cardiovascular risk factors compared runners who averaged at least 15 miles per week with a control group described as "generally sedentary." Both men and women were included in the study. The data set was constructed based on information provided in P. D. Wood et al., "Plasma lipoprotein distributions in male and female runners," in P. Milvey (ed.), *The Marathon: Physiological, Medical, Epidemiological, and Psychological Studies*, New York Academy of Sciences, 1977.

The study design is a 2 × 2 ANOVA with the factors group and gender. There were 200 subjects in each of the four combinations. The variables are Id, a numeric subject identifier; Group, with values "Control" and "Runners"; Gender, with values Female and Male; and HeartRate, heart rate after the subject ran for 6 minutes on a treadmill. Here are the data for the first five subjects:

id	group	sex	beats
1	Control	Female	159
2	Control	Female	183
3	Control	Female	140
4	Control	Female	140
5	Control	Female	125

13 | WORKERS

Text Reference:
Exercises 1.62–1.64,
1.146, 3.71, and 3.79

Each March, the Bureau of Labor Statistics carries out an Annual Demographic Supplement to its monthly Current Population Survey. The data file *workers.dat* contains data about 71,076 people from one of these surveys. We included all people between the ages of 25 and 64 who have worked but whose main work experience is not in agriculture. Moreover, we combined the 16 levels of education in the BLS survey to form 6 levels.

There are four variables in the data set. Age is coded as follows: 6 = 25–29 years, 7 = 30–34 years, 8 = 35–39 years, 9 = 40–44 years, 10 = 45–49 years, 11 = 50–54 years, 12 = 55–59 years, 13 = 60–61 years, 14 = 62–64 years. Education is the highest level of education a person has reached, with the following values: 1 = did not reach high school; 2 = some high school but no high school diploma; 3 = high school diploma; 4 = some college but no bachelor's degree (this includes people with an associate degree); 5 = bachelor's degree; 6 = postgraduate degree (master's, professional, or doctorate). Sex is gender coded as 1 = male and 2 = female. Total income is income from all sources. Note that income can be less than zero in some cases. Here are the first five cases:

Age	Education	Sex	Total income
6	1	1	19,760
6	1	1	20,800
6	1	1	24,700
6	1	1	10,000
6	1	1	36,060

The first worker in the file is a male between 25 and 29 years old who did not graduate from high school and who earned $19,760.

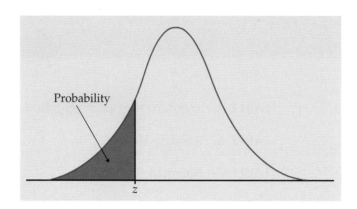

Table entry for z is the area under the standard normal curve to the left of z.

Probability

z

TABLE A	Standard normal probabilities								
z	.00	.01	.02	.03	.04	.05	.06	.07	.08	.09
−3.4	.0003	.0003	.0003	.0003	.0003	.0003	.0003	.0003	.0003	.0002
−3.3	.0005	.0005	.0005	.0004	.0004	.0004	.0004	.0004	.0004	.0003
−3.2	.0007	.0007	.0006	.0006	.0006	.0006	.0006	.0005	.0005	.0005
−3.1	.0010	.0009	.0009	.0009	.0008	.0008	.0008	.0008	.0007	.0007
−3.0	.0013	.0013	.0013	.0012	.0012	.0011	.0011	.0011	.0010	.0010
−2.9	.0019	.0018	.0018	.0017	.0016	.0016	.0015	.0015	.0014	.0014
−2.8	.0026	.0025	.0024	.0023	.0023	.0022	.0021	.0021	.0020	.0019
−2.7	.0035	.0034	.0033	.0032	.0031	.0030	.0029	.0028	.0027	.0026
−2.6	.0047	.0045	.0044	.0043	.0041	.0040	.0039	.0038	.0037	.0036
−2.5	.0062	.0060	.0059	.0057	.0055	.0054	.0052	.0051	.0049	.0048
−2.4	.0082	.0080	.0078	.0075	.0073	.0071	.0069	.0068	.0066	.0064
−2.3	.0107	.0104	.0102	.0099	.0096	.0094	.0091	.0089	.0087	.0084
−2.2	.0139	.0136	.0132	.0129	.0125	.0122	.0119	.0116	.0113	.0110
−2.1	.0179	.0174	.0170	.0166	.0162	.0158	.0154	.0150	.0146	.0143
−2.0	.0228	.0222	.0217	.0212	.0207	.0202	.0197	.0192	.0188	.0183
−1.9	.0287	.0281	.0274	.0268	.0262	.0256	.0250	.0244	.0239	.0233
−1.8	.0359	.0351	.0344	.0336	.0329	.0322	.0314	.0307	.0301	.0294
−1.7	.0446	.0436	.0427	.0418	.0409	.0401	.0392	.0384	.0375	.0367
−1.6	.0548	.0537	.0526	.0516	.0505	.0495	.0485	.0475	.0465	.0455
−1.5	.0668	.0655	.0643	.0630	.0618	.0606	.0594	.0582	.0571	.0559
−1.4	.0808	.0793	.0778	.0764	.0749	.0735	.0721	.0708	.0694	.0681
−1.3	.0968	.0951	.0934	.0918	.0901	.0885	.0869	.0853	.0838	.0823
−1.2	.1151	.1131	.1112	.1093	.1075	.1056	.1038	.1020	.1003	.0985
−1.1	.1357	.1335	.1314	.1292	.1271	.1251	.1230	.1210	.1190	.1170
−1.0	.1587	.1562	.1539	.1515	.1492	.1469	.1446	.1423	.1401	.1379
−0.9	.1841	.1814	.1788	.1762	.1736	.1711	.1685	.1660	.1635	.1611
−0.8	.2119	.2090	.2061	.2033	.2005	.1977	.1949	.1922	.1894	.1867
−0.7	.2420	.2389	.2358	.2327	.2296	.2266	.2236	.2206	.2177	.2148
−0.6	.2743	.2709	.2676	.2643	.2611	.2578	.2546	.2514	.2483	.2451
−0.5	.3085	.3050	.3015	.2981	.2946	.2912	.2877	.2843	.2810	.2776
−0.4	.3446	.3409	.3372	.3336	.3300	.3264	.3228	.3192	.3156	.3121
−0.3	.3821	.3783	.3745	.3707	.3669	.3632	.3594	.3557	.3520	.3483
−0.2	.4207	.4168	.4129	.4090	.4052	.4013	.3974	.3936	.3897	.3859
−0.1	.4602	.4562	.4522	.4483	.4443	.4404	.4364	.4325	.4286	.4247
0.0	.5000	.4960	.4920	.4880	.4840	.4801	.4761	.4721	.4681	.4641

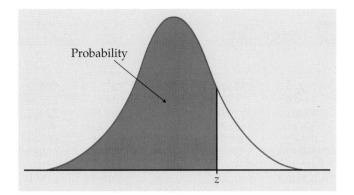

Probability

Table entry for z is the area under the standard normal curve to the left of z.

TABLE A	Standard normal probabilities (continued)								
z	.00	.01	.02	.03	.04	.05	.06	.07	.08	.09
0.0	.5000	.5040	.5080	.5120	.5160	.5199	.5239	.5279	.5319	.5359
0.1	.5398	.5438	.5478	.5517	.5557	.5596	.5636	.5675	.5714	.5753
0.2	.5793	.5832	.5871	.5910	.5948	.5987	.6026	.6064	.6103	.6141
0.3	.6179	.6217	.6255	.6293	.6331	.6368	.6406	.6443	.6480	.6517
0.4	.6554	.6591	.6628	.6664	.6700	.6736	.6772	.6808	.6844	.6879
0.5	.6915	.6950	.6985	.7019	.7054	.7088	.7123	.7157	.7190	.7224
0.6	.7257	.7291	.7324	.7357	.7389	.7422	.7454	.7486	.7517	.7549
0.7	.7580	.7611	.7642	.7673	.7704	.7734	.7764	.7794	.7823	.7852
0.8	.7881	.7910	.7939	.7967	.7995	.8023	.8051	.8078	.8106	.8133
0.9	.8159	.8186	.8212	.8238	.8264	.8289	.8315	.8340	.8365	.8389
1.0	.8413	.8438	.8461	.8485	.8508	.8531	.8554	.8577	.8599	.8621
1.1	.8643	.8665	.8686	.8708	.8729	.8749	.8770	.8790	.8810	.8830
1.2	.8849	.8869	.8888	.8907	.8925	.8944	.8962	.8980	.8997	.9015
1.3	.9032	.9049	.9066	.9082	.9099	.9115	.9131	.9147	.9162	.9177
1.4	.9192	.9207	.9222	.9236	.9251	.9265	.9279	.9292	.9306	.9319
1.5	.9332	.9345	.9357	.9370	.9382	.9394	.9406	.9418	.9429	.9441
1.6	.9452	.9463	.9474	.9484	.9495	.9505	.9515	.9525	.9535	.9545
1.7	.9554	.9564	.9573	.9582	.9591	.9599	.9608	.9616	.9625	.9633
1.8	.9641	.9649	.9656	.9664	.9671	.9678	.9686	.9693	.9699	.9706
1.9	.9713	.9719	.9726	.9732	.9738	.9744	.9750	.9756	.9761	.9767
2.0	.9772	.9778	.9783	.9788	.9793	.9798	.9803	.9808	.9812	.9817
2.1	.9821	.9826	.9830	.9834	.9838	.9842	.9846	.9850	.9854	.9857
2.2	.9861	.9864	.9868	.9871	.9875	.9878	.9881	.9884	.9887	.9890
2.3	.9893	.9896	.9898	.9901	.9904	.9906	.9909	.9911	.9913	.9916
2.4	.9918	.9920	.9922	.9925	.9927	.9929	.9931	.9932	.9934	.9936
2.5	.9938	.9940	.9941	.9943	.9945	.9946	.9948	.9949	.9951	.9952
2.6	.9953	.9955	.9956	.9957	.9959	.9960	.9961	.9962	.9963	.9964
2.7	.9965	.9966	.9967	.9968	.9969	.9970	.9971	.9972	.9973	.9974
2.8	.9974	.9975	.9976	.9977	.9977	.9978	.9979	.9979	.9980	.9981
2.9	.9981	.9982	.9982	.9983	.9984	.9984	.9985	.9985	.9986	.9986
3.0	.9987	.9987	.9987	.9988	.9988	.9989	.9989	.9989	.9990	.9990
3.1	.9990	.9991	.9991	.9991	.9992	.9992	.9992	.9992	.9993	.9993
3.2	.9993	.9993	.9994	.9994	.9994	.9994	.9994	.9995	.9995	.9995
3.3	.9995	.9995	.9995	.9996	.9996	.9996	.9996	.9996	.9996	.9997
3.4	.9997	.9997	.9997	.9997	.9997	.9997	.9997	.9997	.9997	.9998

TABLE B Random digits

Line

101	19223	95034	05756	28713	96409	12531	42544	82853
102	73676	47150	99400	01927	27754	42648	82425	36290
103	45467	71709	77558	00095	32863	29485	82226	90056
104	52711	38889	93074	60227	40011	85848	48767	52573
105	95592	94007	69971	91481	60779	53791	17297	59335
106	68417	35013	15529	72765	85089	57067	50211	47487
107	82739	57890	20807	47511	81676	55300	94383	14893
108	60940	72024	17868	24943	61790	90656	87964	18883
109	36009	19365	15412	39638	85453	46816	83485	41979
110	38448	48789	18338	24697	39364	42006	76688	08708
111	81486	69487	60513	09297	00412	71238	27649	39950
112	59636	88804	04634	71197	19352	73089	84898	45785
113	62568	70206	40325	03699	71080	22553	11486	11776
114	45149	32992	75730	66280	03819	56202	02938	70915
115	61041	77684	94322	24709	73698	14526	31893	32592
116	14459	26056	31424	80371	65103	62253	50490	61181
117	38167	98532	62183	70632	23417	26185	41448	75532
118	73190	32533	04470	29669	84407	90785	65956	86382
119	95857	07118	87664	92099	58806	66979	98624	84826
120	35476	55972	39421	65850	04266	35435	43742	11937
121	71487	09984	29077	14863	61683	47052	62224	51025
122	13873	81598	95052	90908	73592	75186	87136	95761
123	54580	81507	27102	56027	55892	33063	41842	81868
124	71035	09001	43367	49497	72719	96758	27611	91596
125	96746	12149	37823	71868	18442	35119	62103	39244
126	96927	19931	36089	74192	77567	88741	48409	41903
127	43909	99477	25330	64359	40085	16925	85117	36071
128	15689	14227	06565	14374	13352	49367	81982	87209
129	36759	58984	68288	22913	18638	54303	00795	08727
130	69051	64817	87174	09517	84534	06489	87201	97245
131	05007	16632	81194	14873	04197	85576	45195	96565
132	68732	55259	84292	08796	43165	93739	31685	97150
133	45740	41807	65561	33302	07051	93623	18132	09547
134	27816	78416	18329	21337	35213	37741	04312	68508
135	66925	55658	39100	78458	11206	19876	87151	31260
136	08421	44753	77377	28744	75592	08563	79140	92454
137	53645	66812	61421	47836	12609	15373	98481	14592
138	66831	68908	40772	21558	47781	33586	79177	06928
139	55588	99404	70708	41098	43563	56934	48394	51719
140	12975	13258	13048	45144	72321	81940	00360	02428
141	96767	35964	23822	96012	94591	65194	50842	53372
142	72829	50232	97892	63408	77919	44575	24870	04178
143	88565	42628	17797	49376	61762	16953	88604	12724
144	62964	88145	83083	69453	46109	59505	69680	00900
145	19687	12633	57857	95806	09931	02150	43163	58636
146	37609	59057	66967	83401	60705	02384	90597	93600
147	54973	86278	88737	74351	47500	84552	19909	67181
148	00694	05977	19664	65441	20903	62371	22725	53340
149	71546	05233	53946	68743	72460	27601	45403	88692
150	07511	88915	41267	16853	84569	79367	32337	03316

TABLE B   Random digits (continued)

Line								
151	03802	29341	29264	80198	12371	13121	54969	43912
152	77320	35030	77519	41109	98296	18984	60869	12349
153	07886	56866	39648	69290	03600	05376	58958	22720
154	87065	74133	21117	70595	22791	67306	28420	52067
155	42090	09628	54035	93879	98441	04606	27381	82637
156	55494	67690	88131	81800	11188	28552	25752	21953
157	16698	30406	96587	65985	07165	50148	16201	86792
158	16297	07626	68683	45335	34377	72941	41764	77038
159	22897	17467	17638	70043	36243	13008	83993	22869
160	98163	45944	34210	64158	76971	27689	82926	75957
161	43400	25831	06283	22138	16043	15706	73345	26238
162	97341	46254	88153	62336	21112	35574	99271	45297
163	64578	67197	28310	90341	37531	63890	52630	76315
164	11022	79124	49525	63078	17229	32165	01343	21394
165	81232	43939	23840	05995	84589	06788	76358	26622
166	36843	84798	51167	44728	20554	55538	27647	32708
167	84329	80081	69516	78934	14293	92478	16479	26974
168	27788	85789	41592	74472	96773	27090	24954	41474
169	99224	00850	43737	75202	44753	63236	14260	73686
170	38075	73239	52555	46342	13365	02182	30443	53229
171	87368	49451	55771	48343	51236	18522	73670	23212
172	40512	00681	44282	47178	08139	78693	34715	75606
173	81636	57578	54286	27216	58758	80358	84115	84568
174	26411	94292	06340	97762	37033	85968	94165	46514
175	80011	09937	57195	33906	94831	10056	42211	65491
176	92813	87503	63494	71379	76550	45984	05481	50830
177	70348	72871	63419	57363	29685	43090	18763	31714
178	24005	52114	26224	39078	80798	15220	43186	00976
179	85063	55810	10470	08029	30025	29734	61181	72090
180	11532	73186	92541	06915	72954	10167	12142	26492
181	59618	03914	05208	84088	20426	39004	84582	87317
182	92965	50837	39921	84661	82514	81899	24565	60874
183	85116	27684	14597	85747	01596	25889	41998	15635
184	15106	10411	90221	49377	44369	28185	80959	76355
185	03638	31589	07871	25792	85823	55400	56026	12193
186	97971	48932	45792	63993	95635	28753	46069	84635
187	49345	18305	76213	82390	77412	97401	50650	71755
188	87370	88099	89695	87633	76987	85503	26257	51736
189	88296	95670	74932	65317	93848	43988	47597	83044
190	79485	92200	99401	54473	34336	82786	05457	60343
191	40830	24979	23333	37619	56227	95941	59494	86539
192	32006	76302	81221	00693	95197	75044	46596	11628
193	37569	85187	44692	50706	53161	69027	88389	60313
194	56680	79003	23361	67094	15019	63261	24543	52884
195	05172	08100	22316	54495	60005	29532	18433	18057
196	74782	27005	03894	98038	20627	40307	47317	92759
197	85288	93264	61409	03404	09649	55937	60843	66167
198	68309	12060	14762	58002	03716	81968	57934	32624
199	26461	88346	52430	60906	74216	96263	69296	90107
200	42672	67680	42376	95023	82744	03971	96560	55148

TABLE C   Binomial probabilities

Entry is $P(X = k) = \binom{n}{k} p^k (1 - p)^{n-k}$

n	k	.01	.02	.03	.04	.05	.06	.07	.08	.09
						p				
2	0	.9801	.9604	.9409	.9216	.9025	.8836	.8649	.8464	.8281
	1	.0198	.0392	.0582	.0768	.0950	.1128	.1302	.1472	.1638
	2	.0001	.0004	.0009	.0016	.0025	.0036	.0049	.0064	.0081
3	0	.9703	.9412	.9127	.8847	.8574	.8306	.8044	.7787	.7536
	1	.0294	.0576	.0847	.1106	.1354	.1590	.1816	.2031	.2236
	2	.0003	.0012	.0026	.0046	.0071	.0102	.0137	.0177	.0221
	3				.0001	.0001	.0002	.0003	.0005	.0007
4	0	.9606	.9224	.8853	.8493	.8145	.7807	.7481	.7164	.6857
	1	.0388	.0753	.1095	.1416	.1715	.1993	.2252	.2492	.2713
	2	.0006	.0023	.0051	.0088	.0135	.0191	.0254	.0325	.0402
	3			.0001	.0002	.0005	.0008	.0013	.0019	.0027
	4									.0001
5	0	.9510	.9039	.8587	.8154	.7738	.7339	.6957	.6591	.6240
	1	.0480	.0922	.1328	.1699	.2036	.2342	.2618	.2866	.3086
	2	.0010	.0038	.0082	.0142	.0214	.0299	.0394	.0498	.0610
	3		.0001	.0003	.0006	.0011	.0019	.0030	.0043	.0060
	4						.0001	.0001	.0002	.0003
	5									
6	0	.9415	.8858	.8330	.7828	.7351	.6899	.6470	.6064	.5679
	1	.0571	.1085	.1546	.1957	.2321	.2642	.2922	.3164	.3370
	2	.0014	.0055	.0120	.0204	.0305	.0422	.0550	.0688	.0833
	3		.0002	.0005	.0011	.0021	.0036	.0055	.0080	.0110
	4					.0001	.0002	.0003	.0005	.0008
	5									
	6									
7	0	.9321	.8681	.8080	.7514	.6983	.6485	.6017	.5578	.5168
	1	.0659	.1240	.1749	.2192	.2573	.2897	.3170	.3396	.3578
	2	.0020	.0076	.0162	.0274	.0406	.0555	.0716	.0886	.1061
	3		.0003	.0008	.0019	.0036	.0059	.0090	.0128	.0175
	4				.0001	.0002	.0004	.0007	.0011	.0017
	5								.0001	.0001
	6									
	7									
8	0	.9227	.8508	.7837	.7214	.6634	.6096	.5596	.5132	.4703
	1	.0746	.1389	.1939	.2405	.2793	.3113	.3370	.3570	.3721
	2	.0026	.0099	.0210	.0351	.0515	.0695	.0888	.1087	.1288
	3	.0001	.0004	.0013	.0029	.0054	.0089	.0134	.0189	.0255
	4			.0001	.0002	.0004	.0007	.0013	.0021	.0031
	5							.0001	.0001	.0002
	6									
	7									
	8									

TABLE C Binomial probabilities (continued)

Entry is $P(X = k) = \binom{n}{k} p^k (1-p)^{n-k}$

n	k	.10	.15	.20	.25	.30	.35	.40	.45	.50
2	0	.8100	.7225	.6400	.5625	.4900	.4225	.3600	.3025	.2500
	1	.1800	.2550	.3200	.3750	.4200	.4550	.4800	.4950	.5000
	2	.0100	.0225	.0400	.0625	.0900	.1225	.1600	.2025	.2500
3	0	.7290	.6141	.5120	.4219	.3430	.2746	.2160	.1664	.1250
	1	.2430	.3251	.3840	.4219	.4410	.4436	.4320	.4084	.3750
	2	.0270	.0574	.0960	.1406	.1890	.2389	.2880	.3341	.3750
	3	.0010	.0034	.0080	.0156	.0270	.0429	.0640	.0911	.1250
4	0	.6561	.5220	.4096	.3164	.2401	.1785	.1296	.0915	.0625
	1	.2916	.3685	.4096	.4219	.4116	.3845	.3456	.2995	.2500
	2	.0486	.0975	.1536	.2109	.2646	.3105	.3456	.3675	.3750
	3	.0036	.0115	.0256	.0469	.0756	.1115	.1536	.2005	.2500
	4	.0001	.0005	.0016	.0039	.0081	.0150	.0256	.0410	.0625
5	0	.5905	.4437	.3277	.2373	.1681	.1160	.0778	.0503	.0313
	1	.3280	.3915	.4096	.3955	.3602	.3124	.2592	.2059	.1563
	2	.0729	.1382	.2048	.2637	.3087	.3364	.3456	.3369	.3125
	3	.0081	.0244	.0512	.0879	.1323	.1811	.2304	.2757	.3125
	4	.0004	.0022	.0064	.0146	.0284	.0488	.0768	.1128	.1562
	5	.0001	.0003	.0010	.0024	.0053	.0102	.0185	.0312	
6	0	.5314	.3771	.2621	.1780	.1176	.0754	.0467	.0277	.0156
	1	.3543	.3993	.3932	.3560	.3025	.2437	.1866	.1359	.0938
	2	.0984	.1762	.2458	.2966	.3241	.3280	.3110	.2780	.2344
	3	.0146	.0415	.0819	.1318	.1852	.2355	.2765	.3032	.3125
	4	.0012	.0055	.0154	.0330	.0595	.0951	.1382	.1861	.2344
	5	.0001	.0004	.0015	.0044	.0102	.0205	.0369	.0609	.0937
	6	.0001	.0002	.0007	.0018	.0041	.0083	.0156		
7	0	.4783	.3206	.2097	.1335	.0824	.0490	.0280	.0152	.0078
	1	.3720	.3960	.3670	.3115	.2471	.1848	.1306	.0872	.0547
	2	.1240	.2097	.2753	.3115	.3177	.2985	.2613	.2140	.1641
	3	.0230	.0617	.1147	.1730	.2269	.2679	.2903	.2918	.2734
	4	.0026	.0109	.0287	.0577	.0972	.1442	.1935	.2388	.2734
	5	.0002	.0012	.0043	.0115	.0250	.0466	.0774	.1172	.1641
	6	.0001	.0004	.0013	.0036	.0084	.0172	.0320	.0547	
	7	.0001	.0002	.0006	.0016	.0037	.0078			
8	0	.4305	.2725	.1678	.1001	.0576	.0319	.0168	.0084	.0039
	1	.3826	.3847	.3355	.2670	.1977	.1373	.0896	.0548	.0313
	2	.1488	.2376	.2936	.3115	.2965	.2587	.2090	.1569	.1094
	3	.0331	.0839	.1468	.2076	.2541	.2786	.2787	.2568	.2188
	4	.0046	.0185	.0459	.0865	.1361	.1875	.2322	.2627	.2734
	5	.0004	.0026	.0092	.0231	.0467	.0808	.1239	.1719	.2188
	6		.0002	.0011	.0038	.0100	.0217	.0413	.0703	.1094
	7			.0001	.0004	.0012	.0033	.0079	.0164	.0312
	8					.0001	.0002	.0007	.0017	.0039

TABLE C   Binomial probabilities (continued)

$$\text{Entry is } P(X = k) = \binom{n}{k} p^k (1-p)^{n-k}$$

n	k	.01	.02	.03	.04	.05	.06	.07	.08	.09
9	0	.9135	.8337	.7602	.6925	.6302	.5730	.5204	.4722	.4279
	1	.0830	.1531	.2116	.2597	.2985	.3292	.3525	.3695	.3809
	2	.0034	.0125	.0262	.0433	.0629	.0840	.1061	.1285	.1507
	3	.0001	.0006	.0019	.0042	.0077	.0125	.0186	.0261	.0348
	4			.0001	.0003	.0006	.0012	.0021	.0034	.0052
	5						.0001	.0002	.0003	.0005
	6									
	7									
	8									
	9									
10	0	.9044	.8171	.7374	.6648	.5987	.5386	.4840	.4344	.3894
	1	.0914	.1667	.2281	.2770	.3151	.3438	.3643	.3777	.3851
	2	.0042	.0153	.0317	.0519	.0746	.0988	.1234	.1478	.1714
	3	.0001	.0008	.0026	.0058	.0105	.0168	.0248	.0343	.0452
	4			.0001	.0004	.0010	.0019	.0033	.0052	.0078
	5					.0001	.0001	.0003	.0005	.0009
	6									.0001
	7									
	8									
	9									
	10									
12	0	.8864	.7847	.6938	.6127	.5404	.4759	.4186	.3677	.3225
	1	.1074	.1922	.2575	.3064	.3413	.3645	.3781	.3837	.3827
	2	.0060	.0216	.0438	.0702	.0988	.1280	.1565	.1835	.2082
	3	.0002	.0015	.0045	.0098	.0173	.0272	.0393	.0532	.0686
	4		.0001	.0003	.0009	.0021	.0039	.0067	.0104	.0153
	5				.0001	.0002	.0004	.0008	.0014	.0024
	6							.0001	.0001	.0003
	7									
	8									
	9									
	10									
	11									
	12									
15	0	.8601	.7386	.6333	.5421	.4633	.3953	.3367	.2863	.2430
	1	.1303	.2261	.2938	.3388	.3658	.3785	.3801	.3734	.3605
	2	.0092	.0323	.0636	.0988	.1348	.1691	.2003	.2273	.2496
	3	.0004	.0029	.0085	.0178	.0307	.0468	.0653	.0857	.1070
	4		.0002	.0008	.0022	.0049	.0090	.0148	.0223	.0317
	5			.0001	.0002	.0006	.0013	.0024	.0043	.0069
	6						.0001	.0003	.0006	.0011
	7								.0001	.0001
	8									
	9									
	10									
	11									
	12									
	13									
	14									
	15									

TABLE C   Binomial probabilities (continued)

n	k	.10	.15	.20	.25	.30	.35	.40	.45	.50
9	0	.3874	.2316	.1342	.0751	.0404	.0207	.0101	.0046	.0020
	1	.3874	.3679	.3020	.2253	.1556	.1004	.0605	.0339	.0176
	2	.1722	.2597	.3020	.3003	.2668	.2162	.1612	.1110	.0703
	3	.0446	.1069	.1762	.2336	.2668	.2716	.2508	.2119	.1641
	4	.0074	.0283	.0661	.1168	.1715	.2194	.2508	.2600	.2461
	5	.0008	.0050	.0165	.0389	.0735	.1181	.1672	.2128	.2461
	6	.0001	.0006	.0028	.0087	.0210	.0424	.0743	.1160	.1641
	7			.0003	.0012	.0039	.0098	.0212	.0407	.0703
	8				.0001	.0004	.0013	.0035	.0083	.0176
	9						.0001	.0003	.0008	.0020
10	0	.3487	.1969	.1074	.0563	.0282	.0135	.0060	.0025	.0010
	1	.3874	.3474	.2684	.1877	.1211	.0725	.0403	.0207	.0098
	2	.1937	.2759	.3020	.2816	.2335	.1757	.1209	.0763	.0439
	3	.0574	.1298	.2013	.2503	.2668	.2522	.2150	.1665	.1172
	4	.0112	.0401	.0881	.1460	.2001	.2377	.2508	.2384	.2051
	5	.0015	.0085	.0264	.0584	.1029	.1536	.2007	.2340	.2461
	6	.0001	.0012	.0055	.0162	.0368	.0689	.1115	.1596	.2051
	7		.0001	.0008	.0031	.0090	.0212	.0425	.0746	.1172
	8			.0001	.0004	.0014	.0043	.0106	.0229	.0439
	9					.0001	.0005	.0016	.0042	.0098
	10							.0001	.0003	.0010
12	0	.2824	.1422	.0687	.0317	.0138	.0057	.0022	.0008	.0002
	1	.3766	.3012	.2062	.1267	.0712	.0368	.0174	.0075	.0029
	2	.2301	.2924	.2835	.2323	.1678	.1088	.0639	.0339	.0161
	3	.0852	.1720	.2362	.2581	.2397	.1954	.1419	.0923	.0537
	4	.0213	.0683	.1329	.1936	.2311	.2367	.2128	.1700	.1208
	5	.0038	.0193	.0532	.1032	.1585	.2039	.2270	.2225	.1934
	6	.0005	.0040	.0155	.0401	.0792	.1281	.1766	.2124	.2256
	7		.0006	.0033	.0115	.0291	.0591	.1009	.1489	.1934
	8		.0001	.0005	.0024	.0078	.0199	.0420	.0762	.1208
	9			.0001	.0004	.0015	.0048	.0125	.0277	.0537
	10					.0002	.0008	.0025	.0068	.0161
	11						.0001	.0003	.0010	.0029
	12								:0001	.0002
15	0	.2059	.0874	.0352	.0134	.0047	.0016	.0005	.0001	.0000
	1	.3432	.2312	.1319	.0668	.0305	.0126	.0047	.0016	.0005
	2	.2669	.2856	.2309	.1559	.0916	.0476	.0219	.0090	.0032
	3	.1285	.2184	.2501	.2252	.1700	.1110	.0634	.0318	.0139
	4	.0428	.1156	.1876	.2252	.2186	.1792	.1268	.0780	.0417
	5	.0105	.0449	.1032	.1651	.2061	.2123	.1859	.1404	.0916
	6	.0019	.0132	.0430	.0917	.1472	.1906	.2066	.1914	.1527
	7	.0003	.0030	.0138	.0393	.0811	.1319	.1771	.2013	.1964
	8		.0005	.0035	.0131	.0348	.0710	.1181	.1647	.1964
	9		.0001	.0007	.0034	.0116	.0298	.0612	.1048	.1527
	10			.0001	.0007	.0030	.0096	.0245	.0515	.0916
	11				.0001	.0006	.0024	.0074	.0191	.0417
	12					.0001	.0004	.0016	.0052	.0139
	13						.0001	.0003	.0010	.0032
	14								.0001	.0005
	15									

TABLE C — Binomial probabilities (continued)

n	k	.01	.02	.03	.04	.05	.06	.07	.08	.09
20	0	.8179	.6676	.5438	.4420	.3585	.2901	.2342	.1887	.1516
	1	.1652	.2725	.3364	.3683	.3774	.3703	.3526	.3282	.3000
	2	.0159	.0528	.0988	.1458	.1887	.2246	.2521	.2711	.2818
	3	.0010	.0065	.0183	.0364	.0596	.0860	.1139	.1414	.1672
	4		.0006	.0024	.0065	.0133	.0233	.0364	.0523	.0703
	5			.0002	.0009	.0022	.0048	.0088	.0145	.0222
	6				.0001	.0003	.0008	.0017	.0032	.0055
	7						.0001	.0002	.0005	.0011
	8								.0001	.0002
	9									
	10									
	11									
	12									
	13									
	14									
	15									
	16									
	17									
	18									
	19									
	20									

n	k	.10	.15	.20	.25	.30	.35	.40	.45	.50
20	0	.1216	.0388	.0115	.0032	.0008	.0002	.0000	.0000	.0000
	1	.2702	.1368	.0576	.0211	.0068	.0020	.0005	.0001	.0000
	2	.2852	.2293	.1369	.0669	.0278	.0100	.0031	.0008	.0002
	3	.1901	.2428	.2054	.1339	.0716	.0323	.0123	.0040	.0011
	4	.0898	.1821	.2182	.1897	.1304	.0738	.0350	.0139	.0046
	5	.0319	.1028	.1746	.2023	.1789	.1272	.0746	.0365	.0148
	6	.0089	.0454	.1091	.1686	.1916	.1712	.1244	.0746	.0370
	7	.0020	.0160	.0545	.1124	.1643	.1844	.1659	.1221	.0739
	8	.0004	.0046	.0222	.0609	.1144	.1614	.1797	.1623	.1201
	9	.0001	.0011	.0074	.0271	.0654	.1158	.1597	.1771	.1602
	10		.0002	.0020	.0099	.0308	.0686	.1171	.1593	.1762
	11			.0005	.0030	.0120	.0336	.0710	.1185	.1602
	12			.0001	.0008	.0039	.0136	.0355	.0727	.1201
	13				.0002	.0010	.0045	.0146	.0366	.0739
	14					.0002	.0012	.0049	.0150	.0370
	15						.0003	.0013	.0049	.0148
	16							.0003	.0013	.0046
	17								.0002	.0011
	18									.0002
	19									
	20									

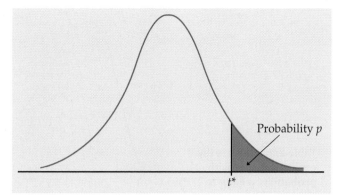

Table entry for p and C is the critical value t^* with probability p lying to its right and probability C lying between $-t^*$ and t^*.

Probability p

t^*

TABLE D t distribution critical values

df	\.25	.20	.15	.10	.05	.025	.02	.01	.005	.0025	.001	.0005
							Upper-tail probability p					
1	1.000	1.376	1.963	3.078	6.314	12.71	15.89	31.82	63.66	127.3	318.3	636.6
2	0.816	1.061	1.386	1.886	2.920	4.303	4.849	6.965	9.925	14.09	22.33	31.60
3	0.765	0.978	1.250	1.638	2.353	3.182	3.482	4.541	5.841	7.453	10.21	12.92
4	0.741	0.941	1.190	1.533	2.132	2.776	2.999	3.747	4.604	5.598	7.173	8.610
5	0.727	0.920	1.156	1.476	2.015	2.571	2.757	3.365	4.032	4.773	5.893	6.869
6	0.718	0.906	1.134	1.440	1.943	2.447	2.612	3.143	3.707	4.317	5.208	5.959
7	0.711	0.896	1.119	1.415	1.895	2.365	2.517	2.998	3.499	4.029	4.785	5.408
8	0.706	0.889	1.108	1.397	1.860	2.306	2.449	2.896	3.355	3.833	4.501	5.041
9	0.703	0.883	1.100	1.383	1.833	2.262	2.398	2.821	3.250	3.690	4.297	4.781
10	0.700	0.879	1.093	1.372	1.812	2.228	2.359	2.764	3.169	3.581	4.144	4.587
11	0.697	0.876	1.088	1.363	1.796	2.201	2.328	2.718	3.106	3.497	4.025	4.437
12	0.695	0.873	1.083	1.356	1.782	2.179	2.303	2.681	3.055	3.428	3.930	4.318
13	0.694	0.870	1.079	1.350	1.771	2.160	2.282	2.650	3.012	3.372	3.852	4.221
14	0.692	0.868	1.076	1.345	1.761	2.145	2.264	2.624	2.977	3.326	3.787	4.140
15	0.691	0.866	1.074	1.341	1.753	2.131	2.249	2.602	2.947	3.286	3.733	4.073
16	0.690	0.865	1.071	1.337	1.746	2.120	2.235	2.583	2.921	3.252	3.686	4.015
17	0.689	0.863	1.069	1.333	1.740	2.110	2.224	2.567	2.898	3.222	3.646	3.965
18	0.688	0.862	1.067	1.330	1.734	2.101	2.214	2.552	2.878	3.197	3.611	3.922
19	0.688	0.861	1.066	1.328	1.729	2.093	2.205	2.539	2.861	3.174	3.579	3.883
20	0.687	0.860	1.064	1.325	1.725	2.086	2.197	2.528	2.845	3.153	3.552	3.850
21	0.686	0.859	1.063	1.323	1.721	2.080	2.189	2.518	2.831	3.135	3.527	3.819
22	0.686	0.858	1.061	1.321	1.717	2.074	2.183	2.508	2.819	3.119	3.505	3.792
23	0.685	0.858	1.060	1.319	1.714	2.069	2.177	2.500	2.807	3.104	3.485	3.768
24	0.685	0.857	1.059	1.318	1.711	2.064	2.172	2.492	2.797	3.091	3.467	3.745
25	0.684	0.856	1.058	1.316	1.708	2.060	2.167	2.485	2.787	3.078	3.450	3.725
26	0.684	0.856	1.058	1.315	1.706	2.056	2.162	2.479	2.779	3.067	3.435	3.707
27	0.684	0.855	1.057	1.314	1.703	2.052	2.158	2.473	2.771	3.057	3.421	3.690
28	0.683	0.855	1.056	1.313	1.701	2.048	2.154	2.467	2.763	3.047	3.408	3.674
29	0.683	0.854	1.055	1.311	1.699	2.045	2.150	2.462	2.756	3.038	3.396	3.659
30	0.683	0.854	1.055	1.310	1.697	2.042	2.147	2.457	2.750	3.030	3.385	3.646
40	0.681	0.851	1.050	1.303	1.684	2.021	2.123	2.423	2.704	2.971	3.307	3.551
50	0.679	0.849	1.047	1.299	1.676	2.009	2.109	2.403	2.678	2.937	3.261	3.496
60	0.679	0.848	1.045	1.296	1.671	2.000	2.099	2.390	2.660	2.915	3.232	3.460
80	0.678	0.846	1.043	1.292	1.664	1.990	2.088	2.374	2.639	2.887	3.195	3.416
100	0.677	0.845	1.042	1.290	1.660	1.984	2.081	2.364	2.626	2.871	3.174	3.390
1000	0.675	0.842	1.037	1.282	1.646	1.962	2.056	2.330	2.581	2.813	3.098	3.300
z^*	0.674	0.841	1.036	1.282	1.645	1.960	2.054	2.326	2.576	2.807	3.091	3.291
	50%	60%	70%	80%	90%	95%	96%	98%	99%	99.5%	99.8%	99.9%
							Confidence level C					

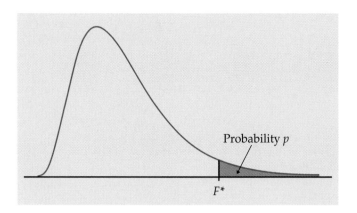

Table entry for p is the critical value F* with probability p lying to its right.

Probability p

F*

TABLE E F critical values

		Degrees of freedom in the numerator								
	p	1	2	3	4	5	6	7	8	9
1	.100	39.86	49.50	53.59	55.83	57.24	58.20	58.91	59.44	59.86
	.050	161.45	199.50	215.71	224.58	230.16	233.99	236.77	238.88	240.54
	.025	647.79	799.50	864.16	899.58	921.85	937.11	948.22	956.66	963.28
	.010	4052.2	4999.5	5403.4	5624.6	5763.6	5859.0	5928.4	5981.1	6022.5
	.001	405284	500000	540379	562500	576405	585937	592873	598144	602284
2	.100	8.53	9.00	9.16	9.24	9.29	9.33	9.35	9.37	9.38
	.050	18.51	19.00	19.16	19.25	19.30	19.33	19.35	19.37	19.38
	.025	38.51	39.00	39.17	39.25	39.30	39.33	39.36	39.37	39.39
	.010	98.50	99.00	99.17	99.25	99.30	99.33	99.36	99.37	99.39
	.001	998.50	999.00	999.17	999.25	999.30	999.33	999.36	999.37	999.39
3	.100	5.54	5.46	5.39	5.34	5.31	5.28	5.27	5.25	5.24
	.050	10.13	9.55	9.28	9.12	9.01	8.94	8.89	8.85	8.81
	.025	17.44	16.04	15.44	15.10	14.88	14.73	14.62	14.54	14.47
	.010	34.12	30.82	29.46	28.71	28.24	27.91	27.67	27.49	27.35
	.001	167.03	148.50	141.11	137.10	134.58	132.85	131.58	130.62	129.86
4	.100	4.54	4.32	4.19	4.11	4.05	4.01	3.98	3.95	3.94
	.050	7.71	6.94	6.59	6.39	6.26	6.16	6.09	6.04	6.00
	.025	12.22	10.65	9.98	9.60	9.36	9.20	9.07	8.98	8.90
	.010	21.20	18.00	16.69	15.98	15.52	15.21	14.98	14.80	14.66
	.001	74.14	61.25	56.18	53.44	51.71	50.53	49.66	49.00	48.47
5	.100	4.06	3.78	3.62	3.52	3.45	3.40	3.37	3.34	3.32
	.050	6.61	5.79	5.41	5.19	5.05	4.95	4.88	4.82	4.77
	.025	10.01	8.43	7.76	7.39	7.15	6.98	6.85	6.76	6.68
	.010	16.26	13.27	12.06	11.39	10.97	10.67	10.46	10.29	10.16
	.001	47.18	37.12	33.20	31.09	29.75	28.83	28.16	27.65	27.24
6	.100	3.78	3.46	3.29	3.18	3.11	3.05	3.01	2.98	2.96
	.050	5.99	5.14	4.76	4.53	4.39	4.28	4.21	4.15	4.10
	.025	8.81	7.26	6.60	6.23	5.99	5.82	5.70	5.60	5.52
	.010	13.75	10.92	9.78	9.15	8.75	8.47	8.26	8.10	7.98
	.001	35.51	27.00	23.70	21.92	20.80	20.03	19.46	19.03	18.69
7	.100	3.59	3.26	3.07	2.96	2.88	2.83	2.78	2.75	2.72
	.050	5.59	4.74	4.35	4.12	3.97	3.87	3.79	3.73	3.68
	.025	8.07	6.54	5.89	5.52	5.29	5.12	4.99	4.90	4.82
	.010	12.25	9.55	8.45	7.85	7.46	7.19	6.99	6.84	6.72
	.001	29.25	21.69	18.77	17.20	16.21	15.52	15.02	14.63	14.33

Degrees of freedom in the denominator

Table entry for *p* is the critical value *F** with probability *p* lying to its right.

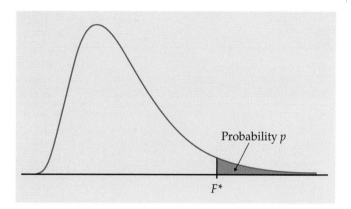

Probability *p*

F*

TABLE E   *F* critical values (continued)

			Degrees of freedom in the numerator							
10	12	15	20	25	30	40	50	60	120	1000
60.19	60.71	61.22	61.74	62.05	62.26	62.53	62.69	62.79	63.06	63.30
241.88	243.91	245.95	248.01	249.26	250.10	251.14	251.77	252.20	253.25	254.19
968.63	976.71	984.87	993.10	998.08	1001.4	1005.6	1008.1	1009.8	1014.0	1017.7
6055.8	6106.3	6157.3	6208.7	6239.8	6260.6	6286.8	6302.5	6313.0	6339.4	6362.7
605621	610668	615764	620908	624017	626099	628712	630285	631337	633972	636301
9.39	9.41	9.42	9.44	9.45	9.46	9.47	9.47	9.47	9.48	9.49
19.40	19.41	19.43	19.45	19.46	19.46	19.47	19.48	19.48	19.49	19.49
39.40	39.41	39.43	39.45	39.46	39.46	39.47	39.48	39.48	39.49	39.50
99.40	99.42	99.43	99.45	99.46	99.47	99.47	99.48	99.48	99.49	99.50
999.40	999.42	999.43	999.45	999.46	999.47	999.47	999.48	999.48	999.49	999.50
5.23	5.22	5.20	5.18	5.17	5.17	5.16	5.15	5.15	5.14	5.13
8.79	8.74	8.70	8.66	8.63	8.62	8.59	8.58	8.57	8.55	8.53
14.42	14.34	14.25	14.17	14.12	14.08	14.04	14.01	13.99	13.95	13.91
27.23	27.05	26.87	26.69	26.58	26.50	26.41	26.35	26.32	26.22	26.14
129.25	128.32	127.37	126.42	125.84	125.45	124.96	124.66	124.47	123.97	123.53
3.92	3.90	3.87	3.84	3.83	3.82	3.80	3.80	3.79	3.78	3.76
5.96	5.91	5.86	5.80	5.77	5.75	5.72	5.70	5.69	5.66	5.63
8.84	8.75	8.66	8.56	8.50	8.46	8.41	8.38	8.36	8.31	8.26
14.55	14.37	14.20	14.02	13.91	13.84	13.75	13.69	13.65	13.56	13.47
48.05	47.41	46.76	46.10	45.70	45.43	45.09	44.88	44.75	44.40	44.09
3.30	3.27	3.24	3.21	3.19	3.17	3.16	3.15	3.14	3.12	3.11
4.74	4.68	4.62	4.56	4.52	4.50	4.46	4.44	4.43	4.40	4.37
6.62	6.52	6.43	6.33	6.27	6.23	6.18	6.14	6.12	6.07	6.02
10.05	9.89	9.72	9.55	9.45	9.38	9.29	9.24	9.20	9.11	9.03
26.92	26.42	25.91	25.39	25.08	24.87	24.60	24.44	24.33	24.06	23.82
2.94	2.90	2.87	2.84	2.81	2.80	2.78	2.77	2.76	2.74	2.72
4.06	4.00	3.94	3.87	3.83	3.81	3.77	3.75	3.74	3.70	3.67
5.46	5.37	5.27	5.17	5.11	5.07	5.01	4.98	4.96	4.90	4.86
7.87	7.72	7.56	7.40	7.30	7.23	7.14	7.09	7.06	6.97	6.89
18.41	17.99	17.56	17.12	16.85	16.67	16.44	16.31	16.21	15.98	15.77
2.70	2.67	2.63	2.59	2.57	2.56	2.54	2.52	2.51	2.49	2.47
3.64	3.57	3.51	3.44	3.40	3.38	3.34	3.32	3.30	3.27	3.23
4.76	4.67	4.57	4.47	4.40	4.36	4.31	4.28	4.25	4.20	4.15
6.62	6.47	6.31	6.16	6.06	5.99	5.91	5.86	5.82	5.74	5.66
14.08	13.71	13.32	12.93	12.69	12.53	12.33	12.20	12.12	11.91	11.72

TABLE E F critical values (continued)

| | | Degrees of freedom in the numerator | | | | | | | | |
	p	1	2	3	4	5	6	7	8	9
8	.100	3.46	3.11	2.92	2.81	2.73	2.67	2.62	2.59	2.56
	.050	5.32	4.46	4.07	3.84	3.69	3.58	3.50	3.44	3.39
	.025	7.57	6.06	5.42	5.05	4.82	4.65	4.53	4.43	4.36
	.010	11.26	8.65	7.59	7.01	6.63	6.37	6.18	6.03	5.91
	.001	25.41	18.49	15.83	14.39	13.48	12.86	12.40	12.05	11.77
9	.100	3.36	3.01	2.81	2.69	2.61	2.55	2.51	2.47	2.44
	.050	5.12	4.26	3.86	3.63	3.48	3.37	3.29	3.23	3.18
	.025	7.21	5.71	5.08	4.72	4.48	4.32	4.20	4.10	4.03
	.010	10.56	8.02	6.99	6.42	6.06	5.80	5.61	5.47	5.35
	.001	22.86	16.39	13.90	12.56	11.71	11.13	10.70	10.37	10.11
10	.100	3.29	2.92	2.73	2.61	2.52	2.46	2.41	2.38	2.35
	.050	4.96	4.10	3.71	3.48	3.33	3.22	3.14	3.07	3.02
	.025	6.94	5.46	4.83	4.47	4.24	4.07	3.95	3.85	3.78
	.010	10.04	7.56	6.55	5.99	5.64	5.39	5.20	5.06	4.94
	.001	21.04	14.91	12.55	11.28	10.48	9.93	9.52	9.20	8.96
11	.100	3.23	2.86	2.66	2.54	2.45	2.39	2.34	2.30	2.27
	.050	4.84	3.98	3.59	3.36	3.20	3.09	3.01	2.95	2.90
	.025	6.72	5.26	4.63	4.28	4.04	3.88	3.76	3.66	3.59
	.010	9.65	7.21	6.22	5.67	5.32	5.07	4.89	4.74	4.63
	.001	19.69	13.81	11.56	10.35	9.58	9.05	8.66	8.35	8.12
12	.100	3.18	2.81	2.61	2.48	2.39	2.33	2.28	2.24	2.21
	.050	4.75	3.89	3.49	3.26	3.11	3.00	2.91	2.85	2.80
	.025	6.55	5.10	4.47	4.12	3.89	3.73	3.61	3.51	3.44
	.010	9.33	6.93	5.95	5.41	5.06	4.82	4.64	4.50	4.39
	.001	18.64	12.97	10.80	9.63	8.89	8.38	8.00	7.71	7.48
13	.100	3.14	2.76	2.56	2.43	2.35	2.28	2.23	2.20	2.16
	.050	4.67	3.81	3.41	3.18	3.03	2.92	2.83	2.77	2.71
	.025	6.41	4.97	4.35	4.00	3.77	3.60	3.48	3.39	3.31
	.010	9.07	6.70	5.74	5.21	4.86	4.62	4.44	4.30	4.19
	.001	17.82	12.31	10.21	9.07	8.35	7.86	7.49	7.21	6.98
14	.100	3.10	2.73	2.52	2.39	2.31	2.24	2.19	2.15	2.12
	.050	4.60	3.74	3.34	3.11	2.96	2.85	2.76	2.70	2.65
	.025	6.30	4.86	4.24	3.89	3.66	3.50	3.38	3.29	3.21
	.010	8.86	6.51	5.56	5.04	4.69	4.46	4.28	4.14	4.03
	.001	17.14	11.78	9.73	8.62	7.92	7.44	7.08	6.80	6.58
15	.100	3.07	2.70	2.49	2.36	2.27	2.21	2.16	2.12	2.09
	.050	4.54	3.68	3.29	3.06	2.90	2.79	2.71	2.64	2.59
	.025	6.20	4.77	4.15	3.80	3.58	3.41	3.29	3.20	3.12
	.010	8.68	6.36	5.42	4.89	4.56	4.32	4.14	4.00	3.89
	.001	16.59	11.34	9.34	8.25	7.57	7.09	6.74	6.47	6.26
16	.100	3.05	2.67	2.46	2.33	2.24	2.18	2.13	2.09	2.06
	.050	4.49	3.63	3.24	3.01	2.85	2.74	2.66	2.59	2.54
	.025	6.12	4.69	4.08	3.73	3.50	3.34	3.22	3.12	3.05
	.010	8.53	6.23	5.29	4.77	4.44	4.20	4.03	3.89	3.78
	.001	16.12	10.97	9.01	7.94	7.27	6.80	6.46	6.19	5.98
17	.100	3.03	2.64	2.44	2.31	2.22	2.15	2.10	2.06	2.03
	.050	4.45	3.59	3.20	2.96	2.81	2.70	2.61	2.55	2.49
	.025	6.04	4.62	4.01	3.66	3.44	3.28	3.16	3.06	2.98
	.010	8.40	6.11	5.19	4.67	4.34	4.10	3.93	3.79	3.68
	.001	15.72	10.66	8.73	7.68	7.02	6.56	6.22	5.96	5.75

Degrees of freedom in the denominator

TABLE E — F critical values (continued)

Degrees of freedom in the numerator

10	12	15	20	25	30	40	50	60	120	1000
2.54	2.50	2.46	2.42	2.40	2.38	2.36	2.35	2.34	2.32	2.30
3.35	3.28	3.22	3.15	3.11	3.08	3.04	3.02	3.01	2.97	2.93
4.30	4.20	4.10	4.00	3.94	3.89	3.84	3.81	3.78	3.73	3.68
5.81	5.67	5.52	5.36	5.26	5.20	5.12	5.07	5.03	4.95	4.87
11.54	11.19	10.84	10.48	10.26	10.11	9.92	9.80	9.73	9.53	9.36
2.42	2.38	2.34	2.30	2.27	2.25	2.23	2.22	2.21	2.18	2.16
3.14	3.07	3.01	2.94	2.89	2.86	2.83	2.80	2.79	2.75	2.71
3.96	3.87	3.77	3.67	3.60	3.56	3.51	3.47	3.45	3.39	3.34
5.26	5.11	4.96	4.81	4.71	4.65	4.57	4.52	4.48	4.40	4.32
9.89	9.57	9.24	8.90	8.69	8.55	8.37	8.26	8.19	8.00	7.84
2.32	2.28	2.24	2.20	2.17	2.16	2.13	2.12	2.11	2.08	2.06
2.98	2.91	2.85	2.77	2.73	2.70	2.66	2.64	2.62	2.58	2.54
3.72	3.62	3.52	3.42	3.35	3.31	3.26	3.22	3.20	3.14	3.09
4.85	4.71	4.56	4.41	4.31	4.25	4.17	4.12	4.08	4.00	3.92
8.75	8.45	8.13	7.80	7.60	7.47	7.30	7.19	7.12	6.94	6.78
2.25	2.21	2.17	2.12	2.10	2.08	2.05	2.04	2.03	2.00	1.98
2.85	2.79	2.72	2.65	2.60	2.57	2.53	2.51	2.49	2.45	2.41
3.53	3.43	3.33	3.23	3.16	3.12	3.06	3.03	3.00	2.94	2.89
4.54	4.40	4.25	4.10	4.01	3.94	3.86	3.81	3.78	3.69	3.61
7.92	7.63	7.32	7.01	6.81	6.68	6.52	6.42	6.35	6.18	6.02
2.19	2.15	2.10	2.06	2.03	2.01	1.99	1.97	1.96	1.93	1.91
2.75	2.69	2.62	2.54	2.50	2.47	2.43	2.40	2.38	2.34	2.30
3.37	3.28	3.18	3.07	3.01	2.96	2.91	2.87	2.85	2.79	2.73
4.30	4.16	4.01	3.86	3.76	3.70	3.62	3.57	3.54	3.45	3.37
7.29	7.00	6.71	6.40	6.22	6.09	5.93	5.83	5.76	5.59	5.44
2.14	2.10	2.05	2.01	1.98	1.96	1.93	1.92	1.90	1.88	1.85
2.67	2.60	2.53	2.46	2.41	2.38	2.34	2.31	2.30	2.25	2.21
3.25	3.15	3.05	2.95	2.88	2.84	2.78	2.74	2.72	2.66	2.60
4.10	3.96	3.82	3.66	3.57	3.51	3.43	3.38	3.34	3.25	3.18
6.80	6.52	6.23	5.93	5.75	5.63	5.47	5.37	5.30	5.14	4.99
2.10	2.05	2.01	1.96	1.93	1.91	1.89	1.87	1.86	1.83	1.80
2.60	2.53	2.46	2.39	2.34	2.31	2.27	2.24	2.22	2.18	2.14
3.15	3.05	2.95	2.84	2.78	2.73	2.67	2.64	2.61	2.55	2.50
3.94	3.80	3.66	3.51	3.41	3.35	3.27	3.22	3.18	3.09	3.02
6.40	6.13	5.85	5.56	5.38	5.25	5.10	5.00	4.94	4.77	4.62
2.06	2.02	1.97	1.92	1.89	1.87	1.85	1.83	1.82	1.79	1.76
2.54	2.48	2.40	2.33	2.28	2.25	2.20	2.18	2.16	2.11	2.07
3.06	2.96	2.86	2.76	2.69	2.64	2.59	2.55	2.52	2.46	2.40
3.80	3.67	3.52	3.37	3.28	3.21	3.13	3.08	3.05	2.96	2.88
6.08	5.81	5.54	5.25	5.07	4.95	4.80	4.70	4.64	4.47	4.33
2.03	1.99	1.94	1.89	1.86	1.84	1.81	1.79	1.78	1.75	1.72
2.49	2.42	2.35	2.28	2.23	2.19	2.15	2.12	2.11	2.06	2.02
2.99	2.89	2.79	2.68	2.61	2.57	2.51	2.47	2.45	2.38	2.32
3.69	3.55	3.41	3.26	3.16	3.10	3.02	2.97	2.93	2.84	2.76
5.81	5.55	5.27	4.99	4.82	4.70	4.54	4.45	4.39	4.23	4.08
2.00	1.96	1.91	1.86	1.83	1.81	1.78	1.76	1.75	1.72	1.69
2.45	2.38	2.31	2.23	2.18	2.15	2.10	2.08	2.06	2.01	1.97
2.92	2.82	2.72	2.62	2.55	2.50	2.44	2.41	2.38	2.32	2.26
3.59	3.46	3.31	3.16	3.07	3.00	2.92	2.87	2.83	2.75	2.66
5.58	5.32	5.05	4.78	4.60	4.48	4.33	4.24	4.18	4.02	3.87

TABLE E *F* critical values (continued)

		Degrees of freedom in the numerator								
	p	1	2	3	4	5	6	7	8	9
18	.100	3.01	2.62	2.42	2.29	2.20	2.13	2.08	2.04	2.00
	.050	4.41	3.55	3.16	2.93	2.77	2.66	2.58	2.51	2.46
	.025	5.98	4.56	3.95	3.61	3.38	3.22	3.10	3.01	2.93
	.010	8.29	6.01	5.09	4.58	4.25	4.01	3.84	3.71	3.60
	.001	15.38	10.39	8.49	7.46	6.81	6.35	6.02	5.76	5.56
19	.100	2.99	2.61	2.40	2.27	2.18	2.11	2.06	2.02	1.98
	.050	4.38	3.52	3.13	2.90	2.74	2.63	2.54	2.48	2.42
	.025	5.92	4.51	3.90	3.56	3.33	3.17	3.05	2.96	2.88
	.010	8.18	5.93	5.01	4.50	4.17	3.94	3.77	3.63	3.52
	.001	15.08	10.16	8.28	7.27	6.62	6.18	5.85	5.59	5.39
20	.100	2.97	2.59	2.38	2.25	2.16	2.09	2.04	2.00	1.96
	.050	4.35	3.49	3.10	2.87	2.71	2.60	2.51	2.45	2.39
	.025	5.87	4.46	3.86	3.51	3.29	3.13	3.01	2.91	2.84
	.010	8.10	5.85	4.94	4.43	4.10	3.87	3.70	3.56	3.46
	.001	14.82	9.95	8.10	7.10	6.46	6.02	5.69	5.44	5.24
21	.100	2.96	2.57	2.36	2.23	2.14	2.08	2.02	1.98	1.95
	.050	4.32	3.47	3.07	2.84	2.68	2.57	2.49	2.42	2.37
	.025	5.83	4.42	3.82	3.48	3.25	3.09	2.97	2.87	2.80
	.010	8.02	5.78	4.87	4.37	4.04	3.81	3.64	3.51	3.40
	.001	14.59	9.77	7.94	6.95	6.32	5.88	5.56	5.31	5.11
22	.100	2.95	2.56	2.35	2.22	2.13	2.06	2.01	1.97	1.93
	.050	4.30	3.44	3.05	2.82	2.66	2.55	2.46	2.40	2.34
	.025	5.79	4.38	3.78	3.44	3.22	3.05	2.93	2.84	2.76
	.010	7.95	5.72	4.82	4.31	3.99	3.76	3.59	3.45	3.35
	.001	14.38	9.61	7.80	6.81	6.19	5.76	5.44	5.19	4.99
23	.100	2.94	2.55	2.34	2.21	2.11	2.05	1.99	1.95	1.92
	.050	4.28	3.42	3.03	2.80	2.64	2.53	2.44	2.37	2.32
	.025	5.75	4.35	3.75	3.41	3.18	3.02	2.90	2.81	2.73
	.010	7.88	5.66	4.76	4.26	3.94	3.71	3.54	3.41	3.30
	.001	14.20	9.47	7.67	6.70	6.08	5.65	5.33	5.09	4.89
24	.100	2.93	2.54	2.33	2.19	2.10	2.04	1.98	1.94	1.91
	.050	4.26	3.40	3.01	2.78	2.62	2.51	2.42	2.36	2.30
	.025	5.72	4.32	3.72	3.38	3.15	2.99	2.87	2.78	2.70
	.010	7.82	5.61	4.72	4.22	3.90	3.67	3.50	3.36	3.26
	.001	14.03	9.34	7.55	6.59	5.98	5.55	5.23	4.99	4.80
25	.100	2.92	2.53	2.32	2.18	2.09	2.02	1.97	1.93	1.89
	.050	4.24	3.39	2.99	2.76	2.60	2.49	2.40	2.34	2.28
	.025	5.69	4.29	3.69	3.35	3.13	2.97	2.85	2.75	2.68
	.010	7.77	5.57	4.68	4.18	3.85	3.63	3.46	3.32	3.22
	.001	13.88	9.22	7.45	6.49	5.89	5.46	5.15	4.91	4.71
26	.100	2.91	2.52	2.31	2.17	2.08	2.01	1.96	1.92	1.88
	.050	4.23	3.37	2.98	2.74	2.59	2.47	2.39	2.32	2.27
	.025	5.66	4.27	3.67	3.33	3.10	2.94	2.82	2.73	2.65
	.010	7.72	5.53	4.64	4.14	3.82	3.59	3.42	3.29	3.18
	.001	13.74	9.12	7.36	6.41	5.80	5.38	5.07	4.83	4.64
27	.100	2.90	2.51	2.30	2.17	2.07	2.00	1.95	1.91	1.87
	.050	4.21	3.35	2.96	2.73	2.57	2.46	2.37	2.31	2.25
	.025	5.63	4.24	3.65	3.31	3.08	2.92	2.80	2.71	2.63
	.010	7.68	5.49	4.60	4.11	3.78	3.56	3.39	3.26	3.15
	.001	13.61	9.02	7.27	6.33	5.73	5.31	5.00	4.76	4.57

Degrees of freedom in the denominator

TABLE E *F* critical values (continued)

Degrees of freedom in the numerator

10	12	15	20	25	30	40	50	60	120	1000
1.98	1.93	1.89	1.84	1.80	1.78	1.75	1.74	1.72	1.69	1.66
2.41	2.34	2.27	2.19	2.14	2.11	2.06	2.04	2.02	1.97	1.92
2.87	2.77	2.67	2.56	2.49	2.44	2.38	2.35	2.32	2.26	2.20
3.51	3.37	3.23	3.08	2.98	2.92	2.84	2.78	2.75	2.66	2.58
5.39	5.13	4.87	4.59	4.42	4.30	4.15	4.06	4.00	3.84	3.69
1.96	1.91	1.86	1.81	1.78	1.76	1.73	1.71	1.70	1.67	1.64
2.38	2.31	2.23	2.16	2.11	2.07	2.03	2.00	1.98	1.93	1.88
2.82	2.72	2.62	2.51	2.44	2.39	2.33	2.30	2.27	2.20	2.14
3.43	3.30	3.15	3.00	2.91	2.84	2.76	2.71	2.67	2.58	2.50
5.22	4.97	4.70	4.43	4.26	4.14	3.99	3.90	3.84	3.68	3.53
1.94	1.89	1.84	1.79	1.76	1.74	1.71	1.69	1.68	1.64	1.61
2.35	2.28	2.20	2.12	2.07	2.04	1.99	1.97	1.95	1.90	1.85
2.77	2.68	2.57	2.46	2.40	2.35	2.29	2.25	2.22	2.16	2.09
3.37	3.23	3.09	2.94	2.84	2.78	2.69	2.64	2.61	2.52	2.43
5.08	4.82	4.56	4.29	4.12	4.00	3.86	3.77	3.70	3.54	3.40
1.92	1.87	1.83	1.78	1.74	1.72	1.69	1.67	1.66	1.62	1.59
2.32	2.25	2.18	2.10	2.05	2.01	1.96	1.94	1.92	1.87	1.82
2.73	2.64	2.53	2.42	2.36	2.31	2.25	2.21	2.18	2.11	2.05
3.31	3.17	3.03	2.88	2.79	2.72	2.64	2.58	2.55	2.46	2.37
4.95	4.70	4.44	4.17	4.00	3.88	3.74	3.64	3.58	3.42	3.28
1.90	1.86	1.81	1.76	1.73	1.70	1.67	1.65	1.64	1.60	1.57
2.30	2.23	2.15	2.07	2.02	1.98	1.94	1.91	1.89	1.84	1.79
2.70	2.60	2.50	2.39	2.32	2.27	2.21	2.17	2.14	2.08	2.01
3.26	3.12	2.98	2.83	2.73	2.67	2.58	2.53	2.50	2.40	2.32
4.83	4.58	4.33	4.06	3.89	3.78	3.63	3.54	3.48	3.32	3.17
1.89	1.84	1.80	1.74	1.71	1.69	1.66	1.64	1.62	1.59	1.55
2.27	2.20	2.13	2.05	2.00	1.96	1.91	1.88	1.86	1.81	1.76
2.67	2.57	2.47	2.36	2.29	2.24	2.18	2.14	2.11	2.04	1.98
3.21	3.07	2.93	2.78	2.69	2.62	2.54	2.48	2.45	2.35	2.27
4.73	4.48	4.23	3.96	3.79	3.68	3.53	3.44	3.38	3.22	3.08
1.88	1.83	1.78	1.73	1.70	1.67	1.64	1.62	1.61	1.57	1.54
2.25	2.18	2.11	2.03	1.97	1.94	1.89	1.86	1.84	1.79	1.74
2.64	2.54	2.44	2.33	2.26	2.21	2.15	2.11	2.08	2.01	1.94
3.17	3.03	2.89	2.74	2.64	2.58	2.49	2.44	2.40	2.31	2.22
4.64	4.39	4.14	3.87	3.71	3.59	3.45	3.36	3.29	3.14	2.99
1.87	1.82	1.77	1.72	1.68	1.66	1.63	1.61	1.59	1.56	1.52
2.24	2.16	2.09	2.01	1.96	1.92	1.87	1.84	1.82	1.77	1.72
2.61	2.51	2.41	2.30	2.23	2.18	2.12	2.08	2.05	1.98	1.91
3.13	2.99	2.85	2.70	2.60	2.54	2.45	2.40	2.36	2.27	2.18
4.56	4.31	4.06	3.79	3.63	3.52	3.37	3.28	3.22	3.06	2.91
1.86	1.81	1.76	1.71	1.67	1.65	1.61	1.59	1.58	1.54	1.51
2.22	2.15	2.07	1.99	1.94	1.90	1.85	1.82	1.80	1.75	1.70
2.59	2.49	2.39	2.28	2.21	2.16	2.09	2.05	2.03	1.95	1.89
3.09	2.96	2.81	2.66	2.57	2.50	2.42	2.36	2.33	2.23	2.14
4.48	4.24	3.99	3.72	3.56	3.44	3.30	3.21	3.15	2.99	2.84
1.85	1.80	1.75	1.70	1.66	1.64	1.60	1.58	1.57	1.53	1.50
2.20	2.13	2.06	1.97	1.92	1.88	1.84	1.81	1.79	1.73	1.68
2.57	2.47	2.36	2.25	2.18	2.13	2.07	2.03	2.00	1.93	1.86
3.06	2.93	2.78	2.63	2.54	2.47	2.38	2.33	2.29	2.20	2.11
4.41	4.17	3.92	3.66	3.49	3.38	3.23	3.14	3.08	2.92	2.78

TABLE E   F critical values (continued)

	p	1	2	3	4	5	6	7	8	9
		\multicolumn{9}{c}{Degrees of freedom in the numerator}								
28	.100	2.89	2.50	2.29	2.16	2.06	2.00	1.94	1.90	1.87
	.050	4.20	3.34	2.95	2.71	2.56	2.45	2.36	2.29	2.24
	.025	5.61	4.22	3.63	3.29	3.06	2.90	2.78	2.69	2.61
	.010	7.64	5.45	4.57	4.07	3.75	3.53	3.36	3.23	3.12
	.001	13.50	8.93	7.19	6.25	5.66	5.24	4.93	4.69	4.50
29	.100	2.89	2.50	2.28	2.15	2.06	1.99	1.93	1.89	1.86
	.050	4.18	3.33	2.93	2.70	2.55	2.43	2.35	2.28	2.22
	.025	5.59	4.20	3.61	3.27	3.04	2.88	2.76	2.67	2.59
	.010	7.60	5.42	4.54	4.04	3.73	3.50	3.33	3.20	3.09
	.001	13.39	8.85	7.12	6.19	5.59	5.18	4.87	4.64	4.45
30	.100	2.88	2.49	2.28	2.14	2.05	1.98	1.93	1.88	1.85
	.050	4.17	3.32	2.92	2.69	2.53	2.42	2.33	2.27	2.21
	.025	5.57	4.18	3.59	3.25	3.03	2.87	2.75	2.65	2.57
	.010	7.56	5.39	4.51	4.02	3.70	3.47	3.30	3.17	3.07
	.001	13.29	8.77	7.05	6.12	5.53	5.12	4.82	4.58	4.39
40	.100	2.84	2.44	2.23	2.09	2.00	1.93	1.87	1.83	1.79
	.050	4.08	3.23	2.84	2.61	2.45	2.34	2.25	2.18	2.12
	.025	5.42	4.05	3.46	3.13	2.90	2.74	2.62	2.53	2.45
	.010	7.31	5.18	4.31	3.83	3.51	3.29	3.12	2.99	2.89
	.001	12.61	8.25	6.59	5.70	5.13	4.73	4.44	4.21	4.02
50	.100	2.81	2.41	2.20	2.06	1.97	1.90	1.84	1.80	1.76
	.050	4.03	3.18	2.79	2.56	2.40	2.29	2.20	2.13	2.07
	.025	5.34	3.97	3.39	3.05	2.83	2.67	2.55	2.46	2.38
	.010	7.17	5.06	4.20	3.72	3.41	3.19	3.02	2.89	2.78
	.001	12.22	7.96	6.34	5.46	4.90	4.51	4.22	4.00	3.82
60	.100	2.79	2.39	2.18	2.04	1.95	1.87	1.82	1.77	1.74
	.050	4.00	3.15	2.76	2.53	2.37	2.25	2.17	2.10	2.04
	.025	5.29	3.93	3.34	3.01	2.79	2.63	2.51	2.41	2.33
	.010	7.08	4.98	4.13	3.65	3.34	3.12	2.95	2.82	2.72
	.001	11.97	7.77	6.17	5.31	4.76	4.37	4.09	3.86	3.69
100	.100	2.76	2.36	2.14	2.00	1.91	1.83	1.78	1.73	1.69
	.050	3.94	3.09	2.70	2.46	2.31	2.19	2.10	2.03	1.97
	.025	5.18	3.83	3.25	2.92	2.70	2.54	2.42	2.32	2.24
	.010	6.90	4.82	3.98	3.51	3.21	2.99	2.82	2.69	2.59
	.001	11.50	7.41	5.86	5.02	4.48	4.11	3.83	3.61	3.44
200	.100	2.73	2.33	2.11	1.97	1.88	1.80	1.75	1.70	1.66
	.050	3.89	3.04	2.65	2.42	2.26	2.14	2.06	1.98	1.93
	.025	5.10	3.76	3.18	2.85	2.63	2.47	2.35	2.26	2.18
	.010	6.76	4.71	3.88	3.41	3.11	2.89	2.73	2.60	2.50
	.001	11.15	7.15	5.63	4.81	4.29	3.92	3.65	3.43	3.26
1000	.100	2.71	2.31	2.09	1.95	1.85	1.78	1.72	1.68	1.64
	.050	3.85	3.00	2.61	2.38	2.22	2.11	2.02	1.95	1.89
	.025	5.04	3.70	3.13	2.80	2.58	2.42	2.30	2.20	2.13
	.010	6.66	4.63	3.80	3.34	3.04	2.82	2.66	2.53	2.43
	.001	10.89	6.96	5.46	4.65	4.14	3.78	3.51	3.30	3.13

Degrees of freedom in the denominator

TABLE E   *F* critical values (continued)

Degrees of freedom in the numerator

10	12	15	20	25	30	40	50	60	120	1000
1.84	1.79	1.74	1.69	1.65	1.63	1.59	1.57	1.56	1.52	1.48
2.19	2.12	2.04	1.96	1.91	1.87	1.82	1.79	1.77	1.71	1.66
2.55	2.45	2.34	2.23	2.16	2.11	2.05	2.01	1.98	1.91	1.84
3.03	2.90	2.75	2.60	2.51	2.44	2.35	2.30	2.26	2.17	2.08
4.35	4.11	3.86	3.60	3.43	3.32	3.18	3.09	3.02	2.86	2.72
1.83	1.78	1.73	1.68	1.64	1.62	1.58	1.56	1.55	1.51	1.47
2.18	2.10	2.03	1.94	1.89	1.85	1.81	1.77	1.75	1.70	1.65
2.53	2.43	2.32	2.21	2.14	2.09	2.03	1.99	1.96	1.89	1.82
3.00	2.87	2.73	2.57	2.48	2.41	2.33	2.27	2.23	2.14	2.05
4.29	4.05	3.80	3.54	3.38	3.27	3.12	3.03	2.97	2.81	2.66
1.82	1.77	1.72	1.67	1.63	1.61	1.57	1.55	1.54	1.50	1.46
2.16	2.09	2.01	1.93	1.88	1.84	1.79	1.76	1.74	1.68	1.63
2.51	2.41	2.31	2.20	2.12	2.07	2.01	1.97	1.94	1.87	1.80
2.98	2.84	2.70	2.55	2.45	2.39	2.30	2.25	2.21	2.11	2.02
4.24	4.00	3.75	3.49	3.33	3.22	3.07	2.98	2.92	2.76	2.61
1.76	1.71	1.66	1.61	1.57	1.54	1.51	1.48	1.47	1.42	1.38
2.08	2.00	1.92	1.84	1.78	1.74	1.69	1.66	1.64	1.58	1.52
2.39	2.29	2.18	2.07	1.99	1.94	1.88	1.83	1.80	1.72	1.65
2.80	2.66	2.52	2.37	2.27	2.20	2.11	2.06	2.02	1.92	1.82
3.87	3.64	3.40	3.14	2.98	2.87	2.73	2.64	2.57	2.41	2.25
1.73	1.68	1.63	1.57	1.53	1.50	1.46	1.44	1.42	1.38	1.33
2.03	1.95	1.87	1.78	1.73	1.69	1.63	1.60	1.58	1.51	1.45
2.32	2.22	2.11	1.99	1.92	1.87	1.80	1.75	1.72	1.64	1.56
2.70	2.56	2.42	2.27	2.17	2.10	2.01	1.95	1.91	1.80	1.70
3.67	3.44	3.20	2.95	2.79	2.68	2.53	2.44	2.38	2.21	2.05
1.71	1.66	1.60	1.54	1.50	1.48	1.44	1.41	1.40	1.35	1.30
1.99	1.92	1.84	1.75	1.69	1.65	1.59	1.56	1.53	1.47	1.40
2.27	2.17	2.06	1.94	1.87	1.82	1.74	1.70	1.67	1.58	1.49
2.63	2.50	2.35	2.20	2.10	2.03	1.94	1.88	1.84	1.73	1.62
3.54	3.32	3.08	2.83	2.67	2.55	2.41	2.32	2.25	2.08	1.92
1.66	1.61	1.56	1.49	1.45	1.42	1.38	1.35	1.34	1.28	1.22
1.93	1.85	1.77	1.68	1.62	1.57	1.52	1.48	1.45	1.38	1.30
2.18	2.08	1.97	1.85	1.77	1.71	1.64	1.59	1.56	1.46	1.36
2.50	2.37	2.22	2.07	1.97	1.89	1.80	1.74	1.69	1.57	1.45
3.30	3.07	2.84	2.59	2.43	2.32	2.17	2.08	2.01	1.83	1.64
1.63	1.58	1.52	1.46	1.41	1.38	1.34	1.31	1.29	1.23	1.16
1.88	1.80	1.72	1.62	1.56	1.52	1.46	1.41	1.39	1.30	1.21
2.11	2.01	1.90	1.78	1.70	1.64	1.56	1.51	1.47	1.37	1.25
2.41	2.27	2.13	1.97	1.87	1.79	1.69	1.63	1.58	1.45	1.30
3.12	2.90	2.67	2.42	2.26	2.15	2.00	1.90	1.83	1.64	1.43
1.61	1.55	1.49	1.43	1.38	1.35	1.30	1.27	1.25	1.18	1.08
1.84	1.76	1.68	1.58	1.52	1.47	1.41	1.36	1.33	1.24	1.11
2.06	1.96	1.85	1.72	1.64	1.58	1.50	1.45	1.41	1.29	1.13
2.34	2.20	2.06	1.90	1.79	1.72	1.61	1.54	1.50	1.35	1.16
2.99	2.77	2.54	2.30	2.14	2.02	1.87	1.77	1.69	1.49	1.22

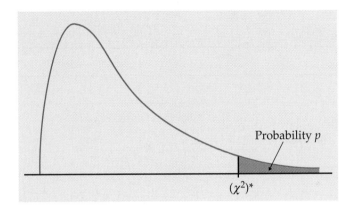

Table entry for *p* is the critical value $(\chi^2)^*$ with probability *p* lying to its right.

Probability *p*

$(\chi^2)^*$

TABLE F	χ^2 distribution critical values

Tail probability *p*

df	.25	.20	.15	.10	.05	.025	.02	.01	.005	.0025	.001	.0005
1	1.32	1.64	2.07	2.71	3.84	5.02	5.41	6.63	7.88	9.14	10.83	12.12
2	2.77	3.22	3.79	4.61	5.99	7.38	7.82	9.21	10.60	11.98	13.82	15.20
3	4.11	4.64	5.32	6.25	7.81	9.35	9.84	11.34	12.84	14.32	16.27	17.73
4	5.39	5.99	6.74	7.78	9.49	11.14	11.67	13.28	14.86	16.42	18.47	20.00
5	6.63	7.29	8.12	9.24	11.07	12.83	13.39	15.09	16.75	18.39	20.51	22.11
6	7.84	8.56	9.45	10.64	12.59	14.45	15.03	16.81	18.55	20.25	22.46	24.10
7	9.04	9.80	10.75	12.02	14.07	16.01	16.62	18.48	20.28	22.04	24.32	26.02
8	10.22	11.03	12.03	13.36	15.51	17.53	18.17	20.09	21.95	23.77	26.12	27.87
9	11.39	12.24	13.29	14.68	16.92	19.02	19.68	21.67	23.59	25.46	27.88	29.67
10	12.55	13.44	14.53	15.99	18.31	20.48	21.16	23.21	25.19	27.11	29.59	31.42
11	13.70	14.63	15.77	17.28	19.68	21.92	22.62	24.72	26.76	28.73	31.26	33.14
12	14.85	15.81	16.99	18.55	21.03	23.34	24.05	26.22	28.30	30.32	32.91	34.82
13	15.98	16.98	18.20	19.81	22.36	24.74	25.47	27.69	29.82	31.88	34.53	36.48
14	17.12	18.15	19.41	21.06	23.68	26.12	26.87	29.14	31.32	33.43	36.12	38.11
15	18.25	19.31	20.60	22.31	25.00	27.49	28.26	30.58	32.80	34.95	37.70	39.72
16	19.37	20.47	21.79	23.54	26.30	28.85	29.63	32.00	34.27	36.46	39.25	41.31
17	20.49	21.61	22.98	24.77	27.59	30.19	31.00	33.41	35.72	37.95	40.79	42.88
18	21.60	22.76	24.16	25.99	28.87	31.53	32.35	34.81	37.16	39.42	42.31	44.43
19	22.72	23.90	25.33	27.20	30.14	32.85	33.69	36.19	38.58	40.88	43.82	45.97
20	23.83	25.04	26.50	28.41	31.41	34.17	35.02	37.57	40.00	42.34	45.31	47.50
21	24.93	26.17	27.66	29.62	32.67	35.48	36.34	38.93	41.40	43.78	46.80	49.01
22	26.04	27.30	28.82	30.81	33.92	36.78	37.66	40.29	42.80	45.20	48.27	50.51
23	27.14	28.43	29.98	32.01	35.17	38.08	38.97	41.64	44.18	46.62	49.73	52.00
24	28.24	29.55	31.13	33.20	36.42	39.36	40.27	42.98	45.56	48.03	51.18	53.48
25	29.34	30.68	32.28	34.38	37.65	40.65	41.57	44.31	46.93	49.44	52.62	54.95
26	30.43	31.79	33.43	35.56	38.89	41.92	42.86	45.64	48.29	50.83	54.05	56.41
27	31.53	32.91	34.57	36.74	40.11	43.19	44.14	46.96	49.64	52.22	55.48	57.86
28	32.62	34.03	35.71	37.92	41.34	44.46	45.42	48.28	50.99	53.59	56.89	59.30
29	33.71	35.14	36.85	39.09	42.56	45.72	46.69	49.59	52.34	54.97	58.30	60.73
30	34.80	36.25	37.99	40.26	43.77	46.98	47.96	50.89	53.67	56.33	59.70	62.16
40	45.62	47.27	49.24	51.81	55.76	59.34	60.44	63.69	66.77	69.70	73.40	76.09
50	56.33	58.16	60.35	63.17	67.50	71.42	72.61	76.15	79.49	82.66	86.66	89.56
60	66.98	68.97	71.34	74.40	79.08	83.30	84.58	88.38	91.95	95.34	99.61	102.7
80	88.13	90.41	93.11	96.58	101.9	106.6	108.1	112.3	116.3	120.1	124.8	128.3
100	109.1	111.7	114.7	118.5	124.3	129.6	131.1	135.8	140.2	144.3	149.4	153.2

CHAPTER | 1

1.1 **(a)** Vehicles. **(b)** Make/model, vehicle type, transmission type (all categorical); number of cylinders, city MPG, highway MPG (all quantitative).

1.3 **(a)** categorical. **(b)** categorical. **(c)** quantitative. **(d)** categorical. **(e)** quantitative.

1.5 Possible answers: time to run a race (instrument: stopwatch), heart rate after exercising (instrument: watch).

1.7 1.75 (1992) and 1.51 (2002) deaths per 100 million miles.

1.9 14.2%; a pie chart could be used.

1.11 Unmet need is greatest at private institutions, especially for-profit ones. A pie chart is incorrect because these numbers do not represent parts of a single whole.

1.15 **(a)** Alaska: 5.7%; Florida: 17.6%. **(b)** Symmetric, centered near 13%, spread from 8.5% to 15.6%.

1.17 359 mg/dl is an outlier; only four are in the desired range.

1.19 Roughly symmetric, centered near 7, spread from 2 to 13.

1.21 There are three peaks; presumably, the lowest group includes public institutions, the highest group exclusive private schools like Harvard, and the middle group other private schools.

1.25 Both plots show random fluctuation. Pasadena temperatures show an upward trend. Redding temperatures are initially similar to Pasadena's but dropped in the mid-1980s.

1.29 Right-skewed, centered near 5 or 6, spread from 0 to 18, no outliers.

1.31 From the top left histogram: 4, 2, 1, 3.

1.33 Use a stemplot or a histogram. Skewed to the right, center near 40 or 50 thousand barrels, with two or three high outliers; apart from these, the numbers are spread from 0 to 118.2 thousand barrels.

1.35 **(a)** Most people will "round" their answers when asked to give an estimate like this. The values 360 and 300 are suspicious. **(b)** The stemplots and midpoints (175 for women, 120 for men) suggest that women (claim to) study more than men.

1.37 **(a)** Four variables: GPA, IQ, and self-concept are quantitative, gender is categorical. **(c)** Skewed to the left, centered around 7.8, spread from 0.5 to 10.8. **(d)** There is more variability among the boys; in fact, there seem to be two groups of boys: those with GPAs below 5, and those with GPAs above 5.

1.39 Skewed to the left, center around 59.5; most scores are between 35 and 73, with a few below that, and one high score of 80 (probably not quite an outlier).

1.41 Five-number summary: $4123, $15,717, $20,072, $27,957.5, $29,875. This and the boxplot do not reveal the three groups of schools visible in the histogram.

1.43 **(a)** Five-number summary (1999 dollars): 0, 2.14, 10.64, 40.96, 88.6. The large gaps between the higher numbers show the skew. **(b)** The *IQR* is 38.82; no observations fall below −56.09 or above 99.19. **(c)** The mean is 21.95 (1999 dollars); the right skew makes it much larger than the median.

1.45 This distribution would almost surely be strongly skewed to the right.

1.47 $\bar{x} = \$62,500$; seven of the eight employees earned less than the mean. $M = \$25,000$.

1.49 The mean rises to $87,500, while the median is unchanged.

1.51   Means are not the appropriate measure of center for skewed distributions.

1.53   (a) pH: $\bar{x} \doteq 5.4256$ and $s \doteq 0.5379$. Density: $\bar{x} \doteq 5.4479$ and $s \doteq 0.2209$. (b) Medians: 5.44 and 5.46.

1.55   The mean and median always agree for two observations.

1.57   (a) Place the new point at the current median.

1.59   (a) *Bihai*: $\bar{x} \doteq 47.5975$, $s \doteq 1.2129$. Red: $\bar{x} \doteq 39.7113$, $s \doteq 1.7988$. Yellow: $\bar{x} \doteq 36.1800$, $s \doteq 0.9753$ (all in mm). (b) *Bihai* and red appear to be right-skewed (although it is difficult to tell with such small samples).

1.61   All four distributions are symmetric. The comparison should include at least the four means: 147, 130, 116, and 103 beats per minute (bpm). The average heart rate for runners is about 30 bpm less than the average sedentary rate.

1.63   Positions 748 and 14,211; about $13,000 and $137,000.

1.65   The five-number summary is 1, 3, 4, 5, and 12 letters.

1.67   Take six or more numbers, with the largest number much larger than Q_3.

1.69   (a) Any set of four identical numbers works. (b) 0, 0, 10, 10 is the only possible answer.

1.71   Five-number summary: 43, 82.5, 102.5, 151.5, and 598 days. Typical guinea pig lifetimes were 82.5 to 151.5 days, but some live quite a bit longer.

1.73   Multiply each value by 0.03937.

1.75   80, 99, 115, and 178 days.

1.77   Full data set: $\bar{x} = 141.8$ and $M = 102.5$ days. 10% trimmed mean: $\bar{x}^* = 118.16$ days; 20% trimmed mean: $\bar{x}^{**} = 111.68$ days.

1.79   The peak at 0 should be "tall and skinny." Near 1, the curve should be "short and fat."

1.81   (a) 1/2. (b) 0.5. (c) 0.4.

1.83   (a) Mean C, median B. (b) Mean A, median A. (c) Mean A, median B.

1.85   (a) 0.6826; the 68–95–99.7 rule gives 0.68. (b) 0.9544 (compared to 0.95); 0.9974 (compared to 0.997).

1.87   (a) 327 to 345 days. (b) 16%.

1.89   Women: $z = 2.96$; men: $z = 0.96$.

1.91 $\bar{x} = 5.4256$ and $s = 0.5379$; 67.62% within $\bar{x} - s$ and $\bar{x} + s$, 95.24% within $\bar{x} \pm 2s$, and all within $\bar{x} \pm 3s$.

1.93   (a) 0.0228. (b) 0.9772. (c) 0.0475. (d) 0.9297.

1.95   (a) 0.8416. (b) 0.1257.

1.97   About 2.5%.

1.99   Jacob's score ($z = -1$) is higher than Emily's ($z = -1.7033$).

1.101   About 1340.

1.103   16th percentile.

1.105   About 0.1%.

1.107   24.8 (round to 25) or above.

1.109   850, 973, 1079, and 1202.

1.111   Above 240 mg/dl: about 31%. Between 200 and 240 mg/dl: about 41%.

1.113   (a) About 3.3%. (b) About 1.2%.

1.115   (a) About 5.2%. (b) About 54.7%. (c) More than 279 days.

1.117   (a) About 1.35. (b) 1.35.

1.119   The plot is nearly linear. Because heart rate is measured in whole numbers, there is a slight "step" appearance to the graph.

1.121   The plot is reasonably close to a line, apart from the stair-step appearance, presumably due to limited accuracy of the measuring instrument and the way that the data are recorded.

1.123 The plot suggests no major deviations from normality, although the three lowest measurements don't quite fall in line with the other points.

1.125 The first plot is nearly linear; the other two each show a low value.

1.127 Histograms will suggest (but not exactly match) Figure 1.35. The uniform distribution does not extend as low or as high as a normal distribution.

1.129 Slightly right-skewed, with one (or more) high outliers. Five-number summary: 22, 23.735, 24.31, 24.845, and 28.55 hours.

1.131 Many (but less than half) of students were 19.

1.133 Women's weights are right-skewed.

1.135 **(a)** About 20% of low-income and 33% of high-income households. **(b)** The majority of low-income households, but only about 7% of high-income households, consist of one person. One-person households often have less income because they would include many young people who have have no job or have only recently started working.

1.137 No to both questions; no summary can exactly describe a distribution that can include any number of values.

1.139 **(a)** The number of home runs declines sharply and steadily. **(b)** Ruth led for a longer time, and appears to have set a new standard for other players.

1.141 **(a)** $x_{new} = -50 + 2x$. **(b)** $x_{new} = -49.0909 + 1.8182x$. **(c)** David (106) is higher than Nancy (92.7). **(d)** About 62% of third-graders and 36% of sixth-graders.

1.143 **(a)** The plot would be linear except for the two extremes. The outliers show up as a point in the lower left below the line and a point in the upper right above the line.

1.147 Men seem to have higher SATM scores than women; women generally have higher GPAs. All four distributions are close to normal; female SATM and both GPA sets have low outliers.

CHAPTER | 2

2.1 **(a)** Explanatory: time spent studying; response: grade. **(b)** Explore the relationship. **(c)** Explanatory: rainfall; response: crop yield. **(d)** Explore the relationship. **(e)** Explanatory: income; response: years of education.

2.3 **(a)** In general, we expect more intelligent children to be better readers, and less intelligent children to be weaker. The plot does show this positive association. **(b)** These four have moderate IQs but poor reading scores. **(c)** Roughly linear, but weak (much scatter).

2.5 **(a)** The response variable (estimated level) can only take on the values 1, 2, 3, 4, 5. **(b)** The association is (weakly) positive. **(c)** The estimate is 4, which is an overestimate; that child had the lowest score on the test.

2.7 **(a)** Areas with many breeding pairs would correspondingly have more males that might potentially return. **(c)** The theory suggests a negative association; the scatterplot shows this.

2.9 A fairly strong, positive, linear association; social exclusion does appear to trigger a pain response.

2.11 **(a)** The Lakers and the Knicks are high in both variables (but fit the pattern). The Grizzlies, Cavaliers, and Rockets have slightly higher values than their revenues would suggest. The association is positive and linear. **(b)** The Lakers and Knicks still stand out, as do the Bulls and Trailblazers, but the association is quite weak. Revenue is a better predictor of value.

2.13 **(b)** The association is linear and positive, stronger for women. Males typically have larger values for both variables.

2.15 The plot shows a fairly steady rate of improvement until the mid-1980s, with much slower progress after that.

2.17 (a) Both show fairly steady improvement. Women have made more rapid progress, but their progress seems to have slowed, while men's records may be dropping more rapidly in recent years. (b) The data supports the first claim but do not seem to support the second.

2.19 (b) Technology. (c) No; positive/negative association makes sense only when both variables are quantitative.

2.21 (a) Price is explanatory; the plot shows a positive linear association. (b) $r = 0.955$.

2.23 Closest to 0.6.

2.25 No; units do not affect correlation.

2.27 (a) Value and revenue: $r_1 = 0.9265$; value and income: $r_2 = 0.2107$. This agrees with conclusions from the scatterplots. (b) Without Portland, $r_2 = 0.3469$. The removal of this point makes the scatterplot appear (slightly) more linear.

2.29 (a) Positive, but not close to 1. (b) $r = 0.5653$. (c) r would not change; it does not tell us that the men were generally taller. (d) r would not change. (e) 1.

2.31 (a) $r = \pm 1$ for a line. (c) Leave some space above your vertical stack. (d) The curve must be higher at the right than at the left.

2.33 (a) The Insight seems to fit the line. (b) Without: 0.9757; with: 0.9934. The Insight increases the strength of the association.

2.35 $r = 1$ for a positively sloped line.

2.37 There is little linear association between research and teaching—for example, knowing a professor is a good researcher gives little information about whether she is a good or bad teacher.

2.39 Both relationships are somewhat linear; GPA/IQ ($r = 0.6337$) is stronger than GPA/self-concept ($r = 0.5418$). The two students with the lowest GPAs stand out in both plots; a few others stand out in at least one plot. Generally speaking, removing these points raises r, except for the lower-left point in the self-concept plot.

2.41 (a) The predicted scores for $x = 90$ and $x = 140$ are about 46 and 90. (b) Reading score $= IQ - 50$.

2.43 (a) 4.2255 cubic km per year. (b) −271 cubic km. (c) 617 cubic km; the prediction error is about 63 cubic km. (d) There are high spikes in the time plot in these two years.

2.45 (b) $\bar{x} = 1987.7778$, $\bar{y} = 304.1111$, $s_x = 8.7003$, $s_y = 3.4440$, $r = 0.7268$, and $\hat{y} = 0.2877x - 267.78$. This line explains about 53% of the variation.

2.47 (a) $\hat{y} = 0.06078x - 0.1261$. (b) $\hat{y} \doteq 0$ (the formula gives −0.0045). (c) 77%.

2.49 For the slower flow rate (8903), icicles grow at 0.158 cm/min; for 8905, the growth rate is 0.0911 cm/min.

2.51 No; the scatterplot shows little or no association, and regression explains only 1.3%.

2.53 For metabolic rate on body mass, the slope is 26.9 (cal/day)/kg. For body mass on metabolic rate, the slope is 0.0278 kg/(cal/day).

2.55 $\hat{y} = 33.67 + 0.54x$. The predicted height is 69.85 inches.

2.57 (a) $\bar{x} = 95$ and $s_x = 53.3854$ minutes; $\bar{y} = 12.6611$ and $s_y = 8.4967$ cm; $r = 0.9958$ (no units). (b) $\bar{y} = 4.9847$ and $s_y = 3.3452$ inches; r is unchanged. (c) 0.0624 inches/minute.

2.59 $r = 0.40$.

2.61 (b) The slope is 0.000051; the plot suggests a horizontal line (slope 0). (c) Storing the oil doesn't help, as the total toxin level does not change over time.

2.63 (b) No, because the pattern is not linear. (c) The sum is 0.01. The first two and last four residuals are negative, and those in the middle are positive.

2.65 The variation of individual data increases the scatter, thus decreasing the strength of the relationship.

2.67   For example, a student who in the past might have received a grade of B (and a lower SAT score) now receives an A (but has a lower SAT score than an A student in the past).

2.69 $r = 0.08795$ and $b = 0.000811$ kg/cal.

2.71   **(a)** Player 7's point is influential. **(b)** The line of the first equation omits Player 7.

2.73   **(b)** $\hat{y} = 6.47 + 1.01x$. **(c)** The largest residuals are the Porsche Boxster (2.365) and Lamborghini Murcielago (-2.545). **(d)** The Insight pulls the line toward its point.

2.75   Without the Insight, $\hat{y} = 4.87 + 1.11x$. For city mileages between 10 and 30 MPG, the difference in predicted highway mileage (with or without the Insight) is no more than 1.4 MPG, so the Insight is not very influential; it falls near the line suggested by the other points.

2.77   **(a)** Drawing the "best line" by eye is a very inaccurate process.

2.79   The plot should show positive association when either set of circles is viewed separately and should show a large number of bachelor's degree economists in business and graduate degree economists in academia.

2.81   **(a)** There is some weak support (the slope of the line is negative). **(b)** There seems to be some suggestion of curvature and also more scatter on the left side; this weakens the conclusion.

2.83   **(a)** Regression more accurately predicts low salaries. **(b)** The plot is curved; salaries are typically overestimated for new players and for those with 15 or more years in the Majors; salaries of players who have been in for 8 years are generally underestimated.

2.85   Age is one lurking variable: married men are generally older.

2.87   No; self-confidence and improving fitness could be a common response to some other personality trait, or high self-confidence could make a person more likely to join the exercise program.

2.89   Students with music experience may have other advantages (wealthier parents, better school systems, etc.).

2.91   The diagram should show that chemical exposure, time standing up, both, or neither affect miscarriages.

2.93   Spending more time watching TV means that less time is spent on other activities; this may suggest lurking variables.

2.95   **(a)** Given two groups of the same age, where one group walks and the other does not, the walkers are half as likely to die in (say) the next year. **(b)** Men who choose to walk might also choose (or have chosen, earlier in life) other habits and behaviors that reduce mortality.

2.97   These results support the idea (the slope is negative), but the relationship is only moderately strong ($r^2 = 0.34$).

2.99   **(a)** One possible measure is mean response: 106.2 spikes/second for pure tones, 176.6 spikes/second for monkey calls. **(b)** $\hat{y} = 93.9 + 0.778x$. The third point has the largest residual. The first point is an outlier in the x direction. **(c)** The correlation drops only slightly (from 0.6386 to 0.6101) when the third point is removed; it drops more drastically (to 0.4793) without the first point. **(d)** Without the first point, the line is $\hat{y} = 101 + 0.693x$; without the third point, it is $\hat{y} = 98.4 + 0.679x$.

2.101   Based on the quantile plot, the distribution is close to normal.

2.103   **(a)** Lines appear to fit the data well; there do not appear to be any outliers or influential points. **(b)** 18.9 cubic ft before, 15.7 cubic ft after. **(c)** 770.4 cubic ft before, 634.8 cubic ft after. **(d)** About $50.44.

2.105   **(a)** Weeds per meter is explanatory; corn yield is the response. **(b)** Yield decreases when there are more weeds. **(c)** $\hat{y} = 166 - 1.10x$. Corn yield decreases by about 1.1 bushels/acre for each additional weed per meter. **(d)** 159.4 bushels/acre.

2.107   **(a)** $\hat{y} = 259.58 - 19.464x$; the relationship appears to be curved. **(b)** Either $\hat{y} = 5.9732 - 0.2184x$ or $\hat{y} = 2.5941 - 0.09486x$; the relationship appears to be linear.

2.109   It is now more common for these stocks to rise and fall together.

2.111   Number of firefighters and amount of damage are common responses to the seriousness of the fire.

2.113 $\hat{y} = 1.28 + 0.00227x$, $r = 0.252$, and $r^2 = 6.3\%$. By itself, SATM does not give reliable predictions of GPA.

CHAPTER | 3

3.1   Jamie and his friends do not constitute a random sample.

3.3   This is an observational study. Explanatory variable: cell phone usage; response variable: presence/absence of brain cancer.

3.5   An experiment: each subject is (presumably randomly) assigned to a treatment group. Explanatory variable: teaching method; response variable: change in each student's test score.

3.7   **(a)** For an observational study, classify each subject as a wine drinker or a beer drinker. For an experiment, assign each subject to drink either wine or beer. In either case, we would then observe the health of the subjects over time. **(b)** Wine drinkers might be wealthier, be better educated, have white-collar jobs, and have different dietary habits.

3.9   Experimental units: pine tree seedlings. Factor: amount of light. Treatments: full light or shaded to 5% of normal. Response variable: dry weight at end of study.

3.11   Subjects: adults from selected households. Factors: level of identification and offer of survey results. Six treatments: interviewer's name/university name/both names, with or without results. Response variable: whether or not the interview is completed.

3.13   Assign nine subjects to each treatment. The first three groups are 05, 16, 17, 20, 19, 32, 04, 25, 29; 31, 18, 07, 13, 33, 02, 36, 23, 27; 35, 21, 26, 08, 10, 11, 15, 12, 14.

3.15   **(a)** Randomly assign six rats to each group. **(b)** Group 1 includes rats 02, 08, 17, 10, 05, and 09. Group 2 is 06, 16, 01, 07, 18, and 15.

3.17   Assign six schools to each treatment group. Choose 07, 19, 14, 17, 13, 15 for Group 1; 08, 21, 20, 11, 24, 09 for Group 2; 06, 23, 16, 18, 12, 04 for Group 3; and the rest for Group 4.

3.19   **(a)** There are three factors (roller type, dyeing cycle time, and temperature), yielding eight treatments, and requiring 24 fabric specimens.

3.21   **(a)** Population = 1 to 150, sample size = 25, then click "Reset" and "Sample."
(b) Without resetting, click "Sample" again.

3.23   The first design is an observational study, while the second is an experiment; with the second, any difference in colon health between the two groups could be attributed to the treatment (bee pollen or not).

3.25   **(a)** See the definitions in this chapter.

3.27   There are nine treatments. Choose the number of letters in each group, and send the letters out at random times over several weeks.

3.29   Each player runs through the sequence (100 yards, four times) once with oxygen and once without (on different days to allow full recovery); observe the difference in times on the final run. We choose 07, 13, 14, 02, 12, 20, 06, 08, 16, 03 for the oxygen-first group.

3.31   **(a)** Randomly assign half the girls to get high-level calcium and the other half to get low-level calcium, then observe how each group processes the calcium. **(b)** Half receive high-calcium punch first; the rest get low-calcium punch first. For each subject, compute the difference in the response variable for each level. The matched pairs design deals with person-to-person variation. **(c)** The first five subjects are 16, 34, 59, 44, and 21.

3.33   A completely randomized design is one possibility; a block design is a better choice.

3.35 (a) False. (b) True. (c) False.

3.37 (a) Population: adult residents of the United States. Sample size: 1002. (b) If the order has any effect on how people respond, rotating the order cancels it out.

3.39 (a) Adult residents of the United States. (b) 37.2%. (c) Response bias (giving inaccurate numbers).

3.41 From line 139 we choose 04, 10, 17, 19, 12, 32, 13, 18, 24, 28.

3.43 With the applet: Population = 1 to 371, sample size = 25, then click "Reset" and "Sample." Table B gives 353, 246, 027, 038, and 207.

3.45 The results would not be random.

3.47 (a) This gives a total of 9000/45 = 200 names. (b) Start with 19.

3.49 (a) A stratified random sample. (b) 12 (559), 04 (209), 11 (805), 19 (562), 02 (707), 06 (925), 08 (650), 25 (619), 17 (626), and 14 (661).

3.51 Label the students $01, \ldots, 29$ and choose four students, then label the faculty $0, \ldots, 9$ and use the table again.

3.53 Label midsize accounts from 001 to 500 and small accounts from 0001 to 4400. Midsize: 417, 494, 322, 247, and 097. Small: 3698, 1452, 2605, 2480, and 3716.

3.55 (a) Households without telephones or with unlisted numbers. Such households would likely be made up of poor individuals, those who choose not to have phones, and those who do not wish to have their phone number published. (b) Those with unlisted numbers.

3.57 (a) Slanted (and based on a questionable conclusion). (b) Slanted in favor of national health insurance. (c) Confusing and vague.

3.59 Children from larger families will be overrepresented in such a sample.

3.61 Slanted questions can produce opposing results.

3.63 Statistic, parameter.

3.65 Statistic, statistic.

3.67 (a) Smaller sample sizes give less information about the population. (b) The margin of error was so large that the results could not be viewed as an accurate reflection of the population of Cubans.

3.69 (a) The variability will be the same for all states. (b) There would be less variability for states with larger samples.

3.71 (b) Size 25: $25,000 to $80,000; size 100: $35,000 to $65,000. Sampling variability decreases as sample size increases. (c) Between $45,000 and $50,000.

3.73 The histograms should be centered at about 0.6 and 0.2.

3.77 (a) Mean GPA 2.6352, standard deviation 0.7794.

3.79 (a) 00001 through 14959. (b) 03638, 07871, 12193.

3.81 (a) A matched pairs design. (b) A stratified sample survey. (c) A block design.

3.83 This is an experiment because treatments are assigned. Explanatory variable: price history (steady or fluctuating); response variable: price the subject expects to pay.

3.87 Randomly choose the order in which the treatments (gear and steepness combination) are tried.

3.89 (a) One possibility: full-time undergraduate students in the fall term on a list provided by the registrar. (b) One possibility: a stratified sample with 125 students from each year. (c) Nonresponse might be higher with mailed (or emailed) questionnaires; telephone interviews exclude some students and may require repeated calling for those who are not home; face-to-face interviews might be too costly. The topic might also be subject to response bias.

3.91 (a) Factors: storage method (fresh, room temperature for one month, refrigerated for one month) and preparation method (cooked immediately or after one hour). This makes six treatments (storage/preparation combinations). Response variables: tasters' color and flavor ratings. (b) Randomly allocate n potatoes to each of the

6 groups, then compare ratings. **(c)** For each taster, randomly choose the order in which the fries are tasted.

3.93 Strong feelings, promotional campaigns, and multiple responses can mean that voluntary response samples do not represent the population of interest.

3.95 CASI will typically produce more honest responses to embarrassing questions.

CHAPTER | 4

4.1 Long trials of this experiment suggest about 40% heads.

4.7 Results will vary; the exact probability is close to 0.5.

4.9 **(c)** Both distributions should be roughly normal with center near 0.3; the second should be much less variable.

4.11 **(a)** {right, left}. **(b)** Upper and lower limits will vary, e.g., (120, 214). **(c)** {0, 0.01, 0.02, 0.03, ...}, e.g., (10, 480). (d) Upper limits will vary.

4.13 **(a)** 0.04, so the sum equals 1. **(b)** 0.56.

4.15 **(a)** Legitimate. **(b)** Not legitimate: the total is more than 1. **(c)** Legitimate.

4.17 **(a)** 0.28. **(b)** 0.88.

4.19 0.15 for Red Sox and Angels, 0.05 for Athletics and White Sox.

4.21 2/15 or 0.1333 for 1, 1/6 for 2 through 5, and 0.2 for 6.

4.23 Take each blood-type probability and multiply by 0.84 and by 0.16. This gives O+ (0.3780), A+ (0.3360), B+ (0.0924), AB+ (0.0336), O− (0.0720), A− (0.0640), B− (0.0176), AB− (0.0064).

4.25 **(a)** 1/38 (all are equally likely). **(b)** 18/38. **(c)** 12/38.

4.27 **(a)** 10,000. **(b)** 0.3439.

4.29 About 0.38 (0.3773).

4.31 The two events (being 75 or older and being a woman) are probably not independent.

4.33 **(a)** 0.2746. **(b)** 0.35. **(c)** 0.545.

4.35 Observe that $P(A \text{ and } B^c) = P(A) - P(A \text{ and } B) = P(A) - P(A)P(B)$.

4.37 **(a)** Either B or O. **(b)** $P(B) = 0.75$, $P(O) = 0.25$.

4.39 **(a)** 0.25. **(b)** 0.015625; 0.140625.

4.41 **(a)** $P(X = 1) = 0.9$. **(b)** "No more than two nonword errors." $P(X = 2) = 0.6$; $P(X < 2) = 0.3$.

4.43 Rented housing typically has fewer rooms and has a sharply skewed distribution.

4.45 **(a)** $P(X \geq 5) = 0.868$. **(b)** "The unit has more than 5 rooms." $P(X > 5) = 0.658$. (c) Pay attention to whether you have "greater than" or "greater than or equal to."

4.47 **(a)** $P(T = 2) = 0.63$, $P(T = 3) = 0.2331$. **(b)** $P(T \leq 4)$ is the probability that one of the first three people breaks the chain; $P(T \leq 4) \doteq 0.9493$.

4.49 Y can be $1, 2, 3, \ldots, 12$, each with probability 1/12.

4.51 **(a)** 0, 1, 2, or 3. **(b)** DDD (probability 0.3890); DDF, DFD, and FDD (probability 0.1439); FFD, FDF, and DFF (probability 0.0532); FFF (probability 0.0197). **(c)** The probabilities for W are 0.3890, 0.4316, 0.1597, and 0.0197.

4.53 **(a)** 0.5. **(b)** 0.5. **(c)** "Equal to" has no effect on the probability.

4.55 **(a)** The height should be 1/2. **(b)** 1/2. **(c)** 0.4. **(d)** 0.6.

4.57 Very close to 1.

4.59 2.78.

4.61 For owner-occupied units, the mean is 6.284 rooms; for rented units, it is 4.187 rooms.

4.63 The owned-unit distribution is more spread out; $\sigma_o \doteq 1.6399$ and $\sigma_r \doteq 1.3077$ rooms.

4.65 **(a)** $562.50 (probability 0.25), $975 (0.5), and $1690 (0.25). **(b)** $1050.625.

4.67 **(a)** $\mu_1 = \sigma_1 = 0.5$. **(b)** $\mu_4 = 2$ and $\sigma_4 = 1$.

4.69   Mean 14 cm and standard deviation 0.0052 cm.

4.71   **(a)** Not independent. **(b)** Independent.

4.73   **(a)** 31 seconds. **(b)** The mean does not change. **(c)** The mean does not change.

4.75   Independent: $\sigma \doteq 8.2462$ seconds. Correlation 0.3: $\sigma \doteq 8.8091$ seconds.

4.77   Show that $\sigma_{X+Y}^2 = (\sigma_X + \sigma_Y)^2$.

4.79   If 1 of the 12 homes were lost, it would cost more than the collected premiums. For many policies, the average claim should be close to $250.

4.81   **(a)** 0.99749. **(b)** $623.22.

4.83 $\mu_R = 11.116\%$ and $\sigma_R \doteq 15.9291\%$.

4.85 $\mu_R = 10.184\%$ and $\sigma_R \doteq 12.3442\%$.

4.87 $P(A|B) \doteq 0.3142$. If A and B were independent, then $P(A|B)$ would equal $P(A)$.

4.89   **(a)** $P(A^c) = 0.31$. **(b)** $P(A^c \text{ and } B) = 0.08$.

4.91   **(a)** 0.3636. **(b)** Not independent.

4.93   1.

4.95 $P(B|C) = 0.25$. $P(C|B) = 0.125$.

4.97   **(a)** $P(M) \doteq 0.4244$. **(b)** $P(B|M) \doteq 0.6776$. **(c)** $P(M \text{ and } B) \doteq 0.2876$.

4.99   **(a)** $P(L) \doteq 0.6722$. **(b)** $P(L|C) \doteq 0.7837$. **(c)** Not independent, because $P(L) \neq P(L|C)$.

4.101   Retired persons are more likely than other adults to have not completed high school.

4.103   75%.

4.105   0.445.

4.107   About 34% (0.3371).

4.109   **(a)** Her brother has allele type aa, and he got one allele from each parent. **(b)** $P(aa) = 0.25$, $P(Aa) = 0.5$, $P(AA) = 0.25$. **(c)** $P(AA|\text{not } aa) = 1/3$, $P(Aa|\text{not } aa) = 2/3$.

4.111   1.

4.113   0.9333.

4.115   **(a)** All probabilities are greater than or equal to 0, and their sum is 1. **(b)** 0.62. **(c)** Both probabilities are 0.39.

4.117   **(a)** 0.9389. **(b)** $1.30. **(c)** $12.9734.

4.119   0.005352.

4.121   0.1622.

4.123   Both $P(B|A)$ and $P(B)$ equal 0.5.

4.125 $P(\text{no}) = 0.35$ when 30% have plagiarized, and 0.40 if only 20% have plagiarized. We estimate that 22% have plagiarized.

CHAPTER | 5

5.1   **(a)** Binomial, $n = 500$, $p = 1/12$. **(b)** Not binomial. **(c)** Not binomial.

5.3   **(a)** Caught: binomial, $n = 20$, $p = 0.7$. Missed: binomial, $n = 20$, $p = 0.3$. **(b)** 0.1133.

5.5   **(a)** 14 errors caught, 6 missed. **(b)** 2.0494 errors. **(c)** With $p = 0.9$, 1.3416 errors; with $p = 0.99$, 0.4450 errors. σ decreases toward 0 as p approaches 1.

5.7 $m = 10$.

5.9   **(a)** 0.4095. **(b)** 4.

5.11   **(a)** 59.044 home runs. **(b)** About 7.6%. **(c)** Almost 0.

5.13   **(a)** $n = 4$ and $p = 0.25$. **(b)** The probabilities are 0.3164, 0.4219, 0.2109, 0.0469, 0.0039. **(c)** $\mu = 1$ child.

5.15   **(a)** 0.5098. **(b)** 0.8198. **(c)** It increases.

5.17   **(a)** 0.62. **(b)** 0.1446.

5.19   **(a)** $p = 0.25$. **(b)** 0.0139. **(c)** $\mu = 5$ and $\sigma = 1.9365$ successes. **(d)** No; the trials would not be independent.

5.21 (a) $\mu = 200$ and $\sigma = 12.6491$ successes. (b) For $\hat{p}$, $\mu = 0.2$ and $\sigma = 0.01265$. (c) About 0.0008. (d) 0.2294 or higher.

5.23 (a) 0.1137. (b) 170.5 blacks. (c) About 0.5.

5.25 (a) 0.1251. (b) 0.0336. (c) 400. (d) Yes.

5.27 Possible values: $1, 2, 3, \ldots$. $P(Y = k) = pq^{k-1}$.

5.29 40.125 mm and 0.001 mm.

5.31 (b) About 0.05. (c) Nearly 0.

5.33 (a) $\bar{x}$ is not systematically higher than or lower than μ. (b) With large samples, $\bar{x}$ is more likely to be close to μ.

5.35 (a) 0.5 and 0.099 moths. (b) About 0.16.

5.37 (a) 0.0668. (b) 0.0013.

5.39 About 133.2 mg/dl.

5.41 About 0.0052.

5.43 0.0294, 0.0853.

5.45 (a) Nearly 1. (b) About 0.9916.

5.47 (a) $\bar{y}$ is $N(\mu_Y, \sigma_Y/\sqrt{m})$ and $\bar{x}$ is $N(\mu_X, \sigma_X/\sqrt{n})$. (b) $N\left(\mu_Y - \mu_X, \sqrt{\sigma_Y^2/m + \sigma_X^2/n}\right)$.

5.49 $N(32, 0.2)$; 0.0124.

5.51 (a) Yes; this rule works for any random variables X and Y. (b) No; this rule requires that X and Y be independent.

5.53 0.0989; these six students were not randomly chosen from the population of all drivers.

5.55 (a) 0.8497. (b) Nearly 1.

5.57 0.0062.

5.59 0.9535.

5.61 (a) No; among other things, X cannot be 0. (b) No; a count assumes only whole-number values. (c) $N(1.5, 0.02835)$. (d) 0.1038.

5.63 0.0375.

5.65 4.0496 to 4.3904.

5.67 (a) $\hat{p}_F$: $N(0.82, 0.01718)$; $\hat{p}_M$: $N(0.88, 0.01453)$. (b) $N(0.06, 0.02250)$. (c) About 0.0038.

CHAPTER | 6

6.1 64 to 88.

6.3 Margins of error: 3.0990, 2.1913, 1.5495, and 0.9800; interval width decreases with increasing sample size.

6.5 Margin of error: 2.2882 U/l; interval: 10.91 to 15.49 U/l.

6.7 Margin of error: 3.0073 U/l; interval: 10.19 to 16.21 U/l.

6.9 Scenario B has a smaller margin of error; less variability in a single grade level.

6.11 No; this is a range of values for the mean rent, not for individual rents.

6.13 (a) 66.3 to 93.7 minutes. (b) No; this is a range of values for the mean study time, not for individual study times.

6.15 (a) 2.1920 cal/day. (b) 4.3841 cal/day. (c) 4.3841 (or 4.2694) cal/day.

6.17 26.2858 to 35.3142 MPH.

6.19 31.8406 to 38.3412.

6.21 We expect a "hit" about 80% of the time (or 40 out of 50).

6.23 Smaller; $n = 487$.

6.25 $n = 41$.

6.27 No; it gives the range of all incomes, not an estimate of the mean income.

6.29   (a) 0.8145. (b) 0.9860.

6.31   (a) The sample will not perfectly represent the population. (b) The method used gives correct results 95% of the time. (c) 34% to 40%. (d) Yes; some fans would change their responses based on what they are currently watching.

6.33   (a) H_0 and H_a have been switched. (b) H_0 should concern μ, not $\bar{x}$. (c) A P-value of 0.95 is not significant.

6.35   (a) $H_0: p = 0.3$; $H_a: p \neq 0.3$. (b) $H_0: \mu = 20$ seconds; $H_a: \mu < 20$ seconds. (c) $H_0: \mu = 1.3$ g/cm^2; $H_a: \mu \neq 1.3$ g/cm^2.

6.37   (a) $H_0: \mu = \$72{,}500$; $H_a: \mu > \$72{,}500$. (b) $H_0: \mu = 1.8$ hr; $H_a: \mu \neq 1.8$ hr.

6.39 $P = 0.1616$.

6.41   (a) Yes. (b) No. (c) To reject, we need $P < \alpha$.

6.43   (a) $P = 0.9452$. (b) $P = 0.0548$. (c) $P = 0.1096$.

6.45   (a) No. (b) No.

6.47   (a) No, because 13 is in the interval. (b) Yes, because 10 is not in the interval.

6.49   The difference was large enough that it would rarely arise by chance. Health issues related to alcohol use are probably discussed in the health and safety class.

6.51   If the mean score for all Boston students were 287, then by chance we might expect to see a sample mean of 262.

6.53 $H_0: \mu = 100$; $H_a: \mu \neq 100$; $z = 5.75$; significant ($P < 0.0001$).

6.55   (a) $z = 2.87$, $P = 0.0021$. (b) The important assumption is that this is an SRS. We also assume a normal distribution, but this is not crucial provided there are no outliers and little skewness.

6.57   (a) $H_0: \mu = 32$; $H_a: \mu > 32$. (b) $z = 1.86$, $P = 0.0314$.

6.59 $\bar{x} = 0.8$ is significant, but 0.7 is not. Smaller a means that $\bar{x}$ must be farther away.

6.61   (a) Yes. (b) No.

6.63   Something that occurs "less than once in 100 repetitions" also occurs "less than 5 times in 100 repetitions."

6.65   Any z with $2.576 < |z| < 2.807$.

6.67 $P > 0.25$.

6.69 $0.1 < P < 0.2$; $P = 0.1802$.

6.71   (a) 99.041 to 109.225 pCi/l. (b) $H_0: \mu = 105$ pCi/l; $H_a: \mu \neq 105$ pCi/l; not significant at $\alpha = 0.05$ because 105 is in the 95% confidence interval.

6.73   The first test was barely significant at $\alpha = 0.05$, while the second was significant at any reasonable α.

6.75 n should be about 100,000.

6.77   With a larger sample, we might have significant results.

6.81   When many variables are examined, "significant" results will show up by chance.

6.83   The intervals are 457.24 to 508.76, 474.85 to 491.15, and 480.42 to 485.58.

6.87   Reject the fourth ($P = 0.004$), fifth ($P = 0.001$), and ninth ($P = 0.002$).

6.91   Larger samples give more power.

6.93   Higher; larger differences are easier to detect.

6.95   Power: about 0.99.

6.97   Power: 0.6406.

6.99   (a) H_0: Patient is ill; H_a: Patient is healthy. Type I error: clearing a patient who is ill. Type II error: sending a healthy patient to the doctor. (Either statement could be taken as the null hypothesis.)

6.101   (a) 64.45 to 70.01; 67.59 to 73.15; 72.05 to 77.61. (c) 63.83 to 70.63; 66.97 to 73.77; 71.43 to 78.23. (d) With the larger margin of error, the intervals overlap more.

6.105   (a) 4.6132 to 6.0534 mg/dl. (b) $z = 1.45$, $P = 0.0735$; not significant.

6.107   (b) 26.2614 to 34.9386 μg/l. (c) $z = 2.53$, $P = 0.0057$.

6.109 **(a)** Under H_0, $\bar{x}$ has a $N(0\%, 5.1971\%)$ distribution. **(b)** $z = 1.31$, $P = 0.0951$. **(c)** Not significant.

6.111 It is essentially correct.

6.113 The number of rejections will depend on the chosen significance level. For example, we expect to reject H_0 five times if $\alpha = 0.05$.

CHAPTER | 7

7.1 **(a)** Distribution is not normal; it has two peaks and one large value. **(b)** We have a large sample, but a small population. **(c)** 27.29 ± 5.717, or 21.57 to 33.01 cm. **(d)** One could argue for either answer.

7.3 **(a)** Right-skewed, but less than the original data, and no clear outliers. **(b)** No; for skewed distributions, the median is more appropriate. **(c)** 1.495 ± 0.449, or 1.05 to 1.94. After undoing the transformation, this is about 1.85 to 5.99 mg/l.

7.5 **(a)** Not normal (lots of 1s and 10s) but no outliers. **(b)** 4.92 to 6.88. **(c)** Because this is not a random sample, it may not represent other children well.

7.7 **(a)** Distribution is not normal but has no outliers. **(b)** t methods should be acceptable. **(c)** $\bar{x} = 30.8$, $s = 11.2$, SE $= 2.505$, and m.e. $= 5.244$ MPH. Interval: 25.6 to 36.0 MPH. **(d)** No; driving speed probably depends on the driver more than on the car.

7.9 90%: 2.1182 to 2.3218. 95%: 2.0983 to 2.3417. Width increases with confidence level.

7.11 **(a)** $t = 5.13$, df $= 15$, $P < 0.001$. **(b)** With 95% confidence, the mean NEAT increase is between 192 and 464 calories.

7.13 **(a)** Not significant ($P = 0.0704$). **(b)** Significant at 5% ($P = 0.0368$). The larger sample makes this t-value significant.

7.15 About ± 2.330.

7.17 For the alternative $\mu < 0$, the answer would be the same ($P = 0.04$). For the alternative $\mu > 0$, the answer would be $P = 0.96$.

7.19 2.262, 2.093, 2.045, 2.023, 2.010, 1.984, 1.972, 1.965. As sample size increases, t^* approaches 1.960.

7.21 **(a)** 29. **(b)** t is between $t^* = 1.311$ ($p = 0.10$) and $t^* = 1.699$ ($p = 0.05$). **(c)** $0.10 < P < 0.20$. **(d)** Not significant at 10% or 5%. **(f)** $P = 0.1875$.

7.23 It depends on whether $\bar{x}$ is on the appropriate side of μ_0.

7.25 **(a)** H_0: $\mu = 1000$ mg; H_a: $\mu \neq 1000$ mg. **(b)** $t = -1.327$, df $= 39$, $P = 0.1912$. **(d)** $P = 0.0166$. This rejects H_0 (the median equals 1000), while the t test does not reject its H_0 because of the high outlier in the data.

7.27 **(a)** The differences are roughly symmetric about zero and spread from -0.018 to 0.020 g. **(b)** $t = -0.347$, df $= 7$, $P = 0.7388$. **(c)** -0.0117 to 0.0087 g. **(d)** They may be representative of future subjects, but the results are suspect because this is not a random sample.

7.29 **(a)** Spread from -3 to 9, left-skewed. **(b)** $\bar{x} = 3.62$, $s = 3.055$, $\text{SE}_{\bar{x}} = 0.524$. **(c)** 2.55 to 4.69 points.

7.31 H_0: $\mu = 0$; H_a: $\mu < 0$; $t = -31.24$, df $= 4$, $P = 0.0001$.

7.33 **(a)** $t = 47.32$, df $= 499$, P is tiny. **(b)** $541.31 to $588.69. **(c)** We have an SRS with a large sample size and no outliers. **(d)** Compare mean increases for a control group (no offer) and a treatment group.

7.35 H_0: $\mu = 4.8$ mg/dl; H_a: $\mu > 4.8$ mg/dl; $t = 2.086$, df $= 5$, $P = 0.046$.

7.37 **(b)** $t = -0.32$, df $= 11$, $P = 0.756$.

7.39 **(a)** Factory: $\bar{x} = 42.85$, $s = 4.793$. Haiti: $\bar{x} = 37.53$, $s = 2.440$. Change: $\bar{x} = -5.33$, $s = 5.59$. **(b)** H_0: $\mu = 0$ mg/100 g; H_a: $\mu < 0$ mg/100 g. **(c)** $t = -4.96$, df $= 26$, $P = 0.00002$. **(d)** Factory: 40.96 to 44.75. Haiti: 36.55 to 38.48. Change: -7.54 to -3.12.

7.41 (a) $H_0: \mu = 0; H_a: \mu > 0$. (b) Slightly left-skewed; $\bar{x} = 2.5$ and $s = 2.893$. (c) $t = 3.865$, df $= 19$, $P = 0.00052$. (d) 1.146 to 3.854.

7.43 $\bar{x} = 114.98, s = 14.80$; 111.14 to 118.82. This might adequately describe the mean IQ at this school, but the sample could not be considered representative of all fifth-graders.

7.45 H_0: median $= 0$; H_a: median $\neq 0$; $P = 0.7266$. This is similar to the t test P-value.

7.47 (a) H_0: median $= 0$; H_a: median > 0; or $H_0: p = 1/2$; $H_a: p > 1/2$. (b) With 28 positive differences out of 34, $z = 3.77$ and $P < 0.0001$.

7.49 H_0: median $= 0$; H_a: median > 0; or $H_0: p = 1/2$; $H_a: p > 1/2$. With 14 positive differences out of 20, $P = 0.0577$.

7.51 Reject H_0 if $|x| = 0.00677$. The power is about 7%.

7.53 (a) Not normal, because all numbers are integers. (b) Yes; we have two large samples, with no outliers. (c) $H_0: \mu_1 = \mu_2; H_a: \mu_1 > (\text{or} \neq) \mu_2$. (d) $t = 6.258$, df $= 354$ or 164, $P < 0.0001$. (e) 0.51 to 0.99 (regardless of df). (f) This may not generalize well to other areas of the country.

7.55 (a) This may be near enough to an SRS if this company's working conditions were similar to those of other companies. (b) 9.99 to 13.01 mg.y/m^3. (c) $t = 15.08$, $P < 0.0001$ with either df $= 137$ or 114. (d) The sample sizes are large enough that skewness should not matter.

7.57 You need either sample sizes and standard deviations or degrees of freedom and a more accurate value for the P-value. The confidence interval will give us useful information about the magnitude of the difference.

7.59 (a) Hypotheses should involve μ_1 and μ_2. (b) The samples are not independent. (c) We need P to be small (for example, less than 0.05) to reject H_0.

7.61 (a) Yes (in fact, $P < 0.005$). (b) No ($P = 0.9975$).

7.63 (a) Use a two-sided alternative. (b) df $= 29$. (c) $0.04 < P < 0.05$.

7.65 This is a matched pairs design.

7.67 The next 10 employees who need screens might not be an independent group— perhaps they all come from the same department, for example.

7.69 (a) This is now a matched pairs design. (b) $t = -49.83$, df $= 1$, $P = 0.0064$.

7.71 Small samples may lead to rejection of H_0 if the evidence is very strong.

7.73 (a) The alternative hypothesis should be chosen without looking at the data. (b) $P = 0.12$.

7.75 (a) The north distribution (five-number summary: 2.2, 10.2, 17.05, 39.1, 58.8 cm) is right-skewed, while the south distribution (2.6, 26.1, 37.70, 44.6, 52.9) is left-skewed. (b) The methods of this section seem to be appropriate. (c) $H_0: \mu_n = \mu_s; H_a: \mu_n \neq \mu_s$. (d) $t = -2.63$ with df $= 55.7$ ($P = 0.011$) or df $= 29$ ($P = 0.014$). (e) Either -19.09 to -2.57 or -19.26 to -2.40 cm.

7.77 (a) Control scores are symmetrical; piano scores are generally higher and slightly left-skewed. (b) Piano: $n = 34$, $\bar{x} = 3.618$, $s = 3.055$, SE $= 0.524$. Control: $n = 44$, $\bar{x} = 0.386$, $s = 2.423$, SE $= 0.365$. (c) $H_0: \mu_p = \mu_c; H_a: \mu_p > \mu_c; t = 5.06$, df $= 61.7$ or df $= 33$, $P < 0.0001$.

7.81 (a) Either -3.06 to 9.06 units (df $= 50$) or -2.98 to 8.98 units (df $= 104.6$). (b) Random fluctuation may account for the difference in the two averages.

7.83 (a) $H_0: \mu_B = \mu_F; H_a: \mu_B > \mu_F; t = 1.654$, $P = 0.053$ (df $= 37.6$) or $P = 0.058$ (df $= 18$). (b) -0.2 to 2.0. (c) We need two independent SRSs from normal populations.

7.85 $s_p = 1.1556; t = 6.251$ (df $= 375$); $P < 0.0001$; 0.51 to 0.99. All results are nearly the same as in Exercise 7.53.

7.87 $s_p = 15.96; t = -2.629$ (df $= 58$); $P = 0.0110$; -19.08 to -2.58 cm. All results are nearly the same as in Exercise 7.75.

7.89 df $= 55.725$.

7.91 (a) df $= 137.066$. (b) $s_p = 5.332$ (slightly closer to s_2, from the larger sample). (c) With no assumption, SE$_1 = 0.7626$; with the pooled method, SE$_2 = 0.6136$. (d) $t = 18.74$, df = 333, $P < 0.0001$. t and df are larger, so the evidence is stronger (although it was quite strong before). **(e)** df = 121.503; $s_p = 1.734$; $SE_1 = 0.2653$ and $SE_2 = 0.1995$. $t = 24.56$, df = 333, $P < 0.0001$.

7.93 **(a)** 9.60. **(b)** 4.03. **(c)** 2.23. **(d)** The ratio decreases with increasing sample size.

7.95 $F = 1.017$ with df = 211 and 164; $P = 0.9114$. The distributions are not normal, so this test may not be reliable (although the conclusion is reasonable). To reject at the 5% level, s_2 would need to be at least 1.39 (based on Table E) or 1.33 (software).

7.97 $F = 5.263$ with df = 114 and 219; $P < 0.0001$. The authors described the distributions as somewhat skewed, so the normality assumption may be violated.

7.99 $F = 1.506$ with df = 29 and 29; $P = 0.2757$. The stemplots in Exercise 7.75 did not appear to be normal.

7.101 **(a)** $H_0: \sigma_m = \sigma_w$; $H_a: \sigma_m > \sigma_w$. **(b)** $F = 1.544$ with df = 19 and 17. **(c)** $P > 0.10$.

7.103 **(a)** Power: about 0.34 (with $\alpha = 0.05$). **(b)** Power: about 0.78. **(c)** We should choose 60 trees per sample.

7.105 Power: about 0.95.

7.107 The approximate margins of error are 416.1, 289.4, 235.0, 203.0, 181.3, 165.3.

7.109 **(a)** Body weight: mean = −0.7 kg, SE = 2.298 kg. Caloric intake: mean = 14 cal, SE = 56.125 cal. **(b)** $t_1 = -0.305$ (body weight) and $t_2 = 0.249$ (caloric intake), both with df = 13; both P-values are about 0.8. **(c)** −5.66 to 4.26 kg and −107.23 to 135.23 cal.

7.111 **(a)** Somewhat right-skewed with no extreme outliers. **(b)** 10.85 to 15.64 U/l.

7.113 **(a)** Slightly left-skewed. **(b)** 2.285 to 2.650. **(c)** 9.82 to 14.15 U/l.

7.115 For north/south differences: $t = 7.15$, df = 575.4 or 283, $P < 0.0001$; the confidence interval is 7.52 to 13.22 cm. For north/south differences: $t = 3.69$, df = 472.7 or 230, $P < 0.0005$; the confidence interval is 2.68 to 8.78 cm. With larger samples, t increases, P decreases, and the intervals shrink.

7.117 78.3% ± 13.8% or 64.5% to 92.1%.

7.119 GPA: $t = -0.91$, df = 74.9 ($P = 0.1811$) or 30 ($0.15 < P < 0.20$). Confidence interval: −1.33 to 0.5. IQ: $t = 1.64$, df = 56.9 ($P = 0.0503$) or 30 ($0.05 < P < 0.10$). Confidence interval: −1.12 to 11.36.

7.121 $t = 3.65$, df = 237.0 or 115, $P < 0.0005$. 95% confidence interval for the difference: 0.78 to 2.60.

7.123 $t = -0.3533$, df = 179, $P = 0.3621$.

7.125 No; what we have is nothing like an SRS.

7.127 **(a)** $\bar{x} = 1.7182$ and $s = 1.3428$. Distribution is right-skewed from 0.068 to 5.417. **(b)** The sample size should be large enough to overcome the skewness. **(c)** 1.396 to 2.041. No; this is an interval for the mean, not for individual observations. **(d)** The transformed distribution is left-skewed, with mean 0.1542 and standard deviation 1.0176. A 95% confidence interval for the mean is −0.090 to 0.399 (0.914 to 1.490, after undoing the logarithms).

7.129 $\bar{x} = 0.8043$ and $s = 0.2765$ g/l; distribution is right-skewed. 95% confidence interval: 0.746 to 0.862 g/l.

7.131 Basal: $\bar{x} = 41.0455$, $s = 5.6356$. DRTA: $\bar{x} = 46.7273$, $s = 7.3884$. Strat: $\bar{x} = 44.2727$, $s = 5.7668$. **(a)** $t \doteq 2.87$, $P < 0.005$. Confidence interval for difference: 1.7 to 9.7 points. **(b)** $t \doteq 1.88$, $P < 0.05$. Confidence interval for difference: −0.24 to 6.7 points.

7.137 As df increases, t^* approaches 1.645.

7.139 Margins of error decrease with increasing sample size.

CHAPTER | 8

8.1 **(a)** Plus four. **(b)** Large-sample (or plus four). **(c)** Plus four. **(d)** Neither. **(e)** Large-sample (or plus four).

8.3   **(a)** Margin of error equals z^* times standard error. **(b)** H_0 should refer to p, not $\hat{p}$. **(c)** Use normal distributions for proportions.

8.5   **(a)** $\hat{p} = 0.6341$; 0.6214 to 0.6467. This interval was found using a procedure that includes the correct proportion 95% of the time. **(b)** We do not know if those who did respond can reliably represent those who did not.

8.7   **(a)** ± 0.0005129. **(b)** Other sources of error are much more significant than sampling error.

8.9   **(a)** $\hat{p} = 0.3275$; 0.3008 to 0.3541. **(b)** $\tilde{p} = 0.3280$; 0.3014 to 0.3547. With large samples, the two methods give similar results. **(c)** The nonresponse rate (3.6%) is quite small, which suggests that the results should be reliable. **(d)** Speakers and listeners probably perceive sermon length differently.

8.11   0.3239 to 0.3761.

8.13   Wider: 0.3157 to 0.3843.

8.15   0.6345 to 0.7455.

8.17   0.2180 to 0.2510.

8.19   **(a)** $\tilde{p} = 0.8571$. **(b)** 0.1833; the interval is 0.6738 to 1. **(c)** No; we would not expect everyone to match the performance of the top salesperson.

8.21   **(a)** No. **(b)** Yes. **(c)** Yes. **(d)** No.

8.23   **(a)** $z = 1.34$, $P = 0.1802$. **(b)** 0.4969 to 0.5165.

8.25 $n = 171$ or 172.

8.27   The sample sizes are 35, 62, 81, 93, 97, 93, 81, 62, and 35; take $n = 97$.

8.29   The margins of error are 0.1859, 0.2479, 0.2840, 0.3036, 0.3099, 0.3036, 0.2840, 0.2479, 0.1859.

8.31   **(a)** Large-sample. **(b)** Plus four. **(c)** Plus four. **(d)** Neither. **(e)** Plus four.

8.33   **(a)** H_0 should refer to p_1 and p_2. **(b)** Only if $n_1 = n_2$. **(c)** Confidence intervals account only for sampling error.

8.35 $z = 20.18$, so P is tiny. Confidence interval: 0.1962 to 0.2377. Nonresponse error could render this interval and test result meaningless.

8.37   Pet owners: 0.4790. Others: 0.5281. Confidence interval: 0.0032 to 0.0950.

8.39   Pet owners: 0.5210. Others: 0.4719. Confidence interval: 0.095 to 0.0032.

8.41 $z = 6.79$ and $P < 0.0001$. Confidence interval: 0.0785 to 0.1415.

8.43   −0.0053 to 0.2335.

8.45   0.0301 to 0.0797.

8.47   **(a)** $\hat{p}_f = 0.8$, SE $\doteq 0.05164$; $\hat{p}_m = 0.3939$, SE $\doteq 0.04253$. **(b)** 0.2960 to 0.5161.

8.49 $z = 2.10$, $P = 0.0360$.

8.51 $z = 5.22$, P is very small.

8.53   Plus four interval: −0.0020 to 0.0489. Although the numbers are smaller than we like for a normal approximation, we can also find $z = 1.82$ and $P = 0.0344$ and the large-sample interval: 0.00175 to 0.04923.

8.55 $\hat{p} = 0.6129$, $z = 4.03$, $P < 0.0001$; confidence interval: 0.5523 to 0.6735.

8.57   **(a)** People have different symptoms; for example, not all who wheeze consult a doctor. **(b)** Sleep: 0.0864; 0.0280 to 0.1448. Number: 0.0307; −0.0361 to 0.0976. Speech: 0.0182; −0.0152 to 0.0515. Activities: 0.0137; −0.0395 to 0.0670. Doctor: −0.0112; −0.0796 to 0.0573. Phlegm: −0.0220; −0.0711 to 0.0271. Cough: −0.0323; −0.0853 to 0.0207. **(c)** It is reasonable to expect that the bypass proportions would be higher. **(d)** In the same order as part (b): $z = 2.64$, $P = 0.0042$; $z = 0.88$, $P = 0.1897$; $z = 0.99$, $P = 0.1600$; $z = 0.50$, $P = 0.3100$; $z = -0.32$, $P = 0.6267$; $z = -0.92$, $P = 0.8217$; $z = -1.25$, $P = 0.8950$. **(e)** 95% confidence interval for sleep improvement: 0.1168 to 0.2023. Part (b) showed improvement relative to control group, which is a better measure of the effect of the bypass.

8.59   **(a)** $z = 6.98$, $P < 0.0001$. **(b)** 0.1145 to 0.2022.

8.61 Education: 1132 users, 852 nonusers. Income: 871 users, 677 nonusers. For users, $\hat{p}_1 = 0.2306$; for nonusers, $\hat{p}_2 = 0.2054$. $z = 1.34$, $P = 0.1802$; -0.0114 to 0.0617. The lack of response about income makes the conclusions for Exercise 8.60 suspect.

8.63 ± 0.0300.

8.65 $z = 8.95$, $P < 0.0001$; 0.3720 to 0.5613.

8.67 All $\hat{p}$-values are greater than 0.5. Texts 3, 7, and 8 have (respectively) $z = 0.82$, $P = 0.4122$; $z = 3.02$, $P = 0.0025$; and $z = 2.10$, $P = 0.0357$. For the other texts, $z = 4.64$ and $P < 0.00005$.

8.69 Margin of error: 0.0306. $z = -2.28$, $P = 0.0226$; confidence interval for difference: 0.0072 to 0.0928.

8.73 z: 0.45, 0.64, 0.90, 1.01, 1.27, 1.42, 3.18, 4.49. P: 0.6527, 0.5222, 0.3681, 0.3125, 0.2041, 0.1556, 0.0015, 0.0000.

8.75 Margins of error: 0.4001 (plus four), 0.3099, 0.2191, 0.1960, 0.1549, 0.1386, 0.0620, 0.0438.

8.77 (a) $p_0 = 0.7911$. (b) $\hat{p} = 0.3897$, $z = -29.1$, P is tiny. (c) $\hat{p}_1 = 0.3897$, $\hat{p}_2 = 0.7930$, $z = -29.2$, P is tiny.

8.79 (a) 0.5278 to 0.5822. (b) 0.5167 to 0.5713. (c) 0.3170 to 0.3690. (d) 0.5620 to 0.6160. (e) 0.5620 to 0.6160. (f) 0.6903 to 0.7397.

CHAPTER | 9

9.1 (a) About 3,388,000. (b) 0.207, 0.024; 0.320, 0.071; 0.104, 0.104; 0.046, 0.125. (c) 0.230, 0.391, 0.208, 0.171. (d) 0.677, 0.323.

9.3 Full-time: 0.305, 0.472, 0.154, 0.069. Part-time: 0.073, 0.220, 0.321, 0.386.

9.5 Two-year FT: 0.479, 0.521. Two-year PT: 0.458, 0.542. Four-year FT: 0.466, 0.534. Four-year PT: 0.394, 0.606. Graduate school: 0.455, 0.545. Vocational school: 0.539, 0.461.

9.7 Start by setting a equal to any number from 0 to 200.

9.9 (a) 51.1%. (b) Small: 41.7%; medium: 51.7%; large: 60.0%. (d) Small: 39.8%; medium: 33.0%; large: 27.3%.

9.11 Success (nonrelapse) rates were 58.3% (desipramine), 25.0% (lithium), and 16.7% (placebo).

9.13 (a) A notably higher percent of women are "strictly voluntary" participants. (b) 40.3% of men and 51.3% of women are participants; the relative risk is 1.273.

9.15 (a) Males: 490 admitted, 310 not; females: 400 admitted, 300 not. (b) Males: 61.25% admitted; females: 57.14% admitted. (c) Business school: 66.67% of males, 66.67% of females; law school: 45% of males, 50% of females. (d) Most male applicants apply to the business school, where admission is easier. A majority of women apply to the law school, which is more selective.

9.17 First- and second-year: A has 8.3% small classes, B has 17.1% small. Upper-level: A has 77.5% small, B has 83.3% small.

9.19 (a) df = 1, $P = 0.2524$. (b) df = 9, $P = 0.0409$. (c) df = 7, $P = 0.0024$. (d) df = 8, $P = 0.1260$.

9.21 (b) $X^2 = 2.591$, df = 1, $P = 0.108$.

9.23 (a) $X^2 = 76.7$, df = 2, $P < 0.0001$. (b) Even with much smaller numbers of students, P is still very small. (c) Our conclusion might not hold for the true percents. (d) Lack of independence could cause the estimated percents to be too large or too small.

9.25 $X^2 = 12.0$, df = 1, $P = 0.001$. The smallest expected count is 6, so the test is valid.

9.27 $X^2 = 23.1$, df = 4, $P < 0.0005$. Dog owners have less education, and cat owners more, than we would expect if there were no relationship between pet ownership and educational level.

9.29 The missing entries are 202, 64, 38, 33. $X^2 = 50.5$, df $= 9$, $P < 0.0005$. The largest contributions to X^2 come from Chemistry/Engineering, Physics/Engineering, and Biology/Liberal arts (more than expected), and Biology/Engineering and Chemistry/Liberal arts (less than expected).

9.31 $X^2 = 56.1$, df $= 2$, $P < 0.0005$. The largest contributions to X^2 arise because younger employees are more likely to have the highest rating, and older employees are more likely to have the lowest rating.

9.33 $X^2 = 3.955$, df $= 4$, $P = 0.413$.

9.35 **(a)** Cats: $X^2 = 6.611$, df $= 2$, $P = 0.037$. Dogs: $X^2 = 26.939$, df $= 2$, $P < 0.0005$. **(b)** Dogs from pet stores are less likely to be taken to a shelter, while "other source" dogs are more likely to be taken. Private-source cats are slightly more likely, and other-source cats slightly less likely, to be taken to a shelter. **(c)** The control group data should be reasonably like an SRS.

9.37 **(a)** $z = -0.5675$, $P = 0.5686$. **(b)** $X^2 = 0.322$, df $= 1$, $P = 0.570$. **(c)** Gastric freezing is no more effective than a placebo.

9.39 $X^2 = 43.487$, df $= 12$, $P < 0.0005$. Science has a large proportion of low-scoring students, while liberal arts/education has a large percent of high-scoring students.

9.41 $X^2 = 852.433$, df $= 1$, $P < 0.0005$.

9.43 **(a)** $X^2 = 2.506$, df $= 2$, $P = 0.286$. **(b)** Divide each echinacea count by 337 and each placebo count by 370. **(c)** The only significant results are for rash ($z = 2.74$, $P = 0.0061$), drowsiness ($z = 2.09$, $P = 0.0366$), and other ($z = 2.09$, $P = 0.0366$). A 10×2 table would not be appropriate because each URI could have multiple adverse events. **(d)** All results are unfavorable to echinacea, so we are not concerned with having detected "false positives." **(e)** We do not have independent observations, but we would expect the dependence to have the same effect on both groups, so our conclusions should be fairly reliable.

9.45 9.11: comparing three populations. 9.23: one could argue for either answer. 9.24: comparing two populations. 9.27: testing independence.

9.47 $X^2 = 5.016$, df $= 3$, $P = 0.1706$.

CHAPTER | 10

10.1 **(a)** -0.5555 to 32.7555. **(b)** -10.5555 to 22.7555. **(c)** 0.1690 to 32.0310.

10.3 **(a)** β_0, β_1, and σ. **(b)** H_0 should refer to β_1. **(c)** The confidence interval will be narrower than the prediction interval.

10.5 **(a)** x (percent forested) is right-skewed; $\bar{x} = 39.3878\%$, $s_x = 32.2043\%$. y (IBI) is left-skewed; $\bar{y} = 65.9388$, $s_y = 18.2796$. **(b)** A very weak positive association, with more scatter in y for small x. **(c)** $y_i = \beta_0 + \beta_1 x_i + \epsilon_i$, $i = 1, 2, \ldots, 49$; ϵ_i are independent $N(0, \sigma)$ variables. **(d)** $H_0: \beta_1 = 0$; $H_a: \beta_1 \neq 0$. **(e)** $\widehat{\text{IBI}} = 59.9 + 0.153\,\text{AREA}$; $s = 17.79$. For testing the hypotheses in (d), $t = 1.92$ and $P = 0.061$. **(f)** Residual plot shows some curvature. **(g)** Residuals are left-skewed.

10.7 **(a)** MOR is fairly symmetric, MOE is slightly right-skewed. For x (MOE), $\bar{x} = 1,799,180$ and $s_x = 329,253$; for y (MOR), $\bar{y} = 11,185$ and $s_y = 1980$. **(b)** Put MOE on the x axis. **(c)** $y_i = \beta_0 + \beta_1 x_i + \epsilon_i$, $i = 1, 2, \ldots, 32$; ϵ_i are independent $N(0, \sigma)$ variables. $\widehat{\text{MOR}} = 2653 + 0.00474\,\text{MOE}$, $s = 1238$, $t = 7.02$, $P < 0.0001$. **(d)** Assumptions appear to be met. There is one unusually small residue (-3308.56).

10.9 **(a)** x (CRP) is sharply right-skewed with high outliers; $\bar{x} = 10.0322$ and $s_x = 16.5632$. y (retinol) is right-skewed; $\bar{y} = 0.7648$ and $s_y = 0.3949$. **(b)** No; no assumption is made about x-values. **(c)** $\widehat{\text{RETINOL}} = 0.843 - 0.00780\,\text{CRP}$, $s = 0.3781$, $t = -2.13$, $P = 0.039$. **(d)** The high outliers in CRP are influential; residuals are right-skewed rather than normal.

10.11 (a) Both distributions are slightly right-skewed. TRAP has $\bar{x} = 13.248$ and $s_x = 6.528$; VO− has $\bar{y} = 889.194$ and $s_y = 427.616$. (b) Put TRAP on the x axis. A moderate positive association. (c) $\hat{y} = 301 + 44.4x$, $s = 319.7$, $t = 4.97$, $P < 0.0001$.

10.13 Both distributions are slightly more symmetric than before. LOGTRAP has $\bar{x} = 2.4674$ and $s_x = 0.4979$; LOGVO− has $\bar{y} = 6.6815$ and $s_y = 0.4832$. A scatterplot shows a moderate positive association, with one low point that might be influential. $\hat{y} = 5.09 + 0.645x$, $s = 0.3674$, $t = 4.78$, $P < 0.0001$.

10.15 (a) Both variables are right-skewed. Pure tones: $\bar{x} = 106.2$ and $s = 91.76$ spikes/second, Monkey calls: $\bar{y} = 176.6$ and $s_y = 111.85$ spikes/second. (b) A moderate positive association; the third point has the largest residual; the first point is an outlier for tone response. (c) $\widehat{CALL} = 93.9 + 0.778\,TONE$, $s = 87.30$, $t = 4.91$, $P < 0.0001$. (d) Without the first point, $\hat{y} = 101 + 0.693x$, $s = 88.14$, $t = 3.18$. Without the third point, $\hat{y} = 98.4 + 0.679x$, $s = 80.69$, $t = 4.49$. With neither, $\hat{y} = 116 + 0.466x$, $s = 79.46$, $t = 2.21$.

10.17 (a) Scatterplot shows a weak negative association; $\widehat{BONDS} = 53.4 - 0.196\,STOCKS$, $s = 59.88$. (b) H_0: $\beta_1 = 0$; H_a: $\beta_1 \neq 0$; $t = -1.27$, $P = 0.226$. (c) The scatterplot shows a lot of variation, so s is large and t is small.

10.19 (a) MA angle is explanatory. (b) A moderate-to-weak positive linear association, with one clear outlier. (c) $y_i = \beta_0 + \beta_1 x_i + \epsilon_i$, $i = 1, 2, \ldots, 38$; ϵ_i are independent $N(0, \sigma)$ variables. (d) H_0: $\beta_1 = 0$; H_a: $\beta_1 > 0$. (e) $t = 1.90$, df = 36, $P = 0.033$.

10.21 (a) Aside from the one high point, there is a moderate positive association. (b) $\widehat{WAGES} = 43.4 + 0.0733\,LOS$, $t = 2.85$, $P = 0.006$. (c) Wages rise an average of 0.0733 wage units per week of service. (d) 0.0218 to 0.1247.

10.23 (a) It appears to be quite linear. (b) $\widehat{LEAN} = -61.1 + 9.32\,YEAR$; $r^2 = 98.8\%$. (c) 8.36 to 10.28 tenths of a millimeter/year.

10.25 (a) $x = 107$. (b) 2.9936 m. (c) Use a prediction interval.

10.27 We cannot reject H_0: $\rho = 0$, because we find $t = -1.83$, $P = 0.089$.

10.29 df = 28, SSE = 10152.4, MSE = 362.6.

10.31 The standard error is 0.2150; the confidence interval is 0.223 to 1.103.

10.33 $n = 20$: $t = 2.45$, df = 18, $P = 0.0248$. $n = 10$: $t = 1.63$, df = 8, $P = 0.1412$.

10.35 (a) Strong positive linear association with one outlier (SAT 420, ACT 21). (b) $\widehat{ACT} = 1.63 + 0.0214\,SAT$, $t = 10.78$, $P < 0.0005$. (c) $r = 0.8167$.

10.37 (a) $a_1 = 0.02617$, $a_0 = -2.7522$. (c) Mean 21.1333 and standard deviation 4.7137—the same as for the ACT scores.

10.39 (a) For squared length: $\widehat{WEIGHT} = -118 + 0.497\,SQLEN$, $r^2 = 0.977$. (b) For squared width: $\widehat{WEIGHT} = -99.0 + 18.7\,SQWID$, $r^2 = 0.965$.

10.41 IBI and area: $r = 0.4459$, $t = 3.42$, $P = 0.001$ (from Exercise 10.4). IBI and percent forested: $r = 0.2698$, $t = 1.92$, $P = 0.061$ (Exercise 10.5). Area and percent forested: $r = -0.2571$, $t = -1.82$, $P = 0.074$.

10.43 $P < 0.001$: All creativity, quickness/analytical, and quickness/verbal. $P < 0.01$: Conscientiousness/analytical, conscientiousness/quantitative, quickness/quantitative, and depth/verbal. $P < 0.05$: Conscientiousness/verbal.

10.45 (a) 95% confidence interval for women: 14.73 to 33.33. For men: −9.47 to 42.97. These intervals overlap quite a bit. (b) For women: 22.78. For men: 16.38. The women's slope standard error is smaller in part because it is divided by a larger number. (c) Choose men with a wider variety of lean body masses.

CHAPTER | 11

11.1 (a) For β_1: −2.83 to 27.03. For β_2: 8.80 to 25.80. (b) For β_1: −2.18 to 26.38. For β_2: −0.83 to 15.43.

11.3 (a) Squared multiple correlation. (b) H_0 should refer to β_2. (c) Only the errors are assumed to be normal.

11.5 **(a)** $H_0: \beta_1 = 0$; $H_a: \beta_1 \neq 0$ ($t = 4.55$, $P < 0.001$). $H_0: \beta_2 = 0$; $H_a: \beta_2 \neq 0$ ($t = 2.69$, $P < 0.01$). $H_0: \beta_3 = 0$; $H_a: \beta_3 \neq 0$ ($t = 2.69$, $P < 0.01$). **(b)** Marijuana use decreases with increasing GPA and increases with popularity and depression. **(c)** The degrees of freedom of the F statistic. **(d)** H_0 ($\beta_1 = \beta_2 = \beta_3 = 0$) is rejected in favor of H_a: at least one β_i is nonzero. **(e)** Students may have lied (or erred) in their responses. **(f)** We cannot assume that students are the same everywhere.

11.7 **(a)** U.S. subjects are less willing to pay more. **(b)** $P < 0.001$ for testing $H_0: \beta_1 = 0$; $H_a: \beta_1 \neq 0$. **(c)** The U.K. response rate is much lower than the U.S. rate, and "don't know" responses are not much better than no response at all.

11.9 **(a)** $y_i = \beta_0 + \beta_1 x_{i1} + \beta_2 x_{i2} + \cdots + \beta_6 x_{i6} + \epsilon_i$, where $i = 1, 2, \ldots, 200$; ϵ_i are independent $N(0, \sigma)$ random variables. **(b)** Model (df = 6), error (df = 193), and total (df = 199).

11.11 $R^2 = 0.1525$.

11.13 **(a)** $\widehat{\text{SCORE}} = 3.33 + 0.82\,\text{UNFAV} + 0.57\,\text{FAV}$. **(b)** We reject $H_0: \beta_1 = \beta_2 = 0$ in favor of H_a: at least one β_i is nonzero. **(c)** The estimates of β_0, β_1, and β_2 are all significantly different from 0. **(d)** df = 149.

11.15 All coefficients are positive, so the associations are positive, as expected. The unfavorable coefficients are larger, so they have a stronger effect.

11.17 The coefficient of the new quadratic term is the same as for the old quadratic term, but the constant and coefficient of accounts have changed (and are now significantly different from 0).

11.19 The log plot appears to be reasonably linear. $\hat{y} = -5.06 + 1.29x$, $t = 6.96$, $P < 0.0005$.

11.21 **(a)** We reject $H_0: \beta_1 = \beta_2 = \cdots = \beta_{13} = 0$ in favor of H_a: at least one $\beta_j \neq 0$ (df = 13 and 2215, $P < 0.001$). **(b)** 29.7%. **(c)** $H_0: \beta_j = 0$; $H_a: \beta_j \neq 0$, df = 2215. Reject H_0 if $|t| > 1.961$. **(d)** Everything except "total payments," "male borrower," and "married." **(e)** Interest rates are lower for larger loans, for longer terms, with larger down payments, when there is a cosigner, when the loan is secured, when the borrower has a higher income, when the credit report is not considered "bad," for older borrowers, when the borrower owns a home, and for borrowers who have lived for a long time at their present address.

11.23 Ten factors significantly influenced the rate for direct loans, while only four had a significant impact for indirect loans.

11.25 $2.1102 to $2.1618, which includes the actual price.

11.27 **(a)** Plot suggests greater variation in VO+ for large OC. $\widehat{\text{VO+}} = 334 + 19.5\,\text{OC}$, $t = 4.73$, $P < 0.0005$. Plot of residuals against OC is slightly curved. Distribution of residuals is skewed to the right. **(b)** $\widehat{\text{VO+}} = 58 + 6.41\,\text{OC} + 53.9\,\text{TRAP}$. Coefficient of OC is not significantly different from 0 ($t = 1.25$, $P = 0.221$), but coefficient of TRAP is ($t = 3.50$, $P = 0.002$). This is consistent with the correlations found in Exercise 11.26.

11.29 The correlations are 0.840 (LVO+ and LVO−), 0.774 (LVO+ and LOC), and 0.755 (LVO+, LTRAP). Regression equations, t statistics, R^2, and s for each model: $\widehat{\text{LVO+}} = 4.38 + 0.706\,\text{LOC}$; $t = 6.58$, $P < 0.0005$; $R^2 = 0.599$, $s = 0.3580$. $\widehat{\text{LVO+}} = 4.26 + 0.430\,\text{LOC} + 0.424\,\text{LTRAP}$; $t = 2.56$, $P = 0.016$; $t = 2.06$, $P = 0.048$; $R^2 = 0.652$, $s = 0.3394$. $\widehat{\text{LVO+}} = 0.872 + 0.392\,\text{LOC} + 0.028\,\text{LTRAP} + 0.672\,\text{LVO−}$; $t = 3.40$, $P = 0.002$; $t = 0.18$, $P = 0.862$; $t = 5.71$, $P < 0.0005$; $R^2 = 0.842$, $s = 0.2326$. As before, this suggests a model without LTRAP: $\widehat{\text{LVO+}} = 0.832 + 0.406\,\text{LOC} + 0.682\,\text{LVO−}$; $t = 4.93$, $P < 0.0005$; $t = 6.57$, $P < 0.0005$; $R^2 = 0.842$, $s = 0.2286$.

11.31 Regression equations, t statistics, R^2, and s for each model: $\widehat{\text{LVO−}} = 5.21 + 0.441\,\text{LOC}$; $t = 3.59$, $P = 0.001$; $R^2 = 0.308$, $s = 0.4089$. $\widehat{\text{LVO−}} = 5.04 + 0.057\,\text{LOC} + 0.590\,\text{LTRAP}$; $t = 0.31$, $P = 0.761$; $t = 2.61$, $P = 0.014$; $R^2 = 0.443$, $s = 0.3732$. $\widehat{\text{LVO−}} = 1.57 - 0.293\,\text{LOC} + 0.245\,\text{LTRAP} + 0.813\,\text{LVO−}$; $t = -2.08$, $P = 0.047$; $t = 1.47$, $P = 0.152$; $t = 5.71$, $P < 0.0005$; $R^2 = 0.748$, $s = 0.2558$. $\widehat{\text{LVO−}} = 1.31 - 0.188\,\text{LOC} + 0.890\,\text{LVO−}$; $t = -1.52$, $P = 0.140$; $t = 6.57$, $P < 0.0005$; $R^2 = 0.728$, $s = 0.2611$.

11.33 **(a)** $y_i = \beta_0 + \beta_1 x_{i1} + \beta_2 x_{i2} + \beta_3 x_{i3} + \beta_4 x_{i4} + \epsilon_i$, where $i = 1, 2, \ldots, 69$; ϵ_i are independent $N(0, \sigma)$ random variables. **(b)** $\widehat{\text{PCB}} = 0.94 + 11.9x_1 + 3.76x_2 + 3.88x_3 + 4.18x_4$. All coefficients are significantly different from 0, although the constant 0.937 is not ($t = 0.76$, $P = 0.449$). $R^2 = 0.989$, $s = 6.382$. **(c)** The residuals appear to be roughly normal, but with two outliers. There are no clear patterns when plotted against the explanatory variables. The plot of the residuals against fitted values shows larger variability for higher fitted values. Residuals against observation order also show increasing variability of the residuals.

11.35 **(a)** $\widehat{PCB} = -1.02 + 12.6x_1 + 0.313x_2 + 8.25x_3$, $R^2 = 0.973$, $s = 9.945$. **(b)** $b_2 = 0.313$, $P = 0.708$. **(c)** In Exercise 11.33, $b_2 = 3.76$, $P < 0.0005$.

11.37 The model is $y_i = \beta_0 + \beta_1 x_{i1} + \beta_2 x_{i2} + \beta_3 x_{i3} + \beta_4 x_{i4} + \epsilon_i$, where $i = 1, 2, \ldots, 69$; ϵ_i are independent $N(0, \sigma)$ random variables. Regression gives $\widehat{TEQ} = 1.06 - 0.097x_1 + 0.306x_2 + 0.106x_3 - 0.0039x_4$, with $R^2 = 0.677$. Only the constant (1.06) and the PCB118 coefficient (0.306) are significantly different from 0. Residuals are slightly right-skewed and show no clear patterns when plotted with the explanatory variables. Plots of residuals against fitted values and against order suggest some problems.

11.39 **(a)** The correlations are all positive, ranging from 0.227 (LPCB28 and LPCB180) to 0.956 (LPCB and LPCB138). LPCB28 has one outlier (specimen 39) when plotted with the other variables; except for that point, all scatterplots appear fairly linear. **(b)** All correlations are higher with the transformed data.

11.41 It appears that a good model is LPCB126 and LPCB28 ($R^2 = 0.768$). If the zero cases of PCB126 are omitted, the model with LPCB126 and LPCB28 has $R^2 = 0.797$.

11.43 $\bar{x}$, M, s, and IQR for each variable are as follows. Taste: 24.53, 20.95, 16.26, 23.9. Acetic: 5.498, 5.425, 0.571, 0.656. H2S: 5.942, 5.329, 2.127, 3.689. Lactic: 1.442, 1.450, 0.3035, 0.430. None of the variables show striking deviations from normality. Taste and H2S are slightly right-skewed, and Acetic has two peaks. There are no outliers.

11.45 $\widehat{Taste} = -61.5 + 15.6\,Acetic$; $t = 3.48$, $P = 0.002$. The residuals seem to have a normal distribution (though slightly left-skewed), but are positively associated with both H2S and Lactic.

11.47 $\widehat{Taste} = -29.9 + 37.7\,Lactic$; $t = 5.25$, $P < 0.0005$. The residuals seem to have a normal distribution; there are no striking patterns for residuals against the other variables.

11.49 $\widehat{Taste} = -26.9 + 3.80\,Acetic + 5.15\,H2S$. For the coefficient of Acetic, $t = 0.84$ and $P = 0.406$. This model is not much better than the model with H2S alone; Acetic and H2S are correlated ($r = 0.618$), so Acetic does not add significant information if H2S is included.

11.51 $\widehat{Taste} = -28.9 + 0.33\,Acetic + 3.91\,H2S + 19.7\,Lactic$. The coefficient of Acetic is not significantly different from 0 ($P = 0.942$). Residuals of this regression appear to be normally distributed and show no patterns in scatterplots with the explanatory variables. It appears that the H2S/Lactic model is best.

CHAPTER | 12

12.1 **(a)** ANOVA is used when the explanatory variable has two (or more) values. **(b)** There is no theoretical limit to the number of means. **(c)** σ is a parameter.

12.3 **(a)** df = 3 and 40; $3.46 < F < 4.31$. **(c)** $0.010 < P < 0.025$. **(d)** We can conclude that at least one mean is different.

12.5 **(a)** df = 4 and 35; $F = 1.14$, $P > 0.100$. **(b)** df = 2 and 18; $F = 4$, $0.025 < P < 0.050$.

12.7 **(a)** Response: rating on five-point scale. Populations: the three groups of students. $I = 3$, $n_1 = 220$, $n_2 = 145$, $n_3 = 76$, $N = 441$. **(b)** Response: change in cholesterol level. Populations: pigs eating different varieties of onions. $I = 3$, $n_i = 5$ ($i = 1, 2, 3$), $N = 15$. **(c)** Response: video game rating. Populations: eighth-graders, third-year high school students, second-year college students. $I = 3$, $n_i = 25$ ($i = 1, 2, 3$), $N = 75$.

12.9 **(a)** DFM = 2, DFE = 438, DFT = 440. $H_0: \mu_1 = \mu_2 = \mu_3$; H_a: at least one mean is different. $F(2, 438)$. **(b)** DFM = 2, DFE = 12, DFT = 14. $H_0: \mu_1 = \mu_2 = \mu_3$; H_a: at least one mean is different. $F(2, 12)$. **(c)** DFM = 2, DFE = 72, DFT = 74. $H_0: \mu_1 = \mu_2 = \mu_3$; H_a: at least one mean is different. $F(2, 72)$.

12.11 (a) Conclusions might not extend to students from schools that are geographically or demographically different from this one, or to students who are not in similar groups (statistics class, health class, co-op housing). (b) The effects on humans might be different from the effects on pigs. (c) The key question is, how well do these groups represent other video game players?

12.13 (a) Yes: $220/190 = 1.16 < 2$. (b) 48,400; 36,100; 40,000. (c) 41,866. (d) 204.6.

12.15 (a) Matched pairs t methods; we examine the change in reaction time for each subject. (b) No: we do not have 4 independent samples.

12.17 (a) The variation in sample size is cause for some concern, but there can be no extreme outliers in a 1-to-7 scale, so ANOVA is probably reliable. (b) Yes: $1.26/1.03 = 1.22 < 2$. (c) $F(4, 405)$, $P = 0.0002$. (d) Hispanic Americans are highest, Japanese are in the middle, and the other three are lowest.

12.19 (a) Immediate: $n = 2$, $\bar{x} = 48.705$, $s = 1.534$, $SE_{\bar{x}} = 1.085$ mg/100 g. One day: $n = 2$, $\bar{x} = 41.955$, $s = 2.128$, $SE_{\bar{x}} = 1.505$ mg/100 g. Three days: $n = 2$, $\bar{x} = 21.795$, $s = 0.771$, $SE_{\bar{x}} = 0.545$ mg/100 g. Five days: $n = 2$, $\bar{x} = 12.415$, $s = 1.082$, $SE_{\bar{x}} = 0.765$ mg/100 g. Seven days: $n = 2$, $\bar{x} = 8.320$, $s = 0.269$, $SE_{\bar{x}} = 0.190$ mg/100 g. (b) H_0: $\mu_1 = \mu_2 = \cdots = \mu_5$; H_a: not all μ_i are equal. $F = 367.74$ with df $= 4$ and 5; $P < 0.0005$, so we reject H_0. (c) Vitamin C content decreases over time.

12.21 Vitamin A—Immediate: $n = 2$, $\bar{x} = 3.350$, $s = 0.01414$, $SE_{\bar{x}} = 0.010$ mg/100 g. One day: $n = 2$, $\bar{x} = 3.240$, $s = 0.05657$, $SE_{\bar{x}} = 0.040$ mg/100 g. Three days: $n = 2$, $\bar{x} = 3.210$, $s = 0.07071$, $SE_{\bar{x}} = 0.050$ mg/100 g. Five days: $n = 2$, $\bar{x} = 3.305$, $s = 0.07778$, $SE_{\bar{x}} = 0.055$ mg/100 g. Seven days: $n = 2$, $\bar{x} = 2.965$, $s = 0.06364$, $SE_{\bar{x}} = 0.045$ mg/100 g. Vitamin E—Immediate: $n = 2$, $\bar{x} = 95.30$, $s = 0.9900$, $SE_{\bar{x}} = 0.700$ mg/100 g. One day: $n = 2$, $\bar{x} = 94.45$, $s = 1.7678$, $SE_{\bar{x}} = 1.250$ mg/100 g. Three days: $n = 2$, $\bar{x} = 95.85$, $s = 2.1920$, $SE_{\bar{x}} = 1.550$ mg/100 g. Five days: $n = 2$, $\bar{x} = 96.35$, $s = 1.9092$, $SE_{\bar{x}} = 1.350$ mg/100 g. Seven days: $n = 2$, $\bar{x} = 93.70$, $s = 1.9799$, $SE_{\bar{x}} = 1.400$ mg/100 g. For vitamin A, $F = 12.09$, $P = 0.009$; we conclude that vitamin A content changes over time. For vitamin E, $F = 0.69$, $P = 0.630$; we cannot reject H_0.

12.25 Assuming the t (ANOVA) test establishes that the means are different, contrasts and multiple comparisons provide no further useful information.

12.27 *H. bihai* and *H. caribaea* red distributions are slightly skewed. *H. bihai*: $n = 16$, $\bar{x} = 47.597$, $s = 1.213$ mm. *H. caribaea* red: $n = 23$, $\bar{x} = 39.711$, $s = 1.799$ mm. *H. caribaea* yellow: $n = 15$, $\bar{x} = 36.180$, $s = 0.975$ mm. This just meets our rule for standard deviations. ANOVA gives $F = 259.12$, df $= 2$ and 51, $P < 0.0005$, so we conclude that the means are different.

12.29 *H. bihai*: $n = 16$, $\bar{x} = 3.8625$, $s = 0.0251$. *H. caribaea* red: $n = 23$, $\bar{x} = 3.6807$, $s = 0.0450$. *H. caribaea* yellow: $n = 15$, $\bar{x} = 3.5882$, $s = 0.0270$. ANOVA gives $F = 244.27$, df $= 2$ and 51, $P < 0.0005$, so we conclude the means are different.

12.31 (a) None of the distributions show any particular skewness. Control: $n = 15$, $\bar{x} = 0.21887$, $s = 0.01159$ g/cm^2. Low dose: $n = 15$, $\bar{x} = 0.21593$, $s = 0.01151$ g/cm^2. High dose: $n = 15$, $\bar{x} = 0.23507$, $s = 0.01877$ g/cm^2. (b) All three distributions appear to be nearly normal. (c) $F = 7.72$, df $= 2$ and 42, $P = 0.001$. (d) For Bonferroni, $t^{**} = 2.49$ and MSD $= 0.0131$. The high-dose mean is significantly different from the other two. (e) High doses increase BMD.

12.33 For Bonferroni, $t^{**} = 2.67$ and MSD $= 0.1476$. All six differences are significant.

12.35 For Bonferroni, $t^{**} = 2.71$ and $s_p \doteq 2.7348$. The Piano mean is significantly higher than the other three, but the other three means are not significantly different from each other.

12.37 (a) Four months: $n = 19$, $\bar{x} = 570.0$, $s = 123.0$ kcal/d. Five months: $n = 18$, $\bar{x} = 483.0$, $s = 112.9$ kcal/d. Six months: $n = 8$, $x = 541.9$, $s = 94.0$ kcal/d. Pooling is reasonable. (b) $F = 2.72$, df $= 2$ and 42, $P = 0.078$. We cannot reject H_0.

12.39 (a) Control: $n = 10$, $\bar{x} = 601.10$, $s = 27.36$ mg/cm^3. Low jump: $n = 10$, $\bar{x} = 612.50$, $s = 19.33$ mg/cm^3. High jump: $n = 10$, $\bar{x} = 638.70$, $s = 16.59$ mg/cm^3. Pooling is reasonable. **(b)** $F = 7.98$, df $= 2$ and 27, $P = 0.002$. We conclude that not all means are equal.

12.41 **(a)** Aluminum: $n = 4$, $\bar{x} = 2.0575$, $s = 0.2520$ mg/100 g. Clay: $n = 4$, $\bar{x} = 2.1775$, $s = 0.6213$ mg/100 g. Iron: $n = 4$, $\bar{x} = 4.6800$, $s = 0.6283$ mg/100 g. Pooling is risky because $0.6283/0.2520 = 2.49 > 2$. **(b)** $F = 31.16$, df $= 2$ and 9, $P < 0.0005$. We cautiously conclude that the means are not the same.

12.43 **(a)** ECM1: $n = 3$, $\bar{x} = 65.0\%$, $s = 8.66\%$. ECM2: $n = 3$, $\bar{x} = 63.33\%$, $s = 2.89\%$. ECM3: $n = 3$, $\bar{x} = 73.33\%$, $s = 2.89\%$. MAT1: $n = 3$, $\bar{x} = 23.33\%$, $s = 2.89\%$. MAT2: $n = 3$, $x = 6.67\%$, $s = 2.89\%$. MAT3: $n = 3$, $\bar{x} = 11.67\%$, $s = 2.89\%$. Pooling is risky because $8.66/2.89 > 2$. **(b)** $F = 137.94$, df $= 5$ and 12, $P < 0.0005$. We conclude that the means are not the same.

12.45 The pooled estimate of the standard deviation is 4.41%. The contrast (average of ECM means minus the average of MAT means) has the value $c = 53.33\%$, with standard error 2.079%. $t = 25.66$ with df $= 12$, and P is tiny.

12.47 **(a)** The plot shows granularity (which varies between groups), but that should not make us question independence; it is due to the fact that the scores are all integers. **(b)** The ratio of the largest to the smallest standard deviations is less than 2. **(c)** Apart from the granularity, the quantile plots are reasonably straight. **(d)** Again, apart from the granularity, the quantile plots look pretty good.

12.49 **(a)** $\psi_1 = (\mu_1 + \mu_2 + \mu_3)/3 - \mu_4$, $\psi_2 = (\mu_1 + \mu_2)/2 - \mu_3$, $\psi_3 = \mu_1 - \mu_2$. **(b)** The pooled standard deviation is $s_p = 1.1958$. $SE_{c_1} \doteq 0.2355$, $SE_{c_2} \doteq 0.1413$, $SE_{c_3} \doteq 0.1609$. **(c)** Testing $H_0: \psi_i = 0$; $H_a: \psi_i \neq 0$ for each contrast, we find $c_1 = -12.51$, $t_1 = -53.17$, $P_1 < 0.0005$; $c_2 = 1.269$, $t_2 = 8.98$, $P_2 < 0.0005$; $c_3 = 0.191$, $t_3 = 1.19$, $P_3 \doteq 0.2359$. The Placebo mean is significantly higher than the average of the other three, while the Keto mean is significantly lower than the average of the two Pyr means. The difference between the Pyr means is not significant (meaning the second application of the shampoo is of little benefit).

12.51 **(a)** Vitamin A—Immediate: $n = 2$, $\bar{x} = 67.0\%$, $s = 0.2828\%$, $SE_{\bar{x}} = 0.2\%$. One day: $n = 2$, $\bar{x} = 64.8\%$, $s = 1.1314\%$, $SE_{\bar{x}} = 0.8\%$. Three days: $n = 2$, $\bar{x} = 64.2\%$, $s = 1.4142\%$, $SE_{\bar{x}} = 1.0\%$. Five days: $n = 2$, $\bar{x} = 66.1\%$, $s = 1.5556\%$, $SE_{\bar{x}} = 1.1\%$. Seven days: $n = 2$, $\bar{x} = 59.3\%$, $s = 1.2728\%$, $SE_{\bar{x}} = 0.9\%$. The transformation has no effect on vitamin E, since the number of milligrams remaining is also the percent of the original 100 mg. Although the SS and MS entries for vitamin A are different from those of Exercise 12.21, everything else is the same: $F = 12.09$, df $= 4$ and 5, $P = 0.009$. So we (again) reject H_0 and conclude that vitamin A content decreases over time. Since the vitamin E numbers are unchanged, we again fail to reject H_0 ($F = 0.69$, df $= 4$ and 5, $P = 0.630$).

12.53 All distributions are reasonably normal, and standard deviations are close enough to justify pooling. For PRE1, $F = 1.13$, df $= 2$ and 63, $P = 0.329$. For PRE2, $F = 0.11$, df $= 2$ and 63, $P = 0.895$. Neither set of pretest scores suggests a difference in means.

12.55 $\hat{y} = 4.36 - 0.116x$. The regression is significant (that is, the slope is significantly different from 0): $t = -13.31$, df $= 158$, $P < 0.0005$. Regression explains $r^2 = 52.9\%$ of the variation in expected price. (This is similar to the ANOVA value: $R^2 = 53.5\%$.) The granularity of the "Number of promotions" observations makes interpreting the plot a bit tricky. For 5 promotions, the residuals seem to be more likely to be negative, while for 3 promotions, the residuals are weighted toward the positive side. This suggests that a linear model may not be appropriate.

12.57 **(a)** F can be made very small (close to 0), and P close to 1. **(b)** F increases, and P decreases.

12.61 **(b)** Answers will vary with choice of H_a and desired power. For example, with $\mu_1 = \mu_2 = 4.4$, $\mu_3 = 5$, $s = 1.2$, three samples of size 75 will produce power 0.89.

CHAPTER | 13

13.1   (a) The RESIDUAL part of the model represents the error. (b) Large F is needed to reject H_0. (c) MS equals SS divided by DF.

13.3   (a) 3 and 32 df; use 3 and 30 df in Table E. (c) $0.05 < P < 0.10$. (d) The lines connecting the means would be roughly parallel, because the interaction term is not significantly different from 0.

13.5   (a) Intervention: 11.6. Control: 9.967. Baseline: 10.0. 3 months: 11.2. 6 months: 11.15. Overall: 10.783. The first two means suggest that the intervention group showed improvement over the control group. (b) Interaction means that the mean number of actions changes differently over time for the two groups.

13.7   There were no significant effects: F_A has df $= 2$ and 24 and $P = 0.3157$. F_B has df $= 1$ and 24 and $P = 0.0688$. F_{AB} has df $= 2$ and 24 and $P = 0.1520$.

13.9   (a) Familiar increases with repetition and has the higher rating for 1 and 3 repetitions, while unfamiliar is higher for 2 repetitions. (b) As repetitions increase, the change in attitude depends on whether the person is familiar with the ad.

13.11 $s_p = 1.308$. Pooling is reasonable, as $1.46/1.16 = 1.26 < 2$.

13.13   For example: Are opinions of university employees (and/or West Coast residents) similar to those of other groups? What do experts consider to be "good" ads? Did the content of the news show affect responses to the ads?

13.15   (b) There seems to be a fairly large difference between the means based on how much the rats were allows to eat, but not very much difference based on the chromium level. There may be an interaction: the NM mean is lower than the LM mean, while the NR mean is higher than the LR mean. (c) L mean: 4.86. N mean: 4.871. M mean: 4.485. R mean: 5.246. LR minus LM: 0.63. NR minus NM: 0.892. Mean GITH levels are lower for M than for R; there is not much difference for L versus N. The difference between M and R is greater among rats who had normal chromium levels in their diets (N).

13.17   The "Other" category had the lowest mean SATM score for both genders; this is apparent from a graph of the means as well as from the marginal means (CS: 605; EO: 624.5; O: 566). Males had higher mean scores in CS and O, while females are slightly higher in EO; this seems to be an interaction. Overall, the marginal means are 611.7 (males) and 585.3 (females).

13.19   (a) $n = 3$ for all combinations. The means and standard deviations are $\bar{x}_{E1,4} = 65.00$, $s_{E1,4} = 8.66$; $\bar{x}_{E1,8} = 63.33$, $s_{E1,8} = 2.89$; $\bar{x}_{E2,4} = 63.33$, $s_{E2,4} = 2.89$; $\bar{x}_{E2,8} = 63.33$, $s_{E2,8} = 5.77$; $\bar{x}_{E3,4} = 73.33$, $s_{E3,4} = 2.89$; $\bar{x}_{E3,8} = 73.33$, $s_{E3,8} = 5.77$; $\bar{x}_{M1,4} = 23.33$, $s_{M1,4} = 2.89$; $\bar{x}_{M1,8} = 21.67$, $s_{M1,8} = 5.77$; $\bar{x}_{M2,4} = 6.67$, $s_{M2,4} = 2.89$; $\bar{x}_{M2,8} = 6.67$, $s_{M2,8} = 2.89$; $\bar{x}_{M3,4} = 11.67$, $s_{M3,4} = 2.89$; $\bar{x}_{M3,8} = 10.00$, $s_{M3,8} = 5.00$. Apart from the first standard deviation, the ratio is below 2 for the rest. (b) ECM means are all higher than MAT means. Time and interaction effects are not clearly suggested. (c) Only Material ($F = 251.26$, df $= 5$ and 24, $P < 0.0005$) is significant. For Time, $F = 0.29$, df $= 1$ and 24, $P = 0.595$. For interaction, $F = 0.06$, df $= 5$ and 24, $P = 0.998$.

13.21   For each time period, there is a significant difference among materials. The pooled standard deviations are 6.236, 4.410, and 4.859. For Bonferroni and $\alpha = 0.05$, $t^{**} = 3.65$, so the MSDs are 18.6, 13.1, and 14.5. The only ECM/MAT difference that is not significant is ECM2 and MAT1 at 2 weeks.

13.23   Yes; the iron-pot means are the highest, and F for testing the effect of the pot type is very large ($F = 92.26$).

13.25   (a) The first six means and standard deviations are $\bar{x}_{11} = 25.0307$, $s_{11} = 0.001155$; $\bar{x}_{12} = 25.0280$, $s_{12} = 0$; $\bar{x}_{13} = 25.0260$, $s_{13} = 0$; $\bar{x}_{21} = 25.0167$, $s_{21} = 0.001155$; $\bar{x}_{22} = 25.0200$, $s_{22} = 0.002$; $\bar{x}_{23} = 25.0160$, $s_{23} = 0$. (b) Except for tool 1, mean diameter is highest at time 2. Tool 1 had the highest mean diameters, followed by tool 2, tool 4, tool 3, and tool 5. (c) $F_A = 412.94$, df $= 4$ and 30, $P < 0.0005$. $F_B = 43.60$, df $= 2$ and 30, $P < 0.0005$. $F_{AB} = 7.65$, df $= 8$ and 30, $P < 0.0005$. (d) There is strong evidence of a difference in mean diameter among the tools (A) and among the times (B). There is also an interaction (specifically, tool 1's mean diameters changed differently over time compared to the other tools).

13.27 **(a)** The first four means and standard deviations are $\bar{x}_{1.40} = 4.423$, $s_{1.40} = 0.1848$; $\bar{x}_{1.30} = 4.225$, $s_{1.30} = 0.3856$; $\bar{x}_{1.20} = 4.689$, $s_{1.20} = 0.2331$; $\bar{x}_{1.10} = 4.920$, $s_{1.10} = 0.1520$. The mean expected price decreases as percent discount increases, and also as the number of promotions increases. **(b)** With A = number of promotions and B = percent discount: $F_A = 47.73$, df = 3 and 144, $P < 0.0005$; $F_B = 47.42$, df = 3 and 144, $P < 0.0005$; $F_{AB} = 0.44$, df = 9 and 144, $P = 0.912$. **(c)** Both main effects are significant, but the interaction is not.

13.29 **(a)** All three F-values have df = 1 and 945, the P-values are < 0.001, < 0.001, and 0.1476. Gender and handedness both have significant effects on mean lifetime, but there is no interaction. **(b)** Women live about 6 years longer than men (on the average), while right-handed people average 9 more years of life than left-handed people. Handedness affects both genders in the same way, and vice versa.

13.31 **(a)** and **(b)** The first three means and standard deviations are $\bar{x}_{1,1} = 3.2543$, $s_{1,1} = 0.2287$; $\bar{x}_{1,2} = 2.7636$, $s_{1,2} = 0.0666$; $\bar{x}_{1,3} = 2.8429$, $s_{1,3} = 0.2333$. The standard deviations range from 0.0666 to 0.3437, for a ratio of 5.16—larger than we like. **(c)** For Plant, $F = 1301.32$, df = 3 and 224, $P < 0.0005$. For Rain, $F = 9.76$, df = 6 and 224, $P < 0.0005$. For interaction, $F = 5.97$, df = 18 and 224, $P < 0.0005$.

13.33 The seven F statistics are 184.05, 115.93, 208.87, 218.37, 220.01, 174.14, and 230.17, all with df = 3 and 32 and $P < 0.0005$.

13.35 Fresh: Plant $F = 81.45$, df = 3 and 84, $P < 0.0005$. Rain $F = 43.71$, df = 6 and 84, $P < 0.0005$. Interaction $F = 1.79$, df = 18 and 84, $P = 0.040$. Dry: Plant $F = 79.93$, df = 3 and 84, $P < 0.0005$. Rain $F = 44.79$, df = 6 and 84, $P < 0.0005$. Interaction $F = 2.22$, df = 18 and 84, $P = 0.008$.

13.37 Fresh: The seven F statistics are 15.88, 11.81, 62.08, 10.83, 22.62, 8.20, and 10.81, all with df = 3 and 12, and $P = 0.003$. Dry: The seven F statistics are 8.14, 26.26, 22.58, 11.86, 21.38, 14.77, and 8.66, all with df = 3 and 12, and $P = 0.003$.

13.39 Men in CS: $n = 39$, $\bar{x} = 526.949$, $s = 100.937$. Men in EOS: $n = 39$, $\bar{x} = 507.846$, $s = 57.213$. Men in Other: $n = 39$, $\bar{x} = 487.564$, $s = 108.779$. Women in CS: $n = 39$, $\bar{x} = 543.385$, $s = 77.654$. Women in EOS: $n = 39$, $\bar{x} = 538.205$, $s = 102.209$. Women in Other: $n = 39$, $\bar{x} = 465.026$, $s = 82.184$. The means suggest that students who stay in the sciences have higher mean SATV scores than those who end up in the "Other" group. Female CS and EO students have higher scores than males in those majors, but males have the higher mean in the Other group. Normal quantile plots suggests some right skewness in the "Women in CS" group and also some nonnormality in the tails of the "Women in EO" group. Other groups look reasonably normal. In the ANOVA, only the effect of major is significant ($F = 9.32$, df = 2 and 228, $P < 0.0005$).

13.41 Men in CS: $n = 39$, $\bar{x} = 7.7949$, $s = 1.5075$. Men in EOS: $n = 39$, $\bar{x} = 7.4873$, $s = 2.1505$. Men in Other: $n = 39$, $\bar{x} = 7.4103$, $s = 1.5681$. Women in CS: $n = 39$, $\bar{x} = 8.8462$, $s = 1.1364$. Women in EOS: $n = 39$, $\bar{x} = 9.2564$, $s = 0.7511$. Women in Other: $n = 39$, $\bar{x} = 8.6154$, $s = 1.1611$. The means suggest that females have higher HSE grades than males. For a given gender, there is not too much difference among majors. Normal quantile plots show no great deviations from normality, apart from the granularity of the grades (most evident among Women in EO). In the ANOVA, only the effect of gender is significant ($F = 50.32$, df = 1 and 228, $P < 0.0005$).

TO TEACHERS | About This Book

1. The committee's report is George Cobb, "Teaching statistics," in L. A. Steen (ed.), *Heeding the Call for Change: Suggestions for Curricular Action*, Mathematical Association of America, 1990, pp. 3–43. A summary has been endorsed by the ASA Board of Directors.

2. See, for example, the evaluation of the current state of the first course in statistics, with discussion by leading statisticians from industry as well as academia, in David S. Moore and discussants, "New pedagogy and new content: the case of statistics," *International Statistical Review*, 65 (1997), pp. 123–165.

3. A. Agresti and B. A. Coull, "Approximate is better than 'exact' for interval estimation of binomial proportions," *The American Statistician*, 52 (1998), pp. 119–126. Alan Agresti and Brian Caffo, "Simple and effective confidence intervals for proportions and differences of proportions result from adding two successes and two failures," *The American Statistician*, 45 (2000), pp. 280–288.

4. Lawrence D. Brown, Tony Cai, and Anirban DasGupta, "Confidence intervals for a binomial proportion and asymptotic expansions," *Annals of Statistics*, 30 (2002), pp. 160–201. For an overview and compelling evidence of the weakness of the traditional intervals see by the same authors "Interval estimation for a binomial proportion," *Statistical Science*, 16 (2001), pp. 101–133.

5. A detailed discussion appears in George Cobb and David S. Moore, "Mathematics, statistics, and teaching," *American Mathematical Monthly*, 104 (1997), pp. 801–823. Readers interested in relations between mathematics and statistics as fields might also look at David S. Moore and George Cobb, "Statistics and mathematics: tension and cooperation," *American Mathematical Monthly*, 107 (2000), pp. 615–630.

TO STUDENTS | What Is Statistics?

1. Federal Interagency Forum on Child and Family Studies, *America's Children in Brief: Key National Indicators of Well-Being*, childstats.gov/americaschildren, 2004.

2. The rise of statistics from the physical, life, and behavioral sciences is described in detail by S. M. Stigler, *The History of Statistics: The Measurement of Uncertainty before 1900*, Harvard-Belknap, 1986.

CHAPTER 1 | Notes

1. Mary Williams Walsh, "Failed pensions: a painful lesson in assumptions," *New York Times*, November 12, 2003.

2. From the Fatal Accident Reporting System Web site, www-fars.nhtsa.dot.gov.

3. Data for 2000, collected by the Current Population Survey and reported in the 2002 *Statistical Abstract of the United States*.

4. Haipeng Shen, "Nonparametric regression for problems involving lognormal distributions," PhD thesis, University of Pennsylvania, 2003. Thanks to Haipeng Shen and Larry Brown for sharing the data.

5. United Nations data found at earthtrends.wri.org.

6. James T. Fleming, "The measurement of children's perception of difficulty in reading materials," *Research in the Teaching of English,* 1 (1967), pp. 136–156.

7. Wayne Nelson, "Theory and applications of hazard plotting for censored failure data," *Technometrics,* 14 (1972), pp. 945–966.

8. Read from a graph in Peter A. Raymond and Jonathan J. Cole, "Increase in the export of alkalinity from North America's largest river," *Science,* 301 (2003), pp. 88–91.

9. Monthly gasoline price index from the Consumer Price Index, from the Bureau of Labor Statistics, www.bls.gov, converted into dollars.

10. U.S. Environmental Protection Agency, *Municipal Solid Waste in the United States: 2000 Facts and Figures,* document EPA530-R-02-001, 2002.

11. Press release for the 2003 DuPont Automotive Color Survey, at www.automotive.dupont.com.

12. National Center for Education Statistics, NEDRC Table Library, nces.ed.gov/surveys/npsas/table_library.

13. Robyn Greenspan, "The deadly duo: spam and viruses, October 2003," found online at cyberatlas.internet.com.

14. See Note 12.

15. Debora L. Arsenau, "Comparison of diet management instruction for patients with non-insulin dependent diabetes mellitus: learning activity package vs. group instruction," MS thesis, Purdue University, 1993.

16. Data from Gary Community School Corporation, courtesy of Celeste Foster, Department of Education, Purdue University.

17. National Climactic Data Center, storm events data base. Found online at sciencepolicy.colorado.edu/sourcebook/tornadoes.html.

18. Found online at earthtrends.wri.org.

19. Data from the U.S. Historical Climatology Network, archived at www.co2science.org. (Despite claims made on this site, temperatures at most U.S. locations show a gradual increase over the past century.)

20. National Oceanic and Atmospheric Administration, www.noaa.gov.

21. We thank Heeseung Roh Ryu for supplying the data, from Heeseung Roh Ryu, Roseann M. Lyle, and George P. McCabe, "Factors associated with weight concerns and unhealthy eating patterns among young Korean females," *Eating Disorders,* 11 (2003), pp. 129–141.

22. This example was provided by Nicolas Fisher.

23. J. Marcus Jobe and Hutch Jobe, "A statistical approach for additional infill development," *Energy Exploration and Exploitation,* 18 (2000), pp. 89–103.

24. S. M. Stigler, "Do robust estimators work with real data?" *Annals of Statistics,* 5 (1977), pp. 1055–1078.

25. T. Bjerkedal, "Acquisition of resistance in guinea pigs infected with different doses of virulent tubercle bacilli," *American Journal of Hygiene,* 72 (1960), pp. 130–148.

26. Data provided by Darlene Gordon, Purdue University.

27. U.S. Environmental Protection Agency, *Model Year 2004 Fuel Economy Guide,* found online at www.fueleconomy.gov.

28. Douglas Fore, "Do we have a retirement crisis in America?" TIAA-CREF Institute, *Research Dialogue,* No. 77, 2003. The data are for the year 2001.

29. *Extreme Weather Sourcebook 2001*, found online at `sciencepolicy.colorado.edu/sourcebook`.

30. We thank Ethan J. Temeles of Amherst College for providing the data. His work is described in Ethan J. Temeles and W. John Kress, "Adaptation in a plant-hummingbird association," *Science*, 300 (2003), pp. 630–633.

31. We thank Charles Cannon of Duke University for providing the data. The study report is C. H. Cannon, D. R. Peart, and M. Leighton, "Tree species diversity in commercially logged Bornean rainforest," *Science*, 281 (1998), pp. 1366–1367.

32. These graphs, like most others in this book, were produced by S-Plus, a professional statistical software package. Both graphs in Figure 1.22 used the same algorithm. The ability to fit normal curves is widespread in statistical software, and general "density estimators" are present in most professional software.

33. Results for 1988 to 1991 from a large sample survey, reported in National Center for Health Statistics, *Health, United States, 1995*, 1996.

34. Elroy Dimson, Paul Marsh, and Mike Stanton, "Irrational optimism," London Business School Institute of Finance and Accounting Working Paper IFA397, 2003.

35. Data provided by Charles Hicks, Purdue University.

36. We thank C. Robertson McClung of Dartmouth College for supplying the data. The study is reported in Todd P. Michael et al., "Enhanced fitness conferred by naturally occurring variation in the circadian clock," *Science*, 302 (2003), pp. 1049–1053.

37. Julie Reinhart and Paul Schneider, "Student satisfaction, self-efficacy, and the perception of the two-way audio/visual distance learning environment," *Quarterly Review of Distance Education*, 2 (2001), pp. 357–365.

38. From `www.isp-planet.com`.

39. From the Current Population Survey, 2003 annual demographic supplement, found online at `www.census.gov`.

40. James W. Grier, "Ban of DDT and subsequent recovery of reproduction of bald eagles," *Science*, 218 (1982), pp. 1232–1235.

CHAPTER 2 | Notes

1. Figure 2.1 displays data for 2003, from the College Board Web site, `www.collegeboard.com`.

2. A sophisticated treatment of improvements and additions to scatterplots is W. S. Cleveland and R. McGill, "The many faces of a scatterplot," *Journal of the American Statistical Association*, 79 (1984), pp. 807–822.

3. Data from National Institute of Standards and Technology, *Engineering Statistics Handbook*, `www.itl.nist.gov/div898/handbook`. The analysis there does not comment on the bias of field measurements.

4. Todd W. Anderson, "Predator responses, prey refuges, and density-dependent mortality of a marine fish," *Ecology*, 81 (2001), pp. 245–257.

5. The motorcycle crash test data are often used to test scatterplot smoothers. They appear in, for example, W. Hardle, *Applied Nonparametric Regression*, Cambridge University Press, 1990. The smoother used to produce Figure 2.5 is called "lowess." A full account appears in section 4.6 of J. M. Chambers, W. S. Cleveland, B. Kleiner, and P. A. Tukey, *Graphical Methods for Data Analysis*, Wadsworth, 1983.

6. James T. Fleming, "The measurement of children's perception of difficulty in reading materials," *Research in the Teaching of English*, 1 (1967), pp. 136–156.

7. Christer G. Wiklund, "Food as a mechanism of density-dependent regulation of breeding numbers in the merlin *Falco columbarius*," *Ecology*, 82 (2001), pp. 860–867.

8. Data from a plot in Naomi I. Eisenberger, Matthew D. Lieberman, and Kipling D. Williams, "Does rejection hurt? An fMRI study of social exclusion," *Science*, 302 (2003), pp. 290–292.

9. We thank C. Robertson McClung of Dartmouth College for supplying the data. The study is reported in Todd P. Michael et al., "Enhanced fitness conferred by naturally occurring variation in the circadian clock," *Science*, 302 (2003), pp. 1049–1053.

10. *Forbes*, February 16, 2004.

11. Alan S. Banks et al., "Juvenile hallux abducto valgus association with metatarsus adductus," *Journal of the American Podiatric Medical Association*, 84 (1994), pp. 219–224.

12. Based on T. N. Lam, "Estimating fuel consumption from engine size," *Journal of Transportation Engineering*, 111 (1985), pp. 339–357. The data for 10 to 50 km/h are measured; those for 60 and higher are calculated from a model given in the paper and are therefore smoothed.

13. From the Web site of the International Olympic Committee, `www.olympic.org`.

14. N. Maeno et al., "Growth rates of icicles," *Journal of Glaciology*, 40 (1994), pp. 319–326.

15. Data provided by Matthew Moore.

16. Compiled from Fidelity data in the *Fidelity Insight* newsletter, 20 (2004), No. 1.

17. A careful study of this phenomenon is W. S. Cleveland, P. Diaconis, and R. McGill, "Variables on scatterplots look more highly correlated when the scales are increased," *Science*, 216 (1982), pp. 1138–1141.

18. Data from a plot in Timothy G. O'Brien and Margaret F. Kinnaird, "Caffeine and conservation," *Science*, 300 (2003), p. 587.

19. See Note 16.

20. Data from a plot in James A. Levine, Norman L. Eberhardt, and Michael D. Jensen, "Role of nonexercise activity thermogenesis in resistance to fat gain in humans," *Science*, 283 (1999), pp. 212–214.

21. E. P. Hubble, "A relation between distance and radial velocity among extragalactic nebulae," *Proceedings of the National Academy of Sciences*, 15 (1929), pp. 168–173.

22. Data from G. A. Sacher and E. F. Staffelt, "Relation of gestation time to brain weight for placental mammals: implications for the theory of vertebrate growth," *American Naturalist*, 108 (1974), pp. 593–613. We found these data in F. L. Ramsey and D. W. Schafer, *The Statistical Sleuth: A Course in Methods of Data Analysis*, Duxbury, 1997.

23. From the Web site of the National Center for Education Statistics, `www.nces.ed.gov`.

24. Data from plots in Chaido Lentza-Rizos, Elizabeth J. Avramides, and Rosemary A. Roberts, "Persistence of fenthion residues in olive oil," *Pesticide Science*, 40 (1994), pp. 63–69.

25. Debora L. Arsenau, "Comparison of diet management instruction for patients with non-insulin dependent diabetes mellitus: learning activity package vs. group instruction," MS thesis, Purdue University, 1993.

26. Gannett News Service article appearing in the *Lafayette (Ind.) Journal and Courier,* April 23, 1994.

27. This example is drawn from M. Goldstein, "Preliminary inspection of multivariate data," *The American Statistician,* 36 (1982), pp. 358–362.

28. The facts in Example 2.22 come from Nancy W. Burton and Leonard Ramist, *Predicting Success in College: Classes Graduating since 1980,* Research Report No. 2001-2, College Board, 2001.

29. Zeinab E. M. Afifi,"Principal components analysis of growth of Nahya infants: size, velocity and two physique factors," *Human Biology,* 57 (1985), pp. 659–669.

30. D. A. Kurtz (ed.), *Trace Residue Analysis,* American Chemical Society Symposium Series No. 284, 1985, Appendix.

31. Data from a plot in Feng Sheng Hu et al., "Cyclic variation and solar forcing of Holocene climate in the Alaskan subarctic," *Science,* 301 (2003), pp. 1890–1893.

32. R. C. Nelson, C. M. Brooks, and N. L. Pike, "Biomechanical comparison of male and female distance runners," in P. Milvy (ed.), *The Marathon: Physiological, Medical, Epidemiological, and Psychological Studies,* New York Academy of Sciences, 1977, pp. 793–807.

33. Frank J. Anscombe, "Graphs in statistical analysis," *The American Statistician,* 27 (1973), pp. 17–21.

34. P. Velleman, *ActivStats 2.0,* Addison Wesley Interactive, 1997.

35. This analysis was done by Crystal Richard.

36. Laura L. Calderon et al., "Risk factors for obesity in Mexican-American girls: dietary factors, anthropometric factors, physical activity, and hours of television viewing," *Journal of the American Dietetic Association,* 96 (1996), pp. 1177–1179.

37. Saccharin was on the National Institutes of Health's list of suspected carcinogens from 1977 to 2000 but remained in wide use with a warning label. It was officially cleared in 2000 and the label was removed.

38. A detailed study of this correlation appears in E. M. Remolona, P. Kleinman, and D. Gruenstein, "Market returns and mutual fund flows," *Federal Reserve Bank of New York Economic Policy Review,* 3, No. 2 (1997), pp. 33–52.

39. M. S. Linet et al., "Residential exposure to magnetic fields and acute lymphoblastic leukemia in children," *New England Journal of Medicine,* 337 (1997), pp. 1–7.

40. *The Health Consequences of Smoking: 1983,* U.S. Public Health Service, 1983.

41. Kevin Phillips, *Wealth and Democracy,* Broadway Books, 2002, p. 151.

42. David E. Bloom and David Canning, "The health and wealth of nations," *Science,* 287 (2000), pp. 1207–1208.

43. Contributed by Marigene Arnold of Kalamazoo College.

44. D. E. Powers and D. A. Rock, *Effects of Coaching on SAT I: Reasoning Test Scores,* Educational Testing Service Research Report 98-6, College Entrance Examination Board, 1998.

45. Oskar Kindvall, "Habitat heterogeneity and survival in a bush cricket metapopulation," *Ecology,* 77 (1996), pp. 207–214.

46. H. Bradley Perry, "Analyzing growth stocks: what's a good growth rate?" *AAII Journal,* 13 (1991), pp. 7–10.

47. Data from a plot in Josef P. Rauschecker, Biao Tian, and Marc Hauser, "Processing of complex sounds in the macaque nonprimary auditory cortex," *Science,* 268 (1995), pp. 111–114. The paper states that there are $n = 41$ observations, but only $n = 37$ can be read accurately from the plot.

48. We thank Zhiyong Cai of Texas A&M University for providing the data. The full study is reported in Zhiyong Cai et al., "Predicting strength of matched sets of test specimens," *Wood and Fiber Science,* 30 (1998), pp. 396–404.

49. Data provided by Robert Dale, Purdue University.

50. Data provided by Samuel Phillips, Purdue University.

51. See Note 32.

52. S. Chatterjee and B. Price, *Regression Analysis by Example,* Wiley, 1977.

53. Data provided by Jim Bateman and Michael Hunt, Purdue University.

54. "Dancing in step," *Economist,* March 22, 2001.

55. Gary Smith, "Do statistics test scores regress toward the mean?" *Chance,* 10, No. 4 (1997), pp. 42–45.

CHAPTER 3 | Notes

1. National Institute of Child Health and Human Development, Study of Early Child Care. The article appears in the July 2003 issue of *Child Development.* The quotation is from the summary on the NICHD Web site, www.nichd.nih.gov.

2. Caffeine equivalents from www.geocities.com/SiliconValley/9692/joltcola. html.

3. J. E. Muscat et al., "Handheld cellular telephone use and risk of brain cancer," *Journal of the American Medical Association,* 284 (2000), pp. 3001–3007.

4. See Jeffrey G. Johnson et al., "Television viewing and aggressive behavior during adolescence and adulthood," *Science,* 295 (2002), pp. 2468–2471. The authors use statistical adjustments to control for the effects of a number of lurking variables. The association between TV viewing and aggression remains significant.

5. For a full description of the STAR program and its follow-up studies, go to www. heros-inc.org/star.htm.

6. Simplified from Arno J. Rethans, John L. Swasy, and Lawrence J. Marks, "Effects of television commercial repetition, receiver knowledge, and commercial length: a test of the two-factor model," *Journal of Marketing Research,* 23 (February 1986), pp. 50–61.

7. L. L. Miao, "Gastric freezing: an example of the evaluation of medical therapy by randomized clinical trials," in J. P. Bunker, B. A. Barnes, and F. Mosteller (eds.), *Costs, Risks, and Benefits of Surgery,* Oxford University Press, 1977, pp. 198–211.

8. Simplified from David L. Strayer, Frank A. Drews, and William A. Johnston, "Cell phone–induced failures of visual attention during simulated driving," *Journal of Experimental Psychology: Applied,* 9 (2003), pp. 23–32. See Example 3.12 for more detail.

9. John H. Kagel, Raymond C. Battalio, and C. G. Miles, "Marijuana and work performance: results from an experiment," *Journal of Human Resources,* 15 (1980), pp. 373–395. A general discussion of failures of blinding is Dean Ferguson et al., "Turning a blind eye: the success of blinding reported in a random sample of randomised, placebo controlled trials," *British Medical Journal,* 328 (2004), p. 432.

10. Martin Enserink, "Fickle mice highlight test problems," *Science,* 284 (1999), pp. 1599–1600. There is a full report of the study in the same issue.

11. See the first citation in Note 9.

12. Geetha Thiagarajan et al., "Antioxidant properties of green and black tea, and their potential ability to retard the progression of eye lens cataract," *Experimental Eye Research*, 73 (2001), pp. 393–401.

13. Joel Brockner et al., "Layoffs, equity theory, and work performance: further evidence of the impact of survivor guilt," *Academy of Management Journal*, 29 (1986), pp. 373–384.

14. Simplified from Sanjay K. Dhar, Claudia González-Vallejo, and Dilip Soman, "Modeling the effects of advertised price claims: tensile versus precise pricing," *Marketing Science*, 18 (1999), pp. 154–177.

15. Data from the source in the preceding note. The study also varied the percent of the price cut. The results in Figure 3.6 are for a 60% discount.

16. Carol A. Warfield, "Controlled-release morphine tablets in patients with chronic cancer pain," *Cancer*, 82 (1998), pp. 2299–2306.

17. R. C. Shelton et al., "Effectiveness of St. John's wort in major depression," *Journal of the American Medical Association*, 285 (2001), pp. 1978–1986.

18. K. Wang, Y. Li, and J. Erickson, "A new look at the Monday effect," *Journal of Finance*, 52 (1997), pp. 2171–2186.

19. Based on Evan H. DeLucia et al., "Net primary production of a forest ecosystem with experimental CO_2 enhancement," *Science*, 284 (1999), pp. 1177–1179. The investigators used the block design.

20. Based on Pierre J. Meunier et al., "The effects of strontium renelate on the risk of vertebral fracture in women with postmenopausal osteoporosis," *New England Journal of Medicine*, 350 (2004), pp. 459–468.

21. E. M. Peters et al., "Vitamin C supplementation reduces the incidence of postrace symptoms of upper-respiratory tract infection in ultramarathon runners," *American Journal of Clinical Nutrition*, 57 (1993), pp. 170–174.

22. From the online version of the Bureau of Labor Statistics *Handbook of Methods*, modified April 17, 2003, at `www.bls.gov`. The details of the design are more complicated than the text describes.

23. For more detail on the material of this section and complete references, see P. E. Converse and M. W. Traugott, "Assessing the accuracy of polls and surveys," *Science*, 234 (1986), pp. 1094–1098.

24. The nonresponse rate for the CPS comes from the source cited in Note 22. The GSS reports its response rate on its Web site, `www.norc.org/projects/gensoc.asp`. The Pew study is described in Gregory Flemming and Kimberly Parker, "Race and reluctant respondents: possible consequences of non-response for pre-election surveys," Pew Research Center for the People and the Press, 1997, found at `www.people-press.org`.

25. George Bishop, "Americans' belief in God," *Public Opinion Quarterly*, 63 (1999), pp. 421–434.

26. For more detail on the limits of memory in surveys, see N. M. Bradburn, L. J. Rips, and S. K. Shevell, "Answering autobiographical questions: the impact of memory and inference on surveys," *Science*, 236 (1987), pp. 157–161.

27. Robert F. Belli et al., "Reducing vote overreporting in surveys: social desirability, memory failure, and source monitoring," *Public Opinion Quarterly*, 63 (1999), pp. 90–108.

28. Sex: Tom W. Smith, "The *JAMA* controversy and the meaning of sex," *Public Opinion Quarterly*, 63 (1999), pp. 385–400. Welfare: from a *New York Times*/CBS News poll reported in the *New York Times*, July 5, 1992. Scotland: "All set for independence?"

Economist, September 12, 1998. Many other examples appear in T. W. Smith, "That which we call welfare by any other name would smell sweeter," *Public Opinion Quarterly*, 51 (1987), pp. 75–83.

29. David W. Moore, "Americans retain an 'internationalist' perspective," found online at www.gallup.com, March 1, 2004.

30. Giuliana Coccia, "An overview of non-response in Italian telephone surveys," *Proceedings of the 99th Session of the International Statistical Institute, 1993*, Book 3, pp. 271–272.

31. From *CIS Boletin 9, Spaniards' Economic Awareness*, found online at www.cis.es/ingles/opinion/economia.htm.

32. The survey question and result are reported in Trish Hall, "Shop? Many say 'Only if I must,'" *New York Times*, November 28, 1990.

33. Adam Nagourney and Janet Elder, "New York Times CBS Poll: What Hispanics Believe," found online at www.Hispanic.cc.

34. Warren McIsaac and Vivek Goel, "Is access to physician services in Ontario equitable?" Institute for Clinical Evaluative Sciences in Ontario, October 18, 1993.

35. Based on Stephen A. Woodbury and Robert G. Spiegelman, "Bonuses to workers and employers to reduce unemployment: randomized trials in Illinois," *American Economic Review*, 77 (1987), pp. 513–530.

36. Based on Christopher Anderson, "Measuring what works in health care," *Science*, 263 (1994), pp. 1080–1082.

CHAPTER 4 | Notes

1. An informative and entertaining account of the origins of probability theory is Florence N. David, *Games, Gods and Gambling*, Charles Griffin, London, 1962.

2. Information for the 1999–2000 academic year is from the 2003 *Statistical Abstract of the United States*, Table 286.

3. You can find a mathematical explanation of Benford's law in Ted Hill, "The first-digit phenomenon," *American Scientist*, 86 (1996), pp. 358–363; and Ted Hill, "The difficulty of faking data," *Chance*, 12, No. 3 (1999), pp. 27–31. Applications in fraud detection are discussed in the second paper by Hill and in Mark A. Nigrini, "I've got your number," *Journal of Accountancy*, May 1999, available online at www.aicpa.org/pubs/jofa/joaiss.htm.

4. Royal Statistical Society news release, "Royal Statistical Society concerned by issues raised in Sally Clark case," October 23, 2001, at www.rss.org.uk. For background, see an editorial and article in the *Economist*, January 22, 2004. The editorial is titled "The probability of injustice."

5. Peter Sheridan Dodds, Roby Muhamad, and Duncan J. Watts, "An experimental study of search in global social networks," *Science*, 301 (2003), pp. 827–829.

6. This is one of several tests discussed in Bernard M. Branson, "Rapid HIV testing: 2003 update," a presentation by the Centers for Disease Control and Prevention, found online at www.cdc.gov.

7. From Statistics Canada, www.statcan.ca.

8. Robyn Greenspan, "The deadly duo: spam and viruses, October 2003," found online at cyberatlas.internet.com.

9. Robert P. Dellavalle et al., "Going, going, gone: lost Internet references," *Science*, 302 (2003), pp. 787–788.

10. Corey Kilgannon, "When New York is on the end of the line," *New York Times*, November 7, 1999.

11. A. Tversky and D. Kahneman, "Extensional versus intuitive reasoning: the conjunction fallacy in probability judgment," *Psychological Review*, 90 (1983), pp. 293–315.

12. We use $\bar{x}$ both for the random variable, which takes different values in repeated sampling, and for the numerical value of the random variable in a particular sample. Similarly, s and $\hat{p}$ stand both for random variables and for specific values. This notation is mathematically imprecise but statistically convenient.

13. We will consider only the case in which X takes a finite number of possible values. The same ideas, implemented with more advanced mathematics, apply to random variables with an infinite but still countable collection of values.

14. From the Web site of the North Carolina State University Department of Registration and Records, `www.ncsu.edu/class/grades`.

15. The probability 0.12 is based on a small study at one university: Michele L. Head, "Examining college students' ethical values," Consumer Science and Retailing honors project, Purdue University, 2003.

16. From the Census Bureau's 1998 American Housing Survey.

17. The mean of a continuous random variable X with density function $f(x)$ can be found by integration:

$$\mu_X = \int xf(x)\,dx$$

This integral is a kind of weighted average, analogous to the discrete-case mean

$$\mu_X = \sum xP(X\overline{x})$$

The variance of a continuous random variable X is the average squared deviation of the values of X from their mean, found by the integral

$$\sigma_X^2 = \int (x-\mu)^2 f(x)\,dx$$

18. See A. Tversky and D. Kahneman, "Belief in the law of small numbers," *Psychological Bulletin*, 76 (1971), pp. 105–110, and other writings of these authors for a full account of our misperception of randomness.

19. Probabilities involving runs can be quite difficult to compute. That the probability of a run of three or more heads in 10 independent tosses of a fair coin is $(1/2) + (1/128) = 0.508$ can be found by clever counting. A general treatment using advanced methods appears in Section XIII.7 of William Feller, *An Introduction to Probability Theory and Its Applications*, vol. 1, 3rd ed., Wiley, 1968.

20. R. Vallone and A. Tversky, "The hot hand in basketball: on the misperception of random sequences," *Cognitive Psychology*, 17 (1985), pp. 295–314. A later series of articles that debate the independence question is A. Tversky and T. Gilovich, "The cold facts about the 'hot hand' in basketball," *Chance*, 2, No. 1 (1989), pp. 16–21; P. D. Larkey, R. A. Smith, and J. B. Kadane, "It's OK to believe in the 'hot hand,'" *Chance*, 2, No. 4 (1989), pp. 22–30; and A. Tversky and T. Gilovich, "The 'hot hand': statistical reality or cognitive illusion?" *Chance*, 2, No. 4 (1989), pp. 31–34.

21. Information from the College Board found online at `www.collegeboard.com`.

22. Commissioners Standard Ordinary Task Force final report, 2002, American Academy of Actuaries, found online at `www.actuary.org/life/cso_0702.htm`.

23. Means and standard deviations from Fidelity Investments, `www.fidelity.com`. Correlations from the *Fidelity Insight* newsletter, `www.fidelityinsight.com`. The correlations concern an unspecified period, and other online sources give different correlations, so these should be regarded as approximate at best.

24. Table 4.1 closely follows the grade distributions for these three schools at the University of New Hampshire for the fall of 2000, found in a self-study document at `www.unh.edu/neasc`. The counts of grades mirror the proportions of UNH students in these schools. The table is simplified to refer to a university with only these three schools.

25. Information about Internet users comes from sample surveys carried out by the Pew Internet and American Life Project, found online at `www.pewinternet.org`. The music downloading data were collected in 2003.

26. These probabilities come from studies by the sociologist Harry Edwards, reported in the *New York Times*, February 25, 1986.

27. See Note 25.

28. From the 2003 *Statistical Abstract of the United States*, Table 491.

29. Projections from the 2002 *Digest of Education Statistics*, found online at `nces.ed.gov`.

30. From the Bureau of Labor Statistics, found online at `www.bls.gov/data`.

31. See Note 29.

32. See Note 25.

CHAPTER 5 | Notes

1. S. A. Rahimtoola, "Outcomes 15 years after valve replacement with a mechanical vs. a prosthetic valve, final report of the Veterans Administration randomized trial," American College of Cardiology, found online at `www.acc.org/education/online/trials/acc2000/15yr.htm`.

2. John Schwartz, "Leisure pursuits of today's young men," *New York Times*, March 29, 2004. The source cited is comScore Media Matrix.

3. Lydia Saad, "Small families are most Americans' ideal," Gallup poll press release, March 29, 2004, found online at `www.gallup.com`.

4. Jeffrey M. Jones, "Gambling a common activity for Americans," Gallup poll press release, March 24, 2004, found online at `www.gallup.com`.

5. Henry Wechsler et al., *Binge Drinking on America's Colleges Campuses*, Harvard School of Public Health, 2001.

6. Haipeng Shen, "Nonparametric regression for problems involving lognormal distributions," PhD thesis, University of Pennsylvania, 2003. Thanks to Haipeng Shen and Larry Brown for sharing the data.

7. Statistical methods for dealing with time-to-failure data, including the Weibull model, are presented in Wayne Nelson, *Applied Life Data Analysis*, Wiley, 1982.

8. From the Web site of the North Carolina State University Department of Registration and Records, `www.ncsu.edu/class/grades`.

9. Values based on Sherri A. Buzinski, "The effect of position of methylation on the performance properties of durable press treated fabrics," CSR490 honors paper, Purdue University, 1985.

10. Values based on David Hon-Kuen Chu, "A test of corporate advertising using the elaboration likelihood model," MS thesis, Purdue University, 1993.

CHAPTER 6 | Notes

1. See Noel Cressie, *Statistics for Spatial Data*, Wiley, 1993. The significance test result that we report is one of several that could be used to address this question. See pp. 607–609 of the Cressie book for more details.

2. Based on information reported in Sandy Baum and Marie O'Malley, "College on credit: how borrowers perceive their education debt," Nellie Mae Corporation, 2003, found online at www.nelliemae.com/library/research_10.html.

3. The standard reference here is Bradley Efron and Robert J. Tibshirani, *An Introduction to the Bootstrap*, Chapman Hall, 1993. A less technical overview is in Bradley Efron and Robert J. Tibshirani, "Statistical data analysis in the computer age," *Science*, 253 (1991), pp. 390–395.

4. C. M. Weaver et al., "Quantification of biochemical markers of bone turnover by kinetic measures of bone formation and resorption in young healthy females," *Journal of Bone and Mineral Research*, 12 (1997), pp. 1714–1720.

5. *Dietary Reference Intakes for Energy, Carbohydrate, Fiber, Fat, Fatty Acids, Cholesterol, Protein, and Amino Acids (Macronutrients)*, Food and Nutrition Board, Institute of Medicine, 2002.

6. The vehicle is a 1997 Pontiac Transport Van owned by the second author.

7. Some analyses of the data collected by the Gambian National Inpregnated Bednet Program are reported in M. C. Thompson et al., "A spatial model of malaria risk in the Gambia: predicting the impact of insecticide treated and untreated bednets on malaria infection," *Proceedings of the Spatial Information Research Centre's 10th Colloquium*, University of Otago, New Zealand, 1998, pp. 321–328.

8. These data were collected in connection with the Purdue Police Alcohol Student Awareness Program run by Police Officer D. A. Larson.

9. National Center for Educational Statistics, *Snapshot Report for the District of Columbia*, 2003.

10. National Center for Educational Statistics, *Snapshot Report for Boston*, 2003.

11. Data provided by Diana Schellenberg, Purdue University School of Health Sciences.

12. R. A. Fisher, "The arrangement of field experiments," *Journal of the Ministry of Agriculture of Great Britain*, 33 (1926), p. 504, quoted in Leonard J. Savage, "On rereading R. A. Fisher," *Annals of Statistics*, 4 (1976), p. 471. Fisher's work is described in a biography by his daughter: Joan Fisher Box, *R. A. Fisher: The Life of a Scientist*, Wiley, 1978.

13. The editorial was written by Phil Anderson. See *British Medical Journal*, 328 (2004), pp. 476–477. A letter to the editor on this topic by Doug Altman and J. Martin Bland appeared shortly after. See "Confidence intervals illuminate absence of evidence," *British Medical Journal*, 328 (2004), pp. 1016–1017.

14. A. Kamali et al., "Syndromic management of sexually-transmitted infections and behavior change interventions on transmission of HIV-1 in rural Uganda: a community randomised trial," *Lancet*, 361 (2003), pp. 645–652.

15. T. D. Sterling, "Publication decisions and their possible effects on inferences drawn from tests of significance—or vice versa," *Journal of the American Statistical Association*, 54 (1959), pp. 30–34. Related comments appear in J. K. Skipper, A. L. Guenther, and G. Nass, "The sacredness of 0.05: a note concerning the uses of statistical levels of significance in social science," *American Sociologist*, 1 (1967), pp. 16–18.

16. For a good overview of these issues see the Editor's Invited Article, "Gene expresssion data: the technology and statistical analysis," by Bruce A. Craig, Michael A. Black, and Rebecca W. Doerge, in *Journal of Agricultural, Biological, and Environmental Statistics*, 8 (2003), pp. 1–28.

17. Robert J. Schiller, "The volatility of stock market prices," *Science*, 235 (1987), pp. 33–36.

18. Sabrina Oesterle et al., "Volunteerism during the transition to adulthood: a life course perspective," *Social Forces*, 83 (2004), pp. 1123–1149.

19. Based on A. M. Garcia et al., "Why do workers behave unsafely at work? Determinants of safe work practices in industrial workers," *Occupational and Environmental Medicine*, 61 (2004), pp. 239–246.

20. Data from Joan M. Susic, "Dietary phosphorus intakes, urinary and peritoneal phosphate excretion and clearance in continuous ambulatory peritoneal dialysis patients," MS thesis, Purdue University, 1985.

21. From a study by Mugdha Gore and Joseph Thomas, Purdue University School of Pharmacy.

22. Barbara A. Almanza, Richard Ghiselli, and William Jaffe, "Foodservice design and aging baby boomers: importance and perception of physical amenities in restaurants," *Foodservice Research International*, 12 (2000), pp. 25–40.

CHAPTER 7 | Notes

1. These data are from "Results report on the vitamin C pilot program," prepared by SUSTAIN (Sharing United States Technology to Aid in the Improvement of Nutrition) for the U.S. Agency for International Development. The report was used by the Committee on International Nutrition of the National Academy of Sciences/Institute of Medicine to make recommendations on whether or not the vitamin C content of food commodities used in U.S. food aid programs should be increased. The program was directed by Peter Ranum and Françoise Chomé.

2. C. Don Wiggins, "The legal perils of 'underdiversification'—a case study," *Personal Financial Planning*, 1, No. 6 (1999), pp. 16–18.

3. These data were collected as part of a larger study of dementia patients conducted by Nancy Edwards, School of Nursing, and Alan Beck, School of Veterinary Medicine, Purdue University.

4. These recommendations are based on extensive computer work. See, for example, Harry O. Posten, "The robustness of the one-sample *t*-test over the Pearson system," *Journal of Statistical Computation and Simulation*, 9 (1979), pp. 133–149; and E. S. Pearson and N. W. Please, "Relation between the shape of population distribution and the robustness of four simple test statistics," *Biometrika*, 62 (1975), pp. 223–241.

5. The data and the use of the log transformation come from Thomas J. Lorenzen, "Determining statistical characteristics of a vehicle emissions audit procedure," *Technometrics*, 22 (1980), pp. 483–493.

6. You can find a practical discussion of distribution-free inference in Myles Hollander and Douglas A. Wolfe, *Nonparametric Statistical Methods*, 2nd ed., Wiley, 1999.

7. Data provided by Francisco Rosales of the Department of Nutritional Sciences, Penn State University. See Rosales et al., "Relation of serum retinol to acute phase proteins and malarial morbidity in Papua New Guinea children," *American Journal of Clinical Nutrition*, 71 (2000), pp. 1580–1588.

8. Data provided by Betsy Hoza of the Department of Psychological Sciences, Purdue University.

9. The vehicle is a 1997 Pontiac Transport Van owned by the second author.

10. David R. Pillow et al., "Confirmatory factor analysis examining attention deficit hyperactivity disorder symptoms and other childhood disruptive behaviors," *Journal of Abnormal Child Psychology,* 26 (1998), pp. 293–309.

11. James A. Levine, Norman L. Eberhardt, and Michael D. Jensen, "Role of nonexercise activity thermogenesis in resistance to fat gain in humans," *Science,* 283 (1999), pp. 212–214. Data for this study are available from the *Science* Web site, `http://www.sciencemag.org`.

12. These data were collected in connection with a bone health study at Purdue University and were provided by Linda McCabe.

13. Data from F. H. Rauscher et al., "Music training causes long-term enhancement of preschool children's spatial-temporal reasoning," *Neurological Research,* 19 (1997), pp. 2–8.

14. Data from Joan M. Susic, "Dietary phosphorus intakes, urinary and peritoneal phosphate excretion and clearance in continuous ambulatory peritoneal dialysis patients," MS thesis, Purdue University, 1985.

15. Data provided by Diana Schellenberg, Purdue University School of Health Sciences.

16. Data provided by Joseph A. Wipf, Department of Foreign Languages and Literatures, Purdue University.

17. Data from Wayne Nelson, *Applied Life Data Analysis,* Wiley, 1982, p. 471.

18. Detailed information about the conservative t procedures can be found in Paul Leaverton and John J. Birch, "Small sample power curves for the two sample location problem," *Technometrics,* 11 (1969), pp. 299–307; in Henry Scheffé, "Practical solutions of the Behrens-Fisher problem," *Journal of the American Statistical Association,* 65 (1970), pp. 1501–1508; and in D. J. Best and J. C. W. Rayner, "Welch's approximate solution for the Behrens-Fisher problem," *Technometrics,* 29 (1987), pp. 205–210.

19. This example is adapted from Maribeth C. Schmitt, "The effects of an elaborated directed reading activity on the metacomprehension skills of third graders," PhD dissertation, Purdue University, 1987.

20. See the extensive simulation studies in Harry O. Posten, "The robustness of the two-sample t test over the Pearson system," *Journal of Statistical Computation and Simulation,* 6 (1978), pp. 295–311.

21. Based on R. G. Hood, "Results of the prices received by farmers for grain—quality assurance project," U.S. Department of Agriculture Report SRB–95–07, 1995.

22. This study is reported in Roseann M. Lyle et al., "Blood pressure and metabolic effects of calcium supplementation in normotensive white and black men," *Journal of the American Medical Association,* 257 (1987), pp. 1772–1776. The individual measurements in Table 7.4 were provided by Dr. Lyle.

23. M. K. Campbell et al., "A tailored multimedia nutrition education pilot program for low-income women receiving food assistance," *Health Education Research,* 14 (1999), pp. 257–267.

24. B. Bakke et al., "Cumulative exposure to dust and gases as determinants of lung function decline in tunnel construction workers," *Occupational Environmental Medicine,* 61 (2004), pp. 262–269.

25. Samara Joy Nielsen and Barry M. Popkin, "Patterns and trends in food portion sizes, 1977–1998," *Journal of the American Medical Association,* 289 (2003), pp. 450–453.

26. Gordana Mrdjenovic and David A. Levitsky, "Nutritional and energetic consequences of sweetened drink consumption in 6- to 13-year old children," *Journal of Pediatrics*, 142 (2003), pp. 604–610.

27. The data were provided by Helen Park. See Helen Park et al., "Fortifying bread with each of three antioxidants," *Cereal Chemistry*, 74 (1997), pp. 202–206.

28. David Han-Kuen Chu, "A test of corporate advertising using the elaboration likelihood model," MS thesis, Purdue University, 1993.

29. See Note 13.

30. Based on C. Papoulias and P. Theodossiou, "Analysis and modeling of recent business failures in Greece," *Managerial and Decision Economics*, 13 (1992), pp. 163–169.

31. Data from a study conducted at the Medical University of South Carolina in 1989.

32. M. F. Picciano and R. H. Deering, "The influence of feeding regimens on iron status during infancy," *American Journal of Clinical Nutrition*, 33 (1980), pp. 746–753.

33. The problem of comparing spreads is difficult even with advanced methods. Common distribution-free procedures do not offer a satisfactory alternative to the *F* test, because they are sensitive to unequal shapes when comparing two distributions. A good introduction to the available methods is W. J. Conover, M. E. Johnson, and M. M. Johnson, "A comparative study of tests for homogeneity of variances, with applications to outer continental shelf bidding data," *Technometrics*, 23 (1981), pp. 351–361. Modern resampling procedures often work well. See Dennis D. Boos and Colin Brownie, "Bootstrap methods for testing homogeneity of variances," *Technometrics*, 31 (1989), pp. 69–82.

34. G. E. P. Box, "Non-normality and tests on variances," *Biometrika*, 40 (1953), pp. 318–335. The quote appears on page 333.

35. Based on Loren Cordain et al., "Influence of moderate daily wine consumption on body weight regulation and metabolism in healthy free-living males," *Journal of the American College of Nutrition*, 16 (1997), pp. 134–139.

36. C. M. Weaver et al., "Quantification of biochemical markers of bone turnover by kinetic measures of bone formation and resorption in young healthy females," *Journal of Bone and Mineral Research*, 12 (1997), pp. 1714–1720. Data provided by Linda McCabe.

37. This exercise is based on events that are real. The data and details have been altered to protect the privacy of the individuals involved.

38. Based loosely on D. R. Black et al., "Minimal interventions for weight control: a cost-effective alternative," *Addictive Behaviors*, 9 (1984), pp. 279–285.

39. These data were provided by Professor Sebastian Heath, School of Veterinary Medicine, Purdue University.

40. J. W. Marr and J. A. Heady, "Within- and between-person variation in dietary surveys: number of days needed to classify individuals," *Human Nutrition: Applied Nutrition*, 40A (1986), pp. 347–364.

41. Data taken from the Web site www.assist-2-sell.com on July 9, 2001.

CHAPTER 8 | Notes

1. Details of exact binomial procedures can be found in Myles Hollander and Douglas Wolfe, *Nonparametric Statistical Methods*, 2nd ed., Wiley, 1999.

2. Results of this survey are reported in Henry Wechsler et al., "Health and behavioral consequences of binge drinking in college," *Journal of the American Medical Association*, 272 (1994), pp. 1672–1677.

3. See A. Agresti and B. A. Coull, "Approximate is better than 'exact' for interval estimation of binomial proportions," *The American Statistician*, 52 (1998), pp. 119–126. A detailed theoretical study is Lawrence D. Brown, Tony Cai, and Anirban DasGupta, "Confidence intervals for a binomial proportion and asymptotic expansions," *Annals of Statistics*, 30 (2002), pp. 160–201.

4. See K. D. R. Setchell et al., "The clinical importance of the metabolite equol—a clue to the effectiveness of soy and its isoflavones," *Journal of Nutrition*, 132 (2002), pp. 3577–3584.

5. National Institute for Occupational Safety and Health, *Stress at Work*, 2000, available online at www.cdc.gov/niosh/stresswk.html.

6. Results of this survey were reported in *Restaurant Business*, September 15, 1999, pp. 45–49.

7. See, for example, http://www.pilatesmethodalliance.org/.

8. Based on information in "NCAA 2003 national study of collegiate sports wagering and associated health risks," which can be found at the NCAA Web site, http://www.ncaa.org.

9. From the "National Survey of Student Engagement, the college student report," 2003, available online at http://www.indiana.edu/~nsse/.

10. James Otto et al., "Training rats to search and alert on contraband odors," *Applied Animal Behaviour Science,* 77 (2002), pp. 271–232.

11. Information about the survey can be found online at http://saint-denis.library.arizona.edu/natcong/.

12. Based on information reported in Sandy Baum and Marie O'Malley, "College on credit: how borrowers perceive their education debt," Nellie Mae Corporation, 2003, found online at www.nelliemae.com/library/research_10.html.

13. From Heeseung Roh Ryu, Roseann M. Lyle, and George P. McCabe, "Factors associated with weight concerns and unhealthy eating patterns among young Korean females," *Eating Disorders*, 11 (2003), pp. 129–141.

14. Data from Roland J. Thorpe Jr. and based on his analysis of "Health ABC," a 10-year longitudinal study of older adults supported by the Laboratory of Epidemiology, Demography, and Biometry of the National Institute on Aging. Additional analyses are given in his PhD dissertation, "Relationship between pet ownership, physical activity, and human health in an elderly population," Purdue University, 2004.

15. Data from Guohua Li and Susan P. Baker, "Alcohol in fatally injured bicyclists," *Accident Analysis and Prevention*, 26 (1994), pp. 543–548.

16. See Alan Agresti and Brian Caffo, "Simple and effective confidence intervals for proportions and differences of proportions result from adding two successes and two failures," *The American Statistician*, 45 (2000), pp. 280–288. The plus four interval is a bit conservative (true coverage probability is higher than the confidence level) when p_1 and p_2 are equal and close to 0 or 1, but the traditional interval is much less accurate and has the fatal flaw that the true coverage probability is *less* than the confidence level.

17. J. M. Tanner, "Physical growth and development," in J. O. Forfar and G. C. Arneil, *Textbook of Paediatrics,* 3rd ed., Churchill Livingston, 1984, pp. 1–292.

18. From M. L. Burr et al., "Effects on respiratory health of a reduction in air pollution from vehicle exhaust emissions," *Occupational and Environmental Medicine,* 61 (2004), pp. 212–218.

19. See Note 14.

20. Lee Rainie et al., "The state of music downloading and file-sharing online," *PEW Internet Project and Comscore Media Metrix Data Memo*, 2004. This memo is available online at `http://www.pewinternet.org/`.

21. Based on Greg Clinch, "Employee compensation and firms' research and development activity," *Journal of Accounting Research*, 29 (1991), pp. 59–78.

22. Sara J. Solnick and David Hemenway, "Complaints and disenrollment at a health maintenance organization," *Journal of Consumer Affairs*, 26 (1992), pp. 90–103.

23. From Monica Macaulay and Colleen Brice, "Don't touch my projectile: gender bias and stereotyping in syntactic examples," *Language*, 73, No. 4 (1997), pp. 798–825. The first part of the title is a direct quote from one of the texts.

24. See S. W. Lagakos, B. J. Wessen, and M. Zelen, "An analysis of contaminated well water and health effects in Woburn, Massachusetts," *Journal of the American Statistical Association*, 81 (1986), pp. 583–596, and the following discussion. This case is the basis for the book and movie *A Civil Action*.

25. See Note 20.

26. From Charles T. Clotfelter et al., "State lotteries at the turn of the century: report to the National Gambling Impact Study Commission," 1999.

27. Marsha A. Dickson, "Utility of no sweat labels for apparel customers: profiling label users and predicting their purchases," *Journal of Consumer Affairs*, 35 (2001), pp. 96–119.

28. From Karin Weber and Weley S. Roehl, "Profiling people searching for and purchasing travel products on the World Wide Web," *Journal of Travel Research*, 37 (1999), pp. 291–298.

29. Based on Robert T. Driescher, "A quality swing with Ping," *Quality Progress*, August 2001, pp. 37–41.

30. Based on a Gallup poll conducted February 16–19, 2004.

31. From Dennis N. Bristow and Richard J. Sebastian, "Holy cow! Wait till next year! A closer look at the brand loyalty of Chicago Cubs baseball fans," *Journal of Consumer Marketing*, 18 (2001), pp. 256–275.

32. Based on Deborah Roedder John and Ramnath Lakshmi-Ratan, "Age differences in children's choice behavior: the impact of available alternatives," *Journal of Marketing Research*, 29 (1992), pp. 216–226.

33. This case is discussed in D. H. Kaye and M. Aickin (eds.), *Statistical Methods in Discrimination Litigation*, Marcel Dekker, 1986; and D. C. Baldus and J. W. L. Cole, *Statistical Proof of Discrimination*, McGraw-Hill, 1980.

34. See Note 12.

CHAPTER 9 | Notes

1. You can find a clear and comprehensive discussion of numerical measures of association for categorical data in Chapter 3 of A. M. Liebetrau, *Measures of Association*, Sage Publications, 1983.

2. These data, from reports submitted by airlines to the Department of Transportation, appear in A. Barnett, "How numbers can trick you," *Technology Review*, October 1994, pp. 38–45.

3. National Science Board, *Science and Engineering Indicators, 1991*, U.S. Government Printing Office, 1991. The detailed data appear in Appendix Table 3-5, p. 274.

4. From P. Azoulay and S. Shane, "Entrepreneurs, contracts, and the failure of young firms," *Management Science*, 47 (2001), pp. 337–358.

5. When the expected cell counts are small, it is best to use a test based on the exact distribution rather than the chi-square approximation, particularly for 2 × 2 tables. Many statistical software systems offer an "exact" test as well as the chi-square test for 2 × 2 tables.

6. The full report of the study appeared in George H. Beaton et al., "Effectiveness of vitamin A supplementation in the control of young child morbidity and mortality in developing countries," United Nations ACC/SCN State-of-the-Art Series, Nutrition Policy Discussion Paper No. 13, 1993.

7. C. M. Ryan, C. A. Northrup-Clewes, B. Knox, and D. I. Thurnham, "The effect of in-store music on consumer choice of wine," *Proceedings of the Nutrition Society*, 57 (1998), p. 1069A.

8. An alternative formula that can be used for hand or calculator computations is

$$X^2 = \sum \frac{(\text{observed})^2}{\text{expected}} - n$$

9. See, for example, Alan Agresti, *Categorical Data Analysis*, Wiley, 1990.

10. From D. A. Redelmeier and R. J. Tibshirani, "Association between cellular-telephone calls and motor vehicle collisions," *New England Journal of Medicine*, 336 (1997), pp. 453–458.

11. See Note 9.

12. See the U.S. Bureau of the Census Web site at http://www.census.gov/population/socdemo/school/ for these and similar data.

13. See Note 12.

14. From F. D. Blau and M. A. Ferber, "Career plans and expectations of young women and men," *Journal of Human Resources*, 26 (1991), pp. 581–607.

15. Data from D. M. Barnes, "Breaking the cycle of addiction," *Science*, 241 (1988), pp. 1029–1030.

16. Data from National Science Foundation Science Resources Studies Division, *Data Brief*, 12 (1996), p. 1.

17. Based on data in Mike Planty et al., "Volunteer service by young people from high school through early adulthood," National Center for Educational Statistics Report NCES 2004–365.

18. Although these data are fictitious, similar though less simple situations occur. See P. J. Bickel and J. W. O'Connell, "Is there a sex bias in graduate admissions?" *Science*, 187 (1975), pp. 398–404.

19. Tom Reichert, "The prevalence of sexual imagery in ads targeted to young adults," *Journal of Consumer Affairs*, 37 (2003), pp. 403–412.

20. See George R. Milne, "How well do consumers protect themselves from identity theft?" *Journal of Consumer Affairs*, 37 (2003), pp. 388–402.

21. Based on information in "NCAA 2003 national study of collegiate sports wagering and associated health risks," which can be found at the NCAA Web site, http://www.ncaa.org.

22. Christopher Somers et al., "Air pollution induces heritable DNA mutations," *Proceedings of the National Academy of Sciences*, 99 (2002), pp. 15,904–15,907.

23. Ethan J. Temeles and W. John Kress, "Adaption in a plant-hummingbird association," *Science,* 300 (2003), pp. 630–633.

24. Dellavalle et al., "Going, going, gone: lost Internet references," *Science,* 302 (2003), pp. 787–788.

25. Data from Roland J. Thorpe Jr. and based on his analysis of "Health ABC," a 10-year longitudinal study of older adults supported by the Laboratory of Epidemiology, Demography, and Biometry of the National Institute on Aging. Additional analyses are given in his PhD dissertation, "Relationship between pet ownership, physical activity, and human health in an elderly population," 2004, Purdue University.

26. These data are from an Undergraduate Task Force study of student retention in the Purdue University College of Science.

27. William D. Darley, "Store-choice behavior for pre-owned merchandise," *Journal of Business Research,* 27 (1993), pp. 17–31.

28. Daniel Cassens et al., "Face check development in veneered furniture panels," *Forest Products Journal,* 53 (2003), pp. 79–86.

29. Gary J. Patronek et al., "Risk factors for relinquishment of cats to an animal shelter," *Journal of the American Veterinary Medical Association,* 209 (1996), pp. 582–588.

30. Gary J. Patronek et al., "Risk factors for relinquishment of dogs to an animal shelter," *Journal of the American Veterinary Medical Association,* 209 (1996), pp. 572–581.

31. Data from James E. Stafford, "Influence of preliminary contact on mail returns," *Journal of Marketing Research,* 3 (1966), pp. 410–411.

32. These data are from Patricia H. Shiono and Mark A. Klebanoff, "The effect of two mailing strategies on the response to a survey of physicians," *American Journal of Epidemiology,* 134 (1991), pp. 539–542.

33. Data provided by Susan Prohofsky, from her PhD dissertation, "Selection of undergraduate major: the influence of expected costs and expected benefits," Purdue University, 1991.

34. Marsha A. Dickson, "Utility of no sweat labels for apparel customers: profiling label users and predicting their purchases," *Journal of Consumer Affairs,* 35 (2001), pp. 96–119.

35. See James A. Taylor et al., "Efficacy and safety of echinacea in treating upper respiratory tract infections in children," *Journal of the American Medical Association,* 290 (2003), pp. 2824–2830.

36. Karen L. Dean, "Herbalists respond to *JAMA* echinacea study," *Alternative & Complementary Therapies,* 10 (2004), pp. 11–12.

CHAPTER 10 | Notes

1. The vehicle is a 1997 Pontiac Transport Van owned by the second author.

2. The data were provided by Linda McCabe and were collected as part of a large study of women's bone health and another study of calcium kinetics, both directed by Professor Connie Weaver of the Department of Foods and Nutrition, Purdue University. Results related to this example are reported in Dorothy Teegarden et al., "Peak bone mass in young women," *Journal of Bone and Mineral Research,* 10 (1995), pp. 711–715.

3. These data were provided by Professor Wayne Campbell of the Purdue University Department of Foods and Nutrition.

4. This quantity is the estimated standard deviation of $\hat{y} - y$, not the estimated standard deviation of $\hat{y}$ alone.

5. For more information about nutrient requirements, see the Institute of Medicine publications on Dietary Reference Intakes available at www.nap.edu.

6. The method is described in Chapter 15 of J. Neter et al., *Applied Linear Statistical Models*, 4th ed., Irwin, 1996.

7. Based on Dan Dauwakter's master's thesis in the Department of Forestry and Natural Resources at Purdue University. More information is available in Daniel C. Dauwalter et al., "An index of biotic integrity for fish assemblages in Ozark Highland streams of Arkansas," *Southeastern Naturalist*, 2 (2003), pp. 447–468. These data were provided by Emmanuel Frimpong.

8. Net cash flow data from Sean Collins, *Mutual Fund Assets and Flows in 2000*, Investment Company Institute, 2001. Found online at www.ici.org. The raw data were converted to real dollars using annual average values of the Consumer Price Index.

9. Alan S. Banks et al., "Juvenile hallux abducto valgus association with metatarsus adductus," *Journal of the American Podiatric Medical Association*, 84 (1994), pp. 219–224.

10. The data were provided by Professor Shelly MacDermid, Department of Child Development and Family Studies, Purdue University, from a study reported in S. M. MacDermid et al., "Is small beautiful? Work-family tension, work conditions, and organizational size," *Family Relations*, 44 (1994), pp. 159–167.

11. These data are from G. Geri and B. Palla, "Considerazioni sulle più recenti osservazioni ottiche alla Torre Pendente di Pisa," *Estratto dal Bollettino della Società Italiana di Topografia e Fotogrammetria*, 2 (1988), pp. 121–135. Professor Julia Mortera of the University of Rome provided valuable assistance with the translation.

12. Data for the first four years from Manuel Castells, *The Rise of the Network Society*, 2nd ed., Blackwell, 2000, p. 41.

13. This study is reported in S. Lau and P. C. Cheung, "Relations between Chinese adolescents' perception of parental control and organization and their perception of parental warmth," *Developmental Psychology*, 23 (1987), pp. 726–729.

14. Donald A. Mahler and John I. Mackowiak, "Evaluation of the short-form 36-item questionnaire to measure health-related quality of life in patients with COLD," *Chest*, 107 (1995), pp. 1585–1589.

15. Data on a sample of 12 of 56 perch in a data set contributed to the *Journal of Statistics Education* data archive www.amstat.org/publications/jse/ by Juha Puranen of the University of Helsinki.

16. L. Cooke et al., "Relationship between parental report of food neophobia and everyday food consumption in 2–6-year-old children," *Appetite*, 41 (2003), pp. 205–206.

17. D. E. Powers and J. K. Kaufman, "Do standardized multiple choice tests penalize deep-thinking or creative students?" Educational Testing Service Research Report RR–02–15 (2002).

CHAPTER 11 | Notes

1. Results of the study are reported in P. F. Campbell and G. P. McCabe, "Predicting the success of freshmen in a computer science major," *Communications of the ACM*, 27 (1984), pp. 1108–1113.

2. Tom Reichert, "The prevalence of sexual imagery in ads targeted to young adults," *Journal of Consumer Affairs*, 37 (2003), pp. 403–412.

3. Miguel A. Diego et al., "Academic performance, popularity, and depression predict adolescent substance abuse," *Adolescence*, 38 (2003), pp. 35–42.

4. Wanki Moon and Siva K. Balasubramanian, "Willingness to pay for non-biotech foods in the U.S. and the U.K.," *Journal of Consumer Affairs,* 37 (2003), pp. 317–339. The authors used a maximum likelihood method to estimate the parameters of their model, but the methods described in this chapter are appropriate for the interpretation of the results given in this exercise.

5. Steven R. Wininger and David Pargman, "Assessment of factors associated with exercise enjoyment," *Journal of Music Therapy,* 40 (2003), pp. 57–73.

6. This study is reported in Scot Burton et al., "Implications of accurate usage of nutrition facts panel information for food product evaluations and purchase intentions," *Journal of the Academy of Marketing Science,* 27 (1999), pp. 470–480.

7. Alan Levinsohn, "Online brokerage, the new core account?" *ABA Banking Journal,* September 1999, pp. 34–42.

8. Michael E. Staten et al., "Information costs and the organization of credit markets: a theory of indirect lending," *Economic Inquiry,* 28 (1990), pp. 508–529.

9. A description of this case as well as other examples of the use of statistics in legal settings is given in Michael O. Finkelstein, *Quantitative Methods in Law,* Free Press, 1978.

CHAPTER 12 | Notes

1. Based on A. M. Garcia et al., "Why do workers behave unsafely at work? Determinants of safe work practices in industrial workers," *Occupational and Environmental Medicine,* 61 (2004), pp. 239–246.

2. This rule is intended to provide a general guideline for deciding when serious errors may result by applying ANOVA procedures. When the sample sizes in each group are very small, the sample variances will tend to vary much more than when the sample sizes are large. In this case, the rule may be a little too conservative. For unequal sample sizes, particular difficulties can arise when a relatively small sample size is associated with a population having a relatively large standard deviation. Careful judgment is needed in all cases. By considering P-values rather than fixed level α testing, judgments in ambiguous cases can more easily be made; for example, if the P-value is very small, say 0.001, then it is probably safe to reject H_0 even if there is a fair amount of variation in the sample standard deviations.

3. This example is based on Amna Kirmani and Peter Wright, "Money talks: perceived advertising expense and expected product quality," *Journal of Consumer Research,* 16 (1989), pp. 344–353.

4. Several different definitions for the noncentrality parameter of the noncentral F distribution are in use. When $I = 2$, the λ defined here is equal to the square of the noncentrality parameter δ that we used for the two-sample t test in Chapter 7. Many authors prefer $\phi = \sqrt{\lambda/I}$. We have chosen to use λ because it is the form needed for the SAS function PROBF.

5. P. Bartel et al., "Attention and working memory in resident anaesthetists after night duty: group and individual effects," *Occupational and Environmental Medicine,* 61 (2004), pp. 167–170.

6. Based on Jack K. Trammel, "The impact of academic accommodations on final grades in a post secondary setting," *Journal of College Reading and Learning,* 34 (2003), pp. 76–90.

7. Christie N. Scollon et al., "Emotions across cultures and methods," *Journal of Cross-cultural Psychology,* 35 (2004), pp. 304–326.

8. These data were provided by Helen Park; see H. Park et al., "Fortifying bread with each of three antioxidants," *Cereal Chemistry,* 74 (1997), pp. 202–206.

9. Michael Wayne Peugh, "Field investigation of ventilation and air quality in duck and turkey slaughter plants," MS thesis, Purdue University, 1996.

10. We thank Ethan J. Temeles of Amherst College for providing the data. His work is described in Ethan J. Temeles and W. John Kress, "Adaptation in a plant-hummingbird association," *Science*, 300 (2003), pp. 630–633.

11. The data were provided by James Kaufman. The study is described in James C. Kaufman, "The cost of the muse: poets die young," *Death Studies*, 27 (2003), pp. 813–821. The quote from Yeats appears in this article.

12. The experiment was performed in Connie Weaver's lab in the Purdue University Department of Foods and Nutrition. The data were provided by Berdine Martin and Yong Jiang.

13. Based on M. U. Kalwani and C. K. Yim, "Consumer price and promotion expectations: an experimental study," *Journal of Marketing Research*, 29 (1992), pp. 90–100.

14. Based on a study by J. E. Stuff and B. L. Nichols reported in Chelsea Lutter, "Recommended length of exclusive breast-feeding, age of introduction of complementary foods and the weaning dilemma," World Health Organization, 1992.

15. Data provided by Jo Welch of the Purdue University Department of Foods and Nutrition.

16. Based on A. A. Adish et al., "Effect of consumption of food cooked in iron pots on iron status and growth of young children: a randomised trial," *The Lancet*, 353 (1999), pp. 712–716.

17. Steve Badylak et al., "Marrow-derived cells populate scaffolds composed of xenogeneic extracellular matrix," *Experimental Hematology*, 29 (2001), pp. 1310–1318.

18. This exercise is based on data provided from a study conducted by Jim Baumann and Leah Jones of the Purdue University School of Education.

CHAPTER 13 | Notes

1. See http://www.malaria.org/ for more information about malaria.

2. This example is based on the work of David I. Thurnham of the Northern Ireland Center for Diet and Health and Delana A. Adelekan of Obafemi Awolowo University in Ile-Ife, Nigeria.

3. We present the two-way ANOVA model and analysis for the general case in which the sample sizes may be unequal. If the sample sizes vary a great deal, serious complications can arise. There is no longer a single standard ANOVA analysis. Most computer packages offer several options for the computation of the ANOVA table when cell counts are unequal. When the counts are approximately equal, all methods give essentially the same results.

4. Samara Joy Nielsen and Barry M. Popkin, "Patterns and trends in food portion sizes, 1977–1998," *JAMA*, 289 (2003), pp. 450–453.

5. The summaries in Examples 13.7 and 13.8 are from Rick Bell and Patricia L. Pliner, "Time to eat: the relationship between the number of people eating and meal duration in three lunch settings," *Appetite*, 41 (2003), pp. 215–218.

6. Karolyn Drake and Jamel Ben El Hine, "Synchronizing with music: intercultural differences," *Annals of the New York Academy of Sciences*, 99 (2003), pp. 429–437.

7. Example 13.11 is based on a study described in P. D. Wood et al., "Plasma lipoprotein distributions in male and female runners," in P. Milvey (ed.), *The Marathon: Physiological, Medical, Epidemiological, and Psychological Studies*, New York Academy of Sciences, 1977.

8. Judith McFarlane et al., "An intervention to increase safety behaviors of abused women," *Nursing Research*, 51 (2002), pp. 347–354.

9. Margaret C. Campbell and Kevin Lane Keller, "Brand familiarity and advertising repetition effects," *Journal of Consumer Research*, 30 (2003), pp. 292–304.

10. Data provided by Julie Hendricks and V. J. K. Liu of the Department of Foods and Nutrition, Purdue University.

11. Results of the study are reported in P. F. Campbell and G. P. McCabe, "Predicting the success of freshmen in a computer science major," *Communications of the ACM*, 27 (1984), pp. 1108–1113.

12. See S. Badylak et al., "Marrow-derived cells populate scaffolds composed of xenogeneic extracellular matrix," *Experimental Hematology*, 29 (2001), pp. 1310–1318.

13. Based on A. A. Adish et al., "Effect of consumption of food cooked in iron pots on iron status and growth of young children: a randomised trial," *Lancet*, 353 (1999), pp. 712–716.

14. Based on a problem from Renée A. Jones and Regina P. Becker, Department of Statistics, Purdue University.

15. Based on M. U. Kalwani and C. K. Yim, "Consumer price and promotion expectations: an experimental study," *Journal of Marketing Research*, 29 (1992), pp. 90–100.

16. For a summary of this study and other research in this area, see Stanley Coren and Diane F. Halpern, "Left-handedness: a marker for decreased survival fitness," *Psychological Bulletin*, 109 (1991), pp. 90–106.

17. Data provided by Neil Zimmerman of the Purdue University School of Health Sciences.

DATA TABLE INDEX

BEYOND THE BASICS INDEX

TABLE B   Random digits

Line								
101	19223	95034	05756	28713	96409	12531	42544	82853
102	73676	47150	99400	01927	27754	42648	82425	36290
103	45467	71709	77558	00095	32863	29485	82226	90056
104	52711	38889	93074	60227	40011	85848	48767	52573
105	95592	94007	69971	91481	60779	53791	17297	59335
106	68417	35013	15529	72765	85089	57067	50211	47487
107	82739	57890	20807	47511	81676	55300	94383	14893
108	60940	72024	17868	24943	61790	90656	87964	18883
109	36009	19365	15412	39638	85453	46816	83485	41979
110	38448	48789	18338	24697	39364	42006	76688	08708
111	81486	69487	60513	09297	00412	71238	27649	39950
112	59636	88804	04634	71197	19352	73089	84898	45785
113	62568	70206	40325	03699	71080	22553	11486	11776
114	45149	32992	75730	66280	03819	56202	02938	70915
115	61041	77684	94322	24709	73698	14526	31893	32592
116	14459	26056	31424	80371	65103	62253	50490	61181
117	38167	98532	62183	70632	23417	26185	41448	75532
118	73190	32533	04470	29669	84407	90785	65956	86382
119	95857	07118	87664	92099	58806	66979	98624	84826
120	35476	55972	39421	65850	04266	35435	43742	11937
121	71487	09984	29077	14863	61683	47052	62224	51025
122	13873	81598	95052	90908	73592	75186	87136	95761
123	54580	81507	27102	56027	55892	33063	41842	81868
124	71035	09001	43367	49497	72719	96758	27611	91596
125	96746	12149	37823	71868	18442	35119	62103	39244
126	96927	19931	36089	74192	77567	88741	48409	41903
127	43909	99477	25330	64359	40085	16925	85117	36071
128	15689	14227	06565	14374	13352	49367	81982	87209
129	36759	58984	68288	22913	18638	54303	00795	08727
130	69051	64817	87174	09517	84534	06489	87201	97245
131	05007	16632	81194	14873	04197	85576	45195	96565
132	68732	55259	84292	08796	43165	93739	31685	97150
133	45740	41807	65561	33302	07051	93623	18132	09547
134	27816	78416	18329	21337	35213	37741	04312	68508
135	66925	55658	39100	78458	11206	19876	87151	31260
136	08421	44753	77377	28744	75592	08563	79140	92454
137	53645	66812	61421	47836	12609	15373	98481	14592
138	66831	68908	40772	21558	47781	33586	79177	06928
139	55588	99404	70708	41098	43563	56934	48394	51719
140	12975	13258	13048	45144	72321	81940	00360	02428
141	96767	35964	23822	96012	94591	65194	50842	53372
142	72829	50232	97892	63408	77919	44575	24870	04178
143	88565	42628	17797	49376	61762	16953	88604	12724
144	62964	88145	83083	69453	46109	59505	69680	00900
145	19687	12633	57857	95806	09931	02150	43163	58636
146	37609	59057	66967	83401	60705	02384	90597	93600
147	54973	86278	88737	74351	47500	84552	19909	67181
148	00694	05977	19664	65441	20903	62371	22725	53340
149	71546	05233	53946	68743	72460	27601	45403	88692
150	07511	88915	41267	16853	84569	79367	32337	03316

Line

151	03802	29341	29264	80198	12371	13121	54969	43912
152	77320	35030	77519	41109	98296	18984	60869	12349
153	07886	56866	39648	69290	03600	05376	58958	22720
154	87065	74133	21117	70595	22791	67306	28420	52067
155	42090	09628	54035	93879	98441	04606	27381	82637
156	55494	67690	88131	81800	11188	28552	25752	21953
157	16698	30406	96587	65985	07165	50148	16201	86792
158	16297	07626	68683	45335	34377	72941	41764	77038
159	22897	17467	17638	70043	36243	13008	83993	22869
160	98163	45944	34210	64158	76971	27689	82926	75957
161	43400	25831	06283	22138	16043	15706	73345	26238
162	97341	46254	88153	62336	21112	35574	99271	45297
163	64578	67197	28310	90341	37531	63890	52630	76315
164	11022	79124	49525	63078	17229	32165	01343	21394
165	81232	43939	23840	05995	84589	06788	76358	26622
166	36843	84798	51167	44728	20554	55538	27647	32708
167	84329	80081	69516	78934	14293	92478	16479	26974
168	27788	85789	41592	74472	96773	27090	24954	41474
169	99224	00850	43737	75202	44753	63236	14260	73686
170	38075	73239	52555	46342	13365	02182	30443	53229
171	87368	49451	55771	48343	51236	18522	73670	23212
172	40512	00681	44282	47178	08139	78693	34715	75606
173	81636	57578	54286	27216	58758	80358	84115	84568
174	26411	94292	06340	97762	37033	85968	94165	46514
175	80011	09937	57195	33906	94831	10056	42211	65491
176	92813	87503	63494	71379	76550	45984	05481	50830
177	70348	72871	63419	57363	29685	43090	18763	31714
178	24005	52114	26224	39078	80798	15220	43186	00976
179	85063	55810	10470	08029	30025	29734	61181	72090
180	11532	73186	92541	06915	72954	10167	12142	26492
181	59618	03914	05208	84088	20426	39004	84582	87317
182	92965	50837	39921	84661	82514	81899	24565	60874
183	85116	27684	14597	85747	01596	25889	41998	15635
184	15106	10411	90221	49377	44369	28185	80959	76355
185	03638	31589	07871	25792	85823	55400	56026	12193
186	97971	48932	45792	63993	95635	28753	46069	84635
187	49345	18305	76213	82390	77412	97401	50650	71755
188	87370	88099	89695	87633	76987	85503	26257	51736
189	88296	95670	74932	65317	93848	43988	47597	83044
190	79485	92200	99401	54473	34336	82786	05457	60343
191	40830	24979	23333	37619	56227	95941	59494	86539
192	32006	76302	81221	00693	95197	75044	46596	11628
193	37569	85187	44692	50706	53161	69027	88389	60313
194	56680	79003	23361	67094	15019	63261	24543	52884
195	05172	08100	22316	54495	60005	29532	18433	18057
196	74782	27005	03894	98038	20627	40307	47317	92759
197	85288	93264	61409	03404	09649	55937	60843	66167
198	68309	12060	14762	58002	03716	81968	57934	32624
199	26461	88346	52430	60906	74216	96263	69296	90107
200	42672	67680	42376	95023	82744	03971	96560	55148

Table entry for *p* and *C* is the critical value *t** with probability *p* lying to its right and probability *C* lying between −*t** and *t**.

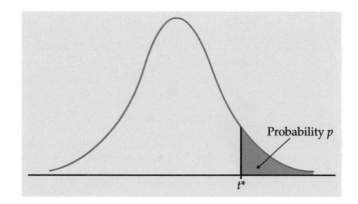

Probability *p*

t*

TABLE D   *t* distribution critical values

df	\multicolumn{12}{c	}{Upper-tail probability *p*}
	.25	.20	.15	.10	.05	.025	.02	.01	.005	.0025	.001	.0005
1	1.000	1.376	1.963	3.078	6.314	12.71	15.89	31.82	63.66	127.3	318.3	636.6
2	0.816	1.061	1.386	1.886	2.920	4.303	4.849	6.965	9.925	14.09	22.33	31.60
3	0.765	0.978	1.250	1.638	2.353	3.182	3.482	4.541	5.841	7.453	10.21	12.92
4	0.741	0.941	1.190	1.533	2.132	2.776	2.999	3.747	4.604	5.598	7.173	8.610
5	0.727	0.920	1.156	1.476	2.015	2.571	2.757	3.365	4.032	4.773	5.893	6.869
6	0.718	0.906	1.134	1.440	1.943	2.447	2.612	3.143	3.707	4.317	5.208	5.959
7	0.711	0.896	1.119	1.415	1.895	2.365	2.517	2.998	3.499	4.029	4.785	5.408
8	0.706	0.889	1.108	1.397	1.860	2.306	2.449	2.896	3.355	3.833	4.501	5.041
9	0.703	0.883	1.100	1.383	1.833	2.262	2.398	2.821	3.250	3.690	4.297	4.781
10	0.700	0.879	1.093	1.372	1.812	2.228	2.359	2.764	3.169	3.581	4.144	4.587
11	0.697	0.876	1.088	1.363	1.796	2.201	2.328	2.718	3.106	3.497	4.025	4.437
12	0.695	0.873	1.083	1.356	1.782	2.179	2.303	2.681	3.055	3.428	3.930	4.318
13	0.694	0.870	1.079	1.350	1.771	2.160	2.282	2.650	3.012	3.372	3.852	4.221
14	0.692	0.868	1.076	1.345	1.761	2.145	2.264	2.624	2.977	3.326	3.787	4.140
15	0.691	0.866	1.074	1.341	1.753	2.131	2.249	2.602	2.947	3.286	3.733	4.073
16	0.690	0.865	1.071	1.337	1.746	2.120	2.235	2.583	2.921	3.252	3.686	4.015
17	0.689	0.863	1.069	1.333	1.740	2.110	2.224	2.567	2.898	3.222	3.646	3.965
18	0.688	0.862	1.067	1.330	1.734	2.101	2.214	2.552	2.878	3.197	3.611	3.922
19	0.688	0.861	1.066	1.328	1.729	2.093	2.205	2.539	2.861	3.174	3.579	3.883
20	0.687	0.860	1.064	1.325	1.725	2.086	2.197	2.528	2.845	3.153	3.552	3.850
21	0.686	0.859	1.063	1.323	1.721	2.080	2.189	2.518	2.831	3.135	3.527	3.819
22	0.686	0.858	1.061	1.321	1.717	2.074	2.183	2.508	2.819	3.119	3.505	3.792
23	0.685	0.858	1.060	1.319	1.714	2.069	2.177	2.500	2.807	3.104	3.485	3.768
24	0.685	0.857	1.059	1.318	1.711	2.064	2.172	2.492	2.797	3.091	3.467	3.745
25	0.684	0.856	1.058	1.316	1.708	2.060	2.167	2.485	2.787	3.078	3.450	3.725
26	0.684	0.856	1.058	1.315	1.706	2.056	2.162	2.479	2.779	3.067	3.435	3.707
27	0.684	0.855	1.057	1.314	1.703	2.052	2.158	2.473	2.771	3.057	3.421	3.690
28	0.683	0.855	1.056	1.313	1.701	2.048	2.154	2.467	2.763	3.047	3.408	3.674
29	0.683	0.854	1.055	1.311	1.699	2.045	2.150	2.462	2.756	3.038	3.396	3.659
30	0.683	0.854	1.055	1.310	1.697	2.042	2.147	2.457	2.750	3.030	3.385	3.646
40	0.681	0.851	1.050	1.303	1.684	2.021	2.123	2.423	2.704	2.971	3.307	3.551
50	0.679	0.849	1.047	1.299	1.676	2.009	2.109	2.403	2.678	2.937	3.261	3.496
60	0.679	0.848	1.045	1.296	1.671	2.000	2.099	2.390	2.660	2.915	3.232	3.460
80	0.678	0.846	1.043	1.292	1.664	1.990	2.088	2.374	2.639	2.887	3.195	3.416
100	0.677	0.845	1.042	1.290	1.660	1.984	2.081	2.364	2.626	2.871	3.174	3.390
1000	0.675	0.842	1.037	1.282	1.646	1.962	2.056	2.330	2.581	2.813	3.098	3.300
z*	0.674	0.841	1.036	1.282	1.645	1.960	2.054	2.326	2.576	2.807	3.091	3.291
	50%	60%	70%	80%	90%	95%	96%	98%	99%	99.5%	99.8%	99.9%

Confidence level *C*